Process Control

Modeling, Design, and Simulation

ISBN 0-13-353640-8
90000
9 790133 536408

PRENTICE HALL INTERNATIONAL SERIES
IN THE PHYSICAL AND CHEMICAL ENGINEERING SCIENCES

NEAL R. AMUNDSON, SERIES EDITOR, *University of Houston*

BALZHISER, SAMUELS, AND ELIASSEN *Chemical Engineering Thermodynamics*
BEQUETTE *Process Control: Modeling, Design, and Simulation*
BEQUETTE *Process Dynamics*
BIEGLER, GROSSMAN, AND WESTERBERG *Systematic Methods of Chemical Process Design*
BROSILOW AND JOSEPH *Techniques of Model-based Control*
CROWL AND LOUVAR *Chemical Process Safety: Fundamentals with Applications, 2nd edition*
CONSTANTINIDES AND MOSTOUFI *Numerical Methods for Chemical Engineers with MATLAB Applications*
CUTLIP AND SHACHAM *Problem Solving in Chemical Engineering with Numerical Methods*
DENN *Process Fluid Mechanics*
DOYLE *Process Control Modules: A Software Laboratory for Control Design*
ELLIOT AND LIRA *Introductory Chemical Engineering Thermodynamics*
FOGLER *Elements of Chemical Reaction Engineering, 3rd edition*
HIMMELBLAU *Basic Principles and Calculations in Chemical Engineering, 6th edition*
HINES AND MADDOX *Mass Transfer*
KYLE *Chemical and Process Thermodynamics, 3rd edition*
PRAUSNITZ, LICHTENTHALER, AND DE AZEVEDO *Molecular Thermodynamics of Fluid-Phase Equilibria, 3rd edition*
PRENTICE *Electrochemical Engineering Principles*
SHULER AND KARGI *Bioprocess Engineering, 2nd edition*
STEPHANOPOULOS *Chemical Process Control*
TESTER AND MODELL *Thermodynamics and Its Applications, 3rd edition*
TURTON, BAILIE, WHITING, AND SHAEIWITZ *Analysis, Synthesis, and Design of Chemical Processes, 2nd edition*
WILKES *Fluid Mechanics for Chemical Engineering*

Process Control

Modeling, Design, and Simulation

B. Wayne Bequette

Rensselaer Polytechnic Institute

Textbook web site:
www.rpi.edu/~bequeb/Control

PRENTICE HALL
Professional Technical Reference
Upper Saddle River, NJ 07458
www.phptr.com

Library of Congress Cataloging-in-Publication Data

Bequette, B. Wayne, 1957–
Process control : modeling, design, and simulation / by B. Wayne Bequette.
p. cm. — (Prentice-Hall international series in the physical and chemical engineering sciences)
ISBN 0-13-353640-8 (paper : alk. paper)
1. Process control—Mathematical models. 2. Manufacturing processes—Mathematical models.
I. Title. II. Series.

TS156.8 .B46 2003
670.42'7—dc21

2002032953

Editorial/production supervision: *Argosy*
Full-service production manager: *Anne R. Garcia*
Cover design director: *Jerry Votta*
Cover design: *Talar Agasyan-Boorujy*
Manufacturing buyer: *Maura Zaldivar*
Publisher: *Bernard Goodwin*
Editorial assistant: *Michelle Vincenti*
Marketing manager: *Dan DePasquale*

PRENTICE HALL PTR

Publishing as Prentice Hall Professional Technical Reference
Upper Saddle River, New Jersey 07458

Prentice Hall books are widely used by corporations and government agencies for training, marketing, and resale.

For information regarding corporate and government bulk discounts please contact:

Corporate and Government Sales (800) 382-3419 or
corpsales@pearsontechgroup.com

For MATLAB product information, please contact:
The MathWorks, Inc.
3 Apple Hill Drive
Natick, MA 01760-2098 USA
Tel: 508-647-7000
Fax: 508-647-7101
E-mail: info@mathworks.com
Web: www.mathworks.com

Printed in the United States of America
10 9 8 7 6 5 4 3 2 1

ISBN 0-13-353640-8

Pearson Education LTD.
Pearson Education Australia PTY, Limited
Pearson Education Singapore, Pte. Ltd.
Pearson Education North Asia Ltd.
Pearson Education Canada, Ltd.
Pearson Educación de Mexico, S.A. de C.V.
Pearson Education—Japan
Pearson Education Malaysia, Pte, Ltd.

To Pat, Brendan, and Eileen
and
my parents, Bill and Ayleen Bequette

About Prentice Hall Professional Technical Reference

With origins reaching back to the industry's first computer science publishing program in the 1960s, Prentice Hall Professional Technical Reference (PH PTR) has developed into the leading provider of technical books in the world today. Formally launched as its own imprint in 1986, our editors now publish over 200 books annually, authored by leaders in the fields of computing, engineering, and business.

Our roots are firmly planted in the soil that gave rise to the technological revolution. Our bookshelf contains many of the industry's computing and engineering classics: Kernighan and Ritchie's *C Programming Language,* Nemeth's *UNIX System Administration Handbook,* Horstmann's *Core Java,* and Johnson's *High-Speed Digital Design.*

PH PTR acknowledges its auspicious beginnings while it looks to the future for inspiration. We continue to evolve and break new ground in publishing by providing today's professionals with tomorrow's solutions.

PRENTICE HALL PTR

Contents

Preface

Background

There are a variety of courses in a standard chemical engineering curriculum, ranging from the introductory material and energy balances course, and culminating with the capstone process design course. The focus of virtually all of these courses is on steady-state behavior; the rare exceptions include the analysis of batch reactors and batch distillation in the reaction engineering and equilibrium stage operations courses, respectively. A concern of a practicing process engineer, on the other hand, is how to best operate a process plant where everything seems to be changing. The process dynamics and control course is where students must gain an appreciation for the dynamic nature of chemical processes and develop strategies to operate these processes.

Textbook Goals

The major goal of this textbook is to teach students to analyze dynamic chemical processes and develop automatic control strategies to operate them safely and economically. My experience is that students learn best with immediate simulation-based reinforcement of basic concepts. Rather than simply present theory topics and develop analytical solutions, this textbook uses "interactive learning" through computer-based simulation exercises (modules). The popular MATLAB software package, including the SIMULINK block-diagram simulation environment, is used. Students, instructors, and practicing process engineers learning new model-based techniques can all benefit from the "feedback" provided by simulation studies.[1]

Depending on the goals of the instructor and the background of the students, roughly one chapter (± 0.5) and one module can be covered each week. Still, it is probably too ambitious to cover the entire text during a typical 15-week semester, so I recommend that instructors carefully choose the topics that best meet their personal objectives. At Rensselaer

[1]It should be noted that I am not a proponent of a solely "simulation-based" control education, where students iteratively adjust parameters in a JavaScript simulation until acceptable responses are obtained. I wish for students to obtain the classic mode of understanding as analyzed so well by Robert Pirsig in *Zen and the Art of Motorcycle Maintenance* (Bantam Books, 1974). This deeper understanding of process control can be obtained by rigorous analysis and by selected simulations where the student plays a direct role in the implementation of an algorithm or strategy of choice.

Polytechnic Institute, we teach the one-semester, 4-credit course in a studio-based format, with students attending two 2-hour sessions and one 2-hour recitation (which also provides plenty of "catch-up" time) each week. During these sessions we typically spend 45 minutes discussing a topic, then have the students spend the remaining hour performing analysis and computer simulation exercises, working in pairs. During the discussion periods the students face the instructor station at the front of the room, and during the simulation periods they swivel in their chairs to the workstations on the countertops behind them. This textbook can also be used in a more traditional lecture-based course, with students working on the modules and solving homework problems on their own.

Chapters

An introduction to process control and instrumentation is presented in Chapter 1. The development and use of models is very important in control systems engineering. Fundamental models are developed in Chapter 2, including the steady-state solution and linearization to form state space models. Chapter 3 focuses on the dynamic behavior of linear systems, starting with state space models and then covering transfer function-based models in detail. Chapter 4 we cover the development of empirical models, including continuous and discrete transfer function models.

Chapter 5 provides a more detailed introduction to feedback control, developing the basic idea of a feedback system, proportional, integral, derivative (PID) controllers, and methods of analyzing closed-loop stability. Chapter 6 presents the Ziegler-Nichols closed-loop oscillation method for controller tuning, since the same basic concept is used in the automatic tuning procedures presented in Chapter 11. Frequency response analysis techniques, important for determining control system robustness, are presented in Chapter 7.

In recent years model-based control has lead to improved control loop performance. One of the clearest model-based techniques is internal model control (IMC), which is presented in Chapter 8. The PID controller remains the most widely used controller in industry; in Chapter 9 we show how to convert internal model controllers to classical feedback (PID) controllers.

In Chapter 10 the widely used cascade and feed-forward strategies are developed. Many control loops suffer from poor performance, either because they were not tuned well originally, or because the process is nonlinear and has changed operating conditions. Two methods of dealing with these problems, automatic tuning and gain scheduling, are presented in Chapter 11. The phenomenon of reset windup and the development of antireset windup strategies are also presented in Chapter 11.

Many control strategies must be able to switch between manipulated inputs or select from several measured outputs. Split-range, selective and override strategies are presented in Chapter 12. Process units contain many control loops that generally do not operate independently. The effects of these control-loop interactions are presented in Chapter 13. The design of multivariable controllers is developed in Chapter 14.

The development of the control instrumentation diagram for an entire chemical process is challenging and remains somewhat of an art. In Chapter 15 recycle systems are shown to cause unique and challenging steady-state and dynamic control problems. In addition, an overview of corporate-wide optimization and control problems is presented. Model predictive control (MPC) is the most widely applied advanced control strategy in industry. The basic step response model-based MPC method is developed in Chapter 16. This is followed by a discussion of the constrained version of MPC, and enhancements to improve disturbance rejection.

Learning Modules

The chapters are followed by a series of learning modules that serve several purposes; some focus on the software tools, while others focus on particular control problems. The first two provide introductions to MATLAB and SIMULINK, the simulation environment for the modules that follow. The third module demonstrates the solution of ordinary differential equations using MATLAB and SIMULINK, while the fourth shows how to use the MATLAB Control Toolbox to create and convert models from one form to another. The modules that follow focus on a particular unit operation, to provide a detailed demonstration of various control system design, analysis or implementation techniques. Module 5 develops a simple isothermal CSTR model that is used in a number of the chapters. Module 6 details the robustness analysis of processes characterized by first-order + deadtime (FODT) models.

Module 7 presents a biochemical reactor with two possible desired operating points; one stable and the other unstable. The controller design and system performance is clearly different at each operating point. The classic jacketed CSTR with an exothermic reaction is studied in Module 8. Issues discussed include recirculation heat transfer dynamics, cascade control, and split-range control. Level control loops can be tuned for two different extremes of closed-loop performance, as shown in Module 9 (steam drum, requiring tight level control) and Module 10 (surge drum, allowing loose level control to minimize outflow variation). Challenges associated with jacketed batch reactors are presented in Module 11. Some motivating biomedical problems are presented in Module 12. Challenges of control loop interaction are demonstrated in the distillation application of Module 13. Module 14 provides an overview of several case study problems in multivariable control.

Here the students can download SIMULINK .mdl files for the textbook web page and perform complete modeling and control system design. These case studies are meant to tie together many concepts presented in the text. Issues particular to flow control are discussed in Module 15, and digital control techniques are presented in Module 16.

Textbook Web Page

MATLAB and SIMULINK files, as well as additional learning material and errata, can be found on the textbook web page:

www.rpi.edu/dept/chem-eng/WWW/faculty/bequette/books/Process_Control/

or

www.rpi.edu/~bequeb/Control

Acknowledgments

A few acknowledgments are in order. First of all, Professor Jim Turpin at the University of Arkansas stimulated my interest in process dynamics and control when I took his course as an undergraduate. As a neophyte process engineer for American Petrofina I had the opportunity to serve as a process operator during two work-stoppages. A newfound respect for control loop interaction led me to graduate study at the University of Texas, where Professor Tom Edgar provided the "degrees of freedom" for me to explore a range of control topics. Collaborations at Merck, Inc., led to the presentation of modeling and control of batch reactors in Module 11. Research sponsored by the Whitaker Foundation and the National Science Foundation resulted in material presented in Modules 12 and 14.

My own graduate students have served as teaching assistants in the dynamics and control courses, and have provided me with valuable feedback on various versions of this textbook. In particular, Lou Russo, now at ExxonMobil, helped me understand what works and what does not work in the classroom and in homework assignments. He certainly had a major positive impact on the education of many Rensselaer undergraduates.

Professor Robert Parker at the University of Pittsburgh classroom tested this textbook, and made a number of valuable suggestions. In addition, Brian Aufderheide (now at the Keck Graduate Institute) critiqued Chapter 16.

My colleagues at Rensselaer have promoted an environment that provides an optimum mix of teaching and research; our department has published four textbooks during the past two years. Various educational initiatives at Rensselaer have allowed me to develop an interactive learning approach to dynamics and control. In particular, the Control Engineering Studio environment gives me immediate feedback on the level of practical understanding on a particular topic and allows me to give immediate feedback to students. A Curriculum Innovation grant from P&G led to the development of experiments and learning modules for the dynamics and control course, and for other courses using the Control Engineering Studio classroom.

Various Troy and Albany establishments have served to "gain schedule" my personal regulatory system and allowed me to obtain a better understanding of the pharmacokinetics and pharmacodynamics of caffeine and ethanol. The Daily Grind (www.dailygrind.com; you won't find a better coffee roaster in Seattle) in Albany provided beans for the many espressos that "kick started" numerous sections of this textbook. Group meetings at the Troy Pub & Uncle Sam Brewery (www.troypub.com; try the Harwood Porter the next time you are in town) led to many interesting education and research[2] discussions (not to mention political and other topics).

Naturally, completing this text would have been a struggle without the support of my wife, Pat Fahy, and the good sleeping habits of my kids, Brendan and Eileen. They have done their best to convince me that not all systems are controllable.

[2]The important interplay of research and education should not be overlooked. Seemingly innocuous problems assigned in the control class have led to interesting graduate research projects. Similarly, graduate research results have been brought into the undergraduate classroom.

CHAPTER 1

Introduction

The goal of this chapter is to provide a motivation for, and an introduction to, process control and instrumentation. After studying this chapter, the reader, given a process, should be able to do the following:

- Determine possible control objectives, input variables (manipulated and disturbance) and output variables (measured and unmeasured), and constraints (hard or soft), as well as classify the process as continuous, batch, or semicontinuous
- Assess the importance of process control from safety, environmental, and economic points of view
- Sketch a process instrumentation and control diagram
- Draw a simplified control block diagram
- Understand the basic idea of feedback control
- Understand basic sensors (measurement devices) and actuators (manipulated inputs)
- Begin to develop intuition about characteristic timescales of dynamic behavior

The major sections of this chapter are as follows:

1.1 Introduction
1.2 Instrumentation
1.3 Process Models and Dynamic Behavior
1.4 Control Textbooks and Journals
1.5 A Look Ahead
1.6 Summary

1.1 Introduction

Process engineers are often responsible for the operation of chemical processes. As these processes become larger scale and/or more complex, the role of process automation becomes more and more important. The objective of this textbook is to teach process engineers how to design and tune feedback controllers for the automated operation of chemical processes.

A conceptual process block diagram for a chemical process is shown in Figure 1–1. Notice that the *inputs* are classified as either manipulated or disturbance variables and the *outputs* are classified as measured or unmeasured in Figure 1–1a. To automate the operation of a process, it is important to use measurements of process outputs or disturbance inputs to make decisions about the proper values of manipulated inputs. This is the purpose of the controller shown in Figure 1–1b; the measurement and control signals are shown as dashed lines. These initial concepts probably seem very vague or abstract to you at this point. Do not worry, because we present a number of examples in this chapter to clarify these ideas.

The development of a control strategy consists of formulating or identifying the following.

1. Control objective(s).
2. Input variables—classify these as (a) manipulated or (b) disturbance variables; inputs may change continuously, or at discrete intervals of time.
3. Output variables—classify these as (a) measured or (b) unmeasured variables; measurements may be made continuously or at discrete intervals of time.
4. Constraints—classify these as (a) hard or (b) soft.
5. Operating characteristics—classify these as (a) continuous, (b) batch, or (c) semicontinuous (or semibatch).
6. Safety, environmental, and economic considerations.
7. Control structure—the controllers can be feedback or feed forward in nature.

Here we discuss each of the steps in formulating a control problem in more detail.

1. The first step of developing a control strategy is to formulate the control *objective(s)*. A chemical-process operating unit often consists of several unit operations. The control of an operating unit is generally reduced to considering the control of each unit operation separately. Even so, each unit operation may have multiple, sometimes conflicting objectives, so the development of control objectives is not a trivial problem.

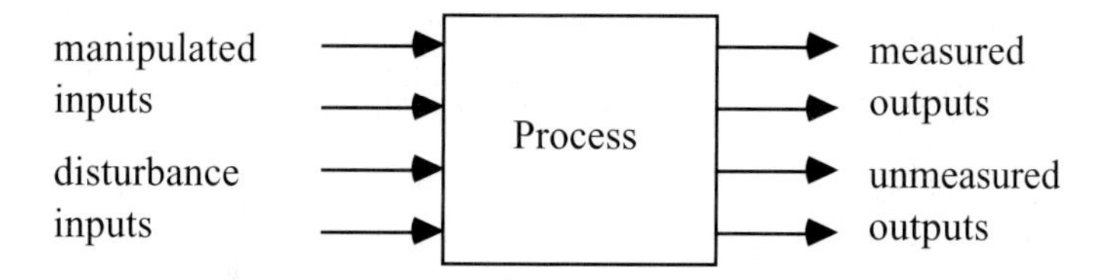

a. Input/Output representation

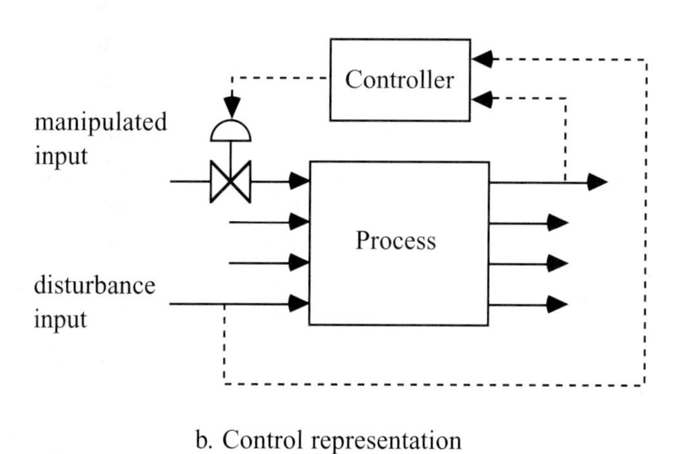

b. Control representation

Figure 1–1 Conceptual process input/output block diagram.

2. Input variables can be classified as *manipulated* or *disturbance* variables. A manipulated input is one that can be adjusted by the control system (or process operator). A disturbance input is a variable that affects the process outputs but that cannot be adjusted by the control system. Inputs may change *continuously* or at *discrete* intervals of time.
3. Output variables can be classified as *measured* or *unmeasured* variables. Measurements may be made *continuously* or at *discrete* intervals of time.
4. Any process has certain operating *constraints,* which are classified as hard or soft. An example of a hard constraint is a minimum or maximum flow rate—a valve operates between the extremes of fully closed or fully open. An example of a soft constraint is a product composition—it may be desirable to specify a composition between certain values to sell a product, but it is possible to violate this specification without posing a safety or environmental hazard.
5. Operating characteristics are usually classified as *continuous, batch,* or *semicontinuous* (*semibatch*). Continuous processes operate for long periods of time

under relatively constant operating conditions before being "shut down" for cleaning, catalyst regeneration, and so forth. For example, some processes in the oil-refining industry operate for 18 months between shutdowns. Batch processes are dynamic in nature—that is, they generally operate for a short period of time and the operating conditions may vary quite a bit during that period of time. Example batch processes include beer or wine fermentation, as well as many specialty chemical processes. For a batch reactor, an initial charge is made to the reactor, and conditions (temperature, pressure) are varied to produce a desired product at the end of the batch time. A typical semibatch process may have an initial charge to the reactor, but feed components may be added to the reactor during the course of the batch run.

Another important consideration is the dominant timescale of a process. For continuous processes this is very often related to the residence time of the vessel. For example, a vessel with a liquid volume of 100 liters and a flow rate of 10 liters/minute would have a residence time of 10 minutes; that is, on the average, an element of fluid is retained in the vessel for 10 minutes.

6. Safety, environmental, and economic considerations are all very important. In a sense, economics is the ultimate driving force—an unsafe or environmentally hazardous process will ultimately cost more to operate, through fines paid, insurance costs, and so forth. In many industries (petroleum refining, for example), it is important to minimize energy costs while producing products that meet certain specifications. Better process automation and control allows processes to operate closer to "optimum" conditions and to produce products where variability specifications are satisfied.

   The concept of "fail-safe" is always important in the selection of instrumentation. For example, a control valve needs an energy source to move the valve stem and change the flow; most often this is a pneumatic signal (usually 3–15 psig). If the signal is lost, then the valve stem will go to the 3-psig limit. If the valve is *air-to-open*, then the loss of instrument air will cause the valve to close; this is known as a *fail-closed* valve. If, on the other hand, a valve is *air-to-close*, when instrument air is lost the valve will go to its fully open state; this is known as a *fail-open* valve.
7. The two standard control types are *feed forward* and *feedback*. A feed-forward controller measures the disturbance variable and sends this value to a controller, which adjusts the *manipulated* variable. A feedback control system measures the output variable, compares that value to the desired output value, and uses this information to adjust the manipulated variable. For the first part of this textbook, we emphasize feedback control of single-input (manipulated) and single-output (measured) systems. Determining the feedback control

structure for these systems consists of deciding which manipulated variable will be adjusted to control which measured variable. The desired value of the measured process output is called the *setpoint.*

A particularly important concept used in control system design is process *gain.* The process gain is the sensitivity of a process output to a change in the process input. If an increase in a process input leads to an increase in the process output, this is known as a positive gain. If, on the other hand, an increase in the process input leads to a decrease in the process output, this is known as a negative gain. The magnitude of the process gain is also important. For example, a change in power (input) of 0.5 kW to a laboratory-scale heater may lead to a fluid temperature (output) change of 10°C; this is a process gain (change in output/change in input) of 20°C/kW. The same input power change of 0.5 kW to a larger scale heater may only yield an output change of 0.5°C, corresponding to a process gain of 1°C/kW.

Once the control structure is determined, it is important to decide on the control *algorithm.* The control algorithm uses measured output variable values (along with desired output values) to change the manipulated input variable. A control algorithm has a number of control *parameters,* which must be "tuned" (adjusted) to have acceptable performance. Often the tuning is done on a simulation model before implementing the control strategy on the actual process. A significant portion of this textbook is on the use of *model-based* control, that is, controllers that have a model of the process "built in."

This approach is best illustrated by way of example. Since many important concepts, such as control instrumentation diagrams and control block diagrams, are introduced in the next examples, it is important that you study them thoroughly.

Example 1.1: Surge Tank

Surge tanks are often used as intermediate storage for fluid streams being transferred between process units. Consider the process flow diagram shown in Figure 1–2, where a fluid stream from process 1 is fed to the surge tank; the effluent from the surge tank is sent to process 2.

There are obvious constraints on the height in this tank. If the tank overflows it may create safety and environmental hazards, which may also have economic significance. Let us analyze this system using a step-by-step procedure.

1. Control objective: The control objective is to maintain the height within certain bounds. If it is too high it will overflow and if it is too low there may be problems with the flow to process 2. Usually, a specific desired height will be selected. This desired height is known as the *setpoint.*

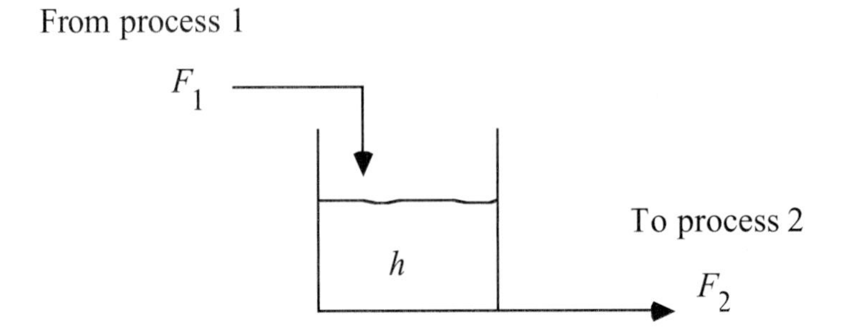

Figure 1–2 Tank level problem.

2. Input variables: The input variables are the flow from process 1 and the flow to process 2. Notice that an outlet flow rate is considered an input to this problem. The question is which input is manipulated and which is a disturbance? That depends. We discuss this problem further in a moment.
3. Output variables: The most important output variable is the liquid level. We assume that it is measured.
4. Constraints: There are a number of constraints in this problem. There is a maximum liquid level; if this is exceeded, the tank will overflow. There are minimum and maximum flow rates through the inlet and outlet valves.
5. Operating characteristics: We assume that this is a continuous process, that is, that there is a continuous flow in and out of the tank. It would be a semicontinuous process if, for example, there was an inlet flow with no outlet flow (if the tank was simply being filled).
6. Safety, environmental, and economic considerations: These aspects depend somewhat on the fluid characteristics. If it is a hazardous chemical, then there is a tremendous incentive from safety and environmental considerations to not allow the tank to overflow. Indeed, this is also an economic consideration, since injuries to employees or environmental cleanup costs money. Even if the substance is water, it has likely been treated by an upstream process unit, so losing water owing to overflow will incur an economic penalty.
   Safety considerations play an important role in the specification of control valves (fail-open or fail-closed). For this particular problem, the control-valve specification will depend on which input is manipulated. This is discussed in detail shortly.
7. Control structure: There are numerous possibilities for control of this system. We discuss first the feedback strategies, then the feed-forward strategies.

Feedback Control

The measured variable for a feedback control strategy is the tank height. Which input variable is manipulated depends on what is happening in process 1 and process 2. Let us consider two different scenarios.

Scenario 1 Process 2 regulates the flow-rate F_2. This could happen, for example, if process 2 is a steam generation system and process 1 is a deionization process. Process 2 varies the flow rate of water (F_2) depending on the steam demand. As far as the tank process is concerned, F_2 is a "wild" (disturbance) stream because the regulation of F_2 is determined by another system. In this case we would use F_1 as the manipulated variable; that is, F_1 is adjusted to maintain a desired tank height.

The control and instrumentation diagram for a feedback control strategy for scenario 1 is shown in Figure 1–3. Notice that the level transmitter (LT) sends the measured height of liquid in the tank (h_m) to the level controller (LC). The LC compares the measured level with the desired level (h_{sp}, the height setpoint) and sends a pressure signal (P_v) to the valve. This valve top pressure moves the valve stem up and down, changing the flow rate through the valve (F_1). If the controller is designed properly, the flow rate changes to bring the tank height close to the desired setpoint. In this process and instrumentation diagram we use dashed lines to indicate signals between different pieces of instrumentation.

A simplified block diagram representing this system is shown in Figure 1–4. Each signal and device (or process) is shown on the block diagram. We use a slightly different

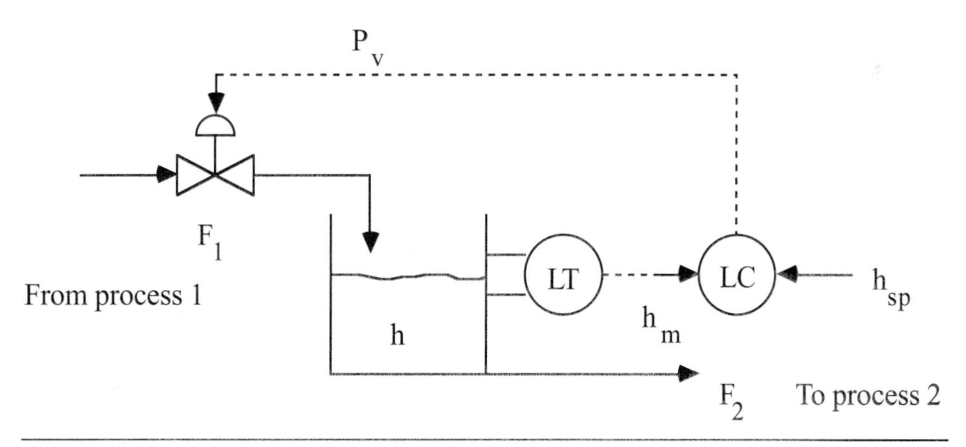

Figure 1–3 Feedback control strategy 1. The level is measured and the inlet flow rate (valve position) is manipulated.

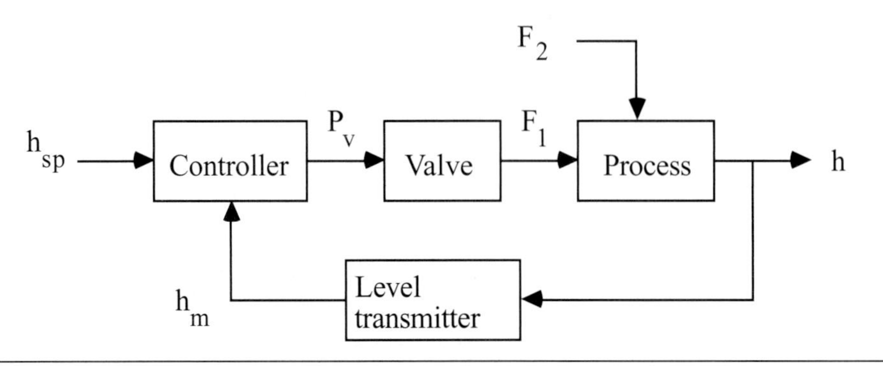

Figure 1–4 Feedback control schematic (block diagram) for scenario 1. F_1 is manipulated and F_2 is a disturbance.

form for block diagrams when we use transfer function notation for control system analysis in Chapter 5. Note that each block represents a dynamic element. We expect that the valve and LT dynamics will be much faster than the process dynamics. We also see clearly from the block diagram why this is known as a *feedback* control "loop." The controller "decides" on the valve position, which affects the inlet flow rate (the manipulated input), which affects the level; the outlet flow rate (the disturbance input) also affects the level. The level is measured, and that value is fed back to the controller [which compares the measured level with the desired level (setpoint)].

Notice that the control valve should be specified as fail-closed or air-to-open, so that the tank will not overflow on loss of instrument air or other valve failure.

Scenario 2 Process 1 regulates flow rate F_1. This could happen, for example, if process 1 is producing a chemical compound that must be processed by process 2. Perhaps process 1 is set to produce F_1 at a certain rate. F_1 is then considered "wild" (a disturbance) by the tank process. In this case we would adjust F_2 to maintain the tank height. Notice that the control valve should be specified as fail-open or air-to-close, so that the tank will not overflow on loss of instrument air or other valve failure.

The process and instrumentation diagram for this scenario is shown in Figure 1–5. The only difference between this and the previous instrumentation diagram (Figure 1–3) is that F_2 rather than F_1 is manipulated.

The simplified block diagram shown in Figure 1–6 differs from the previous case (Figure 1–4) only because F_2 rather than F_1 is manipulated. F_1 is a disturbance input.

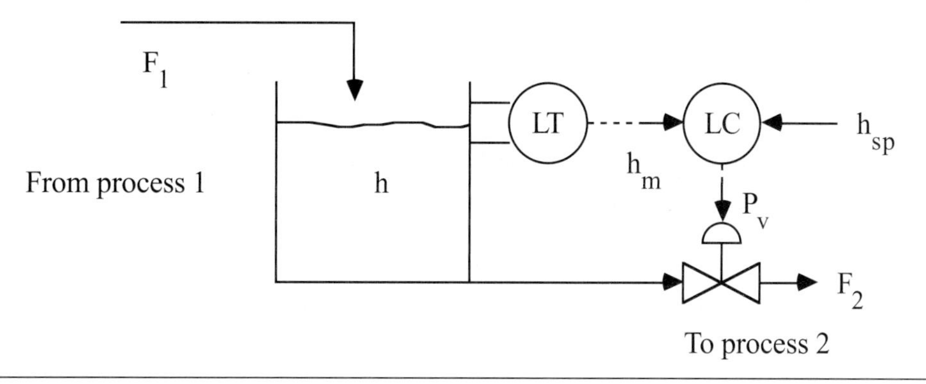

Figure 1–5 Feedback control strategy 2. Outlet flow rate is manipulated.

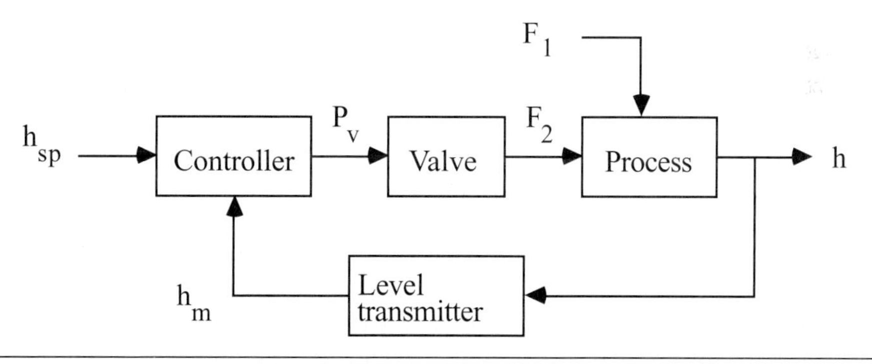

Figure 1–6 Feedback control schematic (block diagram) for scenario 2. F_2 is manipulated and F_1 is a disturbance.

Feed-Forward Control

The previous two feedback control strategies were based on measuring the output (tank height) and manipulating an input (the inlet flow rate in scenario 1 and the outlet flow rate in scenario 2). In each case the manipulated variable is changed after a disturbance affects the output. The advantage to a *feed-forward* control strategy is that a disturbance variable is measured and a manipulated variable is changed before the output is affected. Consider a case where the inlet flow rate can be changed by the upstream process unit and is therefore considered a disturbance variable. If we can measure the inlet flow rate, we can manipulate the outlet flow rate to maintain a constant tank height. This feed-forward control strategy is shown in Figure 1–7, where FM is the flow measurement device and FFC is the feed-forward controller. The corresponding control block diagram is shown in Figure 1–8. F_1 is a disturbance input that directly affects the tank height; the value of F_1 is

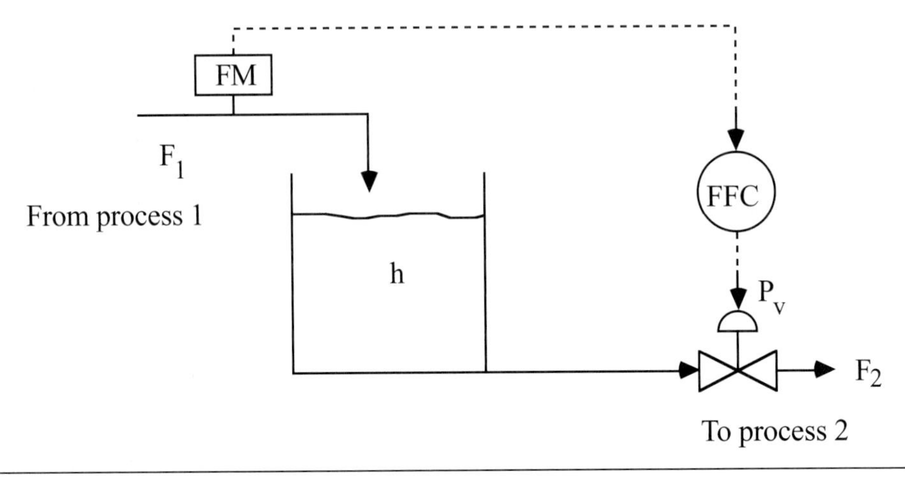

Figure 1–7 Feed-forward control strategy. Inlet flow rate is measured and outlet flow rate is manipulated.

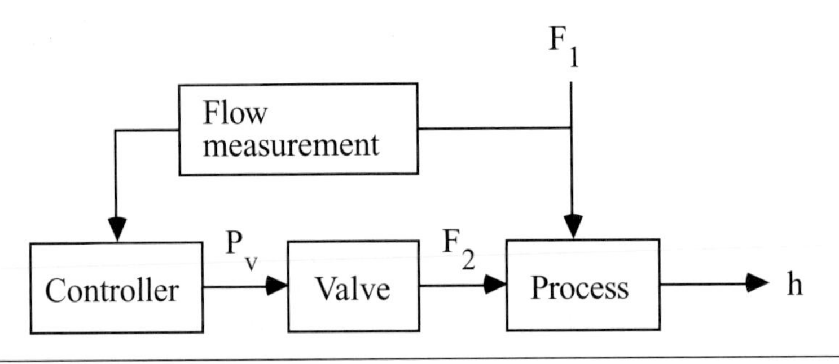

Figure 1–8 Feed-forward control schematic block diagram.

measured by the FM device, and the information is used by an FFC to change the manipulated input, F_2.

The main disadvantage to this approach is sensitivity to uncertainty. If the inlet flow rate is not perfectly measured or if the outlet flow rate cannot be manipulated perfectly, then the tank height will not be perfectly controlled. With any small disturbance or uncertainty, the tank will eventually overflow or run dry. In practice, FFC is combined with feedback control to account for uncertainty. A feed-forward/feedback strategy is shown in Figure 1–9, and the corresponding block diagram is shown in Figure 1–10. Here, the feed-forward portion allows immediate corrective action to be taken before the disturbance (inlet flow rate) actually affects the output measurement (tank height). The feedback controller adjusts the outlet flow rate to maintain the desired tank height, even with errors in the inlet flow-rate measurement.

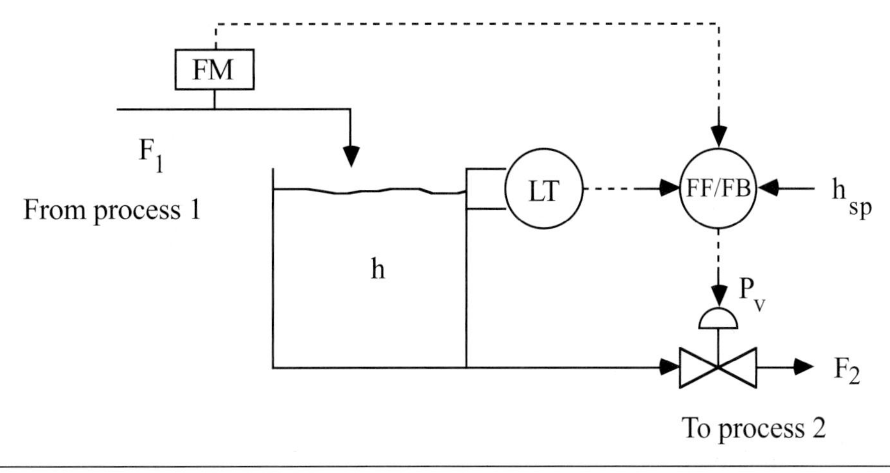

Figure 1–9 Feed-forward/feedback control strategy. The inlet flow rate is the measured disturbance, tank height is the measured output, and outlet flow rate is manipulated.

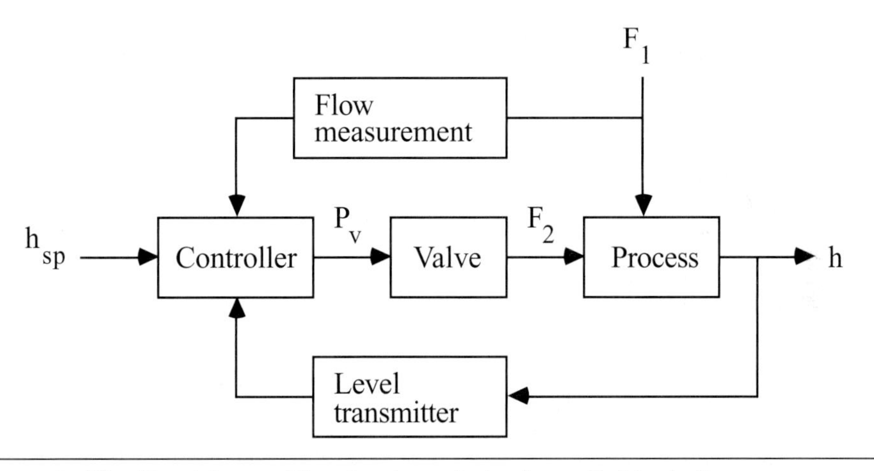

Figure 1–10 Feed-forward/feedback control schematic block diagram.

Discussion of Level Controller Tuning and the Dominant Timescale

Notice that we have not discussed the actual control algorithms; the details of control algorithms and tuning are delayed until Chapter 5. Conceptually, would you prefer to tune level controllers for "fast" or "slow" responses?

When tanks are used as surge vessels it is usually desirable to tune the controllers for a slow return to the setpoint. This is particularly true for scenario 2, where the inlet flow rate is considered a disturbance variable. The outlet flow rate is manipulated but affects another process. In order to not upset the downstream process, we would like to change the outlet flow rate slowly, yet fast enough that the tank does not overflow or go dry.

Related to the controller tuning issue is the importance of the dominant timescale of the process. Consider the case where the maximum tank volume is 200 gallons and the steady-state operating volume is 100 gallons. If the steady-state flow rate is 100 gallons/minute, the "residence time" would be 1 minute. Assume the inlet flow rate is a disturbance and outlet flow rate is manipulated (Figure 1–5). If the feed flow rate increased to 150 gallons/minute and the outlet flow rate did not change, the tank would overflow in 2 minutes. On the other hand, if the same vessel had a steady-state flow rate of 10 gallons/minute and the inlet flow suddenly increased to 15 gallons/minute (with no change in the outlet flow), it would take 20 minutes for the tank to overflow. Clearly, controller tuning and concern about controller failure is different for these two cases.

The first example was fairly easy compared with most control-system synthesis problems in industry. Even for this simple example we found that there were many issues to be considered and a number of decisions (specification of a fail-open or fail-closed valve, etc.) that needed to be made. Often there will be many (and usually conflicting) objectives, many possible manipulated variables, and numerous possible measured variables.

It is helpful to think of common, everyday activities in the context of control, so you will become familiar with the types of control problems that can arise in practice. The following activity is just such an example.

Example 1.2: Taking a Shower

A common multivariable control problem that we face every day is taking a shower. A simplified process schematic is shown in Figure 1–11. We analyze this process step by step.

1. Control objectives: Control objectives when taking a shower include the following:
    - **a.** to become clean
    - **b.** to be comfortable (correct temperature and water velocity as it contacts the body)
    - **c.** to "look good" (clean hair, etc.)
    - **d.** to become refreshed

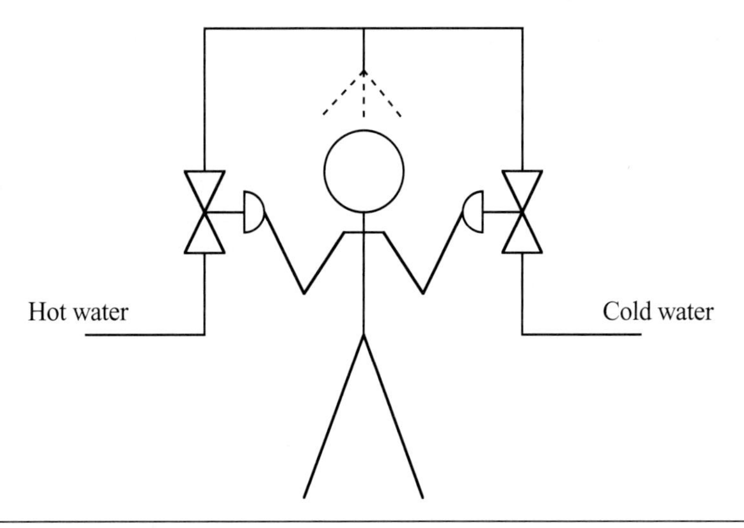

Figure 1–11 Process schematic for taking a shower.

To simplify our analysis, for the rest of the problem we discuss how we can satisfy the second objective (to maintain water temperature and flow rate at comfortable values). Similar analysis can be performed for the other objectives.

2. Input variables: The manipulated input variables are hot-water and cold-water valve positions. Some showers can also vary the velocity by adjustment of the shower head. Another input is body position—you can move into and out of the shower stream. Disturbance inputs include a drop in water pressure (say, owing to a toilet flushing) and changes in hot water temperature owing to "using up the hot water from the heater."
3. Output variables: The "measured" output variables are the temperature and flow rate (or velocity) of the mixed stream as it contacts your body.
4. Constraints: There are minimum and maximum valve positions (and therefore flow rates) on both streams. The maximum mixed temperature is equal to the hot water temperature and the minimum mixed temperature is equal to the cold water temperature. The previous constraints were hard constraints—they cannot be physically violated. An example of a soft constraint is the mixed-stream water temperature—you do not want it to be above a certain value because you may get scalded. This is a soft constraint because it can physically happen, although you do not want it to happen.

5. Operating characteristics: This process is continuous while you are taking a shower but is most likely viewed as a batch process, since it is a small part of your day. It could easily be called a semicontinuous (semibatch) process.
6. Safety, environmental, and economic considerations: Too high of a temperature can scald you—this is certainly a safety consideration. Economically, if your showers are too long, more energy is consumed to heat the water, costing money. Environmentally (and economically), more water consumption means that more water and wastewater must be treated. An economic objective might be to minimize the shower time. However, if the shower time is too short, or not frequent enough, your clothes will become dirty and must be washed more often—increasing your clothes-cleaning bill.
7. Control structure: This is a multivariable control problem because adjusting either valve affects both temperature and flow rate. Control manipulations must be "coordinated," that is, if the hot-water flow rate is increased to increase the temperature, the cold-water flow rate must be decreased to maintain the same total flow rate. The measurement signals are continuous, but the manipulated variable changes are likely to be discrete (unless your hands are continuously varying the valve positions).
   Feedback control: As the body feels the temperature changing, adjustments to one or both valves is made. As the body senses a flow rate or velocity change, one or both valves are adjusted.
   Feed-forward control: If you hear the toilet flush, you move your body out of the stream to avoid the higher temperature that you anticipate. Notice that you are making a manipulated variable change (moving your body) before the effect of an output (temperature or flow rate) change is actually detected.

Some showers may have a relatively large time delay (or dead time) between when a manipulated variable change is made and when the actual output change is measured. This could happen, for example, if there was a large pipe run between the mixing point and the shower head (this would be considered an input time delay). Another type of time delay is measurement dead time, for example if your body takes a while to detect a change in the temperature of the stream contacting your body.

Notice that the control strategy used has more manipulated variables (two valve positions and body movement) than measured outputs (total mixed-stream flow rate and temperature).

In the shower example, the individual taking the shower served as the controller. The measurements and manipulations for this example are somewhat qualitative (you do not know the exact temperature or flow rate, for example). Most of the rest of the textbook consists of quantitative controller design procedures, that is, a mathematical model of the process is used to develop the control algorithm.

This chapter has covered the important first step of control system development—identifying seven basic steps in analyzing a process control problem. We have used simple examples with which you are familiar. As you learn about more chemical and environmental processes, you should get in the habit of thinking about them from a process systems point of view, just as you have with these simple systems.

1.2 Instrumentation

The example level-control problem had three critical pieces of instrumentation: a sensor (measurement device), actuator (manipulated input device), and controller. The sensor measured the tank level, the actuator changed the flow rate, and the controller determined how much to vary the actuator, based on the sensor signal.

There are many common sensors used for chemical processes. These include temperature, level, pressure, flow, composition, and pH. The most common manipulated input is the valve actuator signal (usually pneumatic).

Each device in a control loop must supply or receive a signal from another device. When these signals are continuous, such as electrical current or voltage, we use the term *analog*. If the signals are communicated at discrete intervals of time, we use the term *digital*.

Analog

Analog or continuous signals provided the foundation for control theory and design and analysis. A common measurement device might supply either a 4- to 20-mA or 0- to 5-V signal as a function of time. Pneumatic analog controllers (developed primarily in the 1930s, but used in some plants today) would use instrument air, as well as a bellows-and-springs arrangement to "calculate" a controller output based on an input from a measurement device (typically supplied as a 3- to 15-psig pneumatic signal). The controller output of 3–15 psig would be sent to an actuator, typically a control valve where the pneumatic signal would move the valve stem. For large valves, the 3- to 15-psig signal might be "amplified" to supply enough pressure to move the valve stem.

Electronic analog controllers typically receive a 4- to 20-mA or 0- to 5-V signal from a measurement device, and use an electronic circuit to determine the controller output, which is usually a 4- to 20-mA or 0- to 5-V signal. Again, the controller output is often sent to a control valve that may require a 3- to 15-psig signal for valve stem actuation. In this case the 4- to 20-mA current signal is converted to the 3- to 15-psig signal using an I/P (current-to-pneumatic) converter.

Digital

Many devices and controllers are now based on digital communication technology. A sensor may send a digital signal to a controller, which then does a discrete computation and sends a digital output to the actuator. Very often, the actuator is a valve, so there is usually a D/I (digital-to-electronic analog) converter involved. Indeed, if the valve stem is moved by a pneumatic actuator rather than electronic, then an I/P converter may also be used.

In the past few decades, digital control-system design techniques that explicitly account for the discrete (rather than continuous) nature of the control computations have been developed. If small sample times are used, the tuning and performance of the digital controllers is nearly equal to that of analog controllers.

Techniques Used in This Textbook

Most of the techniques used in this book are based on analog (continuous) control. Although many of the control computations performed on industrial processes are digital, the discrete sample time is usually small enough that virtually identical performance to analog control is obtained. Our understanding of chemical processes is based on ordinary differential equations, so it makes sense to continue to think of control in a continuous fashion. We find that controller tuning is much more intuitive in a continuous, rather than discrete, framework. In Chapter 16 we spend some time discussing techniques that are specific to digital control systems, namely model predictive control (MPC).

Control Valve Placement

In Example 1.1 and in most of the examples given in this textbook, we use a simplified representation for a control valve and signal. It should be noted that virtually all control valves are actually installed in an arrangement similar to that shown in Figure 1–12. When the control valve fails, the adjacent block valves can be closed; the control valve can then be removed and replaced. During the interim, the bypass valve can be adjusted manually to maintain the desired flow rate. Generally, these control valve "stations" are placed at ground level for easy access, even if the pipeline is in a piperack far above the ground.

1.3 Process Models and Dynamic Behavior

Thus far we have mentioned the term *model* a number of times, and you probably have a vague notion of what we mean by model. The following definition of a model is from the McGraw-Hill Dictionary of Scientific and Technical Terms:

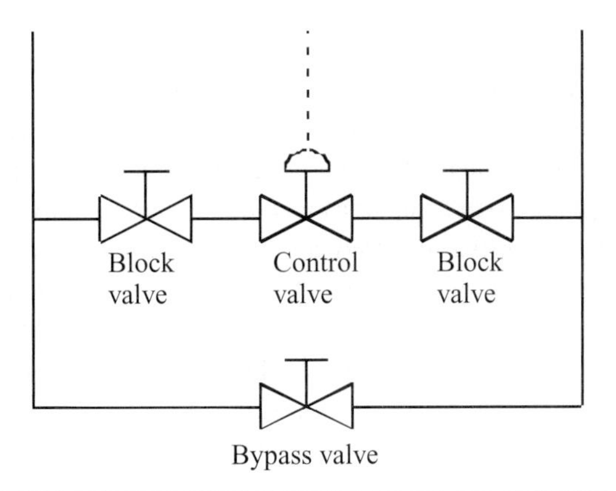

Figure 1–12 Typical control valve arrangement. When the control valve needs to be taken out of service, the two block valves are closed and the control valve is removed. The bypass valve can then be manually adjusted to control the flow.

> "A mathematical or physical system, obeying certain specified conditions, whose behavior is used to understand a physical, biological, or social system to which it is analogous in some way."

In this textbook, model will be taken to mean *mathematical model*. More specifically, we develop process models. A working definition of process model is

> a set of equations (including the necessary input data to solve the equations) that allows us to predict the behavior of a chemical process.

Models play a very important role in control-system design. Models can be used to simulate expected process behavior with a proposed control system. Also, models are often "embedded" in the controller itself; in effect the controller can use a process model to anticipate the effect of a control action. We can see from Example 1.1 that we at least need to know whether an increase in the flow rate will increase or decrease the tank level. For example, an increase in the inlet flow rate increases the tank level (positive gain), while an increase in the outlet flow rate leads to a decrease in the tank level (negative gain). In order to design a controller, then, we need to know whether an increase in the manipulated input increases or decreases the process output variable; that is, we need to know whether the process gain is positive or negative.

An example of a process model is shown next. A number of other examples are developed in Chapter 2.

Example 1.3: Liquid Surge Vessel Model

In the development of a dynamic model, simplifying assumptions are often made. Also, the model requirements are a function of the end-use of the model. In this case, we are ultimately interested in designing a controller and in simulating control-system behavior. Since we have not covered control algorithms in depth, our *objective* here is to develop a model that relates the inputs (manipulated and disturbance) to measured outputs that we wish to regulate.

For this process, we first assume that the density is constant. The model we develop should allow us to determine how the volume of liquid in the vessel varies as a function of the inlet and outlet flow rates. We will list the *state variables*, *parameters*, and the *input* and *output variables*. We must also specify the required information to solve this problem (see Figure 1–2). The system is the liquid in the tank and the liquid surface is the top boundary of the system. The following notation is used in the modeling equations:

F_1 = inlet volumetric flow rate (volume/time);

F_2 = outlet volumetric flow rate;

V = volume of liquid in vessel;

h = height of liquid in vessel;

ρ = liquid density (mass/volume);

A = cross-sectional area of vessel.

Here we write the balance equations based on an instantaneous rate-of-change,

$$\begin{bmatrix} \text{rate of change of} \\ \text{total mass of fluid} \\ \text{inside the vessel} \end{bmatrix} = \begin{bmatrix} \text{mass flow rate} \\ \text{of fluid} \\ \text{into the vessel} \end{bmatrix} - \begin{bmatrix} \text{mass flow rate} \\ \text{of fluid} \\ \text{out of the vessel} \end{bmatrix} \tag{1.1}$$

where the total mass of fluid in the vessel is $V\rho$, the rate of change is $dV\rho/dt$, and the density of the outlet stream is equal to the density of the vessel contents

$$\frac{dV\rho}{dt} = F_1\rho_1 - F_2\rho \tag{1.2}$$

Notice the implicit assumption that the density of fluid in the vessel does not depend on position (the perfect mixing assumption). This assumption allows an ordinary differential equation (ODE) formulation. We refer to any system that can be modeled by ODEs as *lumped parameter systems*. Also notice that the outlet stream density was assumed to be equal to the density of fluid in the tank. Assuming that the density of the inlet stream and fluid in the vessel are equal, this equation is then reduced to[1]

$$\frac{dV}{dt} = F_1 - F_2 \tag{1.3}$$

In Equation (1.3) we refer to V as a *state variable*, and F_1 and F_2 as *input variables* (even though F_2 is an outlet stream flow rate). If density remained in the equation, we would refer to it as a *parameter*.

In order to solve this problem we must specify the inputs $F_1(t)$ and $F_2(t)$ and the initial condition $V(0)$. Direct integration of Equation (1.3) yields

$$V(t) = V(0) + \int_0^t \left[F_i(\sigma) - F(\sigma)\right] d\sigma \tag{1.4}$$

If, for example, the initial volume is 500 liters, the inlet flow rate is 5 liters/second and the outlet flow rate is 4.5 liters/second, we find

$$V(t) = 500 + 0.5 \cdot t$$

Example 1.3 provides an introduction to the notion of states, inputs, and parameters. Consider now the notion of an output. We may consider fluid volume to be a desired output that we wish to control, for example. In that case, volume would not only be a state, it would also be considered an output. On the other hand, we may be concerned about fluid height, rather than volume. Volume and height are related through the constant cross-sectional area, A

$$V = Ah \qquad \text{or} \qquad h = V/A \tag{1.5}$$

Then we have the following modeling equations,

$$\frac{dV}{dt} = F_1 - F_2, \quad h = \frac{V}{A} \tag{1.6}$$

where V is a state, F_1 and F_2 are inputs, h is an output, and A is a parameter. We could also rewrite the state variable equation to find

$$A\frac{dh}{dt} = F_1 - F_2$$

[1]It might be tempting to the reader to begin to directly write a "volume balance" expression, which looks similar to Equation (1.3). We wish to make it clear that there is no such thing as a volume balance and Equation (1.3) is only correct because of the constant density assumption. It is a good idea to always write a mass balance expression, such as Equation (1.2), before making assumptions about the fluid density, which may lead to Equation (1.3).

or

$$\frac{dh}{dt} = \frac{(F_1 - F_2)}{A} \tag{1.7}$$

where fluid height is now the state variable. It should also be noted that inputs can be classified as either *manipulated* inputs (that we may regulate with a control valve, for example) or *disturbance* inputs. If we desired to measure fluid height and manipulate the flow rate of stream 1, for example, then F_1 would be a manipulated input, while F_2 would be a disturbance input.

We have found that a single process can have different modeling equations and variables, depending on assumptions and the objectives used when developing the model.

The liquid level process is an example of an *integrating* process. If the process is initially at steady state, the inlet and outlet flow rates are equal (see Equation 1.3 or 1.7). If the inlet flow rate is suddenly increased while the outlet flow rate remains constant, the liquid level (volume) will increase until the vessel overflows. Similarly, if the outlet flow rate is increased while the inlet flow rate remains constant, the tank level will decrease until the vessel is empty.

In this textbook we first develop process models based on *fundamental* or first-principles analysis, that is, models that are based on known physical-chemical relationships, such as material and energy balances, as well as reaction kinetics, transport phenomena, and thermodynamic relationships. We then develop *empirical* models. An empirical model is usually developed based on applying input changes to a process and observing the response of measured outputs. Model parameters are adjusted so that the model outputs match the observed process outputs. This technique is particularly useful for developing models that can be used for controller design.

1.4 Control Textbooks and Journals

There are a large number of undergraduate control textbooks that focus on control-system design and theory. The following books include an introduction to process modeling and dynamics, in addition to control system design.

Coughanowr, D.R., *Process Systems Analysis and Control*, 2nd ed., McGraw-Hill, New York (1991).

Luyben, M.L., and W.L. Luyben, *Essentials of Process Control*, McGraw-Hill, New York (1997).

Luyben, W.L., *Process Modeling Simulation and Control for Chemical Engineers*, 2nd ed., McGraw-Hill, New York (1990).

Marlin, T.E., *Process Control: Designing Processes and Control Systems for Dynamic Performance*, 2nd ed., McGraw-Hill, New York (2000).

Ogunnaike, B.A., and W.H. Ray, *Process Dynamics, Modeling and Control*, Oxford, New York (1994).

Riggs, J.B., *Chemical Process Control*, Ferret Publishing, Lubbock, Texas (1999).

Seborg, D.E., T.F. Edgar, and D.A. Mellichamp, *Process Dynamics and Control*, Wiley, New York (1989).

Smith, C.A., and A.B. Corripio, *Principles and Practice of Automatic Process Control*, 2nd ed. Wiley, New York (1997).

Stephanopoulos, G., *Chemical Process Control*, Prentice Hall, Englewood Cliffs, NJ (1984).

Svrcek, W.Y., D.P. Mahoney, and B.R. Young, *A Real-Time Approach to Process Control*, Wiley, Chichester (2000).

The following books are generally more applied, with specific control applications detailed.

Levine, W.S. (ed.), *The Control Handbook*, CRC Press, Boca Raton, FL (1996).

Liptak, B.G., and K.Venczel (eds.), *Instrument Engineers Handbook, Process Control Volume*, Chilton Book Company, Radnor, PA (1985).

Luyben, W.L., B.D. Tyreus, and M.L. Luyben, *Plantwide Process Control*, McGraw-Hill, New York (1999).

Schork, F.J., Deshpande, P.B., and K.W. Leffew, *Control of Polymerization Reactors*, Marcel Dekker, New York (1993).

Shinskey, F.G., *Distillation Control*, McGraw-Hill, New York (1977).

Shunta, J.P., *Achieving World Class Manufacturing Through Process Control*, Prentice Hall, Upper Saddle River, NJ (1995).

The following sources often provide interesting process control problems and solutions.

Advances in Instrumentation and Control (ISA Annual Conference)

American Control Conference (ACC) Proceedings—yearly

Chemical Engineering Magazine (McGraw-Hill)—monthly

Chemical Engineering Progress—monthly

Control Engineering Practice (an IFAC Journal)

Hydrocarbon Processing (petroleum refining and petrochemicals)—monthly

Instrumentation Technology (*InTech*, an instrumentation industry magazine)—monthly

IEEE Control Systems Magazine—bimonthly

ISA (Instrument Society of America) Transactions

TAPPI Journal (pulp and paper industry) —monthly

The following sources tend to be more theoretical but often have useful control-related articles.

American Institute of Chemical Engineers (AIChE) Journal

Automatica (Journal of the International Federation of Automatic Control, IFAC)

Canadian Journal of Chemical Engineering

Chemical Engineering Communications

Chemical Engineering Research and Design

Chemical Engineering Science

Computers and Chemical Engineering

Conference on Decision and Control (CDC) Proceedings—yearly

Industrial and Engineering Chemistry Research (I&EC Research)

IEEE Transactions on Automatic Control

IEEE Transactions on Biomedical Engineering

IEEE Transactions on Control System Technology

International Federation of Automatic Control (IFAC) Proceedings

International Journal of Control

International Journal of Systems Sciences

Journal of Process Control

Proceedings of the IEE (part D, Control Theory and Applications)

1.5 A Look Ahead

Chapter 2 develops fundamental models based on material and energy balances, while Chapter 3 covers dynamic analysis. Chapter 4 shows how to develop empirical models from plant tests. Chapter 5 is an introduction to feedback control and provides the first look at quantitative control-system design procedures.

The best way to understand process control is to work many problems. In particular, it is important to use simulation for complex problems. A numerical package that is particularly useful for control-system analysis and simulation is MATLAB; the SIMULINK block-diagram simulator is particularly useful. If you are not familiar with MATLAB/SIMULINK, we recommend that you work through the MATLAB and SIMULINK tutorials (Modules 1 and 2). Simply reading the tutorials will not give you much insight into the use of MATLAB; you must sit at a computer, work through the examples, and try new ideas that you have.

1.6 Summary

You should now be able to formulate a control problem in terms of the following:

- Control objective
- Inputs (manipulated or disturbance)
- Outputs (measured or unmeasured)
- Constraints (hard or soft)
- Operating characteristics (continuous, batch, semibatch)
- Safety, environmental, and economic issues
- Control structure (feedback, feed forward)

You should also be able to sketch control and instrumentation diagrams, and control block diagrams. In addition, you should be able to recommend whether a control valve should be fail-open or fail-closed.

The following terms were introduced in this chapter:

- Actuator
- Air-to-close
- Air-to-open
- Algorithm
- Control block diagram
- Control valve
- Controller
- Deadtime or time-delay
- Digital
- Fail-closed
- Fail-open
- Gain
- Integrating process

- Model
- Process gain
- Process and instrumentation diagram
- Sensor
- Setpoint

The abstract notions of states, inputs, outputs, and parameters were introduced and are covered in more detail in Chapter 2. The examples used were as follows:

1.1 Surge Tank
1.2 Taking a Shower
1.3 Liquid Surge Vessel Model

Student Exercises

1. Discuss the following problems (a–g) in the context of control:

i. Identify control objectives;
ii. Identify input variables and classify as (a) manipulated or (b) disturbance;
iii. Identify output variables and classify these as (a) measured or (b) unmeasured;
iv. Identify constraints and classify as (a) hard or (b) soft;
v. Identify operating characteristics and classify as (a) continuous, (b) batch, (c) semicontinuous (or semibatch);
vi. Discuss safety, environmental, and economic considerations;
vii. Discuss the types of control (feed forward or feedback).

Measurements and manipulated variables can vary continuously or may be sampled discretely.
Select from the following:

a. Driving a car
b. Choose one of your favorite activities (skiing, basketball, making a cappuccino, etc.)
c. A stirred tank heater
d. Beer fermentation
e. An activated sludge process
f. A household thermostat
g. Air traffic control

2. Literature Review. The process control research literature can be challenging to read, with unique notations and rigorous mathematical analyses. Find a paper from one of the magazines or journals listed in Section 1.4 that you would like to understand by the time you have completed this textbook. You will find many articles to choose from, so use some of the following criteria for your selection:

- The process is interesting to you (do not choose mainly a theory paper)
- The modeling equations and parameters are in the paper (make certain the equations are ordinary differential equations and not partial differential equations)
- There are plots that you can verify (eventually) through simulation (the plots should be based on simulation results)
- The control algorithm is clearly written
- The objectives of the paper are reasonably clear to you

Provide the following:

i. A short (one paragraph) summary of the overall objectives of the paper; why are you interested in the paper?

ii. A short list of words and concepts in the paper that are familiar to you.

Suggested Topics (choose one):

a. Fluidized catalytic cracking unit (FCCU)—petroleum refining
b. Reactive ion etching—semiconductor manufacturing
c. Rotary lime kiln—pulp and paper manufacturing
d. Continuous drug infusion—biomedical engineering and anesthesia
e. Anaerobic digester—waste treatment
f. Distillation—petrochemical and many other industries
g. Polymerization reactor—plastics
h. pH—waste treatment
i. Beer production—food and beverage
j. Paper machine headbox—pulp and paper manufacturing
k. Batch chemical reactor—pharmaceutical production

3. Instrumentation Search. Select one of the following measurement devices and use Internet resources to learn more about it. Determine what types of signals are input to or output from the device. For flow meters, what range of flow rates can be handled by a particular flow meter model?

a. Vortex-shedding flow meters
b. Orifice-plate flow meters
c. Mass flow meters
d. Thermocouple-based temperature measurements

   e. Differential pressure (delta P) measurements
   f. Control valves
   g. pH
4. Work through the Module 1: Introduction to MATLAB.
5. A process furnace heats a process stream from near ambient temperature to a desired temperature of 300°C. The process stream outlet temperature is regulated by manipulating the flow rate of fuel gas to the furnace, as shown below.

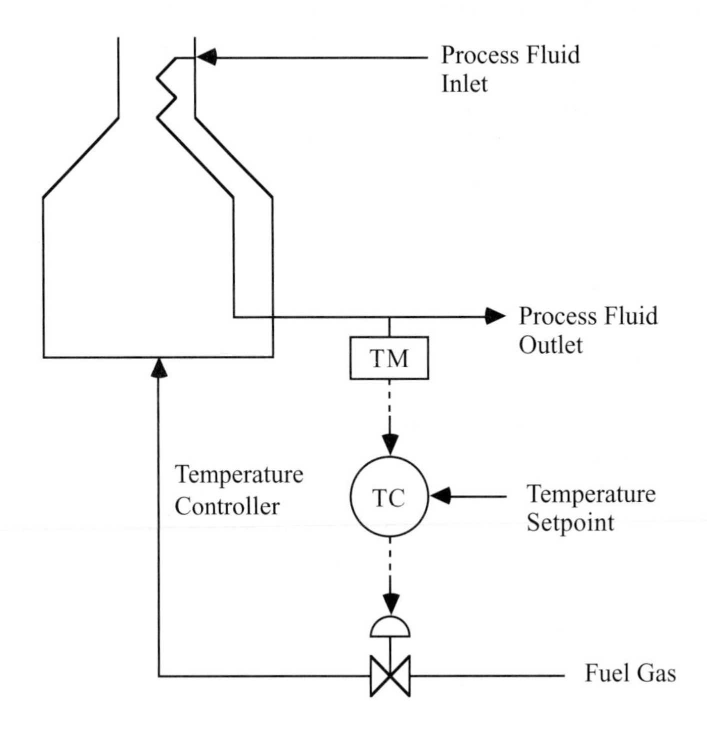

   a. Discuss the objectives of this control strategy.
   b. What is the measured output?
   c. What is the manipulated input?
   d. What are possible disturbances?
   e. Is this a continuous or batch process?
   f. Is this a feed-forward or feedback controller?
   g. Should the control valve fail-open or fail-closed? For the strategy you chose, is the valve gain positive or negative? Why?
   h. Discuss safety, environmental, and economic issues.

i. Draw the control block schematic diagram and label all signals and blocks on the diagram.

6. A fluidized catalytic cracking unit (FCCU) produces a significant portion of the gasoline produced by a typical petroleum refinery. A typical FCCU processes 30,000 Bbl/day (1 Bbl = 42 gallons) of heavy gas oil from a crude oil distillation unit, producing roughly 15,000 Bbl/day of gasoline, along with streams of other components. The value of gasoline alone produced by this unit is on the order of $500,000/day, so any improvement in yield and energy consumption owing to improved control can have a significant economic impact.
Question: A control system revamp for a 30,000 Bbl/day FCCU is estimated to cost $2 million. It is expected that the implementation of advanced control schemes will result in an economic increase of 2% in the value of products produced. Based on the value of gasoline alone, how many days will it take to pay back the control system investment?

7. Furnaces are often used to heat process streams to temperatures above 400°F. A typical fired furnace may have a heat duty of 100×10^6 Btu/hour, requiring roughly 1667 scfm (standard cubic feet per minute) of natural gas (methane has a fuel value of approximately 1000 Btu/scf). The cost of this fuel gas is on the order of $5/1000 scf, yielding an annual fuel cost of $4.4 million/year. Excess combustion air is needed to assure complete combustion; however, too much excess air wastes energy (the heated air simply goes out the exhaust stack). Too little excess air leads to incomplete combustion, wasting energy and polluting the atmosphere with unburned hydrocarbons. It is important, then, to deliver an optimum amount of combustion air to the furnace. With the large flow rates and high temperatures involved, maintaining safe operation is also very important. The control system must be designed so that excess combustion air is maintained, no matter what is happening to the fuel gas flow rate. A fired furnace control system clearly needs to satisfy safety, environmental, and economic criteria.
Question: An advanced control scheme is estimated to save 2% in energy costs, for a fired furnace with a heat duty of 100×10^6 Btu/hour. If it is desired to have a 2-year payback period on this control system investment, what is the maximum investment allowable?

8. Consider the surge vessel process in Example 1.3. If the steady-state volume is 500 liters, and the steady-state inlet and outlet flow rates are 50 liters/minute, find how the liquid volume varies with time if the inlet flow rate is $F_i(t) = 50 + 10\sin(0.1t)$, while the outlet flow rate remains constant at 50 liters/minute.

9. The human body is composed of many innate feedback and feed-forward control loops. For example, insulin is a hormone produced by the pancreas to regulate the blood glucose concentration. The pancreas in a type I (insulin dependent) diabetic has lost the ability to produce significant insulin. An insulin-dependent diabetic must monitor her/his blood glucose (accurate blood glucose strips have been on the market for years) and provide insulin injections several times per day. It is particularly important to use knowledge of the meal characteristics to determine the amount of insulin necessary to compensate for the glucose.

a. Discuss the actions taken by a type I diabetic in terms of the formulation of a control problem. State the objectives and list all variables, etc.

b. It is desirable to form an automated closed-loop system, using a continuous blood glucose measurement and a continuous insulin infusion pump. Draw a "process and instrumentation" diagram and the corresponding control block diagram.

10. Consider the following three heat exchanger control instrumentation diagrams. For each diagram (a, b, and c), the objective is to maintain a desired cold stream outlet temperature. Since the cold stream exiting the exchanger is fed to a reactor, *it is important that the stream temperature never be substantially higher than the setpoint value*. Please answer the two basic questions about each strategy, then the final question (part d).

a. Basic cold stream temperature control strategy.

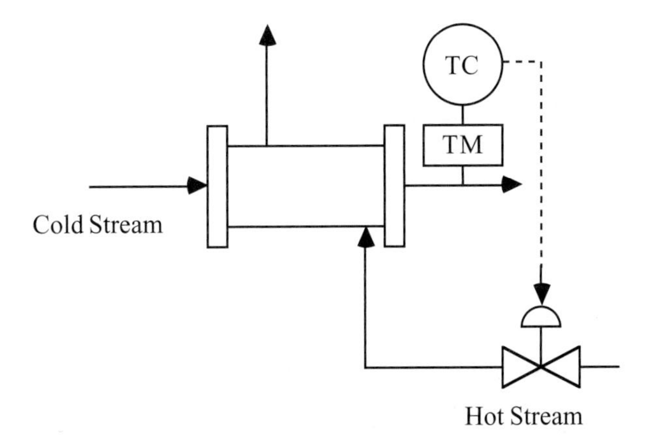

Is the process gain relating the manipulated flow rate to the measured temperature positive or negative?

Should the control valve should be *fail-open* or *fail-closed*? Why?

b. Temperature control using hot stream bypass strategy.

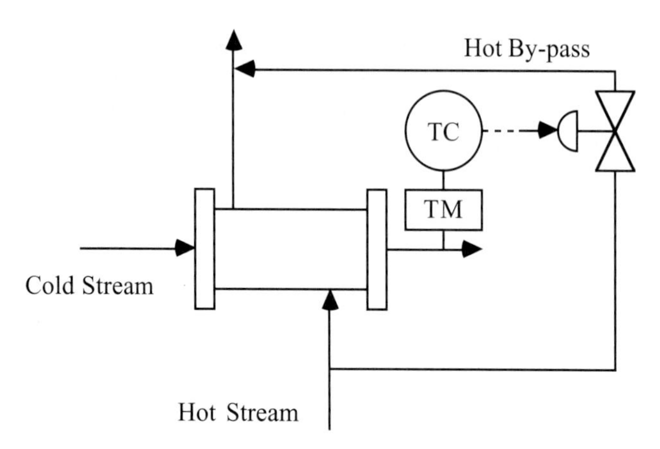

Is the process gain relating the manipulated flow rate to the measured temperature positive or negative?

Should the control valve should be *fail-open* or *fail-closed*? Why?

c. Temperature control using cold stream bypass strategy.

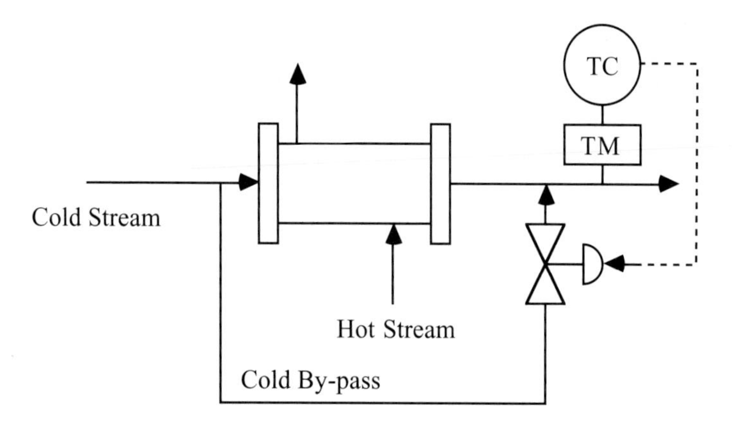

Is the process gain relating the manipulated flow rate to the measured temperature positive or negative?

Should the control valve should be *fail-open* or *fail-closed*? Why?

d. Which strategy (a, b, or c) will have the fastest dynamic behavior? Why?

11. During surgery it is important for an anesthesiologist to regulate a patient's blood pressure to a desired value. She does this by changing the infusion rate

of vasoactive drugs to the patient. In addition to the effect of manipulated vasoactive drugs, blood pressure is affected by the level of anesthetic given to the patient. Discuss actions taken by an anesthesiologist in the context of feedback control. Sketch a control block diagram for an automated system that measures blood pressure and manipulates the infusion rate of a vasoactive drug.

CHAPTER 2

Fundamental Models

In this chapter, a methodology for developing dynamic models of chemical processes is presented. After studying this chapter, the reader should be able to do the following:

- Write balance equations using the integral or instantaneous methods
- Incorporate appropriate constitutive relationships into the equations
- Determine the state, input, and output variables and the parameters for a particular model (set of equations)
- Determine the necessary information to solve a system of dynamic equations
- Linearize a set of nonlinear equations to find the state space model

The major sections of this chapter are as follows:

2.1 Background

Reasons for Modeling

There are many reasons for developing process *models*. Improving or understanding chemical process operation is a major overall objective for developing a dynamic process model. These models are often used for (i) operator training, (ii) process design, (iii) safety system analysis, or (iv) process control.

Operator training: The people responsible for the operation of a chemical manufacturing process are known as *process operators*. A dynamic process model can be used to perform simulations to train process operators, in the same fashion that flight simulators are used to train airplane pilots. Process operators can learn the proper response to upset conditions, before having to experience them on the actual process.

Process design: A dynamic process model can be used to properly design chemical-process equipment for a desired production rate. For example, a model of a batch chemical reactor can be used to determine the appropriate size of the reactor to produce a certain product at a desired rate.

Safety: Dynamic process models can also be used to design safety systems. For example, they can be used to determine how long it will take, after a valve fails, for a system to reach a certain pressure.

Process control: Feedback control systems are used to maintain process variables at desirable values. For example, a control system may measure a product temperature (an output) and adjust the steam flow rate (an input) to maintain that desired temperature. For complex systems, particularly those with many inputs and outputs, it is necessary to base the control-system design on a process model. Also, before a complex control system is implemented on a process, it is normally tested by simulation.

It should be noted that no single model of a process exists, since a model only approximates the process behavior. The desired accuracy and resulting complexity of a process model depends on the final use of the model. Usually more-complex models will require much more data and effort to develop than simplified models, since more model parameters will need to be determined. The focus of this textbook is on process control, so model development is provided with this in mind.

Lumped Parameter System Models

The models developed in this textbook are known as *lumped parameter systems* models. These models consist of initial-value ordinary differential equations, often based on a perfect mixing assumption. The models have the form

$$\dot{x} = f(x, u, p)$$

$$y = g(x, u, p)$$

where x is the vector of state variables, $\dot{x}$ the vector of state variables derivatives with respect to time equal to dx/dt, u the vector of input variables, p the vector of parameters, y the vector of output variables, and, $f(x,u,p)$ and $g(x,u,p)$ the vectors of functions.

State variables are variables that naturally appear in the derivative term of ordinary differential equation models. Common states resulting from overall material balance equations include total mass, volume, level for liquid-phase processes, and pressure for gas-phase processes. Component compositions are the most common states that arise from component material balances. Temperature is the most common state arising from an energy balance modeling equation.

This state-variable representation seems very abstract at this juncture, and it generally takes students some time to become comfortable with it. The easiest way is to work through some simple examples to begin to associate the notion of states, parameters, inputs, and outputs with the physical variables associated with chemical processes. Throughout the text we use matrix and vector notation; you may wish to review any standard linear algebra book to become familiar with this notation. A concise review is also provided in the MATLAB module (Module 1).

2.2 Balance Equations

In this text, we are interested in dynamic balances that have the form

$$\begin{bmatrix} \text{rate of mass or} \\ \text{energy accumulation} \\ \text{in a system} \end{bmatrix} = \begin{bmatrix} \text{rate of mass or} \\ \text{energy entering} \\ \text{a system} \end{bmatrix} - \begin{bmatrix} \text{rate of mass or} \\ \text{energy leaving} \\ \text{a system} \end{bmatrix}$$

This equation is deceptively simple because there may be many in and out terms, particularly for component balances. The in and out terms would then include the generation and conversion of species by chemical reaction, respectively. The rate of mass accumulation in a system has the form dM/dt, where M is the total mass in the system. Similarly, the rate of energy accumulation has the form dE/dt, where E is the total energy in a system. If N_i is used to represent the moles of component i in a system, then dN_i/dt represents the molar rate of accumulation of component i in the system.

When solving a problem, it is important to specify what is meant by system. In some cases the system may be microscopic in nature (a differential element, for example), while in other cases it may be macroscopic in nature (the liquid content of a mixing tank, for example). Also, when developing a dynamic model we can take one of two general viewpoints. One viewpoint is based on an *integral* balance, while the other is based on an *instantaneous* balance. Integral balances are particularly useful when developing models for distributed parameter systems, which result in partial differential equations; the focus in this text is on ordinary differential equation-based models. Another viewpoint is the instantaneous balance where the time rate of change is written directly.

Integral Balances

An integral balance is developed by viewing a system at two different "snapshots" in time. Consider a finite time interval, Δt, and perform a material balance over that time interval,

$$\begin{bmatrix} \text{mass or} \\ \text{energy inside the} \\ \text{system at } t+\Delta t \end{bmatrix} - \begin{bmatrix} \text{mass or} \\ \text{energy inside the} \\ \text{system at } t \end{bmatrix} = \begin{bmatrix} \text{mass or energy} \\ \text{entering the system} \\ \text{from } t \text{ to } t+\Delta t \end{bmatrix} - \begin{bmatrix} \text{mass or energy} \\ \text{leaving the system} \\ \text{from } t \text{ to } t+\Delta t \end{bmatrix}$$

The mean-value theorems of integral and differential calculus are then used to reduce the equations to differential equations. For example, consider the system shown in Figure 2–1, where one boundary represents the mass in the system at time t, while the other boundary represents the mass in the system at $t + \Delta t$.

An integral balance on the total mass in the system is written in the form

$$\begin{bmatrix} \text{mass contained} \\ \text{in the} \\ \text{system at } t+\Delta t \end{bmatrix} - \begin{bmatrix} \text{mass contained} \\ \text{in the} \\ \text{system at } t \end{bmatrix} = \begin{bmatrix} \text{mass entering} \\ \text{the system} \\ \text{from } t \text{ to } t+\Delta t \end{bmatrix} - \begin{bmatrix} \text{mass leaving} \\ \text{the system} \\ \text{from } t \text{ to } t+\Delta t \end{bmatrix}$$

Mathematically this is written as

$$M\big|_{t+\Delta t} - M\big|_t = \int_t^{t+\Delta t} \dot{m}_{in}\,dt - \int_t^{t+\Delta t} \dot{m}_{out}\,dt = \int_t^{t+\Delta t} \left(\dot{m}_{in} - \dot{m}_{out}\right)dt \tag{2.1}$$

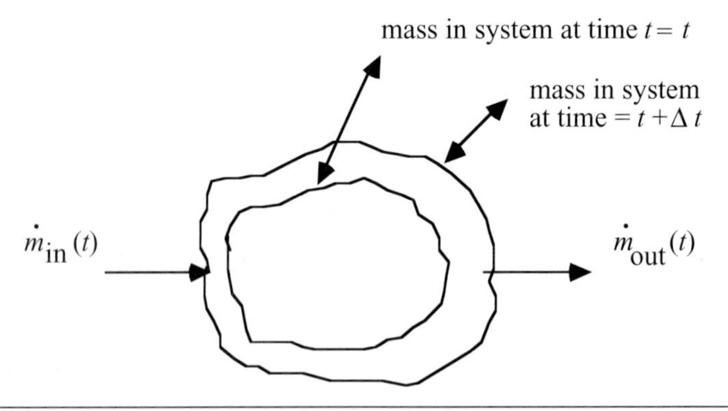

Figure 2–1 Conceptual material balance problem.

where M represents the total mass in the system, while $\dot{m}_{in}$ and $\dot{m}_{out}$ represent the mass rates entering and leaving the system, respectively. We can write the right-hand side of Equation (2.1), using the mean-value theorem of *integral* calculus, as

$$\int_{t}^{t+\Delta t} \left(\dot{m}_{in} - \dot{m}_{out}\right)dt = \left(\dot{m}_{in} - \dot{m}_{out}\right)\Big|_{t+\alpha\Delta t} \cdot \Delta t \tag{2.2}$$

where $0 < \alpha < 1$. Substituting the right-hand side of Equation (2.2) into Equation (2.1), we find

$$M\big|_{t+\Delta t} - M\big|_{t} = \left(\dot{m}_{in} - \dot{m}_{out}\right)\Big|_{t+\alpha\Delta t} \cdot \Delta t \tag{2.3}$$

By dividing Equation (2.3) by Δt, and using the mean value theorem of *differential* calculus ($0 < \beta < 1$) for the left-hand side,

$$\frac{M\big|_{t+\Delta t} - M\big|_{t}}{\Delta t} = \left(\dot{m}_{in} - \dot{m}_{out}\right)\Big|_{t+\alpha\Delta t} \tag{2.4}$$

$$\frac{M\big|_{t+\Delta t} - M\big|_{t}}{\Delta t} = \frac{dM}{dt}\Big|_{t+\beta\Delta t} \tag{2.5}$$

and by substituting Equation (2.5) into Equation (2.4),

$$\frac{dM}{dt}\Big|_{t+\beta\Delta t} = \left(\dot{m}_{in} - \dot{m}_{out}\right)\Big|_{t+\alpha\Delta t} \tag{2.6}$$

and taking the limit as Δt goes to zero, we find

$$\frac{dM}{dt} = \dot{m}_{in} - \dot{m}_{out} \tag{2.7}$$

Representing the total mass as $M = V\rho$, $\dot{m}_{in}$ as $F_{in}\rho_{in}$ and $\dot{m}_{out}$ as $F_{out}\rho$, where V is the volume, ρ is the mass density (mass/volume), and F is a volumetric flow rate (volume/time), we obtain the equation

$$\frac{dV\rho}{dt} = F_{in}\rho_{in} - F_{out}\rho \tag{2.8}$$

Note that we have assumed that the system is *perfectly mixed*, so that the density of material leaving the system is equal to the density of material in the system ($\rho_{out} = \rho$).

Instantaneous Balances

Here we write the dynamic balance equations directly, based on an instantaneous rate-of-change

$$\begin{bmatrix} \text{rate of mass} \\ \text{accumulation} \end{bmatrix} = \begin{bmatrix} \text{rate of mass} \\ \text{in by flow} \end{bmatrix} - \begin{bmatrix} \text{rate of mass} \\ \text{out by flow} \end{bmatrix}$$

$$\frac{dM}{dt} = \dot{m}_{in} - \dot{m}_{out} \tag{2.7}$$

which can also be written as

$$\frac{dV\rho}{dt} = F_{in}\rho_{in} - F_{out}\rho \tag{2.8}$$

This is the same result obtained using an integral balance. Although the integral balance takes longer to arrive at the same result as the instantaneous balance method, the integral balance method is probably clearer when developing distributed parameter (partial differential equation-based) models.

Steady State

At steady state, the derivative with respect to time is zero, by definition, so from Equation (2.7),

$$\dot{m}_{in} = \dot{m}_{out} \tag{2.9}$$

or from Equation (2.8),

$$F_{in}\rho_{in} = F_{out}\rho \tag{2.10}$$

Steady-state relationships are often used for process design and determination of optimal operating conditions.

2.3 Material Balances

The simplest modeling problems consist of material balances. In this section we use two process examples to illustrate the modeling techniques used. Recall that a model for a liquid surge vessel was developed in Chapter 1 (Example 1.3).

Example 2.1: Gas Surge Drum

Surge drums are often used as intermediate storage capacity for gas streams that are transferred between chemical process units. Consider a drum depicted below (Figure 2–2), where q_i is the inlet molar flow rate and q is the outlet molar flow rate. A typical control problem would be to manipulate one flow rate (either in or out) to maintain a desired drum pressure. Here we develop a model that describes how the drum pressure varies with the inlet and outlet flow rates.

Let V = volume of the drum and n = the total amount of gas (moles) contained in the drum.

Assumption: The pressure-volume relationship is characterized by the ideal gas law, $PV = nRT$, where P is pressure, T is temperature (absolute scale), and R is the ideal gas constant.

The rate of accumulation of the mass of gas in the drum is described by the material balance

$$\frac{dnMW}{dt} = q_i MW_i - qMW \tag{2.11}$$

where MW represents the molecular weight. Assuming that the molecular weight is constant, we can write

$$\frac{dn}{dt} = q_i - q \tag{2.12}$$

From the ideal gas law, since V, R, and T are assumed constant,

$$\frac{dn}{dt} = \frac{d(PV/RT)}{dt} = \frac{V}{RT}\frac{dP}{dt}$$

so

$$\frac{V}{RT}\frac{dP}{dt} = q_i - q$$

which can be rewritten

$$\frac{dP}{dt} = \frac{RT}{V}(q_i - q) \tag{2.13}$$

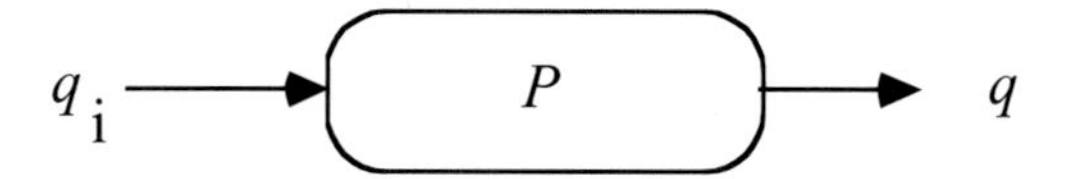

Figure 2–2 Gas surge drum.

To solve this equation for the *state* variable P, we must know the *inputs* q_i and q, the *parameters* R, T, and V, and the *initial condition* $P(0)$. Once again, although q is the molar rate *out* of the drum, we consider it to be an input in terms of solving the model.

It should be noted that just like the liquid level process discussed in Example 1.3, this is an *integrating* system. For example, if the inlet molar flow rate increases while the outlet flow rate stays constant, then the pressure increases without bound. In reality, an increase in pressure would most likely cause an increase in outlet molar flow rate (owing to the increased driving force for flow out of the drum). Indeed, we model that case now.

Outlet Flow as a Function of Gas Drum Pressure

Consider the case where the outlet molar flow rate is proportional to the difference in gas drum pressure and the pressure in the downstream header piping, P_h. Let β represent a flow coefficient. If the flow/pressure difference relationship is linear, then

$$q = \beta(P - P_h) \tag{2.14}$$

So the dynamic modeling equation is

$$\frac{dP}{dt} = \frac{RT}{V}q_i - \frac{RT}{V}\beta(P - P_h) \tag{2.15}$$

At steady state, $dP/dt = 0$, so we find the steady-state relationship

$$q_{is} = \beta(P_s - P_{hs}) \tag{2.16}$$

where we use the subscript s to indicate a steady-state solution. Solving explicitly for P_s, we find

$$P_s = P_{hs} + \frac{q_{is}}{\beta} \tag{2.17}$$

which is a linear relationship. An increase in q_{is} will lead to an increased value of P_s. This type of system is known as *self-regulating*, since a change in an input variable eventually leads to a new steady-state value of the output variable. Contrast self-regulating systems with integrating systems that do not achieve a new steady state (the output "integrates" until a vessel overflows or a tank overpressures).

The modeling equations for Examples 1.3 and 2.1 were based on writing an overall material balance. In the case of a liquid vessel we found that either liquid volume or height could serve as an appropriate state variable. For the gas drum we found that pressure was the most appropriate state variable.

Liquid level and gas pressure vessels represent *inventory* problems, which are integrating by nature. If there is an imbalance in the inlet and outlet flow rates, the inventory material (liquid or gas) can easily increase or decrease beyond desirable limits. It is the independence of the flow rates that can cause this problem. Notice, however, that a feedback controller can be designed to regulate the inventory levels (liquid volume or gas pressure). A feedback controller manipulates a stream flow rate to maintain a desired inventory level.

There are many control loops that a process engineer must consider at the design stage of a process. Because of the critical nature of inventory loops, these must receive the highest level of consideration. In Chapter 15, we find that inventory loops must be closed before other loops are considered.

The next example illustrates the use of modeling for reactor design.

Example 2.2: An Isothermal Chemical Reactor

Ethylene oxide (A) is reacted with water (B) in a continuously stirred tank reactor (CSTR) to form ethylene glycol (P). Assume that the CSTR is maintained at a constant temperature and that the water is in large excess. The stoichiometric equation is

$$A + B \rightarrow P$$

Here we develop a model (Figure 2–3) to find the concentration of each species as a function of time.

Overall Material Balance

The overall mass balance (since the tank is perfectly mixed) is

$$\frac{dV\rho}{dt} = F_i\rho_i - F\rho \tag{2.18}$$

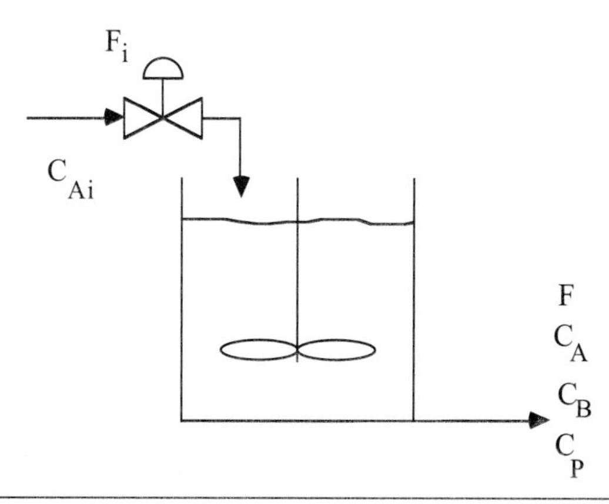

Figure 2–3 Isothermal stirred tank reactor.

Assumption: The liquid-phase density, ρ, is not a function of concentration. The vessel liquid (and outlet) density is then equal to the inlet stream density, so

$$\rho_i = \rho$$

and we can write Equation (2.18) as (notice this is the same result as Example 1.3)

$$\frac{dV}{dt} = F_i - F \tag{2.19}$$

Component Material Balances

It is convenient to work in molar units when writing component balances, particularly if chemical reactions are involved. Let C_A and C_P represent the molar concentrations of A and P (moles/volume). The component material balance equations are

$$\frac{dVC_A}{dt} = F_i C_{Ai} - FC_A + Vr_A \tag{2.20a}$$

$$\frac{dVC_P}{dt} = -FC_P + Vr_P \tag{2.20b}$$

where r_A and r_P represent the rate of *generation* of species A and P per unit volume, and C_{Ai} represents the inlet concentration of species A. Since the water is in large excess its concentration does not change significantly, and the reaction rate is first order with respect to the concentration of ethylene oxide,

$$r_A = -kC_A \tag{2.21}$$

where k is the reaction rate constant and the minus sign indicates that A is consumed in the reaction. Each mole of A reacts with a mole of B (from the stoichiometric equation) and produces one mole of P, so the rate of generation of P (per unit volume) is

$$r_P = kC_A \tag{2.22}$$

Expanding the left-hand side of Equation (2.20a),

$$\frac{dVC_A}{dt} = V\frac{dC_A}{dt} + C_A\frac{dV}{dt} \tag{2.23}$$

Combining Equations (2.19), (2.20a), and (2.23), we find

$$\frac{dC_A}{dt} = \frac{F_i}{V}\left(C_{Ai} - C_A\right) - kC_A \tag{2.24}$$

Similarly, the concentration P can be written as

$$\frac{dC_P}{dt} = -\frac{F_i}{V} C_P + kC_A \tag{2.25}$$

This model consists of three differential equations (2.19, 2.24, 2.25) and, therefore, three state variables (V, C_A, and C_P). To solve these equations we must specify the initial conditions [$V(0)$, $C_A(0)$, and $C_P(0)$], the inputs (F_i, F, C_{Ai}) as a function of time, and the parameter (k).

The state, input, and parameter vectors are

$$x = \begin{bmatrix} x_1 \\ x_2 \\ x_3 \end{bmatrix} = \begin{bmatrix} V \\ C_A \\ C_P \end{bmatrix} \qquad u = \begin{bmatrix} u_1 \\ u_2 \\ u_3 \end{bmatrix} = \begin{bmatrix} F_i \\ F \\ C_{Ai} \end{bmatrix} \qquad p = [p_1] = k \tag{2.26}$$

Using state-variable notation, the model has the form

$$\begin{aligned} \dot{x}_1 &= dx_1/dt = u_1 - u_2 = f_1(x,u,p) \\ \dot{x}_2 &= dx_2/dt = \frac{u_1}{x_1}(u_3 - x_2) - p_1 x_2 = f_2(x,u,p) \\ \dot{x}_3 &= dx_3/dt = -\frac{u_1}{x_1} x_3 + p_1 x_2 = f_3(x,u,p) \end{aligned} \tag{2.27}$$

or,

$$\begin{bmatrix} \dot{x}_1 \\ \dot{x}_2 \\ \dot{x}_3 \end{bmatrix} = \begin{bmatrix} dx_1/dt \\ dx_2/dt \\ dx_3/dt \end{bmatrix} = \begin{bmatrix} f_1(x,u,p) \\ f_2(x,u,p) \\ f_3(x,u,p) \end{bmatrix} = \begin{bmatrix} u_1 - u_2 \\ (u_1/x_1)(u_3 - x_2) - p_1 x_2 \\ -(u_1/x_1)x_3 + p_1 x_2 \end{bmatrix} \tag{2.28}$$

Simplifying Assumptions The reactor model presented in Example 2.2 has three differential equations. Often other simplifying assumptions are made to reduce the number of differential equations to make them easier to analyze and faster to solve. Assuming a constant volume ($dV/dt = 0$), perhaps owing to a feedback controller, reduces the number of equations by one.

The resulting differential equations (since we assumed $dV/dt = 0$, $F = F_i$) are

$$\begin{aligned} \frac{dC_A}{dt} &= \frac{F}{V}(C_{Ai} - C_A) - kC_A \\ \frac{dC_P}{dt} &= -\frac{F}{V} C_P + kC_A \end{aligned} \tag{2.29}$$

Steady-State Solution At steady state, we find the following relationships (where the subscript s represents a steady-state solution):

$$C_{As} = \frac{(F_s/V)C_{Ais}}{(F_s/V)+k} = \frac{C_{Ais}}{(kV/F_s)+1}, \quad C_{Ps} = \frac{kC_{Ais}}{(F_s/V)+k} = \frac{(kV/F_s)C_{Ais}}{(kV/F_s)+1} \tag{2.30}$$

Notice that the concentrations are a function of the *space velocity* (F_s/V), which has units of inverse time. The space velocity can be thought of as the number of reactor volumes that "change over" per unit time. It is inversely related to the fluid residence time (V/F_s), which has units of time and can be thought of as the average time that an element of fluid spends in the reactor.

The concept of conversion is important in chemical-reaction engineering. The conversion of reactant A is defined as the fraction of the feed-stream component that is reacted.

$$X = \frac{C_{Ai} - C_A}{C_{Ai}} \tag{2.31}$$

So, from Equations (2.30) and (2.31), we find that the conversion is related to the space velocity,

$$X = \frac{C_{Ai} - C_A}{C_{Ai}} = \frac{(kV/F_s)}{(kV/F_s)+1} \tag{2.32}$$

Notice that the conversion is a function of the dimensionless term kV/F_s, which is known as the *Damkohler number*. The Damkohler number is the ratio of the characteristic residence time to the characteristic reaction time and is widely used by chemical-reaction engineers to understand reactor behavior. Two different chemical-reaction systems can have the same conversion if their Damkohler numbers are the same. A system with a large rate constant and low residence time can have the same conversion as a system with a small rate constant and high residence time.

Numerical Example Using an Experimentally Determined Rate Constant Laboratory chemists have determined that the reaction rate constant at 55°C is $k = 0.311\ \text{min}^{-1}$. Here we find the steady-state concentrations of ethylene oxide (A) and ethylene glycol (P) as a function of the steady-state space velocity and residence time. The plots in Figure 2–4 all illustrate the same basic concept. On the left-hand plots, the independent variable is the space velocity, while the right-hand plots have residence time as the independent variable. The top plots have concentrations as the dependent variables, while the bottom plots have conversion as the dependent variable. At low space velocities (large residence

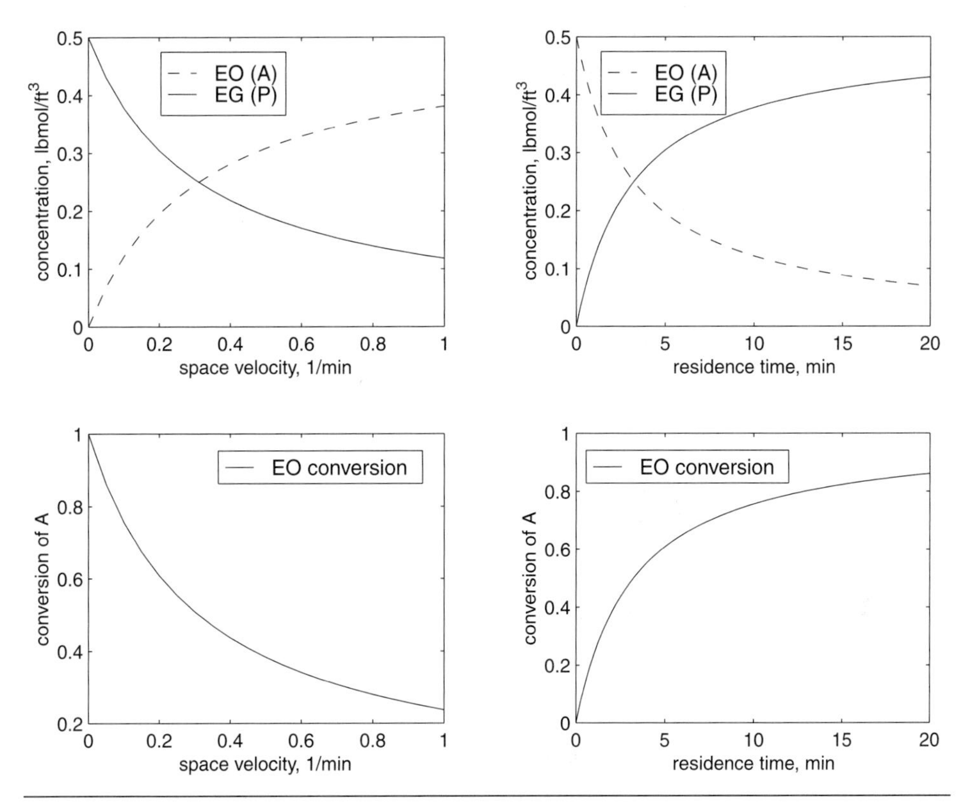

Figure 2–4 Steady-state relationships for ethylene glycol reactor.

times) there is nearly complete conversion of ethylene oxide to ethylene glycol, while at high space velocities (low residence times) there is little conversion.

Design Objective It is desired to produce 100 million pounds per year of ethylene glycol. The feed-stream concentration is 0.5 lbmol/ft^3 and an 80% conversion of ethylene oxide has been determined to be reasonable. What volume of reactor should be specified to meet the production rate requirement? Since process plants often have a shutdown period every 18 months or so, assume 350 days/year of operation.

The design flow-rate calculation is shown below. Since 80% of the ethylene oxide is converted to ethylene glycol, the ethylene glycol concentration is 0.4 lbmol/ft^3 [see Equation (2.32)]. Since the molecular weight is 62 lb/lbmol, the mass concentration is 24.8 lb/ft^3.

The operating flow rate is

$$100 \cdot 10^6 \ \frac{\text{lb}}{\text{yr}} \cdot \frac{\text{ft}^3}{24.8\ \text{lb}} \cdot \frac{\text{yr}}{350\ \text{days}} \cdot \frac{\text{day}}{24\ \text{hr}} \cdot \frac{\text{hr}}{60\ \text{min}} = 8\ \frac{\text{ft}^3}{\text{min}}$$

Solving Equation (2.32) for reactor volume, we find that the required volume is 102.9 ft^3 or 769 gallons. It should be noted that reactors of this size range can be purchased in standard sizes. Most likely the engineer would have a choice of 750-gallon or 1000-gallon models and would choose the 1000-gallon model for expansion capability. Larger scale reactors (greater than roughly 10,000–20,000 gallons) are usually special orders involving on-site construction (or off-site with rail or truck delivery). For the remainder of this problem we assume that the reactor is operated with a volume of 769 gallons, regardless of its maximum capacity.

Dynamic Response Assume that a control strategy will be specified to maintain the desired ethylene glycol concentration in the reactor by manipulating the reactor feed flow rate. In order to design the controller, it is important to understand the dynamic response between an input change and the observed output(s). A step change of 5% in the space velocity (F/V) yields the responses in the ethylene oxide and ethylene glycol concentrations shown in Figure 2–5. An increase in the space velocity (corresponding to a decrease in residence time) results in a decrease in the conversion of A to P. We also see that it takes roughly 10 minutes for the reactor concentrations to achieve new steady-state values. These simulations were performed by integrating differential Equations (2.29) using the techniques presented in Section 2.6.

Examples 2.1 and 2.2 illustrate the use of material balances to develop models. In the gas drum example, the state variable of interest was the drum pressure. In the isothermal ethylene glycol reactor, the state variables of interest were the concentrations of ethylene oxide and ethylene glycol. Material balance equations are rarely adequate to develop most models of interest. In Section 2.5, we review the development of energy balance models, where temperature is often a state variable. First, however, we cover the basic idea of constitutive relationships in Section 2.4.

2.4 Constitutive Relationships

Examples 2.1 and 2.2 required more than simple material balances to define the modeling equations. These required relationships are known as *constitutive* equations; several examples of constitutive equations are shown in this section.

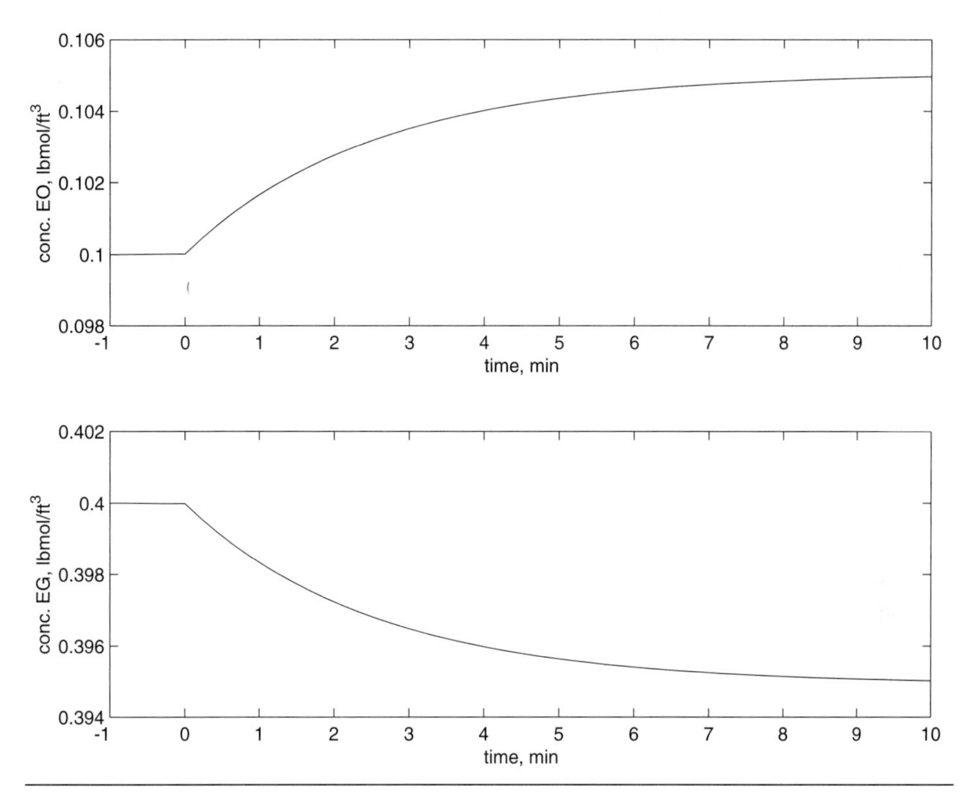

Figure 2–5 Response of ethylene oxide and ethylene glycol concentrations to a step change in space velocity of 5% (from $F/V = 0.0778\text{–}0.0817\ \text{min}^{-1}$).

Gas Law

Process systems containing a gas will often need a gas-law expression in the model. The *ideal gas law* is commonly used to relate pressure (P), molar volume ($\hat{V}$), and temperature (T):

$$P\hat{V} = RT \tag{2.33}$$

The *van der Waal's* $P\hat{V}T$ relationship contains two parameters (a and b) that are system specific:

$$\left(P + \frac{a}{\hat{V}^2}\right)\left(\hat{V} - b\right) = RT \tag{2.34}$$

For other gas laws, see a thermodynamics text, such as Smith, Van Ness, and Abbott (2001).

Chemical Reactions

The rate of reaction per unit volume (mol/volume*time) is usually a function of the concentration of the reacting species. For example, consider the reaction $A + 2B \rightarrow C + 3D$. If the rate of the reaction of A is first order in both A and B, we use the expression

$$r_A = -kC_AC_B \tag{2.35}$$

where r_A is the rate of reaction of A (mol A/volume · time), k the reaction rate constant, C_A the concentration of A (mol A/volume), and C_B the concentration of B (mol B/volume).

Reaction rates are normally expressed in terms of generation of a species. The minus sign indicates that A is consumed in the reaction above. It is good practice to associate the units with all parameters in a model. For consistency in the units for r_A, we find that k has units of (vol/mol B * time). Notice that 2 mol of B react for each 1 mol of A. Then we can write

$$r_B = 2r_A = -2kC_AC_B$$
$$r_C = -r_A = kC_AC_B$$
$$r_D = -3r_A = 3kC_AC_B$$

Usually, the reaction rate coefficient is a function of temperature. The most commonly used representation is the Arrhenius rate law,

$$k(T) = k_0 \exp(-E/RT) \tag{2.36}$$

where $k(T)$ is the reaction rate constant, as a function of temperature, k_0 the frequency factor or preexponential factor, E the activation energy (cal/gmol), R the ideal gas constant (1.987 cal/gmol K), and T the absolute temperature scale (K or R.)

The frequency factor and activation energy can be estimated based on data of the reaction constant as a function of reaction temperature. Taking the natural logarithm of the Arrhenius rate law, we find

$$\ln k = \ln k_0 - \frac{E}{R}\left(\frac{1}{T}\right) \tag{2.37}$$

and we see that k_0 and E can be found from the slope and intercept of a plot of ($\ln k$) vs. ($1/T$).

Equilibrium Relationships

The relationship between the liquid- and vapor-phase compositions of component i, when the phases are in equilibrium, can be represented by

$$y_i = K_i x_i \tag{2.38}$$

where y_i is the vapor-phase mole fraction of component i, x_i the liquid-phase mole fraction of component i, and K_i the vapor/liquid equilibrium constant for component i.

The equilibrium constant is a function of composition and temperature. The simplest assumption for the calculation of an equilibrium constant is to use Raoult's law. Here,

$$K_i = \frac{P_i^{sat}}{P} \tag{2.39}$$

where the pure component vapor (saturation) pressure often has the following form:

$$\ln K_i = A_i - \frac{B_i}{T + C_i} \tag{2.40}$$

Often, we will see a *constant relative volatility* assumption made, to simplify vapor/liquid equilibrium models. In a binary system, the relationship often used between the vapor and liquid phases for the light component is

$$y = \frac{\alpha x}{1 + (\alpha - 1)x} \tag{2.41}$$

where x is the liquid-phase mole fraction of light component, y the vapor-phase mole fraction of light component, and α the relative volatility ($\alpha > 1$).

Heat Transfer

The rate of heat transfer through a vessel wall separating two fluids (a jacketed reactor, for example) can be described by

$$Q = UA\Delta T \tag{2.42}$$

where Q is the rate of heat transfer from hot to cold fluid, U the overall heat transfer coefficient, A the area for heat transfer, and ΔT the difference between hot and cold temperatures.

At the design stage the overall heat transfer coefficient can be estimated from correlations; it is a function of fluid properties and velocities. The individual film heat transfer coefficients (h_i and h_o), the metal conductivity (k, and thickness, Δx), and a fouling factor (f) can be used to determine the overall heat transfer coefficient from the relationship

$$\frac{1}{U} = \frac{1}{h_o} + \frac{1}{h_i} + \frac{\Delta x}{k} + f \tag{2.43}$$

The individual film coefficients are a strong function of fluid properties and velocities. The overall heat transfer coefficient is often estimated from experimental data.

Flow Through Valves

The flow through valve is often described by the relationship

$$F = C_v f(x) \sqrt{\frac{\Delta P_v}{s.g.}} \tag{2.44}$$

where F is the volumetric flow rate, C_v the valve coefficient, x the fraction the valve is open ($0 \leq x \leq 1$), ΔP_v the pressure drop across the valve, $s.g.$ the specific gravity of the fluid, and $f(x)$ the flow characteristic (varies from 0 to 1, as a function of x).

Three common valve characteristics are (i) linear, (ii) equal-percentage, and (iii) quick-opening. For a linear valve, $f(x) = x$. For an equal-percentage valve, $f(x) = \alpha^{x-1}$. For a quick-opening valve, $f(x) = \sqrt{x}$. These flow characteristics are plotted in Figure 2–6.

Notice that for the quick-opening valve, the sensitivity (or "gain") of flow to valve position is high at low openings and low at high openings; the opposite is true for an equal-percentage valve. The sensitivity of a linear valve does not change as a function of valve position. The equal-percentage valve is commonly used in chemical processes because of desirable characteristics when installed in piping systems where a significant piping pressure drop occurs at high flow rates. Knowledge of these characteristics will be important when developing feedback control systems. Flow control is discussed in detail in Module 15.

2.5 Material and Energy Balances

Section 2.3 covered models which consist of material balances only. These are useful if thermal effects are not important, where system properties, reaction rates, etc. do not depend on temperature, or if the system is truly isothermal (constant temperature). Many

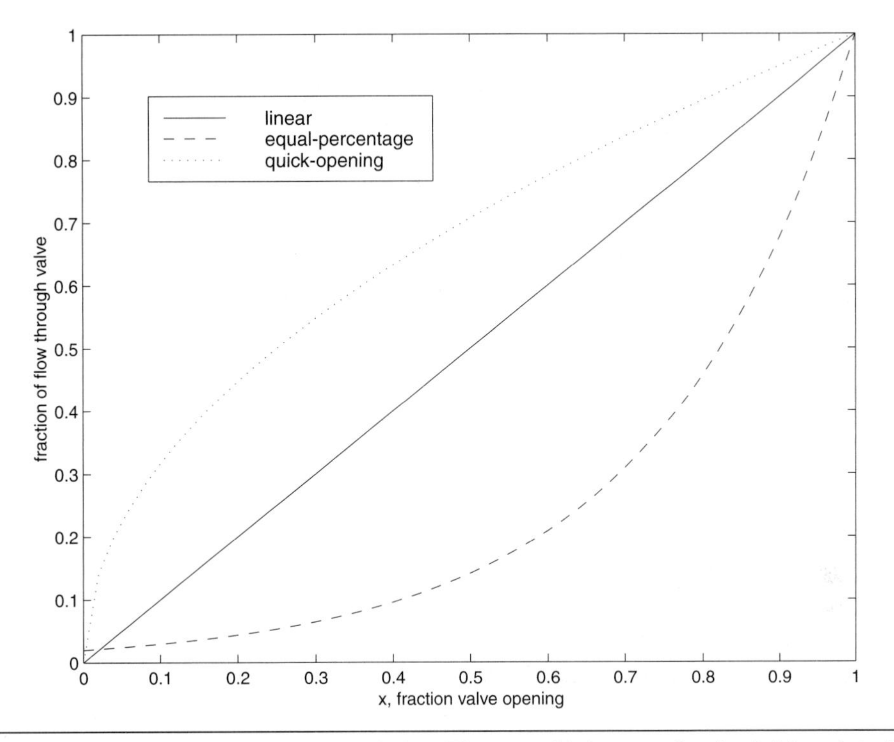

Figure 2–6 Valve flow characteristics. The sensitivity or valve "gain" is related to the slope of the curve.

chemical processes have important thermal effects, so it is necessary to develop material and energy balance models. One key is that a basis must always be selected when evaluating an intensive property such as enthalpy.

Review of Thermodynamics

Developing correct energy balance equations is not trivial and the chemical engineering literature contains many incorrect derivations. Chapter 5 of the book by Denn (1986) points out numerous examples where incorrect energy balances were used to develop process models.

The total energy (TE) of a system consists of internal (U), kinetic (KE), and potential energy (PE),

$$TE = U + KE + PE \tag{2.45}$$

where the kinetic and potential energy terms are

$$KE = \frac{1}{2}mv^2$$

$$PE = mgh$$

For most chemical processes where there are thermal effects, the kinetic and potential energy terms can be neglected, because their contribution is generally at least an order of magnitude less than that of the internal energy term.

When dealing with flowing systems, we usually work with enthalpy. Total enthalpy is defined as

$$H = U + pV \tag{2.46}$$

The heat capacity is defined as the partial derivative of enthalpy with respect to temperature, at constant pressure. The heat capacity, on a unit mass basis, is

$$c_p = \left.\frac{\partial \overline{H}}{\partial T}\right|_p \tag{2.47}$$

where the overbar indicates that the enthalpy is on a unit mass basis. We make use of this relationship in the following example.

The goal of the following example is severalfold:

- develop a model consisting of both material and energy balances
- illustrate the steady-state effect of the input on the output
- illustrate the effect of process "size" (magnitude of flow rate, for example)
- illustrate dynamic behavior

Example 2.3: Heated Mixing Tank

Consider a perfectly mixed stirred-tank heater, with a single feed stream and a single product stream, as shown below. Assuming that the flow rate and temperature of the inlet stream can vary, that the tank is perfectly insulated, and that the rate of heat added per unit time (Q) can vary, develop a model (Figure 2–7) to find the tank liquid temperature as a function of time.

Material Balance

$$\frac{dV\rho}{dt} = F_i\rho_i - F\rho$$

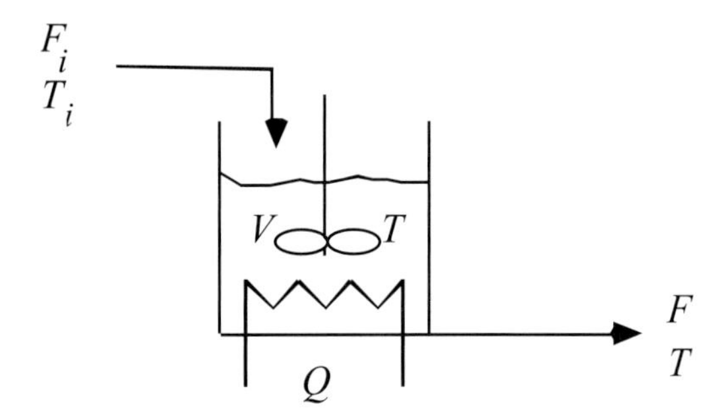

Figure 2–7 Stirred-tank heater.

Neglecting changes in density due to temperature, we find

$$\frac{dV}{dt} = F_i - F \tag{2.48}$$

Energy Balance

Here we neglect the kinetic and potential energy contributions,

$$\frac{dU}{dt} = F_i \rho_i \overline{U}_i - F\rho\overline{U} + Q + W_T \tag{2.49}$$

We write the total work done on the system as a combination of the shaft work (W_S) and the energy added to the system to get the fluid into the tank and the energy that the system performs on the surroundings to force the fluid out.

$$W_T = W_S + F_i p_i - Fp \tag{2.50}$$

This allows us to write Equation (2.49) as

$$\frac{dU}{dt} = F_i \rho_i \left(\overline{U}_i + \frac{p_i}{\rho_i} \right) - F\rho \left(\overline{U} + \frac{p}{\rho} \right) + Q + W_S \tag{2.51}$$

and since $H = U + pV$, and $\overline{H} = \overline{U} + p\overline{V} = \overline{U} + p/\rho$, we can rewrite Equation (2.51) as

$$\frac{dH}{dt} - \frac{dpV}{dt} = F_i \rho_i \overline{H}_i - F\rho\overline{H} + Q + W_S$$

Neglecting pressure*volume changes, we find

$$\frac{dH}{dt} = F_i \rho_i \overline{H}_i - F\rho\overline{H} + Q + W_S \tag{2.52}$$

We must remember the assumptions that went into the development of Equation (2.52).

- The kinetic and potential energy effects were neglected.
- The change in the pV term was neglected.

The total enthalpy term is

$$H = V\rho\overline{H}$$

and assuming no phase change, we select an arbitrary reference temperature (T_{ref}) for enthalpy

$$\overline{H}(T) = \int_{T_{ref}}^{T} c_p dT \tag{2.53}$$

Often we assume that the heat capacity is constant, or calculated at an average temperature, so

$$\overline{H}(T) = c_p\left(T - T_{ref}\right) \tag{2.54a}$$

$$\overline{H}_i(T) = c_p\left(T_i - T_{ref}\right) \tag{2.54b}$$

We now write the energy balance (2.52) in the following fashion,

$$\frac{dV\rho c_p\left(T - T_{ref}\right)}{dt} = F_i\rho_i c_p\left(T_i - T_{ref}\right) - F\rho c_p\left(T - T_{ref}\right) + Q + W_S$$

Expanding the derivative term and assuming that the density is constant, we have

$$V\rho c_p \frac{d\left(T - T_{ref}\right)}{dt} + \rho c_p\left(T - T_{ref}\right)\frac{dV}{dt} = F_i\rho c_p\left(T_i - T_{ref}\right) - F\rho c_p\left(T - T_{ref}\right) + Q + W_S$$

or from Equation (2.48)

$$V\rho c_p \frac{d\left(T - T_{ref}\right)}{dt} + \rho c_p\left(T - T_{ref}\right)\left(F_i - F\right) = F_i\rho c_p\left(T_i - T_{ref}\right) - F\rho c_p\left(T - T_{ref}\right) + Q + W_S$$

Canceling common terms gives

$$V\rho c_p \frac{d\left(T - T_{ref}\right)}{dt} = F_i\rho c_p\left(T_i - T\right) + Q + W_S \tag{2.55}$$

but T_{ref} is a constant, so $d(T - T_{ref})/dt = dT/dt$. Also, neglecting W_S (which is significant only for very viscous fluids), we can write

$$V\rho c_p \frac{dT}{dt} = F_i\rho c_p\left(T_i - T\right) + Q \tag{2.56}$$

which yields the two modeling equations

$$\frac{dV}{dt} = F_i - F \tag{2.57}$$

$$\frac{dT}{dt} = \frac{F_i}{V}(T_i - T) + \frac{Q}{V\rho c_p} \tag{2.58}$$

In order to solve this problem, we must specify the parameters ρ and c_p, the inputs F_i, F, Q, and T_i (as a function of time), and the initial conditions $V(0)$ and $T(0)$.

Steady-State Behavior and the Effect of Scale (Size)

The steady-state solution can be found by setting the derivative terms in Equations (2.57) and (2.58) to 0. The resulting relationship between the manipulated power and the outlet temperature is

$$T_s = T_{is} + \frac{Q_s}{F_s \rho c_p} \tag{2.59}$$

where the subscript s is used to indicate a steady-state value. Notice that for a given steady-state flow rate, the relationship between heater power and outlet temperature is linear. Also, the volume of the vessel has no effect on the steady-state temperature (volume has a solely dynamic effect).

Here we consider a stream of water entering a stirred-tank heater at 20°C, at three possible flow rates: 1 liter/minute (espresso machine), 10 liters/minute (household shower), and 100 liters/minute (small car wash). The outlet temperature as a function of heater power [Equation (2.59)] is plotted in Figure 2–8 for each of the three cases. As expected from Equation (2.59), the curves are linear. The lower flow rate operation has a high sensitivity (slope) of temperature to power, while the higher flow rate operation has a low sensitivity. This makes physical sense, because a given change in power will have a much larger affect on a low flow rate than a high flow rate stream.

This sensitivity is also known as the process gain and is defined as the partial derivative of the output with respect to the input, evaluated at steady state.

$$\frac{\partial T_s}{\partial Q_s} = \frac{1}{F_s \rho c_p} \tag{2.60}$$

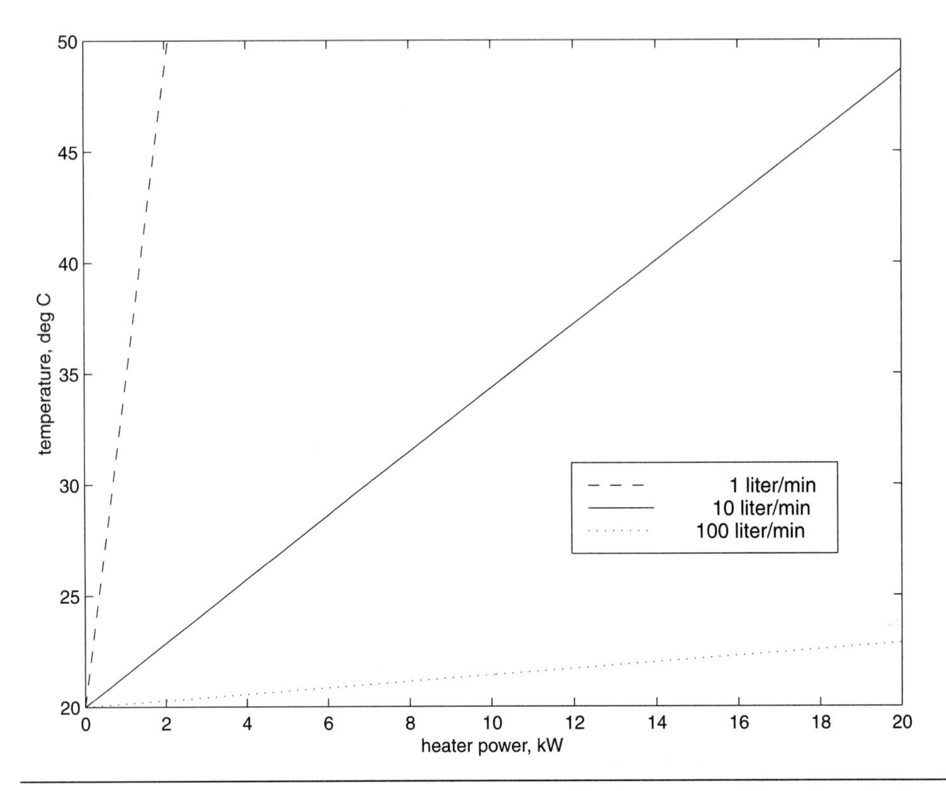

Figure 2–8 Outlet temperature as a function of heater power. The slope is the sensitivity (also known as the "gain") of the output with respect to the input.

It is clear from Equation (2.60) that larger flow rates will have smaller gains (slopes or sensitivities). It is often useful to work with scaled variables. For example, if a scaled steady-state input is defined as

$$Q^{scaled} = \frac{Q_s}{F_s \rho c_p} \tag{2.61}$$

then all three input flow rates have the same steady-state sensitivity of the output to the scaled input. This is shown in Figure 2–9 and the following equation:

$$T_s = T_{is} + Q_s^{scaled} \tag{2.62}$$

The discussion thus far has centered on the steady-state behavior of stirred-tank heaters, and we found that the volume had no effect. The volume has a major impact, however, on the dynamic behavior of a stirred-tank heater.

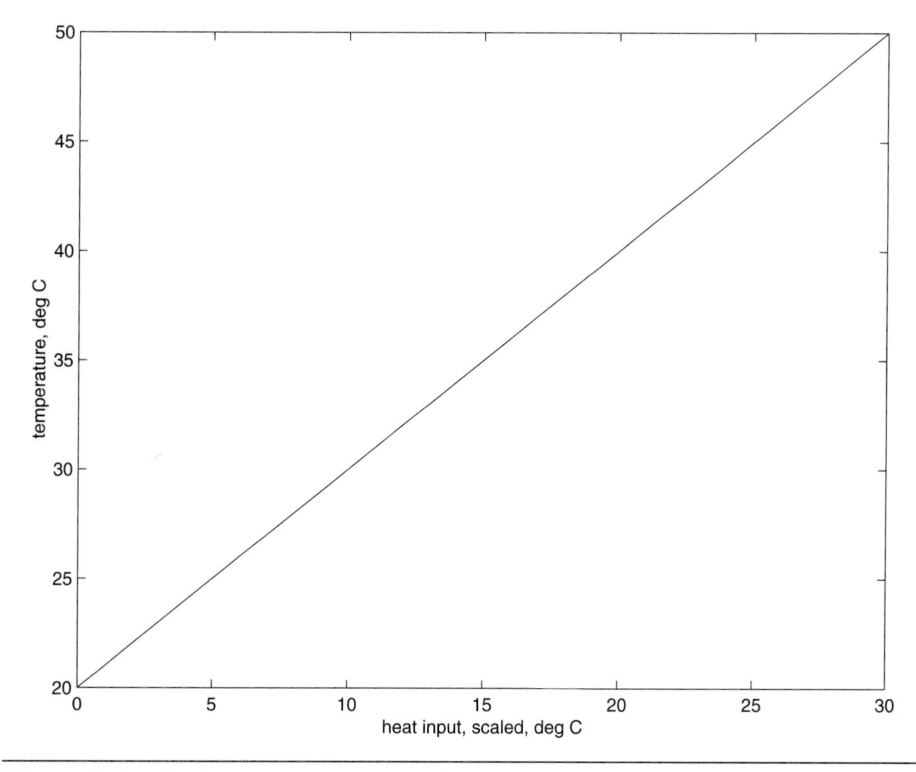

Figure 2–9 Outlet temperature as a function of scaled heater power.

Dynamic Behavior

Volume has a considerable effect on the dynamic behavior of this process. The response of the temperature to a step change in the scaled heat input is shown in Figure 2–10, as a function of the residence time (V/F). As expected, longer residence times have a slower response time than shorter residence-time systems. These curves were obtained by integrating Equation (2.58) for the three different residence times (with V assumed constant). The initial steady-state values are $T = T_i = 20°\text{C}$ and $Q^{\text{scaled}} = 0$. At $t = 0$, Q^{scaled} is stepped from 0° to 1°C.

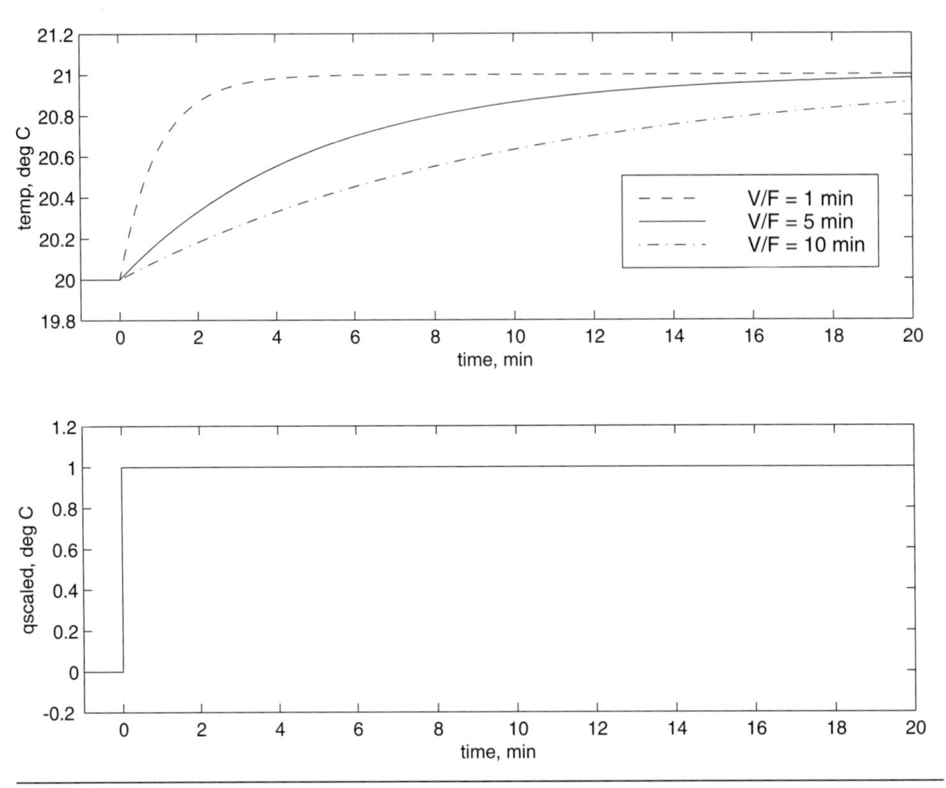

Figure 2–10 Response of temperature to step change in scaled heat addition rate.

2.6 Form of Dynamic Models

The dynamic models derived in this chapter consist of a set of first-order (only first derivatives with respect to time), nonlinear, explicit, initial-value, ordinary differential equations. A representation of a set of first-order differential equations is

$$\begin{aligned}
\dot{x}_1 &= f_1\left(x_1,\ldots,x_{nx},u_1,\ldots,u_{nu},p_1,\ldots,p_{np}\right) \\
\dot{x}_2 &= f_2\left(x_1,\ldots,x_{nx},u_1,\ldots,u_{nu},p_1,\ldots,p_{np}\right) \\
&\vdots \\
\dot{x}_{nx} &= f_{nx}\left(x_1,\ldots,x_{nx},u_1,\ldots,u_{nu},p_1,\ldots,p_{np}\right)
\end{aligned} \tag{2.63}$$

where x_i is a state variable, u_i is an input variable, and p_i is a parameter. The notation $\dot{x}_i$ is used to represent dx_i/dt. Notice that there are nx equations, nx state variables, nu inputs, and np parameters.

Also included in these models is a set of algebraic equations, relating states, inputs, and parameters to output variables.

$$
\begin{aligned}
y_1 &= g_1\left(x_1,\ldots,x_{nx},u_1,\ldots,u_{nu},p_1,\ldots,p_{np}\right)\\
y_2 &= g_2\left(x_1,\ldots,x_{nx},u_1,\ldots,u_{nu},p_1,\ldots,p_{np}\right)\\
&\vdots\\
y_{ny} &= g_{ny}\left(x_1,\ldots,x_{nx},u_1,\ldots,u_{nu},p_1,\ldots,p_{np}\right)
\end{aligned}
\tag{2.64}
$$

State Variables

A state variable is a variable that arises naturally in the accumulation term of a dynamic material or energy balance. A state variable is a measurable (at least conceptually) quantity that indicates the state of a system. For example, temperature is the common state variable that arises from a dynamic energy balance. Concentration is a state variable that arises when dynamic component balances are written.

Input Variables

An input variable is a variable that normally must be specified before a problem can be solved or a process can be operated. Inputs are normally specified by an engineer, based on knowledge of the process being considered. Input variables typically include flow rates of streams entering or leaving a process (notice that the flow rate of an outlet stream might be considered an input variable!). Compositions or temperatures of streams entering a process are also typical input variables. Input variables are often manipulated (by process controllers) in order to achieve desired performance.

Parameters

A parameter is typically a physical or chemical property value that must be specified or known to mathematically solve a problem. Parameters are often fixed by nature, that is, the reaction chemistry, molecular structure, existing vessel configuration, operation, and so forth. Examples include density, viscosity, thermal conductivity, heat transfer coefficient, and mass-transfer coefficient. When designing a process, a parameter might be "adjusted" to achieve some desired performance. For example, reactor volume may be an important design parameter.

Output Variables

An output variable is often a state variable that is measured, particularly for control purposes. Very often the measured outputs are simply a subset of the state variables. Other times the outputs are a nonlinear function of the states (or even inputs).

Vector Notation

The set of differential and algebraic Equations (2.63) and (2.64) can be written more compactly in vector form.

$$\begin{aligned} \dot{x} &= f(x, u, p) \\ y &= g(x, u, p) \end{aligned} \tag{2.65}$$

where x is the vector of state variables, u the vector of input variables, p the vector of parameters, and y the vector of output variables.

Steady-State Solutions

Notice that dynamic models (2.63) can also be used to solve steady-state problems, since

$$\dot{x} = \frac{dx}{dt} = 0$$

That is,

$$f(x, u, p) = 0 \tag{2.66}$$

for steady-state processes, resulting in a set of algebraic equations. In this case, all inputs and parameters would be specified, leaving the nx state values to be solved for; that is, nx equations in nx unknowns must be solved. The focus of this text is not on the development of numerical methods, so we briefly cover the basic idea in Appendix 2.1. Note that differential equation solvers can also be used to solve for the steady state of stable systems, by simply integrating from an initial value for the states for a long period of time, until a steady state is reached.

Numerical Integration

Here we briefly consider numerical methods to integrate ordinary differential equations.

$$\dot{x} = f(x, u, p) \tag{2.67}$$

if the derivative term is approximated (where k represents a time index) as

$$\dot{x} = \frac{dx}{dt} \approx \frac{x(k+1) - x(k)}{t(k+1) - t(k)} = \frac{x(k+1) - x(k)}{\Delta t} \tag{2.68}$$

The explicit *Euler* integration technique involves specifying the integration step size, Δt,

$$x(k+1) = x(k) + \Delta t\, f[x(k)] \tag{2.69}$$

and marching sequentially from one time step to another. This approach is illustrated in Appendix 2.2. In practice, more accurate integration routines using a variable step size are used. For more details on how to use MATLAB integration routines, see Module 3.

2.7 Linear Models and Deviation Variables

Consider the stirred-tank heater model (Example 2.3), when the volume, flow rate, and inlet temperature are constant at their steady-state values (indicated by the subscript s):

$$\frac{dT}{dt} = \frac{F_s}{V_s}(T_{is} - T) + \frac{Q}{V_s \rho c_p} \tag{2.70}$$

Deviation Variable Formulation

Control engineers like to think in terms of "deviation variables," that is, perturbations from a steady-state operating condition. The reader should show that if we define the following deviation variables

$$\begin{aligned} u &= Q - Q_s \\ y &= T - T_s \end{aligned} \tag{2.71}$$

then Equation (2.70) can be written in the form

$$\frac{V_s}{F_s}\frac{dy}{dt} = -y + \frac{1}{F_s \rho c_p} u \tag{2.72}$$

or

$$\tau_p \frac{dy}{dt} = -y + k_p u \tag{2.73}$$

where the new parameters that appear are

$$k_p = \frac{1}{F_s \rho c_p} = \text{process gain}$$

$$\tau_p = \frac{V_s}{F_s} = \text{time constant}$$

Notice that the process gain is the same as the sensitivity shown in Equation (2.60) and the time constant, in this case, is the same as the residence time. Equation (2.73) is one of the most widely used models to describe the dynamic behavior of chemical processes.

Linearization of Nonlinear Models

The material and energy balance models that describe the behavior of chemical processes are generally nonlinear, while commonly used control strategies are based on linear systems theory. It is important, then, to be able to *linearize* nonlinear models for control system design and analysis purposes. The method that we use to form linear models is based on a Taylor series approximation to the nonlinear model. The Taylor series approximation is based on the steady-state operating point of the process.

One State Variable

Consider a single variable function (equation)

$$\frac{dx}{dt} = f(x) \tag{2.74}$$

The value of this function can be approximated using a Taylor series expansion of the form

$$f(x) \approx f(x_s) + \left.\frac{\partial f(x)}{\partial x}\right|_{x_s} (x - x_s) + \frac{1}{2}\left.\frac{\partial^2 f(x)}{\partial x^2}\right|_{x_s} (x - x_s)^2 + \text{higher order terms}$$

where the subscript s is used to indicate the point of linearization (usually the steady-state operating point). The quadratic and higher order terms are neglected, resulting in the following approximate equation:

$$\frac{dx}{dt} = f(x) \approx f(x_s) + \left.\frac{\partial f(x)}{\partial x}\right|_{x_s} (x - x_s) \tag{2.75}$$

Since the steady-state operating point is chosen as the point of linearization, then [by definition of a steady state, $f(x_s) = 0$]

$$\frac{dx}{dt} \approx \left.\frac{\partial f(x)}{\partial x}\right|_{x_s} (x - x_s) \tag{2.76}$$

and since x_s is a constant value, we can write the following form

$$\frac{d(x - x_s)}{dt} \approx \left.\frac{\partial f(x)}{\partial x}\right|_{x_s} (x - x_s) \tag{2.77}$$

or, dropping the "approximately equal" notation

$$\frac{dx'}{dt} = ax'$$

where $x' = x - x_s$ represents a deviation variable, and $a = \partial f/\partial x|_{x_s}$ is the derivative of the function evaluated at the steady-state value.

One State and One Input

Consider now the following single-state, single-input equation,

$$\frac{dx}{dt} = f(x,u) \tag{2.78}$$

The value of this function can be approximated using a Taylor series expansion of the form

$$f(x,u) \approx f(x_s,u_s) + \left.\frac{\partial f(x,u)}{\partial x}\right|_{x_s,u_s} (x - x_s) + \left.\frac{\partial f(x,u)}{\partial u}\right|_{x_s,u_s} (u - u_s) + \text{ higher order terms}$$

where the subscript s is used to indicate the point of linearization (usually the steady-state operating point). The quadratic and higher order terms are neglected, resulting in the following approximate equation

$$\frac{dx}{dt} = f(x,u) \approx f(x_s,u_s) \;+ \left.\frac{\partial f(x,u)}{\partial x}\right|_{x_s} (x - x_s) \;+ \left.\frac{\partial f(x,u)}{\partial u}\right|_{x_s,u_s} (u - u_s) \tag{2.79}$$

Since the steady-state operating point is chosen as the point of linearization, then [by definition of a steady state, $f(x_s, u_s) = 0$]

$$\frac{dx'}{dt} = ax' + bu'$$

where $x' = x - x_s$ represents a deviation variable, and $a = \partial f/\partial x|_{x_s,u_s}$ and $b = \partial f/\partial u|_{x_s,u_s}$ are the derivatives of the function with respect to the state and input, evaluated at the steady-state value.

Output Variable

Consider now the expression for an output variable

$$y = g(x,u) \tag{2.80}$$

A Taylor series expansion about the state and input yields (after neglecting higher order derivatives)

$$y = g(x,u) \approx g(x_s,u_s) + \left.\frac{\partial g(x,u)}{\partial x}\right|_{x_s,u_s}(x - x_s) + \left.\frac{\partial g(x,u)}{\partial u}\right|_{x_s,u_s}(u - u_s) \tag{2.81}$$

and since $y_s = g(x_s,u_s)$

$$y' = cx' + du'$$

where $y' = y - y_s$, $x' = x - x_s$, and $u' = u - u_s$ represent deviation variables, and $c = \partial g/\partial x|_{x_s,u_s}$ and $d = \partial g/\partial u|_{x_s,u_s}$ are the derivatives of the function with respect to the state and input, evaluated at the steady-state value.

These basic ideas are illustrated in the following example.

Example 2.4: A Second-Order Reaction

Consider a CSTR with a single, second-order reaction. The modeling equation, assuming constant volume and density is

$$\frac{dC_A}{dt} = f(C_A,F) = \frac{F}{V}(C_{Ain} - C_A) - kC_A^2$$

Here the state variable is C_A and the input variable is F. A Taylor series expansion performed at the steady-state solution yields

$$a = \left.\frac{\partial f(C_A,F)}{\partial C_A}\right|_{C_{As},F_s} = -\frac{F_s}{V} - 2kC_{As}$$

$$b = \left.\frac{\partial f(C_A,F)}{\partial F}\right|_{C_{As},F_s} = \frac{C_{Ains} - C_{As}}{V}$$

Now, consider the concentration of A to be the output variable

$$y = g(C_A,F) = C_A$$

so

$$c = \frac{\partial g(C_A, F)}{\partial C_A} = 1$$

$$d = \frac{\partial g(C_A, F)}{\partial F} = 0$$

and the state space model is

$$\dot{x}' = ax' + bu'$$
$$y' = cx' + du'$$

where the state, input, and output (in deviation variable form) are

$$x' = C_A - C_{As}$$
$$u' = F - F_s$$
$$y' = x' = C_A - C_{As}$$

For the following parameters,

$$k = 1\frac{\text{liter}}{\text{mol} \cdot \text{min}}, \quad V = 1 \text{ liter}, \; C_{Ains} = 2\frac{\text{mol}}{\text{liter}}$$

A steady-state operating point is

$$F_s = 1\frac{\text{liter}}{\text{min}}, \quad C_{As} = 1\frac{\text{mol}}{\text{liter}}$$

and the partial derivatives are

$$a = \left.\frac{\partial f(C_A, F)}{\partial C_A}\right|_{C_{As}, F_s} = -\frac{F_s}{V} - 2kC_{As} = -1 - 2 = -3 \text{ min}^{-1}$$

$$b = \left.\frac{\partial f(C_A, F)}{\partial F}\right|_{C_{As}, F_s} = \frac{C_{Ains} - C_{As}}{V} = \frac{2-1}{1} = 1\frac{\text{mol/liter}}{\text{liter}} = 1\frac{\text{mol}}{\text{liter}^2}$$

and the linear model is

$$\dot{x}' = -3x' + 1u'$$
$$y' = 1x' + 0u'$$

Generalization

Consider the general nonlinear model with *nx* states, *ny* outputs, *nu* inputs, and *np* parameters

$$\dot{x}_1 = f_1\left(x_1,\ldots,x_{nx},u_1,\ldots,u_{nu},p_1,\ldots,p_{np}\right)$$
$$\dot{x}_2 = f_2\left(x_1,\ldots,x_{nx},u_1,\ldots,u_{nu},p_1,\ldots,p_{np}\right)$$
$$\vdots$$
$$\dot{x}_{nx} = f_{nx}\left(x_1,\ldots,x_{nx},u_1,\ldots,u_{nu},p_1,\ldots,p_{np}\right)$$

$$y_1 = g_1\left(x_1,\ldots,x_{nx},u_1,\ldots,u_{nu},p_1,\ldots,p_{np}\right)$$
$$y_2 = g_2\left(x_1,\ldots,x_{nx},u_1,\ldots,u_{nu},p_1,\ldots,p_{np}\right)$$
$$\vdots$$
$$y_{ny} = g_{ny}\left(x_1,\ldots,x_{nx},u_1,\ldots,u_{nu},p_1,\ldots,p_{np}\right)$$

The elements of the linearization matrices are defined as

$$A_{ij} = \left.\frac{\partial f_i}{\partial x_j}\right|_{x_s,u_s} \qquad B_{ij} = \left.\frac{\partial f_i}{\partial u_j}\right|_{x_s,u_s}$$

$$C_{ij} = \left.\frac{\partial g_i}{\partial x_j}\right|_{x_s,u_s} \qquad D_{ij} = \left.\frac{\partial g_i}{\partial u_j}\right|_{x_s,u_s}$$

where *ij* subscripts refer to the *i*th row and *j*th column of the corresponding matrix. For example, element B_{ij} refers to the effect of the *j*th input on the *i*th state derivative.

The linear *state space* form is

$$\dot{x}' = Ax' + Bu'$$
$$y' = Cx' + Du'$$

where the *deviation variables* are defined as perturbations from their steady-state values

$$x' = x - x_s$$
$$u' = u - u_s$$
$$y' = y - y_s$$

In future chapters we normally drop the prime (′) notation for deviation variables and assume that a state space model is always in deviation variable form.

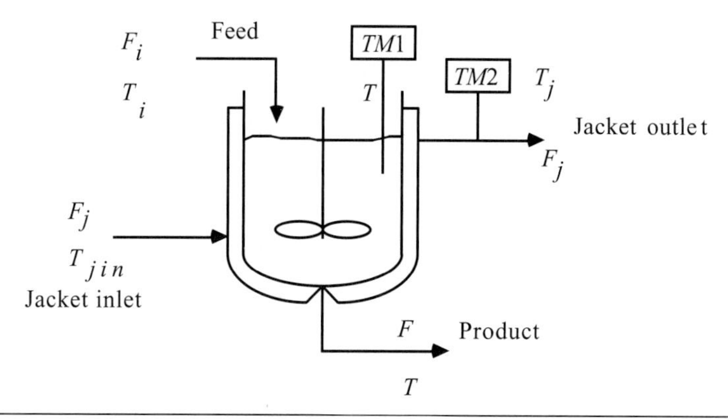

Figure 2–11 Jacketed stirred-tank heater.

Example 2.5: Jacketed Heater

Consider the jacketed stirred-tank heater shown in Figure 2–11. A hot fluid circulated through the jacket (which is assumed to be perfectly mixed), and heat flow between the jacket and vessel increases the energy content of the vessel fluid. The rate of heat transfer from the jacket fluid to the vessel fluid is

$$Q = UA[T_j - T]$$

where U is the overall heat transfer coefficient and A is the area for heat transfer. Assuming that the volume and density are constant, $F_i = F$. Energy balances on the vessel and jacket fluids result in the following equations.

$$\frac{dT}{dt} = \frac{F}{V}(T_i - T) + \frac{UA}{V\rho c_p}(T_j - T) = f_1(T, T_j, F_j, F, T_i, T_{jin})$$

$$\frac{dT_j}{dt} = \frac{F_j}{V_j}(T_{jin} - T_j) - \frac{UA}{V_j \rho_j c_{pj}}(T_j - T) = f_2(T, T_j, F_j, F, T_i, T_{jin})$$

Here the outputs are the vessel and jacket temperatures, which are also the states; the inputs are the jacket flow rate, feed flow rate, feed temperature, and jacket inlet temperature. If the outputs, states, and inputs, in deviation variable form, are

$$y = \begin{bmatrix} y_1 \\ y_2 \end{bmatrix} = x = \begin{bmatrix} x_1 \\ x_2 \end{bmatrix} = \begin{bmatrix} T - T_s \\ T_j - T_{js} \end{bmatrix}$$

$$u = \begin{bmatrix} u_1 \\ u_2 \\ u_3 \\ u_4 \end{bmatrix} = \begin{bmatrix} F_j - F_{js} \\ F - F_s \\ T_i - T_{is} \\ T_{jin} - T_{jins} \end{bmatrix}$$

Then, the linearized model is

$$A_{11} = \frac{\partial f_1}{\partial x_1} = \frac{\partial f_1}{\partial (T - T_s)} = \frac{\partial f_1}{\partial T} = -\frac{F_s}{V} - \frac{UA}{V\rho c_p}$$

$$A_{12} = \frac{\partial f_1}{\partial x_2} = \frac{\partial f_1}{\partial (T_j - T_{js})} = \frac{\partial f_1}{\partial T_j} = \frac{UA}{V\rho c_p}$$

$$A_{21} = \frac{\partial f_2}{\partial x_1} = \frac{\partial f_2}{\partial (T - T_s)} = \frac{\partial f_2}{\partial T} = \frac{UA}{V_j \rho_j c_{pj}}$$

$$A_{22} = \frac{\partial f_2}{\partial x_2} = \frac{\partial f_2}{\partial (T_j - T_{js})} = \frac{\partial f_2}{\partial T_j} = -\frac{F_{js}}{V_j} - \frac{UA}{V_j \rho_j c_{pj}}$$

Similarly, the reader should show that

$$B = \begin{bmatrix} \frac{\partial f_1}{\partial u_1} & \frac{\partial f_1}{\partial u_2} & \frac{\partial f_1}{\partial u_3} & \frac{\partial f_1}{\partial u_4} \\ \frac{\partial f_2}{\partial u_1} & \frac{\partial f_2}{\partial u_2} & \frac{\partial f_2}{\partial u_3} & \frac{\partial f_2}{\partial u_4} \end{bmatrix} = \begin{bmatrix} 0 & \frac{T_{is} - T_s}{V} & \frac{F_s}{V} & 0 \\ \frac{T_{jins} - T_{js}}{V_j} & 0 & 0 & \frac{F_{js}}{V_j} \end{bmatrix}$$

$$C = \begin{bmatrix} 1 & 0 \\ 0 & 1 \end{bmatrix}$$

$$D = \begin{bmatrix} 0 & 0 & 0 & 0 \\ 0 & 0 & 0 & 0 \end{bmatrix}$$

Exercise 8 is a numerical example for this problem.

2.8 Summary

We have focused primarily on the development of ordinary differential equation models that describe the dynamic behavior of processes where perfect mixing can be assumed. The models have the form

$$\dot{x} = f(x,u,p)$$
$$y = g(x,u,p)$$

where the states, inputs, outputs, and parameters are x, u, y, and p, respectively.

States appear in the "accumulation" (derivative with respect to time) term, inputs can be either manipulated or disturbance variables, outputs are often a subset of the states, and parameters are often physical intensive variables, such as density or kinetic rate parameters. The modeling equations are based on material (component or total) and energy balances. States are often concentrations (from the component balances), volume (total material balance on liquids), pressure (total material balances on gases), or temperature (energy balances).

The steady-state solution is $\dot{x} = 0$, or $f(x,u,p) = 0$ which can be found numerically, using MATLAB functions such as `fzero` and `fsolve` (which require the optimization toolbox) as shown in the MATLAB tutorial module.

To integrate the differential equations numerically, we use a variant of Euler integration, which is

$$x(k+1) = x(k) + \Delta t\, f(x(k))$$

The ODE module provides a tutorial on the use of MATLAB to integrate ordinary differential equations.

An understanding of dynamic behavior is obtained by using linear state-space models, which have the form

$$\begin{aligned} \dot{x}' &= Ax' + Bu' \\ y' &= Cx' + Du' \end{aligned}$$

where the prime notation ($'$) is used to indicate a deviation variable. In the next chapter we further analyze linear state space models to obtain an understanding of dynamic behavior. These state space models are converted to Laplace transfer function models, which are used later for control system design.

For the processes studied, some important characteristics were discussed. For example, *residence time* (volume/flow rate) is often a good indicator of the relative "speed" of the process dynamics, particularly for nonreacting systems. For chemical reactors, the *Damkohler* number is an important parameter, since it is the ratio of a characteristic reaction time to residence time.

The concept of *process gain* is extremely important for process control design. Process gain is the sensitivity of the process output to a manipulated input. That is, process gain is the ratio of the long-term (steady-state) change in process output to the change in process input.

Many process systems can be represented as single-state models, where the output is the state. The resulting first-order model has the form

$$\tau_p \frac{dy}{dt} = -y + k_p u$$

where k_p is the process gain and τ_p is the process time constant. The dynamic behavior of first-order processes will be discussed in the next chapter.

The examples used in this chapter were

2.1 Gas Surge Drum
2.2 An Isothermal Chemical Reactor
2.3 Heated Mixing Tank
2.4 A Second-Order Reaction
2.5 Jacketed Heater

Suggested Reading

A nice introduction to chemical engineering calculations is provided by Felder, R. M., and R. Rousseau, *Elementary Principles of Chemical Processes*, 2nd ed. Wiley, New York (1986).

Excellent discussions of the issues involved in modeling a mixing tank, incorporating density effects and energy balances, is provided in the following two books: Denn, M. M., *Process Modeling*, Longman, New York (1986); Russell, T. R. F., and M. M. Denn, *Introduction to Chemical Engineering Analysis*, Wiley, New York (1971).

An introduction to chemical reaction engineering is Fogler, H. S., *Elements of Chemical Reaction Engineering*, 3rd ed., Prentice Hall, Englewood Cliffs, NJ (1999).

An excellent textbook for an introduction to chemical engineering thermodynamics is Smith, J. M., H. C. Van Ness, and M. M. Abbott, *Chemical Engineering Thermodynamics*, 6th ed., McGraw-Hill, New York (2001).

A more detailed discussion of process dynamics is provided in the textbook Bequette, B.W., *Process Dynamics: Modeling, Analysis and Simulation*, Prentice Hall, Upper Saddle River, NJ (1998).

The reaction of ethylene oxide and water to form ethylene glycol is discussed in Fogler, H. S., *Elements of Chemical Reaction Engineering*, 2nd ed., Prentice Hall, Upper Saddle River, NJ (1992).

Student Exercises

1. Consider the gas drum in Example 2.1. Often the outlet flow relationship is actually nonlinear, with the form

$$q = \beta\sqrt{P - P_h}$$

so the modeling equation is

$$\frac{dP}{dt} = \frac{RT}{V} q_i - \frac{RT}{V} \beta\sqrt{P - P_h}$$

Discuss whether this is now a self-regulating process. Also, sketch the steady-state input (q_{is}) -output (P_s) curve, based on the flow coefficient, $\beta = 1 \text{ mol} \cdot \text{s}^{-1} \cdot \text{atm}^{-1/2}$ and a constant header pressure, $P_h = 1$ atm.

2. Consider the heated mixing tank example, which had the modeling equations

$$\frac{dV}{dt} = F_i - F$$

$$\frac{dT}{dt} = \frac{F_i}{V}(T_i - T) + Q$$

For steady-state inlet and outlet flow rates of 100 liters/minute, a liquid volume of 500 liters and inlet and outlet temperatures of 20° and 40°C, respectively:

a. Find the steady-state heating rate, Q.

b. Consider a step inlet temperature change from 20° to 22°C. Use Euler integration with a integration step size of 0.5 minutes to find the vessel temperature response for the first 2 minutes. Compare this with MATLAB ode45 (see the ODE module).

3. Often liquid surge tanks (particularly those containing hydrocarbons) will have a gas "blanket" of nitrogen or carbon dioxide to prevent the accumulation of explosive vapors above the liquid, as depicted below.

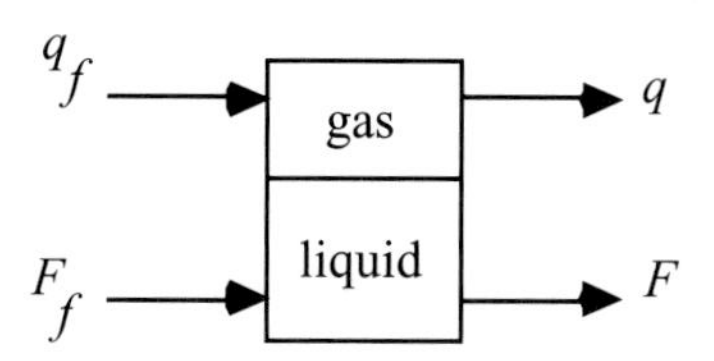

Develop the modeling equations with gas pressure and liquid volume as the state variables. Let q_f and q represent the inlet and outlet gas molar flow rates, F_f and F the liquid volumetric flow rates, V the constant (total) volume, V_l the liquid volume, and P the gas pressure. Assume the ideal gas law. Show that the modeling equations are

$$\frac{dV_l}{dt} = F_f - F$$

$$\frac{dP}{dt} = \frac{P}{V - V_l}(F_f - F) + \frac{RT}{V - V_l}(q_f - q)$$

and state any other assumptions.

4. Consider the ethylene glycol problem (Example 2.2). If two 800-gallon reactors are placed in series, what volumetric flow rate is necessary to produce 100 million lb/year of ethylene glycol? What percentage savings is this compared with using a single 800-gallon reactor?

5. Consider the ethylene glycol problem (Example 2.2). Solve the dynamic equations using ode45 (see the ODE module), for a step change in space velocity from 0.0778 to 0.0817 min^{-1}. Compare your plots with those shown in Figure 2–4.

6. Semibatch reactors are operated as a cross between batch and continuous reactors. A semibatch reactor is initially charged with a volume of material, and a continous feed of reactant is started. There is, however, no outlet stream. Develop the modeling equations for a single first-order reaction. The state variables should be volume and concentration of reactant *A*.

7. A stream contains a waste chemical, *W*, with a concentration of 1 mol/liter. To meet Environmental Protection Agency and state standards, at least 90% of the chemical must be removed by reaction. The chemical decomposes by a second-order reaction with a rate constant of 1.5 liter/(mol hr). The stream flow rate is 100 liter/hour and two available reactors (400 and 2000 liters) have been placed in series (the smaller reactor is placed before the larger one).

a. Write the modeling equations for the concentration of the waste chemical. Assume constant volume and constant density. Let

C_{w1} = concentration in reactor 1, mol/liter
C_{w2} = concentration in reactor 2, mol/liter
F = volumetric flow rate, liter/hr
V_1 = liquid volume in reactor 1, liters
V_2 = liquid volume in reactor 2, liters
k = second-order rate constant, liter/(mol hr)

b. Show that the steady-state concentrations are 0.33333 mol/liter (reactor 1) and 0.09005 mol/liter (reactor 2), so the specification is met. (*Hint*: You need to solve quadratic equations to obtain the concentrations.)

c. Linearize at steady state and develop the state space model (analytical) of the form

$$\dot{x} = Ax + Bu$$

where

$$x = \begin{bmatrix} C_{w1} - C_{w1s} \\ C_{w2} - C_{w2s} \end{bmatrix}, \quad u = \begin{bmatrix} F - F_s \\ C_{win} - C_{wins} \end{bmatrix}$$

d. Show that the A and B matrices are

$$A = \begin{bmatrix} -1.25 & 0 \\ 0.05 & -0.32015 \end{bmatrix}, \quad B = \begin{bmatrix} 0.001667 & 0.25 \\ 0.0001216 & 0 \end{bmatrix}$$

(also, show the units associated with each coefficient).

e. Assuming that each state is an output, show that the C and D matrices are

$$C = \begin{bmatrix} 1 & 0 \\ 0 & 1 \end{bmatrix}, \quad D = \begin{bmatrix} 0 & 0 \\ 0 & 0 \end{bmatrix}$$

f. Find the eigenvalues of A using the MATLAB `eig` function, and find the eigenvalues by hand, by solving det $(\lambda I - A) = 0$.

g. Solve **a** for the nonlinear differential equations, using ode45. Compare the linear and nonlinear variables on the same plots (make certain you convert from deviation to physical variables for the linear results) for a step change in the flow rate from 100 liters/hour to 110 liters/hour. Assume the initial concentrations are the steady-state values (0.3333 and 0.09005). Compare the linear and nonlinear responses of the reactor concentrations. Is the removal specification still obtained?

h. Would better steady-state removal of W be obtained if the order of the reaction vessels was reversed? Why or why not? (Show your calculations.)

8. Consider Example 2.5, the stirred-tank heater example. Read and work through the example. Use the following parameters and steady-state values:

$F_s = 1 \text{ ft}^3/\text{min}$	$\rho c_p = 61.3 \text{ Btu/(°F} \cdot \text{ft}^3)$	$\rho_j c_{pj} = 61.3 \text{ Btu/(°F} \cdot \text{ft}^3)$
$T_{is} = 50°\text{F}$	$T_s = 125°\text{F}$	$V = 10 \text{ ft}^3$
$T_{jis} = 200°\text{F}$	$T_{js} = 150°\text{F}$	$V_j = 2.5 \text{ ft}^3$

a. By solving the steady-state equations, verify that the following values obtained for UA (overall heat transfer coefficient * area for heat transfer) and F_{js} (steady-state jacket flow rate) are correct:

$$UA = 183.9 \text{ Btu/(°F} \cdot \text{min)}$$
$$F_{js} = 1.5 \text{ ft}^3/\text{min}$$

b. Find the values of the matrices in the state space model.

c. Write a function file, `heater.m` (example shown below), to be used with ode45 (Module 3) to solve the two nonlinear ordinary differential equations.

d. First, verify that the steady-state state variable values are correct by simulating the process with no change in the jacket flow rate.

e. Now, perform simulations for small and large step changes in the jacket flow rate. Use the MATLAB step command to solve for the linear state space model. Realize that the step results are based on deviation variables and for a unit step change in input one (jacket flow rate), and convert the linear states to physical variable form.

We now consider the important issue of scale-up: Pilot plants are small-scale (intermediate between lab-scale and full-size manufacturing-scale) chemical processes, used to understand process operating behavior before the manufacturing process is designed. Here we discover the effect of vessel scale on the heat transfer removal capability of a vessel.

f. Consider now a vessel that can handle 10 times the throughput of the previous vessel (that is, 10 ft^3/min rather than 1 ft^3/min). Assume that the same residence time is maintained ($V/F = 10$ minutes), so the volume of the new vessel is 100 ft^3. Assume that the heat transfer coefficient (U) remains constant, but that the heat transfer area changes. Assume that the vessel can be modeled as a cylinder, with the height (L) = 2*diameter (D). Find the value of UA for the larger vessel.

g. For the larger vessel, find the new steady-state value of jacket temperature that must be used to maintain the vessel temperature at $T = 125$°F. Also, find the new steady-state jacket flow rate. *Hint*: Solve the two modeling equations at steady state to obtain these values. Discuss the effect of process scale-up on the operating conditions. How large can the vessel become before the jacket temperature is too high (it approaches the inlet jacket temperature)?

h. For the larger vessel, find the new state space model, assuming that the jacket volume is 0.25 times the vessel volume. Calculate the eigenvalues of the A matrix. How do they compare with the smaller vessel?

i. Find the step response for the nonlinear and linear (state space) systems for a step increase of 0.1 ft^3/min in jacket flow rate. How do these compare with the smaller vessel?

j. Find the step response for the nonlinear and linear (state space) systems for a step increase of 10% in the jacket flow rate. How do these compare with the smaller vessel?

Example m-file to be used by ode45

```
function xdot = heater(t,x,flag,delFj);
%
% Dynamics of a stirred tank heater
% (c) 1994 - B.W. Bequette
% 8 July 94
% 23 Jan 01 - revised for jacket step change in argument list
%
% x(1) = T = temperature in tank
% x(2) = Tj = temperature in jacket
% delFj = change in jacket flowrate
% F = Tank flowrate
% Tin = Tank inlet temperature
% Tji = Jacket inlet temperature
% V = Tank volume
% Vj = Jacket volume
% rhocp = density*heat capacity
% rhocpj = density*heat capacity,jacket fluid
%
% parameter and steady-state variable values are:
%
 F = 1;
 Fjs = 1.5;
 Ti = 50;
 Tji = 200;
 V = 10;
 Vj = 2.5;
 rhocp = 61.3;
 rhocpj= 61.3;
 UA = 183.9;
%
% step change in jacket flow
%
 Fj = Fjs + delFj;
 T = x(1);
 Tj= x(2);
%
% odes
%
```

```
 dTdt = (F/V)*(Ti - T) + UA*(Tj - T)/(V*rhocp);
 dTjdt = (Fj/Vj)*(Tji - Tj) - UA*(Tj - T)/(Vj*rhocpj);
 xdot = [dTdt; dTjdt];
%
% generate simulations using
% [t,x] = ode45('heater',tspan,x0,[],delFj);
% where tspan = [0 30], for example
```

To make the runs go faster, you may wish to generate and run the following script file:

```
%
% runs stirred tank heater example - 23 Jan 01
% step changes to stirred tank heater
%
% make certain you enter the step change value (delFj)
% before running this file
% also, generate function file with the following first line:
% function xdot = heater(t,x,flag,delFj);
%
 options = odeset('RelTol',1e-6,'AbsTol',[1e-6 1e-6]);
%
 [t,x] = ode45('heater',tspan,x0,options,delFj);
%
 figure(1)
 subplot(2,1,1),plot(t,x(:,1))
 title('nonlinear, using ode45')
 xlabel('time, min')
 ylabel('temp, deg F')
 subplot(2,1,2),plot(t,x(:,2))
 xlabel('time, min')
 ylabel('jacket temp, deg F')
%
% state space model with unit step change
 a = [-0.4 0.3;1.2 -1.8];
 b = [0;20];
 c = [1 0;0 1];
 d = 0;
```

```
 sys = ss(a,b,c,d);
 [ylin,tlin] = step(sys);
%
 figure(2)
 subplot(2,1,1),plot(tlin,ylin(:,1))
 title('linear state space, unit step, deviation')
 xlabel('time, min')
 subplot(2,1,2),plot(tlin,ylin(:,2))
 xlabel('time, min')
%
% scale deviation by delFj
 ylinscale = ylin*delFj;
% plot in physical variable form
 figure(3)
 subplot(2,1,1),plot(tlin,125+ylinscale(:,1))
 title('linear state space, physical magnitude')
 xlabel('time, min')
 subplot(2,1,2),plot(tlin,150+ylinscale(:,2))
 xlabel('time, min')
% compare linear and nonlinear on same plot
 figure(4)
 subplot(2,1,1),plot(t,x(:,1),tlin,125+ylinscale(:,1),'--')
 title('nonlinear vs. linear')
 xlabel('time, min')
 ylabel('temp, deg F')
 legend('nonlinear','linear')
 subplot(2,1,2),plot(t,x(:,2),tlin,150+ylinscale(:,2),'--')
 xlabel('time, min')
 ylabel('jacket temp, deg F')
 legend('nonlinear','linear')
```

Appendix 2.1: Solving Algebraic Equations

Fortunately, the MATLAB `fsolve` function is easy to use for solving algebraic equations. For a simplified presentation, we use the form

$$f(x) = 0 \tag{A.1}$$

obtained from Equation (2.66) with a fixed p and u. The most commonly used numerical techniques are related to Newton-Raphson iteration. The "guess" for iteration $k + 1$ is determined from the value at iteration k, using

$$x(k + 1) = x(k) - J^{-1}(k) f[x(k)] \tag{A.2}$$

where $f[x(k)]$ is the vector of function evaluations at *iteration* k, and $J(k)$ is the Jacobian matrix

$$J_{ij}(k) = \frac{\partial f_i}{\partial x_j}(k) \tag{A.3}$$

The ij element of the Jacobian represents the partial derivative of equation i with respect to variable j. If analytical derivatives are not available, elements of the Jacobian are obtained by perturbation of the state variable, requiring $n + 1$ function evaluations for an n-equation system of equations. Various quasi-Newton techniques provide approximations to the Jacobian and do not require as many function evaluations, reducing computational time.

In practice, the Jacobian matrix in Equation (A.2) is not inverted. Rather, a set of linear algebraic equations is solved for $x(k+1)$,

$$J(k)[x(k+1) - x(k)] = -f(x(k)) \tag{A.4}$$

In this text we do not focus on the solution of algebraic equations. See the text by Bequette (1998) for more details on these techniques.

Appendix 2.2: Integrating Ordinary Differential Equations

The Euler integration method often requires very small integration step sizes to obtain a desired level of accuracy. Note that x is a vector of n state variables at each time step. For example, if there are two states (two differential equations)

$$\begin{bmatrix} x_1(k+1) \\ x_2(k+1) \end{bmatrix} = \begin{bmatrix} x_1(k) \\ x_2(k) \end{bmatrix} + \Delta t \cdot \begin{bmatrix} f_1(x_1(k), x_2(k)) \\ f_2(x_1(k), x_2(k)) \end{bmatrix}$$

we have left the inputs and parameters out of the function variable list for convenience. An example is shown next.

Example 2.2: An Isothermal Chemical Reactor, continued

Consider the ethylene glycol reactor problem in Example 2.2. We use k_1 to represent the reaction rate constant, so that it is not confused with the time-step index. The Euler integration algorithm results in the following two equations, where k represents the time-step index.

$$C_A(k+1) = C_A(k) + \Delta t \cdot \left[\frac{F(k)}{V}\left(C_{Ai}(k) - C_A(k)\right) - k_1 C_A(k)\right]$$

$$C_P(k+1) = C_P(k) + \Delta t \cdot \left[-\frac{F(k)}{V} C_P(k) + k_1 C_A(k)\right]$$

For $C_A(0) = 0.1$ and $C_P(0) = 0.4$, and a space velocity $(F/V) = 0.0816 \text{ min}^{-1}$ we find, for the first time step,

$$C_A(k=1) = C_A(k=0) + \Delta t \cdot \left[\frac{F(k=0)}{V}\left(C_{Ai}(k=0) - C_A(k=0)\right) - k_1 C_A(k=0)\right]$$

$$C_P(k=1) = C_P(k=0) + \Delta t \cdot \left[-\frac{F(k=0)}{V} C_P(k=0) + k_1 C_A(k=0)\right]$$

Substituting the numerical values for an integration step size of 0.5 minutes, we find

$$C_A(k=1) = C_A(k=0) + \Delta t \cdot \left[\frac{F(k=0)}{V}\left(C_{Ai}(k=0) - C_A(k=0)\right) - k_1 C_A(k=0)\right]$$

$$C_P(k=1) = C_P(k=0) + \Delta t \cdot \left[-\frac{F(k=0)}{V} C_P(k=0) + k_1 C_A(k=0)\right]$$

resulting in the concentration values at $t = 0.5$ minutes

$$C_A\ (t = 0.5 \text{ min}) = 0.10079 \text{ lbmol/ ft}^3$$
$$C_p\ (t = 0.5 \text{ min}) = 0.39921 \text{ lbmol/ ft}^3$$

Values can be obtained at future times by continuing to march forward in time. MATLAB has a suite of routines for integrating differential equations. These are covered in Module 3.

CHAPTER 3

Dynamic Behavior

The goal of this chapter is to understand dynamic behavior. We begin by working with linear state space models, often obtained by linearizing a nonlinear model, such as those developed in Chapter 2. We then introduce Laplace transforms. The main advantage to Laplace transforms is that they allow us to analyze behavior exhibited by linear differential equations by using simple algebraic manipulations. Laplace transforms are used to create transfer function models, which are the basis for many control system design techniques.

After studying this chapter, the reader should be able to:

- Apply the initial and final value theorems of Laplace transforms
- Understand first-order, first-order + dead time and integrating system step responses
- Understand second-order under-damped behavior
- Understand the effect of pole and zero values on step responses
- Convert state space models to transfer functions

The major sections of this chapter are as follows:

3.1 Background

One of the major goals of this chapter is to obtain an understanding of process dynamics. Process engineers tend to think of process dynamics in terms of the response of a process to a step input change. Assume that the process is initially at steady state, then apply a step change to an input variable. The majority of chemical processes will exhibit one of the responses shown in Figure 3–1. In this plot, we assume that a positive step increase has been made to the input variable of interest. The solid curves are examples of "positive gain" processes; that is, the process output increases for an increase in the input. The dashed lines are those of negative gain processes. The curves in Figure 3–1a show a monotonic change in the output; this behavior is generally known as overdamped. The curves in Figure 3–1b are indicative of "integrating" processes; a prime example is a liquid surge vessel, where the level continues to change when there is an imbalance in the inlet and outlet flow rates. The curves in Figure 3–1c are known as underdamped or oscillatory responses. This type of behavior may occur in exothermic chemical reactors or biochemical reactors. It more often occurs in processes that are under feedback control, particularly if the controller is poorly tuned. The behavior shown in Figure 3–1d is known as "inverse response" and is seen in steam drums, distillation columns, and some adiabatic plug flow reactors.

In the sections that follow, we discuss the characteristics of process models that lead to each of the behaviors shown in Figure 3–1.

3.2 Linear State Space Models

In Chapter 2 we developed fundamental models, which were normally nonlinear in nature. We then developed state space models that were based on linearizing the fundamental

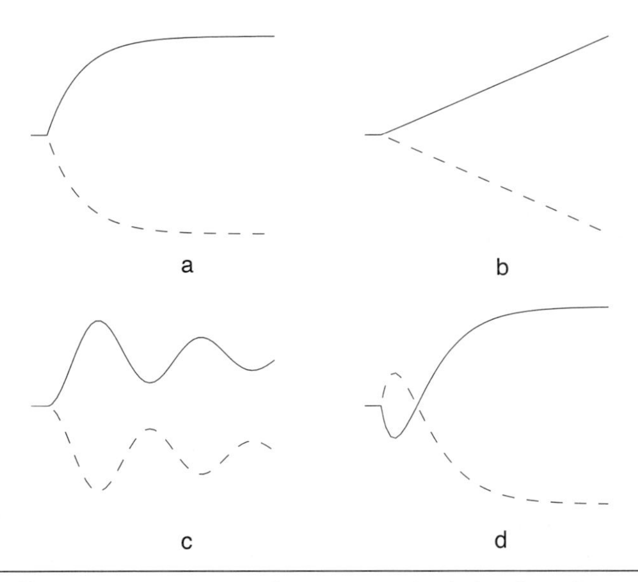

Figure 3–1 Common responses of process outputs to step changes in process inputs. Assuming a positive step change, the solid curves are illustrative of "positive gain" processes, and the dashed curves are indicative of "negative gain" processes. (a) Overdamped or first order, (b) integrating, (c) underdamped or oscillatory, and (d) inverse response.

models at a steady-state solution. This led to the notion of a *perturbation* or *deviation* variable, which is simply the perturbation of a variable from its steady-state value.

State space models have the following form, where the states (x), inputs (u), and outputs (y) are all perturbation or deviation variables

$$\begin{bmatrix} \dot{x}_1 \\ \vdots \\ \dot{x}_n \end{bmatrix} = \begin{bmatrix} a_{11} & \cdots & a_{1n} \\ \vdots & \cdots & \vdots \\ a_{n1} & \cdots & a_{nn} \end{bmatrix} \begin{bmatrix} x_1 \\ \vdots \\ x_n \end{bmatrix} + \begin{bmatrix} b_{11} & \cdots & b_{1m} \\ \vdots & \cdots & \vdots \\ b_{n1} & \cdots & b_{nm} \end{bmatrix} \begin{bmatrix} u_1 \\ \vdots \\ u_m \end{bmatrix}$$
$$\begin{bmatrix} y_1 \\ \vdots \\ y_r \end{bmatrix} = \begin{bmatrix} c_{11} & \cdots & c_{1n} \\ \vdots & \cdots & \vdots \\ c_{r1} & \cdots & c_{rn} \end{bmatrix} \begin{bmatrix} x_1 \\ \vdots \\ x_n \end{bmatrix} + \begin{bmatrix} d_{11} & \cdots & d_{1m} \\ \vdots & \cdots & \vdots \\ d_{r1} & \cdots & d_{rm} \end{bmatrix} \begin{bmatrix} u_1 \\ \vdots \\ u_m \end{bmatrix} \tag{3.1}$$

Recall that in matrix notation, the first subscript refers to the row and the second subscript refers to the column. When matrices multiply vectors, each row corresponds to a particular output of the multiplication, while the column corresponds to a particular input of the operation. Consider the C matrix, which relates the states to the outputs. Element c_{ij} relates the effect of state x_j on output y_i.

The shorthand notation for Equation (3.1) is

$$\begin{aligned} \dot{x} &= Ax + Bu \\ y &= Cx + Du \end{aligned} \tag{3.2}$$

It is important to always check for dimensional consistency in matrix operations. In a matrix-vector operation $y = Cx$, the number of rows in C must be equal to the number of elements in y. Also, the number of columns in C must be equal to the number of elements in x.

Stability

One of the first basic concepts that we need to cover is the notion of stability. Consider a process where one or more states have been perturbed from the steady-state solution or operating point. The process is stable if after a period of time, the variables return to the steady-state values. This means that the state variables, since they are deviation variables, return to zero.

Numerically, we can determine the stability of a state space model by finding the eigenvalues of the state space A matrix. Remember that the A matrix is simply the matrix of derivatives of the dynamic modeling equations with respect to the state variables.

If all of the eigenvalues are negative, then the system is *stable*; if any single eigenvalue is positive, the system is *unstable*. A system with all but one eigenvalues negative and with one eigenvalue equal to zero is called an *integrating* system and is characteristic of processes with liquid levels or gas drum pressures that can vary.

Examples of unstable systems are shown in Figure 3–2. If an eigenvalue is real and positive, the system response is that shown in the top curves. If there are complex conjugate eigenvalues, with positive real portions, the system oscillates (with ever increasing amplitude), as shown at the bottom.

Mathematically, the eigenvalues of the A matrix are found from the roots of the *characteristic polynomial*

$$\det(\lambda I - A) = 0 \tag{3.3}$$

where λ is known as an eigenvalue, and I is the identity matrix. For a state space model with n states, A is an $n \times n$ matrix, and there will be n solutions (eigenvalues) of Equation (3.3). There are analytical solutions for two- and three-state systems; the two-state solution is shown below.

In two-state systems,

$$\lambda I - A = \begin{bmatrix} \lambda & 0 \\ 0 & \lambda \end{bmatrix} - \begin{bmatrix} a_{11} & a_{12} \\ a_{21} & a_{22} \end{bmatrix} = \begin{bmatrix} \lambda - a_{11} & -a_{12} \\ -a_{21} & \lambda - a_{22} \end{bmatrix}$$

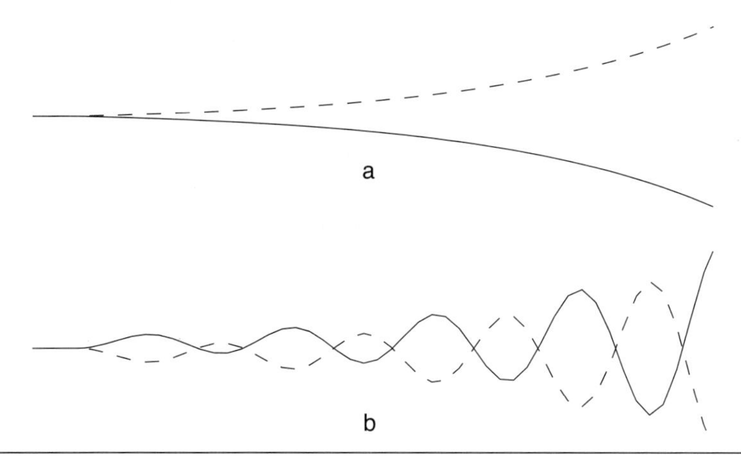

Figure 3–2 Unstable responses. (a) Monotonic and (b) oscillatory.

The determinant can be found by

$$\det\begin{bmatrix} \lambda - a_{11} & -a_{12} \\ -a_{21} & \lambda - a_{22} \end{bmatrix} = (\lambda - a_{11})\cdot(\lambda - a_{22}) - (-a_{12})(-a_{21})$$

$$= \lambda^2 - (a_{11} + a_{22})\lambda + a_{11}a_{22} - a_{12}a_{21}$$

and the eigenvalues are found as the two solutions (roots) to

$$\lambda^2 - (a_{11} + a_{22})\lambda + a_{11}a_{22} - a_{12}a_{21} = 0 \tag{3.4}$$

The roots can be found using the quadratic formula

$$\lambda = \frac{a_{11} + a_{22} \pm \sqrt{(a_{11} + a_{22})^2 - 4(a_{11}a_{22} - a_{12}a_{21})}}{2} \tag{3.5}$$

It is easy to show that if and $a_{11} + a_{22} < 0$ and $a_{11}a_{22} > a_{12}a_{21}$, the roots (eigenvalues) are negative and the system is stable. A more general method of qualitatively checking the stability, known as the *Routh stability criterion*, is shown in Chapter 5.

Example 3.1: Exothermic CSTR

Models for an exothermic, CSTR are detailed in Module 8. For a two-state representation, the first state is the concentration and the second state is the reactor temperature. For a particular reactor with two different operating conditions, the A matrix is (the time unit is hours)

Operating condition 1 *Operating condition 2*

$$A_1 = \begin{bmatrix} -1.1680 & -0.0886 \\ 2.0030 & -0.2443 \end{bmatrix} \qquad A_2 = \begin{bmatrix} -1.8124 & -0.2324 \\ 9.6837 & 1.4697 \end{bmatrix}$$

and the eigenvalues for operating condition 1 can be found using the following steps

$$\lambda I - A_1 = \begin{bmatrix} \lambda + 1.1680 & 0.0886 \\ -2.0030 & \lambda + 0.2443 \end{bmatrix}$$

$$\det(\lambda I - A_1) = (\lambda + 1.1680)(\lambda + 0.2443) - (0.0886)(-2.003) = 0$$

$$= \lambda^2 + 1.4123\lambda + 0.4628 = 0$$

with the solutions [using the quadratic formula (3.5)]

$$\lambda = -0.8955 \text{ hr}^{-1} \quad \text{and} \quad \lambda = -0.5168 \text{ hr}^{-1}$$

Since both eigenvalues are negative, operating condition 1 is *stable*.

The reader should show that the eigenvalues of A_2 are

$$\lambda = -0.8366 \text{ hr}^{-1} \quad \text{and} \quad \lambda = -0.4939 \text{ hr}^{-1}$$

where the positive eigenvalue indicates that operating condition 2 is *unstable*.

MATLAB Eigenvalue Function

The MATLAB `eig` command can be used to quickly find eigenvalues of a matrix. The reader should use the MATLAB command window to verify the following results for the second operating condition:

```
» a2 = [-1.8124 -0.2324;9.6837 1.4697];
» eig(a2)
ans =
   -0.8366
    0.4939
```

Again, the positive eigenvalue indicates that the second operating condition is unstable.

Generalization

Notice that a solution of a second-order polynomial was required to find the eigenvalues of the two-state example; this resulted in two eigenvalues. For the general case of an $n \times n$ matrix, there will be n eigenvalues. It is too complex to find these analytically for all but the simplest low-order systems. The simplest way to find eigenvalues is by using existing numerical analysis software; for example, in MATLAB the `eig` function can be used to find eigenvalues.

The values of the eigenvalues are related to the "speed of response," and the eigenvalue unit is inverse time. If the unit of time used in the differential equations is minutes, for example, then the eigenvalues have min^{-1} as the unit. For stable systems (where all eigenvalues are negative), the larger magnitude (more negative) eigenvalues are faster.

For matrices that are 2×2 or larger, some eigenvalues may occur in complex conjugate pairs. In this case, the stability is determined by the sign of the real portion of the complex number. As long as all real portions are negative, the system is stable.

3.3 Introduction to Laplace Transforms

Most control system analysis and design techniques are based on linear systems theory. Although we could develop these procedures using the state space models, it is generally easier to work with *transfer functions*. Basically, transfer functions allow us to make algebraic manipulations rather than working directly with linear differential equations (state space models). To create transfer functions, we need the notion of the Laplace transform.

The Laplace transform of a time-domain function, $f(t)$, is represented by $L[f(t)]$ and is defined as

$$L[f(t)] = F(s) = \int_0^{\infty} f(t)e^{-st}\,dt \tag{3.6}$$

The Laplace transform is a linear operation, so the Laplace transform of a constant (C) multiplying a time-domain function is just that constant times the Laplace transform of the function,

$$L[C \cdot f(t)] = C \cdot F(s) = C \cdot \int_0^{\infty} f(t)e^{-st}\,dt \tag{3.7}$$

The Laplace transforms of a few common time-domain functions are shown next.

Exponential Function

Exponential functions appear often in the solution of linear differential equations. Here

$$f(t) = e^{-at}$$

The transform is defined for $t > 0$ (we also use the identity that $e^{x+y} = e^x e^y$)

$$L[f(t)] = L\left[e^{-at}\right] = F(s) = \int_0^{\infty} e^{-at}e^{-st}\,dt = \int_0^{\infty} e^{-(s+a)t}\,dt$$

$$= -\frac{1}{s+a}\left[e^{-(s+a)t}\right]_0^{\infty} = -\frac{1}{s+a}[0-1] = \frac{1}{s+a}$$

So we now have the following relationship:

$$L\left[e^{-at}\right] = \frac{1}{s+a} \tag{3.8}$$

Derivatives

This will be important in transforming the derivative term in a dynamic equation to the Laplace domain (using integration by parts),

$$L\left[\frac{df(t)}{dt}\right] = \int_0^\infty \frac{df(t)}{dt} e^{-st} dt = \left[e^{-st} f(t)\right]_0^\infty + \int_0^\infty f(t) s e^{-st} dt$$

$$= \left[0 - f(0)\right] + s\int_0^\infty f(t) e^{-st} dt = -f(0) + sF(s)$$

so we can write

$$L\left[\frac{df(t)}{dt}\right] = sF(s) - f(0) \tag{3.9}$$

For an nth derivative, we can derive

$$L\left[\frac{d^n f(t)}{dt}\right] = s^n F(s) - s^{n-1} f(0) - s^{n-2} f'(0) - \ldots - f^{(n-1)}(0) \tag{3.10}$$

n initial conditions are needed: $f(0),\ldots,f^{(n-1)}(0)$

One reason for using deviation variables is that all of the initial condition terms in Equation (3.10) are 0, if the system is initially at steady-state.

Time Delays (Dead Time)

Time delays often occur owing to fluid transport through pipes, or measurement sample delays. Here we use θ to represent the time delay. If $f(t)$ represents a particular function of time, then $f(t - \theta)$ represents the value of the function θ time units in the past.

$$L\left[f(t-\theta)\right] = \int_0^\infty f(t-\theta) e^{-st} dt = \int_0^\infty f(t-\theta) e^{-s(t-\theta+\theta)} dt = \int_0^\infty f(t-\theta) e^{-s(t-\theta)} e^{-s\theta} dt$$

$$= e^{-s\theta} \int_0^\infty f(t-\theta) e^{-s(t-\theta)} d(t-\theta)$$

We can use a change of variables, $t^* = t - \theta$, to integrate the function. Notice that the lower limit of integration does not change, because the function is defined as $f(t) = 0$ for $t < 0$.

$$L[f(t-\theta)] = e^{-s\theta}\int_0^\infty f(t^*)e^{-st^*}dt^* = e^{-s\theta}F(s) \tag{3.11}$$

So the Laplace transform of a function with a time delay (θ) is simply $e^{-\theta s}$ times the Laplace transform of the nondelayed function.

Step Functions

Step functions are used to simulate the sudden change in an input variable (say a flow rate being rapidly changed from one value to another). A step function is discontinuous at $t = 0$. A "unit" step function is defined as

$$f(t) = 0 \text{ for } t < 0$$
$$f(t) = 1 \text{ for } t \geq 0$$

and using the definition of the Laplace transform,

$$L[f(t)] = \int_0^\infty e^{-st}dt = -\frac{1}{s}\left[e^{-st}\right]_0^\infty = -\frac{1}{s}[0-1] = \frac{1}{s}$$

so

$$L[1] = \frac{1}{s} \tag{3.12}$$

Similarly, the Laplace transform of a constant, C, is

$$L[C] = \frac{C}{s} \tag{3.13}$$

Pulse

Consider a pulse function, where a total integrated input of magnitude P is applied over t_p time units, as shown in Figure 3–3.

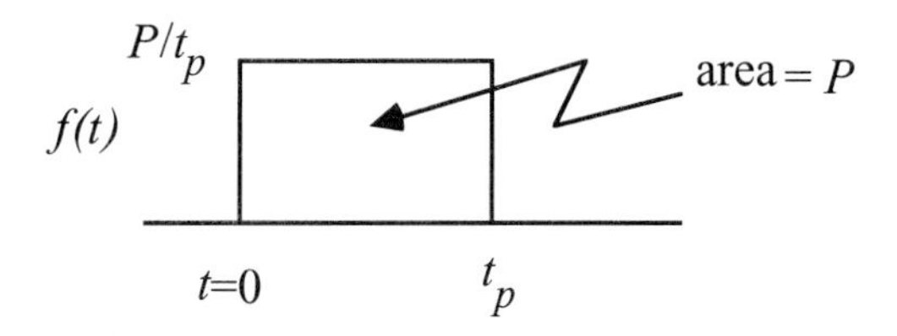

Figure 3–3 Pulse function.

The function is $f(t) = P/t_p$ for $0 < t < t_p$ and $f(t) = 0$ for $t > t_p$. The Laplace transform is

$$F(s) = \int_0^{t_p} \frac{P}{t_p} e^{-st} dt + \int_{t_p}^{\infty} 0 \cdot e^{-st} dt = \frac{P}{t_p} \cdot \frac{\left[1 - e^{-t_p s}\right]}{s} \tag{3.14}$$

Impulse

An impulse function can viewed as a pulse function, where the pulse period is decreased while maintaining the pulse area, as shown in Figure 3–4. In the limit, as t_p approaches 0, the pulse function becomes (using L'Hopital's rule)

$$L[impulse, area = P] = P \tag{3.15}$$

If we denote a unit impulse as $f(t) = \delta$, then the Laplace transform is

$$L\delta = 1 \tag{3.16}$$

Examples of common impulse inputs include a "bolus" (shot or injection) of a drug into a physiological system, or dumping a bucket of fluid or bag of solids into a chemical reactor.

Other Functions

It is rare for one to derive the Laplace transform for a function; rather a table of known transforms (and inverse transforms) can be used to solve most dynamic systems problems.

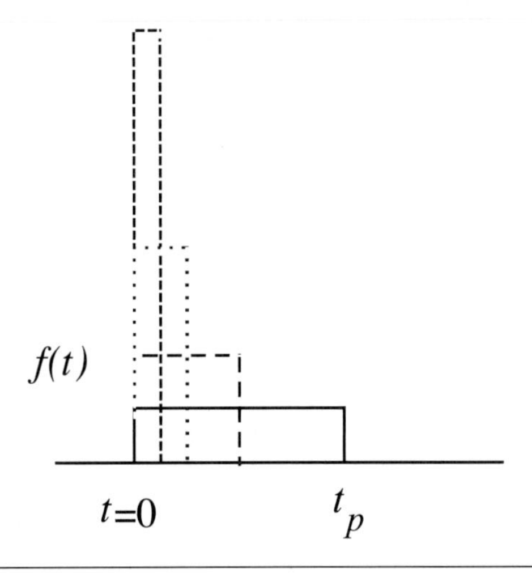

Figure 3–4 Concept of an impulse function.

Table 3–1 presents solutions for most common functions. If you desire to transform a function from the time domain to the Laplace domain, then look for the time-domain function in the first column and write the corresponding Laplace domain function from the second column. Similarly, if your goal is to "invert" a Laplace domain function to the time domain, then look for the Laplace domain function in the second column and write the corresponding time-domain function from the first column. This notion of the inverse Laplace transform can be written

$$L^{-1}[F(s)] = f(t) \tag{3.17}$$

For example

$$L\left[e^{-at}\right] = \frac{1}{s+a} \quad \text{and} \quad L^{-1}\left[\frac{1}{s+a}\right] = e^{-at}$$

Initial- and Final-Value Theorems

The following theorems are very useful for determining limiting values in dynamics and control studies. The long-term behavior of a time-domain function can be found by analyzing the Laplace domain behavior in the limit as the s variable approaches zero. The initial value of a time-domain function can be found by analyzing the Laplace domain behavior in the limit as s approaches infinity.

Table 3–1 Laplace Transforms for Common Time-Domain Functions

Time-domain function	***Laplace domain function***
$f(t)$	$F(s)$
$\delta(t)$ (unit impulse)	1
$S(t)$ (unit step) $\begin{Bmatrix} 0 \text{ for } t < 0 \\ 1 \text{ for } t > 0 \end{Bmatrix}$	$1/s$
C (constant)	C/s
$f(t-\theta)$	$e^{-\theta s}F(s)$
t	$1/s^2$
t^n	$n!/s^{n+1}$
df/dt	$sF(s) - f(0)$
$d^n f/dt^n$	$s^n F(s) - s^{n-1}f(0) - \cdots - sf^{(n-2)}(0) - f^{(n-1)}(0)$

(*continues*)

Table 3–1 *(continued)*

Time-domain function	*Laplace domain function*
e^{-at}	$\frac{1}{s+a}$
te^{-at}	$\frac{1}{(s+a)^2}$
$\frac{t^2}{2}e^{-at}$	$\frac{1}{(s+a)^3}$
$\frac{t^n}{n!}e^{-at}$	$\frac{1}{(s+a)^{n+1}}$
$1-e^{-t/\tau}$	$\frac{1}{s(\tau s+1)}$
$\sin\ \omega t$	$\frac{\omega}{s^2+\omega^2}$
$\cos\ \omega t$	$\frac{s}{s^2+\omega^2}$
$e^{-at}\sin\ \omega t$	$\frac{\omega}{(s+a)^2+\omega^2}$
$e^{-at}\cos\ \omega t$	$\frac{s+a}{(s+a)^2+\omega^2}$
$1+\frac{\left(\tau_1 e^{-t/\tau_1}-\tau_2 e^{-t/\tau_2}\right)}{\tau_2-\tau_1}$	$\frac{1}{s(\tau_1 s+1)(\tau_2 s+1)}$
$1-\left(1-\frac{\tau_n}{\tau_d}\right)e^{-t/\tau_d}$	$\frac{\tau_n s+1}{s(\tau_d s+1)}$
$1+\left(\frac{\tau_3-\tau_1}{\tau_1-\tau_2}\right)e^{-t/\tau_1}+\left(\frac{\tau_3-\tau_2}{\tau_2-\tau_1}\right)e^{-t/\tau_2}$	$\frac{\tau_3 s+1}{s(\tau_1 s+1)(\tau_2 s+1)}$
$1-\frac{1}{\sqrt{1-\zeta^2}}e^{-\zeta t/\tau}\sin(\alpha t+\phi)$ where $\alpha=\frac{\sqrt{1-\zeta^2}}{\tau}$, $\phi=\tan^{-1}\left(\frac{\sqrt{1-\zeta^2}}{\zeta}\right)$	$\frac{1}{s(\tau^2 s^2+2\zeta\tau s+1)}$

The *final-value theorem* is

$$\lim_{t\to\infty} f(t) = \lim_{s\to 0}[sF(s)] \tag{3.18a}$$

The *initial-value theorem* is

$$\lim_{t\to 0} f(t) = \lim_{s\to\infty}[sF(s)] \tag{3.18b}$$

It should be noted that these theorems only hold for stable systems.

Example 3.2: Application of Initial- and Final-Value Theorems

Find the long-term and short-term behavior of the time-domain function, $y(t)$, using the final- and initial-value theorems on the Laplace domain function $Y(s)$ (we see later that this arises from a step input applied to a second-order process):

$$Y(s) = \frac{4}{s(2s+1)(3s+1)}$$

The long-term behavior, $y(t \to \infty)$, is found using the *final-value theorem,*

$$\lim_{t\to\infty} y(t) = \lim_{s\to 0}[sY(s)] = \frac{s4}{s(2s+1)(3s+1)} = \frac{4}{(2s+1)(3s+1)} = 4$$

The short-term behavior, $y(t \to 0)$, is found using the *initial-value theorem,*

$$\lim_{t\to 0} y(t) = \lim_{s\to\infty}[sY(s)] = \frac{s4}{s(2s+1)(3s+1)} = \frac{4}{(2s+1)(3s+1)} = 0$$

The reader should verify that the time-domain function, $y(t)$, can be found by applying Table 3–1 to find

$$y(t) = 4 \cdot [1 - 3e^{-t/3} + 2e^{-t/2}]$$

and that the values of $y(t \to \infty)$, and $y(t \to 0)$, are consistent with the final- and initial-value theorems.

General Solution Procedure

To obtain analytical solutions for differential equation-based models, the general procedure is composed of several steps.

1. Start with an nth order linear differential equation

$$\frac{d^n y}{dt^n} + a_{n-1}\frac{d^{n-1}y}{dt^{n-1}} + \cdots + a_0 y = b_{n-1}\frac{d^{n-1}u}{dt^{n-1}} + \cdots + b_0 u$$

and known initial conditions

$$\left.\frac{d^{n-1}y}{dt^{n-1}}\right|_{t=0}, \cdots, y(0), \left.\frac{d^{n-2}u}{dt^{n-2}}\right|_{t=0}, \cdots, u(0)$$

2. Transform each element of the differential equation to the Laplace domain,

$$L\left[\frac{d^n y}{dt^n}\right] = s^n Y(s) - s^{n-1}y(0) - \cdots - \left.\frac{d^{n-1}y}{dt^{n-1}}\right|_{t=0}, \text{ etc.}$$

3. Use algebraic manipulations to solve for the transformed variable.

$$Y(s) = \frac{N(s)}{D(s)}$$

where $N(s)$ and $D(s)$ are polynomials in s.

4. Perform a partial fraction expansion to isolate the individual elements.

$$Y(s) = \frac{C_1}{s-a} + \frac{C_2}{s-b} + \cdots$$

where a, b, and so forth are the roots of $D(s)$. If roots are repeated the partial fraction expansion has the form (for an example where three roots all have a value of a).

$$Y(s) = \frac{C_1}{s-a} + \frac{C_2}{(s-a)^2} + \frac{C_3}{(s-a)^3} + \frac{C_4}{s-b} + \cdots$$

5. Invert each element back to the time domain to find the final solution for $y(t)$.

$$y(t) = C_1 e^{-at} + C_2 e^{-bt} + \ldots$$

For the case of repeated roots (say 3 that have a value of a), the solution has the form

$$y(t) = C_1 e^{-at} + C_2 t e^{-at} + \frac{C_3}{2} t^2 e^{-at} + C_4 e^{-bt} + \cdots$$

It should be noted that chemical process systems are rarely described by an nth order differential equation. Usually, a set of n first-order differential equations is transformed to a single nth order equation, as shown in Example 3.3.

Example 3.3: Second-Order Differential Equation

Consider the following state space model of an isothermal CSTR (Module 5).

$$\begin{aligned}\frac{dx_1}{dt} &= -2.4048x_1 + 7u \\ \frac{dx_2}{dt} &= 0.8333x_1 - 2.2381x_2 - 1.1170u \\ y &= x_2\end{aligned} \tag{3.19}$$

where x_1 and x_2 represent the concentrations of two components (in deviation variable form) in an isothermal reactor; the initial conditions are $x_1(0) = x_2(0) = 0$. Solve for the output (concentration of component 2) response to a unit step input.

1. The reader should show (see exercise 6) that this can be arranged to a second-order differential equation,

$$\frac{d^2y}{dt^2} + 4.6429\frac{dy}{dt} + 5.3821y = -1.1170\frac{du}{dt} + 3.1472u$$

with the initial conditions $y(0) = dy(0)/dt = 0$, and the input is initially $u(0) = 0$.

2. Taking the Laplace transform of each element,

$$L[d^2y/dt^2] = s^2Y(s) - sy(0) - dy(0)/dt = s^2Y(s)$$
$$L[dy/dt] = sY(s) - y(0) = sY(s)$$
$$L[y] = Y(s)$$
$$L[du/dt] = sU(s) - U(0) = sU(s)$$
$$L[u] = U(s)$$

and substituting back into the differential equation, we find[1]

$$s^2Y(s) + 4.6429(s) + 5.3821Y(s) = -1.1170sU(s) = 3.1472U(s)$$

3. Solving for $Y(s)$, we find

$$Y(s) = \frac{-1.1170s + 3.1472}{s^2 + 4.6429s + 5.3821}U(s)$$

[1]Alternatively, one could take the Laplace transform of each equation in (3.19) and use algebraic substitution to find this second-order equation in s.

Here we consider a unit step input, $U(s) = 1/s$.

$$Y(s) = \frac{-1.1170s + 3.1472}{s^2 + 4.6429s + 5.3821} \cdot \frac{1}{s} = \frac{-1.1170s + 3.1472}{s\left(s^2 + 4.6429s + 5.3821\right)}$$

$$= \frac{-1.1170s + 3.1472}{s(s + 2.4053)(s + 2.2376)}$$

4. Performing a partial fraction expansion using the roots of the denominator polynomial,

$$Y(s) = \frac{-1.1170s + 3.1472}{s(s + 2.4053)(s + 2.2376)} = \frac{C_1}{s} + \frac{C_2}{(s + 2.4053)} + \frac{C_3}{(s + 2.2376)} \tag{3.20}$$

Solve for C_1 by multiplying by s,

$$\frac{-1.1170s + 3.1472}{s(s + 2.4053)(s + 2.2376)} \cdot s = \frac{C_1}{s} \cdot s + \frac{C_2}{(s + 2.4053)} \cdot s + \frac{C_3}{(s + 2.2376)} \cdot s$$

Setting $s = 0$ to find

$$C_1 = \frac{3.1472}{2.4053 \cdot 2.2376} = 0.5848$$

Similarly, solve for C_2 by multiplying Equation (3.20) by $s + 2.4053$ and setting $s = -2.4053$

$$C_2 = 14.4630$$

Also, solve for C_3 by multiplying Equation (3.20) by $s + 2.2376$ and setting $s = -2.2376$

$$C_3 = -15.0478$$

which yields

$$Y(s) = \frac{C_1}{s} + \frac{C_2}{(s + 2.4053)} + \frac{C_3}{(s + 2.2376)} = \frac{0.5848}{s} + \frac{14.4630}{(s + 2.4053)} + \frac{-15.0478}{(s + 2.2376)}$$

5. Inverting each element back to the time domain,

$$y(t) = 0.5848 + 14.4630\exp(-2.4053t) - 15.0478\exp(-2.2376t) \tag{3.21}$$

Although it is nice to have an analytical solution, it is generally more pleasing to engineers to view a plot of a variable as it changes with time, as shown in Figure 3–5. In this plot we notice that the concentration initially decreases, before increasing to a new steady-state value.

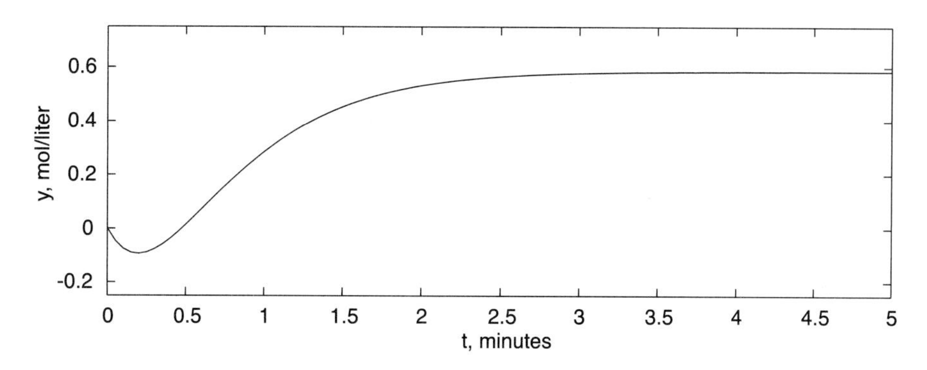

Figure 3–5 Response of the concentration as a function of time. See Equation (3.21).

In Example 3.3 we solved for the coefficients (C_1, C_2, C_3) by selecting values of s to minimize the multiplications performed during each step. An alternative is to solve for three equations in three unknowns by using various values of s (say $s = 1$, 2, and 3, for example) in Equation (3.20).

Differential equations textbooks, such as Boyce and DiPrima (1992) present many examples of applications of Laplace transforms to solve differential equations. In practice, it is rare for process engineers to seek analytical solutions to differential equations; it is far easier to solve these numerically. The primary goal of this section was to provide background material to understand the transfer functions and dynamic responses shown in the next section.

3.4 Transfer Functions

A transfer function relates inputs to outputs in the Laplace domain. In particular, the Laplace domain relationship between a manipulated input and output is called a *process transfer function*.

Consider an nth order differential equation,

$$\frac{d^n y}{dt^n} + a_{n-1}\frac{d^{n-1}y}{dt^{n-1}} + \cdots + a_0 y = b_{n-1}\frac{d^{n-1}u}{dt^{n-1}} + \cdots + b_0 u \tag{3.22a}$$

Since we are assuming that the model is based on deviation variables, and that the system is initially at steady state, the initial conditions are

$$\left.\frac{d^{n-1}y}{dt^{n-1}}\right|_{t=0} = \cdots = y(0) = \left.\frac{d^{n-2}u(0)}{dt^{n-2}}\right|_{t=0} = \cdots = u(0) = 0 \tag{3.22b}$$

Taking the Laplace transform of each term,

$$L\left[\frac{d^n y}{dt^n}\right] = s^n Y(s) - s^{n-1}y(0) - \cdots - \left.\frac{d^{n-1}y}{dt^{n-1}}\right|_{t=0} = s^n Y(s)$$

$$\vdots$$

$$L[y] = Y(s)$$

$$L\left[\frac{d^{n-1} u}{dt^{n-1}}\right] = s^{n-1}U(s) - s^{n-2}u(0) - \cdots - \left.\frac{d^{n-2}u}{dt^{n-2}}\right|_{t=0} = s^{n-1}U(s)$$

$$\vdots$$

$$L[u] = U(s)$$

Solving for $Y(s)$, we find

$$Y(s) = \frac{b_{n-1}s^{n-1} + \cdots + b_0}{s^n + a_{n-1}s^{n-1} + \cdots + a_0} \cdot U(s)$$

The ratio of polynomials is called the *transfer function.* When it relates a manipulated input to an output it is commonly called a process transfer function. In general, we will use $g_p(s)$ to represent the process transfer function.

$$Y(s) = g_p(s)U(s) \tag{3.23}$$

In this case the process transfer function is

$$g_p(s) = \frac{b_{n-1}s^{n-1} + \cdots + b_0}{s^n + a_{n-1}s^{n-1} + \cdots + a_0} \tag{3.24}$$

The roots of the numerator polynomial are known as *zeros*, and the roots of the denominator polynomial are call *poles*. The significance of poles and zeros are discussed in detail in Section 3.9.

We have used capital letters to distinguish Laplace domain variables from the time-domain variables. *In the rest of this text we generally use lowercase letters for all input and output variables.* If the argument is s, then we assume that we are referring to the Laplace domain.

Transfer functions are often used in *block diagrams*. For example, the relationship between an input and output is shown as

$$u(s) \longrightarrow \boxed{g_p(s)} \longrightarrow y(s)$$

In the rest of this chapter we study the dynamic behavior of some commonly used transfer functions. Our focus is on *step responses*, since process engineers often apply step changes to understand dynamic behavior.

3.5 First-Order Behavior

Many chemical processes can be modeled as first-order systems. The differential equation for a linear first-order process is often written in the following form:

$$\tau_p \frac{dy}{dt} + y = k_p u \tag{3.25}$$

This can also be written as

$$\frac{dy}{dt} = -\frac{1}{\tau_p} y + \frac{k_p}{\tau_p} u$$

where the parameters (τ_p and k_p) and variables (y and u) have the following names: τ_p is the process time constant (units of time), k_p the process gain (units of output/input), y the output variable, and u the input variable. Taking the Laplace transform of each term (*notice that we are now using lower-case variables to represent the Laplace domain input and output*), and assuming that the initial condition is $y(0) = 0$,

$$L\left[\tau_p \frac{dy}{dt}\right] = \tau_p L\left[\frac{dy}{dt}\right] = \tau_p [sy(s) - y(0)] = \tau_p \cdot sy(s)$$

$$L[y] = y(s)$$

$$L[k_p u(t)] = k_p \cdot L[u(t)] = k_p \cdot u(s)$$

So the Laplace transform of Equation (3.25) can be written

$$(\tau_p s + 1) y(s) = k_p u(s)$$

or solving for $y(s)$ we find a first-order transfer function

$$y(s) = \frac{k_p}{\tau_p s + 1} u(s) \tag{3.26}$$

Step Response

Consider the case where the output is initially zero (steady state in deviation variable form), and the input is suddenly step changed by an amount Δu. The Laplace transform of the input is

$$u(s) = \frac{\Delta u}{s} \tag{3.27}$$

So Equation (3.26) can be written

$$y(s) = \frac{k_p}{\tau_p s + 1} \cdot \frac{\Delta u}{s} \tag{3.28}$$

Using a partial fraction expansion and inverting to the time domain, you should find (see Exercise 1)

$$y(t) = k_p \Delta u \left(1 - e^{-t/\tau_p}\right) \tag{3.29}$$

Here the notion of a process gain is clear. After a substantial amount of time ($t >> \tau_p$), we find, from Equation (3.29),

$$y(t \to \infty) = k_p \Delta u \tag{3.30}$$

That is,

$$k_p = \frac{y(t \to \infty)}{\Delta u}$$

and, since $y(0) = 0$, we can think of $y(t \to \infty)$ as Δy, so

$$k_p = \frac{\Delta y}{\Delta u} = \frac{\text{long-term output change}}{\text{input change}} \tag{3.31}$$

We can think of the process time constant as the amount of time it takes for 63.2% of the ultimate output change to occur, since when $t = \tau_p$,

$$y\left(t = \tau_p\right) = k_p \Delta u \left(1 - e^{-\tau_p/\tau_p}\right) = k_p \Delta u \left(1 - e^{-1}\right) = 0.632 k_p \Delta u = 0.632 \Delta y$$

Remember that this holds true only for first-order systems.

Impulse Response

Consider now an impulse input of magnitude P, which has units of the input*time; if the input is a volumetric flow rate (volume/time), then the impulse input is a volume. The output response is

$$y(s) = \frac{k_p}{\tau_p s + 1} \cdot P$$

You should find that the time domain solution is

$$y(t) = \frac{P k_p}{\tau_p} e^{-t/\tau_p}$$

which has an immediate response of Pk_p/τ_p followed by a first-order decay with time.

Example 3.4: Stirred-Tank Heater

Recall that an energy balance on a constant-volume stirred-tank heater (Example 2.3) yielded

$$\frac{dT}{dt} = \frac{F_s}{V_s}\left(T_{is} - T\right) + \frac{Q}{V_s \rho c_p}$$

where the subscript s is used to indicate that a particular variable remains at its steady-state value. Defining the following deviation variables

$$u = Q - Q_s$$
$$y = T - T_s$$

The equation can be written in the form

$$\frac{V_s}{F_s}\frac{dy}{dt} = -y + \frac{1}{F_s \rho c_p} u$$

or

$$\tau_p \frac{dy}{dt} = -y + k_p u$$

where the parameters of this first-order model are

$$k_p = \frac{1}{F_s \rho c_p} = \text{ process gain}$$

$$\tau_p = \frac{V_s}{F_s} = \text{ time constant}$$

The gain and time constant are clearly a function of scale. A process with a large steady-state flow rate will have a low gain, compared to a process with a small steady-state flow rate. This makes physical sense, since a given heat power input will have a larger effect on the small process than the large process. Similarly, a process with a large volume-to-flow rate ratio is expected to have a slow response compared to a process with a small volume-to-flow rate ratio.

Consider a heater with a constant liquid volume of V_s = 50 liters and a constant volumetric flow rate of F_s = 10 liters/minute. For liquid water, the other parameters are ρc_p = 1 kcal/ liter°C. The process gain and time constant are then

$$k_p = \frac{1}{F_s \rho c_p} = 1.434 \ \frac{°\text{C}}{kW}$$

$$\tau_p = \frac{V_s}{F_s} = \text{ 5 minutes}$$

Step Response

If a step input change of 10 kW is made, the resulting output change is [from Equation (3.29)]

$$y(t) = k_p \Delta u\left(1 - e^{-t/\tau_p}\right) = 1.434 \cdot 10\left(1 - e^{-t/5}\right) = 14.34\left(1 - e^{-t/5}\right)$$

A plot of the step input and the resulting output are shown in Figure 3–6.

Remember that the inputs and outputs in this expression are in deviation variable form. If the steady-state values are (for an inlet temperature of 20°C)

$$T_s = 20°\text{C}$$
$$Q_s = 0 \text{ kW}$$

Then, the physical temperature response is

$$T = T_s + y = 20 + 14.34(1 - e^{-t/5})$$

Impulse Response

If an impulse input of 30 kJ is made, you should be able to show that the temperature changes immediately by 0.143°C (see Exercise 15).

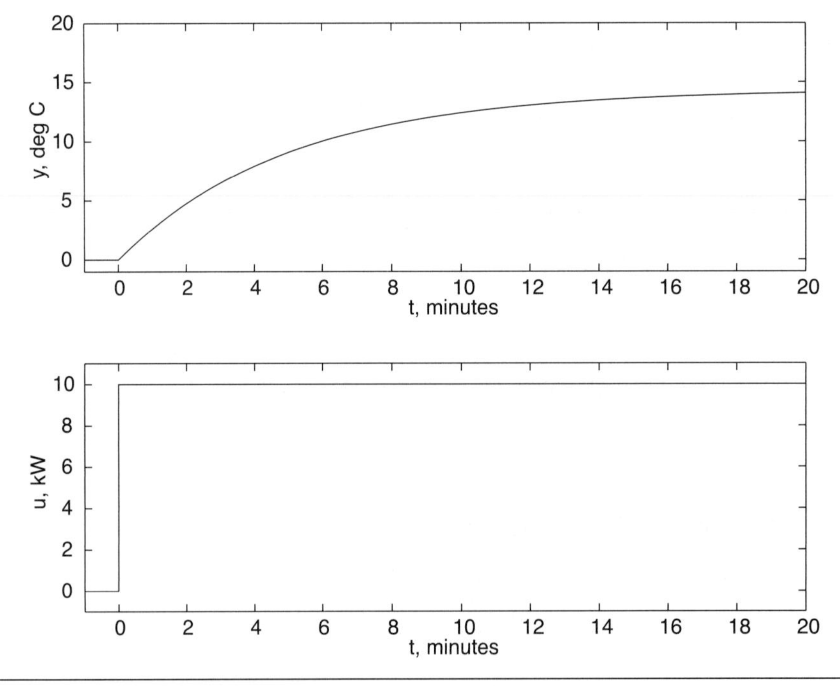

Figure 3–6 Step response of a stirred-tank heater, characteristic of a first-order system. Deviation variables.

3.6 Integrating System

We found in the previous chapter that material balances on liquid surge vessels or gas drums often yielded models with the following form:

$$\frac{dy}{dt} = ku \tag{3.32}$$

In the Laplace domain, this is

$$y(s) = \frac{k}{s}u(s) \tag{3.33}$$

Consider an integrating process initially at steady state, with $y(0) = 0$.

Step Response

If a step input change of Δu is made at $t = 0$,

$$y(s) = \frac{k}{s} \cdot \frac{\Delta u}{s} = \frac{k\Delta u}{s^2}$$

and we find the time-domain value

$$y(t) = k \cdot t \cdot \Delta u \tag{3.34}$$

That is, the output ramps with a constant slope of $k\Delta u$.

Impulse Response

If an impulse input of magnitude P is made at $t = 0$,

$$y(s) = \frac{k}{s} \cdot P$$

then the output immediately changes to a new steady-state value of

$$y(t) = kP$$

Example 3.5: Tank-Height Problem

The mathematical model for a liquid surge tank is (see Example 1.3)

$$\frac{dh}{dt} = \frac{F_1 - F_2}{A}$$

where h is the liquid height, A is the constant cross-sectional area of the tank, F_1 is the inlet flow rate, and F_2 is the outlet flow rate. Assume that the outlet flow rate remains constant at a steady-state value of F_{2s}. Defining the output and input in deviation variable form as

$$y = h - h_s$$
$$u = F_1 - F_{1s}$$

For a constant cross-sectional area of 10 m^2, the model is

$$\frac{dy}{dt} = \frac{1}{10} \cdot u = k \cdot u$$

Step Response

For a step input change of 0.25 m^3/min, the output response is

$$y(t) = k \cdot t \cdot \Delta u = 0.025t$$

which is shown in Figure 3–7. If the steady-state height is 2 meters, then the height as a function of time is

$$h(t) = h_s + y(t) = 2 + 0.025t$$

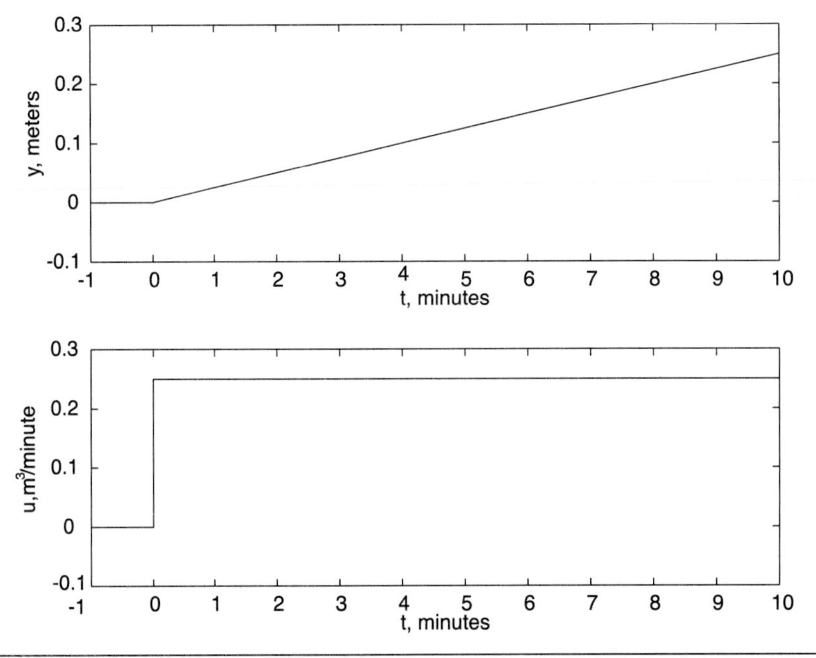

Figure 3–7 Step response of a liquid surge tank. Deviation variables.

Impulse Response

For an impulse input of 1 m^3, the output response is

$$y(t) = 0.1 \cdot 1 = 0.1$$

which makes sense, because the cross-sectional area is 10 m^3.

3.7 Second-Order Behavior

Second-order models arise from systems that are modeled with two differential equations (two states). In this section we separately consider transfer functions that do not have "numerator" dynamics and those that do.

Pure Second-Order Systems

Consider a linear second-order ODE, with constant parameters

$$a_2 \frac{d^2y}{dt^2} + a_1 \frac{dy}{dt} + a_0 y = b_0 u(t) \tag{3.35}$$

This is often written in the form

$$\tau^2 \frac{d^2y}{dt^2} + 2\zeta\tau \frac{dy}{dt} + y = ku(t) \tag{3.36}$$

where (obviously $a_0 \neq 0$)

$$\tau^2 = a_2/a_0,\ 2\zeta\tau = a_1/a_0,\ k = b_0/a_0$$

and the parameters are as follows: k is the gain (units of output/input), ζ the damping factor (dimensionless), and τ the natural period (units of time).

The second-order model shown in Equation (3.35) or (3.36) generally arises by changing a set of two first-order equations (state-space model) to a single second-order equation. For a given second-order ODE, there are an infinite number of sets of two first-order (state-space) models that are equivalent.

Taking the Laplace transform of Equation (3.36),

$$\tau^2[s^2y(s) - sy(0) - \dot{y}(0)] + 2\zeta\tau\,[sy(s) - y(0)] + y(s) = k \cdot u(s) \tag{3.37}$$

where $y(s)$ indicates the Laplace transformed variable.

Assuming initial conditions are zero, that is $dy/dt|_{t=0} = y(0) = 0$, we find

$$y(s) = \frac{k}{\tau^2 s^2 + 2\zeta\tau s + 1} \cdot u(s) \tag{3.38}$$

The *characteristic equation* of the second-order transfer function is $\tau^2 s^2 + 2\zeta\tau s + 1$. We can find the roots (known as the *poles*) by using the quadratic formula

$$p_i = \frac{-2\zeta\tau \pm \sqrt{4\zeta^2\tau^2 - 4\tau^2}}{2\tau^2} = -\frac{\zeta}{\tau} \pm \frac{\sqrt{4\tau^2\left(\zeta^2 - 1\right)}}{2\tau^2} = -\frac{\zeta}{\tau} \pm \frac{\sqrt{\zeta^2 - 1}}{\tau}$$

which yields the following values for the roots:

$$p_1 = -\frac{\zeta}{\tau} + \frac{\sqrt{\zeta^2 - 1}}{\tau}, \; p_2 = -\frac{\zeta}{\tau} - \frac{\sqrt{\zeta^2 - 1}}{\tau} \tag{3.39}$$

The following analysis assumes that $\zeta > 0$ and $\tau > 0$. This implies that the real portions of p_1 and p_2 are negative and, therefore, the system is stable. The three possible cases are shown in Table 3–2.

Step Responses

Now, we consider the dynamic response of second-order systems to step inputs ($u(s) = \Delta u/s$),

$$y(s) = \frac{k}{\tau^2 s^2 + 2\zeta\tau s + 1} \cdot \frac{\Delta u}{s} \tag{3.40}$$

where Δu represents the magnitude of the step change.

Case 1: Overdamped ($\zeta > 1$) For $\zeta > 1$, the denominator polynomial, $\tau^2 s^2 + 2\zeta\tau s + 1$, can be factored into the form

$$\tau^2 s^2 + 2\zeta\tau s + 1 = (\tau_1 s + 1)(\tau_2 s + 1)$$

where the time constants are

$$\tau_1 = \frac{\tau}{\zeta - \sqrt{\zeta^2 - 1}}, \quad \tau_2 = \frac{\tau}{\zeta + \sqrt{\zeta^2 - 1}}$$

Table 3–2 Characteristic Behavior of Second-Order Transfer Functions

Case	***Damping factor***	***Pole location***	***Characteristic behavior***
1	$\zeta > 1$	Two real, distinct roots	Overdamped
2	$\zeta = 1$	Two real, equal roots	Critically damped
3	$\zeta < 1$	Two complex conjugate roots	Underdamped

We can derive the following solution for step responses of overdamped systems,

$$y(t) = k\Delta u\left[1 + \frac{\tau_1 e^{-t/\tau_1} - \tau_2 e^{-t/\tau_2}}{\tau_2 - \tau_1}\right] \tag{3.41}$$

Note that, as in the case of first-order systems, we can divide by $k\Delta u$ to develop a dimensionless output. Also, the dimensionless time is t/τ and we can plot curves for dimensionless output as a function of ζ. This is done in Figure 3–8, which includes the critically damped case, as discussed next. Most chemical processes exhibit overdamped behavior.

Case 2: Critically damped ($\zeta = 1$) The transition between overdamped and underdamped is known as critically damped. We can derive the following for the step response of a critically damped system

$$y(t) = k\Delta u \cdot \left[1 - \left(1 + \frac{t}{\tau}\right)e^{-t/\tau}\right] \tag{3.42}$$

Notice that the main difference between overdamped (or critically damped) step responses and first-order step responses is that the second-order step responses have an *S* shape with a maximum slope at an inflection point, whereas the first-order responses have their maximum slope initially.

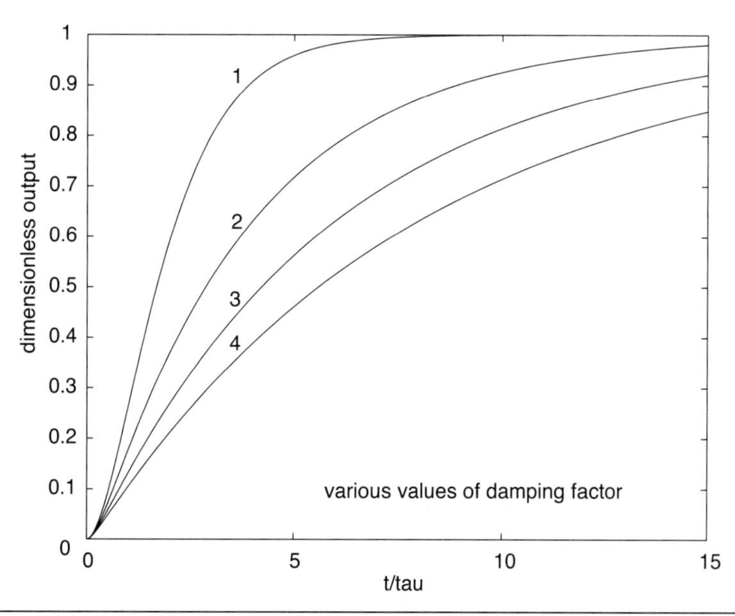

Figure 3–8 Step response of a second-order overdamped system.

The initial behavior for a step change is really dictated by the *relative order* of the system. The relative order is the difference between the orders of the numerator and denominator polynomials in the transfer function. If the relative order is 1, then output response has a nonzero slope at the time of the step input; the step response of a system with a relative order greater than 1 has a zero slope at the time of the step input.

Case 3: Underdamped ($\zeta < 1$) For $\zeta < 1$, we find [from Equation (3.39)] that the poles are complex,

$$p_{1,2} = -\frac{\zeta}{\tau} \pm \frac{\sqrt{\zeta^2 - 1}}{\tau} \quad \text{or} \quad p_{1,2} = \text{Re} \pm j\,\text{Im} \tag{3.43a}$$

where the real and imaginary contributions are

$$\text{Re} = -\frac{\zeta}{\tau} \qquad \text{Im} = \frac{\sqrt{1-\zeta^2}}{\tau} \tag{3.43b}$$

We can derive the following step response for an underdamped system,

$$y(t) = k\Delta u \cdot \left[1 - \frac{1}{\sqrt{1-\zeta^2}} e^{-\zeta t/\tau} \sin(\alpha t + \phi)\right] \tag{3.44a}$$

where

$$\alpha = \frac{\sqrt{1-\zeta^2}}{\tau}, \qquad \phi = \tan^{-1}\left(\frac{\sqrt{1-\zeta^2}}{\zeta}\right) \tag{3.44b}$$

Again, dividing Equation (3.44a) by $k\Delta u$, we can produce the dimensionless plot shown in Figure 3–9.

A number of insights can be obtained from Figure 3–9 and from an analysis of the step response equations. For $\zeta < 1$, the ratio of the imaginary portion to the real portion of the pole [from Equation (3.39)] is

$$\frac{\text{Im}}{\text{Re}} = \frac{\sqrt{1-\zeta^2}}{\zeta}$$

As the imaginary/real ratio gets larger, the response becomes more oscillatory (ζ becomes smaller). We also notice that a decreasing ratio corresponds to a larger negative value for the real portion. As the real portion becomes larger in magnitude (more negative), the response becomes faster. We use these insights to interpret pole/zero plots in Section 3.9.

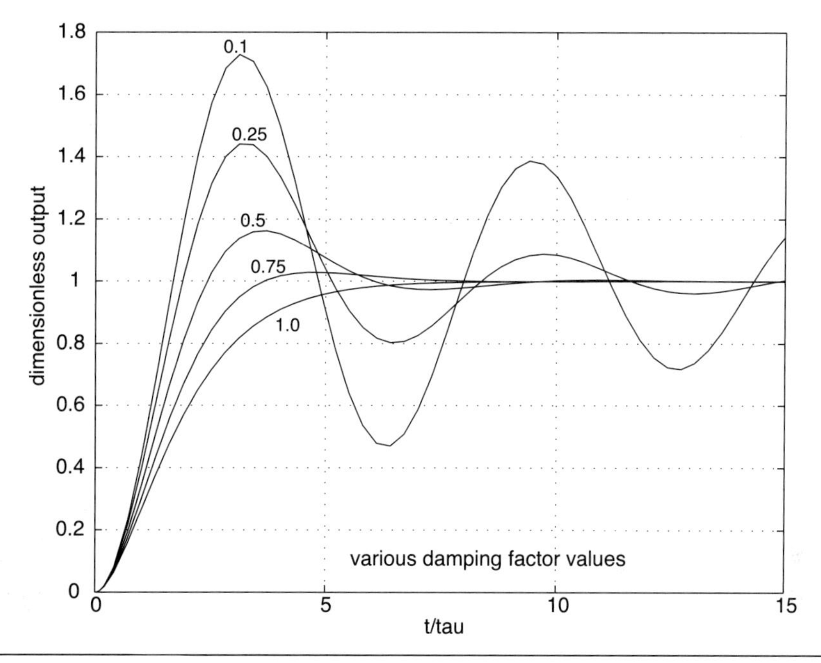

Figure 3–9 Step response of a second-order underdamped system as a function of the damping factor (ζ). Interpolate between the curves for the behavior of other damping factor values.

Underdamped Step Response Characteristics

The following common measures of underdamped second-order step responses are shown in Figure 3–10, and defined below: (1) rise time, (2) time to first peak, (3) overshoot, (4) decay ratio, and (5) period of oscillation.

Rise time is the amount of time it takes to first reach the new steady-state value. *Time to first peak* is the time required to reach the first peak. *Overshoot* is the distance between the first peak and the new steady state. This is usually expressed as the overshoot ratio, as shown in Figure 3–10. *Decay ratio* is a measure of how rapidly the oscillations are decreasing. A b/a ratio of 1/4 is commonly called "quarter wave damping." *Period of oscillation* is the time between successive peaks.

Second-Order Systems with Numerator Dynamics

The previous discussion involved pure second-order systems, where the *relative order* (difference between the denominator and numerator polynomial orders) was two.

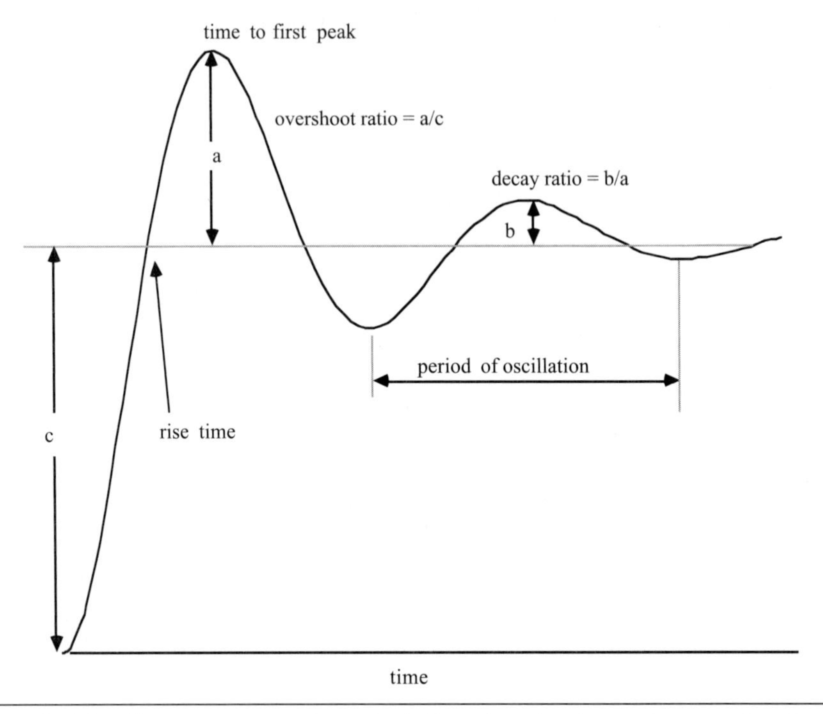

Figure 3–10 Step response characteristics of underdamped second-order processes.

Consider now a second-order system with numerator dynamics with the gain/time constant form

$$g_p(s) = \frac{k_p(\tau_n s + 1)}{(\tau_1 s + 1)(\tau_2 s + 1)} \tag{3.45}$$

which is relative order one. The reader should show that the pole-zero form is

$$g_p(s) = \frac{k_{pz}(s - z_1)}{(s - p_1)(s - p_2)}$$

where the parameters are

$$k_{pz} = \frac{k\tau_n}{\tau_1\tau_2}, \quad p_1 = -\frac{1}{\tau_1}, \quad p_2 = -\frac{1}{\tau_2}, \quad z_1 = \frac{-1}{\tau_n}$$

The gain/time constant form has the following time-domain response to a step input (see Exercise 4):

$$y(t) = k_p \Delta u \cdot \left[1 + \left(\frac{\tau_n - \tau_1}{\tau_1 - \tau_2} \right) e^{-t/\tau_1} + \left(\frac{\tau_n - \tau_2}{\tau_2 - \tau_1} \right) e^{-t/\tau_2} \right] \tag{3.46}$$

The reader should show that, if $\tau_n = \tau_2$, the response is the same as a first-order process.

Example 3.6: Illustration of Numerator Dynamics

Consider the following input-output relationship:

$$y(s) = \frac{(\tau_n s + 1)}{(3s+1)(15s+1)} \cdot u(s)$$

The unit step responses are shown in Figure 3–11. Notice that negative numerator time constants (corresponding to positive zeros) yield a step response which initially decreases before increasing to the final steady state. This type of response is known as *inverse response* and causes tough challenges for process control systems. Positive zeros are often caused by two first-order transfer functions, with gains of opposite sign, acting in parallel (see Exercise 5).

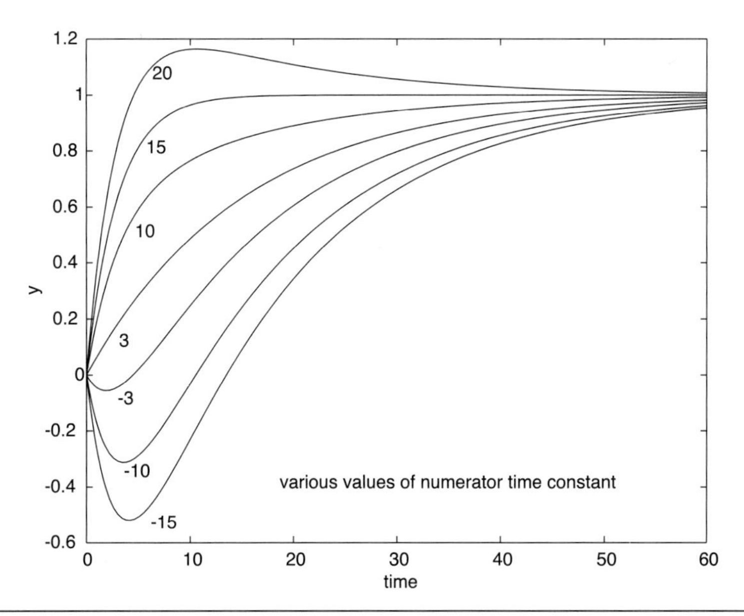

Figure 3–11 Step responses of a second-order system with numerator dynamics.

Notice also that a numerator time constant that is greater than the denominator time constant causes overshoot before settling to the final steady state. Also notice that the inverse response becomes "deeper" as the numerator zero ($-1/\tau_n$) approaches a value of 0 from the positive side.

3.8 Lead-Lag Behavior

Lead-lag transfer functions have the same order numerator polynomial as denominator polynomial. This occurs when the input has a direct effect on the output variable. In terms of the state space model (3.2), this means that $D \neq 0$.

Consider a lead-lag transfer function where the numerator and denominator polynomials are first order:

$$g_p(s) = k_p \cdot \frac{\tau_n s + 1}{\tau_p s + 1} \tag{3.47}$$

For a step input of magnitude Δu, the output response is

$$y(t) = k_p \Delta u \cdot \left[1 - \left(1 - \frac{\tau_n}{\tau_p} \right) e^{-t/\tau_p} \right] \tag{3.48}$$

A dimensionless output can be defined by dividing Equation (3.48) by $k_p \Delta u$, and t/τ_p is a natural dimensionless time. The responses to a step input at $t = 0$ are shown in Figure 3–12. Notice that there is an immediate response that is equal to the τ_n/τ_p ratio. This can also be found by applying the initial value theorem to Equation (3.47) for a step input change (see Exercise 3).

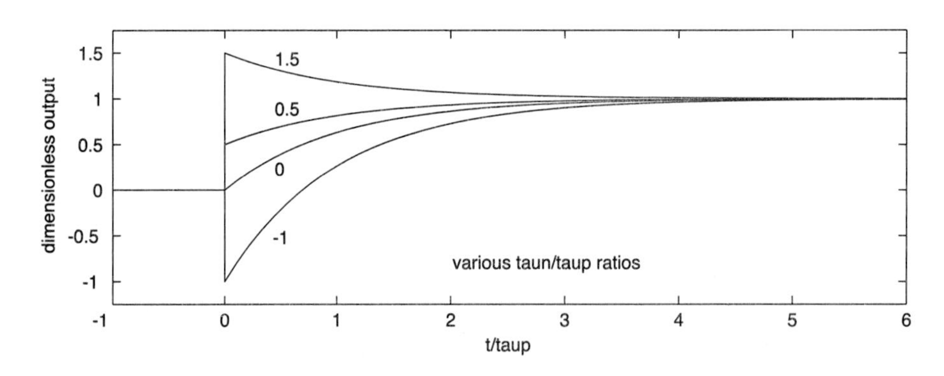

Figure 3–12 Step responses of the lead-lag example.

It is rare for processes to exhibit lead-lag behavior, but many controllers exhibit such behavior.

3.9 Poles and Zeros

There are a number of different ways to represent process transfer functions. The "polynomial" form is

$$g(s) = \frac{b_m s^m + b_{m-1} s^{m-1} + \cdots + b_1 s + b_0}{a_n s^n + a_{n-1} s^{n-1} + \cdots + a_1 s + a_0} \tag{3.49}$$

The values of s that cause the numerator of Equation (3.49) to equal zero are known as the "zeros" of the transfer function. The values of s that cause the denominator of Equation (3.49) to equal zero are known as the "poles" of the transfer function.

The "pole-zero" form is

$$g(s) = \frac{k_{pz}(s - z_1)(s - z_2)\cdots(s - z_m)}{(s - p_1)(s - p_2)\cdots(s - p_n)} \tag{3.50}$$

and complex poles (or zeros) must occur in complex conjugate pairs.

The "gain-time constant" form is the one that we use most often for control system design.

$$g(s) = \frac{k_p(\tau_{n1} s + 1)(\tau_{n2} s + 1)\cdots(\tau_{nm} s + 1)}{(\tau_{p1} s + 1)(\tau_{p2} s + 1)\cdots(\tau_{pn} s + 1)} \tag{3.51}$$

where τ_{ni} is a numerator time constant and τ_{pi} is a denominator time constant. This form is normally used when the roots (poles) of the denominator polynomial are real.

Example 3.7: Comparison of Various Transfer Function Forms

Consider a transfer function with the following gain-time constant form

$$g(s) = \frac{2(-10s + 1)}{(3s + 1)(15s + 1)}$$

The polynomial form is

$$g(s) = \frac{-20s + 2}{45s^2 + 18s + 1}$$

The gain-polynomial form is

$$g(s) = \frac{2(-10s+1)}{45s^2+18s+1}$$

and the pole-zero form is

$$g(s) = \left(-\frac{4}{9}\right)\frac{\left(s-\frac{1}{10}\right)}{\left(s+\frac{1}{3}\right)\left(s+\frac{1}{15}\right)}$$

The zero is 1/10, and the poles are –1/3 and –1/15.

Notice that the zero for Example 3.7 is positive. A positive zero is called a right-half-plane (RHP) zero, because it appears in the right half of the complex plane (with real and imaginary axes). RHP zeros have a characteristic *inverse response*, as shown in Figure 3–11 for $\tau_n = -10$ (which corresponds to a zero of $+0.1$).

Also notice that the poles are negative (left-half-plane), indicating a stable process. RHP (positive) poles are unstable. Recall that complex poles will yield an oscillatory response. A pole-zero plot of the transfer function in Example 3.7 is shown in Figure 3–13 [the pole locations are (-1/3,0) and (-1/15,0) and the zero location is (1/10,0), with the coordinates (real,imaginary)]. For this system, there is no imaginary component and the poles and zeros lie on the real axis.

As poles move further to the left they yield a faster response, and increasing the magnitude of the imaginary portion makes the response more oscillatory. This behavior is summarized in Figure 3–14. Recall also that a process with a pole at the origin (and none in the RHP) is known as an *integrating* system; that is, the system never settles to a steady state when a step input change is made.

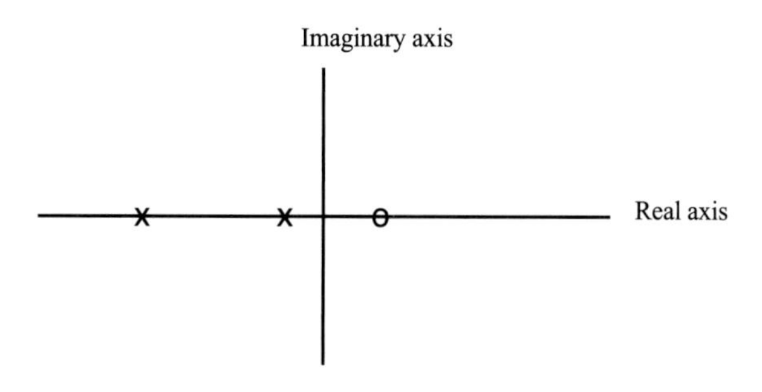

Figure 3–13 Pole-zero location plot for Example 3.7 (x, poles; o, zero).

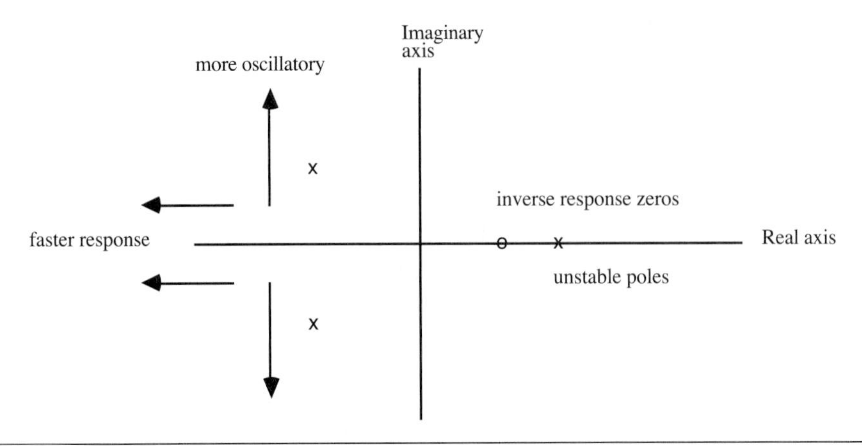

Figure 3–14 Effect of pole-zero location on dynamic behavior (x, poles; o, zero). As poles become more negative, the response is faster. As the imaginary/real ratio increases, the response becomes more oscillatory.

Multiple RHP zeros cause multiple "changes in direction"; for example, with two RHP zeros, the step response, initially going in one direction, switches direction, then switches back to the initial direction.

3.10 Processes with Dead Time

Many processes have a delayed response to a process input, either due to transport lags (such as flow through pipes) or high-order effects. In Section 3.3.4 we found that a time-delay of θ time units had a transfer function of $e^{-\theta s}$.

A first-order process combined with a time-delay has the transfer function

$$g_p(s) = \frac{k_p e^{-\theta s}}{\tau_p s + 1}$$

Consider a process with $k_p = 1$, $\tau_p = 10$, and $\theta = 5$. A unit step input change at $t = 0$ yields the response shown in Figure 3–15. We see that the time-delay shifts the response of the output.

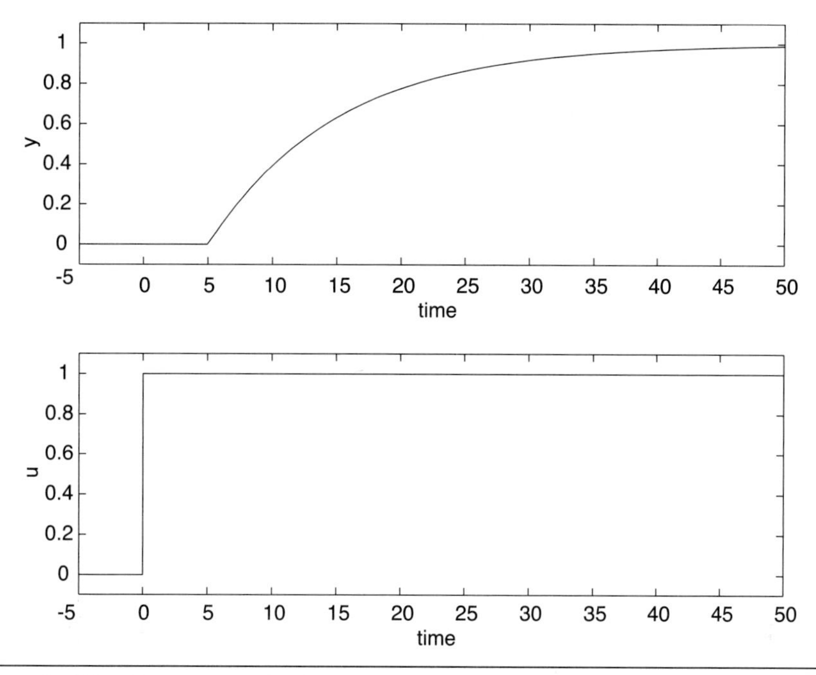

Figure 3–15 Response of a first-order + dead time process.

3.11 Padé Approximation for Dead Time

As discussed in the previous section the transfer function for a pure time delay is $e^{-\theta s}$, where θ is the time delay. Some control system design techniques require a rational transfer function; the Padé approximation for dead time is often used in this case.

A first-order Padé approximation is

$$e^{-\theta s} \approx \frac{-\frac{\theta}{2}s+1}{\frac{\theta}{2}s+1}$$

A second-order Padé approximation is

$$e^{-\theta s} \approx \frac{\frac{\theta^2}{12}s^2-\frac{\theta}{2}s+1}{\frac{\theta^2}{12}s^2+\frac{\theta}{2}s+1}$$

Example 3.8: Application of the Padé Approximations for Dead Time

Consider the first-order + dead time transfer function, where the time-delay dominates the time constant

$$g(s) = \frac{e^{-10s}}{5s+1}$$

The first-order Padé approximation yields the transfer function

$$g_1(s) = \frac{1}{5s+1} \cdot \frac{-5s+1}{5s+1} = \frac{-5s+1}{25s^2+10s+1}$$

and the second-order Padé approximation yields

$$g_2(s) = \frac{1}{5s+1} \cdot \frac{\frac{100}{12}s^2 - \frac{10}{2}s+1}{\frac{100}{12}s^2 + \frac{10}{2}s+1} = \frac{8.3333s^2 - 5s+1}{41.667s^3 + 33.333s^2 + 10s+1}$$

A comparison of the step responses of $g(s)$, $g_1(s)$, and $g_2(s)$ is shown in Figure 3–16. Notice that the first-order approximation has an inverse response, while the second-order approximation has a "double inverse response." The reader should find that there is a single positive zero for $g_1(s)$, and there are two positive, complex-conjugate zeros of the numerator transfer function of $g_2(s)$.

Most ordinary differential equation numerical integrators require pure differential equations (with no time delays). If you have a system of differential equations that has time delays, the Padé approximation can be used to convert them to delay-free differential equations, which can then be numerically integrated.

One of the many advantages to using SIMULINK is that time delays are easily handled so that no approximation is required.

3.12 Converting State Space Models to Transfer Functions

A general state space model can be converted to transfer function form, using the following steps. Starting with the state space model

$$\dot{x} = Ax + Bu$$
$$y = Cx + Du$$

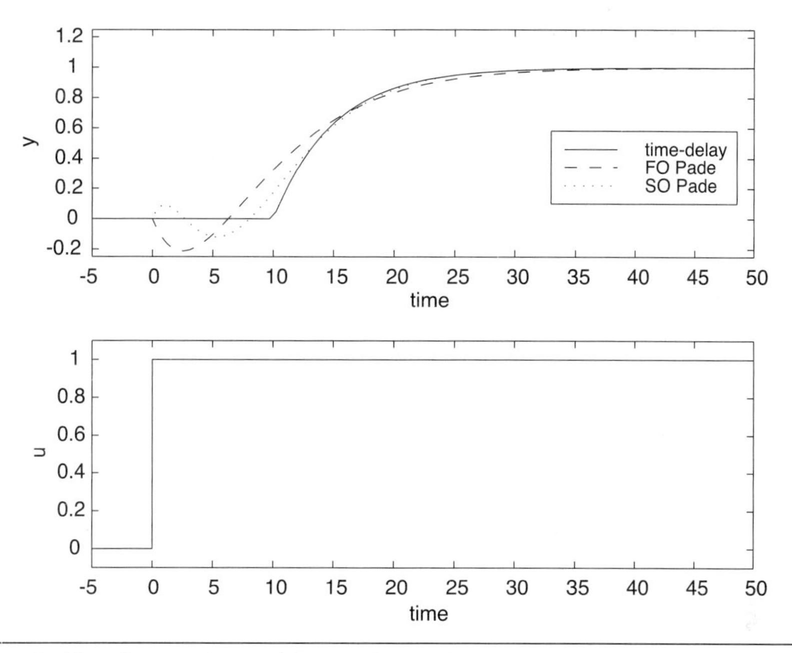

Figure 3–16 Comparison of first-order + dead time response with first and second-order Padé approximations for dead time.

Take the Laplace transform of each term, assuming zero initial conditions

$$sx(s) = Ax(s) + Bu(s)$$
$$y(s) = Cx(s) + Du(s)$$

Solving for $x(s)$, then $y(s)$ (it should be noted that often $D = 0$)

$$x(s) = (sI - A)^{-1}Bu(s)$$
$$y(s) = [C(sI - A)^{-1} B + D]u(s)$$
$$y(s) = G(s)u(s)$$

where $G(s)$ is a transfer function matrix. For example, the transfer function relating input j and output i is

$$y_i(s) = g_{ij}(s)u_j(s)$$

or in matrix form (with m inputs and r outputs)

$$\begin{bmatrix} y_1(s) \\ \vdots \\ y_r(s) \end{bmatrix} = \begin{bmatrix} g_{11}(s) & \cdots & g_{1m}(s) \\ \vdots & & \\ g_{r1}(s) & \cdots & g_{rm}(s) \end{bmatrix} \begin{bmatrix} u_1(s) \\ \vdots \\ u_m(s) \end{bmatrix}$$

Example 3.9: Isothermal CSTR

Consider the isothermal CSTR shown in Example 3.3 and Module 5. The state space model is

$$A = \begin{bmatrix} -2.4048 & 0 \\ 0.83333 & -2.2381 \end{bmatrix} \quad B = \begin{bmatrix} 7 & 0.5714 \\ -1.117 & 0 \end{bmatrix}$$

$$C = \begin{bmatrix} 0 & 1 \end{bmatrix} \qquad D = \begin{bmatrix} 0 & 0 \end{bmatrix}$$

The first input (u_1, manipulated) is the dilution rate (F/V), the second input (u_2, disturbance) is the feed concentration (C_{Af}), and the output is the concentration of the intermediate component, B. The eigenvalues of A (obtained by solving $\det(\lambda I - A) = 0$) are –2.4048 and –2.2381 min^{-1}.

The sequence of steps used to find the transfer function matrix is

$$sI - A = \begin{bmatrix} s + 2.4048 & 0 \\ -0.83333 & s + 2.2381 \end{bmatrix}$$

$$(sI - A)^{-1} = \begin{bmatrix} s + 2.4048 & 0 \\ -0.83333 & s + 2.2381 \end{bmatrix}^{-1} = \begin{bmatrix} s + 2.2381 & 0 \\ 0.83333 & s + 2.4048 \end{bmatrix} \cdot \frac{1}{\det(sI - A)}$$

$$= \begin{bmatrix} s + 2.2381 & 0 \\ 0.83333 & s + 2.4048 \end{bmatrix} \cdot \frac{1}{(s + 2.2381)(s + 2.4048) - (0)(0.8333)}$$

$$= \begin{bmatrix} \dfrac{1}{s + 2.4048} & 0 \\ \dfrac{0.83333}{(s + 2.2381)(s + 2.4048)} & \dfrac{1}{s + 2.2381} \end{bmatrix}$$

and multiplying,

$$C(sI - A)^{-1} = \begin{bmatrix} 0 & 1 \end{bmatrix} \begin{bmatrix} \dfrac{1}{s + 2.4048} & 0 \\ \dfrac{0.83333}{(s + 2.2381)(s + 2.4048)} & \dfrac{1}{s + 2.2381} \end{bmatrix} =$$

$$= \begin{bmatrix} \dfrac{0.83333}{(s + 2.2381)(s + 2.4048)} & \dfrac{1}{s + 2.2381} \end{bmatrix}$$

Also,

$$C(sI - A)^{-1} B = \begin{bmatrix} \dfrac{0.83333}{(s + 2.2381)(s + 2.4048)} & \dfrac{1}{s + 2.2381} \end{bmatrix} \begin{bmatrix} 7 & 0.5714 \\ -1.117 & 0 \end{bmatrix}$$

$$= \begin{bmatrix} \dfrac{0.83333 \cdot 7}{(s + 2.2381)(s + 2.4048)} - \dfrac{1.117}{s + 2.2381} & \dfrac{0.83333 \cdot 0.5714}{(s + 2.2381)(s + 2.4048)} \end{bmatrix}$$

The manipulated input-output process transfer function for the reactor is

$$g_{11}(s) = \frac{5.8333 - 1.117(s+2.4048)}{(s+2.2381)(s+2.4048)} = \frac{-1.1170s + 3.1472}{s^2 + 4.6429s + 5.3821} = \frac{0.5848(-0.3549s+1)}{0.1828s^2 + 0.8627s + 1}$$

and the disturbance input-output transfer function is

$$g_{12}(s) = \frac{0.4762}{(s+2.2381)(s+2.4048)} = \frac{0.4762}{s^2 + 4.6429s + 5.3821} = \frac{0.0885}{0.1828s^2 + 0.8627s + 1}$$

The transfer function poles (–2.2381 and –2.4048) are equal to the eigenvalues of the A matrix. Also, the positive zero (1/0.3549) in $g_{11}(s)$ yields the inverse response shown in Figure 3–5.

We see that it is straightforward to convert state space models to transfer function models. An n-state system results in transfer functions that have a denominator polynomial that is nth order in s, that is, with n poles. Sometimes the resulting transfer functions can be factored into lower order transfer functions because of pole-zero cancellation (a value of a pole is equal to a value of a zero). An example of pole-zero cancellation is shown in Exercise 13.

3.13 MATLAB and SIMULINK

MATLAB and SIMULINK are ideal for analyzing and simulating dynamic behavior. The Control Toolbox uses LTI (linear, time invariant) *objects* to represent dynamic models. These objects can be state space or transfer function models; details on how to use these are presented in Module 4.

Here we first present some useful MATLAB functions not covered in other modules.

`conv` (Convolution) and `roots`

Often it is necessary to multiply polynomials in s together. Although the following multiplication is quite simple: $(5s + 1)(3s + 1) = 15s^2 + 8s + 1$, a more complex multiplication, such as $(2s^2 + 1.5s + 1)(3s^3 + 8s^2 + 4s + 1)$ could be quite time consuming. The following `conv` (convolve) command makes this straightforward. Also, the roots of the resulting polynomial are easily calculated using the `roots` command.

```
» g = conv([2 1.5 1],[3 8 4 1])
g =
 6.0000 20.5000 23.0000 16.0000 5.5000 1.0000
» p = roots(g)
```

```
p =
 -2.1095
 -0.3750 + 0.5995i
 -0.3750 - 0.5995i
 -0.2786 + 0.2836i
 -0.2786 - 0.2836i
```

step

First, define an object, such one of the transfer functions in Example 3.6.

```
» num = [-10 1]; den = [45 18 1];
» sysc = tf(num,den)

Transfer function:
 -10 s + 1
-----------------
45 s^2 + 18 s + 1
```

then plot the step response (use `help step` to view other options)

```
» step(sysc)
```

A number of other useful functions, such as `pole` and `tzero` are shown in Module 4.

SIMULINK

It is highly recommended that you interactively work through the initial sections of Module 2 to understand how SIMULINK can be used to simulate dynamic systems.

3.14 Summary

The main objective of this chapter was to develop an understanding of dynamic process behavior. Although Laplace transforms can be used to obtain analytical solutions to differential equations, we will not be using them for this purpose during the rest of this textbook. The concept of a transfer function is very useful for control system design and analysis. In future chapters we find that transfer functions allow the use of algebra rather

than differential equations when analyzing control systems composed of a number of components (controller, valve, process, sensor, etc.).

The stability of a process is determined by the eigenvalues of the state space *A* matrix, or the poles of the Laplace transfer function: if all are negative, the system is stable. Complex eigenvalues (*poles*) correspond to underdamped (oscillatory) behavior, characterized by damping factors less than 1. Dynamic responses are also affected by the values of the *zeros* (roots of the numerator polynomial in s) of a transfer function. If zeros are positive (in the right-half-plane), step responses have a characteristic *inverse response.*

A first-order + dead time model is the most common process representation. *Time-delays* are represented by an irrational term ($e^{-\theta s}$) in the Laplace domain. First and second-order Padé approximations are sometimes used for controller design; the approximations lead to right-half-plane zeros and inverse response.

The *initial* and *final value* theorems of Laplace transforms can be used to find the limiting behavior of transfer functions without having to obtain a full solution. The *process gain* is a measure of the long-term change in a process output for a given input change.

State space models are easily converted to Laplace transfer function models. The use of LTI objects (see Module 4) allows easy interconversion of models in MATLAB.

References

For more details on analytical mathematical methods (including Laplace transforms) to solve differential equations, see: Boyce, W., and R. DiPrima, *Ordinary Differential Equations and Boundary Value Problems*, 5th ed., Wiley, New York (1992).

Student Exercises

1. Solve for the time-domain output of a first-order transfer function to a step input change.
2. A second-order process with one pole at the origin has the transfer function

$$\frac{3}{s(2s+1)}$$

Find the output as a function of time, for a unit step input change. Sketch the expected behavior.

3. Use the initial value theorem to find the immediate response of a lead-lag transfer function to a step input change at $t = 0$.

$$g_p(s) = k_p \cdot \frac{\tau_n s + 1}{\tau_p s + 1}$$

Also, use the final value theorem to find the long-term response of a lead-lag transfer function to a step input change.

4. For the following second-order process with numerator dynamics, solve for the time-domain output response to a step input change of magnitude Δu at $t = 0$.

$$g_p(s) = \frac{k_p(\tau_n s + 1)}{(\tau_1 s + 1)(\tau_2 s + 1)}$$

For $k_p = 1$°C/Lpm, $\tau_1 = 3$ min, $\tau_2 = 15$ min, $\tau_n = 20$ min find the peak temperature and the time that it occurs.

5. Consider an input-output transfer function that mimics two first-order processes in parallel

$$y(s) = g_p(s)u(s)$$

where

$$g_p(s) = g_1(s) + g_2(s)$$

If the gain of g_1 is positive and the gain of g_2 is negative, find the conditions (relationship between gains and time constants for the two transfer functions) that cause a right-half-plane zero (resulting in inverse response to a step input change) in $g_p(s)$.

6. Consider the state space model

$$\frac{dx_1}{dt} = -2.4048x_1 + 7u$$

$$\frac{dx_2}{dt} = 0.8333x_1 - 2.2381x_2 - 1.1170u$$

$$y = x_2$$

Find the second-order differential equation in y. *Hint:* first solve for x_1 from the second equation, then take the derivative and substitute into the first equation.

7. Consider the following state-space model

$$\begin{bmatrix} \dfrac{dx_1}{dt} \\ \dfrac{dx_2}{dt} \end{bmatrix} = \begin{bmatrix} -1.5 & 1 \\ -0.5 & 0 \end{bmatrix} \begin{bmatrix} x_1 \\ x_2 \end{bmatrix} + \begin{bmatrix} 0.5 \\ 0 \end{bmatrix} u$$

$$y = \begin{bmatrix} 1 & 0 \end{bmatrix} \begin{bmatrix} x_1 \\ x_2 \end{bmatrix}$$

Which has the following input-output transfer function relationship

$$y(s) = \frac{s}{(s+1)(2s+1)} u(s)$$

For a unit step change in the input, $u(s) = 1/s$:

i. Find the output at $t=0$ and as t approaches infinity, using the initial and final value theorems.

ii. Find the time domain solution, $y(t)$

iii. Sketch the time domain behavior of $y(t)$

iv. Are your results for i, ii, and iii consistent?

8. As a process engineer with the Complex Pole Corporation, you are assigned a unit with an exothermic chemical reactor. In order to learn more about the dynamics of the process, you decide to make a step change in the input variable, the coolant temperature, from 10°C to 15°C. Assume that the reactor was initially at a steady state. You obtain the following plot for the output variable, which is reactor temperature (notice that the reactor temperature is in °F). Use Figure 3–9 to help answer the following questions.

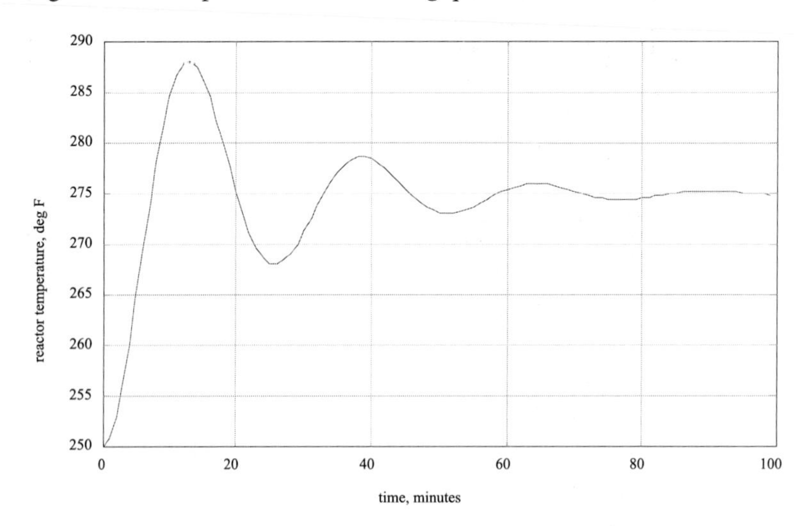

i. What is the value of the process gain? (show units)
ii. What is the value of τ? (show units)
iii. What is the value of ζ? (show units)
iv. What is the decay ratio?
v. What is the period of oscillation? (show units)
vi. Write the second-order transfer function.

Hint: Use Figure 3–9 to assist you.

9. Match the transfer functions with the responses to a unit step input, shown in the figure.

i. $\dfrac{3(-2s+1)}{s^2+0.5s+1}$

ii. $\dfrac{-2e^{-3s}}{3s+1}$

iii. $\dfrac{2}{-5s+1}$

iv. $\dfrac{1}{s(5s+1)}$

v. $\dfrac{4s+1}{2s+1}$

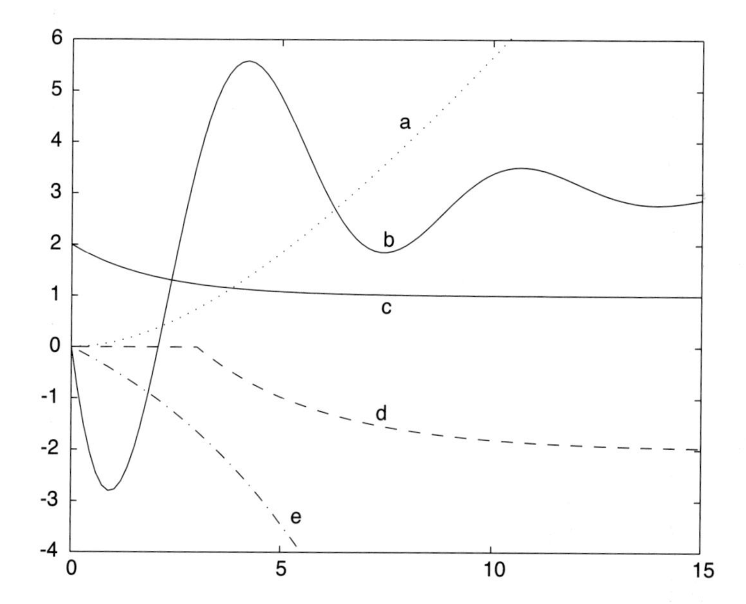

10. Consider the following state space model:

$$\begin{bmatrix} \dot{x}_1 \\ \dot{x}_2 \end{bmatrix} = \begin{bmatrix} -1/3 & 0 \\ 1/2 & 0 \end{bmatrix} \begin{bmatrix} x_1 \\ x_2 \end{bmatrix} + \begin{bmatrix} 1 \\ 0 \end{bmatrix} u$$

$$y = \begin{bmatrix} -2/3 & 2/3 \end{bmatrix} \begin{bmatrix} x_1 \\ x_2 \end{bmatrix}$$

i. Find the transfer function model.

ii. Find (analytically) the time-domain output response to a unit step input change. Sketch the expected response to a unit step input change.

11. As a process engineer, you decide to develop a first-order + time delay model of a process using a step test. The process is initially at steady state, with an input flow rate of 5 gpm and an output of 0.75 mol/L. You make a step increase of 0.5 gpm at 3:00 p.m. and do not observe any changes until 3:07 p.m. At 3:20 p.m., the value of the output is 0.8 mol/L. You become distracted and do not have a chance to look at the output variable again, until you leave for happy hour at a local watering hole at 6:30 p.m. You note that the output has ceased to change and has achieved a new steady-state value of 0.85 mol/L. What are the values of the process parameters, with units? Show your work.

12. Use the initial and final value theorems of Laplace transforms to determine the initial and final values of the process output for a unit step input change to the following transfer functions.

i. $\dfrac{5s+12}{7s+4}$

ii. $\dfrac{(7s^2+4s+2)(6s+4)}{(4s^2+4s+1)(16s^2+4s+1)}$

iii. $\dfrac{4s^2+2s+1}{8s^2+4s+0.5}$

13. Consider the following state space model for a biochemical reactor. Since there are two states (the A matrix is 2×2) we expect that the process transfer function will be second-order. Show that pole-zero cancellation occurs, resulting in a first-order transfer function. Find the values of the gain and time constant.

$$A = \begin{bmatrix} 0 & 0.9056 \\ -0.75 & -2.5640 \end{bmatrix} \qquad B = \begin{bmatrix} -1.5301 \\ 3.8255 \end{bmatrix}$$

$$C = \begin{bmatrix} 1 & 0 \end{bmatrix} \qquad D = \begin{bmatrix} 0 \end{bmatrix}$$

14. Match the transfer functions with the responses to a unit step input, shown in the figure.

i. $\dfrac{-2.5(-4s+1)}{4s^2+4s+1}$

ii. $\dfrac{-2e^{-10s}}{10s+1}$

iii. $\dfrac{-5}{-20s+1}$

iv. $\dfrac{-0.1}{s}$

v. $\dfrac{4s+3}{2s+1}$

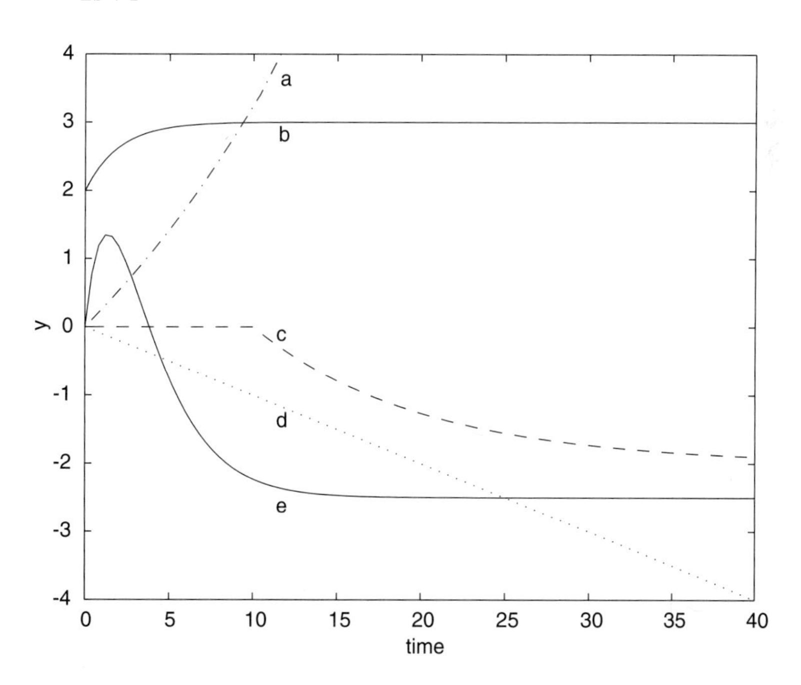

15. Consider Example 3.4. For an impulse input of 30 kJ, find the value of the vessel temperature immediately after the impulse input is applied.

16. Consider the following second-order transfer function

$$g_p(s) = \frac{k_p}{(\tau_1 s + 1)(\tau_2 s + 1)}$$

For a unit *impulse* input, find the output response as a function of time. What is the peak change and when does it occur?

CHAPTER 4

Empirical Models

The previous two chapters have focused on (i) the development of fundamental models based on material and energy balances (Chapter 2), and (ii) understanding dynamic behavior, with an emphasis on linear systems (Chapter 3). In this chapter we discuss the development of empirical models, that is, developing models based on plant tests. For continuous control-system design (most of the focus of this text), a study of Sections 4.1–4.3 will be sufficient. The discrete models developed in Sections 4.4 and 4.5 will be useful when model predictive control (MPC) techniques are presented in Chapter 16, and when digital control is studied in Module 16.

After studying this chapter the reader should be able to:

- Develop continuous first-order and integrator + dead time models from step tests
- Estimate parameters for discrete-time autoregressive models based on input-output data
- Calculate poles and zeros of discrete-time models
- Develop finite step and impulse response models

The major sections of this chapter are as follows:

4.1 Introduction

For many processes there is not enough time, or it is not worth the effort, to develop fundamental process models. Particularly if your main interest is tuning a specific control loop, it is likely that you will develop a transfer function-based model by performing a plant test. The most common plant test is to make a step change in the manipulated input (controller output) and observe the measured process output response. Then a model is developed to provide the best match between the model output and the observed plant output.

There are a number of important issues in developing an input-output model. Foremost is the selection of the proper input and output variables. For many processes this is not trivial, because a particular manipulated input may affect a number of measured outputs. Similarly, a measured output may be affected by a number of manipulated inputs. In this chapter we assume that the manipulated input and measured output have already been selected. The important discussion of the selection of a particular input to be "paired" with a particular output is postponed until Chapter 13.

It is common to base an input-output model on step responses. In this procedure we first bring the process to a consistent and desirable steady-state operating point, then make a step change in the input variable. An important decision is the magnitude of the step change to make.

- If the step change in input is too small, the measured output may not change enough to develop a reliable model. This is particularly true if the measured output is "noisy." Clearly, the magnitude of step input must be enough so that the output "signal-to-noise" ratio is high enough to obtain a good model.
- If the step change in the input is too large, the output variable may change too much and produce product that is "off-specification." This is not desirable because of the severe economic penalty (i.e., the plant loses too much money while the step test is being performed). Also, if the step input change is too large, nonlinear effects may dominate. That is, the operating condition may become significantly different than the desired condition.

Clearly there is a trade-off here. The input must be changed enough to observe a change in the output variable (it must increase above the noise level), yet not so much that the output variable change is too great (causing an economic penalty).

As a process engineer conducting a plant step test, you will usually have some basic knowledge of (or experience with) the input-output pair under consideration. Observations of the measured output with time will provide an estimate of the standard deviation or variance of the measurement noise. A rough estimate of the process gain will enable you to select an input change magnitude so that the output change is "above the noise level."

An estimate of the process gain can often be obtained through steady-state material and energy balances.

In the following sections we show how to estimate parameters for some common simple models. By far the most commonly used model, for control-system design purposes, is the first-order + time-delay model discussed in the next section.

4.2 First-Order + Dead Time

Recall that a first-order + dead-time process, represented by the transfer function relationship

$$y(s) = \frac{k_p e^{-\theta s}}{\tau_p s + 1} u(s) \tag{4.1}$$

has the following output response to a step input change,

$$\begin{aligned} t < \theta,\ & y(t) = 0 \\ t \geq \theta,\ & y(t) = k_p \Delta u \{1 - \exp(-(t - \theta)/\tau_p)\} \end{aligned} \tag{4.2}$$

where the measured output is in deviation variable form. The three process parameters can be estimated by performing a single step test on the process input.

The gain is found as simply the long-term change in process output divided by the change in process input. Also the time delay is the amount of time, after the input change, before a significant output response is observed. There are several easy ways to estimate the time constant for this type of model.

Time for 63.2% Approach to New Steady State

In Equation (4.2), if we set $t = \tau_p + \theta$, we find $y(t = \tau_p + \theta) = k_p \Delta u(1 - \exp(-1)) = 0.632 k_p \Delta u$. Now, since $k_p \Delta u$ is simply the long-term change in the process output, then $t_{63.2\%} = \tau_p + \theta$ is the amount of time it takes for the output to make 63.2% of its ultimate change. This method of estimating the time constant is illustrated in Figure 4–1.

Example 4.1: Numerical Application of 63.2% Method

Consider the response to a step input change at $t = 1$ minute, shown in Figure 4–2. The measured output is shown as the noisy solid curve, and a "best fit" first-order + time-delay model is shown as the dashed curve. Here we find the process gain from

$$k_p = \frac{\Delta y}{\Delta u} = \frac{23 - 25°\text{C}}{10.5 - 10\,\text{gpm}} = -4\,°\text{C/gpm}$$

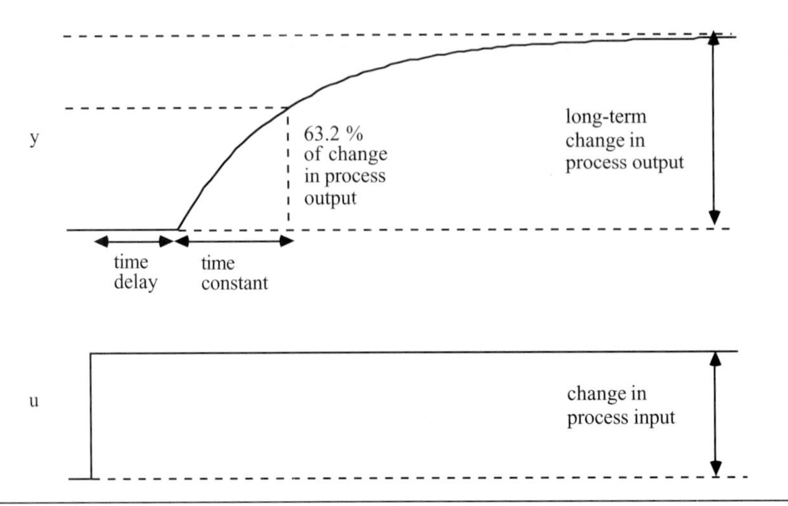

Figure 4–1 Estimating the time constant from a response to a step input.

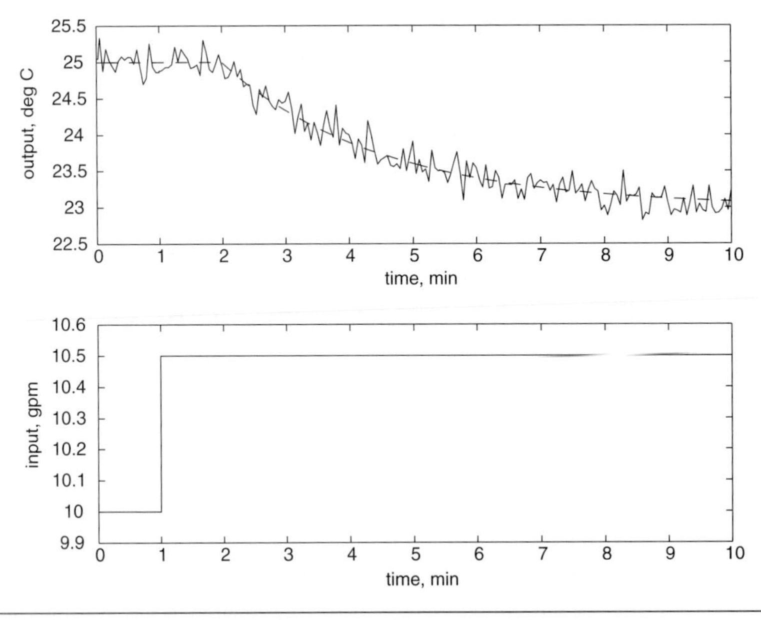

Figure 4–2 Example of the step response with measurement noise.

The time delay is observed to be 1 minute, and the time constant is the time it takes (after the time delay) for the output to change by 0.632(-2) = -1.3°C. That is, the time when y = 23.7°C. In this case,

$$\tau_p = 2.5 \text{ minutes}$$

and the process transfer function is

$$g_p(s) = \frac{-4e^{-s}}{2.5s+1}$$

Maximum Slope Method

We can also find (see Exercise 1) that the maximum slope of the output response to a step input change at $t = 0$ occurs at $t = \theta$ and is

$$slope = \frac{k_p \Delta u}{\tau_p} = \frac{\Delta y}{\tau_p}$$

So the time constant can be estimated from

$$\tau_p = \frac{\Delta y}{slope}$$

A major disadvantage is that the slope estimate may not be accurate if there is significant measurement noise. This approach is particularly useful for higher-order responses that are approximated as a first-order + time-delay response, as shown in the following example.

Example 4.2: Maximum Slope Method

The response of a high order process is shown in Figure 4–3. Here, the dashed line is drawn through the maximum slope of the output response. The slope is

$$slope = \frac{16-10°\text{C}}{30-8 \text{ min}} = 0.273°\text{C/min}$$

$$\tau_p = \frac{\Delta y}{slope} = \frac{6°\text{C}}{0.273°\text{C/min}} = 22 \text{ min}$$

Since the process gain is $k_p = \Delta y/\Delta u = 6°\text{C}/2 \text{ gpm} = 3°\text{C/gpm}$, and the time-delay is $8-1 = 7$ min, the first-order + time-delay transfer function is estimated as

$$g_p(s) = \frac{3e^{-7}}{22s+1}$$

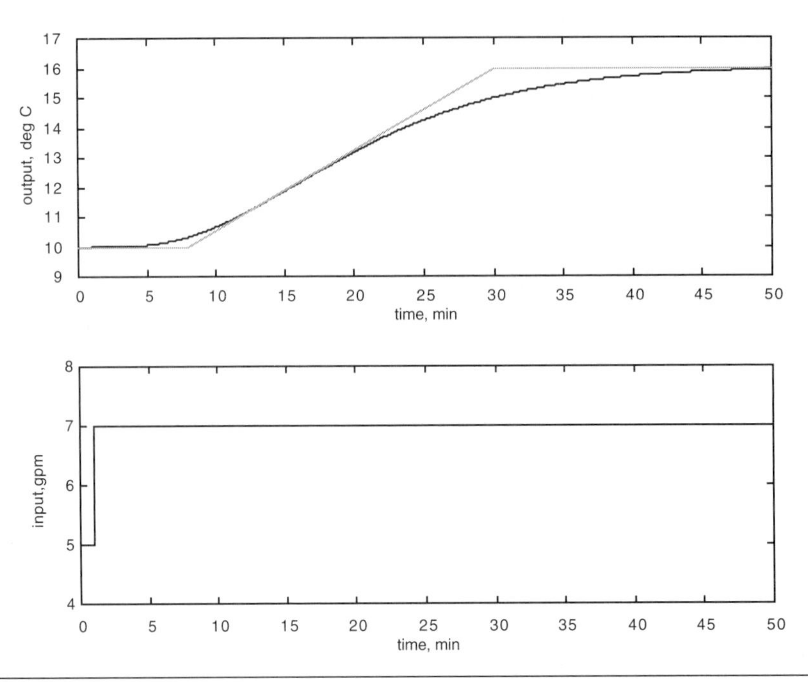

Figure 4–3 Example of the slope method for estimation of time constant.

Two-Point Method for Estimating Time Constant

A technique similar to the first one, but using two points on the output response, is shown in Figure 4–4. Here the time required for the process output to make 28.3% and 63.2% of the long-term change is denoted by $t_{28.3\%}$ and $t_{63.2\%}$, respectively. The time constant and time delay can be estimated from (see Exercise 2)

$$\tau_p = 1.5(t_{63.2\%} - t_{28.3\%})$$
$$\theta = t_{63.2\%} - \tau_p$$

Limitation to FODT Models

The primary limitation to using step responses to identify first-order + dead-time transfer functions is the amount of time required to assure that the process is approaching a new steady state. That is, the major limit is the time required to determine the gain of the process. For large time constant processes it is often desirable to use a simpler model that does not require a long step test time. The integrator + dead-time model shown next will often be good enough for control-system design.

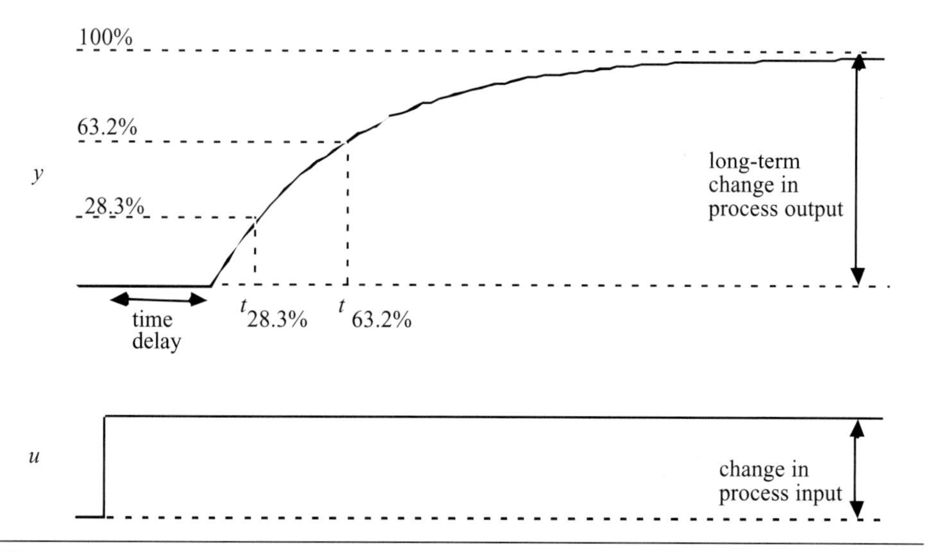

Figure 4–4 Example of the two-point method for estimation of time constant.

4.3 Integrator + Dead Time

An integrator + dead-time process has the input-output transfer function relationship

$$y(s) = \frac{ke^{-\theta s}}{s} \tag{4.3}$$

and the output response to a step input

$$t < \theta, \; y(t) = 0$$
$$t > \theta, \; y(t) = k\Delta u \cdot (t - \theta) \tag{4.4}$$

with the response shown in Figure 4–5. The gain, k, can be found from the slope by solving

$$k = \frac{\text{slope}}{\Delta u}$$

and the time delay is clearly identified by the time required for a change in output.

Integrator + dead-time models are good for describing the behavior of "integrating processes," such as vessel liquid levels or gas drum pressures. They can provide a good short-term approximation to the step response behavior of a first-order + dead-time process. Consider the step response of a distillation column, which has a fairly long time constant. We see from Figure 4–6 that it takes roughly 5–6 hours to obtain good estimates for a FODT model.

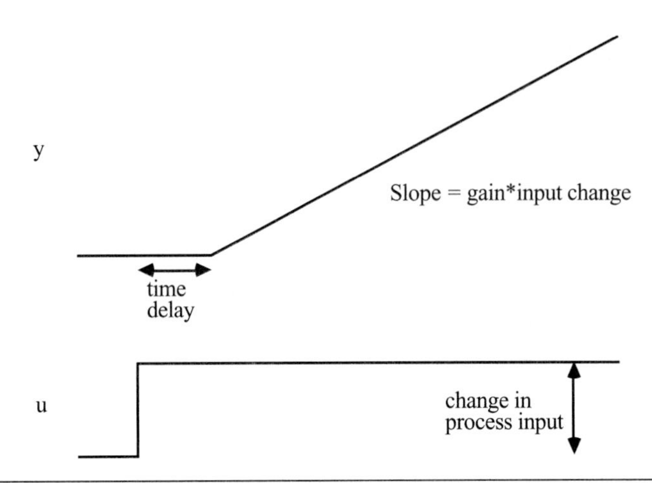

Figure 4–5 Estimating parameters for an integrator + time-delay model from a step input test.

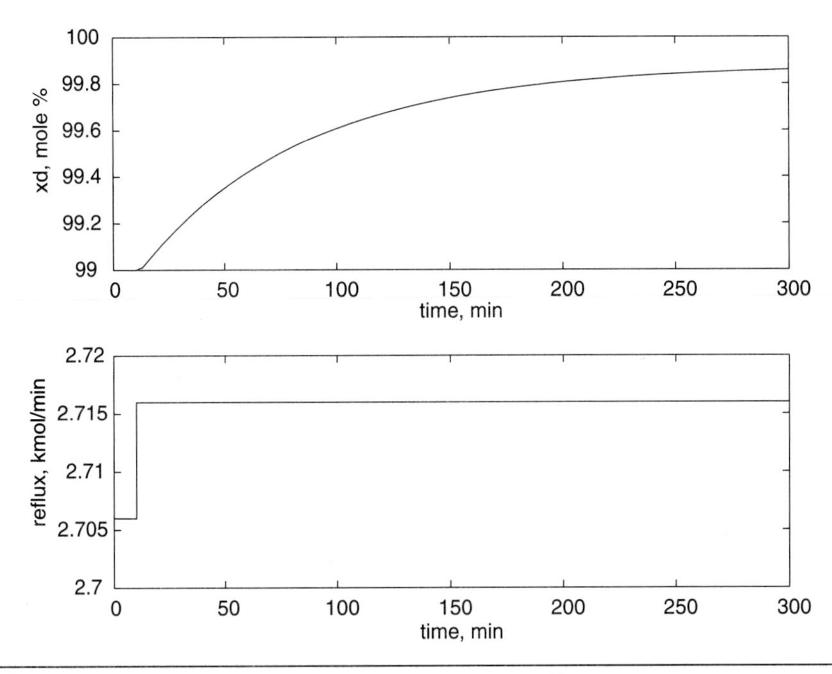

Figure 4–6 Distillation column overhead composition response to a step change of 0.01 kmol/min in reflux rate at $t = 10$ minutes.

The reader should be able to show that the transfer function is approximately

$$g_p(s) = \frac{84e^{-2s}}{75s+1}$$

where the gain has units of mol%/(kmol/min) and the time unit is minutes.

The initial response of Figure 4–6 is "blown-up" in Figure 4–7. Notice that fewer than 20 minutes (the step change is made at t = 10 minutes) is required to obtain a satisfactory integrator + time-delay model. Clearly, plant operators (and managers) will be much more willing to tolerate a 20-minute test than a 4- to 5-hour test.

The reader should be able to show that the estimated transfer function is approximately

$$g_p(s) = \frac{1e^{-2s}}{s}$$

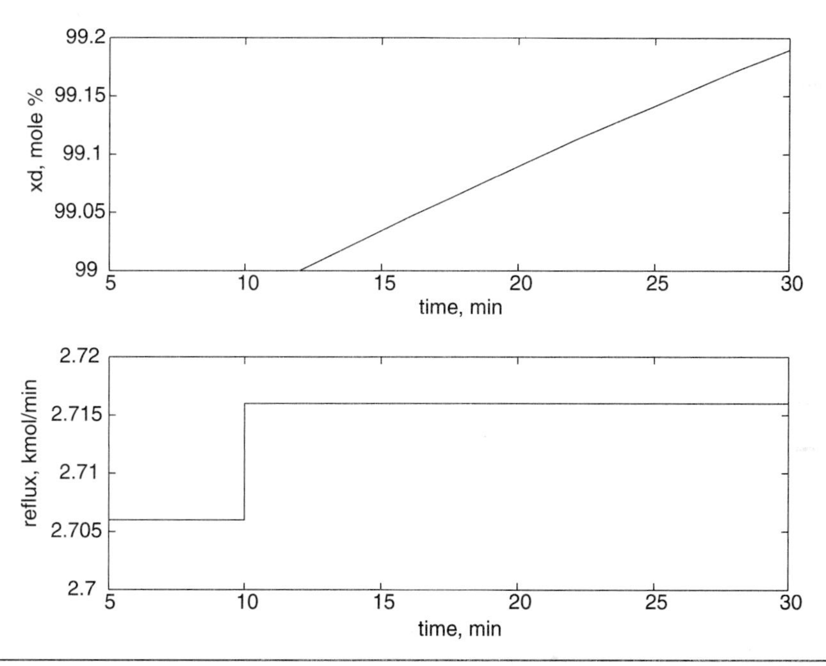

Figure 4–7 Distillation column overhead composition response to a step change of 0.01 kmol/min in reflux rate at t = 10 minutes. Focus is on the short-term response.

4.4 Discrete-Time Autoregressive Models

Most of the models developed and used in this text are based on continuous time. In the most general case of fundamental models, these are nonlinear ordinary differential equations. State space models consist of sets of linear ordinary differential equations, where the states and inputs are in deviation variable form. Transfer functions are used to denote a linear relationship between inputs and outputs. All these previous models assume that the manipulated input and the measured output signals are continuously available.

In practice, manipulated input changes are made at discrete time intervals and measured outputs are available at discrete sample times. The physical inputs and outputs are continuous; only the changes and measurements occur at discrete times. *Since the focus of this textbook is on controller design based on continuous models, the topic of discrete models may be skipped on a first reading*. Discrete models are primarily useful for model predictive control (Chapter 16) and digital control design and analysis (Module 16).

In this section we cover discrete autoregressive models, in Section 4.5 parameter estimation of discrete autoregressive models is covered, and in Section 4.6 we cover finite step and impulse response models. It should be noted that the MATLAB Control Toolbox has routines that convert continuous-time models to discrete and vice versa. The details of the conversion are beyond the scope of this text, but are summarized in the appendix; read Module 4 for explanations on how to use these routines.

Introduction to Autoregressive Models

Discrete autoregressive models assume that an output at the current time step is a function of the outputs and inputs at previous time steps. The general form is

$$\begin{aligned} y(k) &= -a_1 y(k-1) - a_2 y(k-2) - \ldots - a_n y(k-n) + \\ &\quad b_0 u(k) + b_1 u(k-1) + b_2 u(k-2) + \ldots + b_m u(k-m) \end{aligned} \tag{4.5}$$

where k represents the current time step, $k-1$ the previous time step, and so forth. The notation $y(k)$ is used to indicate the value of the output at step k. If all of the outputs are brought to the left hand side, Equation (4.5) can be written

$$\begin{aligned} &y(k) + a_1 y(k-1) + \ldots + a_{n-1} y(k-n+1) + a_n y(k-n) = \\ &b_0 u(k) + b_1 u(k-1) + \ldots + b_{m-1} u(k-m+1) + b_m u(k-m) \end{aligned} \tag{4.6}$$

This form is primarily used to lead to the notion of discrete input-output transfer functions covered next.

Z-Transforms

Similar to the use of Laplace transforms for continuous systems, *Z-transforms* are used for discrete systems.

The Z-transform of a discrete variable, $y(k)$, is

$$y(z) = Z[y(k)] \tag{4.7}$$

The so-called backwards shift operator (z^{-1}) is represented by

$$Z[y(k-1)] = z^{-1}y(z) \tag{4.8}$$

so $Z[y(k-2)] = z^{-2}y(z)$, and so forth. Taking the Z-transform of each term in Equation (4.6), we find

$$\begin{aligned}(1 + a_1 z^{-1} + \ldots + a_{n-1} z^{-n+1} + a_n z^{-n})y(z) = \\ (b_0 + b_1 z^{-1} + \ldots + b_{m-1} z^{-m+1} + b_m z^{-m})u(z)\end{aligned} \tag{4.9}$$

Solving for $y(z)$ yields the discrete transfer function relationship

$$y(z) = \frac{b_0 + b_1 z^{-1} + \ldots + b_{m-1} z^{-m+1} + b_m z^{-m}}{1 + a_1 z^{-1} + \ldots + a_{n-1} z^{-n+1} + a_n z^{-n}} u(z) \tag{4.10}$$

where the discrete transfer function is

$$g_p(z) = \frac{b_0 + b_1 z^{-1} + \ldots + b_{m-1} z^{-m+1} + b_m z^{-m}}{1 + a_1 z^{-1} + \ldots + a_{n-1} z^{-n+1} + a_n z^{-n}} \tag{4.11}$$

The discrete transfer function is the discrete-time analogy to the continuous-time transfer function based on the Laplace transform. Similar to continuous (Laplace) transfer functions, discrete (z-transform) transfer functions can be written in several forms. Multiplying Equation (4.11) by z^n/z^n, we find

$$g_p(z) = \frac{b_0 z^n + b_1 z^{n-1} + \ldots + b_m z^{n-m}}{z^n + a_1 z^{-1} + \ldots + a_{n-1} z^{-n+1} + a_n} \tag{4.12}$$

For most process systems there is not an immediate effect of the input on the output, so $b_0 = 0$.

Poles/Zeros of Discrete Models

Recall that the stability of a *continuous* input-output (transfer function) model is determined by the values of the poles (the roots of the denominator polynomial in s). If the poles are negative, then the model is stable. Similarly, the stability of a *discrete* input-output model is determined by the poles of the denominator polynomial in z. For a discrete-time model, if all poles have a *magnitude* less than 1 (i.e., they are "inside the unit circle") the model is stable. If any pole has a magnitude greater than 1, the model is unstable.

The polynomial forms of Equations (4.11) and (4.12) can be factored into gain-pole-zero form

$$g_p(z) = K \cdot \frac{(z - z_1)(z - z_2)\cdots(z - z_m)}{(z - p_1)(z - p_2)\cdots(z - p_n)} \tag{4.13}$$

where z_i and p_i are the zeros and poles, respectively.

Example 4.3: Discrete Poles and Stability

Consider the simple first-order model

$$y(k) = -a_1 y(k - 1) + b_1 u(k - 1)$$

which has the transfer function

$$g_p(z) = \frac{b_1 z^{-1}}{1 + a_1 z^{-1}} = \frac{b_1}{z + a_1}$$

The pole is found by solving for the roots of the denominator polynomial. In this case

$$p = -a_1$$

To illustrate the importance of pole values, we will study the following values for a_1:

$$a_1 = 0.5, -0.5, 1.5, -1.5$$

yielding the corresponding pole values

$$p = -0.5, 0.5, -1.5, 1.5$$

Also, realize that the modeling equation can be written

$$y(k + 1) = -a_1 y(k) + b_1 u(k)$$

For simplicity, we will assume no input change, so $u(k) = 0$. Also, let the initial value of the output be $y(0) = 1$. The first few values of $y(k)$ for each value of a_1 are shown in the table below.

a_1	0.5	-0.5	1.5	-1.5
p	-0.5	0.5	-1.5	1.5
$y(1)$	-0.5	0.5	-1.5	1.5
$y(2)$	0.25	0.25	2.25	2.25
$y(3)$	-0.125	0.125	-3.375	3.375
$y(4)$	0.0625	0.0625	5.0625	5.0625
Characteristic behavior	Oscillatory, stable	Monotonic, stable	Oscillatory, unstable	Monotonic, unstable

As expected, when the magnitude of the pole is less than 1.0, the process is stable; when the magnitude is greater than 1.0, the process is unstable. One thing that is unusual for this first-order discrete example is that the output can oscillate even with a constant (in this case, 0) input value. For continuous-time systems there must be at least two states (the process must be at least second-order) for oscillation to occur. This is because the condition for oscillation in continuous-time systems is that the eigenvalues must be complex; for this to happen, there must be at least two eigenvalues, each complex-conjugates of the other. Discrete time systems can oscillate, as long as a pole has a negative value; if the negative pole has a magnitude less than one, it is a stable oscillation.

Final and Initial Values Theorems for Discrete Systems

Recall that the process gain of a continuous system could be determined by setting $s = 0$. Similarly, the gain of a discrete-time system can be found by setting $z = 1$.

Final Value Theorem for Discrete-Time Systems

The formal statement of the final value theorem is (where Δt is the sample time)

$$\lim_{n \to \infty} y(n\Delta t) = \lim_{z \to 1} \left(1 - z^{-1}\right) y(z)$$

For a discrete-time input-output model

$$y(z) = g_p(z)u(z)$$

subject to a unit step input

$$u(z) = \frac{1}{1 - z^{-1}} = \frac{z}{z - 1}$$

we find

$$y(z) = g_p(z)u(z) = g_p(z) \cdot \frac{1}{1 - z^{-1}}$$

so, applying the final value theorem

$$\lim_{t \to \infty} y(t) = \lim_{z \to 1} \left(1 - z^{-1}\right) y(z) = \lim_{z \to 1} \left(1 - z^{-1}\right) g_p(z) \frac{1}{1 - z^{-1}} = \lim_{z \to 1} g_p(z)$$

so we can find the long-term behavior of the process output, subject to a unit step input, simply by setting $z = 1$ in $g_p(z)$.

Initial Value Theorem for Discrete-Time Systems

Similarly, the initial value theorem is

$$\lim_{t\to 0} y(t) = \lim_{z\to\infty} \left(1 - z^{-1}\right) y(z)$$

4.5 Parameter Estimation

Often when discrete linear models are developed, they are based on experimental system responses rather than on converting a continuous model to a discrete model. The estimation of parameters for discrete dynamic models is no different from linear regression (least squares) analysis.

The measured inputs and outputs are the independent variables, and the dependent variables are the outputs. For simplicity, consider the following model where two previous values of the input and output are used to predict the next value. This model has four parameters

$$y(k) = -a_1 y(k-1) - a_2 y(k-2) + b_1 u(k-1) + b_2 u(k-2) \tag{4.14}$$

For the system of N data points we can write the following

$$\begin{aligned}
y(1) &= -a_1 y(0) - a_2 y(-1) + b_1 u(0) + b_2 u(-1) \\
y(2) &= -a_1 y(1) - a_2 y(0) + b_1 u(1) + b_2 u(0) \\
&\vdots \\
y(N) &= -a_1 y(N-1) - a_2 y(N-2) + b_1 u(N-1) + b_2 u(N-2)
\end{aligned}$$

Notice that $k = 1$ is an arbitrary starting point for calculating the output, and that two previous data points for the output and the input are needed to start the algorithm. To make it clear that we are performing a model-based prediction of y_k, we will use the notation $\hat{y}_k$ to indicate a model prediction

$$\begin{aligned}
\hat{y}(1) &= -a_1 y(0) - a_2 y(-1) + b_1 u(0) + b_2 u(-1) \\
\hat{y}(2) &= -a_1 y(1) - a_2 y(0) + b_1 u(1) + b_2 u(0) \\
&\vdots \\
\hat{y}(N) &= -a_1 y(N-1) - a_2 y(N-2) + b_1 u(N-1) + b_2 u(N-2)
\end{aligned} \tag{4.15}$$

or, using matrix-vector notation

$$\hat{Y} = \Phi\Theta \tag{4.16}$$

where

$$\hat{Y} = \begin{bmatrix} \hat{y}(1) \\ \cdot \\ \cdot \\ \hat{y}(N) \end{bmatrix}, \quad \Phi = \begin{bmatrix} y(0) & y(-1) & u(0) & u(-1) \\ \cdot & \cdot & \cdot & \cdot \\ \cdot & \cdot & \cdot & \cdot \\ y(N-1) & y(N-2) & u(N-1) & u(N-2) \end{bmatrix}, \quad \Theta = \begin{bmatrix} -a_1 \\ -a_2 \\ b_1 \\ b_2 \end{bmatrix} \tag{4.17}$$

The objective is to choose a set of parameters (a_1, a_2, b_1, b_2) to minimize the square of the *residuals* (the differences between the model outputs and measured outputs).

$$\min_{a_1, a_2, b_1, b_2} \sum_{i=1}^{N} \left(y_i - \hat{y}_i\right)^2 \tag{4.18}$$

The sum of the squares of the residuals can be written in matrix notation as

$$\sum_{i=1}^{N} \left(y_i - \hat{y}_i\right)^2 = \left(Y - \hat{Y}\right)^T \left(Y - \hat{Y}\right) = \left(Y - \Phi\Theta\right)^T \left(Y - \Phi\Theta\right) \tag{4.19}$$

where Y is the vector of measured outputs, $Y = [y(1) \ldots y(N)]^T$.

The solution to this optimization problem is (Ljung, 1996)

$$\Theta = (\Phi^T\Phi)^{-1}\Phi^T Y \tag{4.20}$$

Example 4.4: Process Identification

Although the response to a step input has often been used for process identification, here we apply a pseudo-random binary sequence of inputs (the input changes randomly between two values) to estimate parameters in a discrete time model. The input and output responses are shown in Figure 4–8; the open circles represent the measured outputs obtained with a sample time of 0.25 minutes. The variables are in deviation form.

The output and input data as a function of the discrete time index are:

k	-1	0	1	2	3	4	5
y	0	-0.0889	0.0137	0.1564	0.4618	0.1771	0.3446
u	1.0000	1.0000	1.0000	-1.0000	1.0000	-1.0000	-1.0000
k	6	7	8	9	10	11	12
y	0.2171	-0.1558	0.0485	-0.1879	-0.1123	0.0463	0.2003
u	1.0000	-1.0000	1.0000	1.0000	1.0000	1.0000	-1.0000
k	13	14	15	16	17	18	19
y	0.5007	0.3846	-0.0172	0.1513	-0.1162	0.1134	0.0502
u	-1.0000	1.0000	-1.0000	1.0000	-1.0000	-1.0000	-1.0000

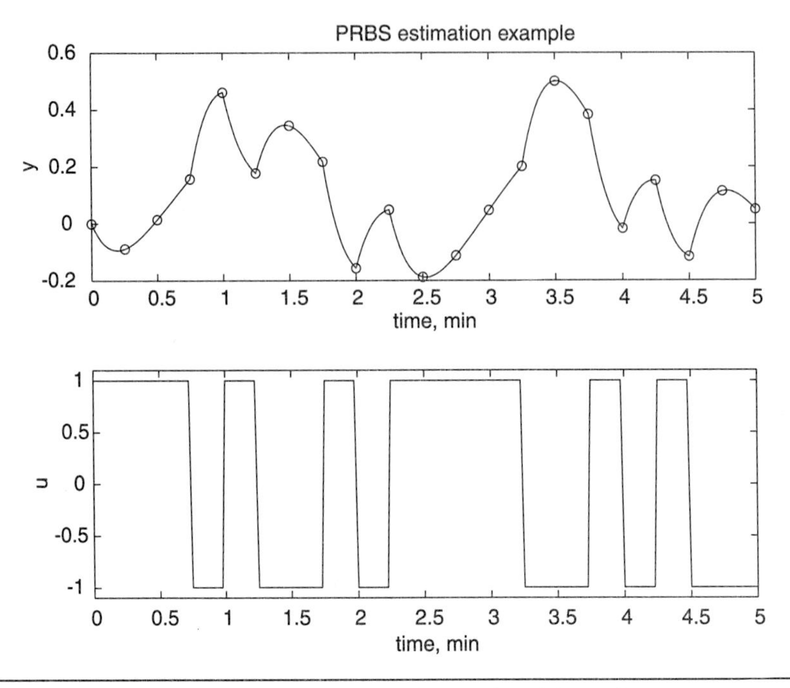

Figure 4–8 Output response to a pseudo-random binary input sequence. Open circles represent output data obtained with a sample time of 0.25 minutes.

The Φ (input-output data) matrix, Θ parameter vector, and discrete transfer functions are shown below.

$$\Phi = \begin{bmatrix} -0.0889 & 0 & 1 & 1 \\ 0.0137 & -0.0889 & 1 & 1 \\ 0.1564 & 0.0137 & -1 & 1 \\ 0.4618 & 0.1564 & 1 & -1 \\ 0.1771 & 0.4618 & -1 & 1 \\ 0.3446 & 0.1771 & -1 & -1 \\ 0.2171 & 0.3446 & 1 & -1 \\ -0.1558 & 0.2171 & -1 & 1 \\ 0.0485 & -0.1558 & 1 & -1 \\ -0.1879 & 0.0485 & 1 & 1 \\ -0.1123 & -0.1879 & 1 & 1 \\ 0.0463 & -0.1123 & 1 & 1 \\ 0.2003 & 0.0463 & -1 & 1 \\ 0.5007 & 0.2003 & -1 & -1 \\ 0.3846 & 0.5007 & 1 & -1 \\ -0.0172 & 0.3846 & -1 & 1 \\ 0.1513 & -0.0172 & 1 & -1 \\ -0.1162 & 0.1513 & -1 & 1 \\ 0.1134 & -0.1162 & -1 & -1 \end{bmatrix}, \quad Y = \begin{bmatrix} 0.0137 \\ 0.1564 \\ 0.4618 \\ 0.1771 \\ 0.3446 \\ 0.2171 \\ -0.1558 \\ 0.0485 \\ -0.1879 \\ -0.1123 \\ 0.0463 \\ 0.2003 \\ 0.5007 \\ 0.3846 \\ -0.0172 \\ 0.1513 \\ -0.1162 \\ 0.1134 \\ 0.0502 \end{bmatrix}$$

$$\Theta = (\Phi^T\Phi)^{-1}\Phi^T Y$$

$$\Theta = \begin{bmatrix} 1.1196 \\ -0.3133 \\ -0.0889 \\ 0.2021 \end{bmatrix} = \begin{bmatrix} -a_1 \\ -a_2 \\ b_1 \\ b_2 \end{bmatrix}$$

$$\begin{aligned} g_p(z) &= \frac{b_1 z + b_2}{z^2 + a_1 z + a_2} = \frac{-0.0889z + 0.2021}{z^2 - 1.1196z + 0.3133} \\ &= \frac{-0.0889z^{-1} + 0.2021z^{-2}}{1 - 1.1196z^{-1} + 0.3133z^{-2}} \\ &= \frac{-0.0889(z - 2.274)}{(z - 0.5716)(z - 0.5481)} \end{aligned}$$

The MATLAB and SIMULINK .m and .mdl files used to generate this example are shown in the Appendix. It should be noted that the discrete transfer function can be converted to a continuous transfer function using d2c to find

$$g_p(s) = \frac{-1.117s + 3.147}{s^2 + 4.642s + 5.373}$$

This is virtually identical to the continuous transfer function used to generate the original data.

This example illustrated the response of a perfectly modeled system (no measurement noise). The approach can also be applied to a system with arbitrary inputs, and with noisy measurements. See Exercise 8 for the same process, with output data corrupted by measurement noise. To verify model parameter estimates it is common to use a portion of experimental data to estimate parameters, then another portion of the data in "simulation" mode to see how well the model predicted outputs match the data.

The data was analyzed in a batchwise fashion, that is, all of the data were collected before the parameter estimation was performed. For on-line (real-time) estimation and control, we would prefer to re-estimate parameters each time we obtain a new data point. This can be done by using a "moving horizon" of past data points (each time a new data point is collected, the oldest one is thrown out), or by using recursive identification techniques. Recursive techniques are commonly applied in adaptive control, where the model is updated at each time step and a control calculation is performed based on the updated model.

4.6 Discrete Step and Impulse Response Models

The common model predictive control techniques presented in Chapter 16 are based on step or impulse responses.

Step Response Models

Finite step response (FSR) models are obtained by making a unit step input change to a process operating at steady state. The model coefficients are simply the output values at each time step, as shown in Figure 4–9. Here, s_i represents the step response coefficient for the ith sample time after the unit step input change. If a nonunit step change is made, the output is scaled accordingly.

The step response model is the vector of step response coefficients,

$$S = [s_1 \;\; s_2 \;\; s_3 \;\; s_4 \;\; s_5 \;\; \ldots \;\; s_n]^T$$

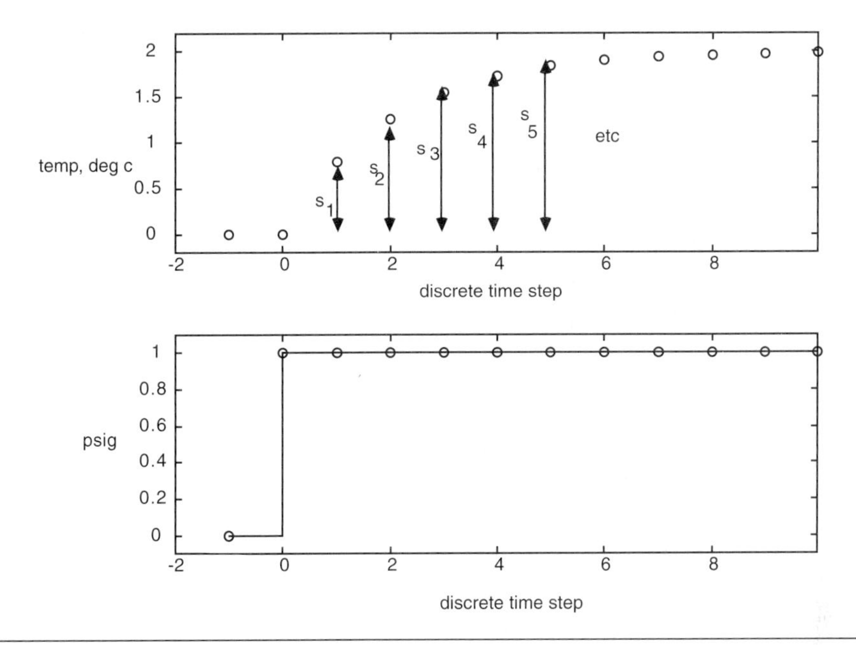

Figure 4–9 Illustration of step response parameter identification.

where the model length n is long enough so that the coefficient values are relatively constant (i.e., the process is close to a new steady state).

Impulse Response Models

Another common form of model is known as a finite impulse response (FIR). Here, a unit pulse is applied to the manipulated input, and the model coefficients are simply the values of the outputs at each time step after the pulse input is applied. As shown in Figure 4–10, h_i represents the ith impulse response coefficient.

There is a direct relationship between step and impulse response models:

$$h_i = s_i - s_{i-1}$$

$$s_i = \sum_{j=1}^{i} h_j$$

Figure 4–11 illustrates how impulse response coefficients can be obtained from step responses. The impulse response coefficients are simply the changes in the step response coefficient at each time step. Similarly, step response coefficients can be found from

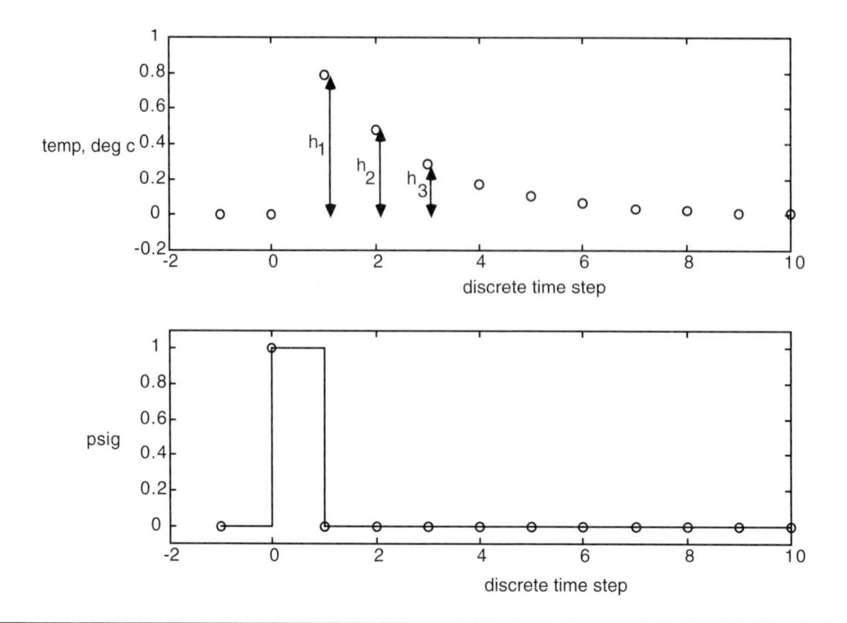

Figure 4–10 Illustration of step response parameter identification.

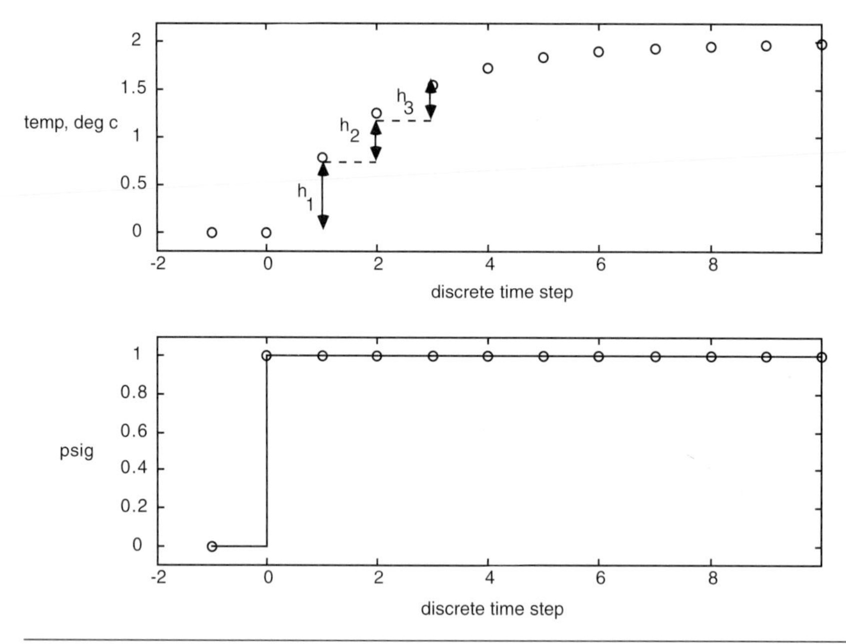

Figure 4–11 Obtaining impulse response models from step response data.

impulse responses; a step response coefficient is the sum of the impulse response coefficients to that point.

4.7 Summary

The first sections of this chapter illustrated the development of first-order + time-delay and integrator + time-delay models based on step input changes. These models will be sufficient for many simplified controller designs. Higher order models can easily be developed by "fitting" (adjusting) parameters to obtain a best match between model predictions and the actual measured process output responses.

Linear regression (least squares) techniques can be used to estimate parameters in discrete time autoregressive models. Usually, parameters are estimated based on a subset of the available experimental data. The model parameters are then verified by applying the model to a different subset of the data. If the model predictions match the measured outputs reasonably well, then the discrete model is usually acceptable for discrete control system design and analysis.

The stability of a discrete transfer function model is determined from the values of the poles of the denominator polynomial. If the poles are less than 1 in magnitude, the discrete model is stable. Module 16 covers classical discrete controller design procedures. Discrete step and impulse response models are often used in model predictive control (Chapter 16).

References

An overview of system identification techniques is provided by Ljung: Ljung, L., System Identification, Chapter 58 (pp. 1033–1054) in *The Control Handbook*, W.S. Levine (ed.), CRC Press (1996).

Student Exercises

1. Consider a first-order + time-delay transfer function

$$y(s) = \frac{k_p e^{-\theta s}}{\tau_p s + 1} u(s)$$

which has the following response to a step input change at $t = 0$:

$$t < \theta,\ y(t) = 0$$
$$t > \theta,\ y(t) = k_p \Delta u(1 - \exp(-(t - \theta)/\tau_p))$$

Show that the maximum rate-of-change (slope) of the output occurs at $t = \theta$. Also, find this slope.

2. Consider the response of a first-order + time-delay transfer function to a step input change. Find the value of the output (y) at the following times, as a fraction of the long-term output change

$$t_1 = \frac{\tau_p}{3} + \theta$$
$$t_2 = \tau_p + \theta$$

Find the time constant and time-delay based on the values of t_1 and t_2.

3. Consider the following step response. Estimate parameters for a first-order + time-delay model using the three techniques shown in Section 4.3. Include the units for each parameter.

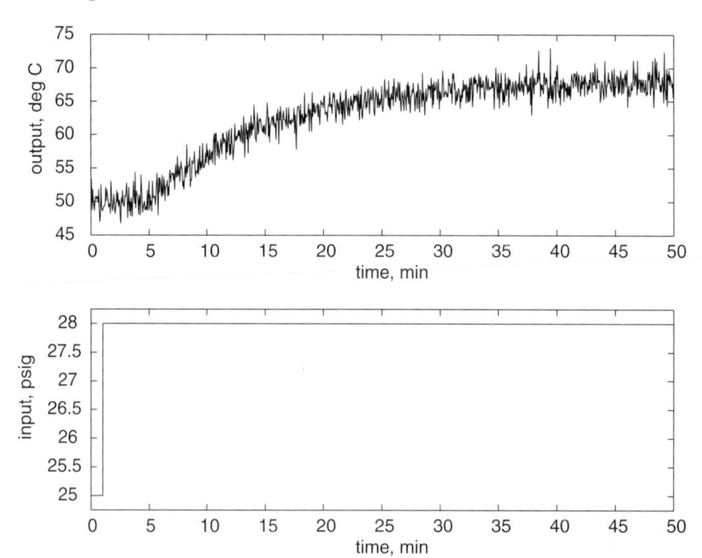

4. Consider the following step response. Estimate the parameters for an integrator + time-delay model, including the units for each parameter.

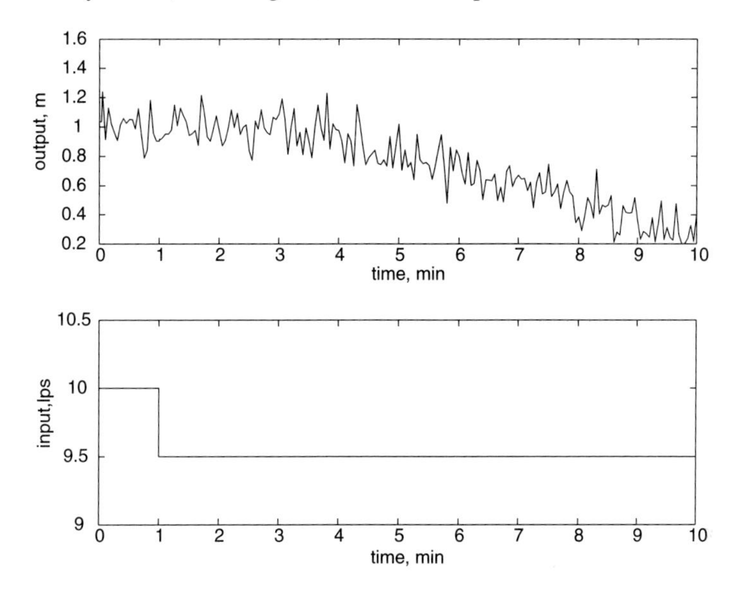

5. Consider the following continuous state space model:

$$A = \begin{bmatrix} -3.6237 & 0 \\ 0.8333 & -2.9588 \end{bmatrix}$$

$$B = \begin{bmatrix} 5.5051 \\ -1.2660 \end{bmatrix}$$

$$C = \begin{bmatrix} 0 & 1 \end{bmatrix}$$

$$D = 0$$

   a. Find the continuous transfer function model (do this analytically).
   b. For a sample time of 0.25, find the discrete state space and transfer function models (use MATLAB; see Module 4).
   c. Compare the step responses of the continuous and discrete models (use MATLAB). What do you observe?

6. Consider the discrete-time model

$$y(k) = 1.425y(k-1) - 0.4966y(k-2) + 0.1194u(k-1) + 0.09456u(k-2)$$

using z-transforms, find the corresponding model represented as equations (4.11), (4.12) and (4.13). Discuss the stability of the process.

7. Consider a unit step input change made at $k = 0$, resulting in the output response shown in the plot and table below.

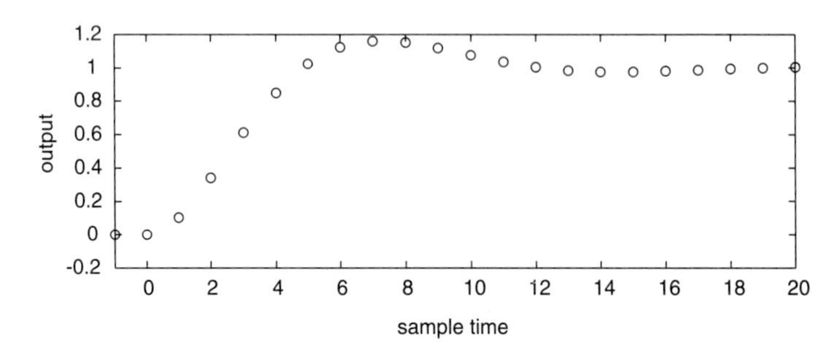

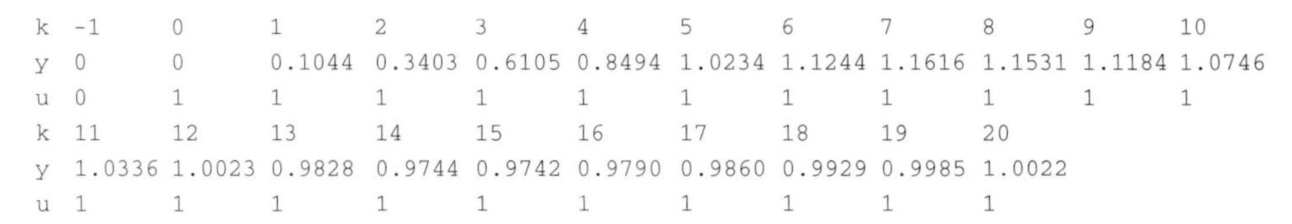

k	-1	0	1	2	3	4	5	6	7	8	9	10
y	0	0	0.1044	0.3403	0.6105	0.8494	1.0234	1.1244	1.1616	1.1531	1.1184	1.0746
u	0	1	1	1	1	1	1	1	1	1	1	1

k	11	12	13	14	15	16	17	18	19	20
y	1.0336	1.0023	0.9828	0.9744	0.9742	0.9790	0.9860	0.9929	0.9985	1.0022
u	1	1	1	1	1	1	1	1	1	1

Estimate the parameters for a discrete linear model with the form

$$g(z) = \frac{b_1 z^{-1} + b_2 z^{-2}}{1 + a_1 z^{-1} + a_2 z^{-2}}$$

Compare the step response of the estimated model with the data. Use MATLAB to convert the discrete model to a continuous model. Compare the step responses of the discrete and continuous models.

8. Consider Example 4.4 with measurement noise on the output variable, as shown below. Estimate the discrete model parameters based on this data. How do the parameters compare with those of Example 4.4? Compare the step responses of the two models.

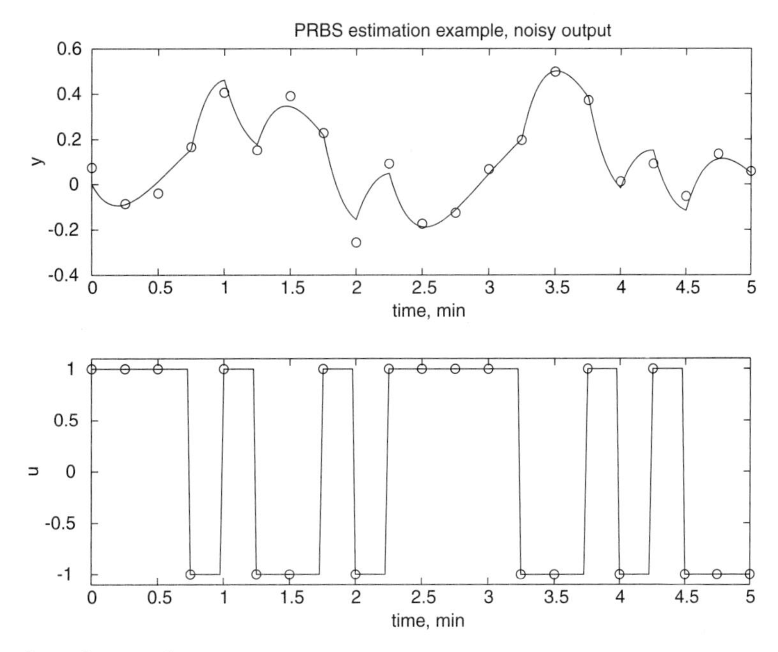

The input/output data are

k	-1	0	1	2	3	4	5
y	0.0741	-0.0857	-0.0399	0.1663	0.4065	0.1521	0.3910
u	1.0000	1.0000	1.0000	-1.0000	1.0000	-1.0000	-1.0000
k	6	7	8	9	10	11	12
y	0.2284	-0.2569	0.0910	-0.1737	-0.1260	0.0668	0.1958
u	1.0000	-1.0000	1.0000	1.0000	1.0000	1.0000	-1.0000
k	13	14	15	16	17	18	19
y	0.4976	0.3724	0.0119	0.0927	-0.0528	0.1357	0.0580
u	-1.0000	1.0000	-1.0000	1.0000	-1.0000	-1.0000	-1.0000

9. Consider the continuous state space model (where the time unit is minutes)

$$\begin{bmatrix} \dot{x}_1 \\ \dot{x}_2 \end{bmatrix} = \begin{bmatrix} -0.1 & 0 \\ 0.04 & -0.04 \end{bmatrix} \begin{bmatrix} x_1 \\ x_2 \end{bmatrix} + \begin{bmatrix} 0.1 \\ 0 \end{bmatrix} u$$

$$y = \begin{bmatrix} 0 & 1 \end{bmatrix} \begin{bmatrix} x_1 \\ x_2 \end{bmatrix}$$

a. Find the eigenvalues and the transfer function (use MATLAB for these calculations, if desired)

b. Using a sample time of 3 minutes, find the discrete state space model and the discrete transfer function. Refer to the Appendix for the form of the discrete state space model, and Module 4 to understand how to use MATLAB for these computations.

Appendix 4.1: Files Used to Generate Example 4.4

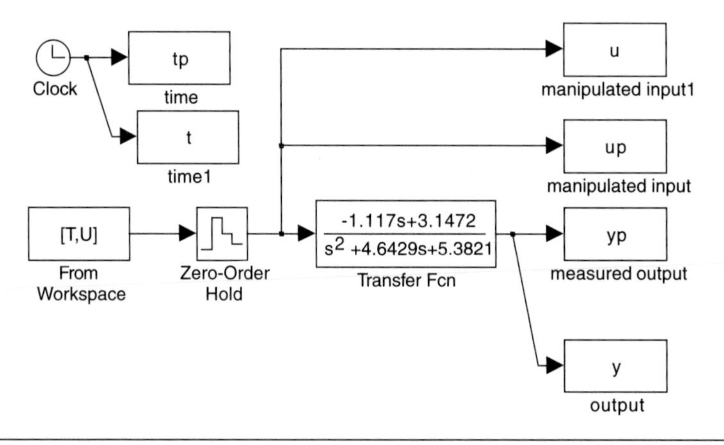

Figure A4–1 SIMULINK File `est_vdv.mdl`.

```
% use PRBS to estimate a discrete model of the vdv reactor
%
  T   = 0:0.25:5;         % time vector (row)
  T   = T';               % time vector (column)
  seed = [3;0.5];         % seed for noise generation
  randn('state',seed); % using the same seed each time
```

```
  uu = 0.5 1 0.25*randn(length(T),1); % mean of 0.5 with variance
%                        of 0.25
  U  = 2*round(uu)-1;  % creates PRBS with -1 and 1 values
%
  sim('est_vdv');      % runs simulation of linear van de vusse
%                        diagram
  figure(1);           % plot input-output data
  subplot(2,1,1),plot(tp,yp,'k',t,y,'ko');
  xlabel('time, min'), ylabel('y')
  title('PRBS estimation example')
  subplot(2,1,2),plot(tp,up,'k'); xlabel('time, min'),ylabel('u')
  axis([0 5 -1.1 1.1])
%                      % generate phi matrix for estimation
  for j = 4:22;
   phi(j-3,:) = [y(j-2) y(j-3) u(j-2) u(j-3)];
  end
%
  theta = inv(phi'*phi)*phi'*y(3:21)  % estimate parameters
%
  num = [theta(3) theta(4)];      % numerator of discrete transfer
                                    function
  den = [1 -theta(1) -theta(2)]; % denominator of discrete
                                    transfer function
  sysd = tf(num,den,0.25)         % create discrete tf object
  tzero(sysd)                     % calculate zeros
  pole(sysd)                      % calculate poles
  syszpk = zpk(sysd)              % zero-pole-k form
```

Appendix 4.2

Discrete time models can be obtained directly from continuous time models, as summarized in this appendix. The MATLAB Control Toolbox can be used for these conversions, as shown in Module 4.

Consider a continuous state space model of the following form (where we have assumed $D = 0$)

$$\begin{aligned} \dot{x} &= Ax + Bu \\ y &= Cx \end{aligned} \tag{A4.1}$$

A similar discrete state space model has the following form

$$x(k+1) = \Phi x(k) + \Gamma u(k)$$
$$y(k) = Cx(k) \tag{A4.2}$$

Assuming that the sample time is Δt, and that the input u is held constant between time t_k and t_{k+1}, (A4.1) can be integrated to yield

$$x(t_k + \Delta t) = e^{A\Delta t} x(t_k) + e^{A\Delta t} \int_{t_k}^{t_k + \Delta t} e^{-A\sigma} d\sigma B u(t_k)$$

which has the solution (although the matrix exponential is not an intuitive concept, the computation is readily performed by MATLAB)

$$x(k+1) = e^{A\Delta t} x(k) + (e^{A\Delta t} - I)A^{-1} B u(k)$$

which is usually written in the following form

$$x(k+1) = \Phi x(k) + \Gamma u(k)$$

where

$$\Phi = e^{A\Delta t}$$
$$\Gamma = (\Phi - I)A^{-1}B$$

The stability of this discrete state space model is determined by the eigenvalues of Φ, which must have a magnitude less that 1.0 for stability.

The discrete transfer function is, similar to the continuous time case

$$G(z) = C(zI - \Phi)^{-1}\,\Gamma$$

CHAPTER 5

Introduction to Feedback Control

The objective of this chapter is to introduce the Laplace domain analysis of feedback control systems. After studying this chapter, the reader should be able to do the following:

- Convert a process and instrumentation diagram (P&ID) to a control block diagram
- Find the controller transfer function for standard feedback controllers
- Find the closed-loop transfer function for a given process and controller transfer function
- Determine the stability of a closed-loop system
- Find the transfer function relationship between any two signals on a block diagram

The major sections of this chapter are as follows:

5.1 Motivation

Consider the level control problem first described in Chapter 1 and shown in Figure 5–1. A level measurement device (usually a differential pressure cell) senses the level and sends a signal to the controller. Notice that the controller is comparing the tank height with the desired setpoint and sending a controller output (pressure signal) to the valve, which changes the valve position and therefore the volumetric flow rate of stream 1, F_1. These signals are shown as dashed lines on the figure.

The question is, what is the control *algorithm*? How does the controller change the flow rate to the process? You can probably think of a number of possible algorithms.

On-Off Control

An *on/off controller* is similar to a controller used in a typical home heating unit. If the liquid level is too high (higher than the desired setpoint), the valve is closed and if the level is too low (lower than the desired setpoint) the valve is fully opened. The problem with an on-off strategy is that the valve will constantly be opening and closing, as shown in Figure 5–2, causing wear and tear on the valve.

Notice that an on-off controller is based on the error between the setpoint (desired height, h_{sp}) and the measured output (measured height). If the error ($h_{sp} - h_m$) is greater than a certain positive value (δ, 1/2 the *dead band*), the valve is turned off; if the error is less than a certain negative value ($-1/2$ the dead band), the valve is turned on. Assume that an increase in valve top pressure increases the flow through the valve [this is known as an *air-to-open* (or *fail-closed*) valve]. A fail-closed valve is used to make certain that the tank does not overflow in the event of a control valve failure. The algorithm could be stated

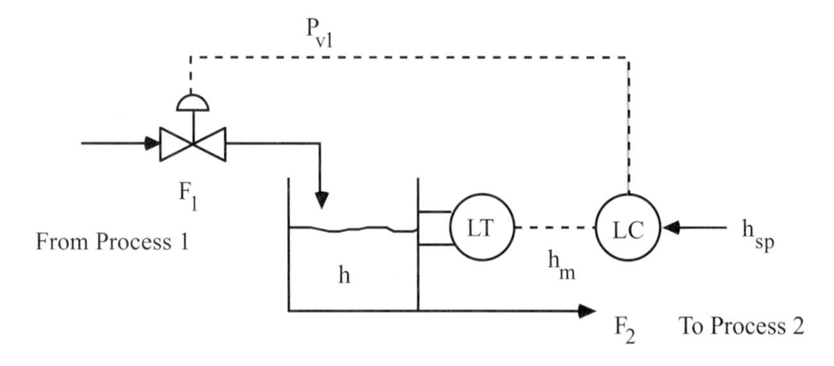

Figure 5–1 Level control, inlet flow rate manipulated. The instrumentation signals are shown as dashed lines. The level transmitter sends a signal to the level control, which sends a signal to the control valve.

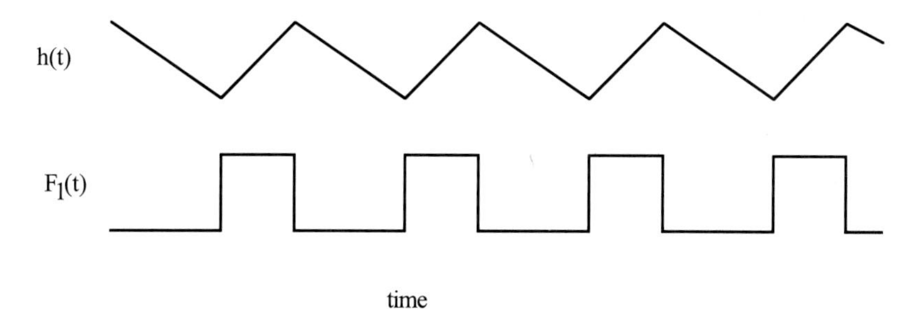

Figure 5–2 On-off control of tank height.

$$
\begin{array}{ll}
\textit{if}\ h \geq h_{sp} + \delta & P_{v1} = P_{v1\,\min} \\
\textit{if}\ h \leq h_{sp} - \delta & P_{v1} = P_{v1\,\max} \\
\textit{if}\ h_{sp} - \delta < h < h_{sp} + \delta & P_{v1} = \text{current value } (P_{v1\,\min} \text{ or } \mathrm{P}_{v1\,\max}).
\end{array}
\tag{5.1}
$$

We see that the controller behavior is periodic, as shown in Figure 5–2.

The next option is to make the signal to the valve *proportional* to the difference between the desired level and the actual level. This is known as a *proportional controller* (or P-controller) and is discussed below.

Proportional Control

The following equation suggests a control action that is proportional to the error (difference between setpoint and measurement),

$$P_{v1} = b + k_c\,(h_{sp} - h_m) \tag{5.2}$$

where b is the bias term, k_c is the *proportional gain* (a tuning parameter), and the difference between the setpoint and the measured output is the *error*

$$e = h_{sp} - h_m \tag{5.3}$$

If the error is 0 at steady state, notice that the bias term is equal to the steady-state valve-top pressure.

Now we discuss the adjustment of the tuning parameter (proportional gain, k_c). Consider the case where the error is positive (setpoint is greater than the measured output). This indicates that the flow rate entering the tank should be higher and therefore the valve-top pressure should increase. The proportional gain, k_c, must then be positive, for the control strategy shown in Figure 5–1. As k_c is increased, more control action is requested for a given amount of error.

Now consider the control strategy shown in Figure 5–3. A similar proportional control algorithm is

$$P_{v2} = b + k_c(h_{sp} - h_m) \tag{5.4}$$

In this case we assume that the valve is *fail-open* (or *air-to-close*), since we desire that the tank not overflow in the event of a control valve (or instrument air) failure. Now we rationalize the adjustment of the tuning parameter. Consider the case where the error is positive (height setpoint is greater than the measured height). This indicates that the flow rate leaving the tank should be lower and that the pressure to the valve should increase. The proportional gain, k_c, must also be positive for the system in Figure 5–3.

Relationship between $k_p k_v$ and k_c

Recall that the process gain is simply the steady-state change in output for a steady-state change in process input.

$$k_p = \frac{\Delta y}{\Delta u}$$

For the *first strategy,* the process gain is positive; that is, an increase in F_1 causes an increase in h.

$$k_{p1} = \frac{\Delta h}{\Delta F_1} = \textit{positive value}$$

Since the input to the valve is the valve-top pressure (P_{v1}) and the output of the valve is flow rate, the valve gain is

$$k_{v1} = \frac{\Delta F_1}{\Delta P_{v1}} = \textit{positive value}$$

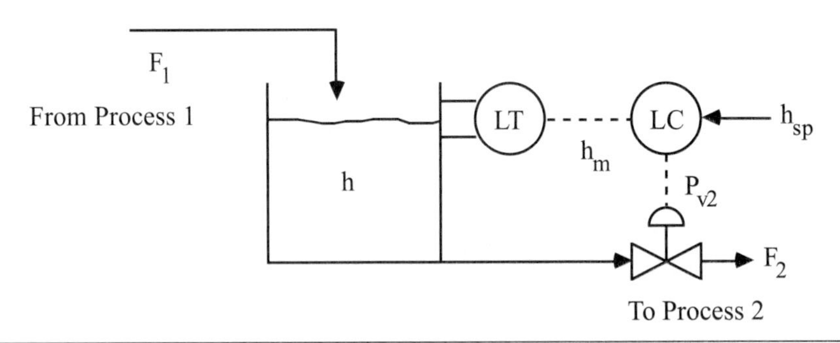

Figure 5–3 Level control with outlet flow rate manipulated.

The overall gain between valve-top pressure and tank height is then

$$k_{p1}k_{v1} = \frac{\Delta h}{\Delta F_1}\frac{\Delta \mathrm{F}_1}{\Delta \mathrm{P}_{\mathrm{v}1}} = \frac{\Delta h}{\Delta P_{v1}} = \textit{positive}$$

For the *second strategy*, the process gain is negative (an increase in F_2 causes a decrease in h),

$$k_{p2} = \frac{\Delta h}{\Delta F_2} = \textit{negative value}$$

and the valve gain is also negative. The gain between valve-top pressure and tank height is

$$k_{p2}k_{v2} = \frac{\Delta h}{\Delta F_2}\frac{\Delta \mathrm{F}_2}{\Delta \mathrm{P}_{\mathrm{v}2}} = \frac{\Delta h}{\Delta P_{v2}} = \textit{positive}$$

Notice that a process with a positive gain ($k_{p1}k_{v1}$ or $k_{p2}k_{v2}$) requires a controller proportional gain that is also positive. Similarly, a process with a negative gain would require a controller proportional gain that is also negative.

Valve Gains

Most control valves are designed to operate with signals between 3 and 15 psig. A *fail-closed* valve will be fully closed at 3 psig and fully open at 15 psig. For example, if the maximum flow rate through the valve is 120 gallons per minute (gpm), then the average valve gain is

$$k_v = \frac{\textit{change in flow rate}}{\textit{change in signal}} = \frac{120 - 0 \text{ gpm}}{15 - 3 \text{ psig}} = 10 \text{ gpm/psig}$$

Similarly, a *fail-open* valve of the same size would have a negative gain (-10 gpm/psig).

5.2 Development of Control Block Diagrams

Control block diagrams are used to analyze the dynamic effect of feedback control loops. All dynamic elements in a control loop are combined, using their Laplace transfer function representation. In this section we use the control strategy from Figure 5–1 (where the inlet flow rate is manipulated) for illustration. Block diagrams are based on Laplace domain signals, which are assumed to be in deviation variable (perturbations from steady state) form. The block diagram for Figure 5–1 has the following components:

- Level controller (relates error to controller output)
- Valve (relates controller output signal to flow-through valve)
- Process (relates manipulated input to process output)
- Disturbance (relates the disturbance input to the process output)
- Sensor (measures tank level)

Controller Transfer Function

Notice first that the level controller compares the measured tank level (h_m) to the desired tank level (h_{sp}). The desired tank level is known as the *setpoint*, and the difference between the setpoint and the measured process output is the *error*.

From the proportional control law (algorithm) for this system [Equation (5.2)], realizing that $b = P_{v1s}$ (steady-state pressure to valve), we find the controller input-output relationship

$$P_{v1} - P_{v1s} = k_c(h_{sp} - h_m)$$

where the controller input signal is the *error*, which is the difference between the *setpoint* and measured process output. We write the equation [where $c(s)$ is the *controller output*, the pressure to the valve]

$$c(s) = k_c e(s) = g_c(s)e(s)$$

which is the transfer function form for a proportional-only controller.

$$g_c(s) = k_c \qquad \text{P control} \tag{5.5}$$

A block diagram for the controller is shown in Figure 5–4, where $g_c(s) = k_c$. For convenience when analyzing block diagrams, the comparator is shown outside the controller transfer function block. The block diagram uses $r(s)$ to represent the setpoint, and $y_m(s)$ to represent the measured process output.

Valve Transfer Function

The block diagram for the valve is shown in Figure 5–5. The input signal is the pressure to the valve top and the output is the flow rate of fluid through the valve.

Process Transfer Function

This process block diagram is shown in Figure 5–6. The input to the block is the flow rate to the tank, and the output is the tank level.

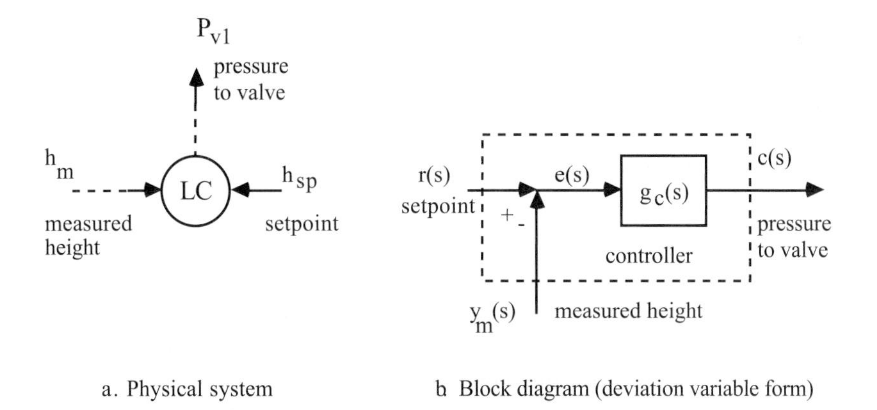

Figure 5–4 Block diagram relationship for controller transfer function. The orientation for the transfer function representation of the controller is slightly different from that of the physical system.

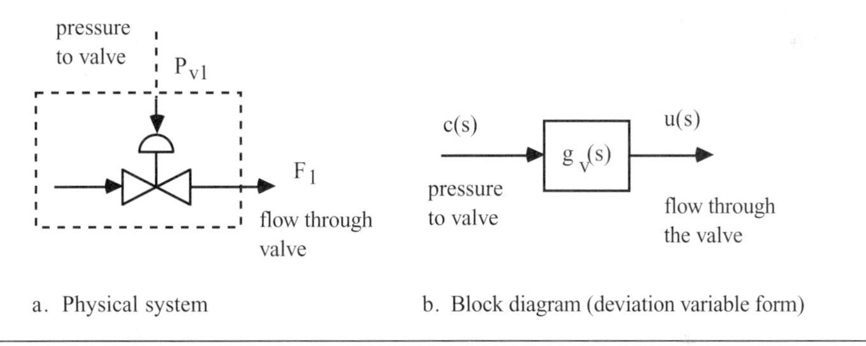

Figure 5–5 Block diagram of valve and the physical system.

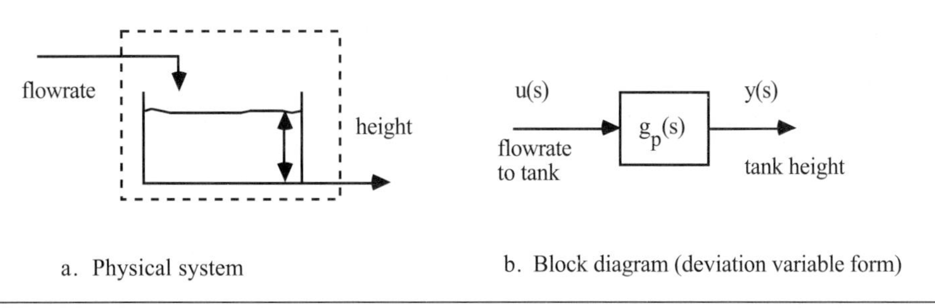

Figure 5–6 Block diagram of the manipulated input effect on the process.

Disturbance Transfer Function

This disturbance (outlet flow rate) block diagram is shown in Figure 5–7. The disturbance input is the flow rate from the tank (which may be due to changes in a downstream control valve not shown), and the output is the tank level.

Measurement (Sensor) Transfer Function

This measurement block diagram is shown in Figure 5–8. The input to the block is the tank level and the output is the actual measurement of tank level.

Control Block Diagram

We can combine Figures 5–4 to 5–8 to obtain the feedback control system block diagram shown in Figure 5–9. This is similar to the conceptual block diagram developed in Chapter 1 (Figure 1–4). There are two externally supplied signals, the setpoint (desired tank height) and the load disturbance (flow rate to downstream process).

We often assume that the output variable can be perfectly measured, and that process input (usually a flow rate) is directly manipulated; in this case we do not include the valve and measurement transfer functions in the closed-loop block diagram. Equivalently, we can "lump" the valve and measurement dynamics into the process transfer function, again allowing us to neglect the valve and measurement transfer functions. Similarly, the measurement device can be lumped into the disturbance transfer function.

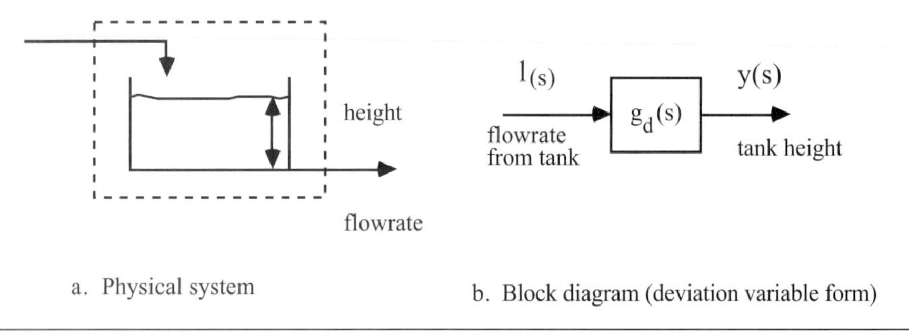

Figure 5–7 Block diagram of the disturbance input effect on the process.

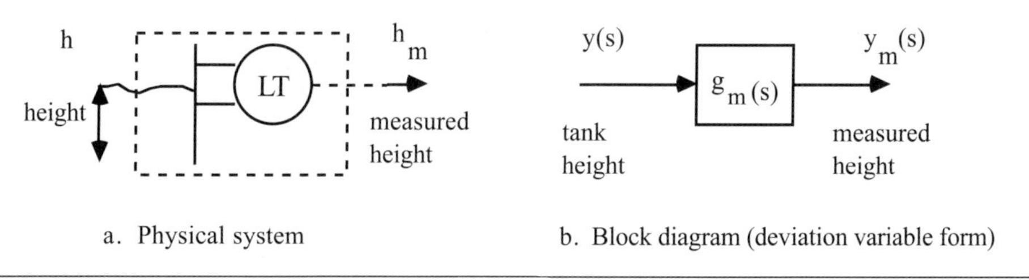

Figure 5–8 Block diagram of sensor.

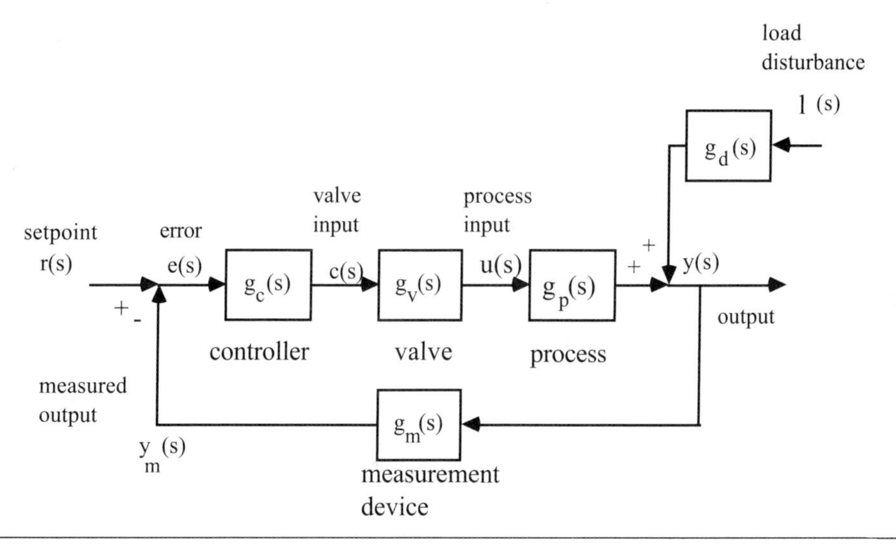

Figure 5–9 Control system block diagram.

5.3 Response to Setpoint Changes

A simplified [assuming $g_m(s) = g_v(s) = 1$] feedback block diagram is shown in Figure 5–10. This simplification is equivalent to lumping the measurement and valve dynamics into the process transfer function. By convention, we denote the output of the controller block as $u(s)$, which is also the manipulated input. Notice that we are focusing on setpoint changes only, so we have not included the disturbance block. In Section 5.7 we will focus on disturbances.

One objective of block diagram analysis is to find the output response to a setpoint change. A critical aspect is to determine the closed-loop stability. The key is to use block diagram manipulation to find the relationship between the setpoint and the output.

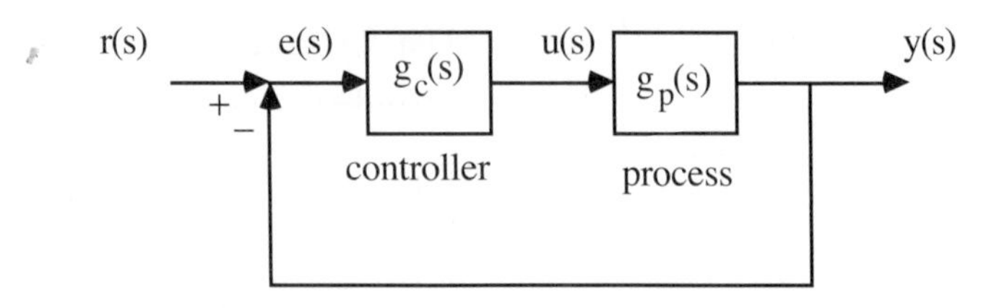

Figure 5–10 Simplified control block diagram. Disturbances are neglected and valve and measurement dynamics are lumped into the process transfer function.

In Figure 5–11, we have absorbed the closed-loop diagram of Figure 5–10 into a single transfer function block, to provide the closed-loop relationship between setpoint and process output.

Our goal is to obtain the output $y(s)$ as a function of the setpoint, $r(s)$. Here, we combine the process input-output relationship, $y(s) = g_p(s)\,u(s)$, with controller relationship, $u(s) = g_c(s)\,e(s)$, to find

$$y(s) = g_p(s)g_c(s)e(s)$$

Also, since the error is defined as $e(s) = r(s) - y(s)$, we can write

$$y(s) = g_p(s)g_c(s)(r(s) - y(s))$$

which can be solved to find

$$y(s) = \frac{g_p(s)g_c(s)}{1 + g_p(s)g_c(s)}r(s)$$

We refer to the relationship between $r(s)$ and $y(s)$ as the *closed-loop transfer function*, $g_{CL}(s)$,

$$g_{CL}(s) = \frac{g_p(s)g_c(s)}{1 + g_p(s)g_c(s)} \tag{5.6}$$

If all of the poles of $g_{CL}(s)$ are stable, then the closed-loop system is stable. The denominator of $g_{CL}(s)$ is also known as the *characteristic equation*.

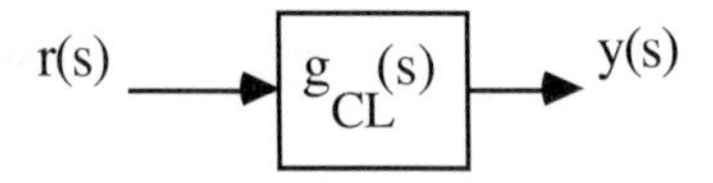

Figure 5–11 External view of Figure 5–10.

Equation (5.6) has been derived for the simple diagram shown in Figure 5–10. Please realize that more-complex block diagrams will have more-complex closed-loop transfer functions. You should be able to derive the closed-loop transfer function for any block diagram. For example the closed-loop transfer function for Figure 5–9 (neglecting disturbances) is

$$g_{CL}(s) = \frac{g_p(s)g_v(s)g_c(s)}{1 + g_p(s)g_v(s)g_c(s)g_m(s)}$$

Possible Problems with *Offset* Using Proportional Controllers

A problem with proportional controllers is that there is generally *offset* when a setpoint change is made. By offset, we mean that the actual process output will not be able to achieve the desired setpoint change. For simplicity, consider a first-order process, $g_p(s) = k_p/(\tau_p s + 1)$, with a proportional controller, $g_c(s) = k_c$, which leads to the following closed-loop transfer function

$$g_{CL}(s) = \frac{\dfrac{k_p k_c}{\tau_p s + 1}}{1 + \dfrac{k_p k_c}{\tau_p s + 1}} = \frac{\dfrac{k_p k_c}{1 + k_p k_c}}{\dfrac{\tau_p}{1 + k_p k_c} s + 1} = \frac{k_{CL}}{\tau_{CL} s + 1}$$

We recognize that this is a first-order transfer function, where $k_{CL} = k_p k_c/(1 + k_p k_c)$and $\tau_{CL} = \tau_p/(1 + k_p k_c)$. For closed-loop *stability*, the requirement is that $\tau_{CL} > 0$ or $\tau_p/(1 + k_p k_c) > 0$. Since $\tau_p > 0$ (the process is open-loop stable), we have the requirement that $1 + k_p k_c > 0$ or

$$k_p k_c > -1$$

for stability. Now consider a step setpoint change of magnitude R. You should be able show that the output response will be

$$y(t) = Rk_{CL}(1 - \exp(-t/\tau_{CL}))$$

Notice that if k_{CL} is not exactly equal to 1, there will be "offset" in the closed-loop response, as shown in Figure 5–12. That is, the process output will not exactly match the desired setpoint, even in the limit as time goes to infinity. Also, as a practical matter, it is important that $k_p k_c > 0$. Notice that if $-1 < k_p k_c < 0$, then k_{CL} is negative; although the closed-loop is stable, the process output actually moves in the opposite direction as the setpoint change.

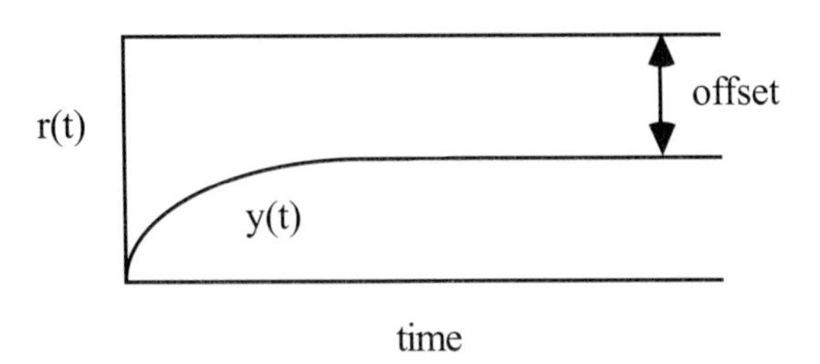

Figure 5–12 Illustration of offset in system with a proportional controller.

Offset can also be shown by the final value theorem of Laplace transforms.

$$\begin{aligned} offset &= \lim_{t\to\infty} e(t) = R - y(t\to\infty) = R[1 - k_{CL}] \\ &= R\left[1 - \frac{k_p k_c}{1 + k_p k_c}\right] = R\left[\frac{1}{1 + k_p k_c}\right], \end{aligned}$$

or the fractional offset is $1/(1 + k_p k_c)$. For a given k_p, the offset will be reduced as k_c gets larger and larger. Also, an increase in k_c decreases τ_{CL}, speeding up the closed-loop response.

Example 5.1: Offset with Proportional (P) Control of a First-Order Process

Consider the process transfer function for a stirred-tank heater, where the output is temperature (°C), the manipulated input is heater power (kW), and the timescale is minutes.

$$y(s) = g_p(s)u(s)$$

$$g_p(s) = \frac{1}{5s + 1}$$

Assume a unit step temperature setpoint change [$r(s) = 1/s$]. The stability condition was $k_p k_c > -1$, so the closed-loop should be stable as long as $k_c > -1$. Closed-loop responses for various values of k_c (-0.5, 1, 5, and 10 kW/°C) are shown in Figure 5–13. As expected, increasing k_c speeds up the response and reduces the offset. Also, it is obvious that the controller gain should be greater than zero (that is, it should be the same sign as the process gain), otherwise the response is in the opposite direction; although this may be stable, it is certainly not desirable.

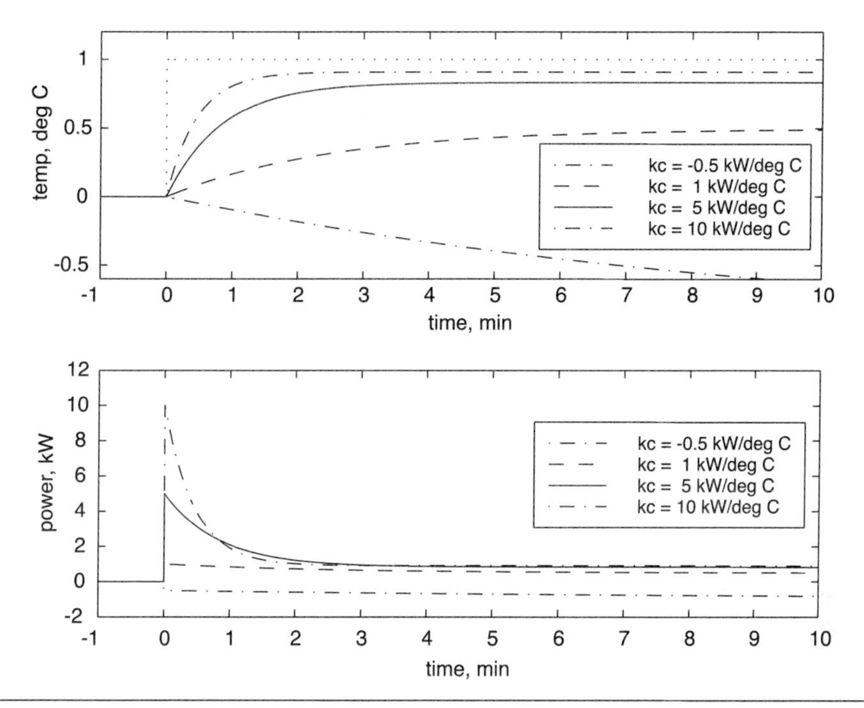

Figure 5–13 Closed-loop response of a first-order system under proportional-only control. Effect of proportional gain on offset, speed of response, and manipulated input action.

Thus far we have not considered the effect of control action. It is easy to derive the following relationship between $r(s)$ and $u(s)$ from Figure 5–10:

$$u(s) = \frac{g_c(s)}{1 + g_p(s)g_c(s)} r(s)$$

For this example,

$$u(s) = \left[\frac{k_c}{1 + k_p k_c}\right]\left[\frac{\tau_p s + 1}{\dfrac{\tau_p}{1 + k_p k_c} s + 1}\right] r(s)$$

which we recognize is a lead-lag response. The control (manipulated) variable response for this example is also shown in Figure 5–13 for various values of k_c. As k_c increases, the magnitude and speed of the manipulated variable response increases. For a particular physical system, too high of a value of k_c may cause the manipulated variable to hit a constraint. Consider a step setpoint change of magnitude R [that is $r(s) = R/s$]. From the definition of a proportional controller, or from the initial value theorem, we can show that

$$u(t = 0^+) = Rk_c$$

If R_{max} represents a maximum setpoint change and u_{max} represents a maximum allowable manipulated variable change, then clearly k_c must be less u_{max}/R_{max} to avoid manipulated variable saturation.

Notice that increasing the controller gain reduces the offset and speeds up the response. In the limit of an infinite controller gain, there is no offset and the dynamic closed-loop response is instantaneous; a clear limitation is that the manipulated input may hit a constraint (in this example, the heater power will be limited to some maximum value). In practice, there are time delays or other model uncertainties and that may cause the closed-loop to be unstable with high controller gains. It is necessary, then, to devise other control strategies to remove offset. These strategies are detailed in Section 5.4.

Please note that there is one type of process where proportional control does not yield offset when a setpoint change is made. These are integrating processes, where one pole of the process has a value of zero; for example,

$$g_p(s) = \frac{1}{s(s+1)}$$

The reader should use the final value theorem of Laplace transforms to show that this process will not have offset when a step setpoint change is made, under proportional control.

5.4 PID Controller Algorithms

Here we continue to work in deviation variable form. Recall that a proportional-only controller could lead to *offset* between the desired setpoint and the actual output. This happened because the controller output (process input) and process output came to new equilibrium values before the error went to zero. The next step is to add a term where the controller output is proportional to the *integral of the error*, in addition to the term that is proportional to the error. This is known as *proportional-integral* (PI) control. As long as there continues to be an error signal to the controller, the controller output (process input) will continue to change. Therefore, the integral of the error forces the error signal to zero. Notice that this type of controller implicitly accounts for the past system response. A proportional-only controller does not account for this "process history."

PI Control

A PI controller has two terms (and two tuning parameters), one proportional to the error (k_c) and the other proportional to the integral of the error (k_I).

$$u(t) = k_c e(t) + k_I \int_o^t e(\sigma) d\sigma$$

This is more commonly written in the form

$$u(t) = k_c \left[e(t) + \frac{1}{\tau_I} \int_o^t e(\sigma) d\sigma \right] \tag{5.7}$$

where τ_I is known as the integral time. Both the proportional gain and integral time are tuning parameters that can be adjusted by the control instrument technician or process operator. Notice that $k_I = k_c/\tau_I$. The transform of an integral is

$$L\left[\int_0^t e(\sigma) d\sigma \right] = \frac{e(s)}{s}$$

Taking the Laplace transform of Equation (5.7), we find as representation for a PI controller

$$u(s) = k_c \left(1 + \frac{1}{\tau_I s} \right) e(s) = k_c \left(\frac{\tau_I s + 1}{\tau_I s} \right) e(s) \tag{5.8}$$

and the controller transfer function for PI control is

$$g_c(s) = k_c \frac{(\tau_I s + 1)}{\tau_I s} \quad \text{PI control} \tag{5.9}$$

The following example is an application of PI control to a first-order process.

Example 5.2: First-Order Process with a PI Controller

A first-order process with a PI controller has the closed-loop transfer function

$$g_{CL}(s) = \frac{g_p(s) g_c(s)}{1 + g_p(s) g_c(s)} = \frac{\frac{k_p k_c}{(\tau_p s + 1)} \frac{(\tau_I s + 1)}{\tau_I s}}{1 + \frac{k_p k_c}{(\tau_p s + 1)} \frac{(\tau_I s + 1)}{\tau_I s}}$$

which can be rearranged to the form

$$g_{CL}(s) = \frac{\tau_I s + 1}{\frac{\tau_I \tau_p}{k_p k_c} s^2 + \frac{\tau_I (1 + k_p k_c)}{k_p k_c} s + 1}$$

Notice that there is *no offset*, because the closed-loop gain is 1. The quadratic formula can be used to show that for a second-order characteristic polynomial, if all the coefficients are positive, the roots are negative (stable). In this case, we have the conditions

$$\frac{\tau_I \tau_p}{k_p k_c} > 0$$

$$\frac{\tau_I (1 + k_p k_c)}{k_p k_c} > 0$$

Assuming that the process is open-loop stable ($\tau_p > 0$) and realizing that $\tau_I > 0$ by definition, we see that from the first condition,

$$k_p k_c > 0$$

that is, k_p and k_c must be the same sign. The second condition requires that $1 + k_p k_c > 0$, or

$$k_p k_c > -1$$

which is less restrictive than the first condition. The first condition, $k_p k_c > 0$, then must be met for closed-loop stability of this system.

We must realize that there are two controller tuning parameters, k_c and τ_I, that can be adjusted to give a desired closed-loop response. The process parameters are not available for adjustment, since they are physical parameters based on the process under consideration.

We have already found the conditions for stability but have not discussed the expected dynamic behavior of a stable closed-loop system. If some poles of the closed-loop system are complex, the response will consist of damped oscillations (underdamped), and if all the poles are real, the response will be critically damped or overdamped. The reader has the opportunity (see Exercise 16) to show that the closed-loop behavior will be underdamped if the following condition is satisfied

$$\tau_I < \frac{4 k_p k_c \tau_p}{(1 + k_p k_c)^2}$$

The reader should also show that if the integral time, τ_I, is equal to the process time constant, τ_p, the closed-loop response will be first order with a closed-loop time constant of

$\tau_p/(k_p k_c)$. These results can be illustrated by simulation for a first-order process with the following parameter values (see Example 5.1)

$$g_p(s) = \frac{1}{5s+1}$$

Simulations for a setpoint change of 1°C are shown in Figure 5–14, for τ_I = 0.25 minutes and τ_I = 5 minutes. In both cases the proportional gain is k_c = 5 kW/°C. For the smaller integral time there is oscillatory performance in both the controlled output (temperature) and manipulated input (heater power). This oscillatory performance is often viewed as unfavorable to process operators. An oscillatory closed-loop system is usually more sensitive to parameter uncertainty; that is, if the process operating conditions changed the closed-loop system could go unstable.

In Example 5.2, a first-order process with PI control, we showed that the controller gain and the process gain must be the same sign for closed-loop stability. This result holds true for any open-loop stable process with integral control. Just remember that the require-

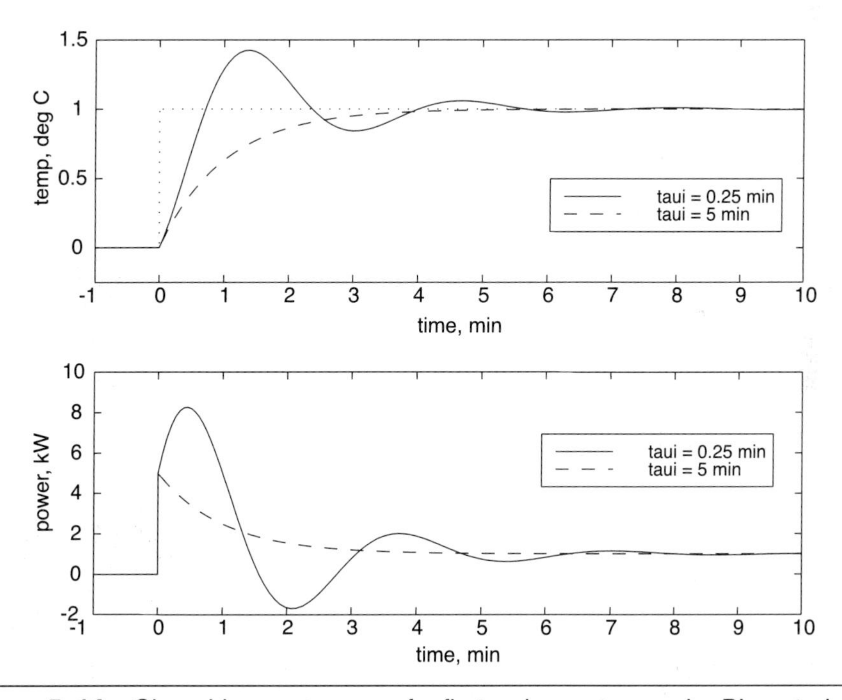

Figure 5–14 Closed-loop response of a first-order system under PI-control. Effect of integral time on offset and speed of response. The proportional gain, k_c, is 5 kW/°C in both cases.

ment $k_p k_c > 0$ is necessary and sufficient for stability of PI control of a *first-order* system only, because the closed-loop transfer function is second-order and positive coefficients for the second-order transfer function are necessary and sufficient for stability. For higher order, open-loop stable systems, $k_p k_c > 0$ is *necessary* for closed-loop stability *but not sufficient. The Routh array* (Section 5.5) must be checked for the sufficient conditions.

PID Control

The next level of controller complexity is to add a term that accounts for the current rate of change (derivative) of the error. Using knowledge of the derivative of the error allows the controller to "predict" where the future error is heading and compensate for it. The time-domain representation of PID control is

$$u(t) = k_c e(t) + k_I \int_o^t e(\sigma)d\sigma + k_D \frac{de(t)}{dt} \tag{5.10}$$

which is often written ($k_D = k_c \tau_D$)

$$u(t) = k_c \left[e(t) + \frac{1}{\tau_I} \int_o^t e(\sigma)d\sigma + \tau_D \frac{de(t)}{dt} \right] \tag{5.11}$$

where τ_D is called the *derivative time*. In the Laplace domain, Equation (5.11) is written

$$u(s) = k_c \left(1 + \frac{1}{\tau_I s} + \tau_D s \right) e(s) = k_c \left(\frac{\tau_D \tau_I s^2 + \tau_I s + 1}{\tau_I s} \right) e(s) \tag{5.12}$$

$$g_c(s) = k_c \left(\frac{\tau_D \tau_I s^2 + \tau_I s + 1}{\tau_I s} \right) \quad \text{"Ideal" PID control} \tag{5.13}$$

It turns out that the "ideal" PID algorithm is not physically realizable, that is, no instrument can take a perfect derivative, so most "practical" PID algorithms have one of the following forms

$$g_c(s) = k_c \left(\frac{\tau_I s + 1}{\tau_I s} \right) \left(\frac{\tau_D s + 1}{\tau_F s + 1} \right) \quad \text{"Real" PID control,} \tag{5.14a}$$

or

$$g_c(s) = k_c \left(\frac{\tau_I s + 1}{\tau_I s} + \frac{\tau_D s}{\tau_F s + 1} \right) \quad \text{"Real" PID control,} \tag{5.14b}$$

where $\tau_F = \alpha \tau_D$. A typical value for α is 0.1 or less. Even if the ideal PID controller of Equation (5.13) could be implemented, it would not be desirable. A step setpoint change, for example, would cause the derivative of the error to be infinity and cause a spike in the manipulated variable; this behavior is known as derivative "kick." For this reason the derivative action is often based on the process output rather than the error. That is, Equation (5.11) is implemented as

$$u(t) = k_c\left[e(t) + \frac{1}{\tau_I}\int_o^t e(\sigma)d\sigma - \tau_D\frac{dy(t)}{dt}\right] \tag{5.15}$$

The transfer function representation is

$$u(s) = k_c\left(\frac{\tau_I s + 1}{\tau_I s}\right)e(s) - k_c\tau_D s y(s) \quad \text{Ideal derivative on process output} \tag{5.16}$$

Again, since an ideal derivative cannot be implemented, the following filtered derivative of the output can be used

$$u(s) = k_c\left(\frac{\tau_I s + 1}{\tau_I s}\right)e(s) - k_c\frac{(\tau_D s)}{(\tau_F s + 1)}y(s) \quad \text{"Real" derivative on process output} \tag{5.17}$$

Proportional Band

We have been using a controller formulation based on proportional gain. Some industrial controllers use proportional band, rather than proportional gain. The proportional band is the range of error that causes the controller output (manipulated input) to change over its full range. The proportional band is related to the proportional gain by

$$PB = \frac{100}{k_c} \tag{5.18}$$

as shown in Module 15.

5.5 Routh Stability Criterion

The stability of the system is determined from the values of the roots (poles) of its characteristic equation. Finding roots is easy for first- and second-order equations (and not too hard for third) since there is an analytical solution for the roots of polynomials through third order. If the polynomial is fourth order or higher, the roots must be determined numerically. There is a method for determining whether any of the roots are positive (unstable) without actually calculating the roots (Routh, 1905). This method involves an analysis of the coefficients of the characteristic polynomial by setting up the *Routh array*. The test of the coefficients in the Routh array is called the *Routh stability criterion*.

The Routh (pronounced like *truth*) stability criterion is based on a polynomial equation that has the following form

$$a_n s^n + a_{n-1} s^{n-1} + \ldots a_1 s + a_0 = 0$$

The *necessary* condition for all the roots to be negative is that all $a_i > 0$ (you can multiply all coefficients by -1 if needed). The *sufficiency* test requires the Routh array

Row				
1	a_n	a_{n-2}	a_{n-4}	$\cdots$
2	a_{n-1}	a_{n-3}	a_{n-5}	$\cdots$
3	b_1	b_2	b_3	$\cdots$
4	c_1	c_2	c_3	$\cdots$
$\vdots$				
$n+1$				

where the elements of rows 3 and higher have the form

$$b_1 = \frac{a_{n-1}a_{n-2} - a_n a_{n-3}}{a_{n-1}}, \qquad b_2 = \frac{a_{n-1}a_{n-4} - a_n a_{n-5}}{a_{n-1}}, \text{ and so forth}$$

$$c_1 = \frac{b_1 a_{n-3} - a_{n-1} b_2}{b_1}, \qquad c_2 = \frac{b_1 a_{n-5} - a_{n-1} b_3}{b_1}, \text{ and so forth}$$

If all the coefficients in the first column of the Routh array are positive, then the sufficient condition for stability is satisfied. If all the coefficients are not positive, then we can determine the number of unstable (positive) roots by the number of changes in the sign of the coefficients as we move down the first column. An example of the use of the Routh array to determine limits of a tuning parameter for closed-loop stability is shown next.

Example 5.3: Third-Order Process with a P-Only Controller

Find the upper bound on proportional gain (for P-only control) for closed-loop stability of the following process

$$g_p(s) = \frac{1}{(3s+1)(2s+1)(s+1)} = \frac{1}{6s^3 + 11s^2 + 6s + 1}$$

The closed-loop transfer function is

$$g_{CL}(s) = \frac{g_p(s)g_c(s)}{1 + g_p(s)g_c(s)} = \frac{k_c}{6s^3 + 11s^2 + 6s + 1 + k_c}$$

We must use the Routh stability criterion on the characteristic polynomial. The coefficients are

$$a_3 = 6, \quad a_2 = 11, \quad a_1 = 6, \quad a_0 = 1 + k_c$$

The Routh array is

Row		
1	a_3	a_1
2	a_2	a_0
3	b_1	
4	a_0	

Inserting our values into the Routh array,

Row		
1	6	6
2	11	$1+k_c$
3	b_1	
4	$1+k_c$	

we see from the necessary condition that $1 + k_c > 0$, or $k_c > 0$. The sufficient condition that must be checked is $b_1 > 0$

$$b_1 = \frac{11 \cdot 6 - 6(1+k_c)}{11} = 6\left[1 - \frac{1+k_c}{11}\right]$$

which leads to $k_c < 10$ for stability. The range of tuning parameters for stability is then

$$-1 < k_c < 10$$

As a practical matter, however, the controller gain should be the same sign as the process gain (recall that in Example 5.1 a controller gain with the wrong sign led to a process output response in the opposite direction of the setpoint change). Simulation results for a unit step setpoint change are shown in Figure 5–15. As k_c is increased, the closed-loop response becomes more oscillatory. The closed-loop system will lose stability at $k_c = 10$.

Indeed, in Table 5–1 we show the relationship between k_c and the closed-loop poles. Notice that one pole stays real and becomes more negative (faster) as k_c increases. The other two poles become complex and the magnitude of the imaginary portion becomes larger as k_c increases, indicating that the magnitude of oscillation is increasing. At $k_c = 10$, the complex poles cross from the left-half plane to the RHP, indicating that the closed-loop system is unstable for $k_c > 10$.

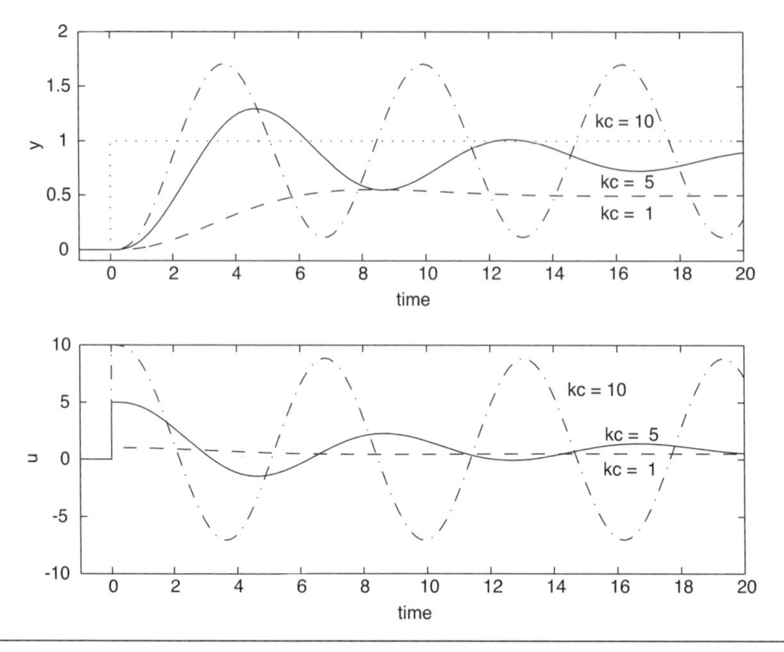

Figure 5–15 Response of the third-order process to a unit setpoint change at $t = 0$. Effect of proportional gain. The system loses stability at $k_c = 10$.

Figure 5–16 is a plot of the roots of the characteristic equation, as a function of k_c. It shows vividly the results presented in Table 5–1. The closed-loop poles start as three real, distinct poles at $k_c = 0$. Two become complex, and move toward the RHP as k_c is increased. When $k_c > 10$, the two complex poles move into the RHP; at $k_c = 10$, a continuous oscillation is formed.

5.6 Effect of Tuning Parameters

We have shown how to use the closed-loop characteristic equation to determine the closed-loop stability of processes. Except for P-only control, we have not discussed the effect of controller tuning parameters on the feedback system performance (closed-loop response).

Our discussion here is based on open-loop stable processes. To better understand the effect of controller tuning parameters, you should experiment with a number of transfer functions. You can also determine the closed-loop poles as a function of the tuning parameters, for a particular process.

Table 5–1 Closed-Loop Poles as a Function of Proportional Gain for Example 5.3[a]

k_c	p_1	p_2	p_3
0	−1.0000	−0.5000	−0.3333
0.1000	−1.0432	−0.3951−0.1402j	−0.3951+0.1402j
0.2500	−1.0926	−0.3704−0.2313j	−0.3704+0.2313j
0.5000	−1.1549	−0.3392−0.3184j	−0.3392+0.3184j
1.0000	−1.2452	−0.2940−0.4257j	−0.2940+0.4257j
3.0000	−1.4612	−0.1861−0.6493j	−0.1861+0.6493j
5.0000	−1.5991	−0.1171−0.7821j	−0.1171+0.7821j
7.0000	−1.7054	−0.0640−0.8819j	−0.0640+0.8819j
9.0000	−1.7938	−0.0198−0.9637j	−0.0198+0.9637j
10.0000	−1.8333	0.0000−1.0000j	0.0000+1.0000j
11.0000	−1.8704	0.0185−1.0339j	0.0185+1.0339j
13.0000	−1.9384	0.0525−1.0959j	0.0525+1.0959j
15.0000	−2.0000	0.0833−1.1517j	0.0833+1.1517j

[a]The poles (p_1,p_2,p_3) are the roots of the characteristic equation, $6s^3 + 11s^2 + 6s + 1 + k_c$.

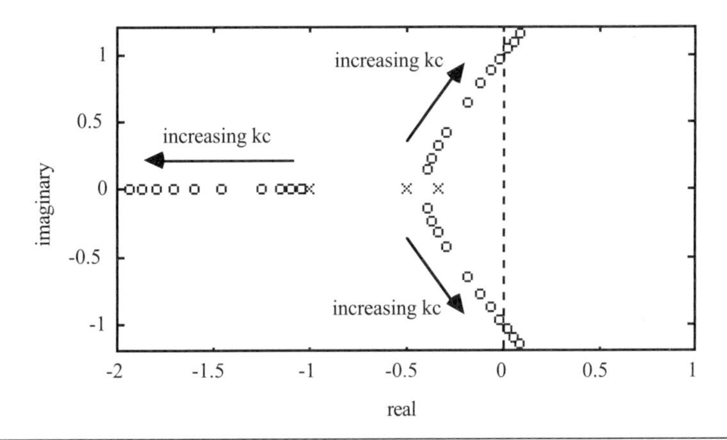

Figure 5–16 Roots of the characteristic equation, as a function of k_c. *x* marks the three poles of the open-loop system ($k_c = 0$). This type of plot is often called a *root-locus* plot.

Effect of Controller Gain

An increase in controller gain (k_c) will speed up the closed-loop response. Except for first- and second-order processes with P-only control, too large a value of controller gain will destabilize the system.

Integral Time

An increase in integral time (τ_I) tends to "slow down" the closed-loop response, while decreasing the integral time speeds up the response. Too small an integral time can cause the closed-loop system to be unstable.

Derivative Time

In this chapter we have not explicitly shown the effect of derivative action—examples are given in the next chapter. An increase in derivative time (τ_D) tends to help stabilize the closed-loop response. A disadvantage of derivative action is that it causes the control system to be sensitive to measurement noise.

The next chapter presents some standard procedures for controller tuning.

5.7 Response to Disturbances

Consider the block diagram in Figure 5–17, which has both setpoint and load disturbance signals. It is easy to derive the following closed-loop relationship:

$$y(s) = \frac{g_p(s)g_c(s)}{1+g_p(s)g_c(s)}r(s) + \frac{g_d(s)}{1+g_p(s)g_c(s)}l(s)$$

In our previous analysis we have been assuming that there is no load disturbance entering the process; that is, our concern has been with setpoint responses. In this section we analyze *disturbance* rejection (or the response to load disturbances) in detail.

Assume that there is no setpoint change, that is, $r(s) = 0$,

$$y(s) = \frac{g_d(s)}{1+g_p(s)g_c(s)}l(s)$$

Notice that a disturbance does not affect the stability of a system (as long as the disturbance transfer function itself is stable), because the closed-loop characteristic equation $[1 + g_p(s)g_c(s)]$ is the same for setpoint tracking or disturbance rejection.

Example 5.4: First-Order Process and Load Transfer Functions with P-Only Control

Consider a first-order process and disturbance with proportional control

$$g_p(s) = \frac{k_p}{\tau_p s + 1}$$

$$g_d(s) = \frac{k_d}{\tau_d s + 1}$$

$$g_c(s) = k_c$$

The effect of the load disturbance on the output is

$$y(s) = \frac{\dfrac{k_d}{\tau_d s + 1}}{1 + \dfrac{k_p k_c}{\tau_p s + 1}} l(s) = \left[\frac{\tau_p s + 1}{\tau_d s + 1}\right]\left[\frac{k_d}{\tau_p s + 1 + k_p k_c}\right] l(s)$$

We see that this system is stable as long as k_c is tuned to satisfy $1 + k_p k_c > 0$. Since we are dealing with deviation variables and we have made no setpoint change, we desire for y to stay as close to 0 as possible.

Consider a step load disturbance of $l(s) = \Delta l/s$. Using the final-value theorem, we find that

$$y(t \to \infty) = \frac{k_d \Delta l}{1 + k_d k_c}$$

which will be a small value as long as the disturbance magnitude (Δl) is low, the disturbance gain (k_d) is low, or the controller gain (k_c) is high. Since it is undesirable to have offset, we would use a PI controller, in practice, to remove the offset (see Exercise 24).

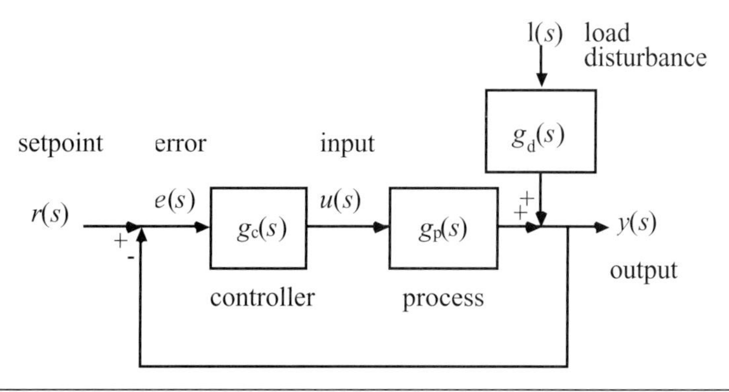

Figure 5–17 Control block diagram including a load disturbance.

5.8 Open-Loop Unstable Systems

The examples presented thus far in this chapter have assumed that the process is open-loop (that is, without control) stable. For these systems we can always "detune" a controller enough so that the closed-loop system is stable. Indeed, if we simply let $k_c = 0$, then we have stability because the closed-loop transfer function is equal to the open-loop transfer function. Open-loop unstable systems are much tougher to control, because the control must be tuned "tightly" enough to stabilize the system, yet not so tightly that the closed-loop system will be unstable. Some exothermic chemical reactors and biochemical reactors are operated at open-loop unstable steady states (that is, at least one eigenvalue of the A matrix is positive; this is equivalent to at least one pole of the process transfer function being positive).

Example 5.5: First-Order Open-Loop Unstable Process with P-Only Control

Consider a first-order open-loop unstable process (see Module 7 on bioreactor control for an example),

$$g_p(s) = \frac{k_p}{-\tau_u s + 1}$$

where $\tau_u > 0$. We use this notation to make it clear that the process is open-loop unstable, since the pole is positive.

Here again we use a proportional controller

$$g_c(s) = k_c$$

The closed-loop transfer function is

$$g_{CL}(s) = \frac{g_p(s)g_c(s)}{1 + g_p(s)g_c(s)} = \frac{k_p k_c/(-\tau_u s + 1)}{1 + k_p k_c/(-\tau_u s + 1)} = \frac{k_p k_c}{-\tau_u s + 1 + k_p k_c}$$

which can be written

$$g_{CL}(s) = \frac{\dfrac{k_p k_c}{1 + k_p k_c}}{\dfrac{-\tau_u}{1 + k_p k_c} s + 1} = \frac{k_{CL}}{\tau_{CL} s + 1}$$

Notice that for stability we require that

$$\tau_{CL} > 0 \quad or \quad \frac{-\tau_u}{1 + k_p k_c} > 0$$

which means that (since $-\tau_u < 0$), $1 + k_p k_c < 0$, or

$$k_p k_c < -1$$

Recall that an open-loop stable first-order process under P-only control (Example 5.1) required that $k_p k_c > -1$ for closed-loop stability (and $k_p k_c > 0$ for reasonable performance). For open-loop stable processes and controllers with integral action, $k_p k_c > 0$ is required for stability; this means that the process gain and controller gain must be the same sign. It is interesting that this open-loop unstable process needs the controller and process gain to be different signs.

Not all open-loop unstable processes can be stabilized by P-only control. For example, one of the processes in Exercise 10 cannot be stabilized by P-only control.

5.9 SIMULINK Block Diagrams

SIMULINK is a natural environment for simulating closed-loop systems. It is recommended that you read Module 2, Introduction to SIMULINK, and reproduce the simulation results shown in Example 5.3. Generate the block diagram shown in Figure 5–18 and run the simulations for various controller proportional gains.

Recall that the closed-loop characteristic equation for Example 5.3 is

$$6s^3 + 11s^2 + 6s + 1 + k_c$$

For $k_c = 10$, the MATLAB `roots` command can be used to find that the process is on the verge of instability, with the following values for the poles:

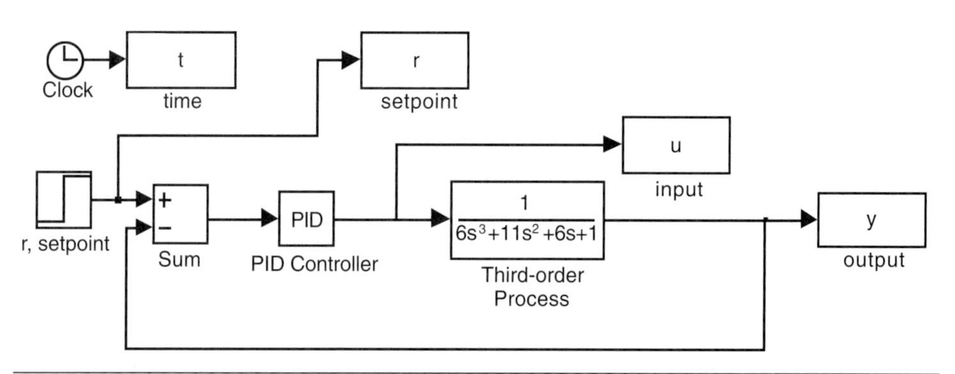

Figure 5–18 SIMULINK diagram for Example 5.3.

```
» roots([6 11 6 11])
ans =
  -1.8333
  -0.0000 + 1.0000i
  -0.0000 - 1.0000i
```

Notice that this analysis can be performed for several values of k_c, to construct Table 5–1.

5.10 Summary

The objective of this chapter was to provide an introduction to feedback control. We showed how to convert an instrumentation and control diagram to a control block diagram. Block diagram manipulations were used to develop a closed-loop transfer function. Transfer function representations for standard PID controllers were developed. The Routh array was used to test for the necessary and sufficient conditions for closed-loop stability based on the characteristic equation of the closed-loop transfer function. We noted that proportional controllers exhibit *offset* for setpoint and disturbance changes, motivating the development of controllers with integral action. Open-loop stable processes generally require that the controller gain be the same sign as the process gain, while open-loop unstable processes often require that the controller gain be the opposite sign of the process gain.

It is suggested that you work some of the problems in the student exercises section. It is particularly important to be able to find the closed-loop relationship (transfer function) between any two signals on a control block diagram.

What has not been discussed in this chapter is the issue of *robustness*. We have presented methods of analysis that assume that the process transfer function is exactly known. In reality, a process model is only an approximate representation of the dynamic behavior of the process. A set of tuning parameters that is stable under one operating condition may be unstable under another. Chapter 7 presents tuning methods that account for uncertainty in process parameters. Also, the model-based methods developed in Chapters 8 and 9 allow for detuning parameters based on model parameter uncertainty.

Common terms used in this chapter are as follows:

Characteristic eqn:	denominator polynomial for closed-loop analysis
Fail-closed:	valve closes on loss of instrument air (air-to-open)
Fail-open:	valve opens on loss of instrument air (air-to-close)
Manipulated input:	also called the controller output (CO)

Necessary condition:	must be satisfied for a chance of stability
Offset:	difference between setpoint and long-term process output
Process output:	also called the process variable (PV)
Routh stability:	stability test based on the closed-loop characteristic equation
Sufficient condition:	guarantees stability (if necessary condition is satisfied)

The **tuning parameters** are as follows:

k_c:	proportional gain	
τ_I:	integral time	
τ_D:	derivative time	
PB:	proportional band	($PB = 100/k_c$)

The **block diagram (Laplace domain) variables** are as follows:

$c(s)$:	controller output
$e(s)$:	error
$r(s)$:	setpoint (reference signal)
$u(s)$:	manipulated variable
$y(s)$:	process output
$y_m(s)$:	measured output

The **transfer functions** are as follows:

$g_c(s)$:	controller
$g_v(s)$:	valve
$g_p(s)$:	process
$g_m(s)$:	measurement
$g_{CL}(s)$:	closed-loop

The **controller types** are as follows:

P:	proportional
PI:	proportional-integral
PID:	proportional-integral-derivative

The **relationships** are as follows:

Relationship	Description
$g_c(s) = k_c$	P control
$g_c(s) = k_c\left(1+\dfrac{1}{\tau_I s}\right) = k_c\dfrac{(\tau_I s+1)}{\tau_I s}$	PI control
$g_c(s) = k_c\left(1+\dfrac{1}{\tau_I s}+\tau_D s\right) = k_c\left(\dfrac{\tau_D\tau_I s^2+\tau_I s+1}{\tau_I s}\right)$	"Ideal" PID control
$g_c(s) = k_c\left(\dfrac{\tau_I s+1}{\tau_I s}\right)\left(\dfrac{\tau_D s+1}{\alpha\tau_D s+1}\right)$	"Real" PID control (series)
$g_c(s) = k_c\left(1+\dfrac{1}{\tau_I s}+\dfrac{\tau_D s}{\alpha\tau_D s+1}\right)$	"Real" PID control (parallel)
$u(s) = k_c\left(\dfrac{\tau_I s+1}{\tau_I s}\right)e(s) - k_c\tau_D s y(s)$	Ideal derivative of output rather than error
$u(s) = k_c\left(\dfrac{\tau_I s+1}{\tau_I s}\right)e(s) - k_c\dfrac{(\tau_D s)}{(\alpha\tau_D s+1)}y(s)$	Real derivative of output rather than error
$g_{CL}(s) = \dfrac{g_p(s)g_v(s)g_c(s)}{1+g_p(s)g_v(s)g_c(s)g_m(s)}$	Closed-loop transfer function (CLTF)
$g_{CL}(s) = \dfrac{g_p(s)g_c(s)}{1+g_p(s)g_c(s)}$	CLTF simplified
$\text{CLCE} = 1 + g_p(s)g_v(s)g_c(s)g_m(s)$	Closed-loop characteristic equation
$\text{CLCE} = 1 + g_p(s)g_c(s)$	Simplified CLCE

References

Routh, E.J., *Dynamics of a System of Rigid Bodies, Part II*, MacMillan, London (1905).

Student Exercises

1. Consider a household heating system with on-off control. Normally there is a dead band of 2°F; that is, the temperature must drop to 1°F below the setpoint before the heater kicks on, and it must go 1°F above the setpoint before the heater kicks off. Clearly the thermostat/heater has periodic behavior with periods where the heater is on, followed by periods where it is off. Discuss the effect of the dead band on this periodic behavior. Sketch the expected heater (on-off) and temperature profiles (as a function of time) as the dead band is changed.
2. **a.** Derive the closed-loop transfer functions relating $L(s)$ and $r(s)$ (the disturbance and setpoint, respectively) to $y_m(s)$ (the measured output) for the following control block diagram. This block diagram is for a system that has significant measurement and valve dynamics.

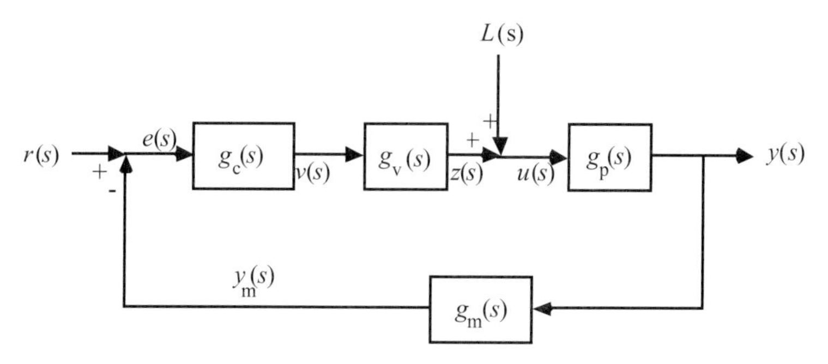

Use the following form: $y_m(s) = g_{CL}(s)\, r(s) + g_{CL1}(s)\, L(s)$ [that is, find $g_{CL}(s)$ and $g_{CL1}(s)$].

b. Also derive the closed-loop transfer functions relating $L(s)$ and $r(s)$ to $u(s)$ (the input to the process).

3. Derive the closed-loop transfer function between $L(s)$ and $y(s)$ for the following control block diagram (this is known as a feed forward/feedback controller).

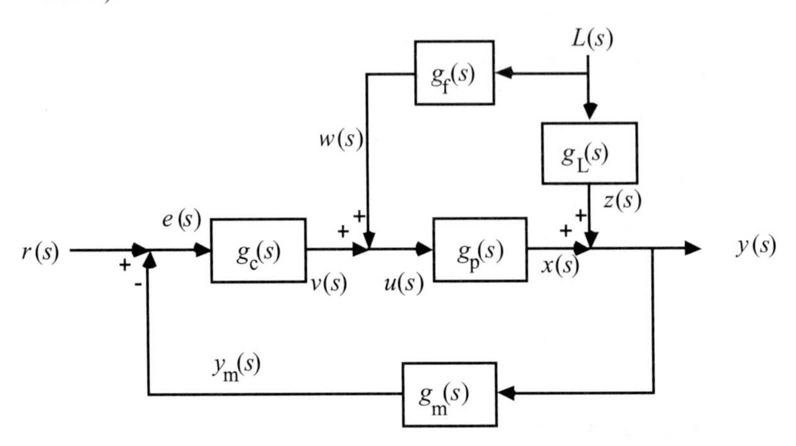

How would you check the stability of the closed-loop system? Will the stability of this system be any different than that of the standard feedback system? Why?

4. Derive the closed-loop transfer function between $r_1(s)$ and $y_1(s)$ for the following control block diagram (this is known as a cascade control system). *Hint*: Find a single block to represent the "inner" feedback loop.

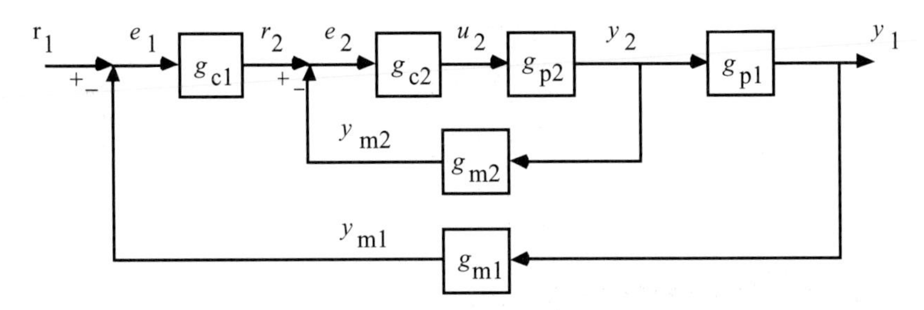

How would you check the stability of the closed-loop system?

5. Consider the following control instrumentation diagram for a heat exchanger.

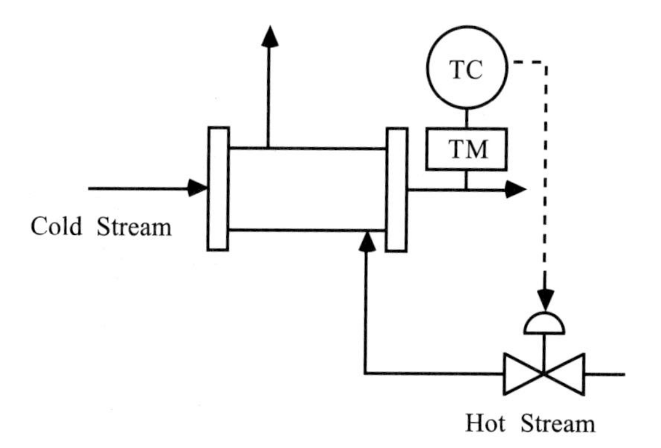

a. Construct a control block diagram, labeling all signals and transfer functions. Make certain that you include valve and measurement sensor dynamics. Also include at least one important disturbance input.

b. Should the control valve should be *fail-open* or *fail-closed*. Why?

c. Based on whether the control valve is fail-open or fail-closed, is the process gain relating the manipulated valve position (or valve top pressure) to the measured temperature positive or negative?

6. Consider a first-order open-loop unstable process that has the following transfer function

$$g_p(s) = \frac{k_p}{-\tau_u s + 1}$$

Find the range of values of parameters for a PI controller that stabilize this process.

7. Calculate the offset to a step setpoint change due to P-only control, for the following process:

$$g_p(s) = \frac{1}{s(2s+1)}$$

Do you have an interesting result? Explain.

8. A process has the following transfer function:

$$g_p(s) = \frac{2(-3s+1)}{(5s+1)}$$

Using a P-only controller, find the range of the controller gain that will yield a stable closed-loop system.

9. Consider the following open-loop unstable process

$$g_p(s) = \frac{3}{-2s+1}$$

For a PI controller, find the range of stabilizing controller proportional gains (k_c) for an integral time constant of $\tau_I = 2$.

10. Consider the two following *unstable* second-order processes:

$$g_{p1}(s) = \frac{1}{(s+1)(s-1)}, \quad g_{p2}(s) = \frac{1}{(s+1)(s-0.1)}$$

Assume that you can apply a P-only feedback controller to each of these processes. What are the bounds on k_c (the controller proportional gain) that will assure stability of the closed-loop system for each process?

11. Consider the open-loop unstable process transfer function

$$g_p(s) = \frac{1}{(s+2)(s-1)}$$

a. Find the range of k_c for a P-only controller that will stabilize this process.

b. It turns out that $k_c = 2$ will yield a stable closed-loop (if this is not in your range from part (**a**), you should check your results) for this system. In practice there is a measurement lag in the feedback loop. Assuming a first-order lag on the measurement, find the maximum measurement time constant which is allowed before the system (with $k_c = 2$) is destabilized.

12. Consider a process with the following transfer function:

$$g_p(s) = \frac{1}{(s-2)(s+1)}$$

Can a PI controller satisfy the *necessary* condition for stability of this process?

13. Consider the following process with a RHP zero:

$$g_p(s) = \frac{2.5(-5s+1)}{(10s+1)(2s+1)}$$

For P-only control, find the bound on the proportional gain to assure closed-loop stability of this process.

14. A PI controller is used on the second-order process

$$g_p(s) = \frac{k_p}{\tau^2 s^2 + 2\zeta\tau s + 1}$$

The process parameters are $k_p = 1$, $\tau = 2$, $\zeta = 0.7$. The tuning parameters used are $k_c = 5$, $\tau_I = 0.2$.

Is the process closed-loop stable?

15. Consider an isothermal CSTR that has the following reaction scheme

$$A \xrightarrow{k_1} B \xrightarrow{k_2} C$$

$$2A \xrightarrow{k_3} D$$

This system exhibits a maximum in the concentration of B vs. dilution rate (F_s/V, where F_s is the steady-state flow rate and V is the constant reactor volume) curve shown in Figure 5–19. In this problem we consider two possible operating points, which we call case 1 and case 2.

Case 1. The steady-state dilution rate of $F_s/V = 4/7\ \text{min}^{-1}$ is on the left side of the peak shown in Figure 5–19. The transfer function relating the input (dilution rate) to the output (concentration of B) is

$$g_p(s) = \frac{-1.1170s + 3.1472}{s^2 + 4.6429s + 5.3821}$$

Case 2. The steady-state dilution rate of $F_s/V = 2.8744\ \text{min}^{-1}$ is on the right side of the peak in Figure 5–19. The transfer function relating the input to the output is

$$g_p(s) = \frac{-1.1170s - 3.1472}{s^2 + 10.2778s + 26.0508}$$

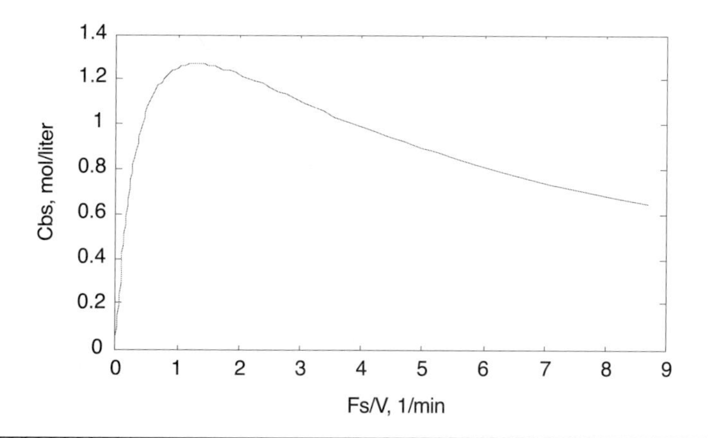

Figure 5–19 Steady-state relationship between input (dilution rate) and output (concentration of *B*).

a. Draw a process and instrumentation diagram for this system.

b. Find the process gain, poles, and zero for the two transfer functions. Are the gains consistent with the steady-state input/output curve?

c. For a P-only controller, find the range of controller gains that will stabilize this system at both operating points (case 1 and case 2).

16. Show that, for a stable first-order process, a PI controller will have an oscillatory response if

$$\tau_I < \frac{4k_p k_c \tau_p}{\left(1 + k_p k_c\right)^2}$$

17. Compare the effect of k_c on the closed-loop stability requirements for the following processes under P-only control:

$$\text{process 1,} \quad \frac{1}{(2s+1)(3s+1)}$$

$$\text{process 2,} \quad \frac{-s+1}{(2s+1)(3s+1)}$$

18. Consider a second order process (without numerator dynamics),

$$g_p(s) = \frac{k_p}{\tau^2 s^2 + 2\zeta\tau s + 1}$$

Determine the conditions (range of values) for the PI tuning parameters to assure closed-loop stability.

19. It is easy to show for the second-order system $g_p(s) = \frac{1}{[(s+1)(s+1)]}$ that there is no upper bound on k_c for P-only control, so that offset can be minimized by making k_c very large. Now consider the more practical case, where there is a measurement lag. How low must τ_m be to assure that the maximum offset for a setpoint change is 5%? (*Hint*: Consider the offset of a P-only control system, but also make certain that the closed-loop system satisfies the Routh stability criterion.)

20. A process is known to have the following input-output transfer function

$$g(s) = \frac{k(-\tau s + 1)}{(2s+1)(s+1)}$$

In the process of tuning a P-only feedback controller for this process, the operator discovers a steady-state offset of 2/5 for a unit setpoint change, when the proportional gain is 1. She also finds that the closed-loop system goes unstable when the proportional gain is greater than 2.

a. What is the value of the process gain, k?

b. What is the value of the numerator time constant, τ?

c. What type of instability occurs when the proportional gain is 2? That is, does the closed-loop system oscillate owing to simultaneous crossing of two complex poles across the imaginary axis? Or does a single real pole cross the imaginary axis?

21. An open-loop unstable chemical reactor is known to have a process transfer function with the following form, where the manipulated input is the coolant flow rate (liters/minute) and the measured process output is the reactor temperature (°C):

$$g_p(s) = \frac{k_p}{(-\tau_u s + 1)(s + 1)}$$

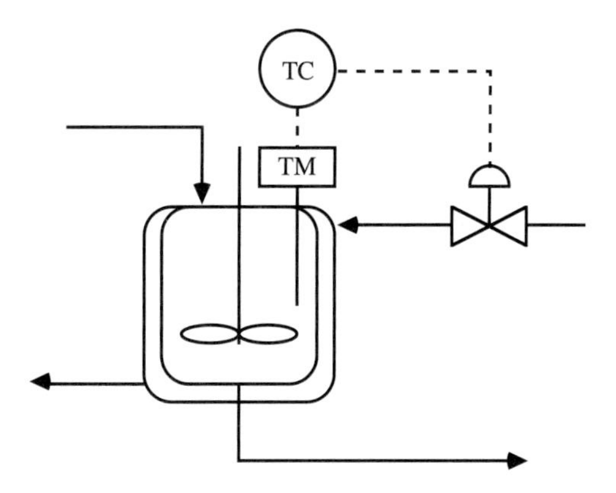

When a P-only controller, with a value k_c = -3 lpm/°C, is used, the response to a unit setpoint change is shown below.

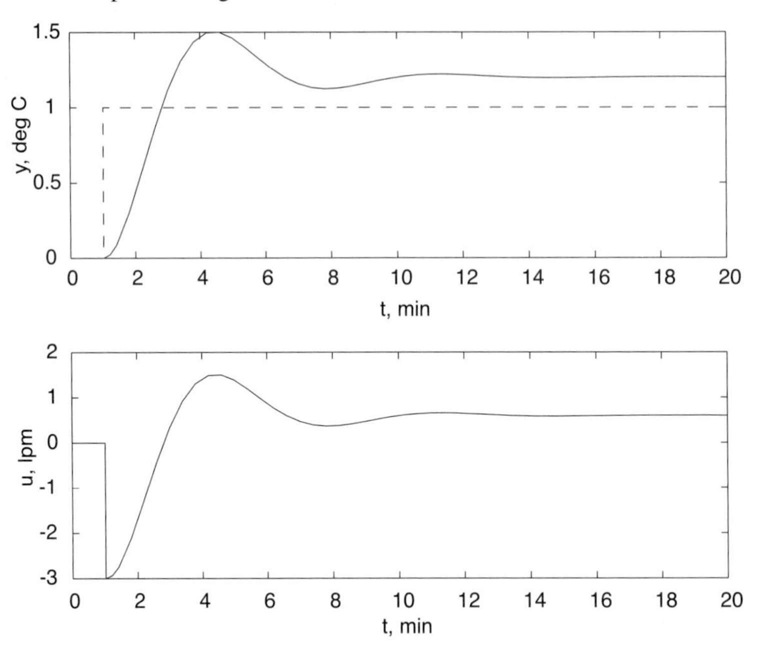

When a PI controller is used, the closed-loop system becomes unstable, as shown below, when the tuning parameter values are k_c = -1.333 lpm/°C and τ_I = 2 minutes.

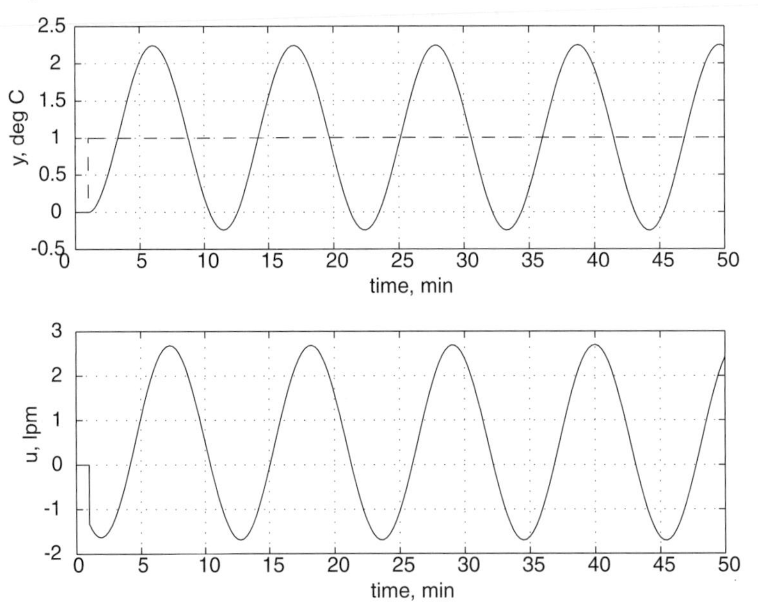

What are the values for the process parameters, k_p and τ_u? *Hints*: Can the final-value theorem be used? Can the Routh stability criterion be used?

22. Consider the following chemical process. The concentration of propylene glycol exiting the adiabatic reactor is measured. The output of the concentration controller is a current signal, which is converted to a pressure signal. The pressure signal actuates a control valve that changes the flow rate of steam to a heat exchanger, which changes the temperature of the reactor inlet stream.

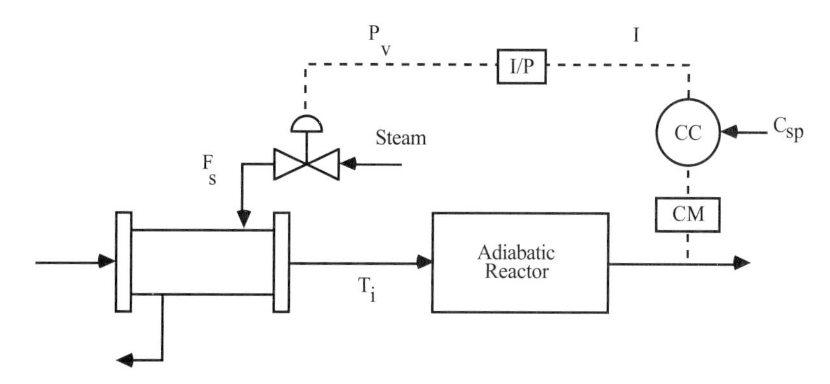

a. Construct the control block diagram, clearly labeling all signals and blocks.

b. Find the closed-loop transfer function from the concentration setpoint to the measured output.

23. Consider a PID controller with a special form that eliminates "derivative kick":

$$u(s) = k_c\left(\frac{\tau_I s + 1}{\tau_I s}\right)e(s) - \frac{k_c \tau_D s}{\alpha \tau_D s + 1}y(s)$$

$$u(s) = g_{c1}(s)e(s) - g_{c2}(s)y(s)$$

This controller is implemented in block diagram form as shown below.

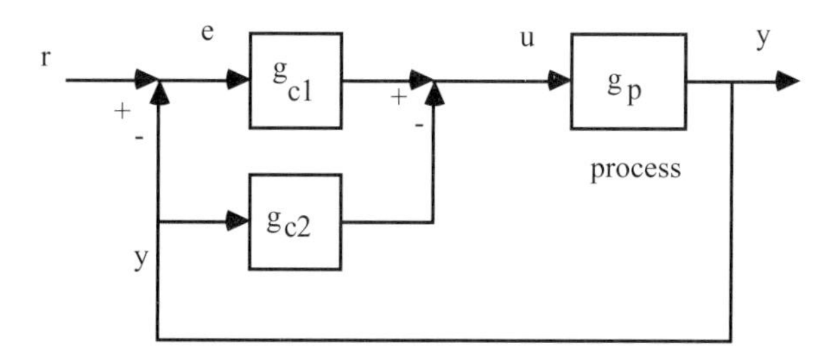

Derive the closed-loop transfer function relating a setpoint change, r, to the process output, y.

24. Consider Example 5.4. If a PI controller is used, rather than P-only, show that there will not be any offset to a step load disturbance. Also, show the conditions on the tuning parameters for closed-loop stability.

CHAPTER 6

PID Controller Tuning

In Chapter 5 we were given a process transfer function and a controller transfer function (P, PI, PID, etc.) and asked to find the closed-loop response (or stability). In this chapter we provide an overview of classical methods for tuning PID controllers. We also provide an introduction to model-based techniques, using the direct synthesis approach.

After studying this chapter, the reader should be able to do the following:

- Understand the different forms of a PID controller
- Tune PID controllers using the classical, Ziegler-Nichols, and Cohen-Coon methods
- Derive controllers based on a process model and desired closed-loop response

The major sections of this chapter are as follows:

6.1 Introduction

Consider the standard feedback block diagram shown in Figure 6–1, where disturbance inputs have been neglected. Transfer functions (and block diagrams) are used to analyze the behavior of control systems, because the algebraic expressions are easy to manipulate.

We noted in Chapter 5 that the closed-loop transfer function could be used to determine, for example, the range of controller gains that assure closed-loop stability.

PID Controller Forms

PID controller algorithms were developed in Chapter 5. Here we provide a concise review of the algorithms in common use.

P-Only Control

The proportional only algorithm is

$$u(t) = k_c e(t) = k_c(r(t) - y(t)) \tag{6.1a}$$

which has the following transfer function relationship between error and controller output:

$$u(s) = k_c e(s) \tag{6.1b}$$

PI-Control

The PI algorithm is

$$u(t) = k_c \left[e(t) + \frac{1}{\tau_I} \int_o^t e(\sigma) d\sigma \right] \tag{6.2a}$$

which has the following transfer function relationship between error and controller output:

$$u(s) = k_c \left(1 + \frac{1}{\tau_I s} \right) e(s) = k_c \left(\frac{\tau_I s + 1}{\tau_I s} \right) e(s) \tag{6.2b}$$

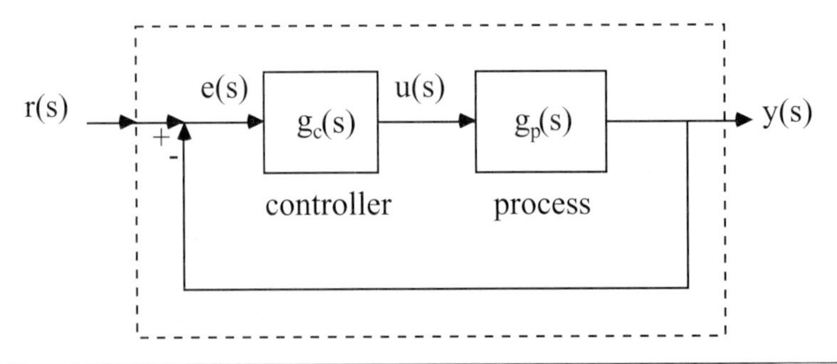

Figure 6–1 Feedback control block diagram.

PID Control

The ideal PID algorithm is

$$u(t) = k_c\left[e(t) + \frac{1}{\tau_I}\int_o^t e(\sigma)d\sigma + \tau_D\frac{de(t)}{dt}\right] \tag{6.3a}$$

which has the following transfer function relationship between error and controller output:

$$u(s) = k_c\left(1 + \frac{1}{\tau_I s} + \tau_D s\right)e(s) = k_c\left(\frac{\tau_D\tau_I s^2 + \tau_I s + 1}{\tau_I s}\right)e(s) \tag{6.3b}$$

In practice it is impossible to perfectly differentiate the error signal, so the following Laplace transfer function approximations are often used for "real" PID control (where $\alpha = 0.1$ is common):

$$g_c(s) = k_c\left(1 + \frac{1}{\tau_I s}\right)\left(\frac{\tau_D s + 1}{\alpha\tau_D s + 1}\right) = k_c\left(\frac{\tau_I s + 1}{\tau_I s}\right)\left(\frac{\tau_D s + 1}{\alpha\tau_D s + 1}\right) \tag{6.4a}$$

or

$$g_c(s) = k_c\left(1 + \frac{1}{\tau_I s} + \frac{\tau_D s}{\alpha\tau_D s + 1}\right) = k_c\left(\frac{\tau_I s + 1}{\tau_I s} + \frac{\tau_D s}{\alpha\tau_D s + 1}\right) \tag{6.4b}$$

A problem with taking the derivative of the error is that step setpoint changes cause the derivative to become unbounded and result in a "spike" in the manipulated variable action. In most practical PID controllers, then, the derivative of the measured process output is used.

$$u(t) = k_c\left[e(t) + \frac{1}{\tau_I}\int_o^t e(\sigma)d\sigma - \tau_D\frac{dy(t)}{dt}\right] \tag{6.5a}$$

The transfer function representation is

$$u(s) = k_c\left(\frac{\tau_I s + 1}{\tau_I s}\right)e(s) - k_c\tau_D s y(s) \tag{6.5b}$$

which is usually implemented in the form of

$$u(s) = k_c\left(\frac{\tau_I s + 1}{\tau_I s}\right)e(s) - \frac{k_c\tau_D s}{\alpha\tau_D s + 1}y(s) \tag{6.5c}$$

where the measured process output has been "filtered" to minimize noise problems.

6.2 Closed-Loop Oscillation-Based Tuning

A PID controller has three tuning parameters. If these are adjusted in an ad hoc fashion, it may take a while for satisfactory performance to be obtained. Also, each tuning technician will end up with a different set of tuning parameters. There is plenty of motivation, then, to develop an algorithmic approach to controller tuning. The first widely used method for PID tuning was published by Ziegler and Nichols in 1942.

Ziegler-Nichols Closed-Loop Method

The Ziegler-Nichols closed-loop tuning technique was perhaps the first rigorous method to tune PID controllers. The technique is not widely used today because the closed-loop behavior tends to be oscillatory and sensitive to uncertainty. We study the technique for historical reasons, and because it is similar to commonly used automatic tuning ("auto-tune") techniques covered in Chapter 11.

The closed-loop Ziegler-Nichols method consists of the following steps.

1. With P-only closed-loop control, increase the magnitude of the proportional gain until the closed-loop is in a continuous oscillation. For slightly larger values of controller gain, the closed-loop system is unstable, while for slightly lower values the system is stable.
2. The value of controller proportional gain that causes the continuous oscillation is called the critical (or ultimate) gain, k_{cu}. The peak-to-peak period (time between successive peaks in the continuously oscillating process output) is called the critical (or ultimate) period, P_u.
3. Depending on the controller chosen, P, PI, or PID, use the values in Table 6–1 for the tuning parameters, based on the critical gain and period.

Tyreus and Luyben have suggested tuning parameter rules that result in less oscillatory responses and that are less sensitive to changes in the process condition. Their rules are shown in Table 6–2.

Table 6–1 Ziegler-Nichols Closed-Loop Oscillation Method Tuning Parameters

Controller type	k_c	τ_I	τ_D
P-only	$0.5\ k_{cu}$	—	—
PI	$0.45\ k_{cu}$	$P_u/1.2$	—
PID	$0.6\ k_{cu}$	$P_u/2$	$P_u/8$

Table 6–2 Tyreus-Luyben Suggested Tuning Parameters Based on the Ziegler-Nichols Closed-Loop Oscillation Tuning Method

Controller type	k_c	τ_I	τ_D
PI	$k_{cu}/3.2$	$2.2\ P_u$	—
PID	$k_{cu}/2.2$	$2.2\ P_u$	$P_u/6.3$

Example 6.1: Third-Order Process

Consider the third-order system used in Example 5.3, with a time unit of minutes

$$g_p(s) = \frac{1}{(3s+1)(2s+1)(s+1)} = \frac{1}{6s^3 + 11s^2 + 6s + 1}$$

Recall that a P-only controller caused the closed-loop to be on the verge of instability (continuous oscillation) when the value of the controller gain was $k_c = 10$. This is the critical proportional gain, k_{cu}. From the response shown in Figure 6–2 we see that the ultimate period (period of oscillation) is $P_u = 6.2$ minutes.

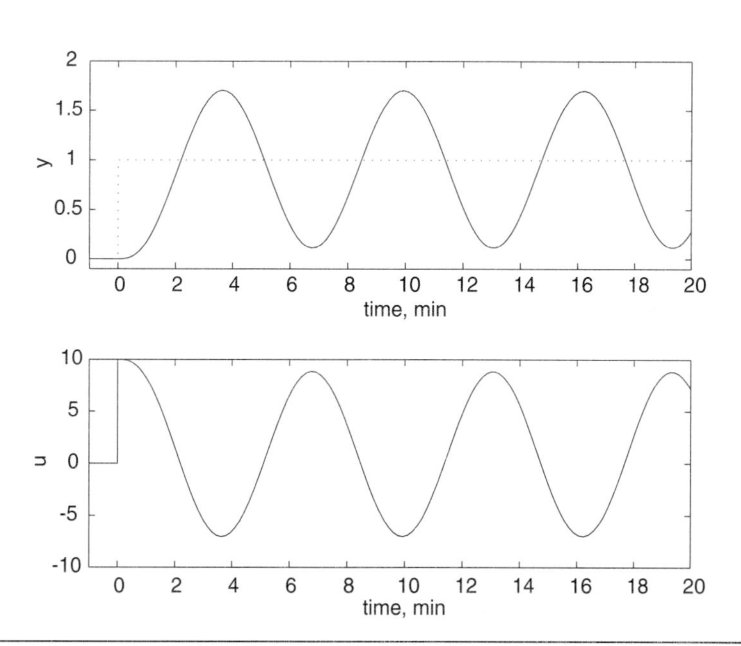

Figure 6–2 Response to unit step setpoint change at $t = 0$; $k_{cu} = 10$. Notice that $P_u = 6.2$ minutes.

The closed-loop responses for Ziegler-Nichols tuning for P, PI, and ideal PID controllers (based on Table 6–1) are shown in Figure 6–3. Notice that a P-only controller has offset, as expected. Also, all the responses are quite oscillatory; this is one of the major disadvantages to the Ziegler-Nichols tuning method. It typically results in more oscillatory behavior than would be allowable in a typical process plant. The tuning parameters are also not very robust, that is, they are very sensitive to process uncertainty. If the process conditions change, then the control system may become unstable.

The closed-loop responses for Tyreus-Luyben tuning for P (assumed to be the Ziegler-Nichols value), PI, and PID controllers are compared in Figure 6–4. The Tyreus-Luyben parameters result in less oscillatory responses and will be less sensitive to uncertainty.

Note that the PID controllers simulated in Figures 6–3 and 6–4 are based on an ideal derivative of the error. A problem is that this results in a "spike" in the manipulated input when a step setpoint change is made. Also, an ideal derivative controller will be more sensitive to measurement noise.

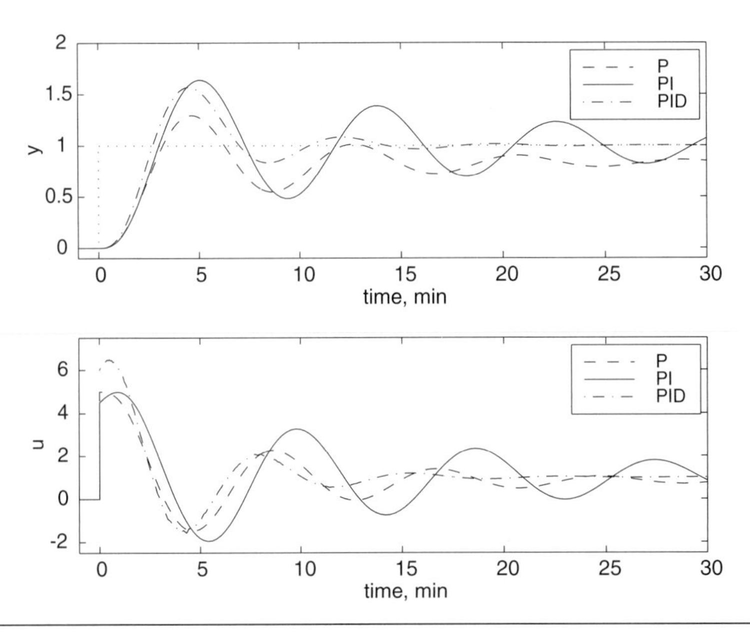

Figure 6–3 Response to unit step setpoint change. Comparison of Ziegler-Nichols P, PI, and PID tuning rules. Notice the "spike" in the manipulated input for the PID controller.

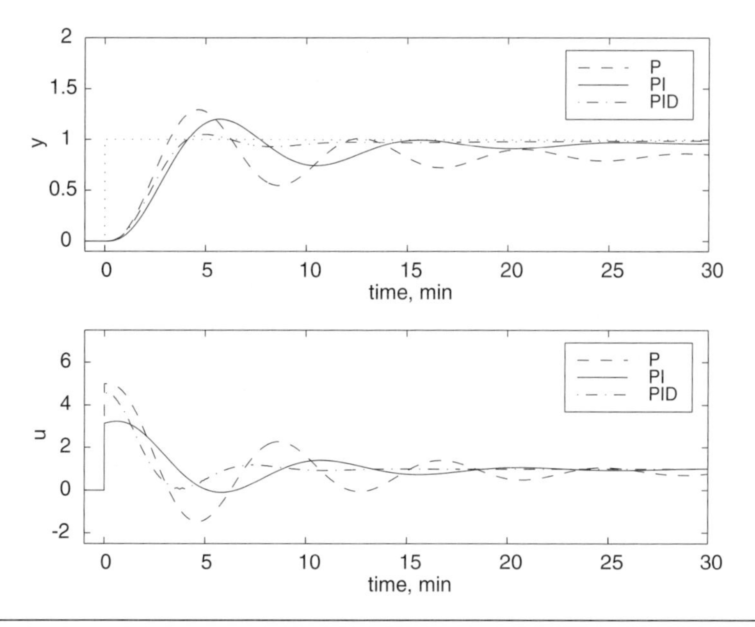

Figure 6–4 Response to unit step setpoint change. Comparison of Tyreus-Luyben P, PI, and PID tuning rules. Notice the "spike" in the manipulated input for the PID controller.

6.3 Tuning Rules for First-Order + Dead Time Processes

The previous tuning rules were based on tests that forced a process into a continuous oscillation. Obvious disadvantages to the techniques are that the system is forced to the edge of instability, and it may take a while to iteratively adjust the controller to obtain a continuous oscillation. In this section we present tuning rules based on process models that have been obtained through the open-loop step tests presented in Chapter 4.

Ziegler-Nichols Open-Loop Method

Ziegler and Nichols also proposed tuning parameters for a process that has been identified as integrator + time-delay based on an open-loop process step response,

$$g_p(s) = \frac{ke^{-\theta s}}{s}$$

Since first-order + time-delay processes have a maximum slope of $k = k_p/\tau_p$ at $t = \theta$ for a unit step input change, these same rules can be used for first-order + time-delay processes,

$$g_p(s) = \frac{k_p e^{-\theta s}}{\tau_p s + 1}$$

Their recommended tuning parameters, which should give roughly quarter-wave damping, are shown in Table 6–3. We see a potential problem for systems with a low time-delay/time-constant ratio, since this causes the proportional gain to become very large. Similarly, the integral time tends to be low, causing oscillatory behavior.

Cohen-Coon Parameters

The method developed by Cohen and Coon (1953) is based on a first-order + time-delay process model. A set of tuning parameters was empirically developed to yield a closed-loop response with a decay ratio of 1/4 (similar to the Ziegler-Nichols methods). The tuning parameters as a function of the model parameters are shown in Table 6–4.

A major problem with the Cohen-Coon parameters is that they tend not to be very robust; that is, a small change in the process parameters can cause the closed-loop system to become unstable.

Table 6–3 Ziegler-Nichols Open-Loop Tuning Parameters

Controller type	k_c	τ_I	τ_D
P-only	$\frac{1}{k\theta}$ or $\frac{\tau_p}{k_p\theta}$	—	—
PI	$\frac{0.9}{k\theta}$ or $\frac{0.9\tau_p}{k_p\theta}$	3.3θ	—
PID	$\frac{1.2}{k\theta}$ or $\frac{1.2\tau_p}{k_p\theta}$	2θ	0.5θ

Table 6–4 Cohen-Coon Tuning Parameters

Controller type	k_c	τ_I	τ_D
P-only	$\frac{\tau_p}{k_p\theta}\left[1+\frac{\theta}{3\tau_p}\right]$	—	—
PI	$\frac{\tau_p}{k_p\theta}\left[0.9+\frac{\theta}{12\tau_p}\right]$	$\frac{\theta\left[30+3\,\theta/\tau_p\right]}{9+20\,\theta/\tau_p}$	—
PID	$\frac{\tau_p}{k_p\theta}\left[\frac{4}{3}+\frac{\theta}{4\tau_p}\right]$	$\frac{\theta\left[32+6\,\theta/\tau_p\right]}{13+8\,\theta/\tau_p}$	$\frac{4\theta}{11+2\,\theta/\tau_p}$

6.4 Direct Synthesis

Consider the standard feedback block diagram shown in Figure 6–1. Recall that we could determine the performance or stability of the closed-loop system from the closed-loop transfer function

$$y(s)=\frac{g_p(s)g_c(s)}{1+g_p(s)g_c(s)}r(s)=g_{CL}(s)r(s) \tag{6.6}$$

In the *direct synthesis* procedure, we select a *desired closed-loop response*, $g_{CL}(s)$, and, based on the known process, $g_p(s)$, find the controller $g_c(s)$ that will yield this response. Solving Equation (6.6) for $g_c(s)$, we obtain

$$g_c(s)=\frac{g_{CL}(s)}{g_p(s)(1-g_{CL}(s))} \tag{6.7}$$

By now you should be able to perform block diagram manipulations to find the relationship between any external signal [such as a setpoint change, r(s)] and any other signal on the control block diagram. For example, it is important to analyze the manipulated variable action required for a setpoint change to make certain that it is not too rapid or that it does not violate constraints. From Figure 6–1 we can easily derive the effect of the setpoint change on the manipulated variable action,

$$u(s)=\frac{g_c(s)}{1+g_p(s)g_c(s)}r(s) \tag{6.8}$$

The direct synthesis procedure then consists of specifying the desired closed-loop transfer function (e.g., first-order response, second-order underdamped, etc.), $g_{CL}(s)$, using Equation (6.7) to find the feedback controller, and considering the manipulated variable response (usually tested by simulation).

The real question is: *How do we specify the desired closed-loop response*?

It turns out that we are not limited by the desired closed-loop response, if the system *is minimum phase* (the process does not have RHP zeros or time delays—the terminology becomes clear in Chapter 7). In the following section we present the direct synthesis method for minimum-phase systems, and cover *non-minimum-phase* systems in the subsequent section.

Direct Synthesis for Minimum-Phase Processes

It seems fairly natural to specify a desired closed-loop response that is first order, since we understand the characteristics of a first-order response.

$$g_{CL}(s) = \frac{1}{\lambda s + 1} \tag{6.9}$$

For a specified first-order response, there is only one tuning parameter, λ, since we desire a closed-loop gain of 1 (we want the process output to equal the setpoint as the closed-loop system goes to a new steady state); small values of λ results in fast responses, while large values result in slow responses. One could also argue for a desired closed-loop response that is second order and underdamped, which would lead to specifying two parameters.

Example 6.2: Direct Synthesis For a First-Order Process

Consider a first-order process

$$g_p(s) = \frac{k_p}{\tau_p s + 1} \tag{6.10}$$

Assume that a first-order closed-loop response is specified. If we desire a fast closed-loop response, we make λ small; for a slower (more "robust") response, we make λ large. Solving for $g_c(s)$ from Equation (6.7), we find

$$g_c(s) = \frac{\dfrac{1}{\lambda s + 1}}{\dfrac{k_p}{\tau_p s + 1}\left(1 - \dfrac{1}{\lambda s + 1}\right)} = \frac{\tau_p s + 1}{k_p \lambda s} \tag{6.11}$$

which can be written (by multiplying by τ_p/τ_p)

$$g_c(s) = \frac{\tau_p}{k_p \lambda} \cdot \frac{\tau_p s + 1}{\tau_p s} \tag{6.12}$$

Recall that the form of a PI controller is

$$g_c(s) = k_c \frac{\tau_I s + 1}{\tau_I s} \tag{6.13}$$

so that our *direct synthesis* controller for a first-order process is simply a *PI controller*, where

$$k_c = \frac{\tau_p}{k_p \lambda}$$

$$\tau_I = \tau_p$$

Notice that there was only one "tuning parameter" for this direct synthesis example. The desired closed-loop time constant, λ, was the only adjustable parameter. Given the first-order transfer function parameters (k_p and τ_p) and the desired closed-loop time constant (λ), we found that the direct synthesis controller was PI, but that we only needed to "adjust" one PI parameter (k_c). This is a nice result, because once we find the process time constant (τ_p), we can set the integral time (τ_I) equal to the time constant, and tune k_c on-line until we achieve a desired response. Tuning a single controller parameter is much easier than tuning two or three.

Numerical Example

Consider the following first-order process, with a time constant of 10 minutes and a process gain of 2 %/%

$$g_p(s) = \frac{2}{10s + 1}$$

The output and manipulated variable responses for $\lambda = 1$, 5, and 10 min ($\tau_I = 10$; $k_c = 5$, 1, and 0.5, respectively) are shown in Figure 6–5. As expected, the manipulated variable response for the faster desired closed-loop time constant is much more aggressive than for the slower closed-loop time constants. Notice that setting the closed-loop time constant equal to the open-loop time constant ($\lambda = \tau_p = 10$) results in a single step change in the manipulated variable action.

The same type of procedure shown in Example 6.2 can be used if a second-order closed-loop response is specified. See Exercise 13.

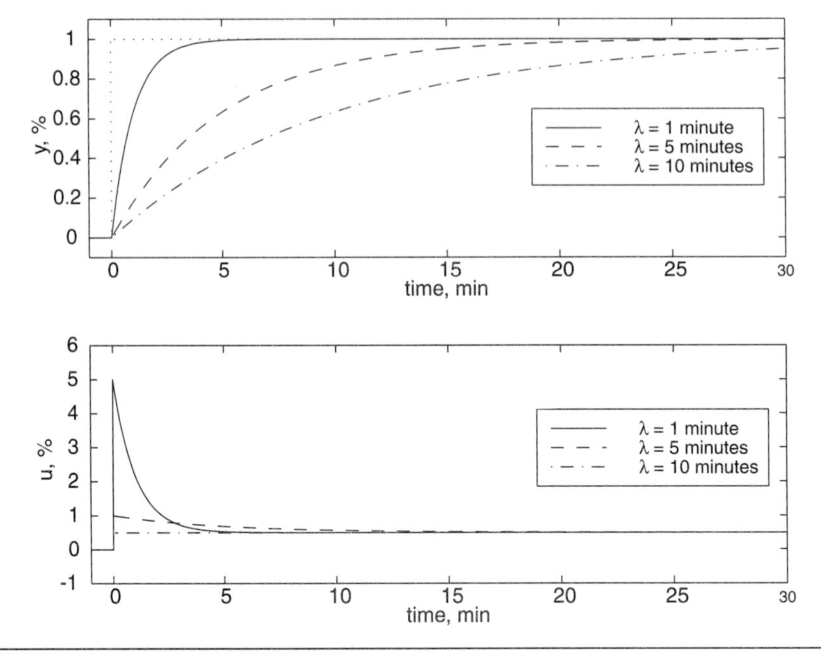

Figure 6–5 Response to a unit step setpoint change.

Direct Synthesis for Nonminimum-Phase Processes

This section presents examples for nonminimum-phase processes, that is, processes that have time delays or RHP zeros. The general technique remains the same for these processes; however, there is a restriction on the type of closed-loop response that can be specified. The next example provides the motivation for specifying different desired closed-loop responses for systems with time delays.

Example 6.3: First-Order + Dead Time Example

Consider the following process transfer function:

$$g_p(s) = \frac{2e^{-3s}}{5s+1}$$

If the desired closed-loop response is first order, $g_{CL}(s) = 1/\lambda s + 1$, the resulting controller is

$$g_c(s) = \frac{2.5}{\lambda}\left(\frac{5s+1}{5s}\right)e^{3s}$$

which is a PI controller with an additional term (e^{3s}). This additional term is not physically realizable because it requires knowledge of future errors to obtain the current control action. This is clearly impossible. Perhaps this is shown more clearly in the time domain

$$u(t) = \frac{2.5}{\lambda}\left[e(t+3) + \frac{1}{5}\int_0^t e(\sigma+3)d\sigma \right]$$

where the control action at time t depends on the error at time $t + 3$, which is clearly impossible to implement.

The next example shows how a system with a *RHP zero* must have a modified desired closed-loop response.

Example 6.4: Process with a RHP Zero

Consider a process with the following transfer function, where the time unit is minutes

$$g_p(s) = \frac{-9s+1}{(15s+1)(3s+1)} \tag{6.14}$$

The direct synthesis procedure with $g_{CL}(s) = 1/\lambda s + 1$ yields the controller

$$g_c(s) = \frac{(15s+1)(3s+1)}{(-9s+1)\lambda s}$$

which is unstable because of the RHP pole. The RHP pole in the controller is due to the *inversion* of the process zero. This *inversion* occurs because of the specification of a first-order closed-loop response. The output and manipulated variable responses are shown in Figure 6–6 for $\lambda = 5$ minutes. Notice that the inverse response does not appear in the output variable but that the manipulated variable is unbounded (unstable). This is often called *internal instability*.

Clearly, the manipulated variable for any physical system will eventually hit a constraint, making the closed-loop system effectively open-loop, since the controller is no longer functioning.

It is easy to understand the behavior shown in Figure 6–6. On a short timescale, the process appears to the controller to have a negative gain, since a step change in the manipulated variable yields an output that initially dips before going in the positive direction. The controller is acting on this effective "negative gain" to continually force the manipulated variable in a negative direction. The reader should verify that once the manipulated variable hits a lower bound, the process output will begin to decrease.

Specifying a desired first-order closed-loop response for a system with a RHP zero resulted in an unstable controller and unbounded manipulated variable action. A stable

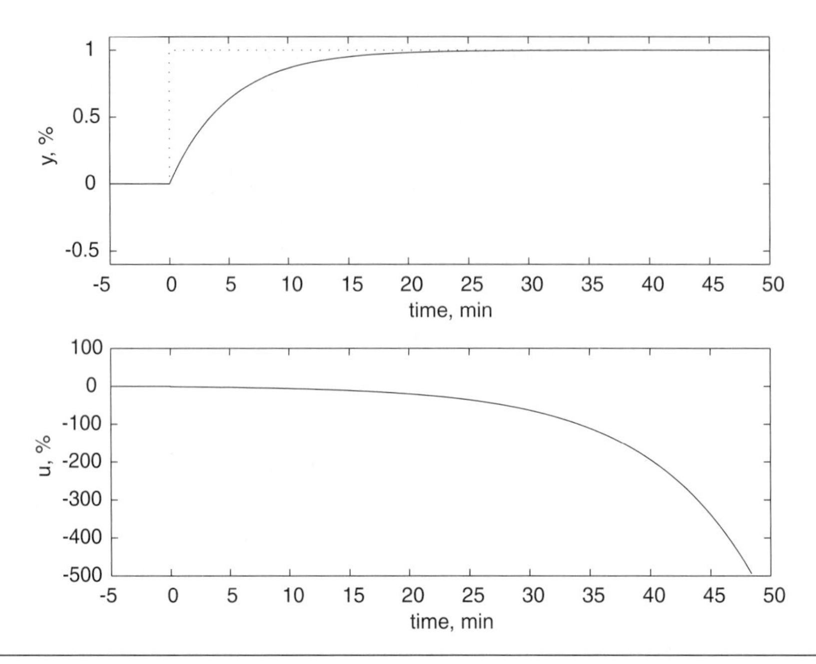

Figure 6–6 Response to a unit step setpoint change when the controller is unstable.

controller can be obtained if the desired closed-loop response has the same RHP zero as the process.

Reformulation of the Desired Response

Including the right-half-plane zero desired closed-loop transfer function specified as

$$g_{CL}(s) = \frac{-9s+1}{(\lambda s+1)^2}$$

using the direct synthesis procedure, you should find the following controller,

$$g_c(s) = \left(\frac{18}{2\lambda+9}\right)\left(\frac{45s^2+18s+1}{18s}\right)\left(\frac{1}{\dfrac{\lambda^2}{2\lambda+9}s+1}\right)$$

which is an ideal PID controller cascaded with a first-order filter. The parameters are

$$k_c = \frac{18}{2\lambda + 9}, \tau_I = 18, \tau_D = 2.5, \tau_F = \frac{\lambda^2}{2\lambda + 9}$$

The output and input responses for a step setpoint change with $\lambda = 5$ minutes are shown in Figure 6–7. The closed-loop system has inverse response (as specified by the closed-loop transfer function) and reasonable manipulated variable action.

6.5 Summary

The *closed-loop* Ziegler-Nichols method (Table 6–1) was shown to lead to oscillatory closed-loop behavior, which is one of the major disadvantages to the approach. A major advantage to the approach is that a process model is not needed. Tyreus-Luyben parameters (Table 6–2) were shown to be less oscillatory and are generally recommended over the Ziegler-Nichols parameters.

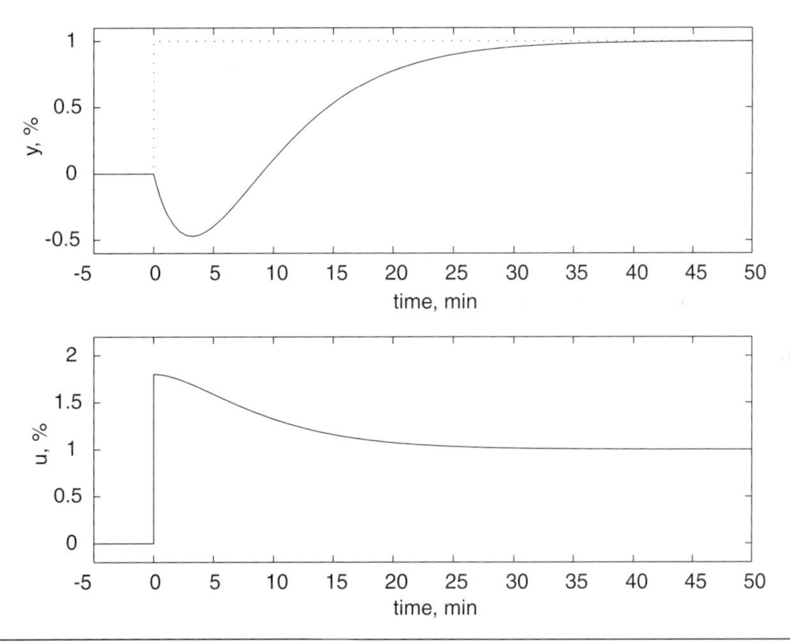

Figure 6–7 Responses to a unit step setpoint change at $t = 0$ minute, when RHP zero is in the desired response ($\lambda = 5$).

The *open-loop* Ziegler-Nichols (Table 6–3) and the Cohen-Coon (Table 6–4) methods also tend to lead to oscillatory closed-loop behavior; we have covered these approaches primarily for historical reasons.

In the direct synthesis design procedure, a desired closed-loop response (or closed-loop transfer function) is specified. A feedback controller is then *synthesized* to obtain that response. We have seen that a PID-type controller often results. Some problems can arise if the process has a time delay or inverse response. In these cases, the desired closed-loop response must also have a time delay or inverse response. We develop a more transparent method for designing controllers for first-order + time-delay processes in Chapters 8 and 9.

References

The original Ziegler-Nichols parameters are developed in the following paper: Ziegler, J. G., and N. B. Nichols, "Optimum Settings for Automatic Controllers," *Trans. ASME*, **64**, 759–768 (1942).

The Cohen-Coon parameters are developed in the following paper: Cohen, G. H., and G. A. Coon, "Theoretical Considerations of Retarded Control," *Trans. ASME*, **75**, 827 (1953).

The Tyreus-Luyben parameters are presented in the following textbook: Luyben, M. L., and W. L. Luyben, *Essentials of Process Control*, McGraw-Hill, New York (1997).

Student Exercises

1. Consider the gas pressure problem shown below. The objective of this problem is to understand (via simulation) how the tuning parameters for a PI controller affect the stability and speed of response for setpoint changes or disturbances.

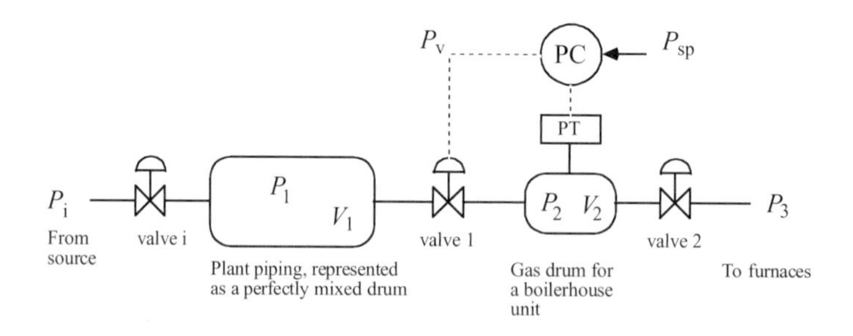

The relationship between the manipulated valve position (u in deviation variables) and the pressure in the second tank (y in deviation variables) is (with a time unit of minutes)

$$y(s) = g_p(s)u(s) = \frac{89.5s + 5.633}{s^2 + 4.744s + 1.4216}u(s) = \frac{3.98(15.8s+1)}{(3.10s+1)(0.226s+1)}u(s)$$

Assume that the dynamic behavior of the pressure measurement/transmitter is characterized by a first-order lag with a time constant of 9 seconds:

$$y_m(s) = g_m(s)y(s) = \frac{1}{0.15s+1}y(s)$$

Also assume that the dynamic behavior of the control valve is also characterized by a first-order lag with a time constant of 6 seconds,

$$u(s) = g_v(s)c(s) = \frac{1}{0.1s+1}c(s)$$

where $c(s)$ is the output from the controller and u(s) is the valve position. For P-only control, find the value of k_c (via simulation) that causes a closed-loop to go unstable. Call this value k_{cu}, and call P_u the period of oscillation (time between peaks) when the system goes unstable. These values are used in the Ziegler-Nichols closed-loop oscillation method.

a. Show that the value of k_c sightly greater than that you obtained causes at least one root of the closed-loop characteristic equation [$g_{CL}(s)$] to be positive. Find the P and PI tuning parameters on the Ziegler-Nichols closed-loop oscillations method.

b. Compare the response of the two different controllers (P vs. PI), for step setpoint changes of 1 psig in the desired output (y).
The closed-loop block diagram is shown below.

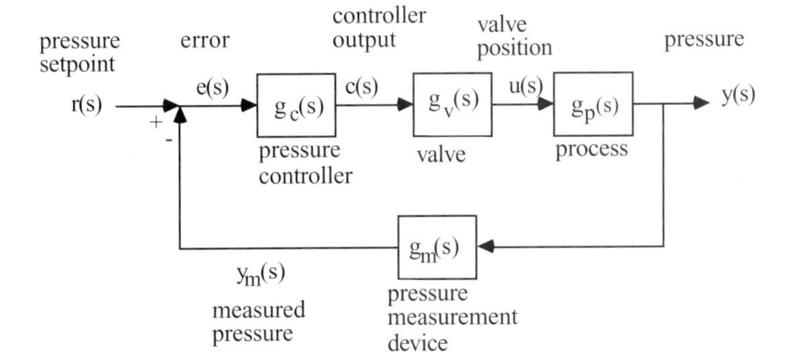

2. Consider the following first-order + time-delay process:

$$g_p(s) = \frac{e^{-5s}}{10s+1}$$

Perform simulations for the process output and manipulated input for unit step changes in the setpoint.

a. Compare the closed-loop step response of this process using P-only control based on (i) Ziegler-Nichols closed-loop oscillations method, (ii) Ziegler-Nichols open-loop method, (iii) Cohen-Coon, and (iv) Tyreus-Luyben tuning.

b. Compare the closed-loop step response of this process using PI control based on (i) Ziegler-Nichols closed-loop method, (ii) Ziegler-Nichols open-loop method, (iii) Cohen-Coon, and (iv) Tyreus-Luyben tuning.

c. Compare the closed-loop step response of this process using PID control based on (i) Ziegler-Nichols closed-loop method, (ii) Ziegler-Nichols open-loop method, (iii) Cohen-Coon, and (iv) Tyreus-Luyben tuning.

Comment on the results for all of these tuning methods.

3. Consider the process transfer function for the Van de Vusse reactor (Module 5).

$$g_p(s) = \frac{0.5848(-0.3549s+1)}{0.1858s^2+0.8627s+1}$$

Find the Ziegler-Nichols controller parameters for P, PI, and PID controllers for this process, based on the closed-loop oscillation method. Compare the responses of all three controllers to a step setpoint change.

4. Most PID controller design procedures assume that a perfect derivative controller is used. For the process transfer function used in problem 3 above, and the Ziegler-Nichols closed-loop method, compare the responses of (i) ideal PID, (ii) real PID, and (iii) PID with ideal derivative action on the process output, rather than the error.

5. Show that the following state space representation of a controller,

$$\frac{dx_{c1}}{dt} = e$$

$$\frac{dx_{c2}}{dt} = \frac{k_c}{\alpha\tau_I}x_{c1} - \frac{1}{\alpha\tau_D}x_{c2} + \frac{k_c}{\alpha}e$$

$$u = \frac{k_c}{\alpha\tau_I}x_{c1} + \frac{1}{\alpha\tau_D}x_{c2} + \frac{k_c}{\alpha}e$$

has the following transfer function representation:

$$u(s) = g_c(s)e(s)$$

where

$$g_c(s) = k_c\left(\frac{\tau_I s+1}{\tau_I s}\right)\left(\frac{\tau_D s+1}{\alpha\tau_D s+1}\right)$$

6. Consider the PID algorithm

$$u(t) = k_c\left[e(t) + \frac{1}{\tau_I}\int_o^t e(\sigma)d\sigma - \tau_D \frac{dy_f}{dt}\right]$$

where y_f is a "filtered" value of the process output. Assuming that a first-order filter is used, with a time constant of τ_f, write the modeling equations (differential and algebraic) to simulate the behavior of this controller.

7. Apply (simulate) the Tyreus-Luyben parameters for PI and PID controllers to the following process

$$g_p(s) = \frac{e^{-5s}}{10s+1}$$

Compare these results with Cohen-Coon. Which do you recommend for implementation on a real process?

8. Find the feedback controller for an integrating process, $g_p(s) = k_p/s$, assuming a desired first-order response using the direct synthesis method.
Answer: It is a P-only controller, with $k_c = 1/k_p\lambda$

9. Find the feedback controller for an integrating process, $g_p(s) = k_p/s$, assuming a desired second-order response $\left[y(s) = \dfrac{1}{\tau^2 s^2 + 2\zeta\tau s + 1} r(s)\right]$
Partial Answer: The controller is a first-order lag.

10. Show that the *direct synthesis* procedure for the following process, assuming a desired first-order response, yields a PID controller

$$g_p(s) = \frac{15}{s^2 + 4.9s + 0.9}$$

Find the PID tuning parameters if a closed-loop time constant of 5 minutes is desired.

11. For a second-order system with numerator dynamics,

$$g_p(s) = \frac{k_p(\tau_n s+1)}{\tau^2 s^2 + 2\zeta\tau s + 1}$$

find a controller that gives a first-order closed-loop response. (*Hint*: It will be a PID with a first-order lag.)

12. Consider the following first-order process:

$$g_p(s) = \frac{2}{3s+1}$$

If the desired closed-loop response to a setpoint change is second order with the following form,

$$g_{CL}(s) = \frac{\alpha s+1}{(\lambda s+1)^2}$$

find the feedback controller required, where α and λ are adjustable tuning parameters (they are both positive). What type of controller is this? If the controller is PID form (perhaps with a lag), find each of the tuning parameters (k_c, τ_I, τ_D, τ_F). Show that $\lambda > 0.5\alpha$ is required for the controller to be stable.

13. Consider a first-order process with a desired closed-loop response that is second order. Use the direct synthesis procedure with the following specified closed-loop transfer function (which is critically damped),

$$g_{CL}(s) = \frac{1}{(\lambda s+1)^2}$$

to derive the controller. Perform simulations for several values of λ and compare and contrast the closed-loop results with those shown in Figure 6–5.

CHAPTER 7

Frequency-Response Analysis

The objective of this chapter is to develop frequency response techniques for control system analysis and design. After studying this chapter, the reader should be able to do the following:

- Substitute $s = j\omega$, and find Bode or Nyquist plots for a transfer function
- Given a Bode plot, construct a Nyquist plot
- Understand the concepts of "all-pass" and "nonminimum phase"
- Understand the effect of gains, time constants, RHP zeros, and time delays on amplitude ratios and phase angles
- Find the gain and phase margins for a given process and controller
- Tune a controller to achieve a given gain or phase margin
- Determine the amount of dead time or gain uncertainty that can be tolerated before a closed-loop system becomes unstable
- Compare the robustness of PI and PID controller tuning techniques

The major sections of this chapter are as follows:

7.1 Motivation

A control system must satisfy desired performance characteristics (closed-loop time constant, minimal overshoot, etc.) for *nominal* operating conditions. By nominal, we normally mean that the plant is perfectly described by the model used for control-system design. In reality, a model is never perfect, so controllers must be designed to be *robust* (to remain stable even when the true plant characteristics are different from the model).

So far we have analyzed the closed-loop characteristic equation ($1 + g_c g_p$) for stability, either by directly calculating the roots or by using the Routh array to verify that the roots (closed-loop poles) are negative. There are two problems with this approach:

1. Although the closed-loop characteristic equation may be stable for the nominal plant, the values of the poles do not indicate how much uncertainty can be tolerated before the system becomes unstable.
2. The closed-loop characteristic equation is irrational if there is a time delay, which is characterized by a $e^{-\theta s}$ term.

The best method for dealing with both of these problems is frequency response analysis. Our motivation for covering frequency response analysis is primarily that it provides a measure of the *robustness* of the controller tuning; that is, it provides a measure of the amount of model uncertainty that can be tolerated before the control system will become unstable. We consider the ability to determine the stability of systems with irrational transfer functions (time delays) to be less important, because a higher order Padé approximation can provide an adequate representation of the time delay, for analysis purposes.

To understand the concept of frequency response, consider the well-insulated, perfectly mixed tank with constant flow in and out (perfect level control), shown in Figure 7–1. Assume that the inlet temperature varies in a sinusoidal fashion between a minimum of 20°C, a maximum of 30°C, with an average of 25°C. The amplitude of the inlet temperature sine wave is then 5°C.

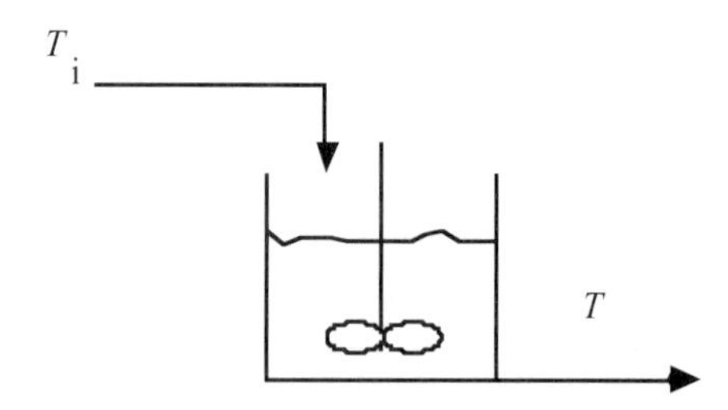

Figure 7–1 Perfectly mixed tank with varying inlet temperature.

This tank has a volume of 100 gallons, with a flow rate of 10 gallons/minute. You can easily show that this system is modeled by a first-order transfer function, with a gain of 1°C/°C and a time constant of 10 minutes. The responses to different frequencies in the inlet temperature are shown in Figure 7–2. Notice that at low-frequency (0.01 rad/minute) forcing (top plot), the outlet temperature (dashed line) has approximately the same amplitude as the inlet temperature (solid line), with very small shift in the peak time. At an intermediate frequency (0.1 rad/minute) forcing, we begin to see a decrease in the amplitude of the outlet temperature, as well as a "lag" in the peak time (center plot). At a high frequency (1 rad/minute), the outlet temperature does not vary substantially (bottom plot).

Please note that radians/time is the common unit used for frequency response analysis in process control. You may be more comfortable thinking in terms of cycles/time. Since there is one cycle for each 2π radians, then a frequency (f) in cycles/time can be converted to a frequency, ω, in radians/time by

$$\omega(radians/time) = 2\pi \cdot f(cycles/time) \tag{7.1}$$

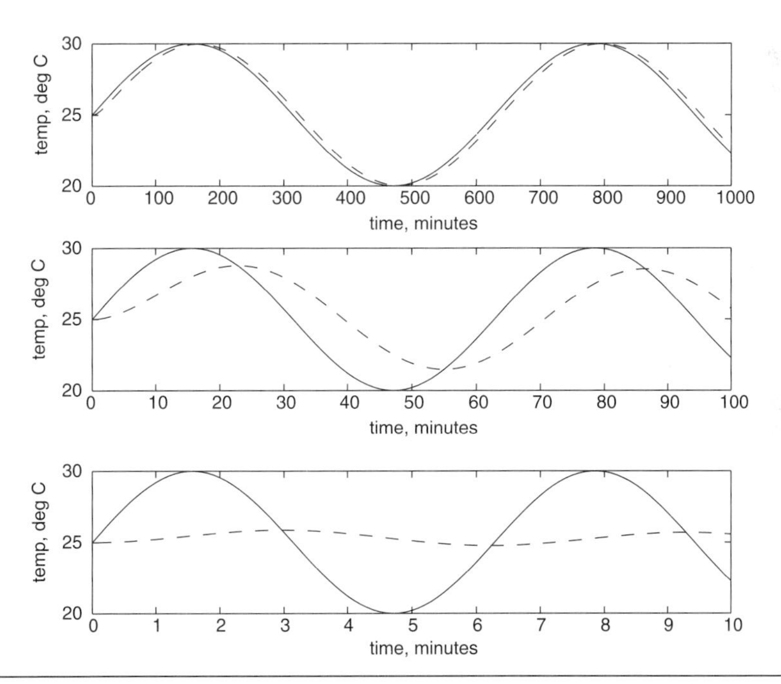

Figure 7–2 Response of outlet temperature to various inlet temperature frequencies. Inlet temperature (solid), outlet temperature (dashed). Top, 0.01 rad/minute; center, 0.1 rad/minute; bottom, 1 rad/minute. Note the different timescales.

Also, the period P ("peak-to-peak time") corresponds to one cycle and can be found by

$$P = 1/f = 2\pi/\omega \tag{7.2}$$

The input-output behavior shown in Figure 7–2 makes physical sense. At low frequencies the inlet temperature changes on a much slower timescale than the characteristic time of the process, so the outlet temperature is virtually identical to the inlet temperature. At high frequencies, the inlet temperature changes so rapidly that the "thermal mass" of the process does not allow the outlet temperature to change significantly.

In terms of dynamic behavior, this type of system is known as a "low-pass filter" that is, low-frequency inlet forcing is directly felt at the output. High input frequencies are effectively filtered by the process and not felt at the output.

There are two important measures that can be obtained from these time-domain plots. One is the *amplitude ratio*, which is the amplitude of the output sine wave divided by the input sine wave. The second measure is the *phase shift* of the output sine wave relative to the input sine wave, as illustrated in Figure 7–3. The phase shift, or phase angle, is

$$\phi = -2\pi(\Delta P/P) = -360°(\Delta P/P) \tag{7.3}$$

where P is the period, and ΔP is the lag between an input peak and an output peak.

For example, if the output were exactly out of phase with the input (output at a minimum when the input is a maximum), the phase angle would be –180° or π radians. If the output lagged the input by a full cycle, the phase angle would be –360°, and it is possible for the phase angle to exceed this value (particularly for processes with a time delay). The phase angle in the bottom plot of Figure 7–2 is approximately –90° (0.5π radians).

Rather than generating numerous time-domain plots like Figure 7–2, each at a different frequency, the equivalent results can be displayed concisely on a *Bode plot*, as shown in Figure 7–4. The amplitude ratio plot is log-log, while the phase angle plot is semilog.

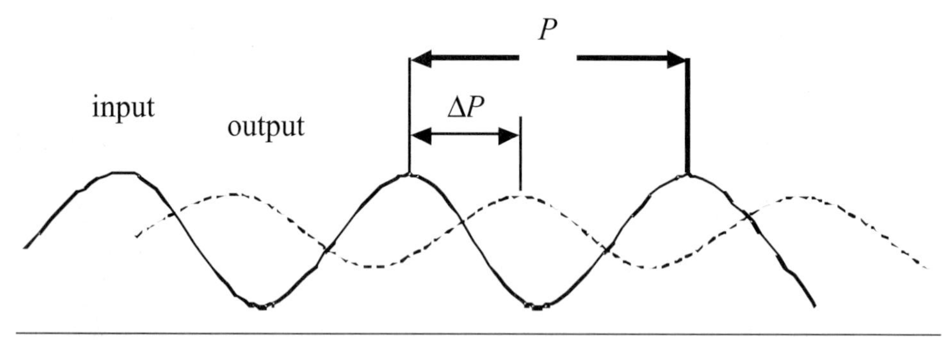

Figure 7–3 Illustration of phase shift or phase angle, $\phi = -360°(\Delta P/P)$.

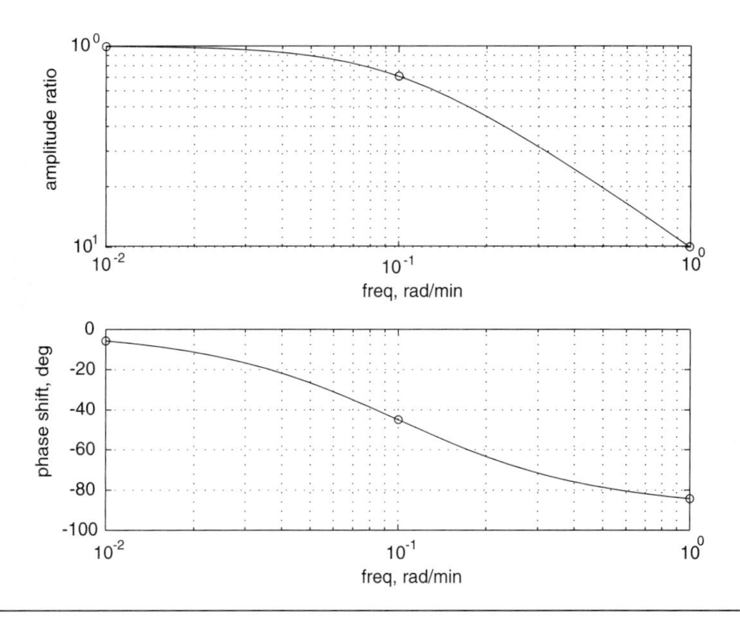

Figure 7–4 Bode plot of the first-order example. Amplitude ratio and phase shift are plotted as a function of frequency. The circles correspond to the three frequencies used on the mixing tank.

7.2 Bode and Nyquist Plots

In the previous section the basic concept of frequency response was presented. For linear systems, an input forcing with a certain frequency results in an output with the same frequency, shifted by the phase angle. There are two ways to perform frequency response analysis: (i) conversion of the input signal to the Laplace domain, partial fraction expansion of the output, and inversion to the time domain, or (ii) substitution of $s = j\omega$ in the transfer function.

Method 2, substitution of $s = j\omega$ in the transfer function, is the easiest approach. There are also two ways to display the frequency response information: Bode plots and Nyquist plots. The following first-order example is used to illustrate these concepts.

Example 7.1: First-Order System

Consider a first-order process transfer function

$$g_p(s) = \frac{k_p}{\tau_p s + 1}$$

We use two methods to find the output, given a sine forcing on the process input.

1. Laplace Transform Method

Consider a first-order system, with an input that is forced with a sine wave of magnitude A and a frequency of ω (rad/time). The time and Laplace domain representations of the input are

$$u(t) = A \sin \omega t$$

$$u(s) = \frac{A\omega}{s^2 + \omega^2}$$

The Laplace domain output is

$$y(s) = \frac{k_p}{\tau_p s + 1} \cdot \frac{A\omega}{s^2 + \omega^2}$$

Using a partial fraction expansion,

$$y(s) = \frac{c_1}{s + \frac{1}{\tau_p}} + \frac{c_2}{s + j\omega} + \frac{c_3}{s - j\omega}$$

Solving for the coefficients and inverting to the time domain (see Exercise 1), we find

$$y(t) = \frac{k_p A \omega \tau_p}{\tau_p^2 \omega^2 + 1} e^{-t/\tau_p} + \frac{k_p A}{\tau_p^2 \omega^2 + 1} \sin \omega t - \frac{k_p A \omega \tau_p}{\tau_p^2 \omega^2 + 1} \cos \omega t$$

The exponential term decays with time, so the first term can be neglected at long times. Also, a trignometric identity can be used to find that

$$y_{ss}(t) = \frac{k_p A}{\sqrt{\tau_p^2 \omega^2 + 1}} \sin(\omega t + \varphi)$$
$$\phi = \tan^{-1}(-\omega \tau_p) \tag{7.4}$$

The *ss* subscript is used to indicate the long-term behavior of the output, after the initial transients have decayed. Notice that the output approaches 0 at high frequencies. This is consistent with the results shown in Figure 7–2. The phase angle also approaches –90° at high frequencies.

The *amplitude ratio* is the magnitude of the output divided by the input.

$$AR = \frac{|y|}{|u|} = \frac{k_p A \Big/ \sqrt{\tau_p^2 \omega^2 + 1}}{A} = \frac{k_p}{\sqrt{\tau_p^2 \omega^2 + 1}} \tag{7.5}$$

Notice that there is a natural dimensionless frequency, $\omega\tau_p$, resulting from this analysis.

An easier way to generate frequency response plots is to simply set $s = j\omega$ in the transfer function. We illustrate this procedure, again with a simple first-order example.

2. *Substitution of $s = j\omega$ Method*

Substitute $s = j\omega$ into the transfer function

$$g_p(s) = \frac{k_p}{\tau_p s + 1}$$

$$g_p(j\omega) = \frac{k_p}{\tau_p j\omega + 1} \cdot \frac{-\tau_p j\omega + 1}{-\tau_p j\omega + 1} = \frac{-k_p \tau_p j\omega + k_p}{-\tau_p^2 j^2 \omega^2 + 1} = \frac{-k_p \tau_p j\omega + k_p}{\tau_p^2 \omega^2 + 1}$$

$$g_p(j\omega) = \frac{k_p}{\tau_p^2 \omega^2 + 1} - j\frac{k_p \tau_p \omega}{\tau_p^2 \omega^2 + 1}$$

Notice that the result is a complex number (real and imaginary portions) that changes as a function of frequency. Since the magnitude of a complex number Re + jIm, can be represented by $\sqrt{\text{Re}^2 + \text{Im}^2}$, then the magnitude (amplitude) of the transfer function (notice that the amplitude at zero frequency is simply the steady-state process gain) is

$$\left|g_p(j\omega)\right| = \sqrt{\frac{k_p^2\left(\tau_p^2 \omega^2 + 1\right)}{\left(\tau_p^2 \omega^2 + 1\right)^2}} = \frac{k_p}{\sqrt{\tau_p^2 \omega^2 + 1}} \tag{7.6a}$$

Also, the phase angle of a complex number can be represented by $\tan^{-1}$(Im/Re), so

$$\phi = \tan^{-1}(-\omega\tau_p) \tag{7.6b}$$

Notice that these are exactly the relationships shown with method 1 [Equations (7.4) and (7.5)].

Also notice that one can simply plot the imaginary values as a function of the real values to obtain a *Nyquist plot*. Such a plot is shown in Figure 7–5 for this system. The curve begins close to $(k_p,0)$ at low frequencies, and approaches (0,0) through the first quadrant at high frequencies. That is, it starts with an amplitude of 1 at a phase angle of 0, and ends with an amplitude of 0 at a phase angle of –90°.

Generalization

To perform a frequency-response analysis, we simply substitute $s = j\omega$ in the transfer function. *Bode plots* consist of amplitude ratio and phase angle plots as a function of frequency. The *amplitude ratio* plots are log-log, while the phase angle plots are semilog. *Nyquist plots* consist of a curve of the real and imaginary components of the transfer function, as a function of frequency.

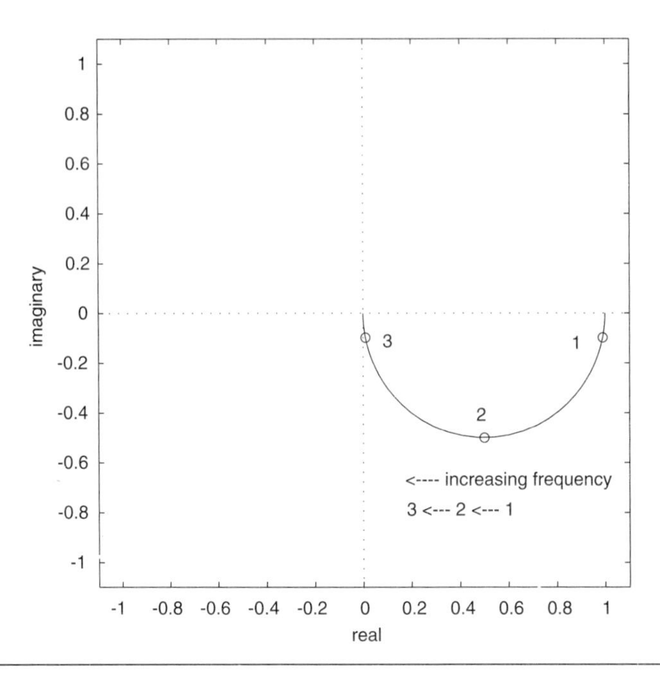

Figure 7–5 Nyquist plot for a first-order system. The circles correspond to the three frequencies (1 = 0.01 rad/minute, 2 = 0.1 rad/minute, 3 = 1 rad/minute) used on the mixing tank.

The steps to generate the plots are as follows.

1. Generate a vector of frequencies, ω, linearly spaced on a logarithmic scale.
2. Substitute $s = j\omega$ into $g_p(s)$.
3. Find the real and imaginary components, $\mathrm{Re}(\omega) + j\mathrm{Im}(\omega)$.
4. For Bode plots, at each frequency point find the amplitude ratio and phase angle

$$AR = |g_p(\omega)| = \sqrt{\mathrm{Re}^2(\omega) + \mathrm{Im}^2(\omega)}$$
$$\phi = \tan^{-1}[\mathrm{Im}(\omega)/\mathrm{Re}(\omega)]$$

   a. Plot amplitude ratio as a function of frequency on a log-log plot.
   b. Plot phase angle as a function of frequency on a semilog plot.
5. For Nyquist plots, plot $\mathrm{Im}(\omega)$ vs. $\mathrm{Re}(\omega)$.

The results for several common transfer functions are shown in Table 7–1.

Table 7–1 Nyquist and Bode Values for Common Transfer Functions

$g(s)$	Re(ω)	Im(ω)	AR	ϕ	$\phi(\omega \to 0)$	$\phi(\omega \to \infty)$
$\frac{k_p}{\tau_p s+1}$	$\frac{k_p}{\tau_p^2\omega^2+1}$	$\frac{-k_p\tau_p\omega}{\tau_p^2\omega^2+1}$	$\frac{k_p}{\sqrt{\tau_p^2\omega^2+1}}$	$\tan^{-1}(-\tau_p\omega)$	0	–90°
$e^{-\theta s}$	$\cos(\theta\omega)$	$-\sin(\theta\omega)$	1	$-\theta\omega$	0	$-\infty$
$\frac{k}{s}$	0	$-\frac{k}{\omega}$	$\frac{k}{\omega}$	–90°	–90°	–90°
$\tau s + 1$	1	$\tau\omega$	$\sqrt{\tau^2\omega^2+1}$	$\tan^{-1}(\tau\omega)$	0	90°
k_p	k_p	0	k_p	0	0	0

Complex Transfer Functions

Consider now a transfer function that is composed of a number of transfer functions

$$g(s) = \frac{g_{n1}(s)g_{n2}(s)...}{g_{d1}(s)g_{d2}(s)...} \tag{7.7}$$

and substituting $s = j\omega$, we can find the magnitude at each frequency

$$|g(\omega)| = \frac{|g_{n1}(\omega)||g_{n2}(\omega)|...}{|g_{d1}(\omega)||g_{d2}(\omega)|...} \tag{7.8}$$

The amplitude ratio and phase angle can be found at each frequency from

$$\log(AR) = \log|g_{n1}(\omega)| + \log|g_{n2}(\omega)| + ... - \log|g_{d1}(\omega)| - \log|g_{d2}(\omega)| - ... \tag{7.9}$$

$$\phi = \phi_{n1} + \phi_{n2} + ... - \phi_{d1} - \phi_{d2} - ... \tag{7.10}$$

That is, at each frequency the amplitude ratios add on a logarithmic scale, while the phase angles add on a linear scale.

Example 7.2: First-Order + Time Delay

Consider a first-order + time-delay process, written as three independent terms

$$g_p(s) = k_p \cdot \frac{1}{\tau_p s+1} \cdot e^{-\theta s}$$

The amplitude ratio and phase angle can be found from Equations (7.9) and (7.10), and Table 7–1

$$\log\left(g_p(\omega)\right) = \log\left|k_p\right| + \log\left|\frac{1}{\sqrt{\tau_p^2\omega^2+1}}\right| + \log\left|\,1\,\right| = \log\left|k_p\right| + \log\left|\frac{1}{\sqrt{\tau_p^2\omega^2+1}}\right|$$

$$\phi(g_p(\omega)) = 0 + \tan^{-1}(-\tau_p\omega) - \theta\omega$$

Notice that the time delay has no effect on the amplitude ratio (compared with the pure first-order process), but it has a great effect on the phase angle, particularly at high frequencies. We can also think of the effect of the changes in each of the parameters.

Gain Change

A change in the gain merely shifts the amplitude ratio up or down (additive on a log scale), with no change in the phase angle.

Time Delay Change

A change in the time delay changes the phase angle, but not the amplitude ratio. An increase in the time delay increases the phase lag (makes it more negative) at any given frequency.

Time Constant Change

A change in the time constant affects both the amplitude ratio and phase angle. An increase in the time constant tends to decrease the amplitude ratio and increase the phase lag (make it more negative) at any given frequency.

7.3 Effect of Process Parameters on Bode and Nyquist Plots

In this section we stress the effect of process parameters and the concepts of "all-pass" and "nonminimum phase."

Effect of Process Order

Consider the following first-, second-, and third-order transfer functions.

$$g_{p1}(s) = \frac{1}{s+1}, \quad g_{p2}(s) = \frac{1}{(s+1)^2}, \quad g_{p3}(s) = \frac{1}{(s+1)^3}$$

The Bode plot is shown below in Figure 7–6. The higher order transfer functions have amplitude ratios that decrease more rapidly with frequency, and the phase angles asymptotically approach larger (more negative) values ($-180°$ and $-270°$ for the second- and third-order processes).

The corresponding Nyquist plot is shown in Figure 7–7. The first-order system stays in the first quadrant, asymptotically approaching the origin along the imaginary axis. The second-order system goes into the second quadrant and asymptotically approaches the origin along the real axis from the negative direction. The third-order system passes into the third quadrant and asymptotically approaches the origin along the positive imaginary axis.

Concepts of "All-Pass" and "Nonminimum Phase"

Consider the following four transfer functions,

$$g_{p1}(s) = 2,\ g_{p2}(s) = 2 \cdot \frac{-0.5s+1}{0.5s+1},\ g_{p3}(s) = 2 \cdot \frac{s^2 - 6s + 12}{s^2 + 6s + 12},\ g_{p4}(s) = 2e^{-s}$$

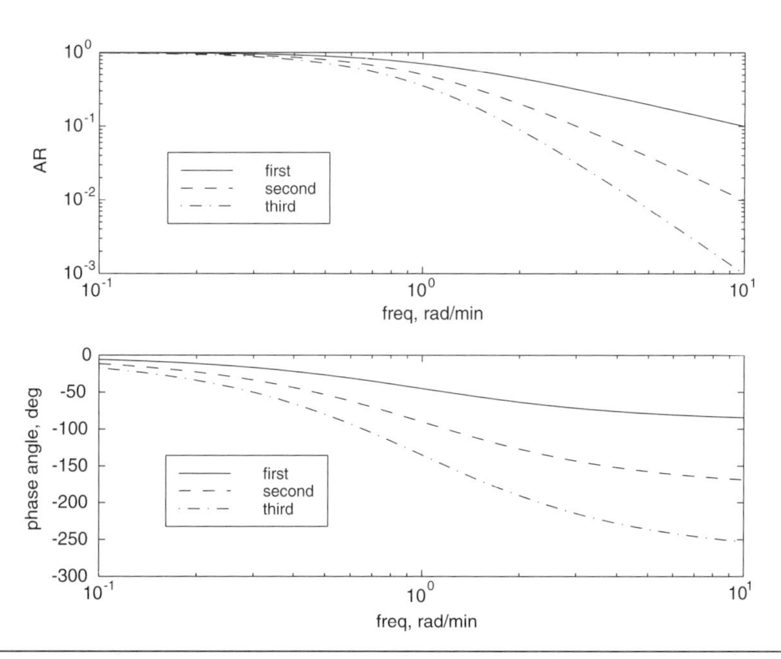

Figure 7–6 Bode plots. Comparison of first-, second-, and third-order systems with a gain of 1 and time constants of 1 minute.

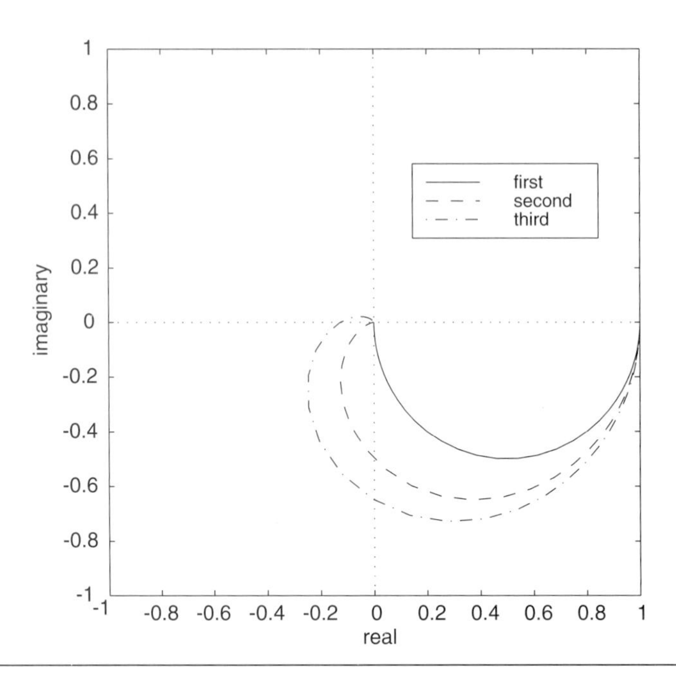

Figure 7–7 Nyquist plots. Comparison of first-, second-, and third-order systems with a gain of 1 and time constants of 1 minute. As the frequency increases, the curves go from (1,0) to (0,0).

where the first transfer function is a pure gain, the second and third are Padé approximations (first and second order, respectively), and the fourth is a pure time delay. The Bode plots are shown in Figure 7–8. Notice that all these transfer functions have an amplitude that is constant (does not vary with frequency). A transfer function with a constant amplitude ratio is called all-pass because all frequencies are directly transferred from input to output with no change in magnitude.

Also, notice that the phase angle is different for each transfer function. A transfer function is "minimum phase" if there is no other transfer function that has the same amplitude ratio, with a larger phase angle. In this case, the first transfer function (gain with no dynamics) is minimum phase, while all of the others are non-minimum phase. The high frequency asymptotes of the first- and second-order Padé approximations are –180° and –360°, respectively. The high-frequency phase angle of the time delay is unbounded. The corresponding Nyquist plots are shown in Figure 7–9; each curve begins at (2,0). The constant gain remains at (2,0). The first-order Padé transfer function ends at $(-2,0)$, which is –180°. The second-order Padé transfer function ends at (2,0), which is –360°. The time delay wraps an infinite number of times.

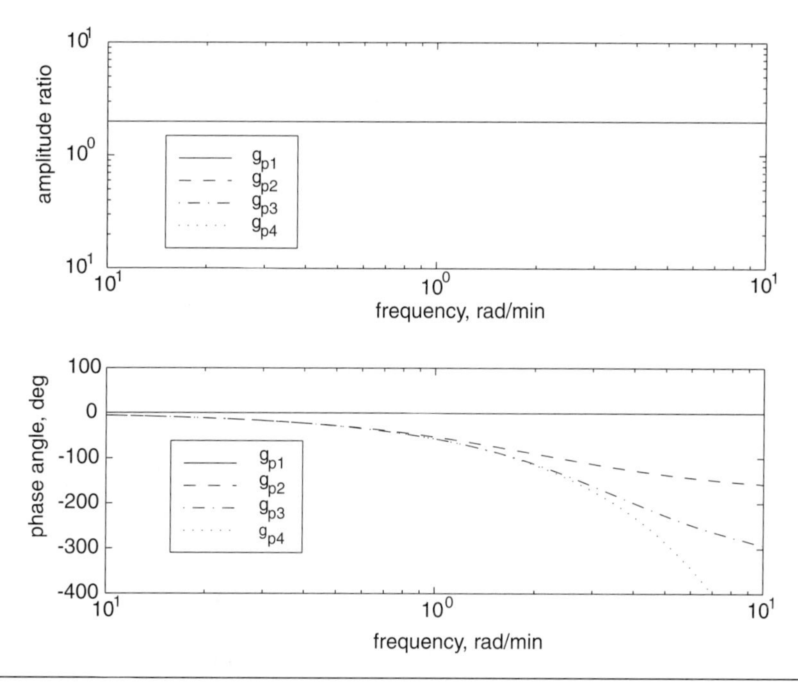

Figure 7–8 Bode plots. Comparison of different processes with an amplitude ratio of 2.

Frequency Response Introductory Summary

We have introduced Bode and Nyquist plots. You should be able to develop Nyquist plots given Bode plots and vice versa. Both analysis techniques are used for control-system tuning, as presented in the next sections.

For stable systems (poles < 0) with no time delays, we can state the following general results. Let m be the order of the numerator polynomial and n the order of the denominator polynomial (the *relative order* is $n–m$).

- If all process zeros are negative (no RHP zeros), the high-frequency phase lag is $(n–m)*(–90°)$
- Each RHP zero adds –90° of phase lag (left-half-plane zeros add +90° of phase lead) at high frequencies
- Gains change the amplitude ratio plot but not the phase angle plot

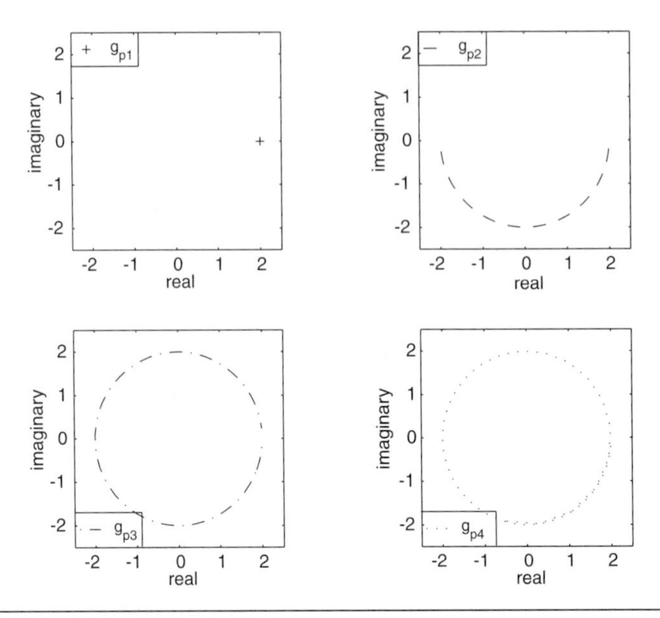

Figure 7–9 Nyquist plots. Comparison of different processes with an amplitude ratio of 2. For each individual Nyquist plot, the frequency increases as the curve moves clockwise from the (2,0) point.

We also can state the following about systems with time delays:

- Time delays change the phase-angle plot but not the amplitude ratio plot
- Processes with time delays have a phase-angle that continuously decreases with frequency; this corresponds to a continuous "wrapping" of the origin on a Nyquist plot

7.4 Closed-Loop Stability Concepts

Sections 7.1–7.3 introduced frequency response analysis, including the construction of Bode and Nyquist plots. In this section we illustrate how Bode and Nyquist plots can be used to determine the stability of a closed-loop system. We also show how to use these plots to tune controllers for robustness, that is, to be able to tolerate a given level of model uncertainty. The tests used involve information about the controller and process in open-loop. The open-loop analysis that we use in this chapter is different from that used in model-based control chapters.

Consider the case where the loop has been opened, as shown in Figure 7–10. We introduce the concept of Bode stability using the following thought experiment. Assume that the setpoint applied to the "open-loop" system in Figure 7–10 is a sine wave. Also assume that the controller has been tuned so that the output lags the setpoint by 180° and has the same amplitude as the setpoint, as shown in Figure 7–11.

Realize that $-y(t)$ is exactly 180° out of phase with $y(t)$, which means that $-y(t)$ is equal to $r(t)$. Now consider the case where the setpoint signal, $r(t)$, is suddenly stopped and simultaneously the loop is closed, as shown in Figure 7–12. This means that the error signal will simply be $-y(t)$, which is identical to $r(t)$. Since it is identical to $r(t)$, then every signal on the control loop diagram remains the same. The output continues to oscillate with the same frequency and magnitude as before the loop was closed. We refer to this control loop as nominally stable.

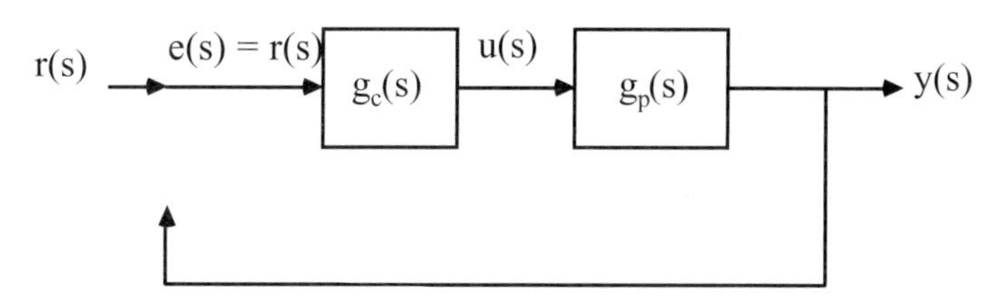

Figure 7–10 Feedback control loop has been "opened."

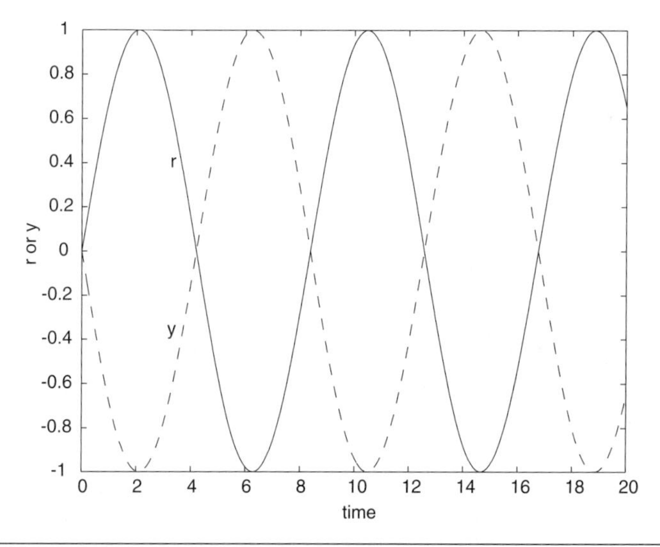

Figure 7–11 Process output (y) lags setpoint (r) by 180°.

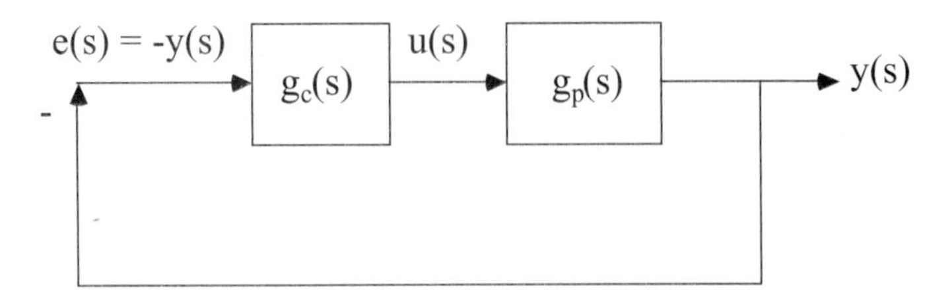

Figure 7–12 Feedback loop has been "closed."

Consider the case where the output is exactly out of phase ($\phi = -180°$) with the setpoint in the open-loop, but the amplitude of the output is less than the setpoint. Therefore the amplitude ratio ($|y|/|r|$) is less than 1. Then, if the loop is closed, the output decreases each time "around the loop" and the system is stable. In contrast, consider the case where the output is exactly out of phase with the setpoint when the control loop is open, but the amplitude of the output is greater than the setpoint. Then, if the loop is then closed, the output increases each time "around the loop" and the system in unstable.

The thought experiment that we have just performed is the basis for the Bode stability criterion stated in the next section.

7.5 Bode and Nyquist Stability

In the previous section we used a thought experiment based on the open-loop (controller + process) response to a sinusoidal setpoint input. We found that if the amplitude ratio of the open-loop transfer function $[g_c(s)g_p(s)]$ is less than 1 when the phase angle is $-180°$, then the system will be stable when the loop is closed. This requirement is also known as the *Bode stability criterion*. Notice that we are assuming no valve and measurement effects [or lumping them into $g_p(s)$]. If these effects are included, then $g_c(s)g_p(s)g_v(s)g_m(s)$ should be used in the analysis.

Bode Stability Criterion

Assume that the process is stable [$g_p(s)$ has no poles in the RHP] and the controller is stable [$g_c(s)$ has no poles in the RHP]. Also assume that the phase angle (ϕ) crosses $-180°$ only once on a Bode plot. The system will be closed-loop stable if and only if the amplitude ratio of $g_c g_p$ is less than 1 at the crossover frequency.

$$AR_{co} = \left|g_c(\omega)g_p(\omega)\right| < 1 \text{ at } \omega_{co} \tag{7.11}$$

where ω_{co} is known as the *crossover frequency*, which is defined as the frequency where the phase angle is –180°.

The requirements placed by the Bode stability criterion are not that restrictive for most chemical processes. For processes where the phase angle crosses –180° more than once, it is necessary to use the Nyquist stability criterion.

Nyquist Stability Criterion

A system will be closed-loop stable if a Nyquist plot of $g_c(s)g_p(s)$ does not encircle the critical point $(-1,0)$. Examples of closed-loop stable and unstable systems are shown in Figure 7–13.

The Bode and Nyquist stability criteria indicate whether a closed-loop system will be stable or not, but they do not indicate how "close to instability" a system is. That is, the Bode and Nyquist stability criteria alone do not indicate the *robustness* of a feedback system to model uncertainty. Gain and phase margins, defined below, will provide an indication of the robustness of a feedback system.

Gain Margin

Let AR_{co} represent the amplitude of $g_c(s)g_p(s)$ at the crossover frequency (ω_{co}, where $\phi = -180°$). The *gain margin* is defined as

$$GM = 1/AR_{co} \tag{7.12}$$

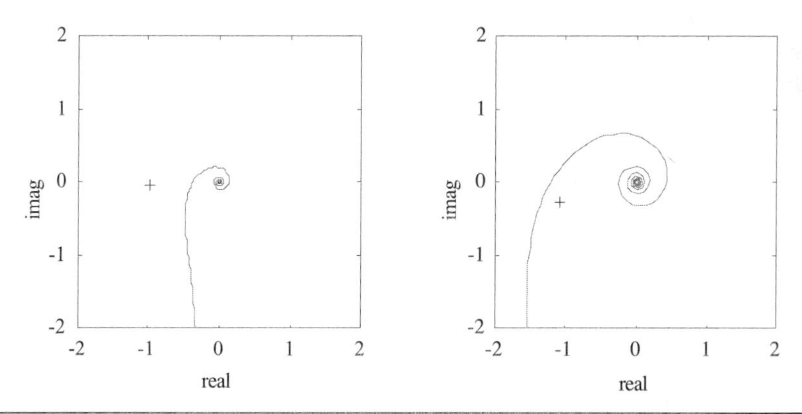

Figure 7–13 Illustration of Nyquist stability criterion. (Left) Stable system, critical point is not encircled. (Right) Unstable system, critical point is encircled.

and the Bode stability requirement is that $AR_{co} < 1$, so for stability

$$GM > 1 \tag{7.13}$$

The gain margin is the multiplicative factor that the gain can be increased by before the system becomes unstable.

Phase Margin

Let ϕ_{pm} represent the phase angle when the amplitude of $g_c(s)g_p(s)$ is 1. This occurs at a frequency ω_{pm}. The phase margin is defined as

$$PM = \phi_{pm} - (-180°) = \phi_{pm} + 180° \tag{7.14}$$

The closed-loop system will be stable if $\phi > -180°$ when AR = 1, so the stability requirement is

$$PM > 1 \tag{7.15}$$

The phase margin is the amount that the phase angle can decrease (in magnitude, at ω_{pm}) before the system becomes unstable.

We have developed the Bode and Nyquist stability criteria and defined the gain and phase margins. Next we use an example to illustrate the techniques.

Example 7.3: Nonminimum-Phase Process

Consider the Van de Vusse reactor (Module 5) controlled using a P-only controller.

$$g_p(s) = \frac{0.5848(-0.3549s + 1)}{\left(0.1858s^2 + 0.8627s + 1\right)}$$

where time unit is minutes. The Bode plots for $g_p(s)g_c(s)$ with $k_c = 2.5$ are shown in Figures 7–14 and 7–15.

We can use the following steps and the Bode plot in Figure 7–14 to calculate the gain margin.

1. Find the frequency where the phase angle for $g_p g_c = -180°$. Here, $\omega_{co} =$ 4.3 rad/min.
2. At this frequency, find the amplitude ratio for $g_p g_c$. Here, AR = 0.60.
3. The gain margin is 1/AR. Here, the gain margin = 1/0.60 = 1.67.

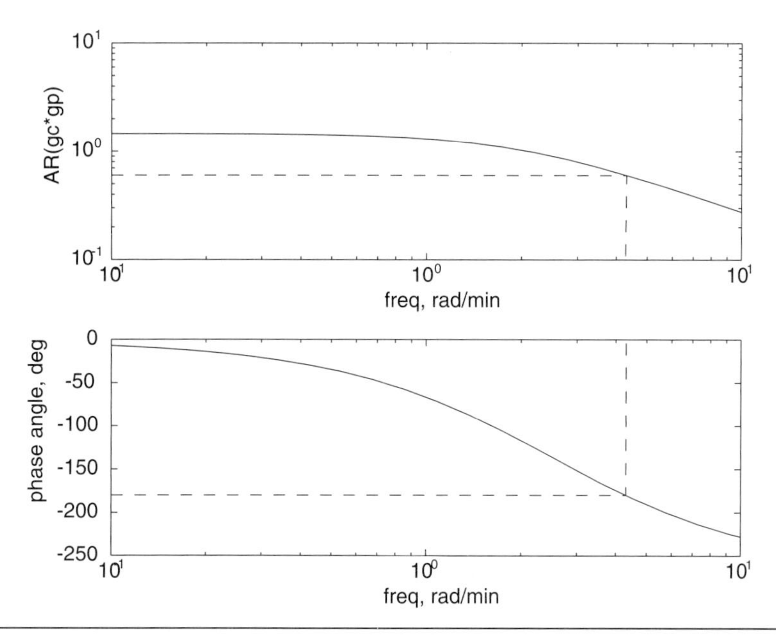

Figure 7–14 Bode plot for Example 7.3. Calculation of the gain margin.

We can use the following steps and the plot in Figure 7–15 to find the phase margin.

1. Find the frequency where the $g_p g_c$ amplitude ratio is 1.0. Here, $\omega_{pm} = 2.1$ rad/minute.
2. At this frequency, find the $g_p g_c$ phase angle. Here, $\phi_{pm} = -121°$.
3. The phase margin is $\phi - (-180)$. Here, the phase margin $= -121 + 180 = 59°$.

The Nyquist plot is shown in Figure 7–16. The unit circle is plotted on this diagram so that it is easy to determine where the amplitude is equal to one. Notice that the Nyquist diagram is consistent with the Bode diagram, since the amplitude ratio is 0.60 when the phase angle is –180°, and the phase angle is –121° when the amplitude ratio is 1.0.

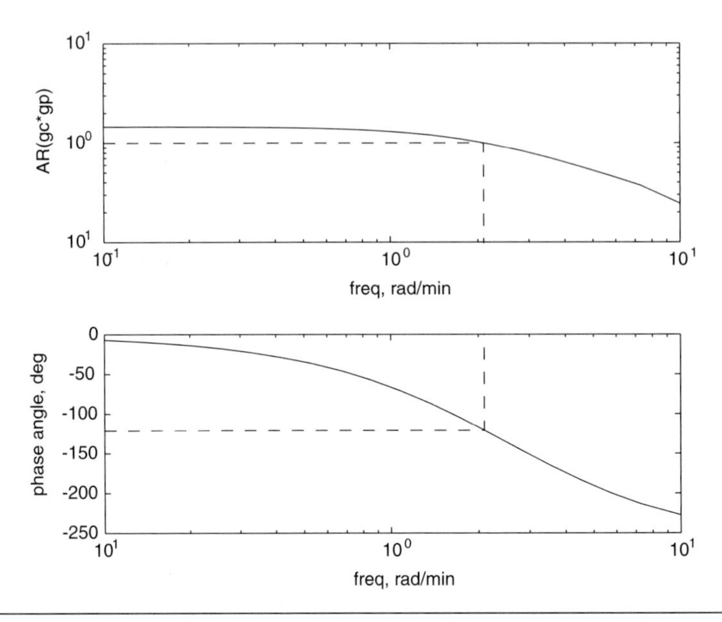

Figure 7–15 Bode plot for Example 7.3. Phase margin calculation.

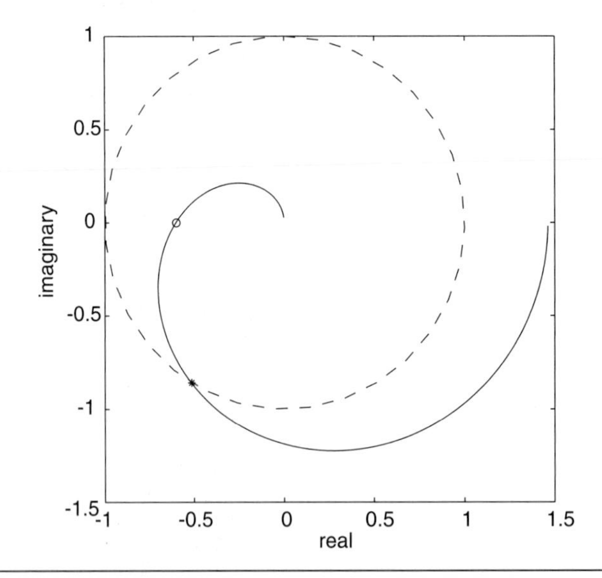

Figure 7–16 Nyquist plot for Example 7.3. The unit circle (dashed line) is included for convenience. The points for the gain margin and phase margin calculations are shown as the open circle and star, respectively.

7.6 Robustness

We have noted throughout this text that it is important to tune controllers to be *robust*. By robust, we mean that a controller can tolerate a certain amount of change in the process parameters without causing the feedback system to go unstable. The gain and phase margins presented in Section 7.5 quantify the amount of uncertainty that can be tolerated.

To illustrate the relationship between gain and phase margins and the model parametric uncertainty that can be tolerated, we consider the following first-order + time-delay process.

The nominal process model is

$$\tilde{g}_p(s) = \frac{\tilde{k}_p e^{-\tilde{\theta}s}}{\tilde{\tau}_p s + 1} \tag{7.16}$$

Let Δk and $\Delta\theta$ represent the additive uncertainty in the process gain and dead time, respectively (*notice that we are neglecting any uncertainty in the time constant*). The actual process is then

$$g_p(s) = \frac{\left(\tilde{k}_p + \Delta k\right) e^{-\left(\tilde{\theta} + \Delta\theta\right)s}}{\tilde{\tau}_p s + 1} \tag{7.17}$$

Notice that we could also use the representation

$$g_p(s) = \frac{\tilde{k}_p (1+\varepsilon) e^{-\tilde{\theta}s} e^{-\Delta\theta s}}{\tilde{\tau}_p s + 1} = \tilde{g}_p(s)(1+\varepsilon)e^{-\Delta\theta s} \tag{7.18}$$

where $\varepsilon = \Delta k / \tilde{k}_p$. The open-loop transfer function used in the frequency response analysis is

$$g_c(s) g_p(s) = g_c(s) \tilde{g}_p(s)(1+\varepsilon)e^{-\Delta\theta s} \tag{7.19}$$

Notice that the gain and time-delay uncertainties are clearly separated from the nominal model and controller ($g_c(s)\tilde{g}_p(s)$) in this equation. Since $1 + \varepsilon$ is a constant term, it will only affect the amplitude ratio. Also notice that $e^{-\Delta\theta s}$ will only affect the phase angle. This means that we can think of the *gain margin* as being related to the amount of gain uncertainty that can be tolerated, while the *phase margin* is related to the amount of dead-time uncertainty that can be tolerated. The change in phase angle, $\Delta\phi$, due to the change in deadtime, $\Delta\theta$, can be found from

$$\Delta\phi = -\Delta\theta\omega_{pm} \tag{7.20}$$

These concepts are shown more clearly in the first-order + dead-time learning module (Module 6).

7.7 MATLAB Control Toolbox: Bode and Nyquist Functions

The MATLAB Control Toolbox has special functions to generate Bode and Nyquist plots, and to calculate gain and phase margins. Here we use Example 7.3 to illustrate the use of the MATLAB `bode`, `nyquist`, and `imargin` functions that are available in the Control Toolbox. For information, simply enter `help bode`, and so forth, in the MATLAB command window.

First, define the process transfer function (g_p).

```
» vdvtf = tf(0.5848*[-0.3549 1],[0.1858 0.8627 1])

Transfer function:
   -0.2075 s + 0.5848
-------------------------
0.1858 s^2 + 0.8627 s + 1
```

Next, generate the open-loop transfer function ($g_c g_p$) for a controller gain (k_c) of 2.5.

```
» kc = 2.5;

» gcgp = kc*vdvtf


Transfer function:
     -0.5189 s + 1.462
-------------------------
0.1858 s^2 + 0.8627 s + 1
```

Now, use the `bode` function to generate magnitude and phase angles as a function of frequency. Also, use `imargin` to perform the gain margin (`Gm`) and phase margin (`Pm`) calculations at their respective frequencies (`Wcg` and `Wcp`).

```
»[mag,phase,w] = bode(gcgp);

[Gm,Pm,Wcg,Wcp] = imargin(squeeze(mag),squeeze(phase),w)

Gm =
     1.6680

Pm =
    59.0905
```

```
Wcg =
    4.3064

Wcp =
    2.0956
```

Figure 7–14 is generated using the following sequence of commands.

```
subplot(2,1,1),loglog(w,squeeze(mag),[0.1 Wcg Wcg],
   [1/Gm 1/Gm min(mag)],'--')
subplot(2,1,2),semilogx(w,squeeze(phase),[min(w) Wcg Wcg],
   [-180 -180 0],'--')
```

Figure 7–15 is generated using the following sequence of commands.

```
subplot(2,1,1),loglog(w,squeeze(mag),[0.1 Wcp Wcp],
   [1 1 min(mag)],'--')
subplot(2,1,2),semilogx(w,squeeze(phase),[0.1 Wcp Wcp],
   [-180+Pm -180+Pm 0],'--')
```

The Nyquist plot, Figure 7–16, is generated using the following sequence of steps. First, define an appropriate frequency range.

```
w1 = logspace(-2,2,300);
```

Then use the nyquist function to generate real and imaginary vectors.

```
[regcgp,imgcgp] = nyquist(gcgp,w1);
```

Generate circles for convenience when interpreting the Nyquist plot.

```
circlex = cos(0:0.2:2*pi);
circley = sin(0:0.2:2*pi);
phaseptx = cos((-180+Pm)*pi/180);
phasepty = sin((-180+Pm)*pi/180);
plot(squeeze(regcgp),squeeze(imgcgp),circlex,circley,'--',
   [-1/Gm],[0],'o', phaseptx,phasepty,'*')
axis('square')
xlabel('real')
ylabel('imaginary')
```

7.8 Summary

In this chapter, we developed the Bode and Nyquist frequency response techniques for control-system design. Gain and phase margins indicate the amount of uncertainty a feedback system can tolerate before becoming unstable. The gain margin is related to the change in gain (process or controller) that can be tolerated, while the phase margin is related to the change in time delay that can be tolerated. From a practical perspective, we recommend a gain margin of at least 2.5 and phase margin of at least 60° to assure robustness.

The basic concepts covered were as follows:

Amplitude ratio (AR): Ratio of the output to input amplitudes

Phase angle (ϕ): Peak-to-peak shift between input and output curves

All-pass: A transfer function with constant amplitude at all frequencies

Nonminimum phase: Generally, transfer functions with time delays and RHP zeros

Bode stability criterion: Amplitude ratio of $g_c g_p < 1$ at –180°.

Nyquist stability criterion: The critical point (–1,0) is not encircled by $g_c g_p$.

Gain margin: Multiplicative factor tolerated by $g_c g_p$ before losing stability

Phase margin: Phase-angle error tolerated by $g_c g_p$ before losing stability

Reference

The following text contains an advanced discussion of control system design for robustness: Morari, M., and E. Zafiriou, *Robust Process Control*, Prentice-Hall, Englewood Cliffs, NJ (1989).

Student Exercises

1. Find the time-domain solution for the output of a first-order process, with an input sine wave of magnitude A at frequency ω.
2. The peak day length (in terms of hours of daylight) occurs on 21 June, while the peak average daily temperature occurs on 21 July (30 days later). Assuming that the relationship between daylight hours and average daily temperature can be modeled by a first-order transfer function, estimate the time constant of Earth's weather system (provide the units). Also, what are the units of the "process gain"?

9. Consider the following Bode plot for a system under PI control.

a. Sketch the Nyquist diagram.

b. How much can the controller gain be increased if a gain margin of 2 is desired?

c. How much can the deadtime increase before the system goes unstable?

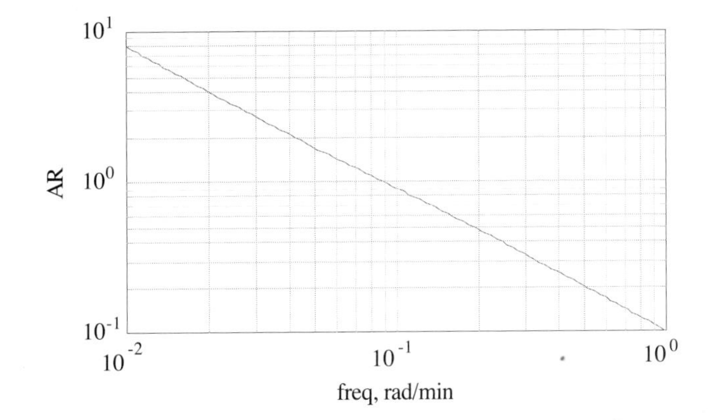

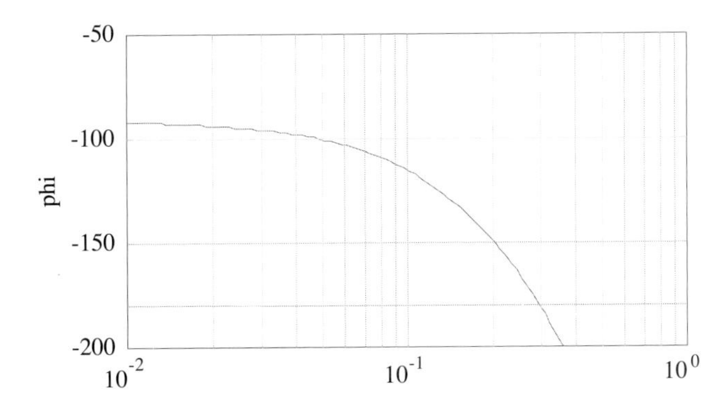

CHAPTER 8

Internal Model Control

In Chapter 6 we presented several methods for tuning PID controllers and developed a model-based procedure (direct synthesis) to synthesize a controller that yields a desired closed-loop response trajectory. In this chapter, we first develop an "open-loop control" design procedure that then leads to the development of an internal model control (IMC) structure. There are a number of advantages to the IMC structure (and controller design procedure), compared with the classical feedback control structure. One is that it becomes very clear how process characteristics such as time delays and RHP zeros affect the inherent controllability of the process. IMCs are much easier to tune than are controllers in a standard feedback control structure.

After studying this chapter, the reader should be able to do the following:

- Design internal model controllers for stable processes (either minimum or non-minimum phase)
- Sketch the closed-loop response if the model is perfect
- Derive the closed-loop transfer functions for IMC
- Design IMC for improved disturbance rejection

The major sections of this chapter are as follows:

8.1 Introduction to Model-Based Control

In the previous chapters we focused on techniques to tune PID controllers. The closed-loop oscillation technique developed by Ziegler and Nichols did not require a model of the process. Direct synthesis, however, was based the use of a process model and a desired closed-loop response to synthesize a control law; often this resulted in a controller with a PID structure. In this chapter we develop a model-based procedure, where a process model is "embedded" in the controller. By explicitly using process knowledge, by virtue of the process model, improved performance can be obtained.

Consider the stirred-tank heater control problem shown in Figure 8–1. We can use a model of the process to decide the heat flow (Q) that needs to be added to the process to obtain a desired temperature (T) trajectory, specified by the setpoint (T_{sp}). A simple steady-state energy balance provides the steady-state heat flow needed to obtain a new steady-state temperature, for example. By using a dynamic model, we can also find the time-dependent heat profile needed to yield a particular time-dependent temperature profile.

Assume that the chemical process is represented by a linear transfer function model, and that it is open-loop stable. The input-output relationship is shown in Figure 8–2, where u is the input variable (heat flow) and y is the output variable (temperature).

When the process is at steady state, and there are no disturbances, then the inputs and outputs are zero (since we are using deviation variables). Consider a desired change

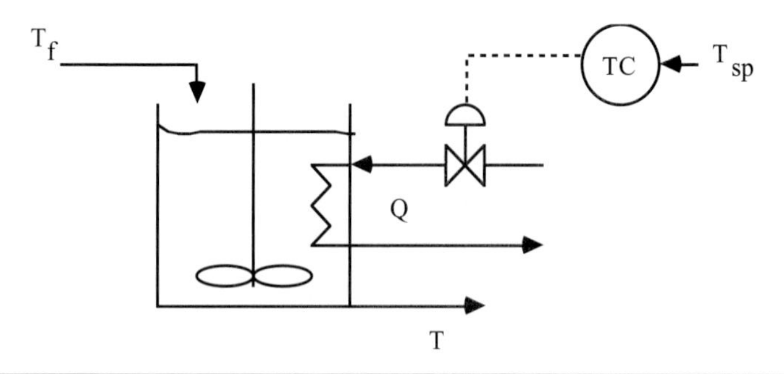

Figure 8–1 Stirred-tank heater.

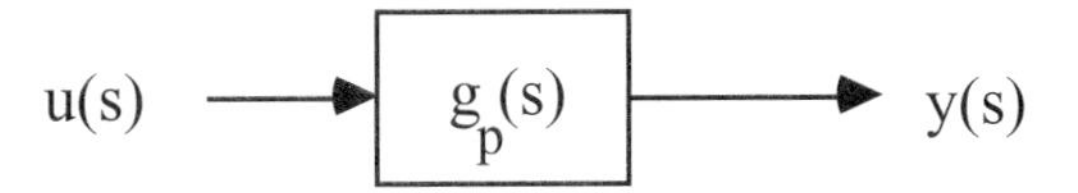

Figure 8–2 Open-loop process system.

in the output y; we refer to the desired value of y as the setpoint, which is represented by r. We wish to design an open-loop controller, $q(s)$, so that the relationship between $r(s)$ and $y(s)$ has *desirable* dynamic characteristics (fast response without much overshoot, no offset, etc.). The open-loop control system is shown in Figure 8–3 (we may also wish to think of this as a feed-forward controller, based on setpoint). We use $q(s)$ to represent the open-loop controller transfer function, to emphasize that it is a different type of controller than the feedback controllers of Chapters 5 and 6.

Using block diagram analysis, we find the following relationship between the setpoint and the output

$$y(s) = g_p(s)q(s)r(s) \tag{8.1}$$

Static Control Law

The simplest controller will result if $q(s)$ is a constant. Let k_q represent this constant. As an example, consider a first-order process, $g_p(s) = k_p/(\tau_p s + 1)$. Then the relationship between $r(s)$ and $y(s)$ is

$$y(s) = \frac{k_p k_q}{\tau_p s + 1} r(s)$$

To obtain a desirable response, $k_q = 1/k_p$; offset will result otherwise. We can see this from the final-value theorem. Consider a step setpoint change, of magnitude R.

$$y(s) = \frac{k_p k_q}{\tau_p s + 1} \cdot \frac{R}{s} \tag{8.2}$$

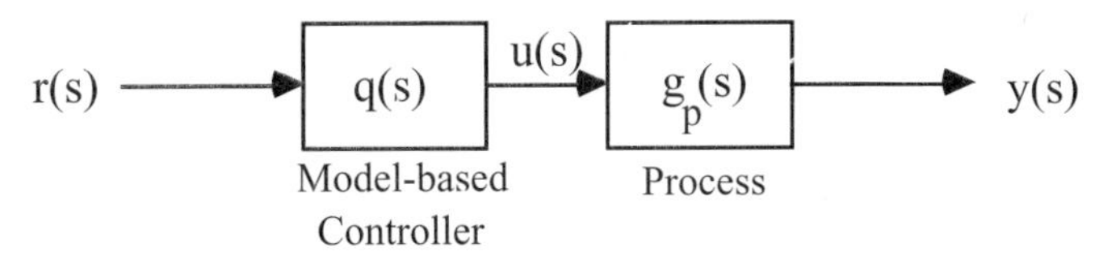

Figure 8–3 Open-loop model-based control system.

From the final-value theorem

$$\lim_{t \to \infty} y(t) = \lim_{s \to 0} sy(s) = k_p k_q R$$

and for no offset, we require that $k_q = 1/k_p$. We can also find the time-domain solution to Equation (8.2)

$$y(t) = k_p k_q R\left(1 - e^{-t/\tau_p}\right)$$

Again, we can see that $k_q = 1/k_p$ is necessary for offset-free performance. Notice also that the speed of response is the same as the time constant of the open-loop process. In order to "speed up" the response, we must use a dynamic control law, as developed in the next section.

Dynamic Control Law

Better control can be obtained if the controller, $q(s)$, is dynamic rather than static. Indeed, we find that if

$$q(s) = \frac{1}{g_p(s)} \tag{8.3}$$

then the relationship between $r(s)$ and $y(s)$ is

$$y(s) = g_p(s)q(s)r(s) = g_p(s)\frac{1}{g_p(s)}r(s) = r(s)$$

That is, we have *perfect control*, since the output perfectly tracks the setpoint! For a first-order process, the controller is

$$q(s) = \frac{1}{g_p(s)} = \frac{\tau_p s + 1}{k_p}$$

Although this is mathematically possible, *perfect control is unachievable* in practical application. Consider the signals in and out of the control block, shown in Figure 8–3. Since the transfer function relationship between $r(s)$ and $u(s)$ is, for this example,

$$u(s) = q(s)r(s) = \frac{\tau_p s + 1}{k_p} r(s) = \frac{1}{k_p}\left(\tau_p s + 1\right)r(s) \tag{8.4}$$

the differential equation that corresponds to Equation (8.4) is

$$u(t) = \frac{1}{k_p}\tau_p \frac{dr}{dt} + \frac{1}{k_p} r(t)$$

From a practical point of view, it is *impossible* to take an exact derivative of $r(t)$, particularly if a discontinuous step setpoint change is made.

Here we use the inverse Laplace transform to solve Equation (8.4) for $u(t)$, when there is a step change in r.

$$u(s) = \frac{\tau_p R}{k_p} + \frac{R}{k_p} \cdot \frac{1}{s}$$

A table of Laplace transforms can be used to find the time-domain solution

$$u(t) = \frac{\tau_p R}{k_p} \delta(t) + \frac{R}{k_p}$$

where $\delta(t)$ is the impulse function, which has infinite height, infinitesimal width, and unit area. Since this is hard to understand conceptually, you probably realize that it is impossible to implement exactly. Think about how you would approximate it.

The Bottom Line. Equation (8.4) is representative of a controller that is not *physically realizable*. Notice that the equation has a numerator polynomial, but not a denominator polynomial. We can state the following general result.

Physical Realizability. For a controller to be *physically realizable*, the order of the denominator of the controller transfer function [$q(s)$] must be at least as great as the order of the numerator.

Definition. A transfer function that satisfies this condition is *proper*. If the order of the denominator is greater than the order of the numerator, then the transfer function is *strictly proper*. If the order of the numerator is equal to the order of the denominator, then the transfer function is often called *semiproper*.

8.2 Practical Open-Loop Controller Design

In order to design a physically realizable controller for the first-order process, there must be a denominator polynomial in the controller. Define a first-order filter as the following transfer function,

$$f(s) = \frac{1}{\lambda s + 1} \tag{8.5}$$

where λ is a filter tuning parameter that has units of time. We see shortly that we can think of λ as a desired time constant for the output response; if a faster response is desired, then

λ is decreased. Now, let the controller be designed in the following fashion to make $q(s)$ proper.

$$q(s) = \frac{f(s)}{g_p(s)} = g_p^{-1}(s)f(s) \tag{8.6}$$

which yields the following controller for a first-order process,

$$q(s) = \frac{\tau_p s + 1}{k_p} \cdot \frac{1}{\lambda s + 1} = \left(\frac{1}{k_p}\right)\frac{\tau_p s + 1}{\lambda s + 1} \tag{8.7}$$

Equation (8.7) is a lead/lag controller. Since the order of the denominator is at least as great as the order of the numerator, the design is physically realizable. The response of the output variable, y, is

$$y(s) = g_p(s)q(s)r(s) = g_p(s)g_p^{-1}(s)f(s)r(s) = f(s)r(s) \tag{8.8}$$

in this case,

$$y(s) = \frac{1}{\lambda s + 1} r(s) \tag{8.9}$$

That is, there will be a first-order response with a time constant of λ. Contrast this with the static control law $[q(s) = k_q = 1/k_p]$, which will yield a first-order response with a time constant of τ_p. As long as $\lambda < \tau_p$, the dynamic controller will have a faster response than the static controller.

Response of Manipulated and Output Variables to Step Setpoint Changes

We have already seen that the output response to a setpoint change is first order [e.g., Equation (8.9)]. We can also find the manipulated variable response to a setpoint change. Here we consider the dynamic controller $[q(s) = f(s)/g_p(s)]$

$$u(s) = \left(\frac{1}{k_p}\right)\frac{\tau_p s + 1}{\lambda s + 1} r(s) \tag{8.10}$$

Notice that this is a lead/lag controller. For a step setpoint change of magnitude R,

$$u(s) = \left(\frac{1}{k_p}\right)\frac{\tau_p s + 1}{\lambda s + 1}\frac{R}{s} \tag{8.11}$$

The time-domain response for the manipulated input is

$$u(t) = \left(\frac{R}{k_p}\right)\left[1 - \left(1 - \frac{\tau_p}{\lambda}\right)e^{-t/\lambda}\right] \tag{8.12}$$

The output response is

$$y(t) = R(1 - e^{-t/\lambda}) \tag{8.13}$$

Figure 8–4 is a plot of $y(t)$ and $u(t)$, as a function of as a function of λ, for a process with $\tau_p = 10$ minutes and $k_p = 1$ %/%. Notice that the manipulated input action occurs immediately at the time of the setpoint change. You should show that the magnitude is consistent with the time-domain solution [Equation (8.12)], and with the Laplace domain solution [Equation (8.11)] using the initial value theorem.

Issues in Dynamic Controller Design

We have already seen that a controller needs to be *proper* to be implemented in a practical situation. We found that a dynamic controller could be designed by using a process model inverse, cascaded with a filter transfer function, to make the controller *proper*.

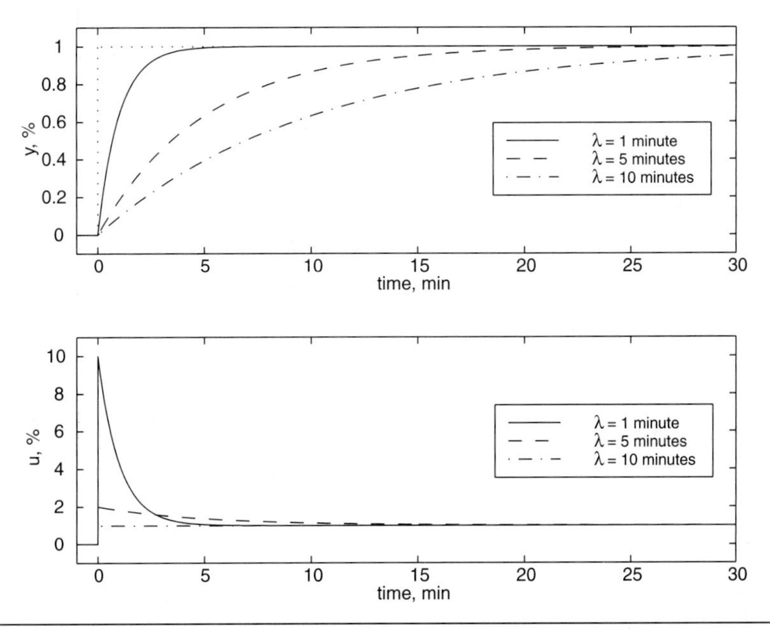

Figure 8–4 Response to step setpoint change, as a function of λ, for a first-order process with a time constant of 10 minutes.

Question 8.2.1: Can the process model inverse always be used for controller design?

The answer to question 8.2.1 is shown clearly in the following example.

Example 8.1: Inverse Response System

Consider the following transfer function,

$$g_p(s) = \frac{k_p(-\beta s + 1)}{(\tau_{p1}s + 1)(\tau_{p2}s + 1)} \tag{8.14}$$

where β is a positive real number, indicating a positive zero (yielding inverse response) in the process transfer function. If we use the previous design procedure for a dynamic open-loop controller, we find

$$q(s) = g_p^{-1}(s)f(s) = \frac{(\tau_{p1}s + 1)(\tau_{p2}s + 1)}{k_p(-\beta s + 1)} f(s) \tag{8.15}$$

and if we let $f(s) = 1/(\lambda s + 1)$, we obtain

$$q(s) = \left(\frac{1}{k_p}\right) \frac{(\tau_{p1}s + 1)(\tau_{p2}s + 1)}{(-\beta s + 1)(\lambda s + 1)} \tag{8.16}$$

Notice that the zeros of the process model become the poles of the controller, when the model inverse is used for control-system design. This creates an unstable controller and the possibility of unbounded, manipulated variable action. Therefore, *if a process has a RHP zero, this zero must be factored out before using the model inverse for the controller design.*

Answer 8.2.1: The process model cannot simply be inverted to form the controller. It must be factored so that the resulting controller is stable and realizable.

Look at Equation (8.16). If we simply take out the unstable pole, we have

$$q(s) = \left(\frac{1}{k_p}\right) \frac{(\tau_{p1}s + 1)(\tau_{p2}s + 1)}{(\lambda s + 1)} \tag{8.17}$$

Notice that this is not acceptable, because $q(s)$ is not *proper*. To make it proper, we can simply increase the order of the filter.

$$q(s) = \left(\frac{1}{k_p}\right) \frac{(\tau_{p1}s + 1)(\tau_{p2}s + 1)}{(\lambda s + 1)(\lambda s + 1)} \tag{8.18}$$

where we have used a second-order filter, $f(s) = 1/(\lambda s + 1)^2$, for the controller design. What kind of output response will be achieved?

$$y(s) = g_p(s)q(s)r(s) = \frac{k_p(-\beta s+1)}{(\tau_{p1}s+1)(\tau_{p2}s+1)}\left(\frac{1}{k_p}\right)\frac{(\tau_{p1}s+1)(\tau_{p2}s+1)}{(\lambda s+1)(\lambda s+1)}r(s)$$
$$= \frac{(-\beta s+1)}{(\lambda s+1)(\lambda s+1)}r(s). \tag{8.19}$$

Notice that the output will exhibit an inverse response when there is a step setpoint change.

This is an important result—process inverse response behavior *cannot* be removed by any *stable* control system.

Example 8.2: Numerical Example of an Inverse Response System

Consider the following transfer function, which has a RHP zero at 1/9 min^{-1}:

$$g_p(s) = \frac{-9s+1}{(15s+1)(3s+1)}$$

Compare the responses (input and output) under open-loop control, for the two cases: (i) when the RHP pole is removed from the controller and (ii) when the RHP pole is not removed from the controller (i.e., the controller is unstable)

(i) *When the unstable controller pole is removed*, we have from Equation (8.18)

$$q(s) = \frac{(15s+1)(3s+1)}{(\lambda s+1)(\lambda s+1)}$$

which yields the following manipulated and output variable response:

$$u(s) = \frac{(15s+1)(3s+1)}{(\lambda s+1)(\lambda s+1)}r(s), \qquad y(s) = \frac{-9s+1}{(\lambda s+1)(\lambda s+1)}r(s)$$

For a unit step change in setpoint, the output and manipulated variable responses are shown in Figure 8–5 for $\lambda = 5$ minutes. Notice that the inverse response appears in the output variable, and that the manipulated input changes immediately when the setpoint change is made. The reader should be able to show that the manipulated input change is consistent with that predicted by the initial value theorem.

(ii) *When the unstable controller pole is not removed*, we have from Equation (8.16)

$$q(s) = \frac{(15s+1)(3s+1)}{(-9s+1)(\lambda s+1)} \tag{8.20}$$

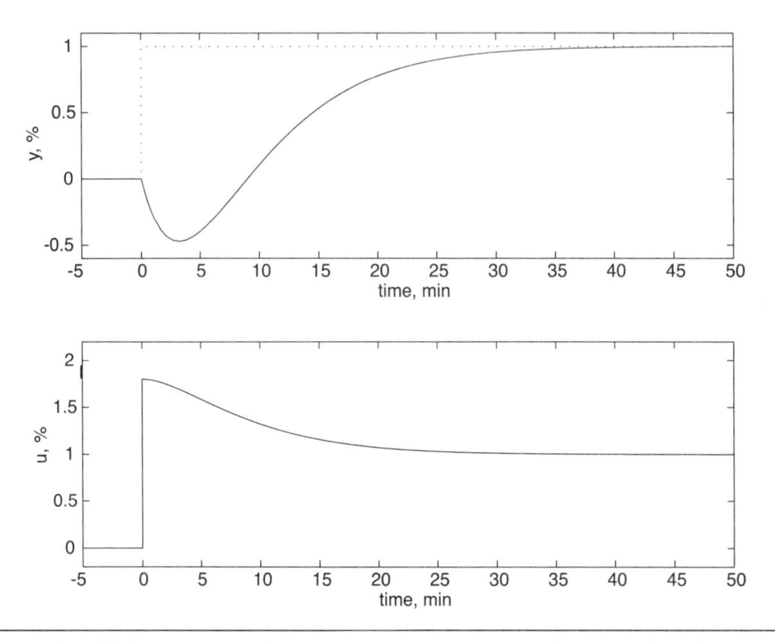

Figure 8–5 Unit step setpoint response when controller is stable ($\lambda = 5$ minutes).

which is an unstable controller. This controller yields the manipulated variable response

$$u(s) = q(s)r(s) = \frac{(15s+1)(3s+1)}{(-9s+1)(\lambda s+1)} r(s) \tag{8.21}$$

The output response is

$$\begin{aligned} y(s) &= q(s)g_p(s)r(s) \\ &= \frac{(15s+1)(3s+1)}{(-9s+1)(\lambda s+1)} \cdot \frac{(-9s+1)}{(15s+1)(3s+1)} r(s) = \frac{1}{\lambda s+1} r(s) \end{aligned} \tag{8.22}$$

The output and manipulated variable responses are shown in Figure 8–6 for $\lambda = 5$ minutes. Notice that the inverse response does not appear in the output variable, but the manipulated variable is unbounded (unstable). This is often called *internal instability*.

This last example illustrated an important point that is not often mentioned in textbooks. If only the process output is calculated or plotted, then the engineer may overlook the fact that a manipulated variable is becoming unbounded. Good output performance does not ensure feasible manipulated variable action.

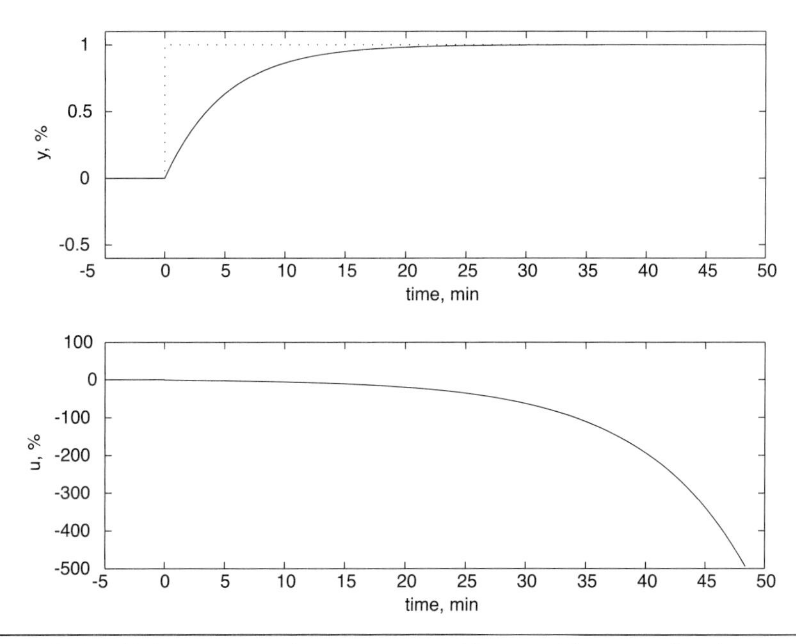

Figure 8–6 Unit step setpoint response with unstable controller (note scale of response).

When designing the open-loop control system of Figure 8–3, it is critical that both the controller and the process be stable. If the process is unstable, an open-loop controller cannot be used. The proper control-system design procedure for open-loop unstable systems is covered in Chapter 9.

8.3 Generalization of the Open-Loop Control Design Procedure

So far we have used two example processes: (i) first-order and (ii) second-order with a RHP zero. Now is the time to generalize our results to any process system.

First of all, we have been using the inverse of the process transfer function, $g_p(s)$, in our control law. We must realize that the actual process transfer function is never known exactly. We now use two transfer function representations of the process. One is considered the *process* (sometimes we use the term *plant*), which is never known exactly. The other is considered the *process model*, which we do know exactly.

Process (or plant) $g_p(s)$ (8.23)

Model $\tilde{g}_p(s)$ (8.24)

Recall that when the process had a RHP zero, we performed a *factorization*, so that the RHP zero did not form a RHP pole in the controller. In general, we factor the process model in the following fashion,

$$\tilde{g}_p(s) = \tilde{g}_{p+}(s)\,\tilde{g}_{p-}(s) \tag{8.25}$$

where $\tilde{g}_{p-}(s)$ contains the invertible elements and $\tilde{g}_{p+}(s)$ contains the noninvertible elements. One easy way to remember this is to recall that if RHP zeros are "inverted" (the RHP is the positive side of the complex plane), they become RHP poles, which are unstable.

Question 8.3.1: Is dead time invertible?

Answer 8.3.1: The reader should verify that dead time is not invertible. If there is dead time in the process, it cannot be removed by any physically realizable controller.

Our controller design will consist of inverting the invertible part of the process model and cascading this with a filter that is of high enough order to make the controller proper.

$$q(s) = \tilde{g}_{p-}^{-1}(s) f(s) \tag{8.26}$$

In the next section, we see that there are a number of ways to factor the same transfer function into invertible and noninvertible parts.

Controller Factorization

There are a number of ways to factor the same transfer function,

$$\tilde{g}_p(s) = \tilde{g}_{p+}(s)\,\tilde{g}_{p-}(s)$$

The most common method, known as an *all-pass* factorization, is shown as method 2 in the example below. The reason for the term *all-pass* was illustrated in Chapter 7, frequency-response techniques.

Example 8.3: Factorization Techniques

Consider a process model with inverse response characteristics (a RHP zero)

$$\tilde{g}_p(s) = \frac{\tilde{k}_p\left(-\tilde{\beta}s+1\right)}{\left(\tilde{\tau}_{p1}s+1\right)\left(\tilde{\tau}_{p2}s+1\right)} \tag{8.27}$$

where $\tilde{\beta}$ is positive.

Method 1: Simple factorization

The simple factorization approach is to simply place the RHP zeros in the noninvertible part of the process model [$\tilde{g}_{p+}(s)$].

$$\begin{aligned} \tilde{g}_p(s) &= \quad \tilde{g}_{p-}(s) \quad \tilde{g}_{p+}(s) \\ \tilde{g}_p(s) &= \frac{\tilde{k}_p}{(\tau_1 s+1)(\tau_2 s+1)}(-\beta s+1) \end{aligned} \tag{8.28}$$

Method 2: All-pass factorization

The all-pass factorization places the RHP zero in the noninvertible part of the process model, but it also places a pole at the reflection of the RHP zero:

$$\begin{aligned} \tilde{g}_p(s) &= \quad \tilde{g}_{p-}(s) \quad \tilde{g}_{p+}(s) \\ \tilde{g}_p(s) &= \frac{\tilde{k}_p(\beta s+1)}{(\tau_1 s+1)(\tau_2 s+1)} \frac{(-\beta s+1)}{(\beta s+1)} \end{aligned} \tag{8.29}$$

Generally, Method 2 (*all-pass*) is used because the resulting controlled system minimizes the integral squared error (ISE).

Comparison of Output Responses for Different Controller Factorizations

Here we compare the different controlled variable responses for methods 1 and 2.

Method 1

$$\begin{aligned} q(s) &= \quad \tilde{g}_{p-}^{-1}(s) \quad f(s) \\ q(s) &= \frac{(\tau_1 s+1)(\tau_2 s+1)}{\tilde{k}_p} \frac{1}{(\lambda s+1)^2} \end{aligned} \tag{8.30}$$

where we have used a second-order filter to make $q(s)$ realizable.

Now, since $y(s) = q(s)\, g_p(s)\, r(s)$,

$$y(s) = \frac{(\tilde{\tau}_1 s+1)(\tilde{\tau}_2 s+1)}{\tilde{k}_p} \frac{1}{(\lambda s+1)^2} \frac{k_p(-\beta s+1)}{(\tau_1 s+1)(\tau_2 s+1)} r(s) \tag{8.31}$$

Assuming a *perfect model* ($\tilde{k}_p = k_p$, etc.),

$$y(s) = \frac{(-\beta s+1)}{(\lambda s+1)^2} r(s) \tag{8.32}$$

which is the same as the result obtained in Equation (8.19).

Method 2

$$q(s) = \tilde{g}_{p-}^{-1}(s) \quad f(s)$$

$$q(s) = \frac{(\tau_1 s+1)(\tau_2 s+1)}{\tilde{k}_p(\tilde{\beta} s+1)} \quad \frac{1}{(\lambda s+1)} \tag{8.33}$$

Again, since $y(s) = q(s)\, g_p(s)\, r(s)$,

$$y(s) = \frac{(\tilde{\tau}_1 s+1)(\tilde{\tau}_2 s+1)}{\tilde{k}_p(\tilde{\beta} s+1)} \frac{1}{(\lambda s+1)} \frac{k_p(-\beta s+1)}{(\tau_1 s+1)(\tau_2 s+1)} r(s) \tag{8.34}$$

and assuming a *perfect model* ($\tilde{k}_p = k_p$, etc.),

$$y(s) = \frac{(-\beta s+1)}{(\beta s+1)(\lambda s+1)} r(s) \tag{8.35}$$

which is slightly different from the response shown in Equation (8.32). The responses will be equivalent only if we set $\lambda = \beta$ (and there is generally no reason to do that).

Summary of Controller Design Procedure

The controller design procedure has been generalized to the following steps.

1. Factor the process model into invertible and noninvertible portions

$$\tilde{g}_p(s) = \tilde{g}_{p+}(s)\, \tilde{g}_{p-}(s)$$

   - Method 1 uses a simple factorization
   - Method 2 uses an all-pass formulation

2. Invert the invertible portion of the process model (the "good stuff") and cascade with a filter that makes the controller $q(s)$ proper.

$$q(s) = \tilde{g}_{p-}^{-1}(s) f(s)$$

The factorization is used only for control-system design. The actual model still contains the RHP zeros and time delays.

Remember that the factorization was performed so that the controller would be stable. For linear systems, *if the controller is stable and the process is stable, then the overall controlled system is stable.* This is true simply because if two transfer functions are stable, then the transfer functions cascaded together (multiplied) are stable. This is a nice result because in a standard feedback control formulation, the controller and the process can each be stable, yet the feedback system may be unstable.

8.4 Model Uncertainty and Disturbances

Thus far we have assumed that the model is perfect when designing the open-loop control system. In practice, it is impossible to have a perfect model. Often, the process gain can be off by 20–50% (or even more, for highly nonlinear processes). The next example shows the problems associated with a purely open-loop control strategy when there are uncertainties or disturbances.

Example 8.4: First-Order Process with Model Uncertainty

The model and process are represented by the following transfer functions:

$$\tilde{g}_p(s) = \frac{\tilde{k}_p}{\tilde{\tau}_p s + 1}$$

$$g_p(s) = \frac{k_p}{\tau_p s + 1}$$

There are no noninvertible elements, so the control design procedure results in the following:

$$q(s) = \tilde{g}_p^{-1}(s) f(s) = \frac{\tilde{\tau}_p s + 1}{\tilde{k}_p} \frac{1}{\lambda s + 1}$$

$$y(s) = q(s) g_p(s) r(s) = \frac{\tilde{\tau}_p s + 1}{\tilde{k}_p} \frac{1}{\lambda s + 1} \frac{k_p}{\tau_p s + 1} r(s)$$

$$y(s) = \frac{\left(k_p / \tilde{k}_p\right)}{\lambda s + 1} \frac{\tilde{\tau}_p s + 1}{\tau_p s + 1} r(s)$$

Notice that the response will be second order with numerator dynamics if the model is not perfect. Also, *there will be offset* when there is a step change in setpoint.

Consider the following numerical example, where the model gain and time constant are 1 %/% and 10 minutes, respectively, but where the actual process has a gain of 1.2 %/% and a time constant of 8 minutes. If the tuning parameter, λ = 4 minutes, the relationship between the setpoint and the output is

$$y(s) = \frac{(1.2)}{4s + 1} \frac{10s + 1}{8s + 1} r(s)$$

Whereas the response with a perfect model would be y(s) = r(s)/(4s+1). The response with model uncertainty is compared with the perfect model response in Figure 8–7 for a unit step change in setpoint. The manipulated input value is the same whether the process model is perfect or in error, since there is no feedback from the measured output to correct for this error.

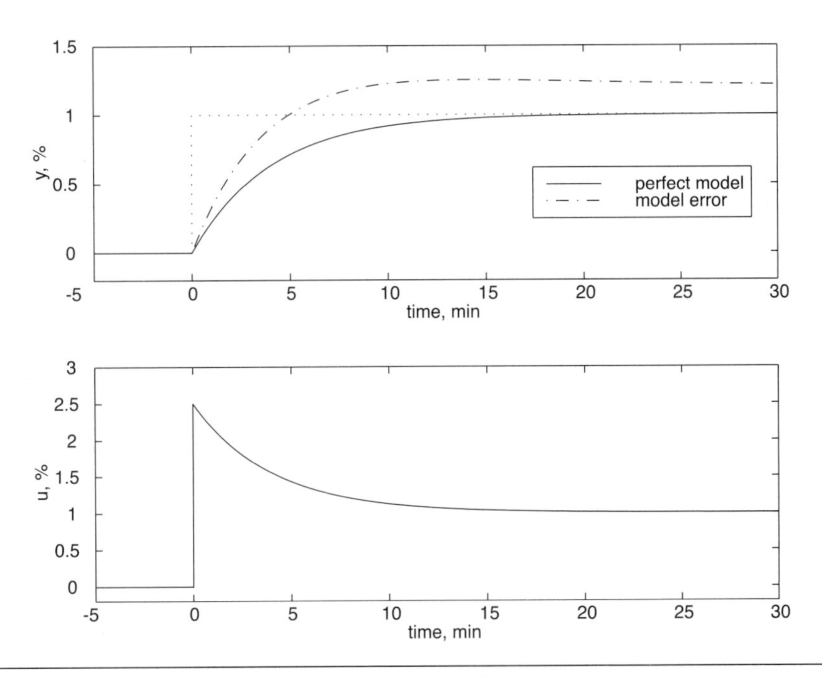

Figure 8–7 Comparison of open-loop controller responses.

Some form of feedback is needed to correct for model uncertainty (as well as any disturbances entering the process). The method that we develop to account for model uncertainty and disturbances is known as internal model control (IMC). In the next section, we begin to develop the IMC structure.

8.5 Development of the IMC Structure

Consider the open-loop control structure shown in Figure 8–3.

We now consider a process model that receives the same manipulated variable signal as the actual process (Figure 8–8). We can now subtract the difference between the process output (actually measured) and the process model output (model predicted) to determine the model error. This is shown in Figure 8–9. We must also realize that disturbances can enter the system, as shown in Figure 8–10.

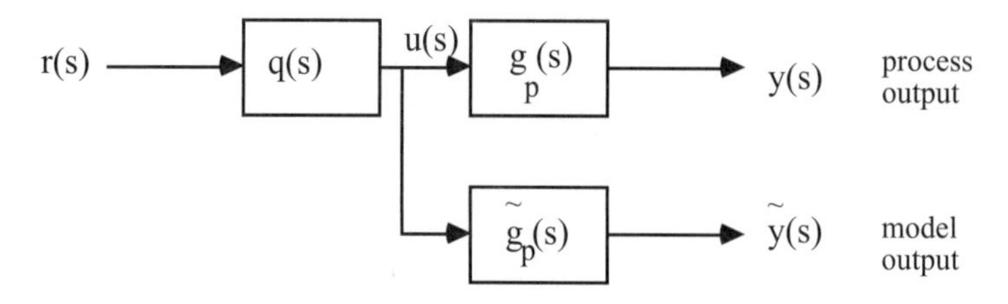

Figure 8–8 Process model in parallel with the actual process.

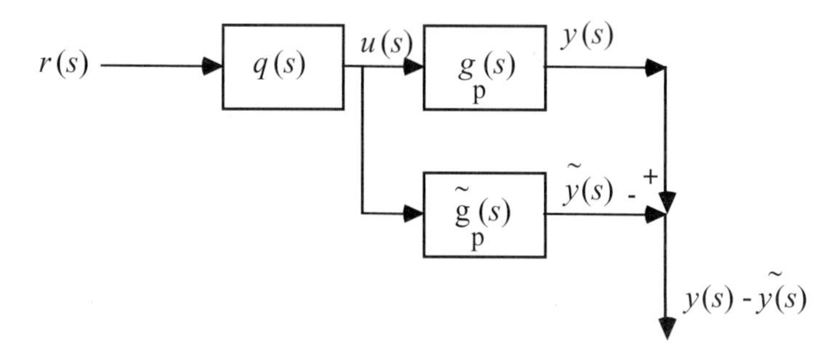

Figure 8–9 Calculating model error.

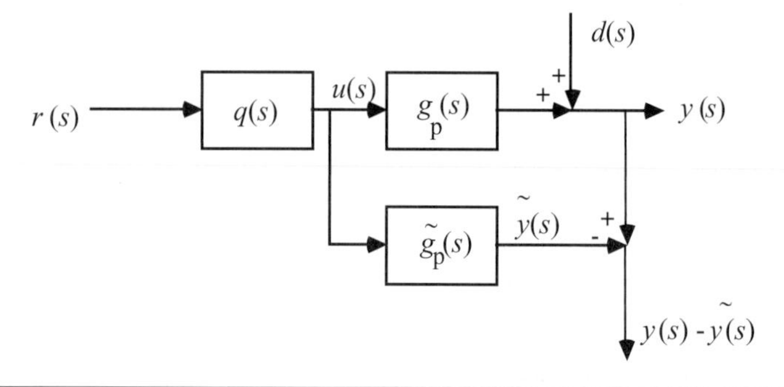

Figure 8–10 Incorporating the process disturbance.

Notice that Figure 8–10 shows the calculation of model uncertainty (which includes unmeasured disturbances). This information can now be used by the controller, to compensate for the model uncertainty. This creates a feedback system, as shown in Figure 8–11.

Figure 8–11 is also known as the IMC structure, which is discussed in depth in the next sections.

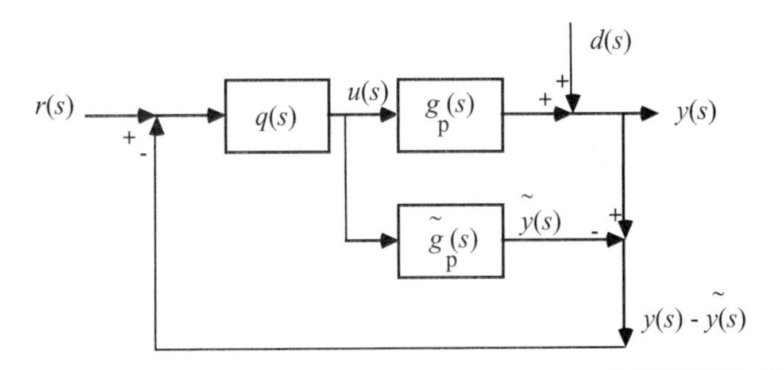

Figure 8–11 Feedback structure derived from the open-loop structure.

8.6 IMC Background

The main advantage to IMC is that it provides a transparent framework for control-system design and tuning. As we show in this chapter, the IMC control structure can be formulated in the standard feedback control structure. For many processes, this standard feedback control structure will result in a PID controller (sometimes cascaded with a first-order lag). This is pleasing because we can use standard equipment and algorithms (i.e., PID controllers) to implement an "advanced" control concept.

The IMC design procedure is exactly that of the open-loop "control" design procedure developed in Section 8.3. Remember that a factorization of the process model was performed so that the resulting controller would be stable. *If the controller is stable and the process is stable, then the overall controlled system is stable.* This is true simply because if two transfer functions are stable, then the transfer functions cascaded together (multiplied) are stable. This is a nice result because in a standard feedback control formulation, the controller and the process can each be stable, yet the feedback system may be unstable. The restriction of this design procedure is that the process must be stable. An extension to unstable processes is covered in the Chapter 9.

Although the IMC design procedure is identical to the open-loop "control" design procedure, the implementation of IMC results in a feedback system. Thus, IMC is able to compensate for disturbances and model uncertainty, while open-loop "control" is not. Note that the internal model controller must be detuned to assure stability if there is model uncertainty.

8.7 The IMC Structure

The IMC structure is shown in Figure 8–12. The distinguishing characteristic of this structure is the process model, which is in parallel with the actual process (plant). Note that (~) is generally used to represent signals associated with the model. Other literature sources may use a subscript (such as *m*) to represent the model. Figure 8–13 illustrates that both the controller and model exist as computer computations; it is convenient to treat them separately for design and analysis.

A list of transfer function variables shown in the IMC block diagram are given below.

$d(s) =$ disturbance

$\tilde{d}(s) =$ estimated disturbance

$g_p(s) =$ process

$\tilde{g}_p(s) =$ process model

$q(s) =$ internal model controller

$r(s) =$ setpoint

$\tilde{r}(s) =$ modified setpoint (corrects for model error and disturbances)

$u(s) =$ manipulated input (controller output)

$y(s) =$ measured process output

$\tilde{y}(s) =$ model output

Notice that the feedback signal is

$$\tilde{d}(s) = (g_p(s) - \tilde{g}_p(s))u(s) + d(s) \tag{8.36}$$

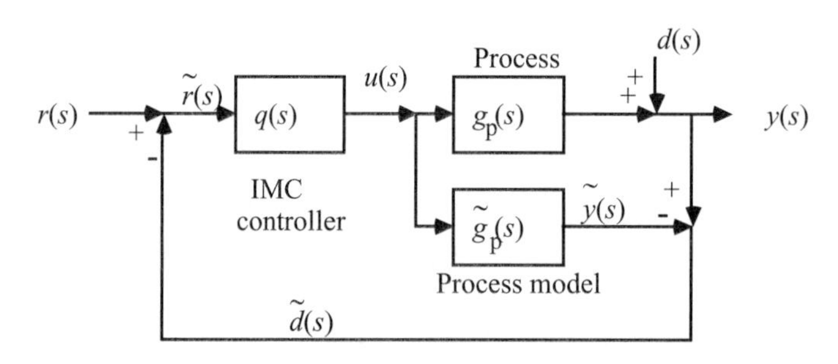

Figure 8–12 The internal model control structure.

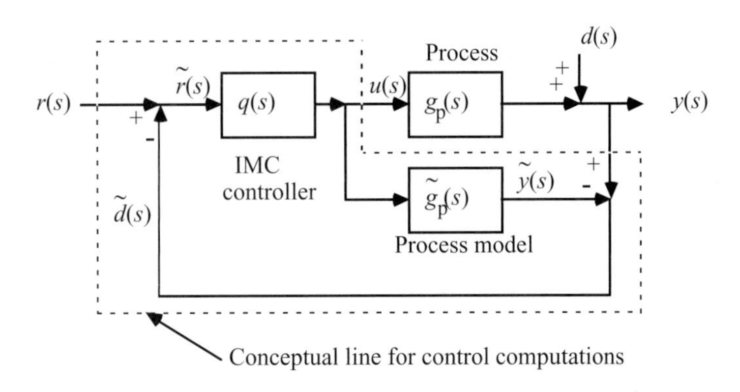

Figure 8–13 The IMC strategy. The dotted line indicates the calculations performed by the model-based controller.

The signal to the controller is

$$\tilde{r}(s) = r(s) - \tilde{d}(s) = r(s) - (g_p(s) - \tilde{g}_p(s))u(s) - d(s) \tag{8.37}$$

Consider now some limiting cases.

Perfect Model, No Disturbances

If the model is perfect ($\tilde{g}_p(s) = g_p(s)$) and there are no disturbances ($d(s) = 0$), then the feedback signal is zero. The relationship between $r(s)$ and $y(s)$ is then

$$y(s) = g_p(s)q(s)r(s) \tag{8.38}$$

Notice that this is the same relationship that we get for an open-loop control system design.

Why is this nice? If the controller, $q(s)$, is stable and the process, $g_p(s)$, is stable, then the closed-loop system is stable. We developed a design procedure that yields a stable, physically realizable controller in Section 8.3.

Recall that a standard feedback controller could actually destabilize a process if we did not correctly choose the tuning parameters. An analysis of the poles of the closed-loop transfer function must be performed to determine the stability of standard feedback controllers.

Perfect Model, Disturbance Effect

If the model is perfect ($\tilde{g}_p(s) = g_p(s)$) and there is a disturbance, then the feedback signal is

$$\tilde{d}(s) = d(s)$$

This illustrates that feedback is needed because of unmeasured disturbances entering a process.

Model Uncertainty, No Disturbances

If there are no disturbances [$d(s) = 0$] but there is model uncertainty ($\tilde{g}_p(s) \neq g_p(s)$), which is always the case in the real world, then the feedback signal is

$$\tilde{d}(s) = (g_p(s) - \tilde{g}_p(s))u(s)$$

This illustrates that feedback is needed because of model uncertainty.

The closed-loop relationship is

$$y(s) = \frac{g_p(s)q(s)}{1 + q(s)\left(g_p(s) - \tilde{g}_p(s)\right)} r(s) + \left[\frac{1 - \tilde{g}_p(s)q(s)}{1 + q(s)\left(g_p(s) - \tilde{g}_p(s)\right)}\right] d(s)$$

Recapitulating, the reasons for feedback control include the following:

- Unmeasured disturbances
- Model uncertainty
- Faster response than the open-loop system (with a static controller)
- Closed-loop stability of open-loop unstable system

The primary disadvantage of IMC is that it does not guarantee stability of open-loop unstable systems. The procedure detailed in Chapter 9 handles these systems.

8.8 The IMC Design Procedure

The IMC design procedure for SISO systems is identical to the design procedure that we developed for open-loop controller design earlier. The assumption we are making is that the model is perfect, so the relationship between the output, y, and the setpoint, r, is given by Equation (8.1). Model uncertainty is handled by adjusting the "filter factor" for *robustness* (tolerance of model uncertainty) and speed of response. The IMC design procedure consists of the following four steps.

1. Factor the process model into invertible ("good stuff") and noninvertible ("bad stuff"—time delays and RHP zeros) elements (generally, an all-pass factorization will be used).

$$\tilde{g}_p(s) = \tilde{g}_{p+}(s)\tilde{g}_{p-}(s). \tag{8.39}$$

This factorization is performed so that the resulting controller will be stable.

2. Form the idealized IMC controller. The ideal internal model controller is the inverse of the invertible portion of the process model ("good stuff").

$$\tilde{q}(s) = \tilde{g}_{p-}^{-1}(s) \tag{8.40}$$

3. Add a filter to make the controller proper. A transfer function is proper if the order of the denominator polynomial is at least as high as the numerator polynomial.

$$q(s) = \tilde{q}(s)f(s) = \tilde{g}_{p-}^{-1}(s)f(s) \tag{8.41}$$

If it is most desirable to track step setpoint changes, the filter transfer function usually has the form

$$f(s) = \frac{1}{(\lambda s + 1)^n} \tag{8.42}$$

and n is chosen to make the controller proper (or semiproper). If it is most desirable to track ramp setpoint changes (often used for batch reactors or transition control problems), then

$$f(s) = \frac{n\lambda s + 1}{(\lambda s + 1)^n}$$

4. Adjust the filter-tuning parameter to vary the speed of response of the closed-loop system. If the λ is "small," the closed loop system is "fast," if λ is "large," the closed-loop system is more *robust* (insensitive to model error).

The student should note that the factorization performed in Equation (8.39) is used only for controller design. The process model that is simulated in parallel with the process is still the full model, $\tilde{g}_p(s)$.

If the process model is *perfect*, then we can easily calculate what the output response to a setpoint change will be. Substituting Equation (8.40) into Equation (8.38), we find

$$y(s) = g_p(s)q(s)r(s) = g_p(s)\tilde{q}(s)f(s)r(s) = g_p(s)\tilde{g}_{p-}^{-1}(s)f(s)r(s) \tag{8.43}$$

If the model is *perfect*, then

$$g_p(s) = \tilde{g}_p(s) = \tilde{g}_{p+}(s)\,\tilde{g}_{p-}(s) \tag{8.44}$$

and we can substitute Equation (8.44) into Equation (8.43) to find

$$y(s) = \tilde{g}_{p+}(s)\tilde{g}_{p-}(s)\tilde{g}_{p-}^{-1}f(s)r(s) \tag{8.45}$$

which yields

$$y(s) = \tilde{g}_{p+}(s)f(s)r(s) \tag{8.46}$$

Equation (8.46) indicates that the bad stuff must appear in the output response. That is, if the open-loop process has a RHP zero (inverse response), then the closed-loop system must exhibit inverse response. Also, if the process has dead time, then dead time must appear in the closed-loop response. *Please remember that Equation (8.46) only holds for the case of a perfect model.*

The most common process model is a first-order plus time-delay transfer function. The design procedure for this system is shown in the next example.

Example 8.5: First-Order + Dead Time Process

Consider a first-order + time-delay model:

$$\tilde{g}_p(s) = \frac{\tilde{k}_p e^{-\tilde{\theta}s}}{\tilde{\tau}_p s + 1}$$

Using the four-step design procedure, first factor out the noninvertible elements,

$$\tilde{g}_p(s) = \tilde{g}_{p+}(s)\tilde{g}_{p-}(s) = e^{-\tilde{\theta}s} \cdot \frac{\tilde{k}_p}{\tilde{\tau}_p s + 1}$$

Then form the idealized IMC controller,

$$\tilde{q}(s) = \tilde{g}_{p-}^{-1}(s) = \frac{\tilde{\tau}_p s + 1}{\tilde{k}_p}$$

and add a filter to make the controller proper

$$q(s) = \tilde{q}(s)f(s) = \tilde{g}_{p-}^{-1}(s)f(s) = \frac{\tilde{\tau}_p s + 1}{\tilde{k}_p} \cdot \frac{1}{\lambda s + 1} = \frac{1}{\tilde{k}_p} \cdot \frac{\tilde{\tau}_p s + 1}{\lambda s + 1}$$

Once again, the controller is of lead-lag form. Finally, adjust λ for response speed and robustness. The closed-loop response (assuming a perfect model) to a setpoint change is

$$y(s) = \tilde{g}_{p+}(s)f(s)r(s) = \frac{e^{-\tilde{\theta}s}}{\lambda s + 1} r(s)$$

For a step setpoint change of magnitude R,

$$y(t) = 0, \qquad 0 \le t < \theta$$

$$y(t) = R\left(1 - e^{-(t-\tilde{\theta})/\lambda}\right), \qquad t \ge \theta$$

Numerical Example

Consider a first-order + time-delay process with a time constant of 10 seconds, a time delay of 5 seconds, and a gain of 1%/%. The output and manipulated variable responses for various values of λ are shown in Figure 8–14.

Notice that the IMC procedure has effectively "compensated" for the time delay. The time delay was not removed, but the controller does not "expect" results from a manipulated variable move until "after" the time delay because the model is integrated "in-parallel" with the process.

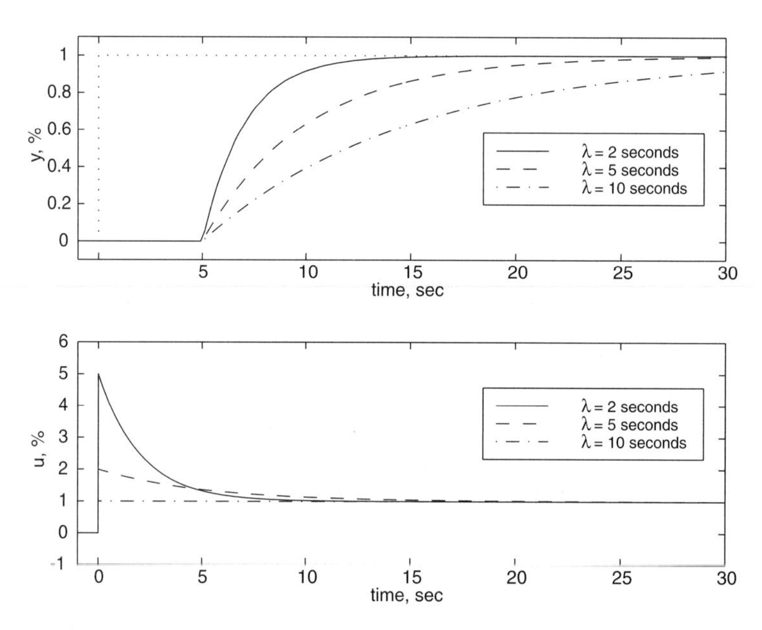

Figure 8–14 Response as a function of λ for IMC of a first-order + time-delay process.

Example 8.6: Second-Order with an RHP Zero

Consider the following transfer function, which has an RHP zero at 1/9 min^{-1}:

$$\tilde{g}_p(s) = \frac{-9s+1}{(15s+1)(3s+1)}$$

Using all-pass factorization of the model, we find

$$\tilde{g}_p(s) = \tilde{g}_{p+}(s)\tilde{g}_{p-}(s) = \frac{-9s+1}{9s+1} \cdot \frac{(9s+1)}{(15s+1)(3s+1)}$$

Forming the idealized controller,

$$\tilde{q}(s) = \tilde{g}_{p-}^{-1}(s) = \frac{(15s+1)(3s+1)}{(9s+1)}$$

Add the filter to make the controller semiproper:

$$q(s) = \tilde{q}(s)f(s) = \tilde{g}_{p-}^{-1}(s)f(s) = \frac{(15s+1)(3s+1)}{(9s+1)} \cdot \frac{1}{\lambda s+1}$$

For a perfect model, the output response is

$$y(s) = \tilde{g}_{p+}(s)f(s)r(s) = \frac{-9s+1}{(9s+1)(\lambda s+1)} r(s)$$

and the manipulated variable response is

$$u(s) = q(s)r(s) = \frac{(15s+1)(3s+1)}{(9s+1)(\lambda s+1)} r(s)$$

The output and manipulated responses for a step setpoint change is shown in Figure 8–15 as a function of λ. Notice that as λ is decreased, the inverse response becomes more pronounced—the inverse response can not be removed by a stable controller.

It should be noted that the initial-value theorem can be used to determine the manipulated variable action that occurs immediately after a step setpoint change.

We have shown how to factor models with time delays and RHP zeros and have used several examples to illustrate the effect of the tuning parameter (or filter time constant), λ, on the output response to setpoint changes if the model is perfect. The IMC design procedure is identical to the open-loop control design procedure presented in Section 8.3. Although the control-system design procedure is the same, the implementation of IMC is much different, since it incorporates feedback to compensate for model uncertainty or disturbances.

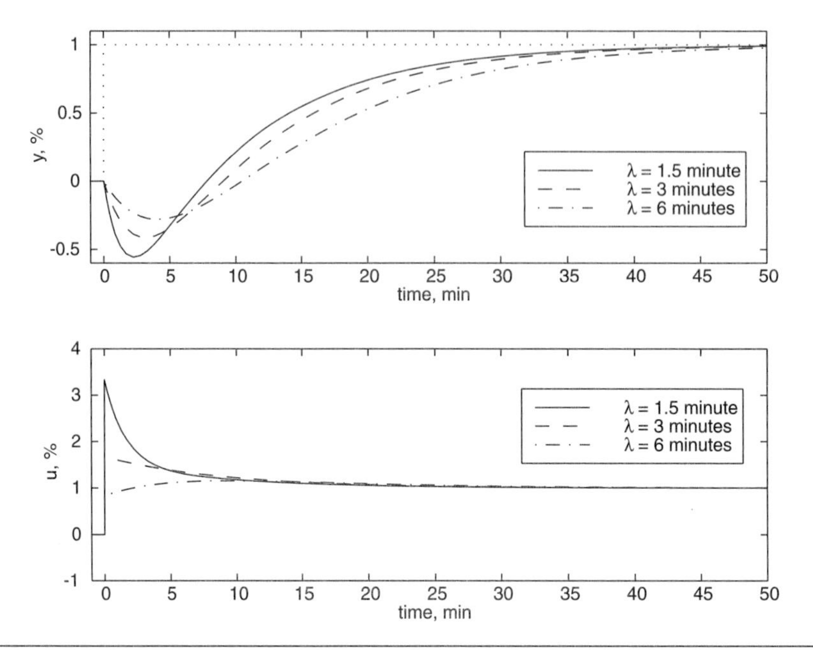

Figure 8–15 Input and output response as a function of λ when all-pass filter is used.

8.9 Effect of Model Uncertainty and Disturbances

For controller design purposes, we assumed that the model was perfect. In practice there is model uncertainty. Model uncertainty will be handled by detuning the filter parameter, λ, for robustness.

The Effect of Model Uncertainty on Setpoint Response

Consider a *setpoint change only*; then

$$y(s) = \frac{g_p(s)q(s)}{1 + q(s)\left(g_p(s) - \tilde{g}_p(s)\right)} r(s)$$

We can see immediately that if the model is perfect [$g_p(s) = \tilde{g}_p(s)$], then we have the open-loop "control" formulation. When the model is not perfect, then the closed-loop response is much more complicated and can even be unstable if the filter [$f(s)$] is not detuned sufficiently.

Disturbance Rejection

Consider a *disturbance only*; then

$$y(s) = \left[\frac{1 - \tilde{g}_p(s)q(s)}{1 + q(s)\left(g_p(s) - \tilde{g}_p(s)\right)} \right] d(s) \tag{8.47}$$

Assume a load disturbance transfer function of the following form

$$d(s) = g_d(s)l(s)$$

Indeed, often the load disturbance is really a disturbance to the process input, so the disturbance transfer function has dynamic behavior similar to the process transfer function. It turns out that a controller that is tightly tuned and works well for setpoint changes may fail miserably on handling disturbance rejection. This is shown by the following example.

Example 8.7: First-Order Process

Here we consider a first-order process model transfer function, with a time unit of minutes

$$\tilde{g}_p(s) = \frac{1}{(10s + 1)}$$

For simplicity, we assume a perfect model ($g_p(s) = \tilde{g}_p(s)$) and that the disturbance transfer function is the same as the process ($g_d(s) = \tilde{g}_p(s)$). The IMC controller design procedure leads to the response for a step setpoint change shown in Figure 8–16, with $\lambda = 2$ min. Although the response is unrealistically fast (λ is too small compared with the process time constant), we wish to show that good setpoint tracking does not necessarily lead to good disturbance rejection.

Disturbance rejection results are shown in Figure 8–17. Notice that there is a very slow response to the step load disturbance. This result is perhaps the greatest criticism of the standard IMC design procedure; it tends to lead to slow responses to load disturbances that occur at the process input. An improved IMC procedure for disturbance rejection is presented in the next section.

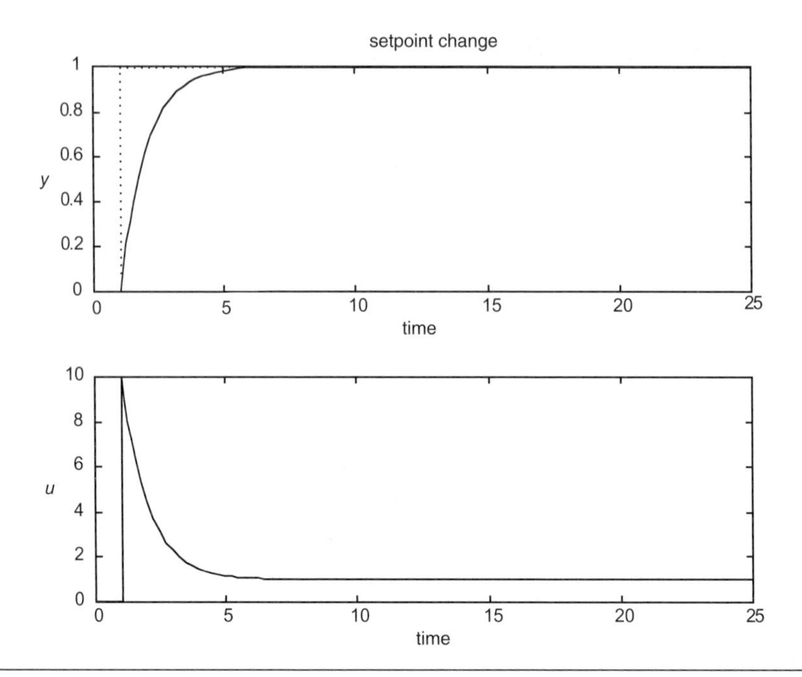

Figure 8–16 Setpoint response for Example 8.7, with $\lambda = 2$ min.

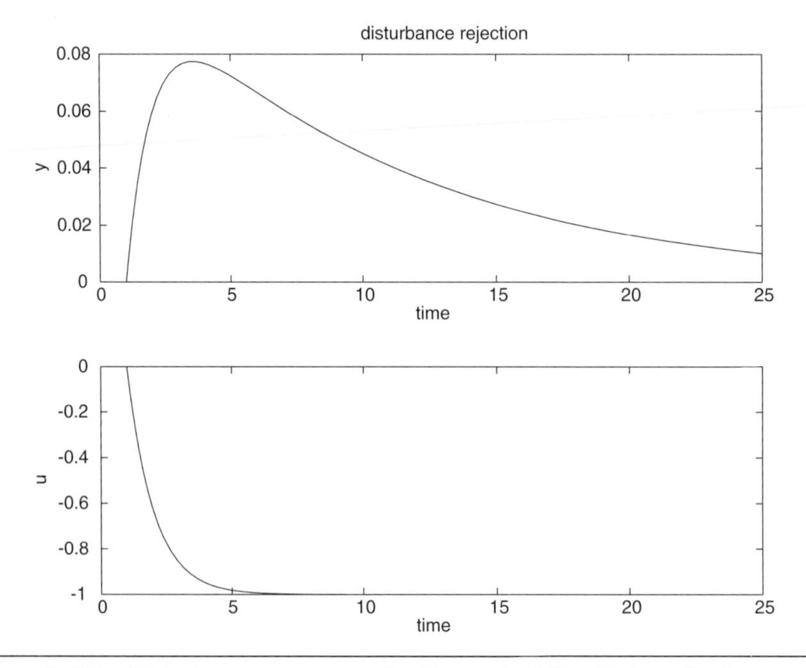

Figure 8–17 Disturbance response for Example 8.7, with $\lambda = 2$ min.

8.10 Improved Disturbance Rejection Design

We can achieve improved disturbance rejection by specifying a different IMC filter design procedure. We perform the same factorization of the process model as before. The difference is in the selection of the IMC filter transfer function. Here, rather than using a filter of the form

$$f(s) = \frac{1}{(\lambda s + 1)^n}$$

we use a filter with the form

$$f(s) = \frac{\gamma s + 1}{(\lambda s + 1)^n} \tag{8.48}$$

where γ is selected to achieve good disturbance rejection. In practice, γ will be selected to cancel a slow disturbance time constant. Consider the closed-loop transfer function for disturbance rejection

$$y(s) = \left[\frac{1 - \tilde{g}_p(s)q(s)}{1 + q(s)\left(g_p(s) - \tilde{g}_p(s)\right)}\right] g_d l(s) \tag{8.49}$$

In the case of a perfect model, this results in

$$y(s) = [1 - \tilde{g}_p(s)q(s)]g_d l(s) \tag{8.50}$$

The controller, using the new filter form, is

$$q(s) = \tilde{g}_{p-}^{-1}(s)f(s) = \tilde{g}_{p-}^{-1}(s)\frac{\gamma s + 1}{(\lambda s + 1)^n} \tag{8.51}$$

so the output response is

$$y(s) = \left[1 - \tilde{g}_p(s)\tilde{g}_{p-}^{-1}(s)\frac{\gamma s + 1}{(\lambda s + 1)^n}\right] g_d(s)l(s)$$

$$y(s) = \left[1 - \tilde{g}_{p+}(s)\frac{\gamma s + 1}{(\lambda s + 1)^n}\right] g_d(s)l(s) \tag{8.52}$$

$$y(s) = \left[\frac{(\lambda s + 1)^n - \tilde{g}_{p+}(s)(\gamma s + 1)}{(\lambda s + 1)^n}\right] g_d(s)l(s)$$

Although it is hard to tell from this general expression, γ should be selected to cancel the slow time constant associated with $g_d(s)$. This is shown by way of an example for a first-order process.

Example 8.8: First-Order Process, Improved Disturbance Rejection Design

For a first-order process model,

$$\tilde{g}_p(s) = \frac{\tilde{k}_p}{(\tilde{\tau}_p s + 1)}$$

The IMC controller, using the proposed filter design, is

$$q(s) = \frac{\tilde{\tau}_p s + 1}{\tilde{k}_p} \frac{(\gamma s + 1)}{(\lambda s + 1)^2}$$

Assuming a perfect model and load disturbance transfer function equal to the process transfer function, Equation (8.52) becomes

$$y(s) = \left[\frac{(\lambda s + 1)^2 - (\gamma s + 1)}{(\lambda s + 1)^2}\right] \frac{\tilde{k}_p}{\tilde{\tau}_p s + 1} l(s)$$

$$y(s) = \left[\frac{\lambda^2 s^2 + 2\lambda s + 1 - (\gamma s + 1)}{(\lambda s + 1)^2}\right] \frac{\tilde{k}_p}{\tilde{\tau}_p s + 1} l(s)$$

$$y(s) = \left[\frac{\lambda^2 s^2 + (2\lambda - \gamma) s}{(\lambda s + 1)^2}\right] \frac{\tilde{k}_p}{\tilde{\tau}_p s + 1} l(s)$$

$$y(s) = \frac{(2\lambda - \gamma) s \left[\dfrac{\lambda^2}{2\lambda - \gamma} s + 1\right]}{(\lambda s + 1)^2} \frac{\tilde{k}_p}{\tilde{\tau}_p s + 1} l(s)$$

If we select $\lambda^2/(2\lambda - \gamma)$ to cancel the process model time constant, $\tilde{\tau}_p$, we find

$$\gamma = \frac{2\lambda\tilde{\tau}_p - \lambda^2}{\tilde{\tau}_p}$$

Numerical Example

Here we consider again the first-order process model transfer function from Example 8.7,

$$\tilde{g}_p(s) = \frac{1}{(10s + 1)}$$

and assume that the disturbance transfer function is equal to the process transfer function (this is equivalent to assuming that the load disturbance occurs at the process input).

$$\gamma = \frac{20\lambda - \lambda^2}{10}$$

which leads to the closed-loop results for a unit step disturbance, shown in Figure 8–18.

It should be noted that the improved disturbance rejection design does not lead to detrimental performance for setpoint changes. In Figure 8–19, the new filter design also leads to faster performing setpoint responses. The major disadvantage is the increased manipulated variable action. In practice, the λ value should be detuned for improved robustness.

8.11 Manipulated Variable Saturation

The reader should note that if the actual manipulated variable signal is used by the model, then manipulated variable *saturation* is not a problem because the system becomes open-loop and the model prediction is the same as the process output (if there are no disturbances and the model is perfect). In conventional PID controllers, special precautions must be

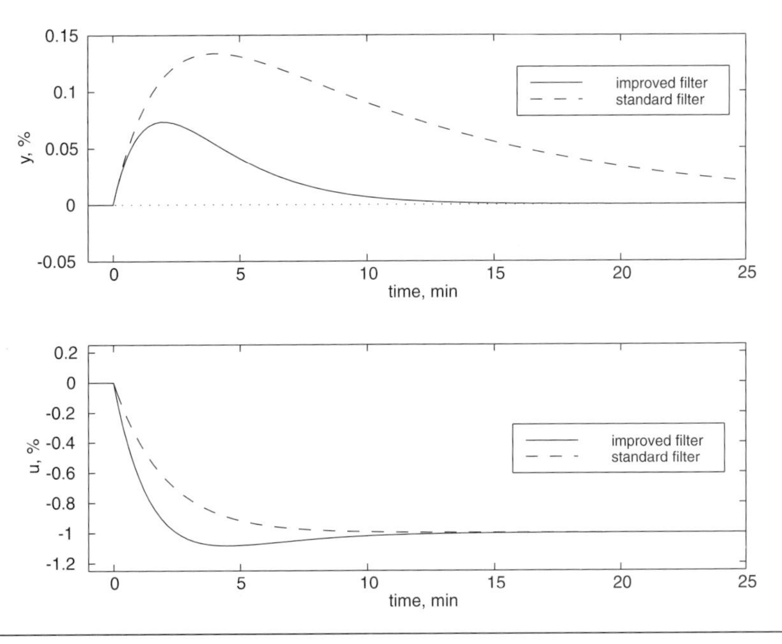

Figure 8–18 Improved disturbance response for Example 8.8, with $\lambda = 2$ min. A comparison with the standard IMC filter design.

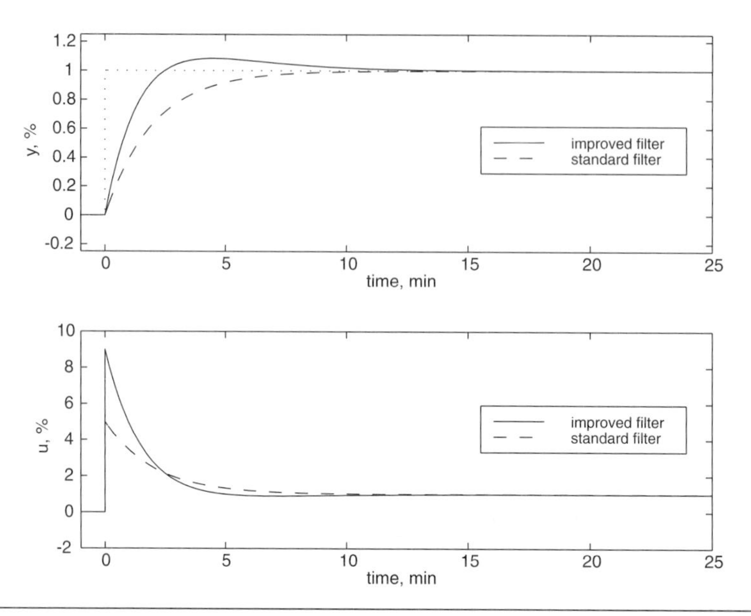

Figure 8–19 Setpoint response for Example 8.8, with $\lambda = 2$ min. A comparison of design for improved disturbance rejection with the standard IMC filter design.

taken to prevent *reset* or *integral windup* from occurring when manipulated variables hit constraints. This phenomenon is discussed more fully in Chapter 11.

The IMC block diagram for the case where the actual manipulated variable value that is implemented on the process is also used by the model is shown in Figure 8–20. Integral "wind-up" will not be a problem with this implementation. Figure 8–21 shows an IMC implementation where the manipulated variable value implemented on the process is constrained, while the process model uses the unconstrained value. Integral wind-up may be a problem in this case. It is very important when implementing IMC to use the actual manipulated input to the model in the control law.

8.12 Summary

The IMC design procedure is exactly the same as the open-loop control design procedure. Unlike open-loop control, the IMC structure compensates for disturbances and model uncertainty. The IMC tuning (filter) factor, λ, is used to detune for model uncertainty. It should be noted that the standard IMC design procedure is focused on setpoint responses, but good setpoint responses do not guarantee good disturbance rejection, particularly for disturbances that occur at process inputs. A modification of the IMC design procedure was

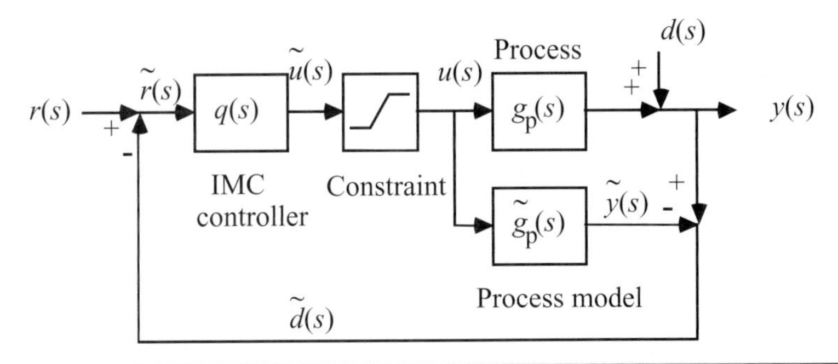

Figure 8–20 The actual manipulated variable value is "measured" and used by the process model. (Correct implementation)

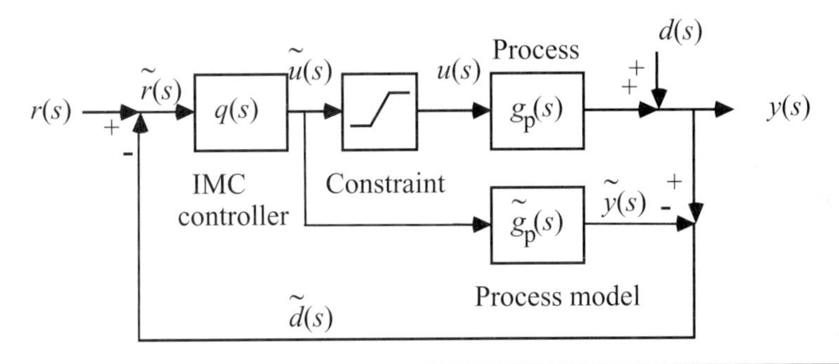

Figure 8–21 The manipulated variable value used by the process model is different from the value actually implemented on the process. (Undesirable implementation)

developed to improve input disturbance rejection. Tolerance of model uncertainty is called *robustness* and was discussed in more detail in Chapter 7.

Like open-loop control, the disadvantage compared with standard feedback control is that IMC does not handle integrating or open-loop unstable systems. In Chapter 9, we develop a procedure, based on IMC, to form a standard feedback control system that can handle open-loop unstable systems. We call this IMC-based PID.

Terms

All-pass: Property of a transfer function (detailed in Chapter 7).

Factorization: Factoring a transfer function into invertible (good stuff) and non-invertible (bad stuff) portions.

Minimum phase: No RHP zeros or time delays.

Nonminimum phase: Contains RHP zeros and/or time delays.

Physical realizability: A controller can be physically implemented if it is proper.

Proper: A transfer function is proper if the order of the denominator is at least as great as the order of the numerator. If they are exactly the same order, the transfer function is *semiproper*. If the order of the denominator is greater than the numerator, the transfer function is *strictly proper*.

Summary of Internal Model Control System Design Procedure

1. Develop a process model.

$$\tilde{g}_p(s)$$

2. Factor the process model into invertible (good) and noninvertible (bad) portions,

$$\tilde{g}_p(s) = \tilde{g}_{p+}(s)\tilde{g}_{p-}(s)$$

usually using an all-pass factorization.

3. Invert the invertible portion of the process model (the good stuff) and cascade with a filter that makes the controller $q(s)$ proper.

$$q(s) = \tilde{g}_{p-}^{-1}(s)f(s)$$

For a focus on step setpoint changes, the following form is often used:

$$f(s) = \frac{1}{(\lambda s + 1)^n}$$

Although not discussed in detail in the chapter, for good tracking of ramp setpoint changes, the following form for the filter should be used:

$$f(s) = \frac{n\lambda s + 1}{(\lambda s + 1)^n}$$

For good rejection of step input load disturbances, the form used is

$$f(s) = \frac{\gamma s + 1}{(\lambda s + 1)^n}$$

where γ is selected to cancel the slow process time constant.

4. Implement in the form of Figure 8–20 to handle constraints on the manipulated input.

References

The improved IMC disturbance rejection procedure was suggested by Braatz and co-workers: Horn, I. G., J. A. Arulandu, C. J. Gombas, J. G. VanAntwerp, and R. D. Braatz, "Improved Filter Design in Internal Model Control," *Ind. Eng. Chem. Res.*, **35**(10), 3437–3441 (1996).

A complete IMC design procedure is presented in the following: Morari, M., and E. Zafiriou, *Robust Process Control*, Prentice-Hall, Upper Saddle River, NJ (1989).

Student Exercises

Problems 1–5 require analytical solutions, Problems 6 and 8 are SIMULINK problems.

1. Design an IMC for the following process model:

$$\tilde{g}_p(s) = \frac{\tilde{k}_p(-\beta s + 1)}{(\tilde{\tau}_{p1}s + 1)(\tilde{\tau}_{p2}s + 1)}$$

Assume that the all-pass factorization technique is used. Assuming the model is perfect, sketch the transient response of $y(t)$ to a step setpoint change in $r(t)$.

2. Consider the following IMC block diagram. Find the closed-loop transfer functions relating the load [$l(s)$] and the setpoint [$r(s)$] to the output [$y(s)$]. Assume that the model is not perfect.

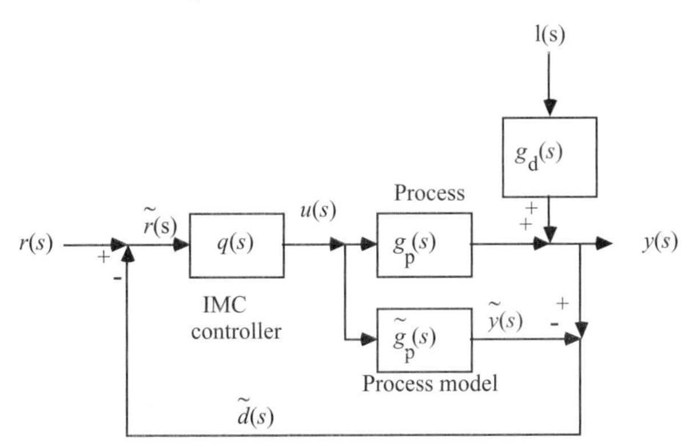

3. A biochemical reactor with several components has the following linearized model, which relates the biomass concentration to the dilution rate:

$$\tilde{g}_p(s) = \frac{1.75(-3s+1)(-5s+1)e^{-1.25s}}{(10s+1)(4s+1)^2}$$

a. Design an IMC, $q(s)$, for this process, assuming that (i) an all-pass filter is used, and (ii) the controller is strictly proper (order of the denominator is greater than the order of the numerator).

b. Qualitatively sketch the type of closed-loop response to a setpoint change that you would expect if the model is perfect.

4. A process has the following relationship between the input and output:

$$\tilde{g}_p(s) = \frac{-10(-3s+1)(-s+1)e^{-4s}}{(12s+1)(3s+1)^2}$$

Design an IMC controller for this process. Let $q(s)$ be *semiproper* (order of denominator is equal to the order of numerator) and use an all-pass factorization. Find the closed-loop transfer function assuming a perfect model. Sketch the closed-loop response to a step setpoint change.

5. Consider the closed-loop response for IMC when the model is *not* perfect. Show that there is no offset for a setpoint change for the following process model and actual process transfer functions. Also, find the minimum value of λ that assures closed-loop stability for this system.

$$\tilde{g}_p(s) = \frac{2}{5s+1}$$

$$g_p(s) = \frac{1.5(-s+1)}{(s+1)(4s+1)}$$

6. Design an IMC for the following process:

$$\tilde{g}_p(s) = \frac{\tilde{k}_p(-\beta s+1)e^{-\tilde{\theta}s}}{(\tilde{\tau}_{p1}s+1)(\tilde{\tau}_{p2}s+1)}$$

where $\tilde{k}_p = 2°\text{F/psig}$, $\tilde{\tau}_{p1} = 10\,\text{min}$, $\tilde{\tau}_{p2} = 3\,\text{min}$, $\tilde{\theta} = 5\,\text{min}$, and $\beta = 6\,\text{min}$.

a. For this part, do not factor out the right-half-plane zero to form the controller. Perform simulations to show that the manipulated variable grows unbounded when a step setpoint change is made, although good performance of the output is achieved.

For the following parts, assume that the "all-pass" factorization technique is used.

b. Assuming the model is perfect, plot the transient response of $y(t)$ to a unit step setpoint change in $r(t)$ as a function of λ. (Show your control loop diagram.)

c. Discuss how dead-time uncertainty degrades the closed-loop performance (show several curves on one plot) and show how the sensitivity depends on the filter constant, λ.

d. Show how gain uncertainty degrades the closed-loop performance (show several curves on one plot) and show how the sensitivity depends on the filter time constant, λ.

e. Now, assume that there is no setpoint change but that there is a step disturbance. Assume that the disturbance transfer function is equal to the process transfer function. Compare and contrast the response of the output with the open-loop control response.

For the following parts, assume that the all-pass factorization technique is not used.

f. Assuming the model is perfect, plot the transient response of $y(t)$ to a unit step setpoint change in $r(t)$ as a function of λ. (Show your control loop diagram.)

g. Discuss how dead-time uncertainty degrades the closed-loop performance (show several curves on one plot) and show how the sensitivity depends on the filter constant, λ.

h. Show how gain uncertainty degrades the closed-loop performance (show several curves on one plot) and show how the sensitivity depends on the filter time constant, λ.

i. Now, assume that there is no setpoint change, but that there is a step disturbance. Assume that the disturbance transfer function is equal to the process transfer function. Compare and contrast the response of the output with the open-loop control response.

7. Packed-bed reactors often exhibit "wrong way" (inverse response) behavior. You are responsible for the control-system design for a packed-bed reactor that has the following step response behavior, where a step decrease in steam valve position was made at $t = 5$ minutes.

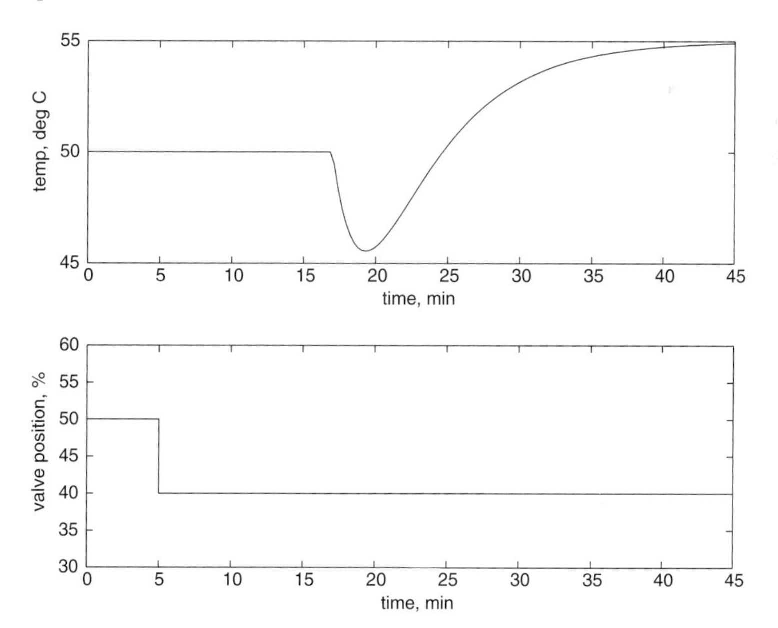

You have developed the following process model (the time unit is minutes):

$$\tilde{g}_p(s) = \frac{-0.5(-10s+1)e^{-12s}}{(5s+1)(3s+1)}$$

a. What are the units for the process gain?
b. Design an IMC for this process. Use the all-pass factorization for the RHP zero, and assume that $q(s)$ is semiproper (numerator order in s is equal to the denominator order in s).
c. Assuming a perfect model, plot qualitatively how the temperature will respond to a step setpoint change of 1°C.
d. It is desirable to make certain that the control valve position, immediately after a 10°C setpoint change, does not move more than 25%. What is the smallest value of λ that you can use? Show your work.

8. Consider the following isothermal chemical reactor, where the dilution rate (feed flow rate per unit volume of reactor) is manipulated to achieve a desired concentration of product.

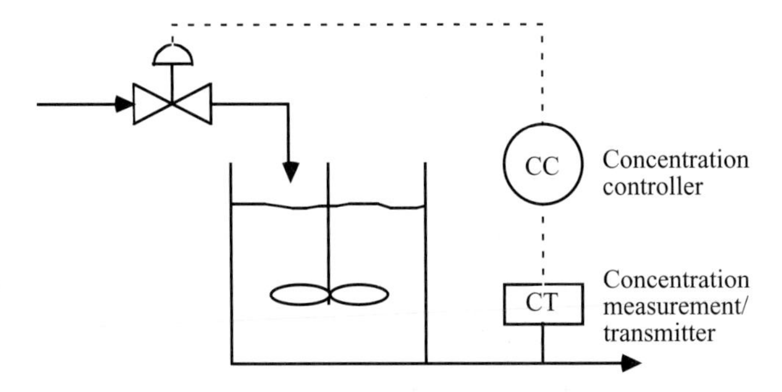

The process model is

$$\tilde{g}_p(s) = \frac{0.5848(-0.3549s+1)e^{-0.5s}}{0.1858s^2+0.8627s+1}$$

which has a RHP zero (inverse response) and a time delay (due to the concentration measurement device).

Design an IMC for this process.

a. *For this part*, do not factor out the RHP zero to form the controller. Perform simulations to show that the manipulated variable grows unbounded when a step setpoint change is made, although good performance of the output is achieved.

For the following parts, assume that the all-pass factorization technique is used.

b. Assuming the model is perfect, plot the transient response of $y(t)$ to a unit step setpoint change in $r(t)$ as a function of λ. (Show your control loop diagram.)

c. Discuss how dead-time uncertainty degrades the closed-loop performance (show several curves on one plot) and show how the sensitivity depends on the filter constant, λ.

d. Show how gain uncertainty degrades the closed-loop performance (show several curves on one plot) and show how the sensitivity depends on the filter time constant, λ.

e. Now, assume that there is no setpoint change but that there is a step disturbance. Assume that the disturbance transfer function is equal to the process transfer function. Compare and contrast the response of the output with the open-loop control response.

For the following parts, assume that the all-pass factorization technique is not used.

f. Assuming the model is perfect, plot the transient response of $y(t)$ to a unit step setpoint change in $r(t)$ as a function of λ. (Show your control loop diagram.)

g. Discuss how dead-time uncertainty degrades the closed-loop performance (show several curves on one plot) and show how the sensitivity depends on the filter constant, λ.

h. Show how gain uncertainty degrades the closed-loop performance (show several curves on one plot) and show how the sensitivity depends on the filter time constant, λ.

i. Now, assume that there is no setpoint change but that there is a step disturbance. Assume that the disturbance transfer function is equal to the process transfer function. Compare and contrast the response of the output with the open-loop control response.

Appendix 8.1: Derivation of Closed-Loop Relationships for IMC

Refer again to Figure 8–12. Recall that the feedback signal is

$$\tilde{d}(s) = (g_p(s) - \tilde{g}_p(s))u(s) + d(s)$$

and the signal to the controller is

$$\tilde{r}(s) = r(s) - \tilde{d}(s) = r(s) - (g_p(s) - \tilde{g}_p(s))u(s) - d(s)$$

The controller output or process input is

$$u(s) = q(s)\tilde{r}(s) = q(s)[r(s) - (g_p(s) - \tilde{g}_p(s))u(s) - d(s)]$$

which can be rearranged to

$$u(s) = \frac{q(s)r(s) - q(s)d(s)}{1 + q(s)\left(g_p(s) - \tilde{g}_p(s)\right)}$$

Now, we realize that

$$y(s) = g_p(s)u(s) + d(s)$$

and we find

$$y(s) = \frac{g_p(s)q(s)r(s) - g_p(s)q(s)d(s)}{1 + q(s)\left(g_p(s) - \tilde{g}_p(s)\right)} + d(s)$$

or

$$y(s) = \frac{g_p(s)q(s)}{1 + q(s)\left(g_p(s) - \tilde{g}_p(s)\right)} r(s) + \left[1 - \frac{g_p(s)q(s)}{1 + q(s)\left(g_p(s) - \tilde{g}_p(s)\right)}\right] d(s)$$

which can be written

$$y(s) = \frac{g_p(s)q(s)}{1 + q(s)\left(g_p(s) - \tilde{g}_p(s)\right)} r(s) + \left[\frac{1 - \tilde{g}_p(s)q(s)}{1 + q(s)\left(g_p(s) - \tilde{g}_p(s)\right)}\right] d(s)$$

CHAPTER 9

The IMC-Based PID Procedure

In Chapter 8 we developed a transparent framework for control-system design: the IMC structure. One nice thing about the IMC procedure is that it results in a controller with a single tuning parameter, the IMC filter (λ). For a system that is "minimum phase," λ is equivalent to a closed-loop time constant (the "speed of response" of the closed-loop system). Although the IMC procedure is clear and IMC is easily implemented, the most common industrial controller is still the PID controller. The purpose of this chapter is to show that the IMC block diagram can be rearranged to the form of a standard feedback control diagram. We find that the IMC law, for a number of common process transfer functions, is equivalent to PID-type feedback controllers.

After studying this chapter, the reader should be able to do the following:

- Design an internal model controller, then find the equivalent feedback controller (IMC-based PID) in standard form; derive and use the results presented in Table 9–1
- Use an approximation for time delays in order to find a PID-type control law; compare the performance of various IMC-based PID controllers for time-delay processes; use Table 9–2
- Use Table 9–3 to find PID-type controllers for unstable processes

The major sections of this chapter are as follows:

9.1 Background

As we show in this chapter, the IMC structure can be rearranged to the standard feedback structure.

Question: Why do we care about IMC, if we can show that it can be rearranged into the standard feedback (PID) structure?

Answer: Because the process model is *explicitly* used in the control-system design procedure. The standard feedback structure uses the process model in an *implicit* fashion; that is, PID tuning parameters are often "tweaked" based on a transfer function model, but it is not always clear how the process model affects the tuning decision. In the IMC formulation, the controller, $q(s)$, is based directly on the "good" part of the process transfer function. The IMC formulation generally results in only one tuning parameter, the closed-loop time constant (λ, the IMC filter factor). The IMC-based PID tuning parameters are then a function of this closed-loop time constant. The selection of the closed-loop time constant is directly related to the robustness (sensitivity to model error) of the closed-loop system. Also, for open-loop unstable processes, it is necessary to implement the IMC strategy in standard feedback form, because the IMC suffers internal stability problems.

The reader should realize that the *IMC-based PID* controller presented in this chapter will not give the same results as the IMC strategy when there are process time delays, because the IMC-based PID procedure uses an approximation for dead time, while the IMC strategy uses the exact representation for dead time.

9.2 The Equivalent Feedback Form to IMC

In this section we derive the standard feedback equivalence to IMC by using block-diagram manipulation. Begin with the IMC structure shown in Figure 9–1; the point of comparison between the model and process output can be moved as shown in Figure 9–2.

Figure 9–2 can be rearranged to the form of Figure 9–3.

The arrangement shown inside the dotted line of Figure 9–3 is shown in Figure 9–4.

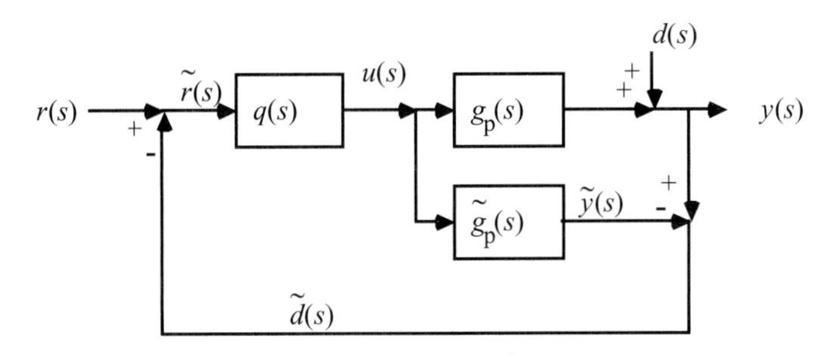

Figure 9–1 IMC structure.

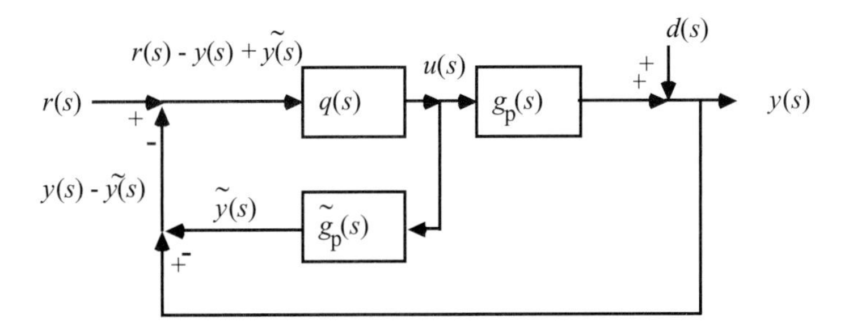

Figure 9–2 Cosmetic change in IMC structure.

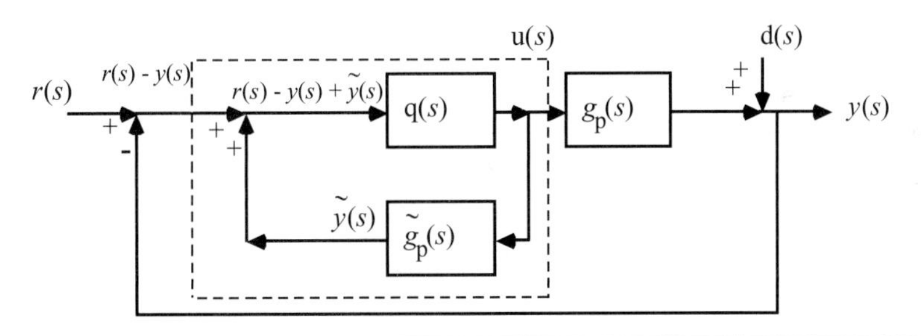

Figure 9–3 Rearrangement of IMC structure.

Figure 9–4 can be rearranged to the form of Figure 9–5.

Notice that $r(s) - y(s)$ is simply the error term used by a standard feedback controller. Therefore, we have found that the IMC structure can be rearranged to the feedback control (FBC) structure, as shown in Figure 9–6. This reformulation is advantageous because we find that a PID controller often results when the IMC design procedure is

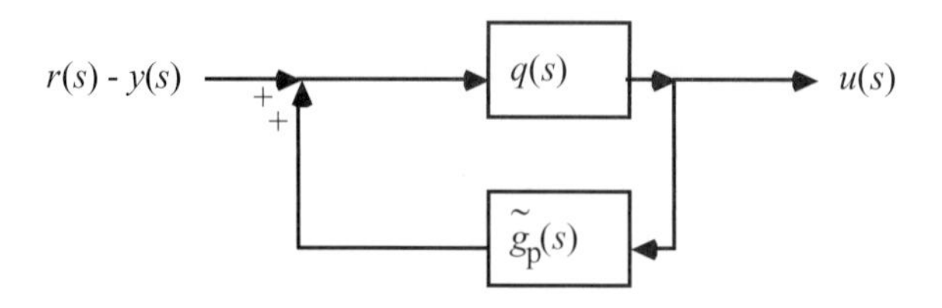

Figure 9–4 Inner loop of the rearranged IMC structure shown in Figure 9–3.

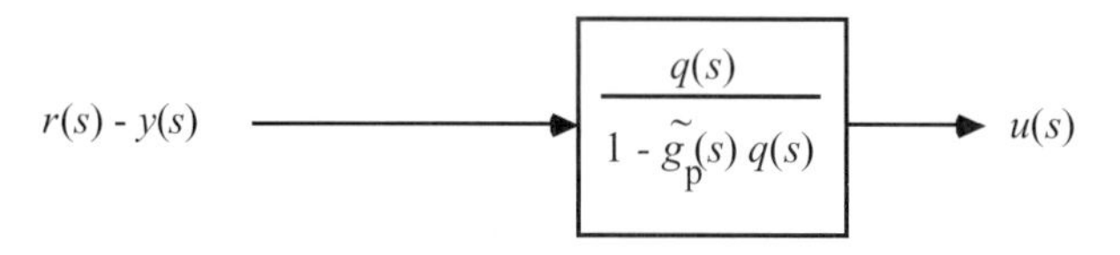

Figure 9–5 Equivalent block to Figure 9–4.

used. Also, the standard IMC block diagram cannot be used for unstable systems, so this feedback form must be used for those cases.

Now, we can use the IMC design procedure to help us design a standard feedback controller. The standard feedback controller is a function of the internal model, $\tilde{g}_p(s)$, and internal model controller, $q(s)$, as shown in Equation (9.1).

The standard feedback controller which is equivalent to IMC is

$$g_c(s) = \frac{q(s)}{1 - \tilde{g}_p(s)q(s)} \tag{9.1}$$

We refer to Equation (9.1) as the *IMC-based PID* relationship because the form of $g_c(s)$ is often that of a PID controller. The *IMC-based PID* procedure is similar to the IMC procedure of the previous chapter, with some additional steps. One major difference is that

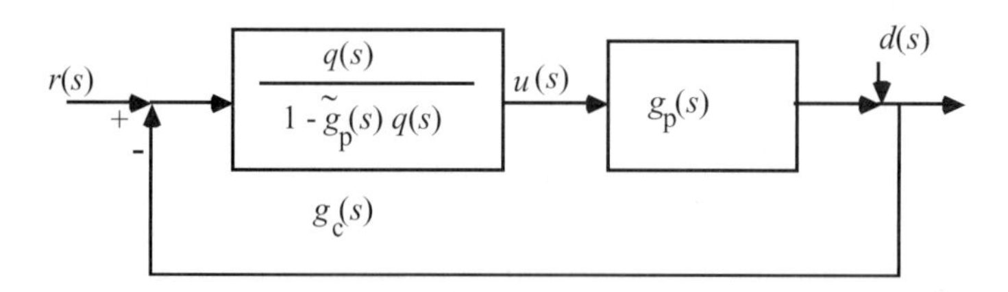

Figure 9–6 Standard feedback diagram illustrating the equivalence with IMC. The feedback controller, $g_c(s)$, contains both the internal model, $\tilde{g}_p(s)$, and the internal model controller, $q(s)$.

the IMC-based procedure will, many times, not require that the controller be *proper*. Also, the process dead time will be approximated using the Padé procedure, in order to arrive at an equivalent PID-type controller. Because of the Padé approximation for dead time, the IMC-based PID controller may not perform as well as IMC for processes with significant time delays.

The IMC-Based PID Control Design Procedure

The following steps are used in the IMC-based PID control system design.

1. Find the IMC controller transfer function, $q(s)$, which includes a filter, $f(s)$, to make $q(s)$ semiproper or to give it derivative action [the order of the numerator of $q(s)$ is one order greater that the denominator of $q(s)$]. Notice that this is a major difference from the IMC procedure. Here, we may allow $q(s)$ to be *improper*, in order to find an equivalent PID controller. For integrating or unstable processes, or for better disturbance rejection, a filter with the following form will often be used

$$f(s) = \frac{\gamma s + 1}{(\lambda s + 1)^n}$$

2. Find the equivalent standard feedback controller using the transformation

$$g_c(s) = \frac{q(s)}{1 - \tilde{g}_p(s) q(s)} \tag{9.1}$$

   Write this in the form of a ratio between two polynomials.

3. Show this in PID form and find k_c, τ_I, τ_D. Sometimes this procedure results in an ideal PID controller cascaded with a first-order filter, with a filter time constant (τ_F):

$$g_c(s) = k_c \left[\frac{\tau_I \tau_D s^2 + \tau_I s + 1}{\tau_I s} \right] \left[\frac{1}{\tau_F s + 1} \right] \tag{9.2}$$

4. Perform closed-loop simulations for both the perfect model case and cases with model mismatch. Adjust λ considering a tradeoff between performance and *robustness* (sensitivity to model error). Initial values for λ will generally be around 1/3 to 1/2 the dominant time constant.

9.3 IMC-Based Feedback Design for Delay-Free Processes

The IMC-based PID procedure, like the IMC procedure from Chapter 8, can be based on either good setpoint tracking or improved disturbance rejection. Here, we first consider design for setpoint tracking.

Focus on Setpoint Tracking

The procedure outlined in Section 9.2 is illustrated by way of two examples: a first-order process and a second-order process. For clarity we drop the (~) notation on all of the process model parameters.

Example 9.1: IMC-Based PID Design for a First-Order Process

Find the PID-equivalent to IMC for a first-order process

$$\tilde{g}_p(s) = \frac{k_p}{\tau_p s + 1}$$

Step 1. Find the IMC controller transfer function, $q(s)$, which includes a filter to make $q(s)$ *semiproper.*

$$q(s) = \tilde{q}(s) f(s) = \tilde{g}_{p-}^{-1}(s) f(s) = \frac{\tau_p s + 1}{k_p} \frac{1}{\lambda s + 1}$$

$$q(s) = \frac{1}{k_p} \frac{\tau_p s + 1}{\lambda s + 1} \tag{9.3}$$

Step 2. Find the equivalent standard feedback controller using the transformation

$$g_c(s) = \frac{q(s)}{1 - \tilde{g}_p(s) q(s)} = \frac{\dfrac{\tau_p s + 1}{k_p(\lambda s + 1)}}{1 - \dfrac{k_p}{\tau_p s + 1} \dfrac{\tau_p s + 1}{k_p(\lambda s + 1)}} = \frac{\tau_p s + 1}{k_p \lambda s} \tag{9.4}$$

Recall that the transfer function for a PI controller is

$$g_c(s) = k_c \frac{\tau_I s + 1}{\tau_I s} \tag{9.5}$$

Step 3. Rearrange Equation (9.4) to fit the form of Equation (9.5), to find how the model parameters and λ are related to the PI controller parameters. Multiplying Equation (9.4) by τ_p/τ_p, we find

$$g_c(s) = \left(\frac{\tau_p}{k_p\lambda}\right)\frac{\tau_p s + 1}{\tau_p s} \tag{9.6}$$

Equating terms in Equations (9.5) and (9.6), we find the following values for the PI tuning parameters

$$k_c = \frac{\tau_p}{k_p\lambda} \qquad \tau_I = \tau_p \tag{9.7}$$

The IMC-based PID design procedure for a first-order process has resulted in a PI control law. The major difference is that there are no longer two degrees of freedom in the tuning parameters (k_c, τ_I)—the IMC-based procedure shows that only the proportional gain needs to be adjusted. The integral time is simply set equal to the process time constant. Notice that the proportional gain is inversely related to λ, which makes sense. If λ is small (closed loop is "fast"), the controller gain must be large. Similarly, if λ is large (closed loop is "slow"), the controller gain must be small. Also notice that the same results were obtained using the *direct synthesis method*—a specified first-order closed-loop response for a first-order process leads to a PI controller with the parameter values shown in Equation (9.7).

This procedure can be used to develop the equivalent PID (plus lag, in some cases) controller for a number of other transfer functions, as shown in Table 9–1. In the next example, we derive the PID controller for a second-order process.

Example 9.2: IMC-Based PID Design for a Second-Order Process

Find the PID equivalent to IMC for a second-order process

$$\tilde{g}_p(s) = \frac{k_p}{(\tau_1 s + 1)(\tau_2 s + 1)}$$

Step 1. Find the IMC controller transfer function, $q(s)$—here we allow $q(s)$ to be *improper* because we wish to end up with an ideal PID controller.

$$q(s) = \tilde{q}(s)f(s) = \tilde{g}_{p-}^{-1}(s)f(s) = \frac{(\tau_1 s + 1)(\tau_2 s + 1)}{k_p}\frac{1}{(\lambda s + 1)} \tag{9.8}$$

Step 2. Find the equivalent standard feedback controller using the transformation

$$g_c(s) = \frac{q(s)}{1 - \tilde{g}_p(s)q(s)} = \frac{\dfrac{(\tau_1 s + 1)(\tau_2 s + 1)}{k_p(\lambda s + 1)}}{1 - \dfrac{k_p}{(\tau_1 s + 1)(\tau_2 s + 1)} \dfrac{(\tau_1 s + 1)(\tau_2 s + 1)}{k_p(\lambda s + 1)}} \tag{9.9}$$

$$= \frac{\tau_1 \tau_2 s^2 + (\tau_1 + \tau_2)s + 1}{k_p \lambda s}$$

Recall that the transfer function for an ideal PID controller is

$$g_c(s) = k_c \left[\frac{\tau_I \tau_D s^2 + \tau_I s + 1}{\tau_I s} \right] \tag{9.10}$$

Step 3. Rearrange Equation (9.9) to fit the form of Equation (9.10), to find how the model parameters and λ are related to the PID controller parameters. Multiplying Equation (9.9) by $(\tau_1 + \tau_2)/(\tau_1 + \tau_2)$, we find

$$g_c(s) = \left(\frac{\tau_1 + \tau_2}{k_p \lambda} \right) \frac{\tau_1 \tau_2 s^2 + (\tau_1 + \tau_2)s + 1}{(\tau_1 + \tau_2)s} \tag{9.11}$$

Equating terms in Equations (9.10) and (9.11), we find the following relationships:

$$k_c = \frac{(\tau_1 + \tau_2)}{k_p \lambda}$$

$$\tau_I = \tau_1 + \tau_2$$

$$\tau_D = \frac{\tau_1 \tau_2}{\tau_1 + \tau_2}$$

which is shown as row C in Table 9–1.

The IMC-based PID controllers for first- and second-order processes, as designed above, have good setpoint tracking characteristics. Although setpoint tracking may be quite good, disturbance rejection can be poor, particularly for input disturbances. Next, we apply the IMC-based PID procedure to develop good disturbance rejection.

Focus on Disturbance Rejection

For improved disturbance rejection, we use an IMC filter with the form

$$f(s) = \frac{\gamma s + 1}{(\lambda s + 1)^n}$$

where γ is selected to achieve good disturbance rejection. In practice, it is selected to cancel a slow disturbance time constant. Consider the closed-loop transfer function for disturbance rejection in the case of a perfect model (as developed in Section 8.10):

$$y(s) = [1 - \tilde{g}_p(s)q(s)]g_d l(s)$$

The internal model controller, using the new filter form, is

$$q(s) = \tilde{g}_{p-}^{-1}(s) f(s) = \tilde{g}_{p-}^{-1}(s)\frac{\gamma s + 1}{(\lambda s + 1)^n}$$

so the output response is

$$y(s) = \left[1 - \tilde{g}_p(s)\tilde{g}_{p-}^{-1}(s)\frac{\gamma s + 1}{(\lambda s + 1)^n}\right] g_d(s)l(s)$$

$$y(s) = \left[1 - \tilde{g}_{p+}(s)\frac{\gamma s + 1}{(\lambda s + 1)^n}\right] g_d(s)l(s) \tag{9.12}$$

$$y(s) = \left[\frac{(\lambda s + 1)^n - \tilde{g}_{p+}(s)(\gamma s + 1)}{(\lambda s + 1)^n}\right] g_d(s)l(s)$$

For a first-order process and an input disturbance, in Chapter 8 (Example 8.8), we found that the internal model controller

$$q(s) = \frac{\tau_p s + 1}{k_p} \cdot \frac{\gamma s + 1}{(\lambda s + 1)^2}$$

Using the procedure in Equation (9.12) results in a numerator filter term

$$\gamma = \frac{2\lambda\tau_p - \lambda^2}{\tau_p}$$

The reader should show that the IMC-based PID procedure leads to a PI controller with the parameter values

$$k_c = \frac{2\tau_p - \lambda}{k_p \lambda} \quad \text{and} \quad \tau_I = \frac{2\tau_p\lambda - \lambda^2}{\tau_p}$$

with the requirement that $\lambda < 2\tau_p$.

Integrating Processes

For integrating processes, a filter with the following form is used:

$$f(s) = \frac{2\lambda s + 1}{(\lambda s + 1)^n}$$

The reader should be able to derive the controllers shown in elements I and K of Table 9–1.

Summary for Delay-Free Processes

The results for the previous examples are shown in Table 9–1. To develop a more complete understanding of this procedure, the reader should derive the parameters for some of the other relationships shown in Table 9–1. The reader should also note that the same results would be obtained using the direct synthesis approach if the proper desired closed-loop transfer function is specified. The proper desired closed-loop transfer function appears clearly in the IMC procedure.

Notice that Table 9–1 is for process transfer functions that do not have a time delay. The following section develops PID tuning relationships for a first-order + time-delay process.

9.4 IMC-Based Feedback Design for Processes with a Time Delay

In order to arrive at a PID equivalent form for processes with a time delay, we must make some approximation to the dead time. (Please note that we do not use an approximation for dead time when implementing the "pure" IMC strategy developed in Chapter 8. We are doing this only for the *IMC-based PID* procedure presented in this chapter; this is only done to yield a PID-type controller.) We use either a *first-order Padé approximation* for dead time or neglect dead time entirely and use the first-order transfer function results.

First-Order + Dead Time

Since first-order + dead time is the most common representation of chemical process dynamics, the PID equivalent form developed here is useful for a large number of process control loops. We develop the procedure by way of example.

Example 9.3: IMC-Based PID Design for a First-Order + Dead Time Process

Find the PID controller which approximates IMC for a first-order + time-delay process

$$\tilde{g}_p(s) = \frac{k_p e^{-\theta s}}{\tau_p s + 1} \tag{9.13}$$

Step 1. Use a first-order Padé approximation for dead time $\left(e^{-\theta s} \approx \frac{-0.5\theta s + 1}{0.5\theta s + 1} \right)$,

$$\tilde{g}_p(s) = \frac{k_p e^{-\theta s}}{\tau_p s + 1} \approx \frac{k_p(-0.5\theta s + 1)}{(\tau_p s + 1)(0.5\theta s + 1)} \tag{9.14}$$

Step 2. Factor out the noninvertible elements (this time do not make the bad part all-pass),

$$\begin{aligned} \tilde{g}_{p-}(s) &= \frac{k_p}{(\tau_p s + 1)(0.5\theta s + 1)} \\ \tilde{g}_{p+}(s) &= -0.5\theta s + 1 \end{aligned} \tag{9.15}$$

Step 3. Form the idealized controller,

$$\tilde{q}(s) = \frac{(\tau_p s + 1)(0.5\theta s + 1)}{k_p} \tag{9.16}$$

Step 4. Add the filter—this time we will not make $q(s)$ proper because a PID controller will not result. We use the "derivative" option, where we allow the numerator of $q(s)$ to be one order higher than the denominator (NOTE: *This is done only so that we will obtain an ideal PID controller*).

$$q(s) = \tilde{q}(s)f(s) = \tilde{g}_{p-}^{-1}(s)f(s) = \frac{(\tau_p s + 1)(0.5\theta s + 1)}{k_p} \frac{1}{\lambda s + 1} \tag{9.17}$$

Now, find the PID equivalent. Recall that

$$\begin{aligned} g_c(s) &= \frac{q(s)}{1 - \tilde{g}_p(s)q(s)} = \frac{\tilde{q}(s)f(s)}{1 - \tilde{g}_p(s)\tilde{q}(s)f(s)} \\ g_c(s) &= \frac{\tilde{q}(s)f(s)}{1 - \tilde{g}_{p-}(s)\tilde{g}_{p+}(s)\tilde{g}_{p-}^{-1}(s)f(s)} = \frac{\tilde{q}(s)f(s)}{1 - \tilde{g}_{p+}(s)f(s)} = \left(\frac{1}{k_p}\right)\frac{(\tau_p s + 1)(0.5\theta s + 1)}{(\lambda + 0.5\theta)s} \end{aligned} \tag{9.18}$$

We can expand the numerator term to find

$$g_c(s) = \left(\frac{1}{k_p}\right)\frac{0.5\tau_p \theta s^2 + (\tau_p + 0.5\theta)s + 1}{(\lambda + 0.5\theta)s} \tag{9.19}$$

We can multiply Equation (9.19) by $(\tau_p + 0.5)/(\theta/\tau_p + 0.5\theta)$ to find the PID parameters

$$\begin{aligned} k_c &= \frac{(\tau_p + 0.5\theta)}{k_p(\lambda + 0.5\theta)} \\ \tau_I &= \tau_p + 0.5\theta \\ \tau_D &= \frac{\tau_p \theta}{2\tau_p + \theta} \end{aligned} \tag{9.20}$$

The IMC-based PID controller design procedure has resulted in a PID controller, when the process is first-order + dead time. Remember that a Padé approximation for dead time was used in this development, meaning that the filter factor (λ) cannot be made arbitrarily small; therefore, there will be performance limitations to the IMC-based PID strategy that do not occur in the IMC strategy. Rivera et al. (1986) recommend that $\lambda > 0.8\theta$ because of the model uncertainty due to the Padé approximation.

In Example 9.3 the all-pass formulation was *not* used. The reader should show that the use of an all-pass in the factorization will lead to a PID controller in series with a first-order lag. The parameters, in this case, are shown as the first entry in Table 9–2. Morari and Zafiriou (1989) recommend $\lambda > 0.25\theta$ for the PID + lag formulation. The third and fourth entries neglect the time delay in forming the PI controller.

Summary of PI(D) Control of First-Order + Time-Delay Processes

Different assumptions are used to derive the PI and PID controllers shown in Table 9–2. A zeroth-order Padé approximation leads to a PI controller while a first-order Padé approximation leads to PID + lag (when an all-pass filter is used) and PID (when the all-pass is not used) controllers. Generally, the PID + lag controller will be easier to tune for robustness and will certainly be less sensitive to noise than the PID controller. The PID + filter perfomance is almost identical to the "pure" IMC. This shows that the powerful IMC framework can be used to design PID-type controllers that can be implemented in industrial processes using existing (PID) control equipment.

Integrator + Dead Time

For processes where the time constant is dominant, the step response behavior can be approximated as integrator + dead time, as characterized by the following transfer function.

$$g_p(s) = \frac{ke^{-\theta s}}{s} \tag{9.21}$$

Here, assume that a Taylor series approximation for dead time is used. Also, the special filter form for integrating systems is used.

$$e^{-\theta s} \approx -\theta s + 1$$

$$\tilde{g}_p(s) \approx \frac{k(-\theta s + 1)}{s} = \frac{k}{s}(-\theta s + 1)$$

$$q(s) = \frac{s}{k} \cdot \frac{\gamma s + 1}{(\lambda s + 1)^2}$$

Using the IMC-based PID procedure,

$$g_c(s) = \frac{q(s)}{1 - \tilde{g}_p(s)q(s)}$$

the reader should show that a PI controller results with the following parameters:

$$k_c = \frac{2\lambda + \theta}{k(\lambda + \theta)^2} \qquad \tau_I = 2\lambda + \theta \tag{9.22}$$

Gain + Dead Time

For processes where the time delay is dominant, the step response behavior can be approximated as gain + dead time, as characterized by the following transfer function.

$$g_p(s) = ke^{-\theta s} \tag{9.23}$$

Using a second-order Padé approximation for the time delay,

$$e^{-\theta s} \approx \frac{\frac{\theta^2}{12}s^2 - \frac{\theta}{2}s + 1}{\frac{\theta^2}{12}s^2 + \frac{\theta}{2}s + 1}$$

$$\tilde{g}_p(s) \approx k \cdot \frac{\frac{\theta^2}{12}s^2 - \frac{\theta}{2}s + 1}{\frac{\theta^2}{12}s^2 + \frac{\theta}{2}s + 1} = \frac{k}{\frac{\theta^2}{12}s^2 + \frac{\theta}{2}s + 1} \cdot \left(\frac{\theta^2}{12}s^2 - \frac{\theta}{2}s + 1\right)$$

$$q(s) = \frac{\frac{\theta^2}{12}s^2 + \frac{\theta}{2}s + 1}{k} \cdot \frac{1}{(\lambda s + 1)^2}$$

the reader should find that a PID + filter controller results, with

$$k_c = \frac{\theta}{k(4\lambda + \theta)} \qquad \tau_I = \frac{\theta}{2} \qquad \tau_D = \frac{\theta}{6} \qquad \tau_F = \frac{2\lambda^2 - \frac{\theta^2}{6}}{4\lambda + \theta} \tag{9.24}$$

with the condition that $\lambda > \frac{\theta}{\sqrt{12}}$.

9.5 Summary of IMC-Based PID Controller Design for Stable Processes

We have shown several examples where the IMC design procedure could be used to develop an equivalent PID-type control law. For stable processes with no time delay, the *IMC-based PID* procedure gives exactly the same feedback performance as does IMC. For stable processes with a time delay, the IMC-based PID procedure will not give exactly the same performance as IMC, because a Padé approximation for dead time is used in the controller design.

We want this following point to be clear to the reader. For process transfer functions without time delays, the IMC-based PID controller will yield exactly the same performance as does IMC. This will occur if no approximation has to be made in the process model to find a feedback form that is equivalent to PID. If an approximation (such as Padé) is made in the IMC-based PID strategy, and this approximation is not made in the IMC strategy, then the performance will not be the same.

It should also be noted that although the standard IMC filter form, $f(s) = 1/(\lambda s + 1)^n$, leads to good setpoint tracking, it generally does not lead to good rejection of disturbances that have dynamics similar to the process transfer function. It is generally better to use a filter of the form $f(s) = (\gamma s + 1)/(\lambda s + 1)^n$.

Table 9–1 provides a summary of the PID tuning parameters for systems without a time delay. Table 9–2 summarizes the PID tuning parameters for stable processes with a time delay. Notice that there are minimum recommended values for λ shown, since there

Table 9–1 Ideal PID Tuning Parameters for Open-Loop Stable and Integrating Processes[a]

	$g_p(s)$	$g_{CL}(s)$	k_c	τ_I	τ_D	τ_F
A	$\frac{k_p}{\tau_p s+1}$	$\frac{1}{\lambda s+1}$	$\frac{\tau_p}{k_p\lambda}$	τ_p		
B[b]	$\frac{k_p}{\tau_p s+1}$	$\frac{\gamma s+1}{(\lambda s+1)^2}$	$\frac{2\tau_p-\lambda}{k_p\lambda}$	$\frac{2\tau_p\lambda-\lambda^2}{\tau_p}$		
C	$\frac{k_p}{(\tau_1 s+1)(\tau_2 s+1)}$	$\frac{1}{\lambda s+1}$	$\frac{\tau_1+\tau_2}{k_p\lambda}$	$\tau_1+\tau_2$	$\frac{\tau_1\tau_2}{\tau_1+\tau_2}$	

(continues)

Table 9–1 *(continued)*

	$g_p(s)$	$g_{CL}(s)$	k_c	τ_I	τ_D	τ_F
D	$\frac{k_p}{\tau^2 s^2 + 2\zeta\tau s + 1}$	$\frac{1}{\lambda s + 1}$	$\frac{2\zeta\tau}{k_p\lambda}$	$2\zeta\tau$	$\frac{\tau}{2\zeta}$	
E[c]	$\frac{k_p}{\tau^2 s^2 + 2\zeta\tau s + 1}$	$\frac{1}{(\lambda s + 1)^2}$	$\frac{\zeta\tau}{k_p\lambda}$	$2\zeta\tau$	$\frac{\tau}{2\zeta}$	$\frac{\lambda}{2}$
F[c,d]	$\frac{k_p(-\beta s + 1)}{\tau^2 s^2 + 2\zeta\tau s + 1}$	$\frac{-\beta s + 1}{(\beta s + 1)(\lambda s + 1)}$	$\frac{2\zeta\tau}{k_p(2\beta + \lambda)}$	$2\zeta\tau$	$\frac{\tau}{2\zeta}$	$\frac{\beta\lambda}{2\beta + \lambda}$
G[d]	$\frac{k_p(-\beta s + 1)}{\tau^2 s^2 + 2\zeta\tau s + 1}$	$\frac{-\beta s + 1}{\lambda s + 1}$	$\frac{2\zeta\tau}{k_p(\beta + \lambda)}$	$2\zeta\tau$	$\frac{\tau}{2\zeta}$	
H	$\frac{k}{s}$	$\frac{1}{\lambda s + 1}$	$\frac{1}{k\lambda}$			
I[e]	$\frac{k}{s}$	$\frac{2\lambda s + 1}{(\lambda s + 1)^2}$	$\frac{2}{k\lambda}$	2λ		
J	$\frac{k}{s(\tau s + 1)}$	$\frac{1}{\lambda s + 1}$	$\frac{1}{k\lambda}$		τ	
K[e]	$\frac{k}{s(\tau s + 1)}$	$\frac{2\lambda s + 1}{(\lambda s + 1)^2}$	$\frac{2\lambda + \tau}{k\lambda^2}$	$2\lambda + \tau$	$\frac{2\lambda\tau}{2\lambda + \tau}$	

[a]Parameters for other process transfer functions are given in Rivera et al. (1986) and Morari and Zafiriou (1989).

[b]The controller is designed for improved input disturbance rejection; $\gamma = \frac{2\tau_p\lambda - \lambda^2}{\tau_p}$. Notice that we desire $\gamma > 0$, which leads to $\lambda < 2\tau_p$.

[c]The controller is PID + lag, $g_c(s) = k_c\left[\frac{\tau_I\tau_D s^2 + \tau_I s + 1}{\tau_I s}\right]\left[\frac{1}{\tau_F s + 1}\right]$.

[d]It is assumed that $\beta > 0$ (inverse response, RHP zeros).

[e]The controllers are designed for ramp setpoint changes. This also generally leads to better input disturbance rejection.

Table 9–2 PID Tuning Parameters for Stable Time-Delay Processes[a]

	$g_p(s)$	k_c	τ_I	τ_D	τ_F	Notes[b]
A	$\frac{k_p e^{-\theta s}}{\tau_p s+1}$	$\frac{\tau_p+\frac{\theta}{2}}{k_p(\theta+\lambda)}$	$\tau_p+\frac{\theta}{2}$	$\frac{\tau_p\theta}{2\tau_p+\theta}$	$\frac{\lambda\theta}{2(\lambda+\theta)}$	1
B	$\frac{k_p e^{-\theta s}}{\tau_p s+1}$	$\frac{\tau_p+\frac{\theta}{2}}{k_p\left(\lambda+\frac{\theta}{2}\right)}$	$\tau_p+\frac{\theta}{2}$	$\frac{\tau_p\theta}{2\tau_p+\theta}$		2
C	$\frac{k_p e^{-\theta s}}{\tau_p s+1}$	$\frac{\tau_p}{k_p\lambda}$	τ_p			3
D	$\frac{k_p e^{-\theta s}}{\tau_p s+1}$	$\frac{\tau_p+\frac{\theta}{2}}{k_p\lambda}$	$\tau_p+\frac{\theta}{2}$			4
E	$\frac{ke^{-\theta s}}{s}$	$\frac{2\lambda+\theta}{k(\lambda+\theta)^2}$	$2\lambda+\theta$			5
F	$\frac{ke^{-\theta s}}{s}$	$\frac{2}{k\left(\lambda+\frac{\theta}{2}\right)}$	$2\lambda+\theta$	$\frac{\lambda\theta+\frac{\theta^2}{4}}{2\lambda+\theta}$		6
G	$ke^{-\theta s}$	$\frac{\theta}{k(2\lambda+\theta)}$	$\frac{\theta}{2}$			7
H	$ke^{-\theta s}$	$\frac{\theta}{k(4\lambda+\theta)}$	$\frac{\theta}{2}$	$\frac{\theta}{6}$	$\frac{2\lambda^2-\frac{\theta^2}{6}}{4\lambda+\theta}$	8

[a]Based on the ideal PID controller transfer function: $g_c(s)=k_c\left[\frac{\tau_I\tau_D s^2+\tau_I s+1}{\tau_I s}\right]\left[\frac{1}{\tau_F s+1}\right]$.

[b]IMC-based PID is based on a first-order Padé approximation, unless otherwise noted. 1, With an all-pass factorization and semiproper $q(s)$ recommended $\lambda > 0.25\theta$. 2, Without an all-pass factorization and improper $q(s)$; recommended $\lambda > 0.8\theta$. 3, Time-delay neglected; recommended $\lambda > 1.7\theta$. 4, Time-delay neglected; effective time constant increased by 0.5θ; recommended $\lambda > 1.7\theta$. 5, Using a Taylor series approximation for time delay $(-\theta s + 1)$. 6, With an improper $q(s)$. 7, Using a first order Padé approximation for time delay. 8, Using a second order Padé approximation for time delay. In all cases it is recommended that $\lambda > 0.2\tau_p$.

is inherent model uncertainty due to the Padé approximation. Selecting λ near this value will put the closed-loop system on the edge of stability, so most often your initial tuning values for λ will be significantly larger than these minimum values.

9.6 IMC-Based PID Controller Design for Unstable Processes

The IMC procedure must be modified for unstable processes. Rotstein and Lewin (1991) have used the procedure developed by Morari and Zafiriou (1989) to find IMC-based PID controllers for unstable processes. The modification to the procedure shown in Sections 9.2 and 9.3 is to use a slightly more complicated filter transfer function.

1. Find the IMC controller transfer function, $q(s)$, which includes a filter, $f(s)$, to make $q(s)$ semiproper. An additional requirement is that the value of $f(s)$ at $s = p_u$ (where p_u is an unstable pole) must be 1. That is,

   $$f(s = p_u) = 1$$

   Morari and Zafiriou (1989) recommend a filter transfer function that has the form

   $$f(s) = \frac{\gamma s + 1}{(\lambda s + 1)^n}$$

   where n is chosen to make $q(s)$ proper (usually *semiproper*). A value of γ is found that satisfies the filter requirement $f(s = p_u) = 1$.
2. Find the equivalent standard feedback controller using the transformation

   $$g_c(s) = \frac{q(s)}{1 - \tilde{g}_p(s)q(s)}$$

   Write this in the form of a ratio between two polynomials.
3. Show this in PID form and find k_c, τ_I, τ_D. Sometimes this procedure results in a PID controller cascaded with a lag term (τ_F):

   $$g_c(s) = k_c\left[\frac{\tau_I\tau_D s^2 + \tau_I s + 1}{\tau_I s}\right]\left[\frac{1}{\tau_F s + 1}\right]$$

4. Perform closed-loop simulations for both the perfect model case and cases with model mismatch. Choose the desired value for λ as a tradeoff between performance and robustness.

Example 9.4 illustrates this procedure for a first-order unstable process.

Example 9.4: IMC-Based PID Design for a First-Order Unstable Process

Find the IMC-based PID controller for a first-order unstable process

$$\tilde{g}_p(s) = \frac{k_p}{-\tau_u s + 1} \tag{9.25}$$

where τ_u is given a positive value. The pole, p_u, is $1/\tau_u$, which is positive, indicating instability.

Step 1. Find the IMC controller transfer function, $q(s)$,

$$q(s) = \tilde{q}(s)f(s) = \tilde{g}_{p-}^{-1}(s)f(s) = \frac{-\tau_u s + 1}{k_p} \frac{\gamma s + 1}{(\lambda s + 1)^2} \tag{9.26}$$

Note that we have selected a second-order polynomial filter to make the controller, $q(s)$, semiproper. Now we solve for γ so that $f(s = p_u) = 1$.

$$f(s) = \frac{\gamma s + 1}{(\lambda s + 1)^2}$$

$$f\left(s = \frac{1}{\tau_u}\right) = \frac{\gamma(1/\tau_u) + 1}{\left(\lambda(1/\tau_u) + 1\right)^2} = 1$$

so

$$\gamma\left(\frac{1}{\tau_u}\right) + 1 = \left(\lambda\left(\frac{1}{\tau_u}\right) + 1\right)^2$$

Solving for γ, we find

$$\gamma = \lambda\left(\frac{\lambda}{\tau_u} + 2\right) \tag{9.27}$$

Step 2. Find the equivalent standard feedback controller using the transformation

$$g_c(s) = \frac{q(s)}{1 - \tilde{g}_p(s)q(s)}$$

After a lengthy bit of algebra,

$$g_c(s) = \frac{\gamma}{k_p(2\lambda - \gamma)} \frac{(\gamma s + 1)}{\gamma s} \tag{9.28}$$

Step 3. This is in the form of a PI controller, where

$$k_c = \frac{\gamma}{k_p(2\lambda - \gamma)} = \frac{-(\lambda + 2\tau_u)}{k_p\lambda}$$
$$\tau_I = \gamma = \lambda\left(\frac{\lambda}{\tau_u} + 2\right) \tag{9.29}$$

As a numerical example, consider $g_p(s) = \frac{1}{-s+1}$. The closed loop output responses for various values of λ are shown in Figure 9–7a, while the manipulated variable responses are shown in Figure 9–7b. Notice that we do not achieve the nice overdamped-type of closed-loop output responses that we were able to obtain with open-loop stable processes. The reader should show that the closed-loop relationship for this system is

$$y(s) = \frac{g_c(s)g_p(s)}{1 + g_c(s)g_p(s)} r(s) = \frac{\gamma s + 1}{(\lambda s + 1)^2} r(s)$$

which has overshoot if $\gamma > \lambda$ [this is always the case for this system; see Equation 9.27)].

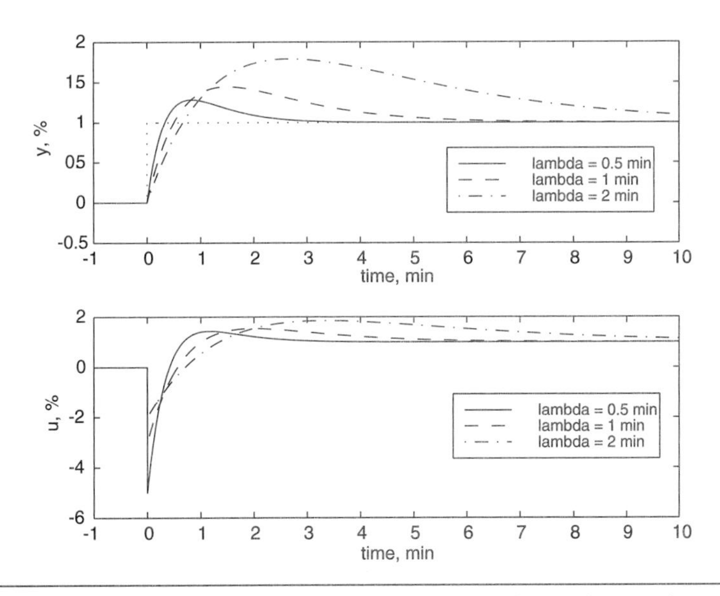

Figure 9–7 Responses for a step setpoint change, for various values of λ. (a) Output; (b) manipulated input.

In Figure 9–7 we notice that the closed-loop response had overshoot. A response to a setpoint change, without overshoot, can be obtained by including a setpoint filter, as shown in Figure 9–8. The setpoint filter is

$$f_{sp}(s) = \frac{1}{\gamma s + 1}$$

which yields the following closed-loop response for a setpoint change.

$$y(s) = \frac{g_c(s)g_p(s)}{1 + g_c(s)g_p(s)} f_{sp}(s)r(s) = \frac{1}{(\lambda s + 1)^2} r(s)$$

The results for several unstable process transfer functions are shown in Table 9–3. See Rotstein and Lewin (1991) for a discussion of the effect of dead time and model uncertainty on the control of unstable processes.

Summary of IMC-Based PID Controller Design for Unstable Processes

A major tuning consideration by the student is that there are both upper and lower bounds on λ to assure stability of an unstable process. This is in contrast to stable processes, where the closed loop is guaranteed to be stable under model uncertainty, simply by increasing λ to a large value (detuning the controller).

9.7 Summary

We have shown how the IMC procedure can be used to design PID-type feedback controllers. If the process has no time delay and the inputs do not hit a constraint, then the IMC-based PID controllers will have the same performance as does IMC. If there is dead

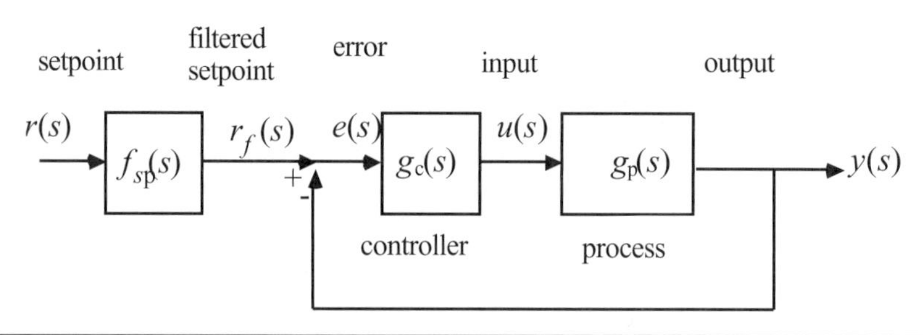

Figure 9–8 FBC with a setpoint filter.

Table 9–3 PID Tuning Parameters for Unstable Processes

	$g_p(s)$	$g_{CL}(s)$	k_c	τ_I	τ_D	Notes[a]
A	$\dfrac{k_p}{-\tau_u s+1}$	$\dfrac{\gamma s+1}{(\lambda s+1)^2}$	$\dfrac{-(\lambda+\tau_u)}{k_p\lambda}$	γ		1
B	$\dfrac{k_p}{(-\tau_u s+1)(\tau_p s+1)}$	$\dfrac{\gamma s+1}{(\lambda s+1)^2}$	$\dfrac{-\tau_u(\gamma+\tau_p)}{k_p\lambda^2}$	$\gamma+\tau_p$	$\dfrac{\gamma\tau_p}{\gamma+\tau_p}$	1
C	$\dfrac{k_p(\tau_n s+1)}{(-\tau_u s+1)(\tau_p s+1)}$	$\dfrac{\gamma s+1}{(\lambda s+1)^2}$	$\dfrac{-1}{k_p}\left(1+\dfrac{2\tau_u}{\lambda}\right)$	γ		1, 2

[a]1, $\gamma=\lambda\left(\dfrac{\lambda}{\tau_u}+2\right)$. 2, PI cascaded with a lead-lag filter, $\dfrac{\tau_p s+1}{\tau_n s+1}$.

time, then the IMC-based PID controllers will not perform as well as IMC because of the Padé approximation for deadtime.

It is interesting to note that the IMC-based PID controllers for all of the transfer functions shown in Table 9–1 could have been designed using the *direct synthesis* method of Chapter 6. The key issue in the direct synthesis method is the specification of the closed-loop response characteristic. If the process has a RHP zero, then the specified closed-loop response must also have a RHP zero. The IMC-based PID procedure provides a clear method for handling this. Also, note that the standard IMC filter design results in good setpoint response performance, but other filter designs must be used for good input disturbance rejection.

The IMC design method of Chapter 8 was modified to handle unstable processes. The standard IMC structure cannot handle unstable processes, so the controller for an unstable process must be implemented in standard PID feedback form.

For a good example application of the IMC-based PID procedure, work through Module 7 on Biochemical Reactors.

IMC-Based PID Procedure Summary

1. Find the IMC controller transfer function, $q(s)$, which includes a filter, $f(s)$. The controller, $q(s)$, may be *semiproper* or even *improper* to give the resulting PID controller derivative action. If a process model has a time delay, use an approximation, such as a first- or second-order Padé.

For good setpoint tracking, a filter with the following form is generally used

$$f(s)=\frac{1}{(\lambda s+1)^n}$$

For improved disturbance rejection, or for integrating and open-loop unstable processes, a filter with the following form is used:

$$f(s) = \frac{\gamma s + 1}{(\lambda s + 1)^n}$$

For disturbance rejection, γ is selected so that the term inside the brackets cancels a slow pole in the distubance transfer function [often $g_d(s) = g_p(s)$ for input disturbances],

$$y(s) = [1 - \tilde{g}_p(s)q(s)]g_d l(s)$$

For unstable processes, a value of γ is found that satisfies the filter requirement $f(s = p_u) = 1$.

2. Find the IMC-based PID controller using the transformation

$$g_c(s) = \frac{q(s)}{1 - \tilde{g}_p(s)q(s)}$$

   Write this in the form of a ratio between two polynomials.

3. Show this in PID form and find k_c, τ_I, τ_D. Sometimes this procedure results in an ideal PID controller cascaded with a first-order filter, with a filter time constant (τ_F):

$$g_c(s) = k_c \left[\frac{\tau_I \tau_D s^2 + \tau_I s + 1}{\tau_I s}\right]\left[\frac{1}{\tau_F s + 1}\right]$$

4. Perform closed-loop simulations for both the perfect model case and the cases with model mismatch. Adjust the value of λ on-line as a tradeoff between performance and *robustness* (sensitivity to model error).

References

The following paper derives IMC-based PID controllers for "real PID" control: Chien, I.-L., and P. S. Fruehauf, "Consider IMC Tuning to Improve Controller Performance," *Chem. Eng. Prog.*, **86**(10), 33–41 (1990).

The IMC procedure is fully developed in the following monograph: Morari, M., and E. Zafiriou, *Robust Process Control*, Prentice-Hall, Englewood Cliffs, NJ (1989).

Initial IMC-based tuning parameters were presented in the following paper: Rivera, D. E., M. Morari, and S. Skogestad, "Internal Model Control. 4. PID Controller Design," *Ind. Eng. Chem. Proc. Des. Dev.*, **25**, 252–265 (1986).

The following paper discusses the IMC-based tuning rules for open-loop unstable systems: Rotstein, G. E., and D. R. Lewin, "Simple PI and PID Tuning for Open-Loop Unstable Systems,"*Ind. Eng. Chem. Res.*, **30**, 1864–1869 (1991).

Student Exercises

Exercises 1, 19, and 20 are SIMULINK-based problems.

1. Use the *IMC-based* PID design procedure to find the PID controller for a second-order transfer function

$$\tilde{g}_p(s) = \frac{k_p}{\tau^2 s^2 + 2\zeta\tau s + 1}$$

Hint: First design the IMC, assuming that you will allow the controller to be improper; that is, the order of the numerator is one higher than the denominator. What order filter do you find?

a. Find the PID parameters, k_c, τ_I, τ_D, as a function of the process model parameters, k_p, τ, ζ, and the filter factor, λ.

b. For a perfect model, plot the closed-loop response of y to a step setpoint change in r (as a function of λ). (Show your control block diagram.)

$$\tau = 2 \text{ minutes}, \ \zeta = 0.8, \ k_p = 5.25 \text{ psig/gpm}.$$

c. Discuss the effect of uncertainty in ζ (show plots) on closed-loop performance, as a function of λ.

2. Compare the response of the following first-order + time-delay process, using IMC and IMC-based PID. Discuss the effect of λ on the closed-loop stability for both systems.

$$\tilde{g}_p(s) = \frac{1e^{-10s}}{5s + 1}$$

Do you find that there is a minimum λ required for the stability of the IMC-based PID strategy? How does this relate to the recommendations in Table 9–2? Is there a minimum λ required for the stability of the IMC strategy?

3. In Example 9.3, the all-pass formulation was not used for the first-order + time-delay process with a Padé approximation for dead time. Show that the use of an all-pass in the factorization [and semi-proper $q(s)$] leads to a PID controller in series with a first order lag.

4. Use the following inverse response process,

$$\tilde{g}_p(s) = \frac{k_p(-\beta s+1)}{(\tau_1 s+1)(\tau_2 s+1)}$$

   a. Find the IMC-based PID controller if no all-pass is used and the controller is improper.
   b. Find the IMC-based PID controller (which must be cascaded with a first-order lag) if an all-pass is used and the IMC controller, $q(s)$, is semiproper.

5. For the following integrating process,

$$\tilde{g}_p(s) = \frac{k_p}{s}$$

show that the IMC-based PID procedure results in a P controller if $f(s) = \frac{1}{\lambda s+1}$ and in a PI controller if $f(s) = \frac{2\lambda s+1}{(\lambda s+1)^2}$.

6. The following process transfer function represents the relationship between boiler feedwater flow rate and steam drum level:

$$\tilde{g}_p(s) = \frac{k_p(-\beta s+1)}{s(\tau_p s+1)}$$

   a. Use the all-pass factorization method to show that the IMC-based PID controller is PD with a first-order lag.
   b. Use $f(s) = \frac{2\lambda s+1}{(\lambda s+1)^2}$ and a non-"all-pass" factorization to find a PID controller.

7. Use the *IMC-based* feedback controller design procedure to design a PID controller for the following process.

$$\tilde{g}_p(s) = \frac{k_p}{s(\tau_p s+1)}$$

Assume that the IMC filter is $f(s) = \frac{2\lambda s+1}{(\lambda s+1)^2}$. Find k_c, τ_I and τ_D—these will be a function of the process parameters and λ (show all work). What is the closed-loop transfer function? Sketch the expected response of the output to a unit step setpoint change.

8. Consider the following first-order + time-delay process

$$\tilde{g}_p(s) = \frac{25e^{-20s}}{15s+1}$$

Find the tuning parameters for the IMC-based PID controller (no lag). What is the maximum value of k_c that you would recommend? Why?

9. Consider Example 9.2, where $q(s)$ was allowed to be improper. This yielded a PID controller. Now assume that $q(s)$ is forced to be proper (actually semi-proper). Find the resulting feedback controller. Elaborate on the control structure.

10. A stack gas scrubber has the following relationship between the fresh feedwater flow rate and the SO_2 concentration in the water leaving the scrubber:

$$y(s) = \frac{-2(6s+1)}{(10s+1)(3s+1)}u(s)$$

Use the IMC-based PID procedure to find the PID controller (plus first-order lag) for this process.

Hint: Let the IMC $q(s)$ be semiproper (order of denominator = order of numerator). Which controller tuning parameters does λ affect?

11. Use the *IMC-based* feedback controller design procedure to design a PID-type controller for the following process. Assume that $q(s)$ is semiproper and $\tau_n > 0$.

$$\tilde{g}_p(s) = \frac{k_p(\tau_n s+1)}{\tau^2 s^2 + 2\zeta\tau s + 1}$$

Find the PID tuning parameters (assuming ideal PID) as a function of the process parameters and λ (show all work). What is the closed-loop transfer function? Sketch the expected response behavior.

12. Consider the closed-loop response for IMC when the model is *not* perfect. Show that there is no offset for a setpoint change for the following model and process transfer functions. Also, find the minimum value of λ that assures closed-loop stability when

process model $\quad \tilde{g}_p(s) = \dfrac{1}{10s+1}$

process $\quad g_p(s) = \dfrac{0.75(-s+1)}{(2s+1)(8s+1)}$

13. Show that the IMC strategy cannot be implemented on an unstable process and must be implemented in standard feedback form.

14. A styrene polymerization reactor is operated at an open-loop unstable point and has the following input-output model

$$\tilde{g}_p(s) = \frac{-2}{-10s+1}$$

Design the IMC-based PI controller for this system. Sketch the expected response for a setpoint change.

15. A vinyl-acetate polymerization reactor is operated at an open-loop unstable point and has the following input-output model, for control of output temperature (°C) by manipulating the jacket temperature (°C). The timescale is minutes.

$$\tilde{g}_p(s) = \frac{-2.5}{(-10s+1)(2s+1)}$$

a. Design the IMC-based PID controller for this system; that is, find k_c, τ_I, and τ_D. Show units for all parameters.

b. What is the closed-loop transfer function? Sketch the expected response for a step setpoint change.

16. Derive γ for the first three elements in Table 9–3. The value of the filter must be one at the location of the unstable pole.

17. A reactor is operated at an open-loop unstable point and has the following input-output model:

$$\tilde{g}_p(s) = \frac{-2(3s+1)}{(-4s+1)(5s+1)}$$

a. Design the IMC-based PID controller (perhaps with a lead-lag filter) for this system, that is, find the tuning parameters.

b. What is the closed-loop transfer function? Sketch the expected response for a step setpoint change.

18. Show that a PI controller cannot stabilize the process

$$\tilde{g}_p(s) = \frac{k_p}{(-\tau_u s+1)(\tau_p s+1)}$$ if $\tau_u < \tau_p$. (*Hint*: Use the Routh stability criterion.)

The IMC-based PID controller (entry B in Table 9–3) can handle this process.

19. Consider a first-order + dead-time process with the following transfer function (use this to represent the process in all analyses and simulations)

$$g_p(s) = \frac{1e^{-\theta s}}{10s+1}$$

Now assume two different PI design procedures for this process: The model is first-order + dead time with the known parameter values; and the model is integrator + dead time where $k = 1/10 = 0.1$.

a. For a small time delay (0.5 minutes), compare the performance of the two PI control strategies for step setpoint changes. Discuss the effect of λ. Also compare the performance of the two PI control strategies for step input load disturbance changes (let the load disturbance transfer function be equal to the process transfer function). Discuss the effect of λ.

b. For a larger time delay (5 minutes), compare the performance of the two PI control strategies for step setpoint changes. Discuss the effect of λ. Also compare the performance of the two PI control strategies for step input load disturbance changes (let the load disturbance transfer function be equal to the process transfer function). Discuss the effect of λ.

c. Summarize your results and make recommendations.

20. Consider a first-order + dead time process with the following transfer function (use this to represent the process in all analysis and simulations):

$$g_p(s) = \frac{1e^{-\theta s}}{10s+1}$$

Compare the setpoint performances of PI, improved PI, PID, and PID + filter tuning rules from Table 9–2 (rows A–D). Discuss the effect of the closed-loop time constant, λ; for what values is unstable or poor performance obtained?

21. Consider the following process model with a RHP zero

$$\tilde{g}_p(s) = \frac{k(-\beta s+1)}{s}$$

Use the IMC-based PID design procedure to find a PI controller. Assume that an all-pass factorization is used for the RHP zero. Also, assume that $q(s)$ is semiproper. Use an IMC filter with the form

$$\frac{\gamma s+1}{(\lambda s+1)^n}$$

and solve for γ to give exactly a PI controller (no other terms).

Find the proportional gain and integral time as a function of the model parameters (k, β) and λ.

CHAPTER 10

Cascade and Feed-Forward Control

The objective of this chapter is to develop feed-forward- and cascade-control system design procedures. The emphasis of both of these strategies is on rejecting disturbances. After studying this chapter, the reader should be able to do the following:

- Given an instrumentation diagram for a cascade-control strategy, develop the corresponding control block diagram using either a series or parallel structure
- Tune cascade controllers; tune the secondary loop first; then tune the primary controller with the secondary loop closed
- Given an instrumentation diagram for feed-forward/feedback control, develop the corresponding control block diagram
- Develop the closed-loop transfer functions for feed-forward/feedback control
- Design feed-forward/feedback controllers. Understand physical realizability limitations due to time delays and RHP zeros

The major sections of this chapter are as follows:

10.1 Background

Thus far in this textbook we have emphasized control-system design for single input–single output (SISO) processes; that is, processes with one output (measured) variable and one manipulated variable. We have also focused on setpoint responses, primarily because it is then easier to tune a feedback controller. In particular, it is easier to tune on-line for setpoint changes because we do not know when a disturbance is going to enter a system. We have developed a number of control-system design procedures that are based on a desired response to a setpoint change.

In practice, *disturbance rejection* is very important. The primary disadvantage to feedback-only control is that a disturbance must be "felt" by the output variable before there is a control-system response. The purpose of this chapter is to show how to use multiple measurements to improve the response to a disturbance. In Sections 10.2 through 10.5, we study *cascade* control. In cascade control, multiple output measurements are used to improve the response of the most important (primary) output to a disturbance. In the last part of the chapter (Sections 10.6 to 10.8) we cover *feed-forward* control, where the measurement of a disturbance is directly used to improve the system response to the disturbance. In Section 10.10 we combine feed-forward and cascade control.

10.2 Introduction to Cascade Control

Cascade control involves the use of multiple measurements and a single manipulated input. As a motivating example, consider the temperature-control problem shown in Figure 10–1, where a fired furnace is used to heat a process fluid stream. The outlet temperature is controlled by manipulating the valve position of the fuel gas control valve. Clearly, disturbances in the fuel gas header pressure (upstream of the valve) will end up changing the fuel gas flow rate, and, therefore, the outlet temperature. Also, any problems with the control valve, such as stiction or hysteresis (see Module 15), will affect the fuel gas flow rate.

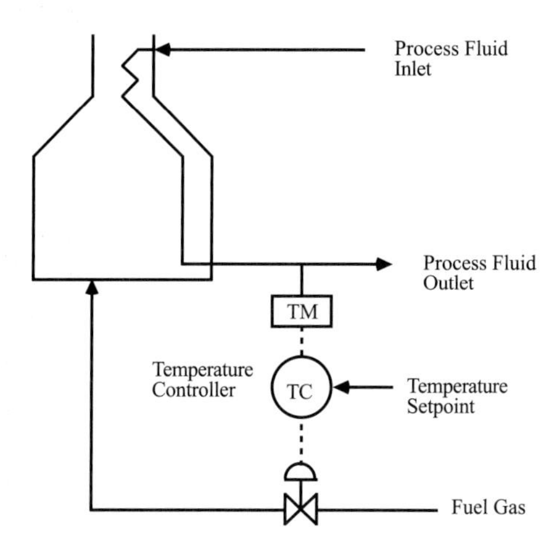

Figure 10–1 Feedback control of process outlet temperature.

Cascade to Flow Control

The best way to compensate for disturbances directly affecting the fuel gas flow rate is to *cascade* the temperature controller to a fuel gas flow controller, as shown in Figure 10–2. Here, the output of the temperature controller is the setpoint to the fuel gas flow controller. The output of the flow controller is the pressure to the control valve, which changes the valve position and, therefore, the flow rate. Any change in the fuel gas header pressure will be "felt" immediately by the flow measurement[1], allowing the flow controller to take immediate corrective action. A block diagram of this strategy is shown in Figure 10–3.

In this strategy, the temperature controller is known as the *primary*, *master,* or *outer-loop* controller, while the flow controller is the *secondary*, *slave,* or *inner-loop* controller. The dynamics of the flow control loop are very fast, making the flow controller easy to tune. The temperature-control loop is much slower, so the primary loop can be effectively tuned as if the flow controller response is instantaneous.

For cascade control strategies where flow control is the inner loop, a control block diagram where the secondary and primary process transfer functions are in *series* (as in Figure 10–3) is natural. The primary disturbance (process inlet temperature) directly affects the primary output (process outlet temperature), with no direct effect on the

[1] Note that the flowmeter has been placed upstream of the control valve. This is common practice because a length of straight-run piping is needed upstream of the flowmeter for reasonable accuracy. Also, the pressure upstream of the valve is relatively constant (making the fluid density relatively constant), compared to the downstream pressure.

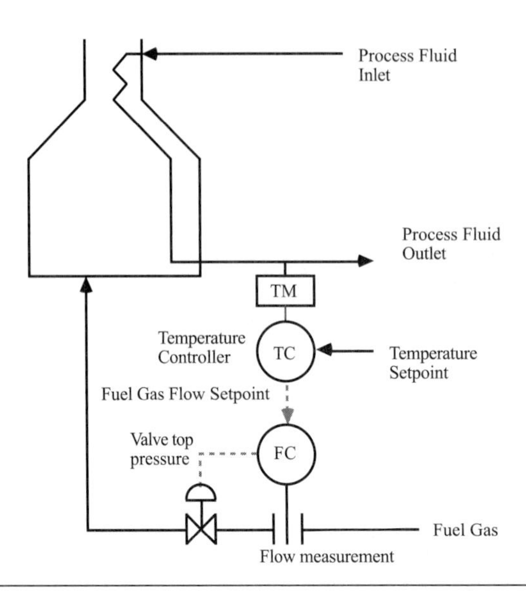

Figure 10–2 Cascade control of process outlet temperature.

secondary output (fuel gas flow rate, which is also the manipulated input for the primary process). The secondary disturbance (fuel gas header pressure) directly affects the secondary output, so the control block diagram shown in Figure 10–3 is relatively straightforward to generate. This is known as a *series* representation for the block diagram that is useful when the "secondary" process (such as the control valve or fuel gas flow rate) are naturally separated from the "primary" process (the heater and process fluid). It should be noted that other primary disturbances, such as process fluid flow rate, could easily be added to the diagram. See Exercise 9 for a numerical study of this process.

An example of the improved performance achievable with cascade control is shown in Figure 10–4, where a 1-psig disturbance in fuel gas header pressure occurs at $t = 0$ minutes. In the cascade-control strategy (Figure 10–2), the flow controller (secondary controller) rejects the header pressure disturbance before it is even "felt" by the process outlet temperature, since there is a primary process time-delay of one minute. The standard feedback controller (Figure 10–1) has poor performance due to the process time delay. It should be noted that setpoint changes and responses to a primary disturbance (process feed temperature) are the same for both control strategies (see Exercise 9 for a numerical study).

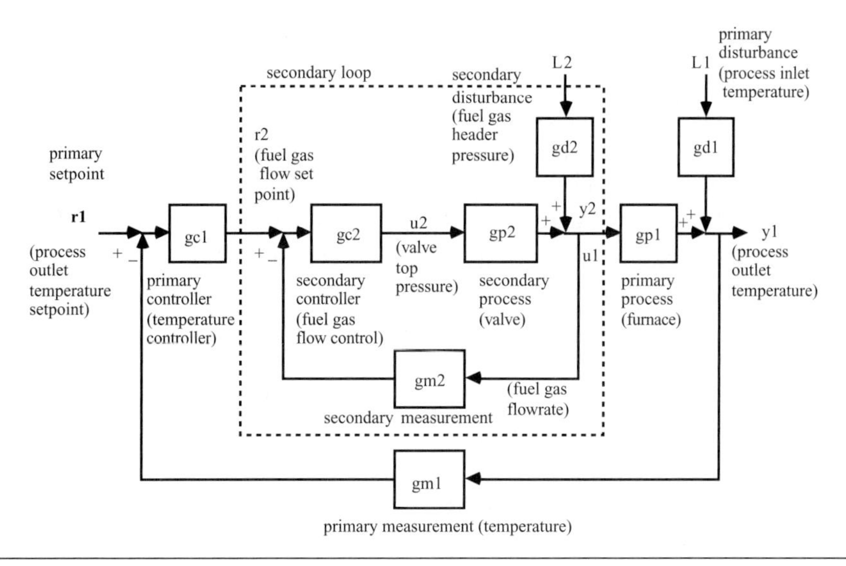

Figure 10–3 Cascade control block diagram for the fired heater. Series transfer function form.

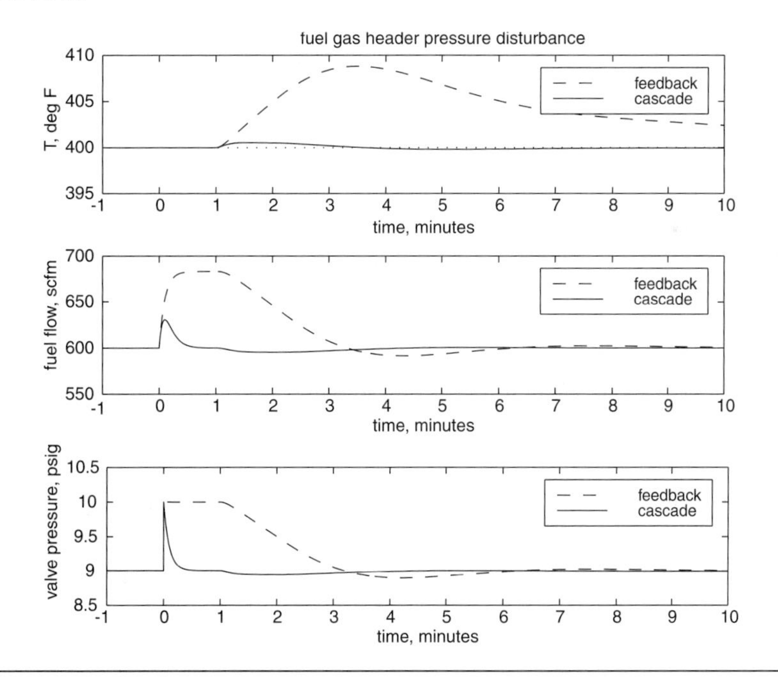

Figure 10–4 Response to a disturbance in fuel gas header pressure of 1 psig. Comparison of cascade control with standard feedback control.

Reactor Temperature Cascade Control

As another example of cascade control, consider the stirred-tank reactor shown in Figure 10–5. In the standard feedback-control strategy the temperature of the reactor is measured and the jacket flow rate is manipulated. If there is a disturbance in the jacket feed temperature, it will affect the jacket temperature, which will affect the reactor temperature. In the cascade-control strategy (Figure 10–6) the temperature of the reactor is measured and compared with the desired reactor temperature. The output of this reactor temperature controller is a setpoint to the jacket temperature controller. The jacket temperature controller manipulates the jacket flow rate. Notice that two measurements (reactor temperature and jacket temperature) are made but only one manipulated variable (jacket flow rate) is ultimately adjusted.

In this strategy, the reactor temperature controller is the primary (master or outer loop) controller, while the jacket temperature controller is the secondary (slave or inner loop) controller. This is effective because the jacket temperature dynamics are normally significantly faster than the reactor temperature dynamics. An inner-loop disturbance, such as jacket feed temperature, will be "felt" by the jacket temperature before it has a significant effect on the reactor temperature. This inner-loop (secondary) controller then adjusts the manipulated variable before a substantial effect on the primary output has occurred.

Notice that an appropriate block diagram representation of cascade control is less obvious here. Jacket flow rate certainly has a direct effect on the jacket temperature, which then affects the reactor temperature; one would think that a series representation would again be appropriate. Notice, however, that the jacket and reactor temperatures interact in a way that is different than the primary and secondary processes of the fired heater. Here, the reactor and jacket temperatures are directly coupled by the heat transfer between them. A change in the reactor temperature causes a change in jacket temperature and vice versa; a simple input-output transfer function block diagram does not truly capture this behavior. A more appropriate representation is shown in Figure 10–7, where a state space model is used to represent the interactions, and the temperatures are simply viewed as two outputs from the process. See Exercise 10 for a numerical study of a similar stirred-tank heater problem.

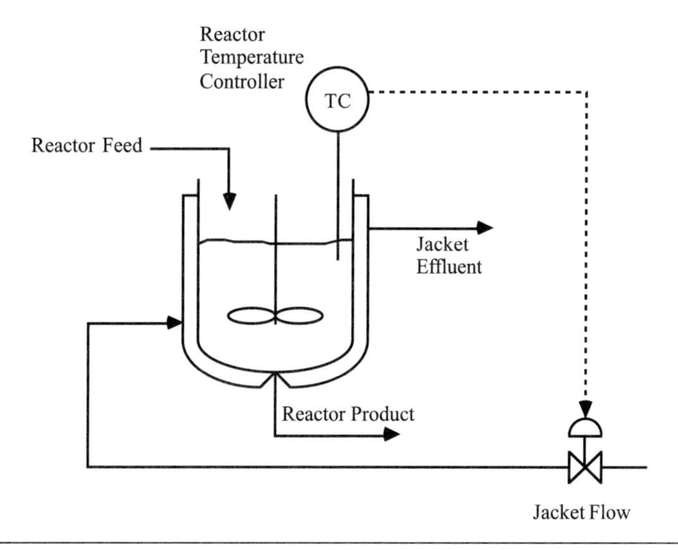

Figure 10–5 Stirred-tank reactor. Standard feedback control.

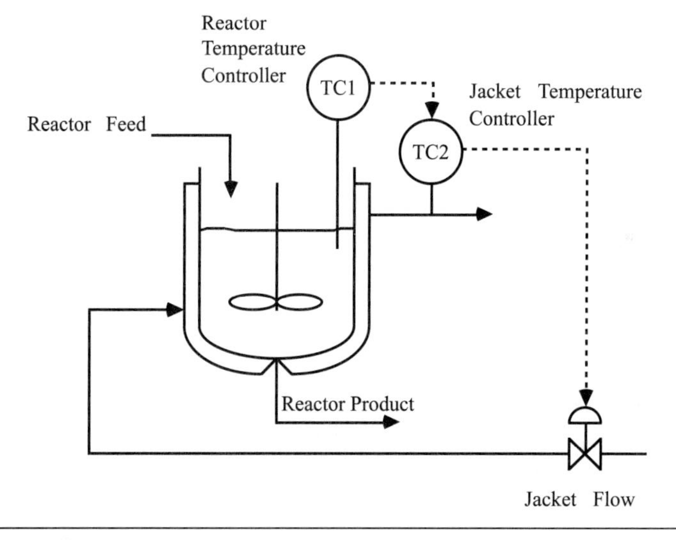

Figure 10–6 Stirred-tank reactor. Cascade control.

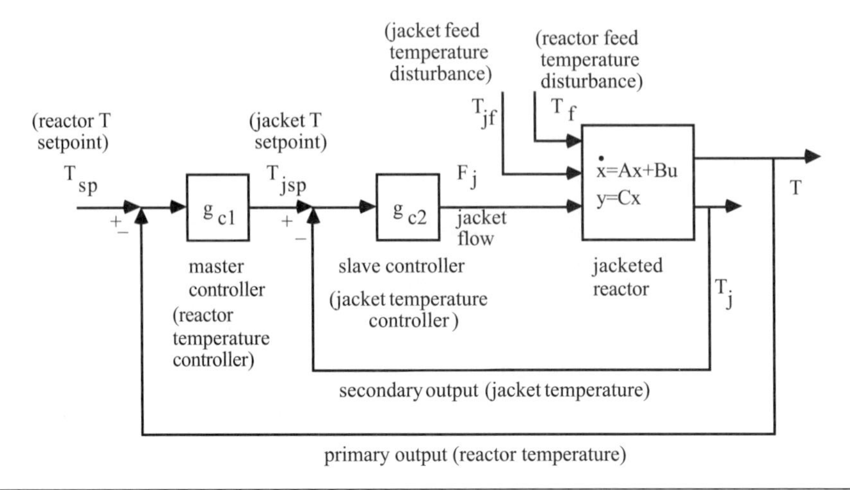

Figure 10–7 Cascade control block diagram for a jacketed chemical reactor. State space form for the reactor.

10.3 Cascade-Control Analysis

There are two common ways to represent a cascade-control system in transfer function form: *series* cascade and *parallel* cascade. The series cascade representation shown in Figure 10–8 is the most common, so it will be used in the analysis performed in this section. Students interested in analysis using the parallel structure can work Exercise 11.

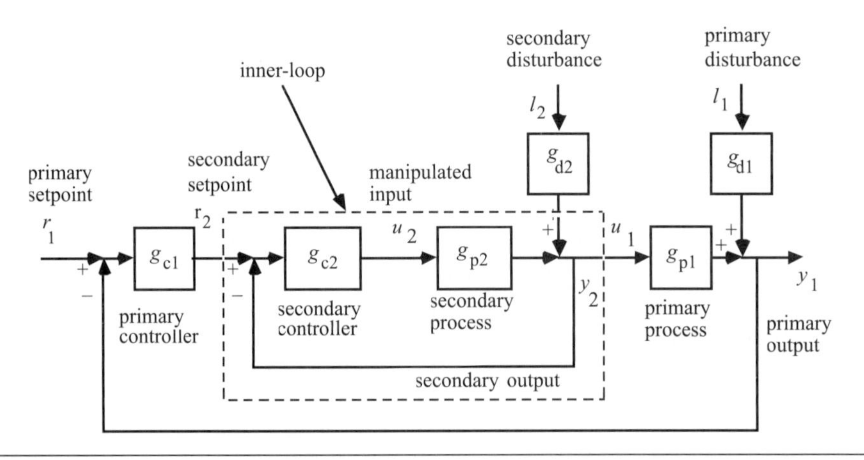

Figure 10–8 Series cascade structure.

We use the following algebraic manipulations to understand the effect of the secondary (inner-loop) on the primary (or outer) loop. Notice that the secondary output can be written

$$y_2(s) = \frac{g_{c2}(s)g_{p2}(s)}{1+g_{c2}(s)g_{p2}(s)} r_2(s) + \frac{g_{d2}(s)}{1+g_{c2}(s)g_{p2}(s)} l_2(s) \tag{10.1}$$

The secondary closed-loop transfer function can be defined as

$$g_{c2cl}(s) = \frac{g_{c2}(s)g_{p2}(s)}{1+g_{c2}(s)g_{p2}(s)} \tag{10.2}$$

Further analysis yields

$$y_1(s) = \frac{g_{c2}(s)g_{p2}(s)g_{p1}(s)}{1+g_{c2}(s)g_{p2}(s)} r_2(s) + \frac{g_{d2}(s)g_{p1}(s)}{1+g_{c2}(s)g_{p2}(s)} l_2(s) + g_{d1}l_1(s) \tag{10.3}$$

After tuning the inner loop, we can use the following transfer function to design the outer-loop controller.

$$g_{p1,eff}(s) = \frac{g_{c2}(s)g_{p2}(s)g_{p1}(s)}{1+g_{c2}(s)g_{p2}(s)} = g_{c2cl}(s)g_{p1}(s) \tag{10.4}$$

and the closed-loop relationship for a primary setpoint change is

$$y_1(s) = \frac{g_{c1}(s)g_{p1,eff}(s)}{1+g_{c1}(s)g_{p1,eff}(s)} r_1(s) = \frac{g_{c1}(s)g_{c2cl}(s)g_{p1}(s)}{1+g_{c1}(s)g_{c2cl}(s)g_{p1}(s)} r_1(s) \tag{10.5}$$

where it is clear that the secondary closed-loop transfer function affects the primary control loop. Notice that if the secondary control loop is much faster than the primary loop, so that $g_{c2CL} = 1$ (on a relative time scale to the primary control loop), then the closed-loop transfer function for the primary loop is

$$y_1(s) \approx \frac{g_{c1}(s)g_{p1}(s)}{1+g_{c1}(s)g_{p1}(s)} r_1(s)$$

10.4 Cascade-Control Design

Tuning the cascade controller consists of two steps. First, tune the inner-loop controller $[g_{c2}(s)]$ based on the secondary process $[g_{p2}(s)]$ and use the inner-loop/closed-loop transfer function [Equation (10.2)] to find an effective process transfer function [Equation(10.4)] for tuning the outer-loop controller $[g_{c1}(s)]$.

The inner-loop control system design can be based on IMC-based PID or any other procedure, such as frequency-response gain and phase margins. Once the inner-loop is

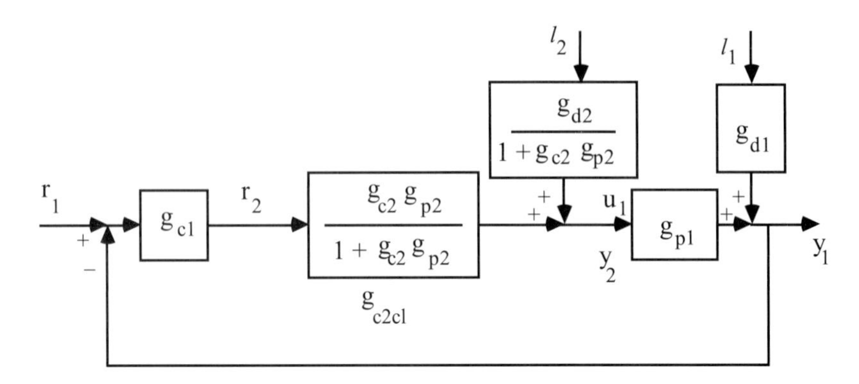

Figure 10–9 Series cascade structure using a closed-loop transfer function for the inner-loop.

tuned, the effective outer-loop transfer function can be used to tune the outer-loop controller. Again, any control design procedure can be used.

Rules of Thumb for Cascade Control

- Cascade control can be successfully used to reject secondary-process disturbances when the primary process has a much larger time constant and a time delay or RHP zero while the secondary process has a small time constant and little or no dead time or nonminimum-phase (RHP zeros) behavior.
- The most common cascade-control loop involves a flow controller as the inner loop. This type of loop easily rejects disturbances in fluid stream pressure, either upstream or downstream of the valve.
- The inner loop (secondary, slave loop) in a cascade-control strategy should be tuned before the outer loop (primary, master loop). After the inner loop is tuned, and closed, the outer loop should be tuned using knowledge of the dynamics of the inner loop.
- Cascade control is widely used on chemical processes, and should not be considered an "advanced control strategy."
- There is little or no advantage to using cascade control if the secondary process is not significantly faster than the primary process dynamics. In particular, if there is much dead time in the secondary process, or if there is an RHP zero, it is unlikely that cascade control will be much better than standard feedback control.
- Cascade control can be easily combined with other forms of control, such as feed-forward control.

- Any of your favorite methods can be used for tuning the inner and outer loops. Indeed, the IMC strategy can be used on either or both of the inner and outer loops.

10.5 Cascade IMC

IMC can also be implemented in a cascade structure. Frequently, the primary controller is IMC, while the secondary controller is PID; this is particularly true when the secondary control loop involves flow control, as shown in Figure 10–10. Notice that the measured flow implemented on the process is also used on the model (recall that the model output is obtained by integrating the model differential equations). If the control valve hits a constraint and cannot achieve the desired flow rate (flow setpoint), then the process model "knows" this and a realistic model output is predicted. This minimizes a problem known as "reset windup," which is sometime associated with the manipulated inputs hitting a constraint (this will be discussed further in Chapter 11).

The internal model controller should be designed based on a model that includes the flow control loop dynamics

$$\tilde{g}_{p\_des}(s) = \tilde{g}_p(s)\tilde{g}_{c2cl}(s)$$

The reader should also be able to develop a control block diagram for cascade control where both the primary and secondary loops use IMC (see Exercise 12).

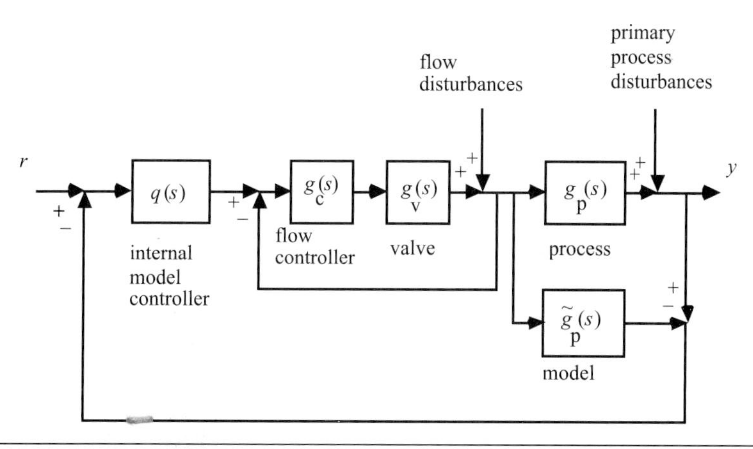

Figure 10–10 Cascade IMC outer-loop, PID flow-control inner-loop.

10.6 Feed-Forward Control

When a disturbance affects a process under feedback control it is necessary for a measured process output to change before corrective action is taken to change the manipulated input. It would be preferable to have a sensor that measures the disturbance and adjusts the manipulated input before the process output changes. Consider the furnace system studied earlier in this chapter and shown in Figure 10–11. One possible disturbance is the process fluid flow rate; if the flow rate increases 20%, then the required heat duty should also increase by 20%. The purpose of the feed-forward control strategy shown is to immediately change the fuel gas flow rate (for simplicity we assume that the fuel gas flow rate is directly manipulated in this diagram) when a change in the process flow rate is sensed.

For this furnace example, the steady-state energy balance can be used to find the fuel gas flow rate

$$\dot{n}_{fg} = F\rho c_p\left(T - T_f\right)\cdot\left(\frac{\varepsilon}{\Delta H^c}\right)$$

where F = process stream flow rate, ρ = density, c_p = heat capacity, T = desired outlet temperature, T_f = feed temperature, ε = furnace efficiency, and ΔH^c = heat of combustion of fuel gas. A steady-state feed-forward controller computes the fuel gas flow rate based on the measured process flow rate (and possibly the process feed temperature). Sometimes it is important to also consider dynamic effects, as shown in several examples that follow.

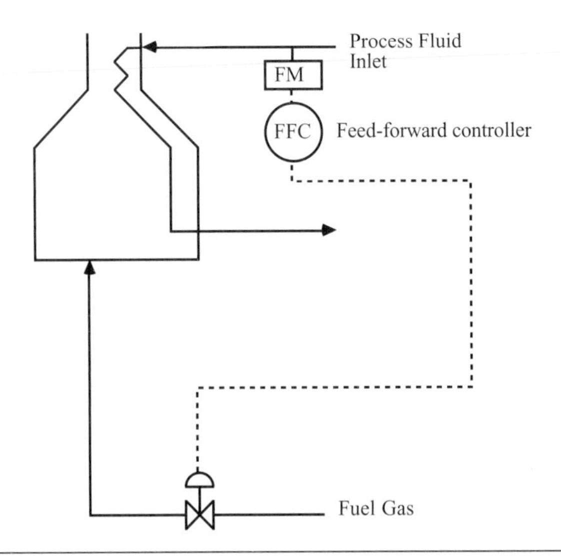

Figure 10–11 Feed-forward control of temperature based on the measured process fluid flow rate (disturbance input).

The performance of feed-forward control is limited by model uncertainty. Without a perfect process model the controller cannot completely compensate for disturbances and there will generally be offset. In practice, feed-forward control is combined with feedback control, as shown in Figure 10–12. Here the feed-forward controller compensates for process flow rate disturbances, while the feedback controller compensates for other disturbances and model uncertainty. The corresponding block diagram is shown in Figure 10–13.

Consider the feed-forward/feedback block diagram shown in Figure 10–13. We can develop the closed-loop transfer function by using the following steps:

$$y(s) = g_d(s)l(s) + g_p(s)u(s) = g_d(s)l(s) + g_p(s)[u_f(s) + u_c(s)]$$
$$y(s) = g_d(s)l(s) + g_p(s)[g_{cf}(s)g_{mf}(s)l(s) + g_c(s)(r(s) - g_m(s)y(s))]$$
$$y(s) = [g_d(s) + g_p(s)g_{cf}(s)g_{mf}(s)]l(s) + g_p(s)g_c(s)r(s) - g_p(s)g_c(s)g_m(s)y(s)$$

Rearranging, to solve for $y(s)$, we find

$$(1 + g_p(s)g_c(s)g_m(s))y(s) = [g_d(s) + g_p(s)g_{cf}(s)g_{mf}(s)]l(s) + g_p(s)g_c(s)r(s)$$

or

$$y(s) = \frac{g_d(s) + g_p(s)g_{cf}(s)g_{mf}(s)}{1 + g_p(s)g_c(s)g_m(s)}l(s) + \frac{g_p(s)g_c(s)}{1 + g_p(s)g_c(s)g_m(s)}r(s) \tag{10.6}$$

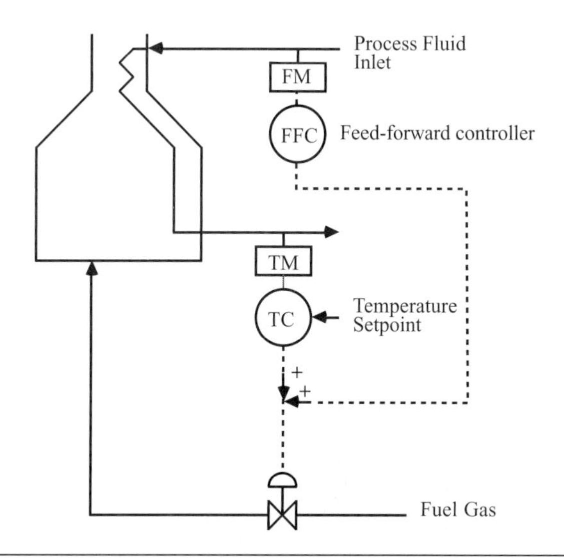

Figure 10–12 Feed-forward/feedback control of temperature based on the measured process fluid flow rate (disturbance input).

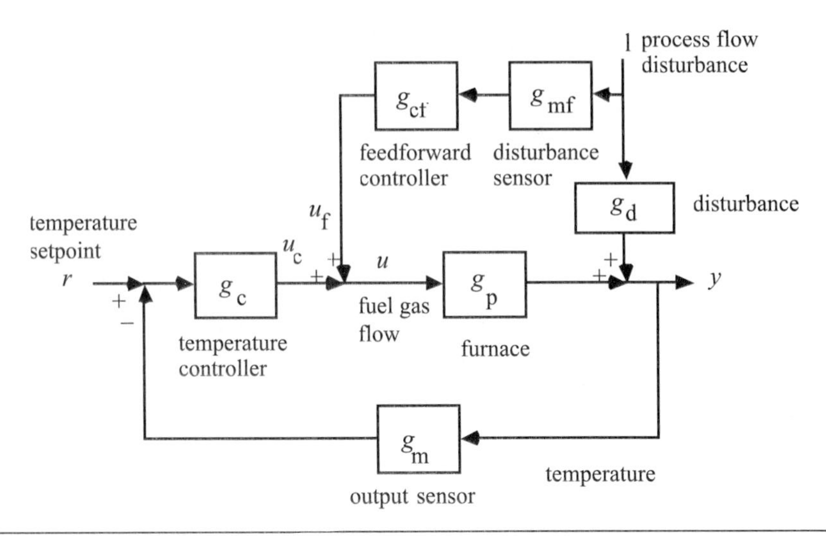

Figure 10–13 Feed-forward/feedback control block diagram for the furnace example.

Equation (10.6) is the closed-loop relationship for feed-forward/feedback control. Notice that a (stable) feed-forward controller does not affect the closed-loop stability, since the feed-forward controller transfer function does not appear in the closed-loop characteristic equation (denominator polynomial).

10.7 Feed-Forward Controller Design

Remember that we are working in deviation variables. Assume that we are not making a setpoint change, so $r(s) = 0$. Since we do not want the output variable to change, that is equivalent to saying that we desire $y(s) = 0$. From Equation (10.6) we can see that the only way to assure $y(s) = 0$ is for the following relationship to hold

$$g_d(s) + g_p(s)g_{cf}(s)g_{mf}(s) = 0 \tag{10.7}$$

In this expression all the transfer functions are determined for a particular system, except for the feed-forward controller, $g_{cf}(s)$. Solving Equation (10.7) for $g_{cf}(s)$, we find

$$g_{cf}(s) = -\frac{g_d(s)}{g_p(s)g_{mf}(s)} \tag{10.8}$$

If we make the simplifying assumption that the disturbance measurement has no dynamics, then

$$g_{cf}(s) = -\frac{g_d(s)}{g_p(s)} \tag{10.9}$$

Notice that Equation (10.9) requires the inverse of the process model. Clearly, there will be problems if the process model has RHP zeros, or if the process time delay is greater than the disturbance time delay. The use of Equation (10.9) as a feed-forward design equation is shown in Examples 10.1–10.4.

Example 10.1: First-Order Process and Disturbance Transfer Functions

Consider the first-order process and disturbance transfer functions

$$g_p(s) = \frac{k_p}{\tau_p s + 1}, \quad g_d(s) = \frac{k_d}{\tau_d s + 1} \tag{10.10}$$

The feed-forward control law, from Equations (10.9) and (10.10), is

$$g_{cf}(s) = -\frac{k_d/\tau_d s + 1}{k_p/\tau_p s + 1} = -\left(\frac{k_d}{k_p}\right)\frac{\tau_p s + 1}{\tau_d s + 1} \tag{10.11}$$

which is simply a lead-lag controller. Most control systems have feed-forward controllers that are of lead-lag form.

Numerical Example

Here we study the case where $k_d = k_p = 1$ and $\tau_d = \tau_p = 5$ minutes. If we use the IMC-based PI procedure for feedback control, we find that $\tau_I = 5$ and $k_c = 2$ for $\lambda = 2.5$ minutes. We compare the disturbance rejection of feedback-only with feed-forward/feedback control in Figure 10–14 for a unit step change in the load disturbance at $t = 0$. The feed-forward controller is a static controller with $k_{cf} = 1$ for this case. Notice that the feed-forward controller implements an immediate change in the manipulated input to counteract the disturbance.

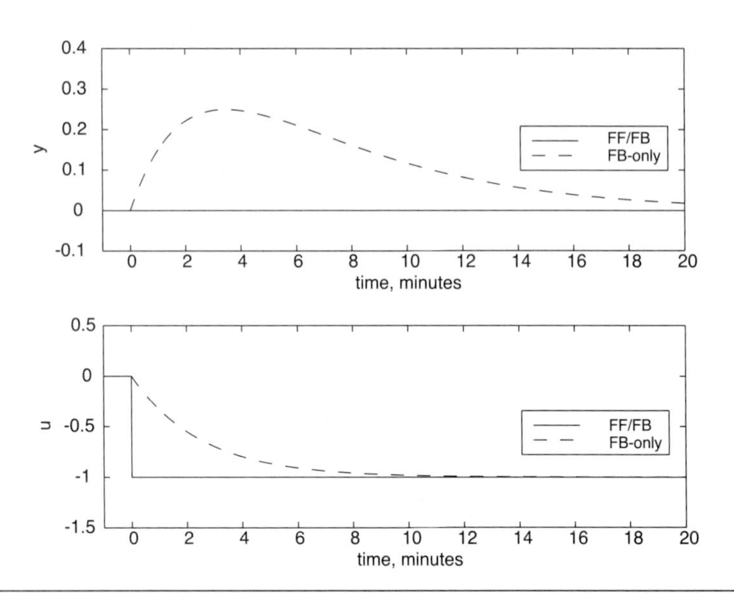

Figure 10–14 Comparison of feed-forward/feedback with feedback-only for numerical example.

Example 10.2: First-Order + Dead Time Process and Disturbance Transfer Functions

Consider the transfer functions

$$g_p(s) = \frac{k_p e^{-\theta_p s}}{\tau_p s + 1}, \quad g_d(s) = \frac{k_d e^{-\theta_d s}}{\tau_d s + 1} \tag{10.12}$$

The feed-forward control law, from Equations (10.9) and (10.12), is

$$g_{cf}(s) = -\frac{k_d e^{-\theta_d s} / \tau_d s + 1}{k_p e^{-\theta_p s} / \tau_p s + 1} = -\left(\frac{k_d}{k_p}\right) \frac{\tau_p s + 1}{\tau_d s + 1} e^{-(\theta_d - \theta_p)} \tag{10.13}$$

which is simply a lead-lag feed-forward controller with a time delay. The main thing that we notice about Equation (10.13) is that $\theta_d \geq \theta_p$ is required for the controller to be realizable. If $\theta_p > \theta_d$, we cannot have a time-delay term in the feed-forward controller and must accept the fact that we cannot have perfect feed-forward control, even if the models are perfect.

Example 10.3: Process Higher Order than Disturbance Transfer Function

Consider the following transfer functions:

$$g_p(s) = \frac{2}{(15s+1)(3s+1)}, \quad g_d(s) = \frac{1.5}{5s+1} \tag{10.14}$$

The feed-forward controller is

$$g_{cf}(s) = -0.75\frac{(15s+1)(3s+1)}{5s+1} \tag{10.15}$$

Since Equation (10.15) is improper, it is not physically realizable. We could simply add a first-order filter, perhaps something like $1/(s + 1)$, to make the feed-forward controller second order over second order. Perhaps the simplest thing to do is to make the numerator term first order. Most likely, we would use 18s + 1 to approximate the numerator to arrive at the lead-lag controller

$$g_{cf}(s) = -0.75\frac{(18s+1)}{5s+1} \tag{10.16}$$

Numerical Simulation

Assume that an IMC-based PID controller is used for feedback control. With $\lambda = 9$, we find (from Table 9–1) that $k_c = 1$, $\tau_I = 18$, and $\tau_D = 2.5$. Figure 10–15 compares the closed-loop response of the simplified feed-forward/feedback [Equation (10.16)] with feedback-only to a unit step change in load disturbance at $t = 0$. Although the feed-forward/feedback control is not perfect, the performance is substantially better than feedback only control. This performance is achieved by the initial manipulated variable action that is characteristic of lead-lag feed-forward controllers with a numerator lead time constant that is greater than the denominator lag time constant (see the lead-lag transfer function behavior in Chapter 3).

Example 10.4: Process Has Inverse Response, Disturbance Does Not

Consider the process and disturbance transfer functions

$$g_p(s) = \frac{-9s+1}{(15s+1)(3s+1)}, \quad g_d(s) = \frac{1.5}{5s+1} \tag{10.17}$$

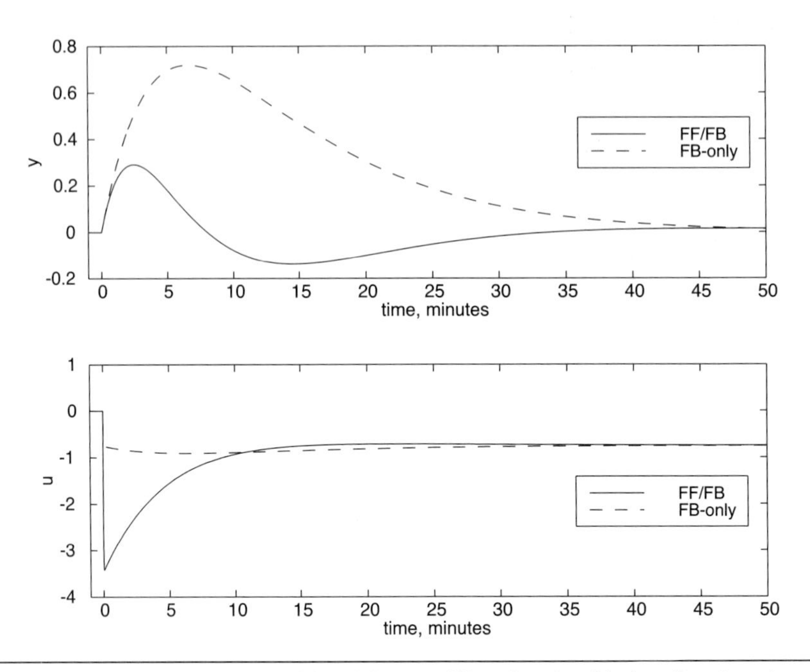

Figure 10–15 Comparison of feed-forward/feedback with feedback-only.

The feed-forward controller is

$$g_{cf}(s) = -1.5\frac{(15s+1)(3s+1)}{(-9s+1)(5s+1)} \tag{10.18}$$

Notice that Equation (10.18) has a RHP pole and is therefore unstable. We must factor RHP zeros from the process transfer function before designing a feed-forward controller. In all likelihood, we would neglect the dynamic part and use only a static feed-forward controller, as shown in Equation (10.19) below.

Static Feed-Forward Control

Often, we will be satisfied to neglect the differences in the dynamics between a process and a disturbance, and to simply design a static feed-forward compensator,

$$g_{cf}(s) = k_{cf} = -\frac{k_d}{k_p} \tag{10.19}$$

This will get us the majority of the performance improvement over feedback-only control.

In the next section, we show feed-forward control in the IMC structure.

10.8 Feed-Forward Control in the IMC Structure

Feed-forward control can also be implemented with an IMC structure, as shown in Figure 10–16, where we have neglected the measurement dynamics for clarity. The feed-forward control design is

$$q_f(s) = -\frac{\tilde{g}_d(s)}{\tilde{g}_p(s)} \tag{10.20}$$

Again, noninvertible elements of $q_f(s)$ must be removed before the feed-forward controller can be implemented.

The closed-loop transfer relationship (the derivation is left as an exercise for the reader) is

$$y(s) = \frac{g_p(s)q(s)}{1+q(s)\left(g_p(s)-\tilde{g}_p(s)\right)}r(s) + \left[\frac{g_p(s)q_f(s) - g_p(s)q(s)\left(g_d(s)-\tilde{g}_d(s)\right)}{1+q(s)\left(g_p(s)-\tilde{g}_p(s)\right)} + g_d(s)\right]l(s)$$

10.9 Summary of Feed-Forward Control

We have developed feed-forward control based on models of the disturbance and process transfer functions. If the disturbance and process models are not available, feed-forward control can still be implemented as a tunable gain plus lead-lag controller. You should consider the following important issues when implementing feed-forward control:

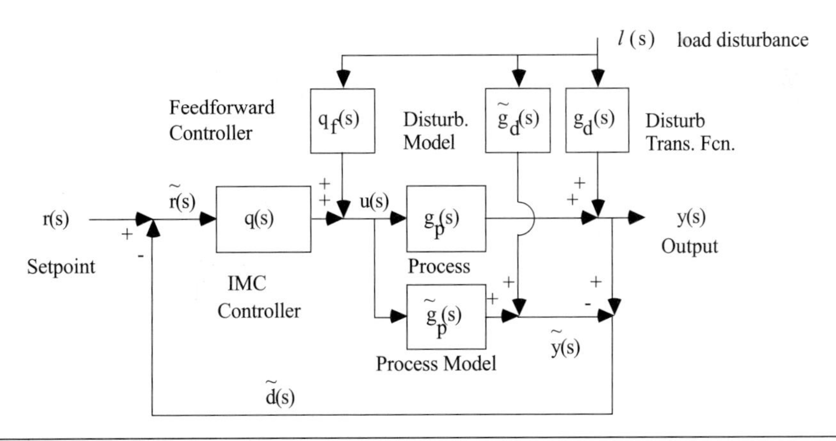

Figure 10–16 Feed-forward control in the IMC structure.

- The feed-forward controller must be physically realizable and stable, as noted by the next three items.
  - The disturbance time delay must be greater than the process time delay for perfect feed-forward compensation. If the process time delay is greater than the disturbance time delay, the feed-forward controller will have no time-delay compensation, and perfect control cannot be achieved.
  - If the process has an RHP zero, it must be factored out before designing the feed-forward controller. Otherwise, the feed-forward controller will be unstable.
  - If the process is higher order than the disturbance, fast time constants probably must be neglected in designing the feed-forward controller.
- Often a static feed-forward controller will have good performance, particularly if the disturbance dynamics are the same timescale as the process dynamics.
- A feed-forward controller does not change the closed-loop stability of the feedback system, assuming the feed-forward controller is stable. Also, a feed-forward controller does not change the setpoint response of a closed-loop system.
- Feed-forward control can be implemented with either the classical feedback (PID-type) or IMC structure.

10.10 Combined Feed-Forward and Cascade

In Section 10.6, a fired heater was used to provide motivation for feed-forward control. In the example it was assumed that the fuel gas flow was directly manipulated. In practice a flow controller would be used, resulting in the feed-forward, feedback, and cascade-control strategy shown in Figure 10–17.

Notice that each "mode" is used to reject a different type of disturbance. The feed-forward controller is used to reject feed flow rate disturbances. The fuel gas flow controller is used to reject disturbances in the fuel gas header pressure. The temperature controller plays a role in compensating for all disturbances. See Exercise 19 for a numerical example.

A common process that combines feed-forward/feedback and cascade control is a steam drum. See Module 9 for a case study of steam drum level control.

10.11 Summary

Disturbance rejection provides the motivation for cascade and feed-forward control. Cascade control uses a secondary process measurement to detect and correct for a disturbance before it affects the primary process output. Feed-forward control uses a direct measurement

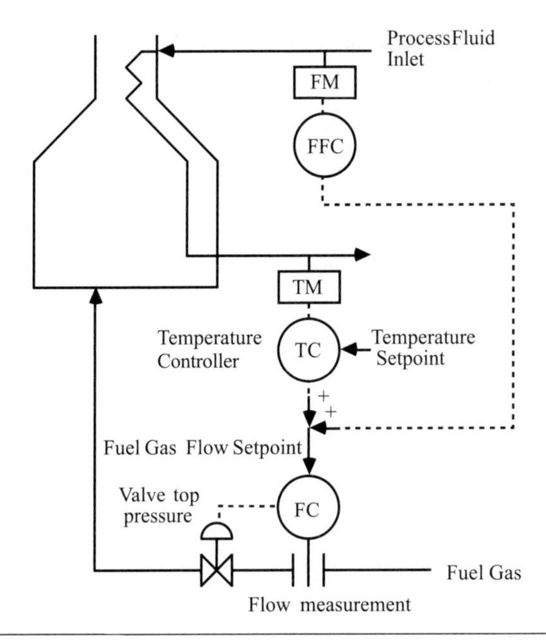

Figure 10–17 Feed-forward, feedback, and cascade control of a fired heater.

of the disturbance to change the manipulated input before the disturbance affects the output. It is common to combine feed-forward/feedback with cascade control, as shown in the furnace control block diagram, Figure 10–17.

Although some texts refer to cascade and feed-forward control as "advanced control," these techniques are commonly used in industry. Almost every control system that has a flow rate as a manipulated variable will have a flow controller as an inner loop in a cascade strategy.

References

A number of feed-forward control examples are presented in the following sources:

Badavas, P.C., "Feedforward Methods for Process Control Systems," *Chemical Engineering*, 103–108 (October 15, 1984).

Stephanopoulos, G., *Chemical Process Control*, Prentice Hall (1984).

Student Exercises—Cascade Control

1. Derive the closed-loop transfer function relating the primary setpoint to the primary process output for a cascade-control system.
2. Derive the closed-loop transfer function relating a secondary disturbance to the primary output for a cascade-control system.
3. For the following process, develop the *double cascade*-control loop diagram, where the jacket flow controller is a "tertiary controller." Use the state space form similar to Figure 10–6, where jacket flow rate is an input to the state space representation of the reactor.

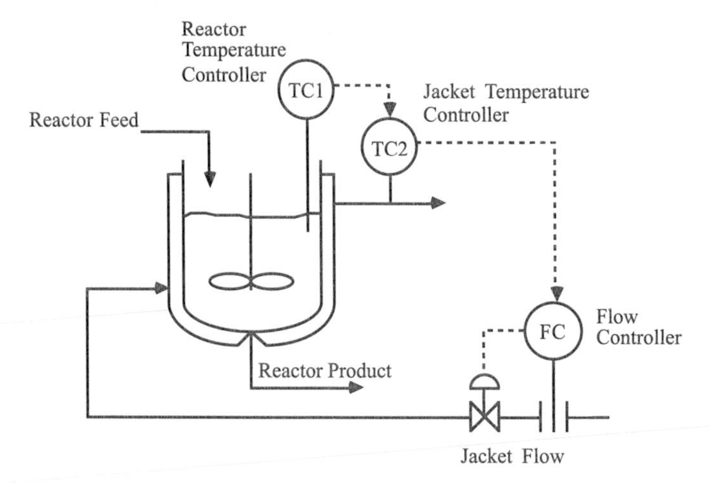

4. Consider the following instrumentation diagram for a chemical reactor. C_{sp} represents a concentration setpoint.

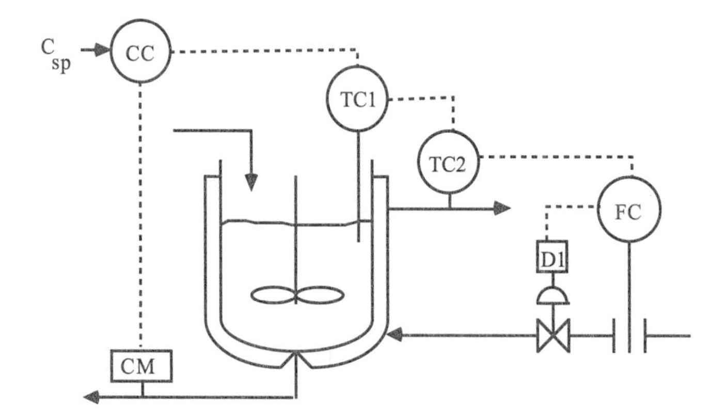

Draw the control block diagram for this system. Label every transfer function and signal on the diagram. What type of control strategy would you call this? What is the primary measured variable for this system? What is the primary control variable for this system? Which controller is probably tuned to be the slowest? Which controller is probably tuned to be the fastest? What is the gain for device D1 if a current signal (4–20 mA) is being converted to a pneumatic signal (3–15 psig) (show units)?

5. For a particular biochemical reactor, the concentration of the microorganism (bug) is controlled by varying the pH in the reactor. The pH is controlled by manipulating the flow of caustic solution to the reactor. The bugs grow by "eating" a waste chemical in the feed stream.
   a. Complete the instrumentation diagram shown below. Draw the control block diagram associated with the bug concentration control strategy.
   b. What type of control strategy is this (for the bug concentration control)?
   c. Identify the control objective and possible disturbance variables. Name the disturbance variables that each controller is meant to reject.

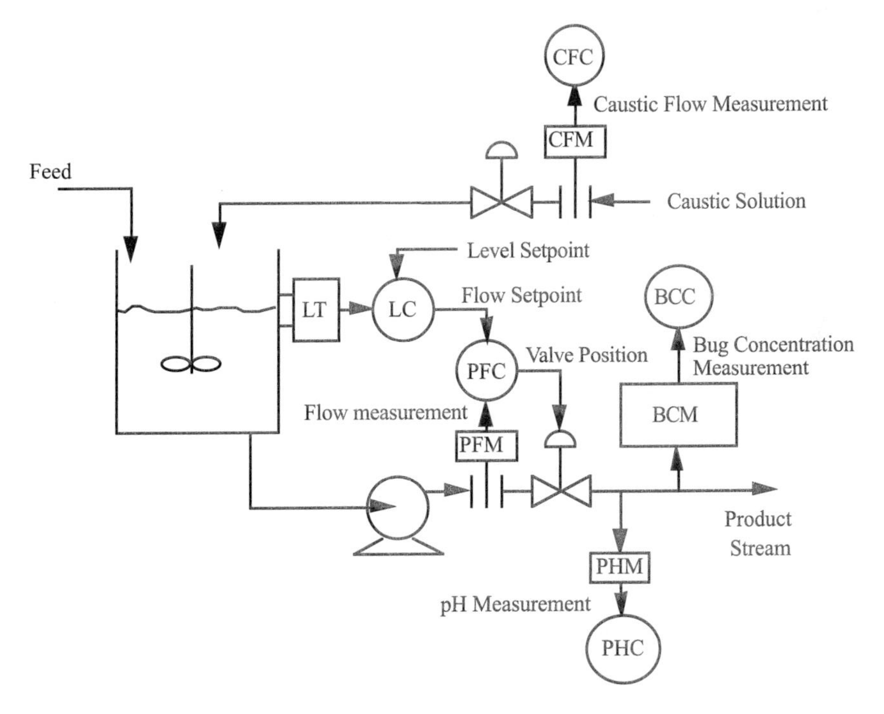

6. As a budding young engineer in the semiconductor device manufacturing industry, your first assignment involves a chemical vapor deposition reactor. Before you obtained this position, the process did not have many feedback

control loops. Your goal is to design a control system that can control the growth rate of a thin film on a wafer. A spectroscopic ellipsometer is available to measure the growth rate of this film. The growth rate is a function of the composition [mole fraction of TMG (trimethyl gallium)] of the inlet feed stream. The composition of the inlet feed stream is a function of the flow rate of hydrogen that bubbles through a bottle of TMG.

Draw the following control instrumentation diagram. Now, *draw the corresponding control block diagram*, in order to be able to analyze the control loops and tune the controllers. Label all controllers, measurement devices, and signals on the block diagram.

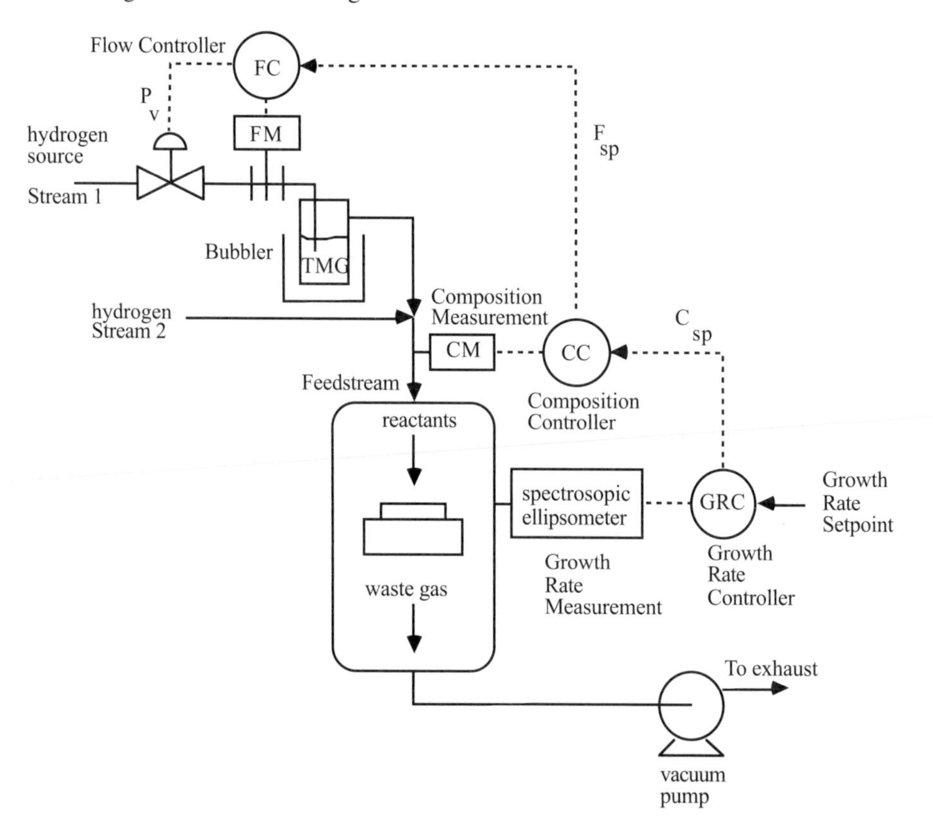

7. Consider the following depiction of the bottom portion of a distillation column. It is desired to control the composition of the bottom stream at a certain value (setpoint) by manipulating the steam to the reboiler. Composition is a slow variable, and the measurement device can have significant delays; it is known that measuring and controlling the temperature of a tray in the column can improve disturbance rejection. The temperature is regulated by the pressure of

the steam in the reboiler, which is a function of the steam flow to the reboiler. Show the proper signals for a cascade type of control system. Also, develop the corresponding block diagram, showing clearly the disturbances that each loop is meant to reject.

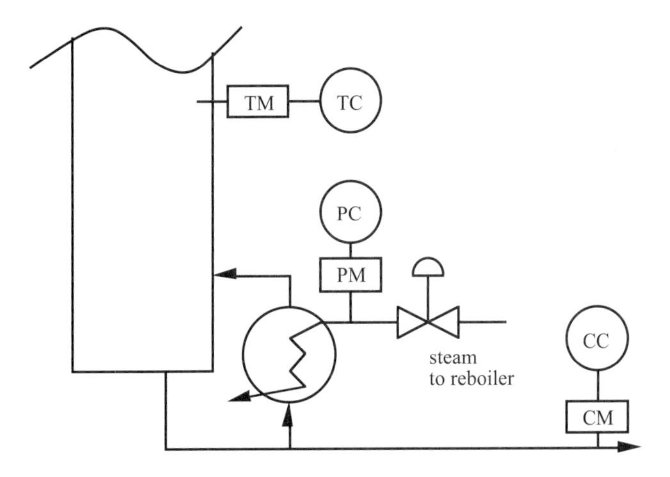

8. A common type of chemical reactor is a packed-bed reactor, which is assumed to operate adiabatically (no heat loss). The objective of the following reactor control strategy is to maintain a desired reactor outlet concentration by manipulating the temperature of the inlet stream to the reactor. The reactor inlet temperature is maintained by manipulating the steam flow to a feed preheat exchanger. Connect the measurements and control devices on the control instrumentation diagram and draw the corresponding control block diagram. Label all controllers and signals on both the control instrumentation and control block diagrams. Draw appropriate disturbances on the control block diagram.

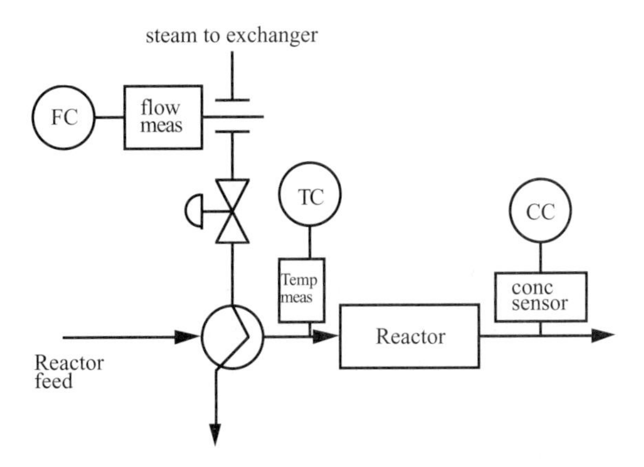

9. As a young process engineer, you have been given the responsibility for a catalytic reforming unit at a refinery that, for some reason, is just now getting involved with "advanced control." The catalytic reforming unit produces roughly 25% of the gasoline produced by the refinery. The throughput of the unit is 25,000 barrels per day (Bbl/day) and the value is roughly $30/Bbl. The gross operating income of the unit is then roughly $1 million/day. You feel certain that tighter control can improve the yield of the unit, increase the value of the product, and reduce operating costs. The reforming unit consists of three fuel gas-fired furnaces and three catalytic reactors with a number of additional heat exchangers and separation columns. You feel that you must prove that advanced control can work on part of the process before you attempt to implement it on the entire process. You decide that the place to start is on the temperature control of the process fluid leaving the first furnace.

A simplified process instrumentation diagram for the furnace that heats the process stream for the first reactor is shown in Figure 10–1. The "old school" operations personnel have been controlling the outlet temperature of the process stream by simply manipulating the fuel gas valve directly. You realize that this method allows fuel gas header pressure changes to affect the fuel gas flow rate, which will affect the process fluid outlet temperature.

You wish to show the benefits of *cascade control* for this system. Use your practical engineering knowledge, a number of discussions with the process operators responsible for the reforming unit, an investigation of the unit log books, and determine the following.

Relevant information:

a. For each increase of 1 psig to the fuel gas valve actuator, the fuel gas flow rate increases by 83.3 scfm (standard cubic feet per minute).

b. For each increase of 1 scfm in fuel gas flow, there is an increase in the outlet reformate stream temperature of 0.36°F.

c. For a change in the upstream fuel gas header pressure of 1 psig, there is a corresponding change of 13.5 scfm in the fuel gas flow rate.

d. For a change in the inlet process stream of 1°F, there is a change of 1°F in the outlet temperature.

e. You assume that the time constant associated with the control valve is roughly 6 seconds (0.1 minutes).

f. The process time constant associated with the furnace is approximately 5 minutes and there is an additional 1 minute time delay.

Design a cascade-control strategy and compare results (using SIMULINK) with the standard feedback-control strategy for two types of disturbances: fuel gas header pressure (±5 psig), and process stream inlet temperature disturbances (±25°F).

10. Consider a stirred-tank heater that has the following process model:

$$\frac{dT}{dt} = \frac{F}{V}\left(T_f - T\right) - \frac{UA}{V\rho c_p}\left(T - T_j\right)$$

$$\frac{dT_j}{dt} = \frac{F_{jf}}{V_j}\left(T_{jf} - T_j\right) + \frac{UA}{V_j\rho_j c_{pj}}\left(T - T_j\right)$$

with the parameter and variable values

$V/F = 30$ minutes
$UA = 24$ kcal/min °C
$V\rho c_p = 850$ kcal/°C
$V_j\rho_j c_{pj} = 120$°C
$T_{jf} = 120$°C
$T = 40$°C
$T_f = 20$°C

a. Find the state space model for these conditions, if the inputs are F_j, T_{jf}, and T_f.

b. Show that the linear transfer function model has the form of the following block diagram and find all of the process and disturbance transfer functions. Is a zero of the secondary process transfer function related to the pole of the primary process transfer function?

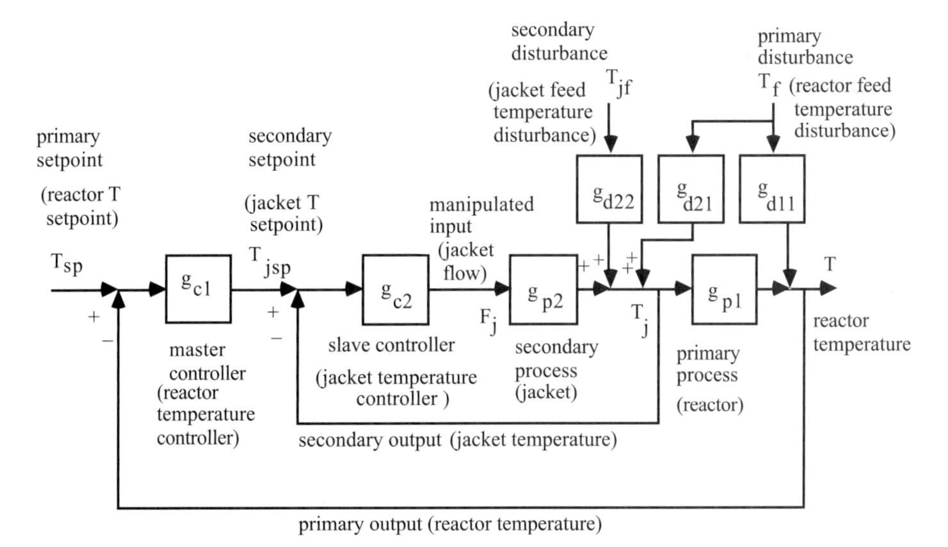

c. If changes in the vessel temperature are slow and can be neglected so that only the second (jacket temperature) equation is considered, find the secondary transfer function, g_{p2} (this is first-order). How does this compare with part b?

11. Consider a cascade control block diagram based on a parallel process representation (note that the disturbances are excluded here). Show that the closed-loop transfer function for the inner-loop is the same as for the series representation. Derive the closed-loop transfer function for the outer-loop as a function of the inner-loop closed-loop transfer function. Also, relate g_p to g_{p1} and g_{p2}.

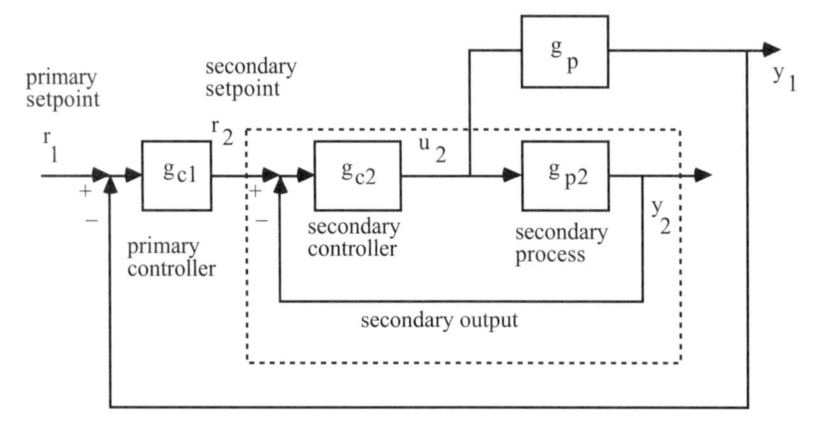

12. Develop a block diagram that has the IMC structure on both the primary and secondary loops.

Student Exercises—Feed-Forward Control

13. Use physical reasoning to discuss why the disturbance dead time must be greater than the process dead time to even have the possibility of perfect feed-forward disturbance compensation.

14. Let $g_d(s) = \dfrac{10.5(-4s+1)e^{-2s}}{(1.5s+1)(12s+1)}$ and $g_p(s) = \dfrac{3.5(-5s+1)e^{-3s}}{(1s+1)(10s+1)}$.

Design a feed-forward controller, $g_{cf}(s)$, for this system. Elaborate on its expected performance.

15. A mixing vessel is used to maintain a desired pH level in a stream flowing to a waste treatment plant. The pressure to the valve on an acid stream is used as the manipulated variable. Most of the variability in pH is due to waste stream 1, which is a caustic stream. It is desirable to implement a feed-forward controller to reject the pH disturbances due to this stream.

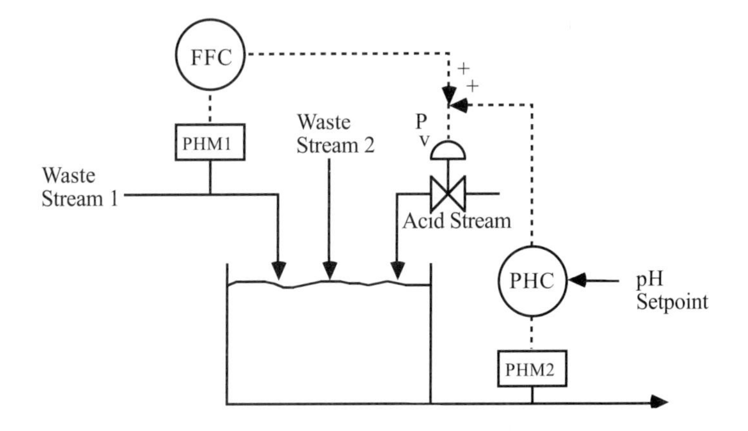

The following data are relevant to this problem. Without control, a change in the inlet pH of 0.5 leads to a change of 0.25 pH in the outlet stream. The time delay is 10 minutes and the time constant is 30 minutes. A change in the acid stream valve-top pressure of 1 psig leads to a change in pH in the outlet stream of 0.4 pH. The time constant is 25 minutes and the time delay is 7.5 minutes.

Design a feed-forward controller for this process. Show units.

16. Derive the following closed-loop relationship for IMC with feed-forward control,

$$y(s) = \frac{g_p(s)q(s)}{1+q(s)\left(g_p(s)-\tilde{g}_p(s)\right)} r(s) + \left[\frac{g_p(s)q_f(s) - g_p(s)q(s)\left(g_d(s)-\tilde{g}_d(s)\right)}{1+q(s)\left(g_p(s)-\tilde{g}_p(s)\right)} + g_d(s)\right] l(s)$$

and show that there is no offset for a step load disturbance.

Student Exercises—Feed-Forward and Cascade

17. Consider again the reactor control system from Exercise 8. If there is also a measurement of the concentration of the reactor feed stream (before it enters the heat exchanger), show how it can be used in a feed-forward-control strategy (using the previously developed cascade type of strategy). Show this on both the control instrumentation diagram below and the control block diagram.

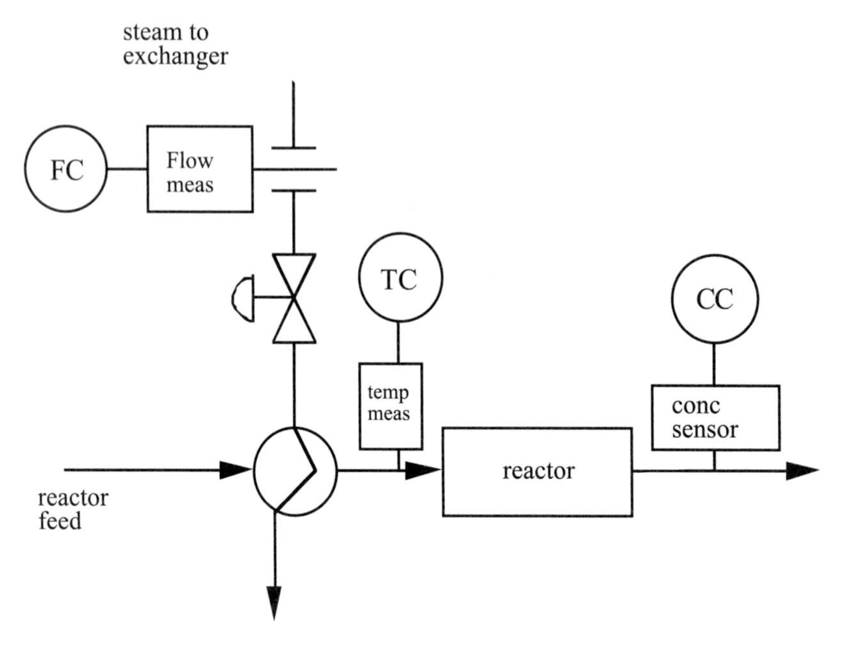

18. Design a feed-forward/cascade control strategy for the following level control problem. Draw the instrumentation directly on the figure. Draw a control block diagram, labeling all the signals on the diagram

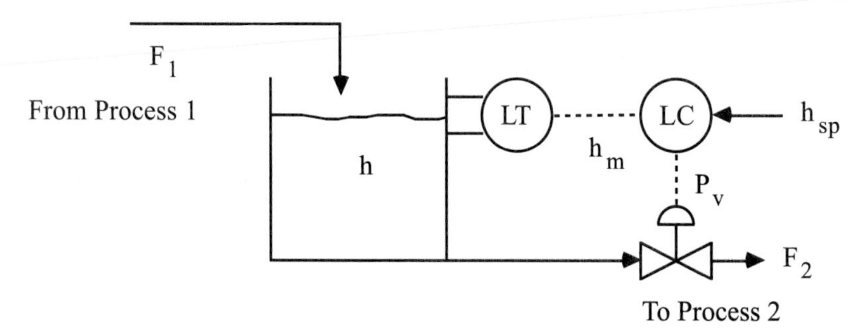

19. Consider the heater shown in Exercise 9. Now consider implementing feed-forward and cascade control when process flow disturbances can occur, as shown below.

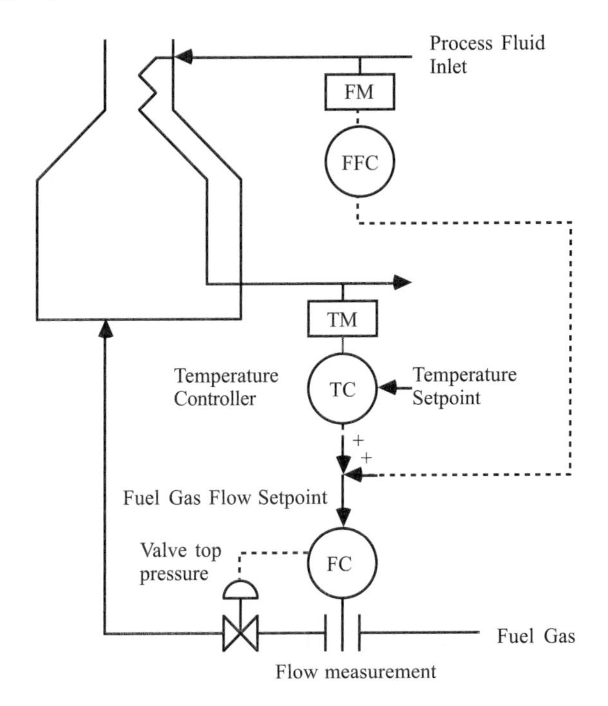

For a change in the inlet process flow rate of 1,000 Bbl/day, there is a change of –10°F in the outlet temperature. Also, the dynamic behavior can be described by a first-order lag of 5 minutes and a one minute time delay. Use the information from Exercise 9 to design a feed-forward/feedback/cascade-control strategy, and perform simulations for a step feed flow rate disturbance of 2,000 Bbl/day.

CHAPTER 11

PID Enhancements

Commercial PID controllers generally have many features beyond the simple control algorithm. All have antireset windup capability and many have automatic tuning and gain scheduling functions. The objective of this chapter is to provide an overview of the common techniques used. The major sections of this chapter are as follows:

11.1 Background

The focus of the techniques developed thus far in this textbook has been on linear systems, although it is recognized that chemical processes are inherently nonlinear. Fortunately, when processes are operated close to a desired operating point, the linear-system approaches generally work well. Also, when the main nonlinearity is due to the control valve or flow-pressure drop characteristics, cascade-control approaches work quite well. The inner-loop flow controller has fast dynamics, and the nonlinear effects will usually not propagate to the outer-loop.

In this chapter we cover several different approaches to handling nonlinear effects. The first nonlinear effect is due to manipulated input saturation. When manipulated inputs hit a constraint, the phenomena of *reset* (or *integral*) *windup* can occur. Approaches for *antireset windup* (ARW) are presented in Section 11.2.

When chemical processes change operating points, the dynamic input-output characteristics often change, requiring a control system retuning. Automatic tuning procedures can be used to retune controllers. So-called autotune procedures are developed in Section 11.3.

Sometimes it is desirable to have a controller gain that is a function of the error. If small errors can be tolerated, then a controller with a proportional gain that is a function of the absolute value of the error can often be used. The basic idea of nonlinear PID is introduced in Section 11.4. If the process gain changes as a function of operating condition, then "gain scheduling" approaches can be used to vary the controller tuning parameters as a function of the operating condition. Gain scheduling is presented in Section 11.5. Sometimes process input-output behavior can be made more linear by proper measurement/actuator selection. Such techniques are discussed in Section 11.6.

11.2 Antireset Windup

Reset (Integral) Windup

The phenomenon of reset windup occurs when a controller with integral action operates with the manipulated variable at a constraint for a period of time. When the manipulated input hits the constraint, the integral term in a PID controller continues to accumulate the error, requesting more and more manipulated variable action; the manipulated input cannot increase, however, since it is already constrained. Owing to this integral windup, the manipulated variable may stay saturated longer than is necessary. A block diagram for a process with manipulated input constraints is shown in Figure 11–1a, where v represents the unconstrained input from the controller, while u represents the actual constrained input implemented on the process. The "saturation" element detailed in Figure 11–1b, has the following mathematical description

$$\begin{aligned} &\text{if } v < u_{\min} \text{ then } u = u_{\min} \\ &\text{if } u_{\min} < v < u_{\max} \text{ then } u = v \\ &\text{if } v > u_{\max} \text{ then } u = u_{\max} \end{aligned} \tag{11.1}$$

The effect of reset windup is illustrated in the following example.

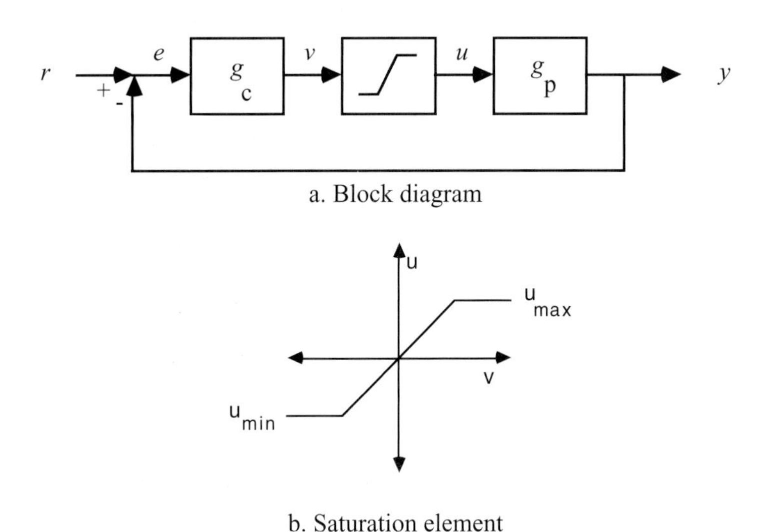

Figure 11–1 Representation of a process with a constrained input.

Example 11.1: Illustration of Reset Windup

Consider the first-order process,

$$g_p(s) = \frac{0.5}{s+1}$$

where the time unit is minutes. Here we compare two processes: where the manipulated input is unconstrained, and where the manipulated input is constrained between –1 and 1. Assuming deviation variable form and an initial steady-state output of 0, we know that the constrained process cannot achieve a steady-state output change of greater than 0.5. Here we tune the PI controller for a closed-loop response time of 0.25 minutes. The controller gain and integral time parameter values are $k_c = 8$, $\tau_I = 1$ min.

Figure 11–2 compares the input and output responses for a setpoint change of 0.4 in both the constrained and unconstrained control systems. The process output for the constrained system exhibits overshoot, while the unconstrained system has the expected first-order closed-loop response. The overshoot is due to reset windup. Notice the constrained manipulated input stays saturated for 2 minutes, although the output is above the setpoint after 1.5 minutes (after the setpoint change).

A SIMULINK block diagram for this problem is shown in Section 11.7. For small setpoint changes (say, ± 0.1) you should find that reset windup is not a problem. You should show that as the setpoint change approaches 0.5 (remember, this is the maximum steady-

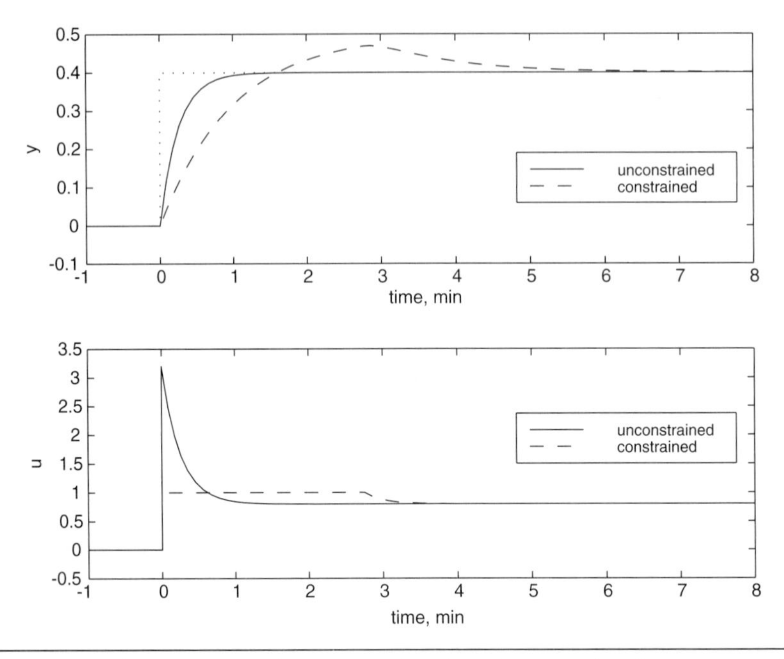

Figure 11–2 Comparison of unconstrained and constrained responses to a step setpoint change.

state change that can be handled by the process, owing to the input constraints), the reset windup effect becomes worse. See Exercise 2.

Antireset Windup (ARW) Techniques

There are a number of ways that have been developed to minimize problems with reset windup. Two ways are compared with the standard PI algorithm in Figure 11–3. Figure 11–3a shows the standard PI algorithm in block diagram form, with a saturation element to represent the manipulated input constraints. It is clear that there is no feedback to the algorithm to prevent the unconstrained input v from changing due to integration of the error signal. The "back-calculating" ARW strategy is shown in Figure 11–3b. Here, the difference between the ideal (unconstrained) input v and the actual (constrained) input u is fed-back to the integrator. The resetting time, τ_r, is an adjustable parameter that is normally set equal to the integral time, τ_I. The first-order feedback method of implementing integral action is shown in Figure 11–3c. For a resetting time equal to the integral time, the two antiwindup strategies (Figures 11–3 b and c) yield the same closed-loop performance.

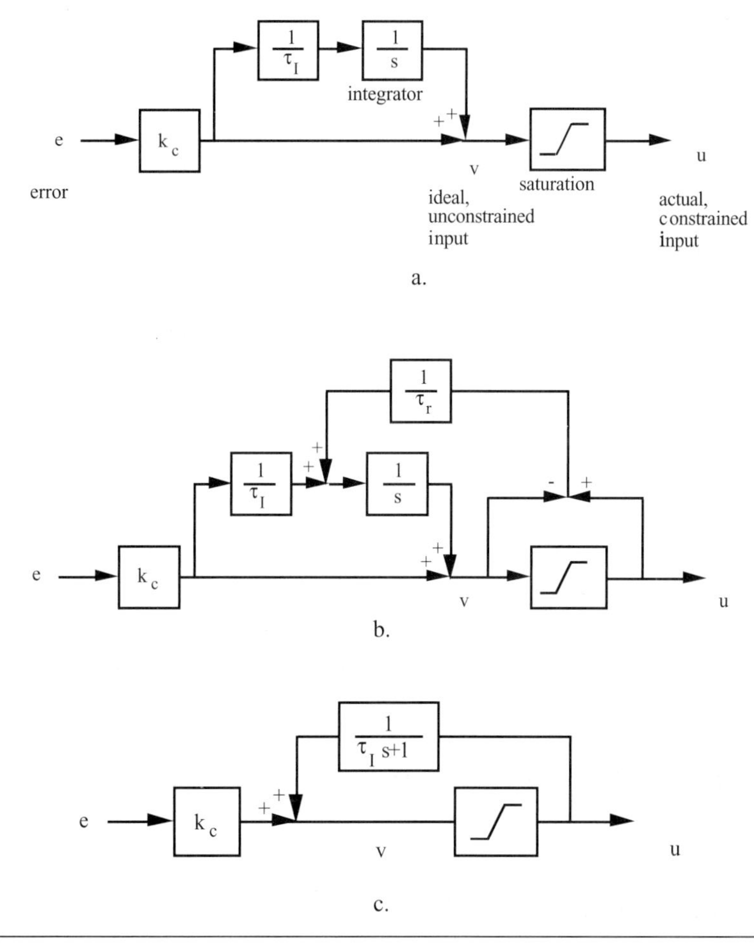

Figure 11–3 Various forms of PI control. (a) Standard PI control without windup compensation. (b) Reset back-calculating. (c) First-order feedback. (b, c) Antiwindup strategies.

Example 11.2: Illustration of Reset Windup Compensation (ARW)

A comparison of the antiwindup approach with the saturating PI strategy is shown in Figure 11–4, for the system given in Example 11.1. Note that the controller with antiwindup remains saturated for a much shorter time. Clearly, the antiwindup strategy is preferred.

Reset windup can be particularly problematic in cascade-control systems, particularly when batch processes are involved. This is illustrated by the cascade-control block

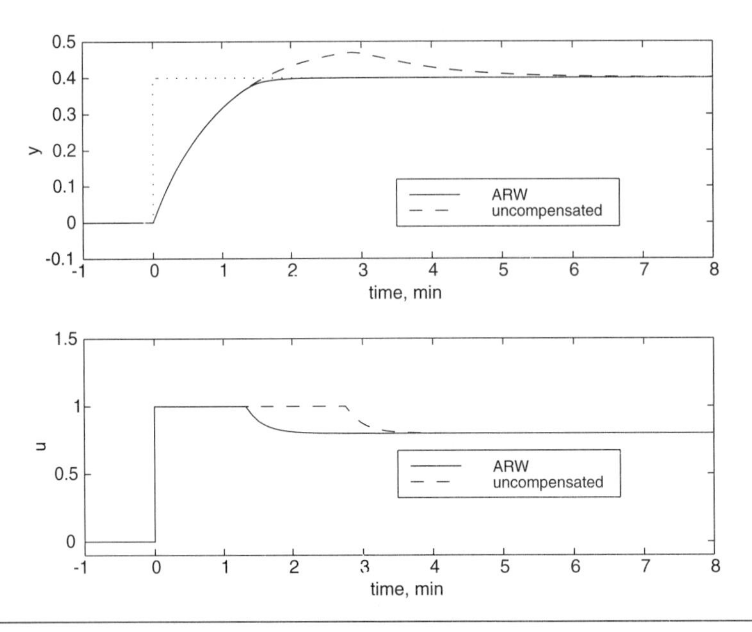

Figure 11–4 Comparison of antiwindup (ARW) and uncompensated responses for system in Example 11.1. The overshoot in the uncompensated system is due to reset (integral) windup.

diagram shown in Figure 11–5. The outer loop (primary control loop) of a cascade strategy is normally tuned for much slower dynamics than is the inner loop (secondary control loop). If the inner loop performs more slowly than expected, due to constraints on the manipulated input for example, the performance of the outer loop controller may suffer. In this case external windup protection can be provided, as shown in the following example.

Example 11.3: Cascade Control of a CSTR

A physical example (temperature control of a jacketed CSTR) is shown in Figure 11–6. The output of the primary controller (reactor temperature) is the setpoint to the inner-loop controller. The manipulated variable for the secondary loop is the jacket flow rate. If the flow rate becomes constrained, or if the jacket temperature controller is not tuned tightly, the jacket temperature will not closely match its setpoint, causing problems with the integral action in the primary loop.

For simplicity, let the primary and secondary processes be represented by the following first-order transfer functions,

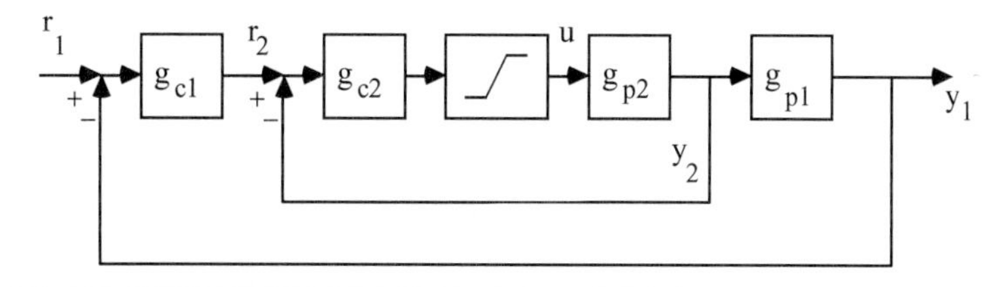

Figure 11–5 Series representation of a cascade-control system with saturation limits on the manipulated input.

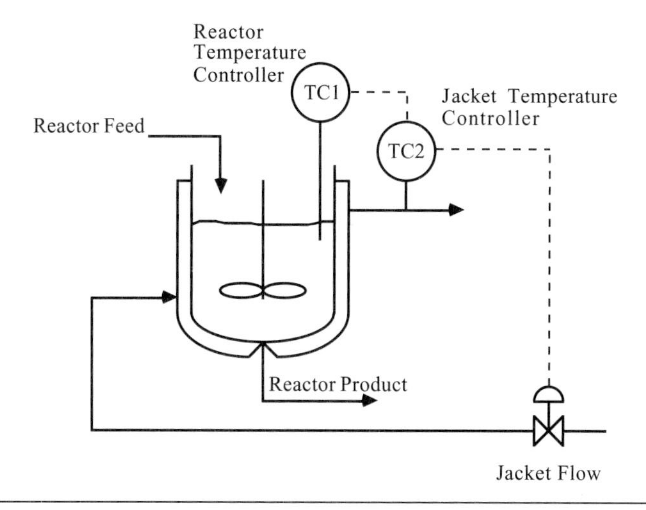

Figure 11–6 Cascade-control applied to a CSTR.

$$g_{p1}(s) = \frac{1}{50s+1}$$

$$g_{p2}(s) = \frac{1}{5s+1}$$

and assume IMC-based PI tuning, where the secondary process dynamics were neglected in the design of the primary controller. Closed-loop time constants of 10 and 2 minutes were used for the primary and secondary loops, respectively. The manipulated input is assumed to saturate at ± 3. The primary and secondary tuning parameters are then

$$k_{c1} = \frac{\tau_{p1}}{k_{p1}\lambda_1} = \frac{50 \text{ min}}{1 \cdot 10 \text{ min}} = 5, \quad \tau_{I1} = 50 \text{ min}$$

$$k_{c2} = \frac{\tau_{p2}}{k_{p2}\lambda_2} = \frac{5 \text{ min}}{1 \cdot 2 \text{ min}} = 2.5, \quad \tau_{I2} = 5 \text{ min}$$

We first perform a simulation with antiwindup on the secondary control loop, since the jacket flow rate is constrained. The response to a unit setpoint change is shown in Figure 11–7. Notice that there is still some windup involved, causing overshoot of the primary process output. This is because the output of the primary controller is a setpoint to the secondary controller. The secondary controller does not track its setpoint very well because of the saturation of the manipulated input. This is shown clearly in the middle plot of Figure 11–7.

Improved performance can be obtained using an external input to the primary controller. The basic idea is illustrated in Figure 11–8 for the reactor temperature (master) controller, where the controller output is the jacket temperature setpoint. Here the jacket temperature (actual manipulated input) is compared with the jacket temperature setpoint

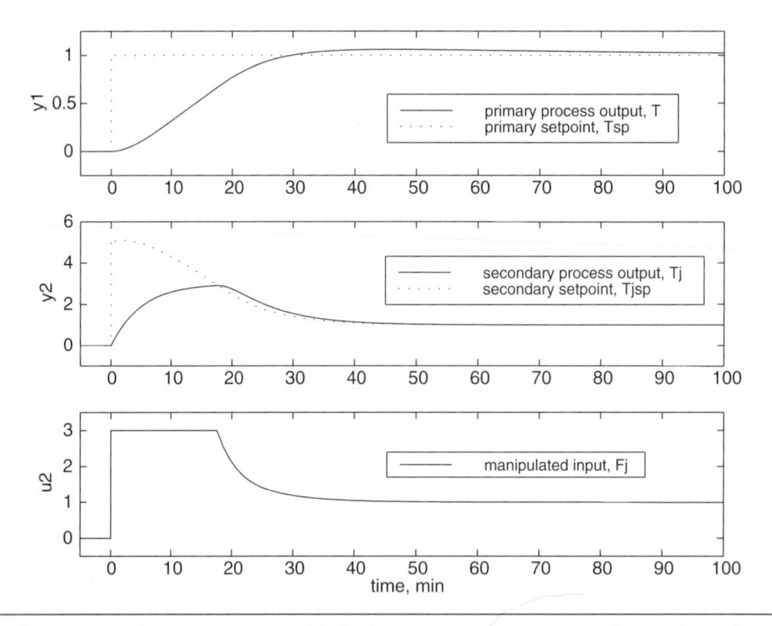

Figure 11–7 Performance of antiwindup secondary controller. The primary controller output (jacket temperature setpoint) differs substantially from the secondary measurement (jacket temperature).

(ideal manipulated input). If the jacket temperature is not equal to its setpoint, some of the integral action is "removed" by corrective feedback action. The improved cascade control, with antiwindup (Figure 11–3c) on the secondary loop and external resetting antiwindup (Figure 11–8) on the primary loop, has excellent performance, as shown in Figure 11–9.

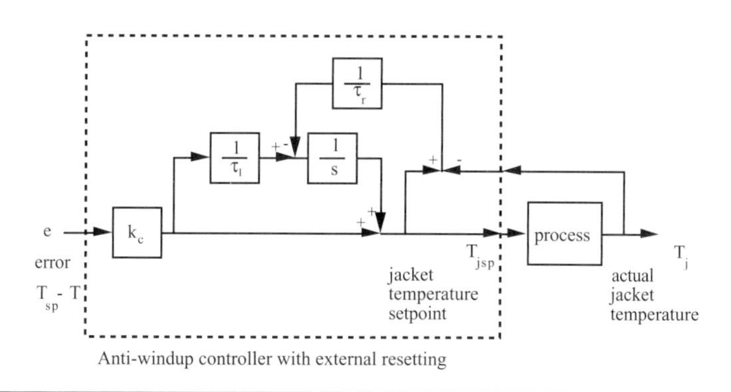

Figure 11–8 Reactor temperature controller with external variable reset.

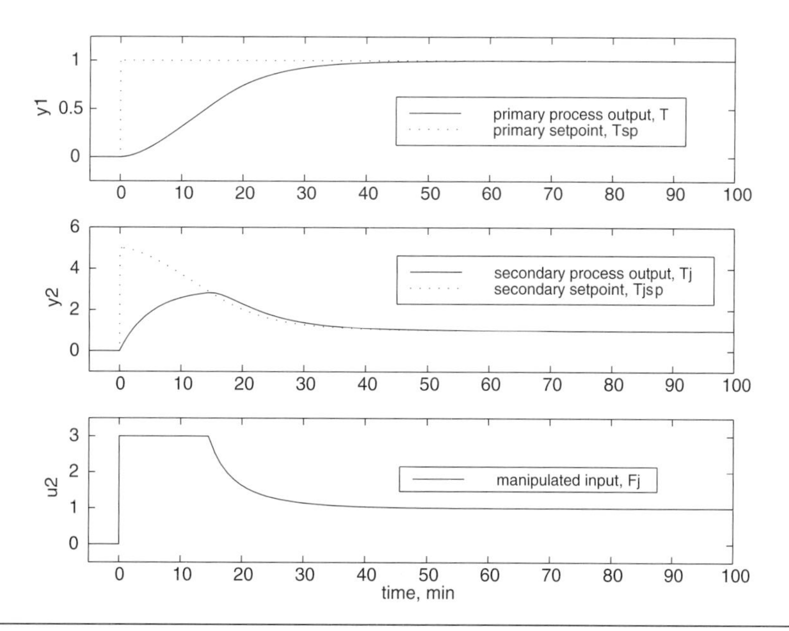

Figure 11–9 Performance of CSTR cascade control with antiwindup protection on both controllers. The external ARW control strategy of Figure 11–8 is implemented on the reactor temperature controller.

There is now no overshoot in the primary process output (reactor temperature). The external reset windup protection is shown as a signal between the measured jacket temperature and the reactor temperature controller, on the control instrumentation diagram in Figure 11–10.

It should be noted that antiwindup protection can also be provided by using the "velocity form" in a digital control system. This is discussed in Module 16.

11.3 Autotuning Techniques

Many process control systems have an automatic tuning (autotune) feature. The operator can simply push the autotune button and have the controller tune itself, that is, determine the values of the tuning parameters. This is done not continuously (an approach known as adaptive control) but only when the operator feels that the current set of tuning parameters is not performing well.

The most common method of automatic tuning uses a relay switch to create what is essentially an on-off controller. The resulting oscillatory behavior is analyzed to determine the proper controller settings.

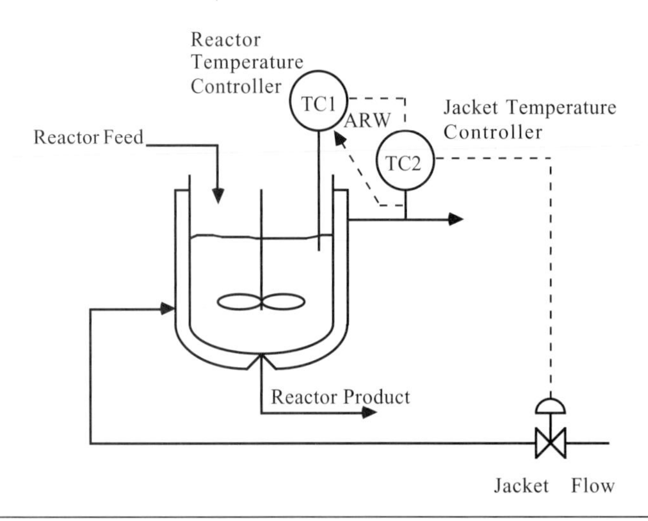

Figure 11–10 Cascade-control applied to a CSTR. The reactor temperature controller has external antireset windup (ARW) protection based on a measurement of the jacket temperature.

Control Block Diagram

The basic control loop for a controller with an autotune relay switch is shown in Figure 11–11. For normal operation, the switch is connected to the PID controller, as shown in Figure 11–11a. When the autotune function is operating, the switch is set to the output of the relay block, as shown in Figure 11–11b.

Notice that the relay block represents a nonlinear function. As drawn, it represents the following:

$$\begin{aligned} &\textit{if } e < 0, \textit{ then } u = u_{\min} \\ &\textit{if } e > 0, \textit{ then } u = u_{\max} \end{aligned} \tag{11.2}$$

Assuming that the input is symmetrical with a steady-state value of 0 (in deviation variable form), the magnitude of the input change allowed is h, so $u_{\max} = h$, and $u_{\min} = -h$. Notice that this assumes that the process gain is positive. If the process gain is negative, then switch the inequalities shown in Equation 11.2.

In autotune mode, the closed-loop system oscillates and the manipulated variable action is on-off (or bang-bang), as shown in Figure 11–12. In this example, $h = 0.05$.

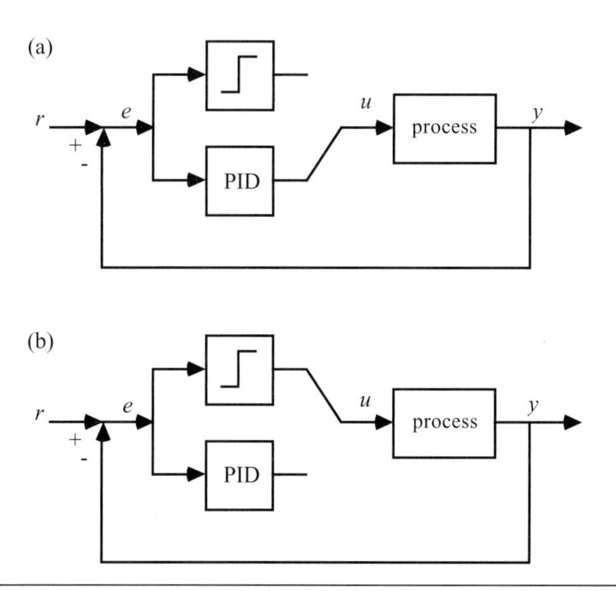

Figure 11–11 Control block diagram for an autotune relay switch. (a) Switch set to PID Control. (b) Switch set to relay function for automatic tuning.

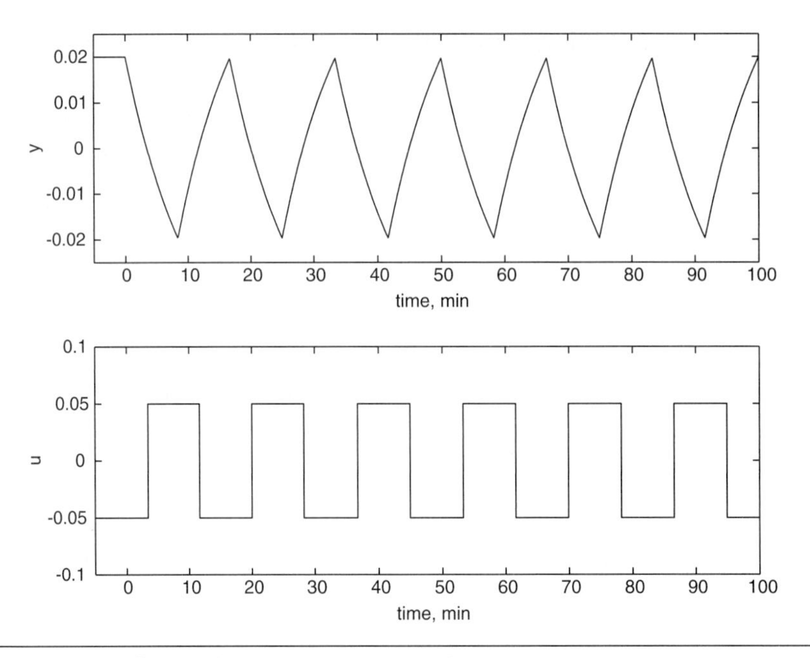

Figure 11–12 Oscillations in autotune mode.

There are two parameters that result from this autotune test. One is the period (time between successive peaks), P, and the other is the amplitude of the process output, a. The period has units of time; the ultimate frequency can be found from

$$\omega_u = \frac{2\pi}{P} \tag{11.3}$$

And the ultimate gain can be found from the amplitude

$$k_{cu} = \frac{4h}{\pi a} \tag{11.4}$$

In the results shown in Figure 11–12, the period is 16.4 min and the process output amplitude is 0.019. The manipulated input magnitude is 0.05, yielding the following values for k_{cu} and ω_u:

$$k_{cu} = \frac{4h}{\pi a} = \frac{4 \cdot 0.05}{\pi \cdot 0.019} = 3.35$$

$$\omega_u = \frac{2\pi}{16.4 \text{ min}} = 0.383 \text{ min}^{-1}$$

The behavior is very similar to that obtained from the Ziegler-Nichols closed-loop cycling method. The Ziegler-Nichols results for the same process are shown in Figure 11–13.

The Ziegler-Nichols test yields an ultimate gain of 3.83 and an ultimate period of 17.13 min; contrast these with the relay values of 3.35 and 16.4 min.

Relay Deadband (Hysteresis)

If the ideal relay is implemented, there can be problems if there is process or measurement noise. To handle this, a deadband (or hysteresis) with a magnitude of ε is added to the relay switch, as shown below (again, a positive process gain is assumed). The magnitude of the deadband is selected to be at least two to three times the standard deviation of the measurement noise:

$$\begin{aligned} &\textit{if } e < -\varepsilon, \textit{ then } u = u_{\min} \\ &\textit{if } e > +\varepsilon, \textit{ then } u = u_{\max} \end{aligned} \tag{11.5}$$

Also, if the error is in between the minimum and maximum limits, then the control action stays at the current value.

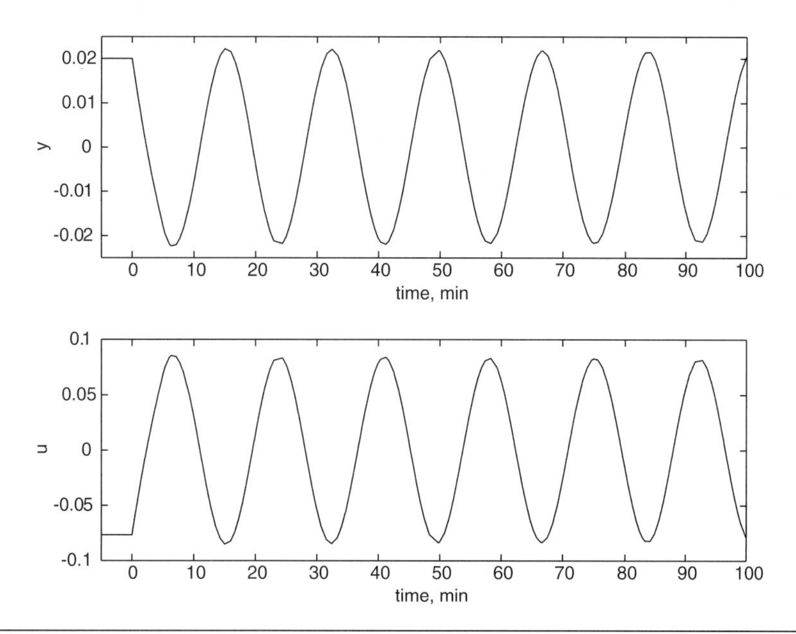

Figure 11–13 Ziegler-Nichols closed-loop test. The controller proportional gain is $k_c = 3.83$.

Controller Tuning

Once the process parameters (ultimate gain and frequency) are identified from the relay test, a controller tuning procedure must be used. An obvious method is to use the Ziegler-Nichols values, since the autotune test is similar to the Ziegler-Nichols test. These controller tuning parameters are not robust, however, and result in closed-loop systems that may go unstable if the process conditions change.

An alternative is to use a model-based procedure, such as IMC-based PID. A problem is that the ultimate gain and frequency must be converted to model parameters. Since only two parameters, ultimate gain and frequency, are obtained from the relay test, a two-parameter process model must be developed. The classical first-order + time-delay model cannot be used because it has three parameters. Two options are integrator + time delay (11.6) and gain + time delay (11.7):

$$g_p(s) = \frac{ke^{-\theta s}}{s} \tag{11.6}$$

$$g_p(s) = ke^{-\theta s} \tag{11.7}$$

The integrator + time-delay model is generally a very good approximation to a first-order + time-delay process if the time constant is much larger than the time delay. The gain + time-delay model is generally a good approximation if the time delay is much larger than the time constant (such as for liquid flows with significant pipe volume delays).

Integrator + Time Delay

For the integrator + time-delay model, the dead time and process gain can be found from the relay parameters in the following fashion:

$$\begin{aligned} \theta &= P/4 \\ k &= 4a/Ph \end{aligned} \tag{11.8}$$

The IMC-based PI tuning parameters can be found from (see Chapter 9)

$$\begin{aligned} k_c &= \frac{2\lambda + \theta}{k(\lambda + \theta)^2} \\ \tau_I &= 2\lambda + \theta \end{aligned} \tag{11.9}$$

Gain + Time Delay

For the gain + time-delay model, the dead time and process gain can be found from the relay parameters in the following fashion

$$\begin{aligned} \theta &= P/2 \\ k &= a/h \end{aligned} \tag{11.10}$$

The IMC-based PI tuning parameters can be found from

$$k_c = \frac{\theta/2}{k\lambda} \qquad \tau_I = \theta/2 \tag{11.11}$$

In both cases, the IMC filter-factor (λ) is adjusted to provide a tradeoff between nominal performance and robustness to model uncertainty.

Comparison of Model-Based PI Controllers

In the example shown in Figure 11–12, the period and output amplitudes were 16.4 min and 0.019 for a control magnitude of 0.05. The IMC-based PI controllers obtained under the two different model-based assumptions are compared in Table 11–1.

The responses of the two model-based PI controllers are shown in Figure 11–14. The λ value was adjusted to achieve 25% overshoot in both cases. For more robust performance, the λ values must be increased.

11.4 Nonlinear PID Control

The fixed-parameter PID controller is ubiquitous in the process industries. The *ideal* PID controller has the form (see Chapter 5)

$$u(t) = k_c\left[e(t) + \frac{1}{\tau_I}\int_o^t e(\sigma)d\sigma + \tau_D\frac{de(t)}{dt}\right] \tag{11.12}$$

where it is understood that any bias associated with steady-state operation is absorbed into the integral term. Other forms for PID control are presented in Chapter 5.

Often it is useful to have a nonlinear controller even if the process is linear. A common example is level control of a surge drum where the primary objective is to maintain a

Table 11–1 Example Parameters for Model-Based Controllers

Model	*Integrator + time delay*	*Gain + time delay*
k	4(0.019)/(16.4)0.05 = 0.093	0.019/0.05 = 0.38
θ	16.4/4 = 4.1 min	16.4/2 = 8.2 min
k_c	$(2\lambda+4.1)/[0.093(\lambda+4.1)^2]$ = 1.831 for λ = 5 min	$(8.2/2)/(0.38\lambda) = 10.79/\lambda$ = 1.269 for λ = 8.5 min
τ_I	$2\lambda + 4.1$ = 14.1 for λ = 5 min	8.2 min

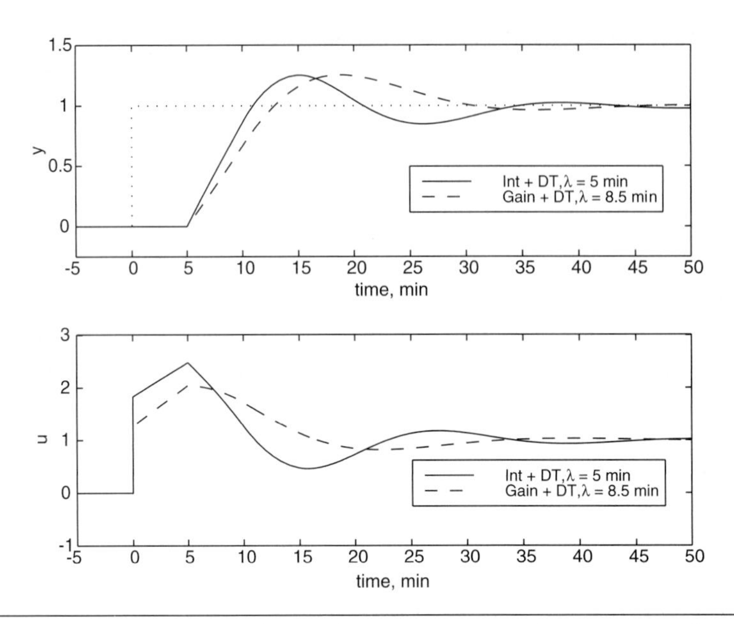

Figure 11–14 Comparison of closed-loop responses for the example process. IMC-based PI tuning parameters based on two different models (integrator + time-delay and gain + time-delay) identified from a relay test.

reasonably consistent outlet flow under disturbances in the inlet flow rate, subject to high and low drum level constraints. In this situation, some small nominal controller gain can be used when the level is close to the setpoint, and as the error increases, the gain can increase. A common choice is to make the proportional gain a function of the absolute value of the error,

$$k_c = k_{c0}\left(1 + a \cdot |e(t)|\right) \tag{11.13}$$

This means that the controller output is effectively proportional to the square of the error, as shown below for a P-only nonlinear controller.

$$u(t) = u_0 + k_c e(t) = u_0 + k_{c0}\left(1 + a \cdot |e(t)|\right)e(t) \tag{11.14}$$

An alternative to making the controller gain a continuous function of the error is to form a piecewise linear controller (sometimes called a gap controller), where (often $k_{c,\text{high}} = k_{c,\text{low}}$)

$$\begin{aligned} k_c &= k_{c,\text{low}} && \text{for } e < e_{low}, \\ k_c &= k_{c0} && \text{for } e_{low} \le e \le e_{high}, \\ k_c &= k_{c,\text{high}} && \text{for } e > e_{high} \end{aligned} \tag{11.15}$$

An example of nonlinear PID control is given in Module 10 (Surge Vessel Level Control).

11.5 Controller Parameter (Gain) Scheduling

Perhaps the most widely mentioned technique for controlling nonlinear processes is gain scheduling. Let α represent the scheduling variable, which will often be the measured output, y. A gain-scheduled PID controller is then represented by

$$u(t) = k_c(\alpha)\left[e(t) + \frac{1}{\tau_I}\int_o^t e(\sigma)d\sigma + \tau_D \frac{de(t)}{dt}\right] \tag{11.16}$$

Notice that the primary difference between this algorithm and the nonlinear PID controllers of the previous section is that α is a general scheduling variable, whereas the controllers in the previous section were based on error as the scheduling variable. Although all of the controller parameters can be scheduled, controller gain is commonly scheduled because processes are often characterized as a changing gain with relatively constant dynamics. Typically, the controller gain, k_c, is varied to keep $k_c k_p$ constant, which then keeps the stability margin constant. If the process gain can be characterized as a function of the scheduling parameter, $k_p(\alpha)$, then the controller gain can be scheduled as

$$k_c(\alpha) = \frac{k_c(\alpha_0)k_p(\alpha_0)}{k_p(\alpha)} \tag{11.17}$$

where the subscript 0 indicates a nominal operating point.

An important step in developing a gain-scheduled controller is determining the proper scheduling variable. Rules of thumb such as scheduling on a “slow variable” or using a variable that “captures the nonlinearities” are often used. It is most common, then, to schedule based on either the setpoint or measured output, since the process input (manipulated variable) will vary more rapidly than the output. When there are additional measurements, then a measured or inferred “auxiliary” variable can be used for scheduling. The four basic steps in developing a parameter-scheduled controller are as follows:

1. Develop a linear process model for a set of operating conditions
2. Design linear controllers for each operating condition (model)
3. Develop a schedule for the controller parameters
4. Implement the parameter-scheduled controller on the nonlinear plant

Note that steps 1 and 2 will often be combined by using some form of closed-loop tuning at discrete operating points. This is particularly true if the autotune feature of commercial controllers is used; the use of three operating points appears to be common.

There are several different options for the scheduling of the controller parameters.

- Switch parameters at discrete values of the scheduling variable
- Interpolate parameters as a function of the scheduling variable
- Vary parameters continuously with the scheduling variable

A major advantage of a parameter-scheduled controller is that linear control system design procedures can be used, and at least the local control system behavior can be understood. A problem can occur in steps three and four, where parameter scheduling is developed and implemented. If the dynamic effect of the scheduling variable is not included in the design process, then even the local linearized closed-loop behavior may not be predicted correctly.

A general representation of a gain-scheduled controller is shown in Figure 11–15, where α represents the scheduling variable. The double lines on the process and controller blocks indicate that these are nonlinear.

An example application of gain scheduling is distillation. Since the process gain is a function of the product composition, it makes sense to schedule the controller gain as a function of composition. This approach was taken by Tsogas and McAvoy (1985) in the control of distillation composition, with distillate flow rate as the manipulated variable. The scheduling algorithm used was

$$\begin{aligned} k_c(x_D) &= k_{c0} \cdot \frac{1 - x_{D0}}{1 - x_D} \qquad \text{for } x_D > x_{D0} \\ k_c(x_D) &= k_{c0} \qquad \text{for } x_D < x_{D0} \end{aligned} \tag{11.18}$$

Notice that this is a "one-way" algorithm. Since the process gain increases with lower purity, maintaining a constant controller gain speeds up the response when the distillate is less pure.

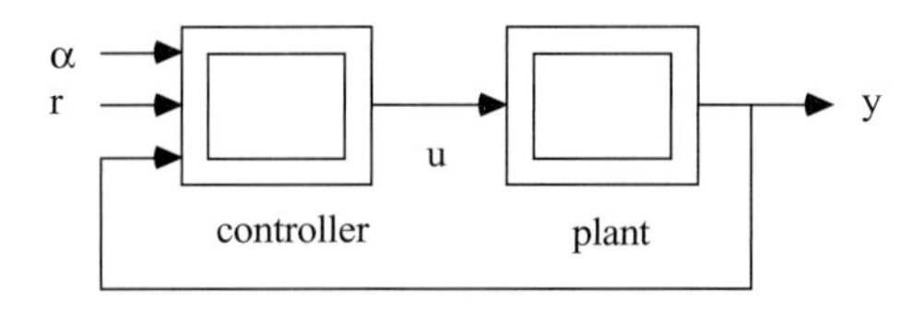

Figure 11–15 Parameter-scheduled control strategy.

11.6 Measurement/Actuator Selection

Perhaps the best way to assure that a nonlinear control system is well maintained is through design of the physical control devices (actuators and measurements). For example, in Module 15 (flow control), we show that the installed characteristic of an equal-percentage valve makes the valve position/flow rate relationship linear over a wider range of operating conditions than other valve types.

In distillation control, defining a process output as the logarithm of the product composition can be used to linearize the input-output relationship. For example,

$$y = \ln(x_D) \tag{11.19}$$

where x_D is the distillate composition. The new setpoint variable, r, can be found as the logarithm of the desired distillate composition, x_{Dsp}, by

$$r = \ln(x_{Dsp}) \tag{11.20}$$

and the controller output (reflux flow rate, for example) would be a function of $r - y$.

11.7 Implementing PID Enhancements in SIMULINK

SIMULINK has a number of standard features that make it easy to perform simulations for the control strategies developed in this chapter. Saturation and relay blocks, for example, can be found in the `Discontinuities` SIMULINK library in MATLAB version 6.5 (Release 13).

Constrained vs. Unconstrained

The simulation results in Figure 11–2 were generated using the SIMULINK `.mdl` diagram shown in Figure 11–16.

ARW

The antiwindup results presented in Figure 11–4 were generated using the SIMULINK `.mdl` diagram shown in Figure 11–17. The cascade results of Figures 11–7, and 11–9 were generated using the SIMULINK `.mdl` diagrams shown in Figures 11–18 and 11–19.

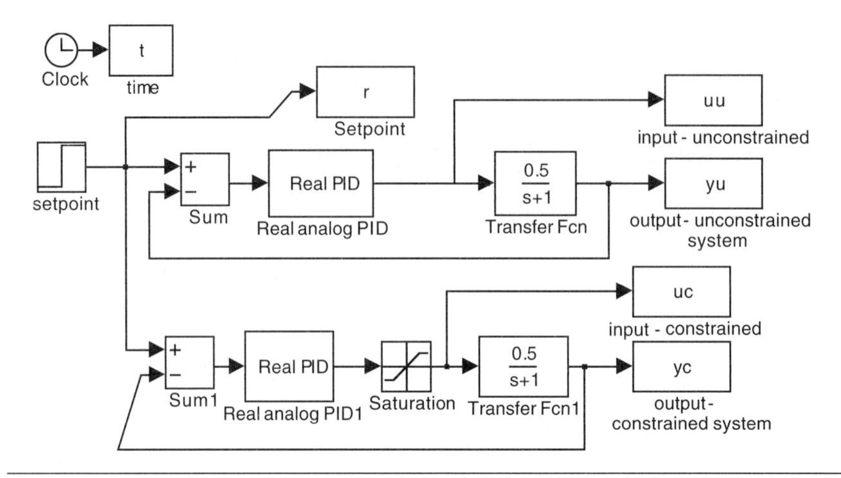

Figure 11–16 Comparison of constrained and unconstrained processes to a step setpoint change.

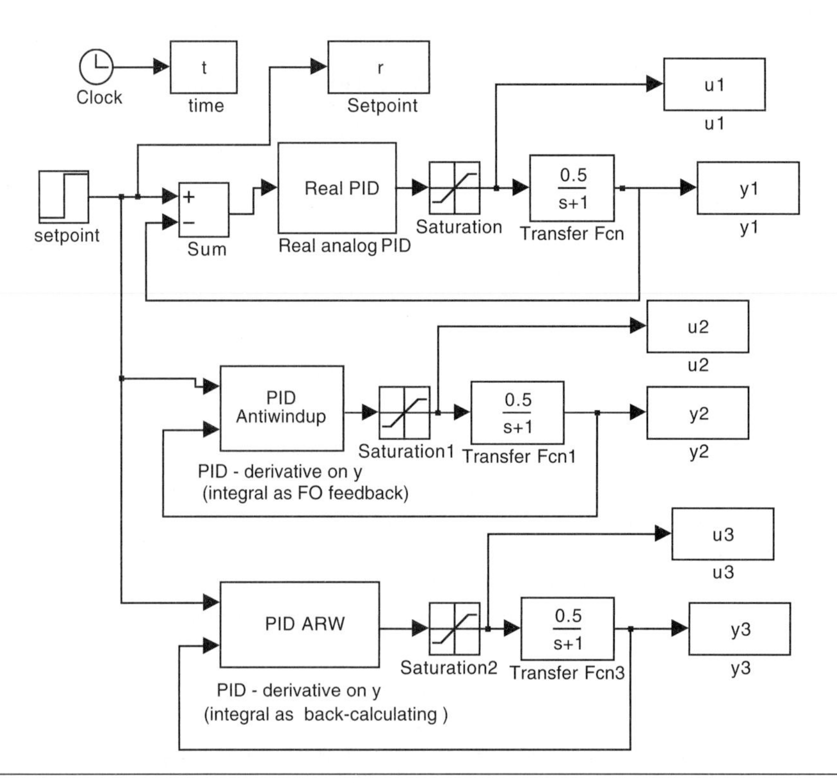

Figure 11–17 Comparison of ARW strategies with “classical” PID.

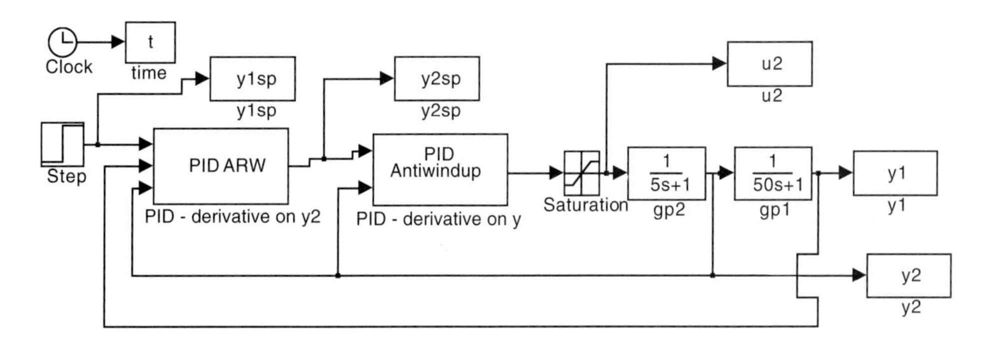

Figure 11–18 SIMULINK cascade-control simulation with antiwindup on both controllers.

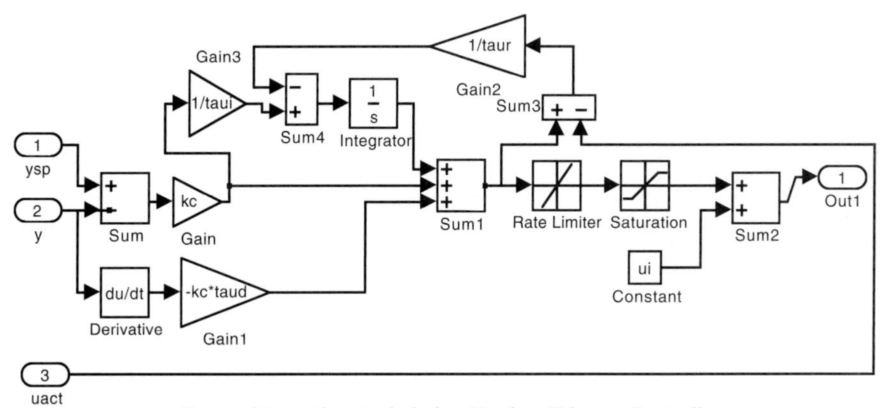

a. External Resetting Antiwindup Used on Primary Controller

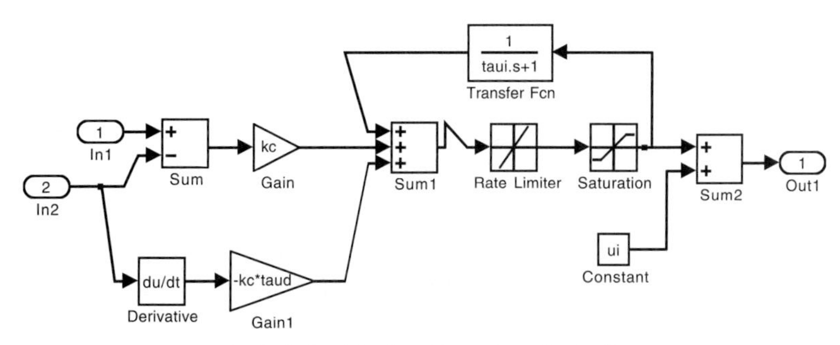

b. Antiwindup Used on Secondary Controller

Figure 11–19 *Unmasked* view of anti-windup strategies used for the cascade-control simulations.

Autotuning

The autotuning simulation shown in Figure 11–12 was generated using Figure 11–20.

11.8 Summary

A number of enhancements to standard PID controllers have been presented in this chapter. All commercial controllers implement ARW methods of one form or another. It is particularly important for cascade controllers to use external reset windup protection.

A number of controllers also have autotune features, but some of these are proprietary. We have presented the approach that is most widely presented in the literature. Using a relay function in the closed loop provides a closed-loop oscillation similar to the Ziegler-Nichols closed-loop oscillation method. There are several of advantages to the relay-based approach: the magnitude of the oscillations is bounded by proper selection of the manipulated input constraints, and the time required to develop several periods of the oscillation is much shorter than the time it takes to laboriously change a proportional gain to create a closed-loop oscillation.

One of the primary ways of creating a nonlinear PID controller is to make the controller gain a function of the absolute value of the error. This approach works well for processes where small deviations from setpoint are acceptable, such as surge vessel level control. Gain scheduling is a approach that can be used when the model and controller gain can be characterized as a function of the operating condition.

The proper selection of measurements and actuators can help linearize the input-output relationship. Equal percentage valves often make flow systems appear linear over a wider range. Similarly, using the logarithm of distillation product compositions can make the input-output behavior more linear.

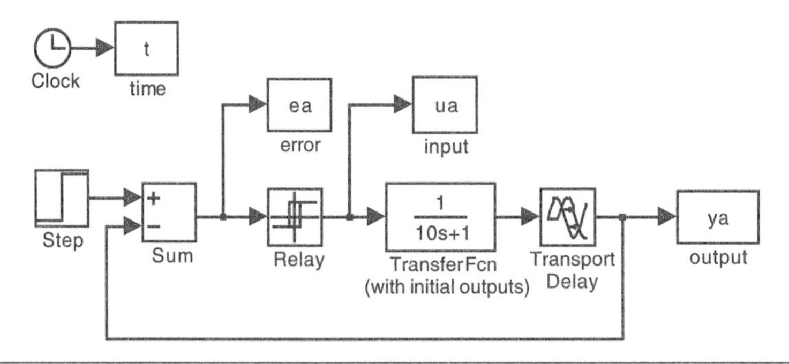

Figure 11–20 SIMULINK `.mdl` diagram for autotuning simulation.

References

A nice presentation of antireset windup is presented in: Astrom, K. J., and L. Rundqwist, "Integrator Windup and How to Avoid It," in *Proceedings of the 1989 American Control Conference*, Pittsburgh, pp. 1693–1698 (1989).

The following paper presents model-based (integrator + time delay and gain + time delay) PI approaches for tuning SISO and two-input, two-output processes: Friman, M., and K. V. Waller, "Autotuning of Multiloop Control Systems," *Ind. Eng. Chem. Res.*, **33**(7), 1708–1717 (1994).

A nice summary of PID design and implementation is presented in the following book: Kiong, T. K., W. Quing-Guo, and H. C. Chieh with T. J. Hagglund, *Advances in PID Control*, Springer Verlag, London (1999).

The following paper reviews gain scheduling approaches to process control: Bequette, B. W., "Practical Approaches to Nonlinear Control: A Review of Process Applications," in *Nonlinear Model-based Process Control*, NATO ASI Series, Ser. E, Vol. 353, pp. 3–32 (R. Berber and C. Kravaris, Eds.), Kluwer, Dordrecht (1998).

Distillation gain-scheduling control has been studied in the following: Tsogas, A., and T. J. McAvoy, "Gain Scheduling for Composition Control of Distillation Columns," *Chem. Eng. Commun.*, **37**, 275–291 (1985).

Student Exercises

1. Verify that the two block diagrams below result in the same PI control algorithm transfer function.

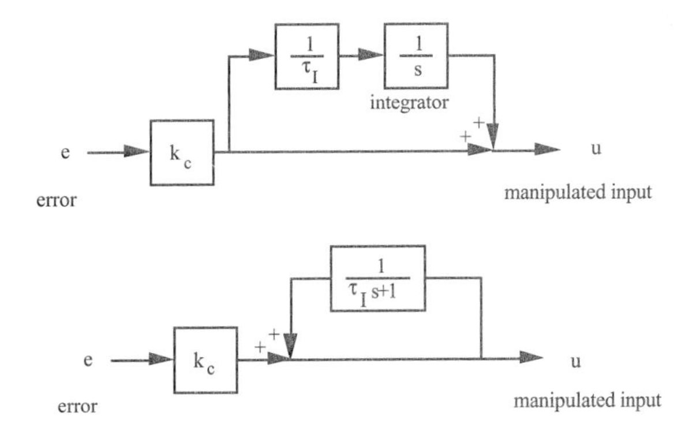

2. Use the SIMULINK diagram presented in Figure 11–16 to verify the results presented in Example 11.1. Show that, as the setpoint change approaches 0.5 the integral windup effect becomes worse.
3. Verify the results of Example 11.2 using a SIMULINK simulation. Also, discuss the effect of the resetting time, τ_r, on the ARW results.
4. Consider the process furnace temperature cascade control system shown in Figure 10–2, where the output of the temperature control loop is a setpoint for the fuel gas flow controller. Draw a signal, similar to Figure 11–10, that shows that an external antireset strategy is used for the temperature controller.
5. Consider Example 11.3. Use SIMULINK to show that cascade control without antireset windup protection yields significant overshoot for a setpoint change.
6. Show that the relay procedure for an integrator + time-delay process will exhibit the following period and output amplitude:

$$P = 4\theta$$
$$a = Phk/4$$

7. Show that the relay procedure for a gain + time-delay process will exhibit the following period and output amplitude:

$$P = 2\theta$$
$$a = hk$$

8. The example process used in this module is first order + time delay, with a gain of 1, a time constant of 10, and a time delay of 5. Compare the Ziegler-Nichols PI control performance when the critical gain and period are obtained using two different methods: relay-based cycling, and Ziegler-Nichols closed-loop cycling. Also, compare the Tyreus-Luyben tuning parameters obtained using the two methods. Does the relay-based method generally yield more conservative results?
9. For an integrator + time-delay model, find the IMC-based PI parameters, assuming an IMC filter with the form

$$f(s) = \frac{\gamma s + 1}{(\lambda s + 1)^2}$$

where γ is selected to yield a PI controller.
10. For a gain + time-delay model, find the IMC-based PI parameters when the all-pass method is not used.
11. For a first-order + time-delay process, with a time constant of 10 minutes and a time delay of 0.5 minutes, use the autotune procedure. How close are the critical gain and ultimate period to the values obtained by the Ziegler-Nichols

closed-loop oscillation method? What are the process parameters for an integrator + time-delay model? Based on these parameters, simulate the closed-loop behavior of a PI controller, with various values of λ. What λ values do you recommend? Why?

12. For a first-order + time-delay process, with a time constant of 0.5 minutes and a time delay of 10 minutes, use the autotune procedure. How close are the critical gain and ultimate period to the values obtained by the Ziegler-Nichols closed-loop oscillation method? What are the process parameters for a gain + time-delay model? Based on these parameters, simulate the closed-loop behavior of a PI controller, with various values of λ. What λ values do you recommend? Why?

CHAPTER 12

Ratio, Selective, and Split-Range Control

After reading this chapter the reader should be able to understand and develop ratio, selective and override control, and split-range control strategies. Ratio control is similar to feed-forward control, while selective and override control choose between alternative outputs or inputs. Split-range control chooses between two or more control valves to implement a manipulated input.

The major sections of this chapter are as follows:

12.1 Motivation

In Chapter 10 we discussed feed-forward and cascade-control strategies. The purpose of these control strategies was to take advantage of additional measurements to improve the disturbance rejection capability. In this chapter, we also develop strategies that use more than one measurement and, in addition, manipulate more than one input. Ratio control, presented in Section 12.2, is very similar to feed-forward control. In Section 12.3, we discuss selective and override control, which involves the use of "selectors" to decide the proper control action to take. In Section 12.4, split-range control is presented as a way to use more than one valve to regulate a process.

12.2 Ratio Control

Ratio control is similar to feed-forward control, since both typically involve the measurement of a stream flow rate. Ratio control is often used in component blending problems. Consider the control strategies shown in Figure 12–1, where stream *A* is a "wild stream" (disturbance) and it is desirable to maintain the flow rate of stream *B* at a constant ratio to stream *A*. In Figure 12–1a the F_B/F_A ratio is calculated and sent to a ratio controller. In Figure 12–1b the flow rate of stream *A* is measured and multiplied by desired ratio to determine the setpoint for the stream *B* flow rate.

Alternative 12–1b is favored over 12–1a since it results in a linear input-output relationship. In Figure 12–1a the steady-state relationship between the manipulated input (valve position, related to F_B) and the measured output (the ratio, F_B/F_A) is

$$y = R = F_B/F_A = (1/F_A)F_B \tag{12.1}$$

and the steady-state process gain is

$$k_p = \partial y/\partial u = 1/F_A \tag{12.2}$$

Clearly, the process gain varies as a function of the flow rate of stream *A*. This means the ratio controller would need to be tuned differently depending the stream *A* flow rate. In contrast, the relationship between the manipulated input (valve position, related to F_B) and the measured output (F_B) for the flow controller in Figure 12–1b is

$$y = F_B = u \tag{12.3}$$

so the gain is

$$k_p = \partial y/\partial u = 1 \tag{12.4}$$

which is constant.

The ratio could be specified to maintain an excess of component *B* in a reaction, for example. A common use of this would be the supply of combustion air to a furnace. Stream *A* could represent the fuel gas, for example, and stream *B* would represent the combustion air.

The blend stream could also have a composition or product property measurement and controller; the output of this controller would be cascaded to the desired ratio of stream *B* flow to stream *A* flow. In Exercise 1 you have the opportunity to sketch this control strategy.

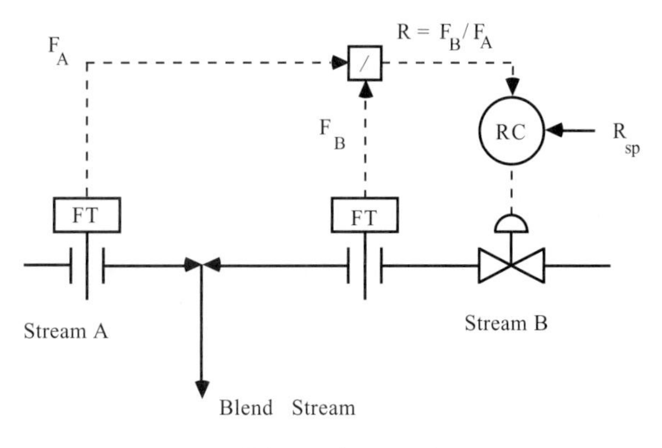

a. Direct ratio control

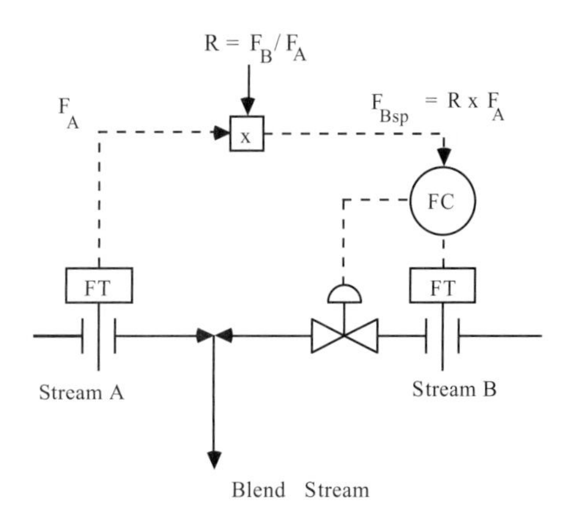

b. Preferred control method

Figure 12–1 Ratio control. Comparison of two alternatives.

12.3 Selective and Override Control

Often a controller needs to select from a set of possible measurements for control. For example, consider the packed-bed reactor shown in Figure 12–2. It is important to maintain the "hot spot" temperature at a certain value to minimize catalyst degradation. In this strategy, the "high selector" (represented as the block with >) chooses the highest of three bed temperatures to send to the temperature controller. In algorithmic form, let T_1, T_2, and T_3 represent the three temperature measurements, and T represent the temperature measurement used by the controller, then

$$T = \max(T_1, T_2, T_3) \tag{12.5}$$

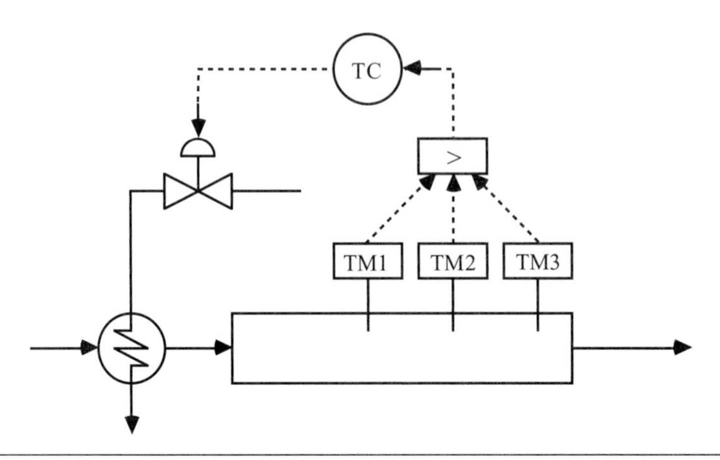

Figure 12–2 High-temperature selector for a packed-bed reactor.

When manipulating the fuel gas flow to a fired heater (furnace), it is important to make certain that a minimum air/fuel ratio is maintained. High/low selectors can be used to assure this, as shown in Figure 12–3; here the *A/F* boxes indicate the air/fuel ratio. To understand how this strategy works, first consider a positive change in the fuel flow setpoint. Notice that the combination of high and low selectors will force the combustion air flow rate to increase before the fuel flow increases. Also, if a decrease in fuel flow setpoint is made, the fuel flow must decrease before the combustion air flow. This strategy makes certain that excess combustion air is always available.

In practice this strategy is just one part of the overall furnace control strategy. Usually, the fuel flow setpoint will be the output of another controller (e.g., process fluid temperature or steam system pressure), and the air/fuel ratio might be the output of a furnace stack gas O_2 or CO measurement and control strategy.

12.4 Split-Range Control

A split-range strategy is often used in situations where one or more valves may be used, depending on the operating scenario. An example application is a batch reactor. Batch reactors must often be heated from ambient temperature to a desired operating temperature. Particularly if the reaction is exothermic, then cooling will be needed to maintain the desired reactor temperature. In the split-range control strategy shown in Figure 12–4, if the jacket temperature controller output is between 0 and 50%, the cold glycol valve is open; if the jacket temperature controller output is between 50 and 100%, the hot glycol valve is open. For safety reasons we wish the cold glycol valve to fail-open and the hot

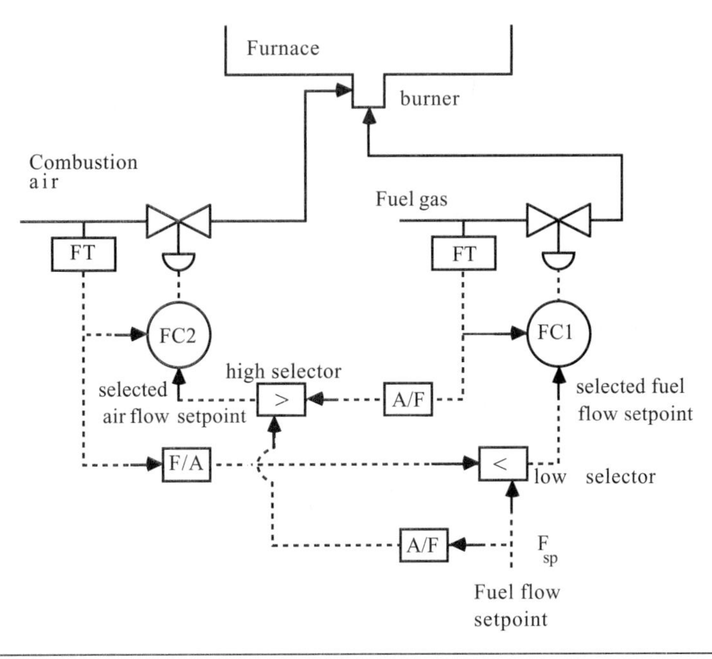

Figure 12–3 Use of high/low selectors for fuel gas and air flow control of a fired heater. Notice the simplified representation for the air flow; ordinarily a damper inside an air duct would be adjusted to vary the air flow.

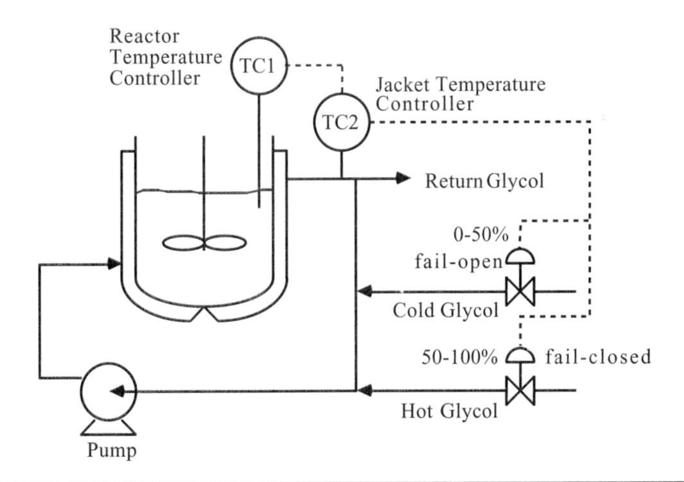

Figure 12–4 Batch reactor temperature control. The jacket temperature controller has a split-range output, where the cold glycol valve is open during "cooling mode" and the hot glycol valve is open during "heating mode."

glycol valve to fail-closed. The diagram in Figure 12–5 more clearly shows the relationship between valve positions and the jacket temperature controller output.

The implementation is as follows. Let u represent the output of the jacket temperature controller (0–100%). Also, let v_c and v_h represent the cold and hot glycol valve positions, respectively. The split-range algorithm can be stated as

$$\begin{aligned} 0 \le u \le 50\% &: v_c = 100 - 2u,\ v_h = 0 \\ 50 < u \le 100\% &; v_c = 0,\ v_h = 2(u - 50) \end{aligned} \tag{12.6}$$

Sometimes there is a small deadband, where the cold glycol is open between 0 and 45%, while the hot glycol is open between 55 and 100%. This adds the equivalent of a small time delay to the control strategy but avoids situations where both cold and hot valves are open owing to small miscalibrations.

Example 12.1: 1000 Liter Stirred-Tank Heater

Here we show results for a 1000 liter jacketed, continuous stirred-tank heater, operated at 85% of capacity. For clarity, we consider only the *jacket temperature controller*, which is split-ranged as shown in Figure 12–5. Ordinarily this jacket temperature controller would be the secondary (inner-loop or slave controller) where the output of a vessel temperature controller (outer-loop or primary controller) is the setpoint to the jacket temperature controller.

The modeling equations are

$$\frac{dT}{dt} = \frac{F}{V}\left(T_i - T\right) + \frac{UA}{V\rho c_p}\left(T_j - T\right)$$

$$\frac{dT_j}{dt} = \frac{F_j}{V_j}\left(T_{jin} - T_j\right) - \frac{UA}{V_j\rho_j c_{pj}}\left(T_j - T\right)$$

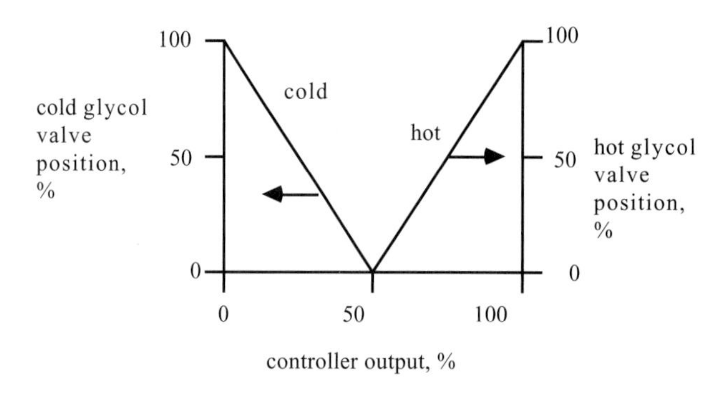

Figure 12–5 Depiction of the split-range controller action.

where the jacket make-up flow rate (F_j; of either cold or hot glycol) is the process input. Numerical values are shown in Exercise 5.

The responses for setpoint changes in jacket temperature are shown in Figure 12–6. A positive setpoint change is made at t = 0 minutes, followed by a negative change at t = 10 minutes. The middle plot shows the jacket temperature controller output and the bottom plot shows the individual valve positions. All variables are in deviation form. When the controller output is positive, the hot valve is activated; when the controller output is negative, the cold valve is activated.

12.5 SIMULINK Functions

SIMULINK has a number of built-in functions to perform ratio, selective, and split-range control. Examples are shown here.

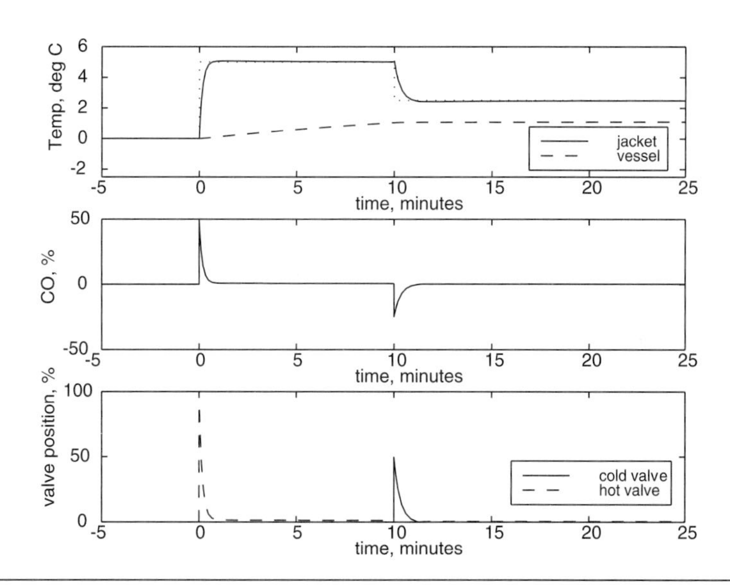

Figure 12–6 Response to step setpoint changes in jacket temperature. All variables are in deviation form.

`Product`	Can be used for ratio control
`MinMax`	Can be used for high or low selection
`Switch`	Can be used for switching between alternatives
`Saturation`	Can be used for split-range control, and controller saturation

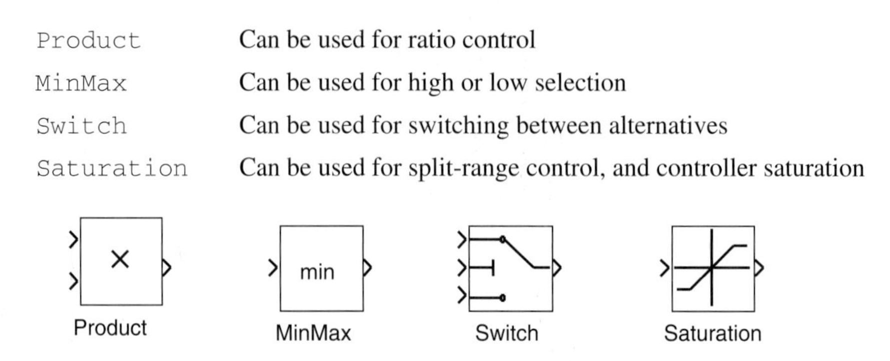

An example SIMULINK block diagram to implement a split-range jacket temperature controller from Example 12.1 is shown in Figure 12–7. Since this has been formulated in deviation variable form, the output of the jacket temperature controller ranges from –50 to 50%. Saturation elements are used for the cold and hot valve positions.

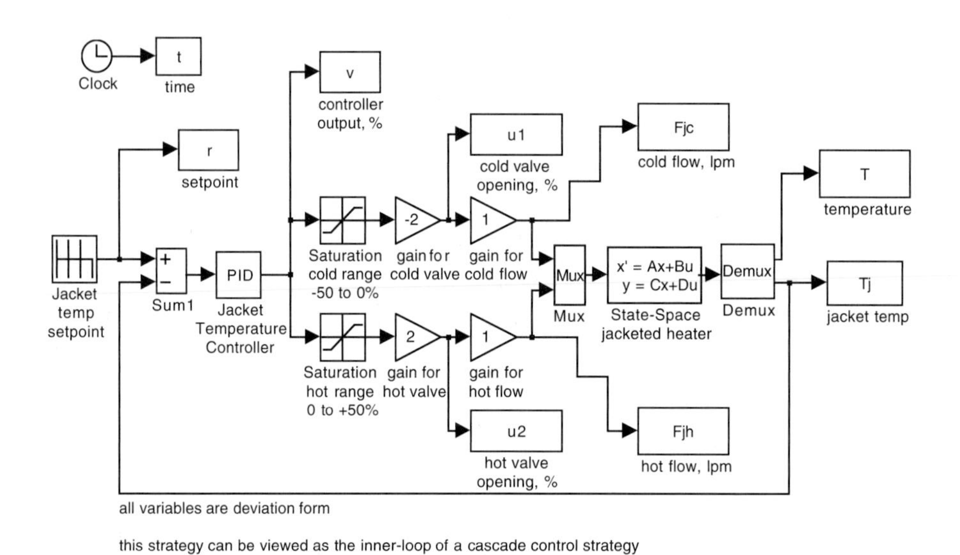

Figure 12–7 SIMULINK implementation of a split-range strategy for a jacket temperature controller.

12.6 Summary

Ratio control has many similarities to feed-forward control. Usually a "wild stream" flow rate is measured and another stream is manipulated to maintain a constant ratio to the wild stream. Selective and override strategies are often used to improve the safety of a process system. A prime example is a fired heater where it is important for combustion air to "lead" the fuel gas flow on increasing loads, but to "lag" fuel gas flow on decreasing loads. Split-range control allows the use of two or more valves as manipulated inputs, depending on the operating condition.

References

Ratio control is discussed in: Myron Jr., T. J., "Feedback Methods for Process Control Systems," *Chemical Engineering*, 233–238 (Nov. 14, 1983).

A number of selective control examples are shown in: Kurth, T. G., "How to Use Feedback Loops to Meet Process Conditions," *Chemical Engineering*, 77–83 (April 30, 1984).

Student Exercises

1. Consider the ratio control strategy shown in Figure 12–1b. If the blend stream has a concentration measurement and controller, construct the control instrumentation diagram where the output of the concentration controller is the desired ratio of stream B to stream A flows.
2. Consider the high-temperature select control problem shown in Figure 12–2. Modify the control instrumentation diagram to include a cascade-control strategy, where the output of the reactor temperature controller is the setpoint to a reactor inlet temperature controller. The reactor inlet temperature controller then manipulates the flow to the preheat exchanger.
3. Consider a gas-phase reactor where a valve on the output stream is normally regulated to maintain a desired outlet composition. Design a selector-based control strategy that switches to pressure control if the pressure is too high.
4. Consider a waste treatment system. A pH controller may need to adjust the pH by manipulating either basic or acidic stream flow rates. Design a split-range control strategy to do this.
5. Consider Example 12.1, where input 1 is the cold glycol make-up flow rate, and input 2 is the hot glycol make-up flow rate. The steady-state temperature is

20°C, which is also the vessel feed temperature. At steady-state the cold and hot glycol make-up flow rates are 0. The parameter values are

```
fov      =  1/30; % minutes^-1
ua       =  24  ; % kcal/min deg C
vrhocp   = 850  ; % kcal/deg C
vrhocpj  = 250  ; % kcal/deg C
Tjfcold  = -20  ; % deg C
Tjfhot   = 100  ; % deg C
Tjs      =  20  ; % deg C
vj       = 250  ; % liters
```

and the state space model matrices are

$$A = \begin{bmatrix} -\frac{F_s}{V} - \frac{UA}{V\rho c_p} & \frac{UA}{V\rho c_p} \\ \frac{UA}{V_j \rho_j c_{pj}} & -\frac{\left(F_{jfcs} + F_{jfhs}\right)}{V_j} - \frac{UA}{V_j \rho_j c_{pj}} \end{bmatrix}$$

$$B = \begin{bmatrix} 0 & 0 \\ \frac{T_{jfcold} - T_{js}}{V_j} & \frac{T_{jfhot} - T_{js}}{V_j} \end{bmatrix}$$

$$C = \begin{bmatrix} 1 & 0 \\ 0 & 1 \end{bmatrix}$$

$$D = \begin{bmatrix} 0 & 0 \\ 0 & 0 \end{bmatrix}$$

Solve for the numerical values of all matrices. Design a jacket temperature controller and implement the SIMULINK diagram shown in Figure 12–7. Discuss your jacket temperature controller design and tuning procedure. Notice that the jacket temperature output has a range of 100%, constrained between –50% and +50%. The cold valve is fully open at –50% and fully closed at 0%. The hot valve is fully closed at 0% and fully open at +50%,

6. Develop a SIMULINK diagram of a cascade control strategy, where Figure 12–7 represents the inner-loop (jacket temperature) control system. The output of the primary (outer-loop) controller that you develop should be the setpoint for the jacket temperature controller. Discuss the design and implementation (tuning) of both controllers. Use the parameter values from Exercise 5.

CHAPTER 13

Control-Loop Interaction

Most process unit operations have a number of manipulated inputs, each of which affects some measured output, and it is not always clear which input should be "paired" with which output for control-system design. This general pairing problem is known as control structure selection and is often more important than the actual controller design and tuning method used. After studying this chapter, the reader should be able to do the following:

- Use the *relative gain array* (RGA) to determine the proper input-output pairings for multiple single input–single output controllers
- Understand that the RGA gives insight about process sensitivity
- Discuss the implications of the RGA for failure sensitivity

The major sections of this chapter are as follows:

13.1 Introduction

Thus far in this textbook we have focused primarily on SISO control systems. Most chemical processes, however, have many manipulated variables affecting many output variables, as shown in Figure 13–1. We generally assume that a desired output variable is also measured. In some cases (composition for example), an output variable cannot be measured and must be "inferred" from other measurements (such as temperatures and flow rates).

Although most processes are multivariable, SISO control loops are formed by selecting a measured output that is most strongly affected by a particular manipulated input. This is done for two reasons: SISO design techniques are easy to understand, and hardware and software are readily available for SISO controllers. We refer to control systems that use several SISO controllers on a multivariable process as MVSISO (multivariable single input–single output) controllers. The selection of which measured output to pair with which manipulated input is known as variable "pairing" and is the focus of much of this chapter.

13.2 Motivation

A specific example of a multivariable control problem is shown in Figure 13–2. Here, two different streams are blended together. The objective is to manipulate the individual stream flow rates to meet specifications on the output flow rate and the output composition. Let the outputs be represented by y_2 (total flow) and y_1 (composition), and the inputs by u_1 (flow rate of stream 1) and u_2 (flow rate of stream 2). How should our outputs and inputs be paired? If we assume that the output-input pairings are y_1-u_1 and y_2-u_2, the instrumentation

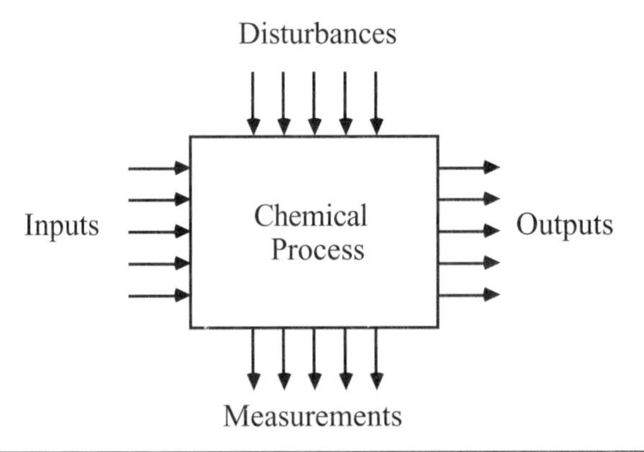

Figure 13–1 General multivariable process.

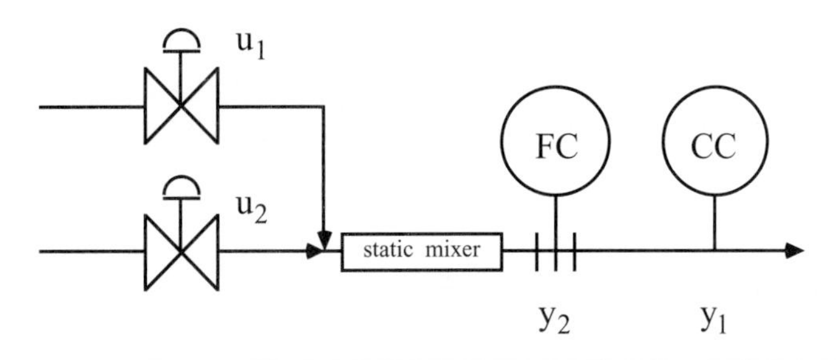

Figure 13–2 Blending system.

diagram is shown in Figure 13–3a. If we assume that the output-input pairings are y_1-u_2 and y_2-u_1, the instrumentation diagram is shown in Figure 13–3b.

We can see potential problems in the operation of the control strategy in either Figure 13–3a or b. For discussion purposes, consider Figure 13–3a. A setpoint change in output 1 (y_1, concentration) will cause the flow controller to change input 1 (u_1, stream 1 flow). The change in input 1 will also affect output 2 (y_2, total flow). Once output 2 is disturbed, the flow controller will cause a change in input 2. Input 2 will disturb the concentration,

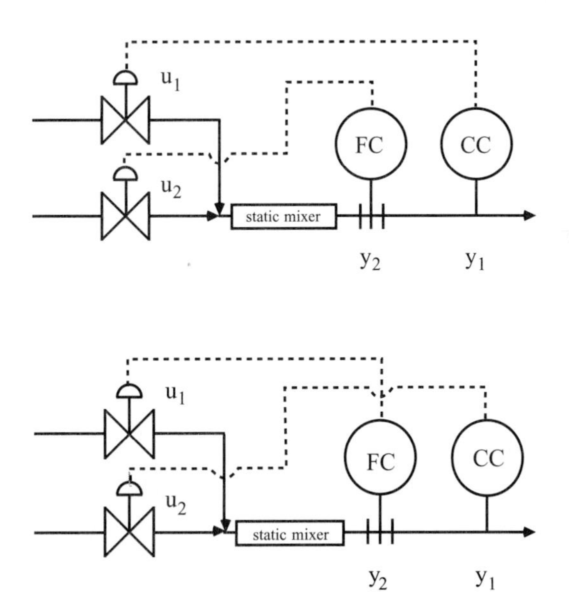

Figure 13–3 Blending system alternative closed-loop pairings. (a) u_1-y_1, u_2-y_2 pairing. (b) u_1-y_2, u_2-y_1 pairing.

causing the concentration controller to change input 1, and we are back to where we started. The effect of one control loop on the other is referred to as *control loop interaction*. Although it is natural to think of these interactions as occurring in a sequential fashion, they actually happen simultaneously.

Example 13.1: Whiskey Blending

Consider now a very specific problem to see whether physical intuition can be used to determine the proper pairing. Assume that an enterprising control student has decided that there is a profit to be made by blending a stream of 80 proof (40% ethanol) whiskey with pure water to produce a product stream that is 60 proof (30% ethanol). Assume that it is important to meet both quality and production-rate requirements, and that you can measure the alcohol content and total flow rate accurately. Assume that the 80 proof stream flow rate is the first manipulated input (u_1) and that the dilution water flow rate is the second manipulated input (u_2). The blended product flow rate is output 2 and the blend composition is output 1.

Your intuition is probably to pair output 1 (alcohol measurement) with input 1 (80 proof whiskey stream), since there is no alcohol in the water stream; this is the strategy shown in Figure 13–3a. We find later that this is not the best pairing, indicating that the best control structure is often counterintuitive. We first see the problem by observing the behavior of control loop 1 (y_1-u_1 pairing) when control loop 2 (y_2-u_2 pairing) is open (or under "manual" control). The response to an ethanol concentration setpoint change is shown in Figure 13–4. Observe that the closed-loop behavior of ethanol concentration is quite good, but that the total flow rate deviates, since the flow control loop (control loop 2) is not closed.

Assume that it is desirable to maintain the total blend rate at a constant value. This requires that loop 2 be closed. Loop 2 has blend flow as the measured process output (y_2) and the dilution water flow rate as the manipulated input (u_2). Assume that it has been tuned independently from loop 1, that is, loop 1 is open. The response to a setpoint change in flow rate is shown in Figure 13–5. Notice that the flow-rate response is quite good but that the ethanol concentration deviates from the desired value, because loop 1 is open.

We have seen that the individual control loops have quite good performance. Since we desire to control both the concentration and the flow rate of the blend stream, we need to have both loops closed simultaneously. The response to a setpoint change in the total blend flow rate with both loops closed is shown in Figure 13–6. Notice that closed-loop system is now unstable! What characteristic of this control strategy caused the instability to occur when both loops were closed? Would this behavior have occurred if a different variable pairing had been used? That is, would there be problems if loop 1 paired the

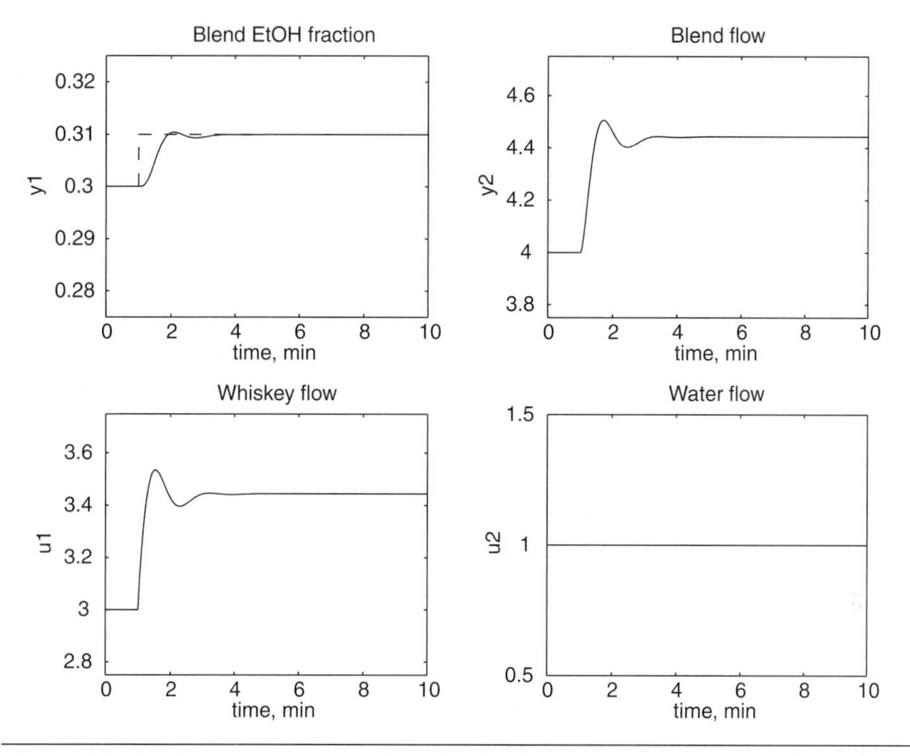

Figure 13–4 Response to an ethanol setpoint change; loop 1 has ethanol concentration as the measured output (y_1) and whiskey flow rate as the manipulated input (u_1). Loop 2 is open.

ethanol measurement (y_1) with dilution water flow (u_2) and loop 2 paired the total blend flow rate (y_2) with the whiskey flow (u_1)? We find the answer to this question in the sections that follow.

13.3 The General Pairing Problem

Given a general multivariable process diagram, how do we select the pairings for a control structure that consists of multiple SISO loops? In the following sections, we use matrix vector notation, where $y(s)$ is a vector of n outputs and $u(s)$ is a vector of m inputs,

$$y(s) = G(s)u(s) \tag{13.1}$$

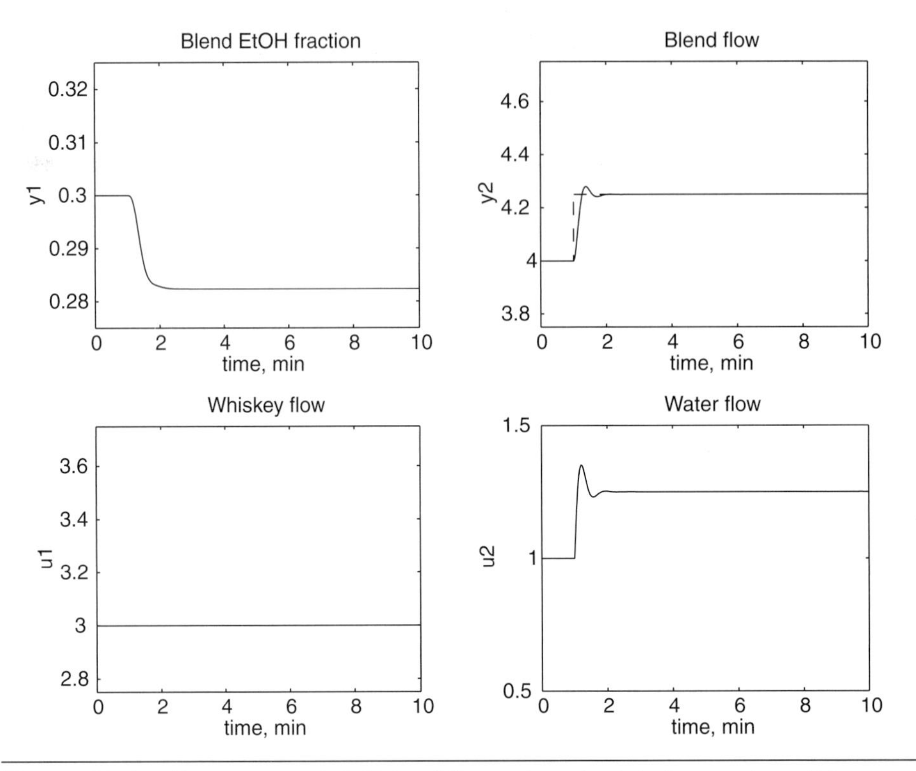

Figure 13–5 Response to total blend flow rate setpoint change; loop 2 has blend flow rate as the measured output (y_2) and dilution water flow rate as the manipulated input (u_2). Loop 1 is open.

$$\underbrace{\begin{bmatrix} y_1(s) \\ \vdots \\ y_n(s) \end{bmatrix}}_{n \text{ outputs}} = \underbrace{\begin{bmatrix} g_{11}(s) & \cdots & g_{1m}(s) \\ \vdots & \cdots & \vdots \\ g_{n1}(s) & \cdots & g_{nm}(s) \end{bmatrix}}_{\substack{n \times m \text{ transfer} \\ \text{function matrix}}} \underbrace{\begin{bmatrix} u_1(s) \\ \vdots \\ u_m(s) \end{bmatrix}}_{m \text{ inputs}} \tag{13.2}$$

where each output can be represented by

$$y_i(s) = \sum_{j=1}^{m} g_{ij}(s) u_j(s) \tag{13.3}$$

To form a MVSISO system, we must ask the following:

Which output y_i should be paired with which input u_j to form an SISO control loop?

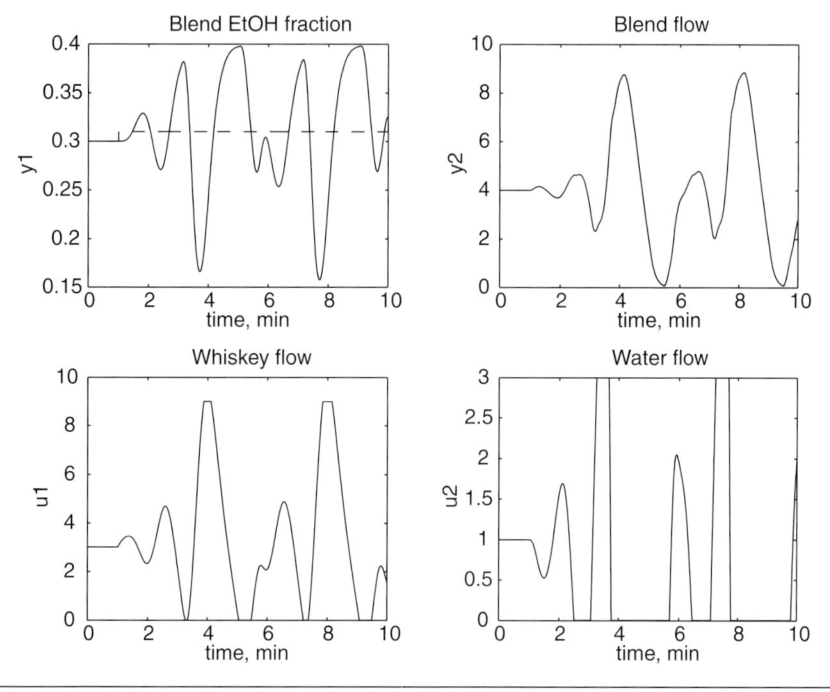

Figure 13–6 Response to a total blend flow rate setpoint change with both loops closed. Loop 2 pairs blend flow rate (y_2) and dilution water flow rate (u_2), while loop 1 pairs ethanol concentration (y_1) with whiskey flow (u_1).

Two Input–Two Output Processes

Let us consider the relationship between y_i and u_j under a number of conditions. For simplicity, we will deal primarily with 2×2 (two input–two output) systems. The open-loop input-output relationships are

$$\begin{aligned} y_1(s) &= g_{11}(s)u_1(s) + g_{12}(s)u_2(s) \\ y_2(s) &= g_{21}(s)u_1(s) + g_{22}(s)u_2(s) \end{aligned} \tag{13.4}$$

which is written in matrix form as

$$\begin{bmatrix} y_1(s) \\ y_2(s) \end{bmatrix} = \begin{bmatrix} g_{11}(s) & g_{12}(s) \\ g_{21}(s) & g_{22}(s) \end{bmatrix} \begin{bmatrix} u_1(s) \\ u_2(s) \end{bmatrix} \tag{13.5}$$

and we can think of the input output relationships as shown in Figure 13–7.

The corresponding feedback control system, if the pairings are u_1-y_1 and u_2-y_2, is shown in Figure 13–8.

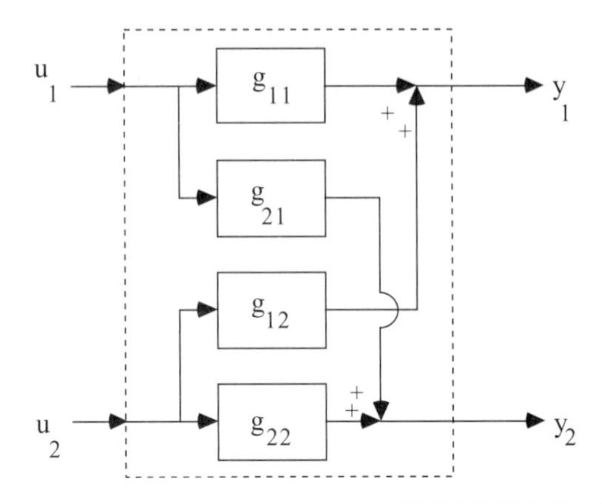

Figure 13–7 Input-output block diagram for a two input–two output process.

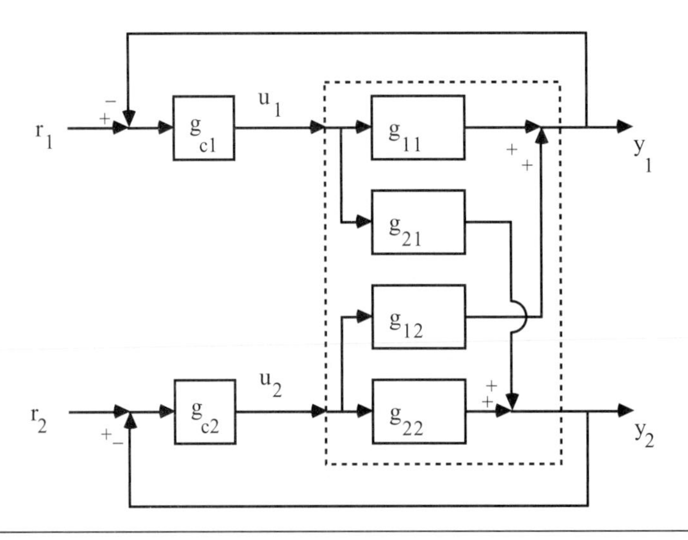

Figure 13–8 Feedback block diagram for a two input–two output process where output y_1 is paired with input u_1.

Input 1–Output 1 Dynamic Behavior

For consistency in notation, we refer to loop 1 as the y_1-u_1 pairing and loop 2 as the y_2-u_2 pairing. We wish to consider how control loop 1 should be designed. We have focused on the use of input-output models for control system design. In the case of loop 1, we need to know how input 1 affects output 1.

Now, consider how u_1 affects y_1 if (a) loop 2 is open and if (b) loop 2 is closed.

(a) *Loop 2 is open* (loop 1 is also open). If the second input (u_2) is constant, then the outputs depend only on the first input (u_1), as shown in Figure 13–9.

The input-output relationship is

$$y_1(s) = g_{11}(s)u_1(s)$$

If we designed a SISO feedback control system for loop 1, we would use $g_{11}(s)$ as the process transfer function for control system design.

(b) *Loop 2 is closed.* If the second loop (y_2-u_2) is closed, then the outputs depend on input 1 (u_1), as shown in Figure 13–10. Notice that there is an additional effect of input 1 on input 2 owing to the action of controller 2. Input 1 affects output 2, which then changes input 2 through the second controller. This change in input 2 then has an additional effect on output 1; this effect can be positive or negative.

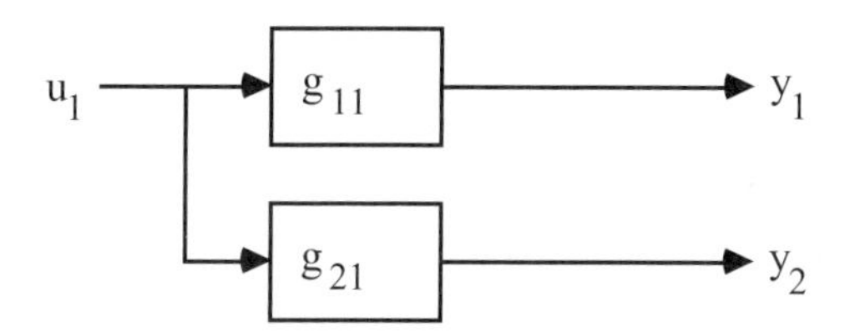

Figure 13–9 Block diagram with loop 2 open (loop 1 is open as well).

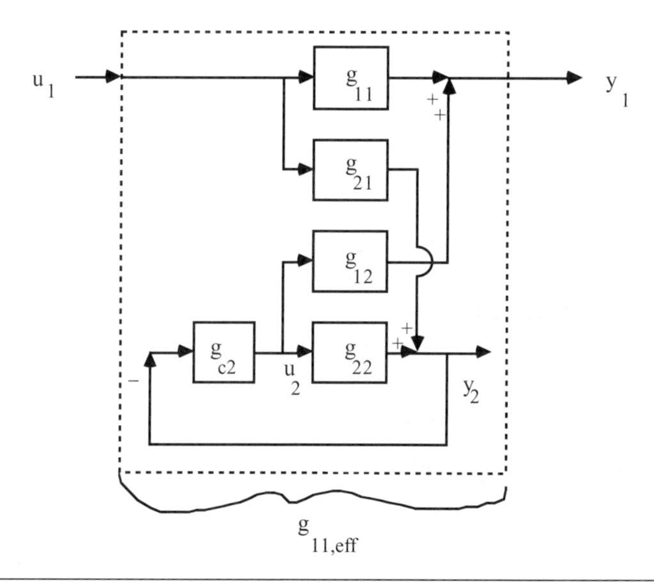

Figure 13–10 Block diagram of input 1–output 1 relationship with loop 2 closed.

Thus, the relationship between $u_1(s)$ and $y_1(s)$ is no longer just $g_{11}(s)$. Indeed, we can derive the relationship in the fashion:

$$y_1(s) = g_{11}(s)u_1(s) + g_{12}(s)u_2(s) \tag{13.6}$$

but (assuming no setpoint change in loop 2, $r_2 = 0$),

$$u_2(s) = g_{c2}(s)e_2(s) = g_{c2}(s)(-y_2(s)) \tag{13.7}$$

also

$$y_2(s) = g_{21}(s)u_1(s) + g_{22}(s)u_2(s) \tag{13.8}$$

Combining Equations (13.7) and (13.8),

$$y_2(s) = g_{21}(s)u_1(s) + g_{22}(s)g_{c2}(s)(-y_2(s)) \tag{13.9}$$

and solving for $y_2(s)$,

$$y_2(s) = \frac{g_{21}(s)u_1(s)}{1 + g_{22}(s)g_{c2}(s)} \tag{13.10}$$

Substituting this into Equation (13.7),

$$u_2(s) = \frac{-g_{21}(s)g_{c2}(s)}{1 + g_{22}(s)g_{c2}(s)}u_1(s) \tag{13.11}$$

and substituting this result into Equation (13.6), we find

$$y_1(s) = \left[g_{11}(s) - \frac{g_{12}(s)g_{21}(s)g_{c2}(s)}{1 + g_{22}(s)g_{c2}(s)}\right]u_1(s) \tag{13.12}$$

That is, $y_1(s) = g_{11,\text{eff}}(s)\ u_1(s)$, where

$$g_{11,eff}(s) = g_{11}(s) - \frac{g_{12}(s)g_{21}(s)g_{c2}(s)}{1 + g_{22}(s)g_{c2}(s)} \tag{13.13}$$

$g_{11,\text{eff}}(s)$ is the effective input-output relationship between u_1 and y_1, with loop 2 closed. This means that $g_{11,\text{eff}}(s)$ should be used as the process transfer function to design controller 1 if loop 2 is closed. Unless $g_{12}(s)$ or $g_{21}(s) = 0$, if the second control loop is closed, then the first control loop must be designed differently than if the second control loop were open. Naturally, the relationship between u_1 and y_1 is a function of the controller g_{c2} when the second loop is closed.

Example 13.1, continued

Consider the expected effect of the second loop in the blending example posed earlier. In Figure 13–11 we compare the response of ethanol concentration due to a change in

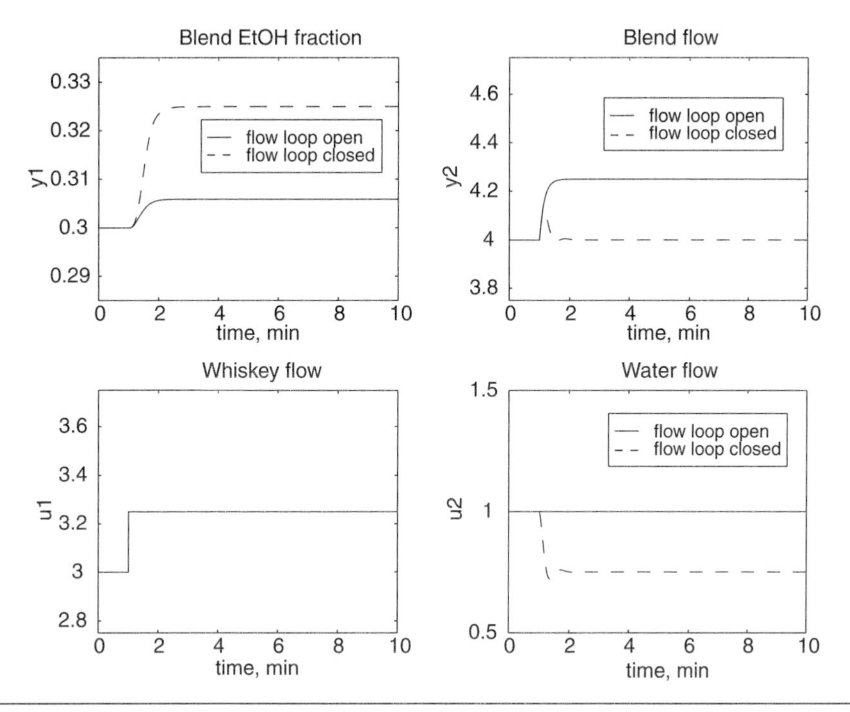

Figure 13–11 Response of ethanol concentration to a step increase in whiskey flow rate; loop 2 open (solid curve) and loop 2 closed (dashed curve).

whiskey flow rate, under two conditions: with the flow controller (loop 2) open (solid), and with the flow controller closed (dashed). Notice the tremendous amplification of the effect of the whiskey flow rate. Here, we rationalize the effect of the flow loop being closed. When the whiskey flow rate is increased, it causes the total blend flow rate to increase. The total blend flow-rate controller then cuts back on the dilution water flow rate, which causes the ethanol concentration to increase even more. The flow rate controller has caused the whiskey stream to have roughly four times the effect on the ethanol concentration, compared to when the flow control loop is open.

Steady-State Effective Gain

Comment:

What we would really like is a method to determine the relationship between u_1 and y_1 with loop 2 closed, without knowing $g_{c2}(s)$.

We can easily determine this relationship for a limiting case, the steady state. For the steady state, we simply let $s \to 0$. Assuming integral action is used in the controller, we find that as $s \to 0$, $g_{c2}(s) \to \infty$. Also, from Equation (13.13) and $g_{c2}(0) \to \infty$ as $s \to 0$, we find the following relationship (where 0 represents $s = 0$, the steady state; do not confuse this with $t = 0$, the initial condition):

$$g_{11,eff}(0) = g_{11}(0) - \frac{g_{12}(0)g_{21}(0)}{g_{22}(0)} = k_{11} - \frac{k_{12}k_{21}}{k_{22}} = k_{11,eff} \tag{13.14}$$

This can also be determined from the steady-state relationship

$$\begin{bmatrix} y_1 \\ y_2 \end{bmatrix} = \begin{bmatrix} k_{11} & k_{12} \\ k_{21} & k_{22} \end{bmatrix} \begin{bmatrix} u_1 \\ u_2 \end{bmatrix}$$

Assuming *perfect* control of y_2, that is, $y_2 = 0$,

$$y_2 = k_{21}u_1 + k_{22}u_2 = 0 \tag{13.15}$$

$$u_2 = \frac{-k_{21}u_1}{k_{22}} \tag{13.16}$$

and we can then find the following relationships:

$$\begin{aligned} y_1 &= k_{11}u_1 + k_{12}u_2 \\ y_1 &= k_{11}u_1 + k_{12}\left(-\frac{k_{21}}{k_{22}}\right)u_1 \\ y_1 &= \left(k_{11} - \frac{k_{12}k_{21}}{k_{22}}\right)u_1 = k_{11eff}u_1 \end{aligned} \tag{13.17}$$

We now have a steady-state effective gain relationship between u_1 and y_1 with the loop between u_2 and y_2 closed.

13.4 The Relative Gain Array

Ed Bristol (a control engineer with Foxboro) developed a heuristic technique to predict possible interactive effects between control loops when multiple SISO loops are used. He was motivated to develop a dimensionless measure of interaction by some boiler control problems he was working on. His simple measure of interaction is known as the *relative gain*, or the Bristol *relative gain array* (RGA). Since then, many papers have been written to provide a more rigorous theoretical basis for the technique.

Two-Inputs and Two-Outputs

The relationship between u_1 and y_1 at steady state ($s = 0$), with no change in u_2, is

$$\left(\frac{\partial y_1}{\partial u_1}\right)_{u_2} = g_{11}(0) = k_{11} \tag{13.18}$$

This is referred to as the gain between u_1 and y_1 with all other loops *open* (that is, u_2 is constant for this 2×2 example).

The steady-state relationship between u_1 and y_1 at steady state, with y_2 maintained constant at its setpoint (using a controller with integral action), is

$$\left(\frac{\partial y_1}{\partial u_1}\right)_{y_2} = g_{11}(0) - \frac{g_{12}(0)g_{21}(0)}{g_{22}(0)} = k_{11} - \frac{k_{12}k_{21}}{k_{22}} = k_{11eff} \tag{13.19}$$

Notice that Equation (13.19) is the gain between input 1 and output 1, assuming output 2 is constant.

Definition of the Relative Gain

The relative gain (λ_{ij}) between input j and output i is defined in the following fashion:

$$\lambda_{ij} = \frac{\textit{gain between input j and output i with all other loops open}}{\textit{gain between input j and output i with all other loops closed}}$$

$$\lambda_{ij} = \frac{\left(\frac{\partial y_i}{\partial u_j}\right)_{u_k \; k \neq j}}{\left(\frac{\partial y_i}{\partial u_j}\right)_{y_k \; k \neq i}} \tag{13.20}$$

By convention, the symbol used for the relative gain is λ_{ij}. Please do not confuse it with the λ that was used for the IMC filter or for the λ that is commonly used to represent the eigenvalues of a matrix. The relative gain will normally be shown with two subscripts, distinguishing it from eigenvalues or the IMC filter factor.

Relative Gain Between Input 1 and Output 1 for a Two Input–Two Output System

The relative gain between input 1 and output 1 is then [see Equation (13.23)]

$$\lambda_{11} = \frac{\text{gain between } u_1 \text{ and } y_1 \text{ with } u_2 \text{ constant}}{\text{gain between } u_1 \text{ and } y_1 \text{ with } y_2 \text{ constant}}$$

We see that the relative gain between u_1 and y_1 for a 2 × 2 system is

$$\lambda_{11} = \frac{\left(\frac{\partial y_1}{\partial u_1}\right)_{u_2}}{\left(\frac{\partial y_1}{\partial u_1}\right)_{y_2}} = \frac{k_{11}}{k_{11eff}} = \frac{k_{11}}{k_{11} - \frac{k_{12}k_{21}}{k_{22}}} = \frac{1}{1 - \frac{k_{12}k_{21}}{k_{11}k_{22}}} \tag{13.21}$$

Remember that Equation (13.21) is only true for the two input–two output case.

The RGA

The RGA is simply the matrix that contains the individual relative gain as elements, that is $\Lambda = \{\lambda_{ij}\}$. For a 2 × 2 system, the RGA is

$$\Lambda = \begin{bmatrix} \lambda_{11} & \lambda_{12} \\ \lambda_{21} & \lambda_{22} \end{bmatrix} \tag{13.22}$$

Question: What value do we desire for the relative gain?

Think of it this way. If we want to use a u_1-y_1 pairing for a SISO controller, we do not want it to matter whether the other loops in the system are closed or not. This tells us that we desire a relative gain close to 1.0. That is,

Answer: We desire $\lambda_{11} \approx 1.0$ if we wish to pair output 1 with input 1.

The λ_{11} calculation for a 2 × 2 system yielded $\lambda_{11} = \frac{k_{11}k_{22}}{k_{11}k_{22} - k_{12}k_{21}}$. The more general result is (see Appendix 13.1 for a derivation)

$$\lambda_{ij} = g_{ij}(0)\hat{g}_{ji}(0) = k_{ij}\hat{k}_{ji} \tag{13.23}$$

where $\hat{k}_{ji} = ji^{th}$ element of $G^{-1}(0)$. Notice that this is not the same as matrix multiplication. Here, the individual elements of the two matrices are being multiplied together.

We check the general results derived above by applying them to a 2 × 2 (two output, two input) system,

$$G(0) = \begin{bmatrix} g_{11}(0) & g_{12}(0) \\ g_{21}(0) & g_{22}(0) \end{bmatrix} = \begin{bmatrix} k_{11} & k_{12} \\ k_{21} & k_{22} \end{bmatrix}$$

and recalling our rules for inverting a 2 × 2 matrix,

$$\hat{G}(0) = G^{-1}(0) = \begin{bmatrix} k_{22} & -k_{12} \\ -k_{21} & k_{11} \end{bmatrix} \cdot \frac{1}{\det K} = \begin{bmatrix} \frac{k_{22}}{k_{11}k_{22} - k_{12}k_{21}} & \frac{-k_{12}}{k_{11}k_{22} - k_{12}k_{21}} \\ \frac{-k_{21}}{k_{11}k_{22} - k_{12}k_{21}} & \frac{k_{11}}{k_{11}k_{22} - k_{12}k_{21}} \end{bmatrix}$$

Multiplying the individual elements of $G(0)$ and $\hat{G}(0)$, as in Equation (13.23), we find

$$\Lambda = \begin{bmatrix} \dfrac{k_{11}k_{22}}{k_{11}k_{22} - k_{12}k_{21}} & \dfrac{-k_{12}k_{21}}{k_{11}k_{22} - k_{12}k_{21}} \\ \dfrac{-k_{21}k_{12}}{k_{11}k_{22} - k_{12}k_{21}} & \dfrac{k_{11}k_{22}}{k_{11}k_{22} - k_{12}k_{21}} \end{bmatrix} = \begin{bmatrix} \lambda_{11} & \lambda_{12} \\ \lambda_{21} & \lambda_{22} \end{bmatrix} \tag{13.24}$$

13.5 Properties and Application of the RGA

Sum of Rows and Columns Property of the RGA

Notice that Equation (13.24) yields the following relationships

$$\lambda_{11} + \lambda_{12} = 1 \qquad \lambda_{11} + \lambda_{21} = 1$$
$$\lambda_{12} + \lambda_{22} = 1 \qquad \lambda_{21} + \lambda_{22} = 1$$

This property for 2×2 systems is general for $n \times n$ systems. That is, each row of the RGA sums to 1.0 and each column of the RGA sums to 1.0.

Then, for a 2×2 system, only one relative gain must be calculated to find the entire array,

$$\Lambda = \begin{bmatrix} \lambda_{11} & 1-\lambda_{11} \\ 1-\lambda_{11} & \lambda_{11} \end{bmatrix} \tag{13.25}$$

For a 3×3 system, only four of the nine elements would need to be calculated. In practice, they are all calculated simultaneously using (13.23).

Use of RGA to Determine Variable Pairing

It is desirable to pair output i and input j such that λ_{ij} is as close to 1 as possible.

Example 13.2: RGA for Variable Pairing

Consider a relative gain array, $\Lambda = \begin{bmatrix} 0.95 & 0.05 \\ 0.05 & 0.95 \end{bmatrix}$

We would pair y_1 with u_1 and y_2 with u_2 in this case.

However, if $\Lambda = \begin{bmatrix} 0.05 & 0.95 \\ 0.95 & 0.05 \end{bmatrix}$

we would pair y_1 with u_2 and y_2 with u_1.

Two Input–Two Output Systems

It is easy to show the two possible cases for the relative gain array of a 2×2 system. If there are an *odd* number of positive elements in $\mathbf{G}(0)$, then

$$\lambda_{ij} \in (0,1)$$

That is, for an odd number of positive elements in the process gain matrix, all the relative gains will be between 0 and 1. If there are an *even* number of positive elements in $\mathbf{G}(0)$, then

$$\lambda_{ij} \in (-\infty,0) \cup (1,\infty)$$

This means that for an even number of positive elements in the process gain array, all the relative gains will be outside the 0–1 range.

Implications for the Sign of a Relative Gain

First of all, we can state that if λ_{11} was negative, we would not want to pair input 1 with output 1. Let us use the following reasoning. Assume that the open-loop gain between input 1 and output 1 (g_{11}) is positive; this implies that the controller gain is also positive. A negative relative gain tells us that the gain between input 1 and output 1 with output 2 perfectly controlled (that is, loop 2 closed) is negative. This means that if loop 2 is closed, we need a negative controller gain on loop 1. However, we noted before that if loop 2 is open, we need a positive controller gain on loop 1. It is not desirable to have a control system where you have to change the sign of the controller gain depending on whether the other loops are open or closed, so the bottom line is the following:

We do not want to pair output i and input j if λ_{ij} is negative.

The results for a 2×2 system indicate that if $\lambda_{11} > 1$ then $\lambda_{12} = \lambda_{21} < 0$. Also, if $0 < \lambda_{11} < 1$, then $0 < \lambda_{12} < 1$. Let us think about the differences between having a λ_{11} greater than 1 and having a λ_{11} less then 1.

If $\lambda_{11} > 1$, then $\left.\dfrac{\partial y_1}{\partial u_1}\right|_{u_2} > \left.\dfrac{\partial y_1}{\partial u_1}\right|_{y_2}$.

This tells us that the gain between u_1 and y_1 is larger with loop 2 open than with loop 2 closed.

- If we design the u_1-y_1 controller with loop 2 open, then close loop 2, we expect somewhat sluggish behavior, since the effective process gain will be lower than the u_1-y_1 gain.

However,

- If we design the u_1-y_1 controller with loop 2 closed, then if we have to open loop 2 for some reason, we could have destabilizing behavior, since the u_1-y_1 gain will be higher than when the controller was tuned.

If $0 < \lambda_{11} < 1$, then $\left.\frac{\partial y_1}{\partial u_1}\right|_{u_2} < \left.\frac{\partial y_1}{\partial u_1}\right|_{y_2}$.

This tells us that the gain between u_1 and y_1 is smaller with loop 2 open than with loop 2 closed.

- If we design the u_1-y_1 controller with loop 2 open, then close loop 2, we expect more aggressive behavior, since the effective process gain will be higher than the u_1-y_1 gain.

However,

- If we design the u_1-y_1 controller with loop 2 closed, then if we have to open loop 2 for some reason, we expect sluggish behavior, since the u_1-y_1 gain will be lower than when the controller was tuned.

Midchapter Summarizing Remarks

You must be careful when tuning SISO loops in a multiple input–multiple output (MIMO) system, because the other loops in the system can greatly affect the gain between the input-output variables under consideration.

Also, note that we have only been concerned with static (steady-state) effects. *Dynamic* interaction has a major effect when control systems are tuned. Our experience is that one rarely reverses a pairing decision based on the steady-state relative gain, when one considers the dynamic effects.

The primary result from a dynamic interaction analysis is to recognize additional interaction affects above the steady-state ones and either accept the degradation in control system performance, or use a "true" multivariable control system design technique rather than separate SISO controllers. Sometimes the interaction and other problems will be so severe that only one loop can be closed.

13.6 Return to the Motivating Example

Consider Example 13–1, where 40% ethanol (stream 1) is being blended with pure water (stream 2) to produce a product of 30% ethanol. Here we wish to use the RGA to help us develop the control structure (pair inputs and outputs).

We first develop the steady-state model, then linearize to find the process gain matrix, and finally calculate the RGA to decide on variable pairing. For simplicity we neglect density differences between the streams and components.

Total Material Balance

Neglecting density differences between the streams and components, the total blend product volumetric flow rate is equal the sum of the volumetric flow rates of the two feed streams,

$$F = F_1 + F_2$$

Component Material Balance on Ethanol

Assume a binary system with z_1 = volume fraction of ethanol in stream 1. Let z = volume fraction of ethanol in the blend stream. Since there is no ethanol in the water stream, we find

$$Fz = F_1 z_1$$

Solving these equations for the volume fraction of ethanol in the blend stream,

$$z = \frac{F_1 z_1}{F_1 + F_2}$$

Let the volume fraction of ethanol be output 1 and total flow rate output 2. Also, inputs 1 and 2 are the flow rates of the whiskey and water feedstreams, respectively. The following steady-state values were used in Example 13.1: $F_1 = 3$ gpm, $F_2 = 1$ gpm, $z = 0.3$ mole fraction ethanol, and $F = 4$ gpm.

The steady-state input-output gains are

$$k_{11} = \frac{\partial z}{\partial F_1} = \frac{F_2 z_1}{(F_1 + F_2)^2} = \frac{1 \cdot 0.4}{4^2} = 0.025 \text{ mol frac/gpm}$$

$$k_{12} = \frac{\partial x}{\partial F_1} = \frac{-F_1 z_1}{(F_1 + F_2)^2} = \frac{-3 \cdot 0.4}{4^2} = -0.075 \text{ mol frac/gpm}$$

$$k_{21} = \frac{\partial F}{\partial F_1} = 1\,\text{gpm/gpm}$$

$$k_{22} = \frac{\partial F}{\partial F_2} = 1\ \text{gpm/gpm}$$

which yields the process gain matrix

$$K = \begin{bmatrix} 0.025 & -0.075 \\ 1 & 1 \end{bmatrix}$$

The relative gain relating input 1 to output 1 is

$$\lambda_{11} = \frac{1}{1 - \dfrac{k_{12}k_{21}}{k_{11}k_{22}}} = \frac{1}{1 - \dfrac{(-0.075)\cdot 1}{0.025 \cdot 1}} = 0.25$$

This value of 0.25 indicates that the effective gain relating input 1 to output 1 with loop 2 closed is four times the value of the gain with loop 2 open. This explains the result shown in Figure 13–11.

The RGA is then

$$\Lambda = \begin{bmatrix} 0.25 & 0.75 \\ 0.75 & 0.25 \end{bmatrix}$$

which indicates that the blend total flow (y_2) should be paired with the whiskey flow rate (u_1) and the blend ethanol concentration (y_1) should be paired with the water flow (u_2). Notice that the largest component stream flow rate then controls the total flow.

13.7 RGA and Sensitivity

The previous sections have introduced the concept of the RGA, to help determine which input should be paired with which output to form a multivariable control system composed of a number of single input–single output controllers (we refer to this as a MVSISO system). The purpose of this section is to discuss the relationship between relative gain and controller tuning. It should become clear that the RGA gives valuable insight into the failure sensitivity of a control system. We also present numerous examples to illustrate the important concepts.

Failure Sensitivity

When tuning a set of SISO controllers to form a MVSISO strategy, it is important to consider the *failure sensitivity* of the system. For example, consider a two input–two output system. If loop 1 is tuned to operate well when loop 2 is closed, we must also consider what happens when loop 2 is opened. It is possible for loop 2 to be opened in a number of ways. The most obvious way is for an operator to put loop 2 on manual control (i.e., open loop). Another way is if the second manipulated variable hits a constraint, say if the valve goes all the way open or closed—this means that the second input can no longer affect the second output, which is equivalent to being open loop.

Relative Gain as a Perturbation of the Nominal Process

Consider a system where loop one has output 1 paired with input 1. The relative gain is defined as

$$\lambda_{11} = \frac{\text{gain between input 1 and output 1 with all loops } \textit{open}}{\text{gain between input 1 and output 1 with all other loops } \textit{closed}} \tag{13.26}$$

Using k_{11} to represent the steady-state gain between input 1 and output 1,

$$\lambda_{11} = \frac{k_{11}}{k_{11,eff}} \tag{13.27}$$

From Equation (13.27), we can represent the effective gain as

$$k_{11,eff} = \frac{k_{11}}{\lambda_{11}} \tag{13.28}$$

where

$$k_{11,eff} = k_{11} - \frac{k_{12}k_{21}}{k_{22}}$$

We can really represent the effect of the relative gain by comparing the two block diagrams shown in Figures 13–12a and b.

Comparing Figures 13–12a and b, we see that the effective process gain between input 1 and output 1 with loop 2 is closed is simply a perturbation of the open-loop gain between input 1 and output 1 by a multiplicative factor of $1/\lambda_{11}$.

Relative Gains Between 0 and 1

If a relative gain is less than 1, this means that $k_{11,\text{eff}} > k_{11}$. If a controller for loop 1 is based on k_{11} (implicitly assuming that loop 2 is open), then when loop 2 is closed, the controller gain for loop 1 (k_{c1}) should be detuned approximately by a factor of λ_{11}, otherwise the control system will respond too rapidly and there will be a chance of instability. If, on the

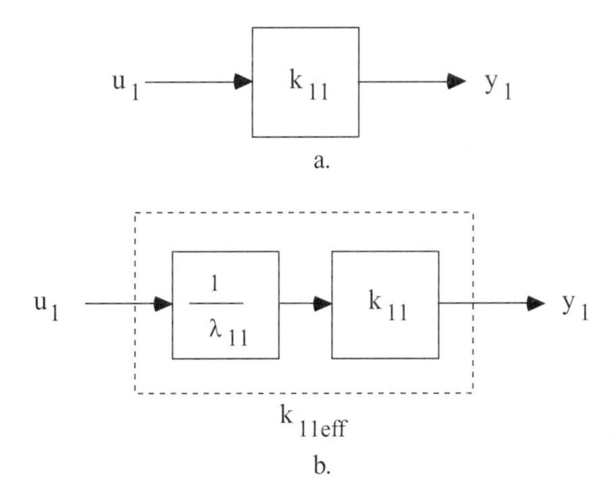

Figure 13–12 Process gain between input 1 and output 1 when loop 2 is open versus when loop 2 is closed. (a) Process gain between input 1 and output 1 when loop 2 is open. (b) Effective process gain between input 1 and output 1 when loop 2 is closed.

other hand, the controller for loop 1 is tuned based on loop 2 being closed, then k_{c1} can probably be increased by a factor of $1/\lambda_{11}$ when loop 2 is opened. The problem is, if the k_{c1} is kept at this new larger value when loop 2 is closed again, there is again a potential for instability.

This analysis indicates that for safety purposes, if the relative gain is less than 1, then loop 1 should be tuned based on loop 2 closed.

Relative Gains Greater than 1

If a relative gain is greater than 1, this means that $k_{11,\text{eff}} < k_{11}$. If a controller for loop 1 is based on k_{11} (implicitly assuming that loop 2 is open), then when loop 2 is closed, the controller gain for loop 1 (k_{c1}) can be increased approximately by a factor of λ_{11}. The problem is, if loop 2 is then opened again, the controller gain will be too high, and there is a chance of instability. If, on the other hand, the controller for loop 1 is tuned based on loop 2 being closed, then k_{c1} should be decreased by a factor of $1/\lambda_{11}$ when loop 2 is opened.

This analysis indicates that for safety purposes, if the relative gain is greater than 1, then loop 1 should be tuned based on loop 2 open.

Negative Relative Gains

Think about the ramifications of having a negative relative gain. If λ_{11} is negative, this means that the gain between u_1 and y_1 with loop 2 closed ($k_{11,\text{eff}}$) has the opposite sign compared with the gain between u_1 and y_1 with loop 2 open (k_{11}). This has the following ramifications. First of all, assume that the process is open-loop stable. For open-loop stable processes that have a controller with integral action, the controller gain must have the same sign as the process gain. If the controller for loop 1 is designed with loop 2 open, it will have a controller gain with a certain sign. Then, when loop 2 is closed, it must have an opposite sign in order to be stable! This is certainly not a desirable situation and is unacceptable in practice. A controller for loop 1 designed based on g_{11} (loop 2 open) will be destabilized when loop 2 is closed. If the controller for loop 1 is designed based on $g_{11,\text{eff}}$ (loop 2 closed), the controller will be destabilized when loop 2 is opened. This leads to the following statement regarding failure sensitivity.

Failure Sensitivity

A control strategy that is based on pairing on a negative relative gain is *failure sensitive* and is *unacceptable*. Never select input-output pairings based on a negative relative gain.

Implication for Subsystems

Many control strategies have more than two control loops. Consider a case where there are three control loops, for example. If the variable pairings have been performed so that the corresponding relative gains are positive (and hopefully close to 1.0), then we need to consider the failure sensitivity of all possible controller failures. For example, if loop 1 fails, it is important then to make certain that the RGA of the subsystem with loop 1 removed still satisfies the requirement of positive relative gains on all paired inputs and outputs.

Sensitivity to Model Uncertainty

Large Relative Gains (>>1)

Large relative gains correspond to a high sensitivity to uncertainty. This is shown in the following example.

Example 13.3: Model Uncertainty and the RGA

Consider a steady-state process *model* with the corresponding RGA:

$$K_m = \begin{bmatrix} 1.00 & 1.05 \\ 0.95 & 1.00 \end{bmatrix}, \quad \Lambda = \begin{bmatrix} 400 & -399 \\ -399 & 400 \end{bmatrix}$$

The actual *process* could very easily have a minor error of 5% in the gain relating u_1 to y_2, as shown in the actual process gain matrix and RGA:

$$K_p = \begin{bmatrix} 1.00 & 1.05 \\ 1.00 & 1.00 \end{bmatrix}, \quad \Lambda = \begin{bmatrix} -20 & 21 \\ 21 & -20 \end{bmatrix}$$

Notice that the model requires that input 1 be paired with output 1, while the relative gain for the actual plant (model + uncertainty) indicates that input 1 must be paired with output 2. In this case, we have no choice but to use only one control loop. In this case we would use physical insight and dynamic considerations to select the loop to close.

Sensitivity to Model Uncertainty

A large relative gain (the order of 25 and higher) corresponds to an extreme sensitivity to model uncertainty and indicates that some loop should be opened; that is, not all outputs should be under feedback control. One process that can have multivariable control, even with a large relative gain, is distillation, where nonlinear effects dominate.

13.8 Using the RGA to Determine Variable Pairings

In the previous section, we showed how to tune controllers, assuming that a decision on the pairing had already been established. The purpose of this section is to give some examples to illustrate how inputs and outputs should be paired to form MVSISO control loops.

Example 13.1, continued

Consider the whiskey-blending problem, which had the steady-state process gain matrix and RGA:

$$\begin{bmatrix} y_1 \\ y_2 \end{bmatrix} = \begin{bmatrix} 0.025 & -0.075 \\ 1 & 1 \end{bmatrix} \begin{bmatrix} u_1 \\ u_2 \end{bmatrix}, \quad K = \begin{bmatrix} 0.025 & -0.075 \\ 1 & 1 \end{bmatrix}, \quad \Lambda = \begin{bmatrix} 0.25 & 0.75 \\ 0.75 & 0.25 \end{bmatrix}$$

indicating that the output-input pairings should be y_1-u_2 and y_2-u_1. In order to achieve this pairing, we could use the block diagram shown in Figure 13–13.

Notice that the difference between r_2 and y_2 (the error in output 2) is used to adjust u_1, using a PID controller (g_{c1}), hence, we refer to this pairing as y_2-u_1. Similarly, the difference between r_1 and y_1 (the error in output 1) is used to adjust u_2, using a PID controller (g_{c1}); hence, we refer to this pairing as y_1-u_2. This corresponds to the physical diagram shown in Figure 13–3b.

An alternative is to rearrange (renumber) our outputs or inputs so that the natural pairing is y_1-u_1 and y_2-u_2 (in the renumbered variables). For example, we could renumber the outputs. Let $y_1^* = y_2$ and $y_2^* = y_1$, where * indicates the newly redefined variable. This corresponds to the control and instrumentation diagram shown in Figure 13–14.

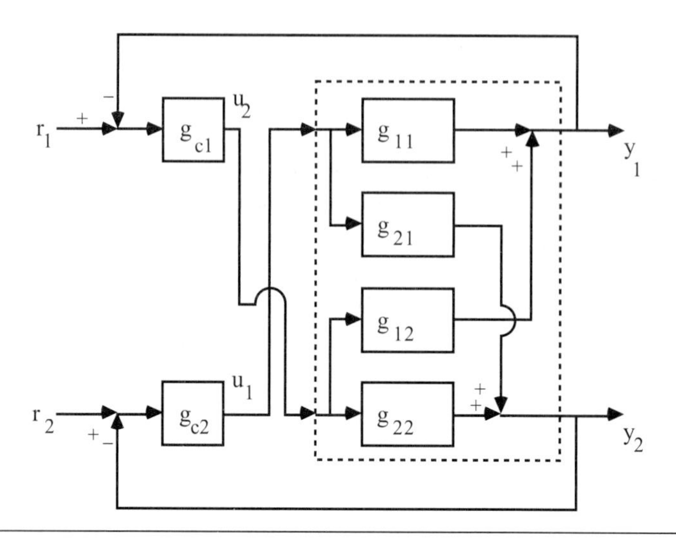

Figure 13–13 Control block diagram for 1-2/2-1 pairing.

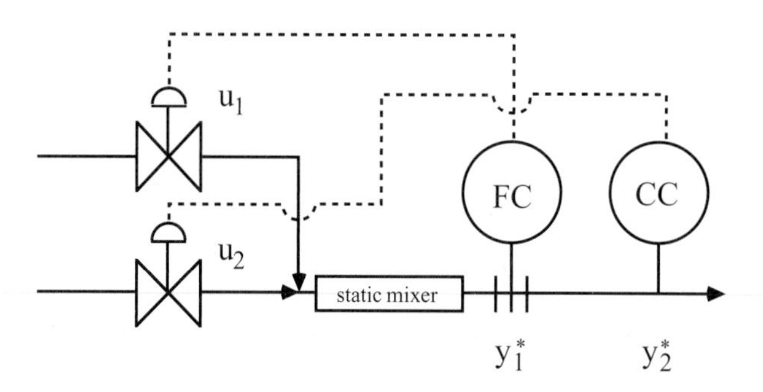

Figure 13–14 Control instrumentation diagram for redefined variables.

We can write

$$\begin{bmatrix} y_2^* \\ y_1^* \end{bmatrix} = \begin{bmatrix} 1 & 1 \\ 0.025 & -0.075 \end{bmatrix} \begin{bmatrix} u_1 \\ u_2 \end{bmatrix}, \quad K = \begin{bmatrix} 1 & 1 \\ 0.025 & -0.075 \end{bmatrix}$$

or

$$\begin{bmatrix} y_1^* \\ y_2^* \end{bmatrix} = \begin{bmatrix} 0.025 & -0.075 \\ 1 & 1 \end{bmatrix} \begin{bmatrix} u_1 \\ u_2 \end{bmatrix}, \quad K^* = \begin{bmatrix} 0.025 & -0.075 \\ 1 & 1 \end{bmatrix}$$

The reader should verify that the new RGA is

$$\Lambda^* = \begin{bmatrix} 0.75 & 0.25 \\ 0.25 & 0.75 \end{bmatrix}$$

Notice that redefining the *outputs* is equivalent to switching the *rows* of the process gain matrix related to the corresponding outputs (in this case, switching rows 1 and 2 is equivalent to exchanging outputs 1 and 2). The new RGA is found by making the same transformation of rows in the RGA, as was performed on the process gain matrix.

Similarly, redefining the *inputs* is equivalent to switching the *columns* of the process gain matrix related to the corresponding inputs. The new RGA is found by making the same transformation of columns in the RGA, as was performed on the process gain matrix.

Generalization

We have shown, by way of example, that outputs can be renumbered by switching the rows in the process gain matrix, and inputs can be renumbered by switching columns in the process gain matrix. Each time we perform this exchange of variables, we do not need to recalculate the relative gain array—we simply find the new RGA by performing the same row and column exchanges on the RGA that we performed on the process gain matrix. By performing this exchange (renumbering) of variables, we can always rearrange our inputs and outputs such that loop 1 has a pairing of y_1 and u_1 (in the newly defined variables). For this reason, we will often assume that a y_1-u_1 pairing has been made and discuss the effect of λ_{11} on controller tuning.

Example 13.4: A Three Input–Three Output System

Consider the following RGA for a system with three inputs and three outputs:

$$\Lambda = \begin{bmatrix} 1 & 1 & -1 \\ 3 & 4 & 2 \\ -3 & 4 & 0 \end{bmatrix}$$

How would you choose input-output pairings for this process?

Solution:

1. First of all, we know that we do not want to pair on a negative relative gain.
2. Second, we do not want to pair with a relative gain of 0 because that means that the particular input does not have an effect on the particular output, when all of the other loops are open.

Look at row 3 in the relative gain array; this corresponds to output 3. We would not pair y_3 with u_3 because of the 0 term. We also would not pair y_3 with u_1 because of the –3 term. This means that y_3 must be paired with u_2. Let us indicate this choice in our RGA,

$$\Lambda = \begin{bmatrix} 1 & 1 & -1 \\ 3 & 4 & 2 \\ -3 & \{4\} & 0 \end{bmatrix}$$

Now we have eliminated the third output (row) and the second input (column) from our selection process. From the first row (output), we see that we would not pair y_1 with u_3 because of the –1 term. We also cannot pair y_1 with u_2 because we have already paired u_2 with y_3 (hence, the circle). Our only choice is to pair y_1 with u_1. Let us indicate this choice in our RGA,

$$\Lambda = \begin{bmatrix} \{1\} & 1 & -1 \\ 3 & 4 & 2 \\ -3 & \{4\} & 0 \end{bmatrix}$$

Now, we have eliminated outputs 1 and 3 (rows 1 and 3) as well as inputs 1 and 2 (columns 1 and 2). We have no choice but to pair y_2 with u_3, and we make that choice below:

$$\Lambda = \begin{bmatrix} \{1\} & 1 & -1 \\ 3 & 4 & \{2\} \\ -3 & \{4\} & 0 \end{bmatrix}$$

Notice also, that we can renumber our outputs, such that pairings of the new variables occur on the diagonal. For example, if we define $y_2^* = y_3$ and $y_3^* = y_2$, this is equivalent to switching the second and third rows to obtain

$$\Lambda = \begin{bmatrix} \{1\} & 1 & -1 \\ -3 & \{4\} & 0 \\ 3 & 4 & \{2\} \end{bmatrix}$$

which is based on the new output vector

$$y^* = \begin{bmatrix} y_1^* \\ y_2^* \\ y_3^* \end{bmatrix} = \begin{bmatrix} y_1 \\ y_3 \\ y_2 \end{bmatrix}$$

Example 13.5: A Four Input–Four Output Distillation Column (Alatiqi and Luyben, 1986)

The steady-state gain input–output relationship is

$$\begin{bmatrix} y_1 \\ y_2 \\ y_3 \\ y_4 \end{bmatrix} = \begin{bmatrix} -11.18 & 14.04 & -0.10 & 4.49 \\ -1.73 & 5.11 & 4.61 & -5.49 \\ -4.17 & 6.93 & -0.05 & 1.53 \\ 4.09 & -6.36 & -0.25 & -0.49 \end{bmatrix} \begin{bmatrix} u_1 \\ u_2 \\ u_3 \\ u_4 \end{bmatrix}$$

The RGA (using the m-file `rga.m` shown in Appendix 13.2) is

$$\Lambda = \begin{bmatrix} 3.0078 & -2.8276 & -0.0349 & 0.8546 \\ -0.0839 & 0.0544 & 1.5507 & -0.5212 \\ -5.0327 & 4.6749 & -0.0396 & 1.3974 \\ 3.1087 & -0.9017 & -0.4761 & -0.7309 \end{bmatrix}$$

Now, let us systematically choose our pairings. Looking at *column 3,* we see only one relative gain that is not negative, therefore, we must pair y_2 with u_3. Looking at *row 4*, we notice that there is only one relative gain that is not negative, therefore, we must pair y_4 with u_1. Looking at *column 2*, there are 2 nonnegative relative gains, but we have already used output 2, so we must pair y_3 with u_2. This leaves y_1 and u_4, which, fortunately, has a favorable relative gain (0.8546). If the y_1-u_4 relative gain had not been favorable, we would have been forced to drop output 1 and input 4 and simply have three control loops for our system.

13.9 MATLAB RGA Function File

The MATLAB function file, `rga.m`, given in Appendix 13.2 can be used to the find the RGA of any square (number of inputs = number of outputs) gain matrix. For example, consider the wet grinding circuit of Exercise 10.

First, the gain matrix is entered

```
» g = [119 153 -21; 370 767 -50; 903 -667 -1033]
g =
        119          153          -21
        370          767          -50
        903         -667        -1033
```

Then the RGA is calculated using the call to the function file:

```
» lambda = rga(g)
lambda =
     3.6449   -1.9131   -0.7318
    -2.3616    2.9581    0.4035
    -0.2833   -0.0450    1.3283
```

13.10 Summary

The purpose of this chapter was to develop a technique (the RGA) to assist us with the decision of how to pair input and output variables to form multiple SISO control loops. The relative gain between output i and input j is defined as

$$\lambda_{ij} = \frac{\textit{gain between input } j \textit{ and output } i \textit{ with all other loops open}}{\textit{gain between input } j \textit{ and output } i \textit{ with all other loops closed}}$$

$$\lambda_{ij} = \frac{\left(\dfrac{\partial y_i}{\partial u_j}\right)_{u_k \;\; k \neq j}}{\left(\dfrac{\partial y_i}{\partial u_j}\right)_{y_k \;\; k \neq i}}$$

and the RGA is a matrix of relative gains. Some of the important things that we learned about the RGA are summarized below.

- The rows and columns of the RGA sum to 1
- Do not pair on a negative or zero relative gain
- Try to pair on relative gains close to 1
- If you cannot pair on positive relative gains, or if the relative gains are high (>25), then the number of closed-loop control loops must generally be reduced; that is, one or more loops must be placed on manual control
- Switching rows on the process gain and relative gain matrices is equivalent to renumbering the outputs; switching columns on the process gain and relative gain matrices is equivalent to renumbering the inputs
- After selecting variable pairings, check the relative gain array for all subsystems to make certain that they are not sensitive to controller failure

References

A nice summary of the features of the relative gain array are presented in the following: Grosdidier, P., M. Morari, and and B. Holt, "Closed-Loop Properties from Steady-State Gain Information," *Ind. Eng. Chem. Fund.*, **24**(2), 221–235 (1985).

The following sources were used for some of the RGA examples in this chapter: Alatiqi, I. M., and W. L. Luyben, "Control of a Complex Sidestream Column/Stripper Distillation Configuration," *Ind. Eng. Chem. Proc. Des. Dev.*, **25**(3), 762–767 (1986).

Doukas, N., and W. L. Luyben, "Control of Sidestream Separating Ternary Mixtures," *Instrum. Technol. (INTECH)*, pp. 43–48 (June, 1978).

Exercise 12 is presented in the following: Reeves, D. E., and Y. Arkun, "Interaction Measures for Nonsquare Decentralized Control Structures," *AIChE J.*, **35**(4), 603–613 (1989).

The grinding circuit problem (Exercise 10) was presented in the following: Hulbert, D. G., and E. T. Woodburn, "Multivariable Control of a Wet Grinding Circuit," *AIChE J.*, **29**, 186 (1983).

Student Exercises

1. A 3 × 3 system has the following steady-state gain matrix and RGA:

$$K = \begin{bmatrix} 1 & 1 & 5/3 \\ 1/3 & 1 & 1 \\ 1 & 1/3 & 1 \end{bmatrix}, \quad \Lambda = \begin{bmatrix} -4.5 & -4.5 & 10 \\ 1 & 4.5 & -4.5 \\ 4.5 & 1 & -4.5 \end{bmatrix}$$

   How would you pair the variables for an MVSISO strategy and why?

2. A two input–two output process has the following RGA:

$$\Lambda = \begin{bmatrix} -1.5 & 2.5 \\ 2.5 & -1.5 \end{bmatrix}$$

   Assume that the MVSISO system is tuned for stable feedback control with both loops closed.
   Case 1—u_1 is paired with y_1 (loop 1) and u_2 is paired with y_2 (loop 2). What happens if loop 2 is opened? Why? Must loop 1 be retuned? How?
   Case 2—u_1 is paired with y_2 (loop 1) and u_2 is paired with y_1 (loop 2). What happens if loop 2 is opened? Why? Must loop 1 be retuned? How?

3. A system has the following process gain matrix:

$$K = \begin{bmatrix} 2.5 & 5 \\ 1.25 & 2.55 \end{bmatrix}$$

   How would you pair variables for your control loops? Why?

4. A three input–three output process has the following RGA:

$$\Lambda = \begin{bmatrix} -2 & 1.05 & 1.95 \\ 2.05 & -1.95 & 0.9 \\ 0.95 & 1.9 & -1.85 \end{bmatrix}$$

How would you pair variables for your control loops? Why?

5. Consider the following process gain matrix and RGA:

$$K = \begin{bmatrix} 1 & -2 & -2 \\ 0 & 1 & 1.5 \\ -1 & -2 & -1.5 \end{bmatrix}, \quad \Lambda = \begin{bmatrix} 0.6 & 1.2 & -0.8 \\ 0 & -1.4 & 2.4 \\ 0.4 & 1.2 & -0.6 \end{bmatrix}$$

a. Can three SISO control loops be used?

b1. If your answer to **a** is yes, what are the recommended pairings [will these pairings assure stability if one of the loops must be opened (think about the 2×2 subsystems)]? Reorder the inputs and outputs such that the paired variables are on the diagonal of the new $G(0)$ matrix. Show the new $G(0)$ matrix and the new Λ matrix.

b2. If your answer to **a** is no, do you recommend a reduction to two loops? If so, what are your recommendations for input-output pairings for this system?

6. Consider the following steady-state gain matrix for a process with three inputs and four outputs. [This is for a Benzene-Toluene distillation column, with a side draw—see Doukas and Luyben (1978) for details.] Assuming all the outputs are of equal importance, which input-output pairings would you use for a MVSISO control system?

$$K = \begin{bmatrix} -9.811 & 0.374 & -11.3 \\ 5.984 & -1.986 & 5.24 \\ 2.380 & 0.0204 & -0.33 \\ -11.672 & -0.176 & 4.48 \end{bmatrix}$$

7. For a 2×2 system, show that a y_1-u_2 and y_2-u_1 pairing yields the following result for λ_{12}.

$$\lambda_{12} = \frac{k_{12}}{k_{12,eff}}$$

where $k_{12,eff} = k_{12} - \dfrac{k_{11}k_{22}}{k_{21}}$

8. Consider the steady-state process gain matrix

$$K = \begin{bmatrix} 1 & 1 & -0.1 \\ 0.1 & 2 & -1 \\ -2 & -3 & 1 \end{bmatrix}$$

which yields the following RGA:

$$\Lambda = \begin{bmatrix} -1.9 & 3.6 & -0.7 \\ -0.1 & 3.0 & -1.9 \\ 3.0 & -5.6 & 3.6 \end{bmatrix}$$

Is it possible to pair outputs and inputs to form three separate single-loop controllers? If it can be done, do it. If it cannot be done, explain why and suggest another multivariable control strategy.

9. Your boss has told you to develop a control strategy for the following stirred-tank heater, which has two inputs (rate of heat addition, Q, and flow rate to tank, F_i), and two outputs (tank height and temperature). She also gave you the following transfer function matrix, assuming that you knew which output corresponded to which measured variable and which input corresponded to which manipulated variable.

$$\begin{bmatrix} y_1(s) \\ y_2(s) \end{bmatrix} = \begin{bmatrix} \dfrac{-25}{5s+1} & \dfrac{0.5}{5s+1} \\ \dfrac{1}{10s+1} & 0 \end{bmatrix} \begin{bmatrix} u_1(s) \\ u_2(s) \end{bmatrix}$$

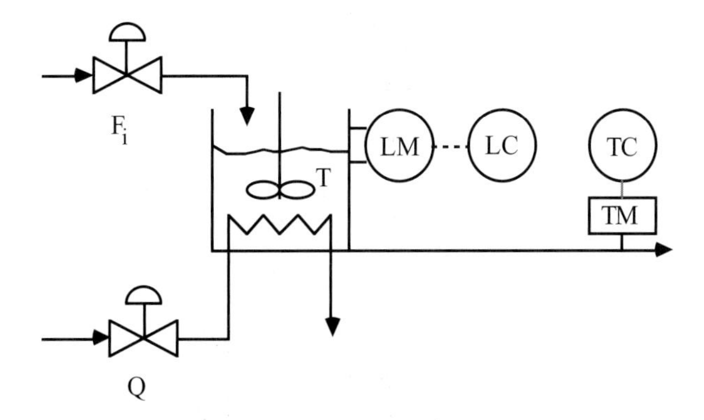

Use physical reasoning to determine which output corresponds to which measurement (i.e., is y_1 the tank height or the tank temperature) and which input corresponds to which manipulated variable (i.e., is u_1 the heat addition rate or the flow rate to the tank?). If two SISO loops are used to control this process, how should the variables be paired? Renumber the outputs such that the output-input pairings are on the diagonal of the transfer function matrix. Show the new transfer function matrix.

10. The objective of a wet grinding circuit is to grind large solid material to form granular material. Solids are fed with a water stream to a ball mill where the solids are ground. The slurry effluent from the ball mill is fed to a sump which is used as surge capacity. The slurry is pumped to a cyclone separator, which separates water from solid material and returns water to the ball mill. A schematic diagram is shown below.

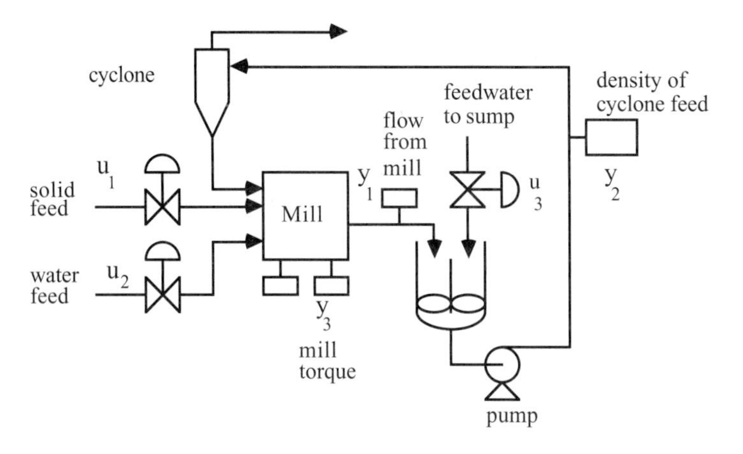

Consider the following process model (Hulbert and Woodburn, 1983)

$$\begin{pmatrix} y_1(s) \\ y_2(s) \\ y_3(s) \end{pmatrix} = \begin{pmatrix} \dfrac{370}{500s+1} & \dfrac{767}{33s+1} & \dfrac{-50}{10s+1} \\ \dfrac{903}{500s+1} & \dfrac{-667e^{-320s}}{(166s+1)} & \dfrac{-1033}{47s+1} \\ \dfrac{119}{217s+1} & \dfrac{153}{337s+1} & \dfrac{-21}{10s+1} \end{pmatrix} \begin{pmatrix} u_1(s) \\ u_2(s) \\ u_3(s) \end{pmatrix}$$

Which has the following RGA

$$\Lambda = \begin{pmatrix} -2.36 & 2.96 & 0.40 \\ -0.28 & -0.05 & 1.33 \\ 3.64 & -1.91 & -0.73 \end{pmatrix}$$

a. What pairing is suggested by the RGA for this system? Why?

b. Connect the appropriate input and output variables with controllers, on the diagram above, to form a control instrumentation diagram.

c. Discuss the failure sensitivity of this three input–three output system. That is, if any of the control loops failed, would the resulting two input–two output system have the correct pairing?

11. A forced circulation evaporator is used to concentrate dilute liquors by evaporating solvent from a feed stream. Feed is mixed with recirculating liquor and pumped into the tube side of a vertical heat exchanger. Steam enters the shell side of the vertical heat exchanger and condenses on the outside of the tubes. The liquor is partially vaporized as it passes through the tube side of the heat exchanger and passes into the separator. Vapor from the separator is condensed by a cooling water exchanger, while a portion of the liquid stream from the separator is withdrawn as product and the rest is recirculated to the heat exchanger. A schematic process and instrumentation diagram is shown below.

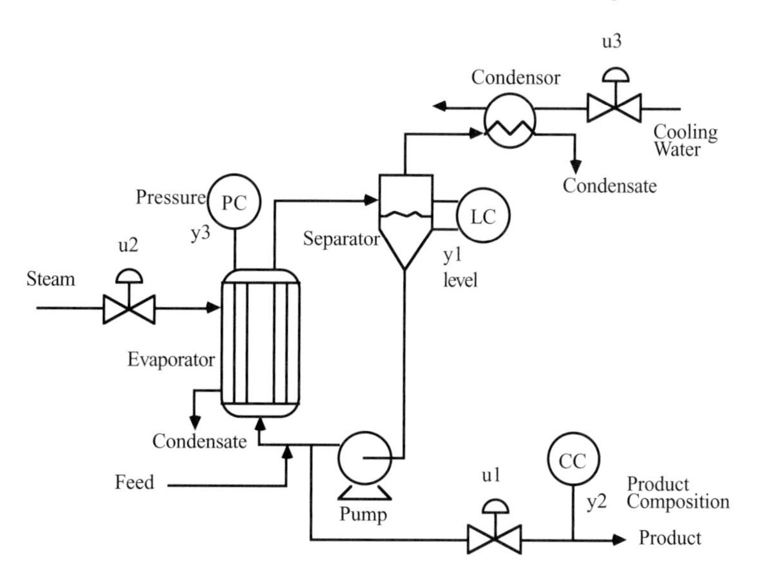

It seems fairly clear that it is desirable to control the level in the separator (y_1) by manipulating the product flow rate (u_1). It is particularly important to keep the level under feedback control, since is has integrating action (a pole is at the origin). With the level controlled by manipulating the product flow rate, the following steady-state input-output relationship is found for the other variables

$$\begin{bmatrix} y_2 \\ y_3 \end{bmatrix} = \begin{bmatrix} -4.7245 & 39.2363 \\ 1.5642 & 21.1487 \end{bmatrix} \begin{bmatrix} u_2 \\ u_3 \end{bmatrix}$$

How should the other loops be paired? Why?

12. Consider the following system with two outputs and three inputs (Reeves and Arkun, 1989). This model is for a mixing tank with three input streams. The outputs are the concentration and tank height, while the three inputs are stream flow rates:

$$G(s) = \begin{bmatrix} \dfrac{4}{20s+1} & \dfrac{4}{20s+1} & \dfrac{4}{20s+1} \\ \dfrac{3}{10s+1} & \dfrac{-3}{10s+1} & \dfrac{5}{10s+1} \end{bmatrix}$$

To use a MVSISO strategy, we must remove one of the inputs (set it to a constant value, rather than using it for feedback control).

13. As a promising young engineer, you are placed in charge of a gasoline blending facility. This facility blends three refinery streams together to form a final product stream that must meet road octane (RO) and Reid vapor pressure (RVP, psi) specifications. Two streams are available for manipulation. You are given the following data

Flow 1	Flow 2	RO	RVP	Total flow
10 Bbl/min	20 Bbl/min	90.000	10.000	60 Bbl/min
11	20	90.164	10.164	61
10	21	90.164	9.918	61

Let Flow 1 be the first input and Flow 2 be the second input. RO is the first output and RVP is the second output.

a. Calculate the RGA.

b. How would you pair the input-output variables for this process? (Why?)

c. Complete the control instrumentation diagram shown below.

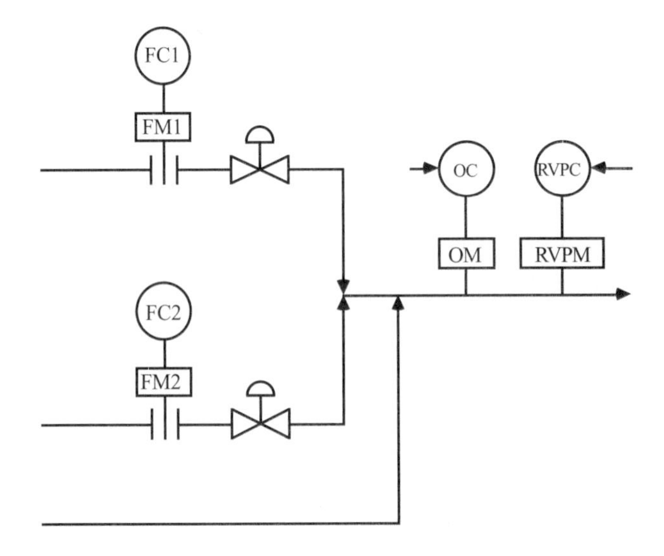

d. Draw a control block diagram for this process (with the proper pairings). (Are there any cascade loops?)

14. Consider the whiskey blending problem. A new market for 70 proof whiskey (0.35 volume fraction ethanol) has developed. Assume that the feedstreams are still 80 proof whiskey and pure water, and that the desired production rate of 70 proof whiskey is 4 gpm. Find the new gain matrix and relative gain array. What are the implications for controller pairing?

Appendix 13.1: Derivation of the Relative Gain for an *n*-Input–*n*-Output System

The relative gain between input j and output i is defined as

$$\lambda_{ij} = \frac{\left(\dfrac{\partial y_i}{\partial u_j}\right)_{u_k\ k\neq j}}{\left(\dfrac{\partial y_i}{\partial u_j}\right)_{y_k\ k\neq i}}$$

Output i is the following steady-state function of all of the inputs:

$$y_i = \sum_{j=1}^{n} g_{ij}(0)u_j = \sum_{j=1}^{n} k_{ij}u_j$$

The relationship between input j and output i with all other inputs constant is

$$\left(\frac{\partial y_i}{\partial u_j}\right)_{u_{k,k\neq i}} = g_{ij}(0) = k_{ij}$$

For any desired output vector, y, we can find the corresponding input vector, u, from

$$u = G^{-1}(0)y = K^{-1}\text{y}$$

Defining $\hat{G}(0) = G^{-1}(0) = K^{-1}$

$$u = \hat{G}(0)y = \hat{K}y$$

The relationship between all outputs and input j is

$$u_j = \sum_{j=1}^{n} \hat{g}_{ji}(0)y_i = \sum_{j=1}^{n} k_{ji}y_i$$

The relationship between output i and input j with all other outputs constant is

$$\left(\frac{\partial u_j}{\partial y_i}\right)_{y_{k,k\neq i}} = \hat{g}_{ji}(0) = \hat{k}_{ji}$$

and

$$\lambda_{ij} = \frac{\left(\frac{\partial y_i}{\partial u_j}\right)_{u_k\ k\neq j}}{\left(\frac{\partial y_i}{\partial u_j}\right)_{y_k\ k\neq i}} = \left(\frac{\partial y_i}{\partial u_j}\right)_{u_{k,k\neq j}} \left(\frac{\partial u_j}{\partial y_i}\right)_{y_{k,k\neq i}}$$

The relative gain between input j and output i is

$$\lambda_{ij} = g_{ij}(0)\hat{g}_{ji}(0) = k_{ij}\hat{k}_{ji} \tag{13.23}$$

where $\hat{k}_{ji} = ji^{th}$ element of $G^{-1}(0)$.

What Equation (13.23) tells us is that the *ij* element of the RGA is found by multiplying the *ij* element of the steady-state process gain matrix by the *ji* element of the inverse of the steady-state process gain matrix.

Appendix 13.2: m-File to Calculate the RGA

```
  function lambda = rga(g)
%
% calculate the relative gain array for a
% square gain matrix of any size
%
% first, check the dimensions to see if the
% process gain matrix is square
%
  [mrow ncol]=size(g);
%
% if the gain matrix is not square, let the
% user know
%
  if mrow ~= ncol;
    disp('needs to be square, buddy')
```

```
    end;
  else
%
% the matrix ghat is simply the inverse of the
% gain matrix.  ghatt is the transpose of ghat
%
    ghat = inv(g);
    ghatt=ghat';
%
%  perform an element by element multiplication
%  of g and ghatt to find the relative gain array
%
    lambda = g.*ghatt;
  end
```

CHAPTER 14

Multivariable Control

After studying this chapter, the reader should be able to do the following:

- Understand the performance limitations of processes with RHP transmission zeros
- Understand the effect of "direction" in a multivariable system
- Perform matrix transfer function block-diagram analyses and understand that the order of multiplication is very important
- Design a decoupling controller
- Understand the basic multivariable IMC design

The major sections of this chapter are as follows:

14.1 Background

Chapter 13 presented the problem of control-loop interaction, and a method (RGA) to help determine the best input-output pairing to form a set of SISO control loops. The goal of this chapter is severalfold. One is to understand the inherent performance limitations of a multivariable system due to RHP *transmission zeros*. Also, multivariable systems exhibit a "directional sensitivity" that can be detected by performing a *singular value decomposition* (SVD). Use of the SVD requires that the gain matrix be properly scaled, so that the scaled inputs and outputs have roughly the same ranges of expected values.

Two forms of multivariable control are presented in this chapter: decoupling and IMC. The more commonly used multivariable technique in practice, model predictive control (MPC), is the topic of Chapter 16.

14.2 Zeros and Performance Limitations

Recall that RHP zeros severely limit the performance of SISO processes. For example, a model inverse can not be implemented as the controller, because this would cause the controller to be unstable. A similar problem exists for multivariable processes; RHP *transmission* zeros limit MIMO performance.

SISO Zeros

For a SISO process transfer function with the form

$$g(s) = \frac{N(s)}{D(s)} \tag{14.1}$$

the zeros are the roots of the numerator polynomial, $N(s)$. The inverse of the process transfer function is

$$g^{-1}(s) = \frac{D(s)}{N(s)} \tag{14.2}$$

That is, the zeros of $g(s)$ are the poles of $g^{-1}(s)$. A RHP zero then results in an inverse transfer function with a RHP pole, which is unstable. Some model-based control strategies, such as IMC, make use of a model inverse for control system design. In Chapter 8, we found that the RHP zeros must be factored out, and that only a portion of the model could be inverted for controller design.

Multivariable Transmission Zeros

The *transmission zeros* of a multivariable transfer function matrix are the values of s that cause the input-output matrix to lose rank. That is, the transmission zeros are the values of s that cause

$$\det[G(s)] = 0 \tag{14.3}$$

Similar to single variable systems, the multivariable transmission zeros become the poles of the inverse system. RHP transmission zeros for a particular transfer function matrix then become RHP poles for the inverse transfer function matrix (that is, the inverse is unstable). Again, a multivariable controller that simply tries to invert the process model will be unstable.

We illustrate these concepts with the following example. Here we use the symbol z to represent the zero.

Example 14.1: Calculation of Transmission Zeros

The calculation of transmission zeros are illustrated for two different systems (A and B).

System A

Consider the transfer function matrix, where the time unit is minutes

$$G(s) = \begin{bmatrix} s+3 & 2 \\ 3 & 1 \end{bmatrix} \cdot \frac{1}{(s+1)(s+2)}$$

The determinant of $G(s)$ is

$$\det[G(s)] = \frac{(s+3)-6}{(s+1)^2(s+2)^2}$$

and solving for $\det[G(s)] = 0$,

$$z = 3 \text{ min}^{-1}.$$

We can also see that the rank of the G matrix drops to 1 when $s = 3 \text{ min}^{-1}$,

$$G(s=3) = \begin{bmatrix} 6 & 2 \\ 3 & 1 \end{bmatrix} \cdot \frac{1}{20}$$

since the first column of the matrix is now a multiplicative factor (3) of the second column. Note that although the zero of the g_{11} element is -3 min^{-1}, the matrix transmission zero is $+3 \text{ min}^{-1}$ (that is, a RHP transmission zero)!

The inverse[1] of the transfer function matrix is

$$G^{-1}(s) = \begin{bmatrix} \dfrac{(s+1)(s+2)}{s-3} & \dfrac{-2(s+1)(s+2)}{s-3} \\ \dfrac{-3(s+1)(s+2)}{s-3} & \dfrac{(s+3)(s+1)(s+2)}{s-3} \end{bmatrix}$$

$$= \begin{bmatrix} \dfrac{-0.667(s+1)(0.5s+1)}{(-0.333s+1)} & \dfrac{0.667(s+1)(0.5s+1)}{(-0.333s+1)} \\ \dfrac{2(s+1)(0.5s+1)}{(-0.333s+1)} & \dfrac{-2(0.333s+1)(s+1)(0.5s+1)}{(-0.333s+1)} \end{bmatrix}$$

which has an unstable pole of $p = +3$, as expected.

It is possible for an individual element to have a RHP (positive) zero while the matrix transmission zero is in the left-half-plane (negative); this is illustrated by the following transfer function matrix.

System B

Consider the transfer function matrix

$$G(s) = \begin{bmatrix} -s+3 & 2 \\ 3 & 1 \end{bmatrix} \cdot \frac{1}{(s+1)(s+2)}$$

which differs from system A by the RHP zero in the $g_{11}(s)$ transfer function. The determinant of $G(s)$ is

$$\det[G(s)] = \frac{(-s+3)-6}{(s+1)^2(s+2)^2}$$

and solving for $\det[G(s)] = 0$,

$$z = -3$$

We can also see that the rank of the G matrix drops to 1 when $s = -3$,

$$G(s=-3) = \begin{bmatrix} 6 & 2 \\ 3 & 1 \end{bmatrix} \cdot \frac{1}{2}$$

---

[1]Recall that the inverse of a 2 × 2 matrix, $G = \begin{bmatrix} g_{11} & g_{12} \\ g_{21} & g_{22} \end{bmatrix}$, is calculated by

$$G^{-1} = \begin{bmatrix} g_{22} & -g_{12} \\ -g_{21} & g_{11} \end{bmatrix} \cdot \frac{1}{\det(G)} = \begin{bmatrix} g_{22} & -g_{12} \\ -g_{21} & g_{11} \end{bmatrix} \cdot \frac{1}{g_{11}g_{22} - g_{12}g_{21}}.$$

since the first column of the matrix is (like system A) a multiplicative factor (3) of the second column. Note that although the zero of the g_{11} element is $+3$ (RHP), the matrix transmission zero is $-3\ \text{min}^{-1}$ (that is, a left-half-plane transmission zero).

It is left to the reader to show that the inverse of this matrix has a stable pole at $-3\ \text{min}^{-1}$.

An example of a physical process with a RHP transmission zero is shown in Figure 14–1. This is a laboratory-scale system developed by Johansson (2000) to study the effect of operating condition on the location of the transmission zeros. The inputs are the voltages to the pumps (v_1, v_2) and the outputs are the voltage measurements of the height of liquid in tanks 1 and 2 (y_1, y_2). The flow from pump 1 is split between tanks 1 and 4, while the flow from pump 2 is split between tanks 2 and 3. Depending on the actual flow splits, it is possible to obtain a positive RHP transmission zero. Numerical details are presented in Example 14.2.

Although this is a "toy" problem, you can see how some processes with recycle might have similar behavior.

Example 14.2: Quadruple Tank Problem (Johansson, 2000)

This tank system can be operating at two operating points. The first one has two transmission zeros that are in the left-half plane (negative), while the second has a RHP transmission zero.

Operating Point 1

Operating point 1 has the following steady-state inputs and outputs:

$$v_1 = 3.00,\ v_2 = 3.00\ \text{V};\ F_1 = 10,\ F_2 = 10.05\ \text{cm}^3/\text{s}$$

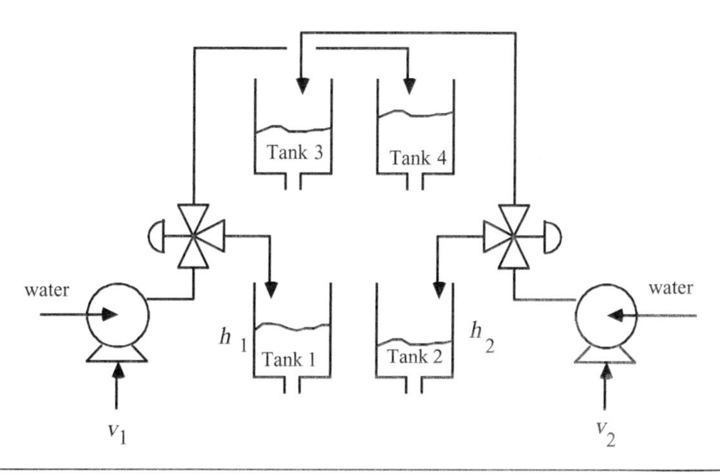

Figure 14–1 Quadruple tank control problem.

$$h_1 = 12.4, h_2 = 12.7, h_3 = 1.8, h_4 = 1.4 \text{ cm};$$
$$y_1 = 6.2, y_2 = 6.35 \text{ V}$$

The result is the process transfer function matrix

$$G_1(s) = \begin{bmatrix} \dfrac{2.6}{62s+1} & \dfrac{1.5}{(23s+1)(62s+1)} \\ \dfrac{1.4}{(30s+1)(90s+1)} & \dfrac{2.8}{(90s+1)} \end{bmatrix}$$

where the gains are volts/volt, and the timescale is seconds. The transmission zeros are

$$z = -0.060 \text{ and } -0.018 \text{ sec}^{-1}$$

which are both in the left-half-plane (so the matrix inverse will be stable).

Operating Point 2

Operating point 2 has the following steady-state inputs and outputs

$$v_1 = 3.15, v_2 = 3.15 \text{ V}; F_1 = 9.89, F_2 = 10.36 \text{ cm}^3/\text{s};$$
$$h_1 = 12.6, h_2 = 13.0, h_3 = 4.8, h_4 = 4.9 \text{ cm};$$
$$y_1 = 6.3, y_2 = 6.5 \text{ V}$$

This operating point has the transfer function matrix

$$G_2(s) = \begin{bmatrix} \dfrac{1.5}{63s+1} & \dfrac{2.5}{(39s+1)(63s+1)} \\ \dfrac{2.5}{(56s+1)(91s+1)} & \dfrac{1.6}{(91s+1)} \end{bmatrix}$$

The transmission zeros are

$$z = -0.057 \text{ and } +0.013 \text{ sec}^{-1}$$

Notice that there is a RHP transmission zero, so the matrix inverse will be unstable. In Exercise 1, you will find that the feedback control performance of this operating point is severely limited compared with the first operating point.

The MATLAB routine for calculating transmission zeros is discussed in Section 14.7.

14.3 Scaling Considerations

To properly use the SVD method presented in Section 14.4, it is necessary to scale the process transfer function matrix such that expected ranges of all scaled inputs and outputs are approximately the same.

An output scaling matrix can be found to assure that the scaled outputs cover the same range. Let (*) represent the scaled variables. A diagonal scaling matrix for the output variable is

$$y^* = S_o y \tag{14.4}$$

Similarly, a diagonal scaling matrix for the input variable is

$$u^* = S_i u \tag{14.5}$$

Since the input-output relationship is

$$y(s) = G(s)u(s) \tag{14.6}$$

we find

$$y^*(s) = S_o y(s) = S_o G(s)u(s). \tag{14.7}$$

Also,

$$u(s) = S_i^{-1} u^*(s) \tag{14.8}$$

so the scaled input-output relationship is

$$y^*(s) = S_o G(s) S_i^{-1} u^*(s) = G^*(s) u^*(s) \tag{14.9}$$

yielding the scaled transfer function matrix

$$G^*(s) = S_o G(s) S_i^{-1} \tag{14.10}$$

This is illustrated by the following example.

Example 14.3: Mixing Tank

The mixing tank shown in Figure 14–2 mixes two streams together (plus a possible disturbance stream). It is desirable to control both the level (y_1) and the temperature (y_2) in the tank. The flow rates of the hot (u_1) and cold (u_2) streams can be manipulated. The flow rate out of the tank is proportional to the square root of the height of liquid in the tank ($F = \beta\sqrt{h}$). The residence time at steady state is 10 minutes. The steady-state flow rates and temperatures are as follows:

Stream	***Temperature***	***Flow, liters/minute***
Hot (manipulated)	60°C	25
Cold (manipulated)	10°C	25
Cold (disturbance)	10°C	0
Outlet	35°C	50

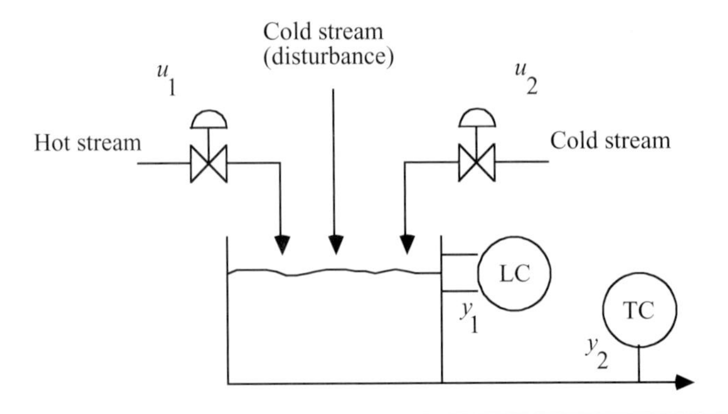

Figure 14–2 Mixing tank.

The cross-sectional area of the tank is 0.581 m^2. The steady-state tank height is then 0.86 m, with a steady-state volume of 0.5 m^3 or 500 liters (assume that the tank is half full at steady state). The input-output transfer function relationship is (see Exercise 3).

$$\begin{bmatrix} y_1(s) \\ y_2(s) \end{bmatrix} = \begin{bmatrix} \dfrac{0.03442}{20s+1} & \dfrac{0.03442}{20s+1} \\ \dfrac{0.5}{10s+1} & \dfrac{-0.5}{10s+1} \end{bmatrix} \begin{bmatrix} u_1(s) \\ u_2(s) \end{bmatrix} \tag{14.11}$$

The time constant unit is minutes, and the units of the gains are

$$k_{11}, k_{12} [\equiv] \frac{\text{m}}{\text{liter/min}} \qquad k_{21}, k_{22} [\equiv] \frac{^\circ\text{C}}{\text{liter/min}}$$

It is interesting that the height time constant is twice the temperature time constant. Notice that if the height unit was centimeters and the temperature unit was degrees Celsius, the transfer function matrix would be

$$\begin{bmatrix} y_1(s) \\ y_2(s) \end{bmatrix} = \begin{bmatrix} \dfrac{3.442}{20s+1} & \dfrac{3.442}{20s+1} \\ \dfrac{0.9}{10s+1} & \dfrac{-0.9}{10s+1} \end{bmatrix} \begin{bmatrix} u_1(s) \\ u_2(s) \end{bmatrix} \tag{14.12}$$

If, in addition, the time units were seconds, the input-output relationship would be

$$\begin{bmatrix} y_1(s) \\ y_2(s) \end{bmatrix} = \begin{bmatrix} \dfrac{3.442}{1200s+1} & \dfrac{3.442}{1200s+1} \\ \dfrac{0.9}{600s+1} & \dfrac{-0.9}{600s+1} \end{bmatrix} \begin{bmatrix} u_1(s) \\ u_2(s) \end{bmatrix} \tag{14.13}$$

The point of this is that it is very important to understand the units of the variables you are dealing with, and that the parameters can vary (numerically) by orders of magnitudes depending on those units.

In this example, the expected range of the manipulated variables is roughly the same, since we assume that the control valves are designed to handled flows roughly twice the nominal steady-state values. That is, we expect the manipulated inputs to be bounded between 0 and 50 liters/minute.

Notice that the outputs do not cover the same range. For one thing, the units are different (m for height, and °C for temperature).

The steady-state value of the tank height is 0.86 m, with a minimum value of 0 m and a maximum height of 1.72 m. We desire, however, for the height to stay between 20 and 80% of the full tank level. That is, the height should range from 0.344 to 1.376 m. Since we work in deviation variables, and it is convenient to cover a range of −1 to 1, then it is natural to scale by 0.516 m (this is found from 0.86 to 0.344). We can then define a scaled y_1 variable as y^*_1, where

$$y_1^* = y_1/0.516$$

Similarly, it is convenient to scale the second output. The approximate range of temperatures is from 10° to 60°C, with a steady-state value of 35°C. Again, since we are working with deviation variables and it is natural to consider variations of −1 to 1 in the scaled second output, we use 25°C (obtained from 35°C − 10°C) as the scaling factor. The second scaled output is then

$$y_2^* = y_2/25$$

And the scaled output vector is related to the dimensional output vector as

$$\begin{bmatrix} y_1^* \\ y_2^* \end{bmatrix} = \begin{bmatrix} 1/0.516 & 0 \\ 0 & 1/25 \end{bmatrix} \begin{bmatrix} y_1 \\ y_2 \end{bmatrix} \tag{14.14}$$

Then from Equation (14.11),

$$\begin{bmatrix} y_1^*(s) \\ y_2^*(s) \end{bmatrix} = \begin{bmatrix} 1/0.516 & 0 \\ 0 & 1/25 \end{bmatrix} \begin{bmatrix} \dfrac{0.03442}{20s+1} & \dfrac{0.03442}{20s+1} \\ \dfrac{0.5}{10s+1} & \dfrac{-0.5}{10s+1} \end{bmatrix} \begin{bmatrix} u_1(s) \\ u_2(s) \end{bmatrix}$$

$$\begin{bmatrix} y_1^*(s) \\ y_2^*(s) \end{bmatrix} = \begin{bmatrix} \dfrac{0.0667}{20s+1} & \dfrac{0.0667}{20s+1} \\ \dfrac{0.02}{10s+1} & \dfrac{-0.02}{10s+1} \end{bmatrix} \begin{bmatrix} u_1(s) \\ u_2(s) \end{bmatrix} \tag{14.15}$$

The gains indicate that the outputs will vary approximately the same order of magnitude (there is roughly a factor of 3 difference in the gains).

Although not important in this particular example, the inputs can be scaled as well. Since the steady-state manipulated input flow rates are each 25 liters/minute, and we can assume that the flows each range from 0 to 50 liters/minute, it is convenient to scale each input by a factor of 25 liters/minute.

$$\begin{bmatrix} u_1^* \\ u_2^* \end{bmatrix} = \begin{bmatrix} 1/25 & 0 \\ 0 & 1/25 \end{bmatrix} \begin{bmatrix} u_1 \\ u_2 \end{bmatrix} \tag{14.16}$$

or

$$\begin{bmatrix} u_1 \\ u_2 \end{bmatrix} = \begin{bmatrix} 25 & 0 \\ 0 & 25 \end{bmatrix} \begin{bmatrix} u_1^* \\ u_2^* \end{bmatrix} \tag{14.17}$$

and we can write [combining Equations (14.15) and (14.17)]

$$\begin{bmatrix} y_1^*(s) \\ y_2^*(s) \end{bmatrix} = \begin{bmatrix} \frac{0.0667}{20s+1} & \frac{0.0667}{20s+1} \\ \frac{0.02}{10s+1} & \frac{-0.02}{10s+1} \end{bmatrix} \begin{bmatrix} 25 & 0 \\ 0 & 25 \end{bmatrix} \begin{bmatrix} u_1^*(s) \\ u_2^*(s) \end{bmatrix}$$
$$\begin{bmatrix} y_1^*(s) \\ y_2^*(s) \end{bmatrix} = \begin{bmatrix} \frac{1.667}{20s+1} & \frac{1.667}{20s+1} \\ \frac{0.5}{10s+1} & \frac{-0.5}{10s+1} \end{bmatrix} \begin{bmatrix} u_1^*(s) \\ u_2^*(s) \end{bmatrix} \tag{14.18}$$

Now it is clear from the gain matrix that a full manipulated variable move will cause the desired tank height range to be violated, while the temperature range will be satisfied.

When dealing with dynamic systems, it is always important to know the units of all parameters and variables, and to understand expected operating ranges for all inputs and outputs. It is convenient to scale the inputs and outputs such that the expected operating range is -1 to $+1$; for one reason, it is easier to compare responses on the same set of plots. Also, some analysis tools (such as the SVD presented in Section 14.4) can be interpreted only on properly scaled systems.

14.4 Directional Sensitivity and Operability

In a number of examples in this textbook, we have noted that the design of a process can limit the ability to control the process. For example, if the "gain" for an input-output pairing is not large enough, then there is a limit to how large a setpoint change can be made before the manipulated input saturates (hits a constraint, such as a valve being fully open or closed). For multivariable systems, the effect of a particular input on a particular output is influenced by the other inputs. For example, consider the mixing tank of Example 14.3. A change in the hot-stream flow rate can significantly affect the temperature if the cold-stream flow rate remains constant. If the cold-stream flow rate changes simultaneously, however, the change in the hot-stream flow rate may not change the temperature. There are two ways of exploring these multivariable effects. One is to apply a SVD to the steady-state process gain matrix; this is an inherently linear procedure. An alternative is to "map out" an operating window or the possible range of process outputs that corresponds to a range of possible process inputs. This can be done with either the linear model or a fundamental nonlinear model.

In this section, we first present the SVD technique, followed by the operating window technique.

SVD

The SVD can be used to predict the *directional sensitivity* of a process. The process gain matrix is decomposed into three matrices,

$$G = U\Sigma V^T \tag{14.19}$$

where U is the left singular vector matrix, Σ the diagonal matrix of singular values, and V the right singular vector matrix. The left and right singular vector matrices are both orthonormal matrices; that is, each column of the matrix is orthogonal to all other columns and the columns each are unit length. The diagonal singular value matrix is ordered so that the largest singular value is in the (1,1) position. *Note that the standard notation for SVD is to use U to represent the left singular vector matrix. Please do not confuse this with the u vector commonly used to represent the vector of manipulated inputs.*

The left singular vector matrix indicates the strongest and weakest *output* directions, while the right singular vector matrix indicates the strongest and weakest *input* directions. The ordered matrix of singular values provides the "magnitude" of the strongest and weakest directions. The ratio of the minimum and maximum singular values is known as the condition number of the matrix. Large condition numbers are indicative of *ill-conditioned* systems, which effectively lose a degree of freedom (not all inputs are truly independent).

This analysis has the following control-related implications. It is relatively easy to make a setpoint change in the strongest output direction (indicated by the first column of the left singular vector matrix), and it is relatively hard to make a setpoint change in the weakest output direction (indicated by the last column of the left singular vector matrix).

Here we use the mixing tank example to illustrate the use of SVD analysis.

Example 14.3, continued

The singular value decomposition of the mixing tank gain matrix (steady state, $s = 0$) is (see Section 14.8 for the MATLAB calculation)

$$G = U\Sigma V^T$$

$$\underbrace{\begin{bmatrix} 1.667 & 1.667 \\ 0.5 & -0.5 \end{bmatrix}}_{G} = \underbrace{\begin{bmatrix} 1 & 0 \\ 0 & -1 \end{bmatrix}}_{U} \underbrace{\begin{bmatrix} 2.3575 & 0 \\ 0 & 0.7071 \end{bmatrix}}_{\Sigma} \underbrace{\begin{bmatrix} 0.7071 & -0.7071 \\ 0.7071 & 0.7071 \end{bmatrix}^T}_{V}$$

The first column of the U (left singular) matrix indicates that output 1 (liquid height) is the *strongest* output direction, while the second column indicates that output 2 (temperature) is the *weakest* input direction. The first column of the V (right singular) matrix indicates that the strongest input direction is to increase (or decrease) both inputs by the same amount, while the second column indicates that changing both inputs in opposite directions (decreasing one, increasing the other) results in the weakest effect. The matrix of ordered singular values (Σ) indicates that the strongest direction has 3.33 times (2.3575/0.7071) the effect of the weakest direction.

These results have the following physical interpretation. A simultaneous increase in both the hot (input 1) and cold (input 2) flow rates will increase the tank height (output 1) but will not change the outlet temperature (output 2). Similarly, a decrease in the hot flow rate simultaneously with an increase in the cold flow rate will not change the tank height but will result in a temperature decrease.

First, consider a simultaneous increase in both flow rates (notice that the "magnitude" of the input vector is $\sqrt{0.7071^2 + 0.7071^2} = 1$ and refer to this as input vector **a**:

$$u_a = \begin{bmatrix} 0.7071 \\ 0.7071 \end{bmatrix}$$

The output for this particular input can be found from

$$y_a = Gu_a = \begin{bmatrix} 1.667 & 1.667 \\ 0.5 & -0.5 \end{bmatrix} \begin{bmatrix} 0.7071 \\ 0.7071 \end{bmatrix}$$

$$y_a = \begin{bmatrix} 2.3575 \\ 0 \end{bmatrix}$$

That is, the scaled height increases by 2.3575, but there is no change in temperature. This can be viewed in Figure 14–3, where the input vector is in the direction $(u_1,u_2) = (0.71,0.71)$ and the output vector is in the direction $(y_1,y_2) = (2.36,0)$.

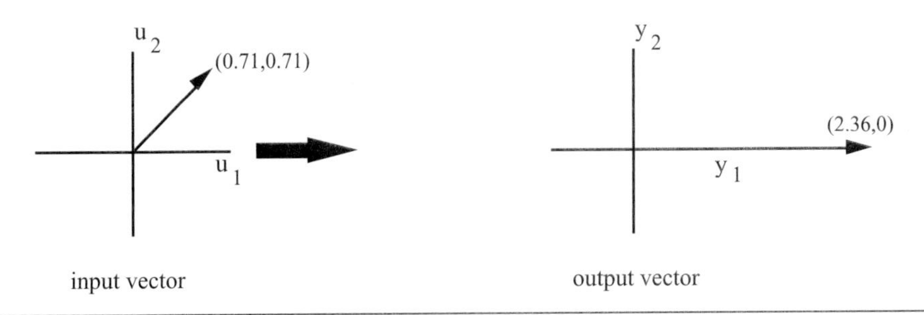

Figure 14–3 Input vector; output vector.

Now, consider a decrease in hot flow rate simultaneously with an increase in cold flow rate (again, notice that the "magnitude" of the input vector is $\sqrt{(-0.7071)^2 + 0.7071^2} = 1$) and refer to this as input vector **b**:

$$u_b = \begin{bmatrix} -0.7071 \\ 0.7071 \end{bmatrix}$$

The output can be found from

$$y_b = Gu_b = \begin{bmatrix} 1.667 & 1.667 \\ 0.5 & -0.5 \end{bmatrix} \begin{bmatrix} -0.7071 \\ 0.7071 \end{bmatrix}$$

$$y_b = \begin{bmatrix} 0 \\ -0.7071 \end{bmatrix}$$

That is, the scaled temperature decreases by 0.7071, but there is no change in height. Notice that we can construct Figure 14–4, where the "unit circle" on the left represents all possible input vectors with a length of 1. The input vectors u_a and u_b are also shown. The ellipse on the right represents the corresponding set of outputs (obtained from $y = Gu$). Notice that **a** is the longest output vector and **b** is the shortest. That is, a unit input in the **a** direction results in an output in the **a** direction shown, with a length of the largest singular value. Similarly, a unit input in the **b** direction results in the output shown, with a length of the smallest singular value. All other inputs result in output vectors with lengths between the smallest and largest singular values.

A minor disadvantage of the SVD analysis is the assumption that all possible input vectors are the same length. For this example, you could argue that it is possible for both the hot and cold flows to be at their maximum values of 1. A slightly more realistic interpretation of the input-output mapping is then shown in Figure 14–5.

The output space, in physical (rather than scaled) variables, is shown in Figure 14–6. Notice that the linear analysis predicts negative height values, which clearly cannot be obtained in practice.

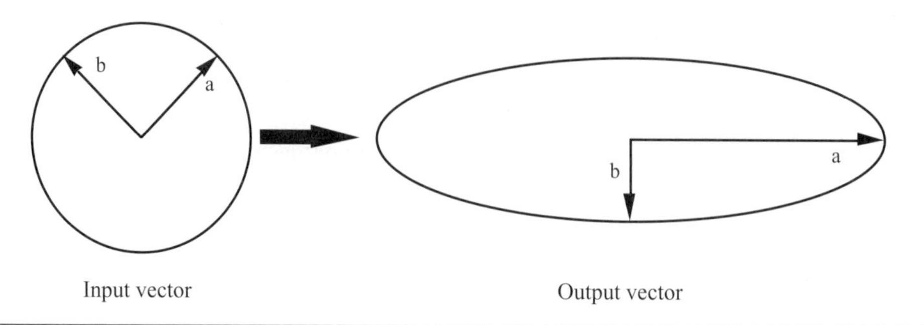

Figure 14–4 Input to output mapping.

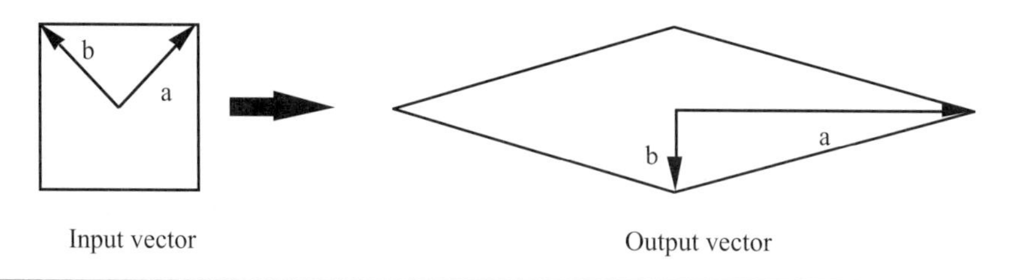

Figure 14–5 Input to output mapping.

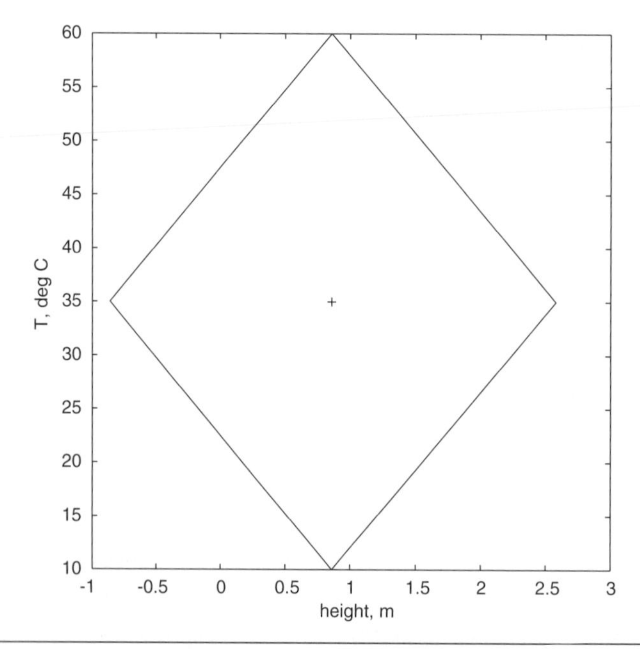

Figure 14–6 Linear output space in physical coordinates.

Operating Window

This approach is similar to the "rectangular" mapping shown in the previous example. The idea is to map the effect of a region of process inputs onto a region of process outputs. Here, simulations are performed using the steady-state nonlinear model, with various values for the inputs.

Example 14.3, continued

The nonlinear steady-state modeling equations result in the two steady-state output equations

$$T = \frac{F_H T_H + F_C T_C}{F_H + F_C}$$

$$h = \left(\frac{F_H + F_C}{\beta} \right)^2$$

Here it is understood that the steady-state values are being used, so the subscript s has been neglected. The flow coefficient is $\beta = 53.9(\text{liter/min})/\sqrt{\text{m}}$ under these conditions. For the flows bounded between 0 and 50 liters/minute each, we find the output region shown in Figure 14–7. The vessel will overflow for tank heights greater than 1.72 m.

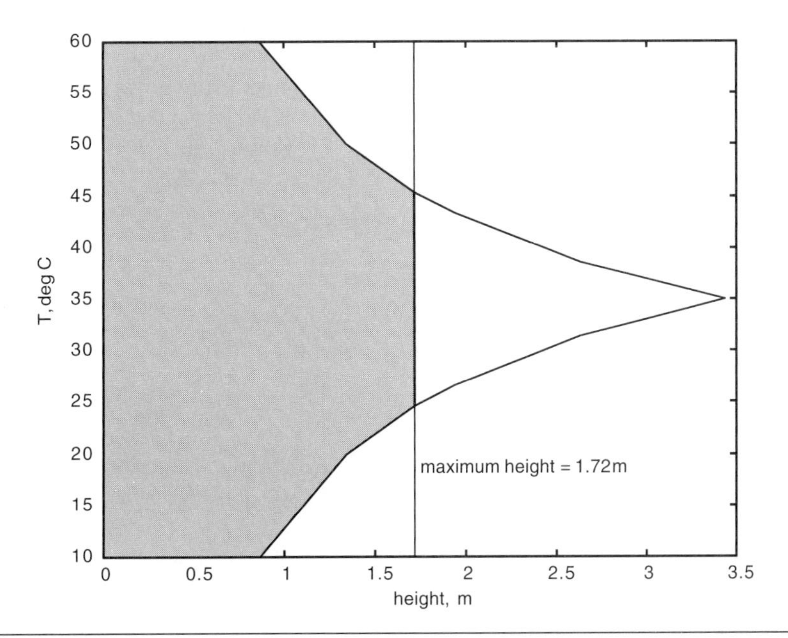

Figure 14–7 Output operating window based on nonlinear model. The input space is a rectangle with flows between 0 and 50 liters/min.

Notice that some high and low temperatures cannot be obtained at heights between 0.86 and 1.72 m; this is also predicted by the "linear rectangle" analysis shown earlier. At low tank heights, however, the entire temperature range is feasible; this is not predicted by the previous linear analysis and is a nonlinear effect.

We have shown that the SVD provides insight about the strong and weak input and output directions. The primary limitation is the assumption of a linear model. Linear models are suitable for perturbations from a nominal operating point and are generally very useful for control system design. After all, the purpose of a controller is to maintain the process close to a desired operating point. However, linear models usually do not predict the behavior of a nonlinear system over the entire range of operating conditions. To explore a "window" of operating conditions, the nonlinear model should be used. If operating the process at a different operating point is desired, the model can be linearized at that point for controller design and analysis. Another limitation to the analysis presented in this section is that, like the RGA, it is based solely on steady-state information. Dynamics certainly play an important role, but this steady-state analysis provides a quick screening tool to understand physical limitations to steady-state operating conditions.

14.5 Block-Diagram Analysis

Consider the multivariable block diagram shown in Figure 14–8. Each of the variables shown on this diagram is a vector and the blocks represent matrix transfer functions. That is, for two process outputs, two manipulated inputs, and one disturbance, the corresponding vectors are given below.

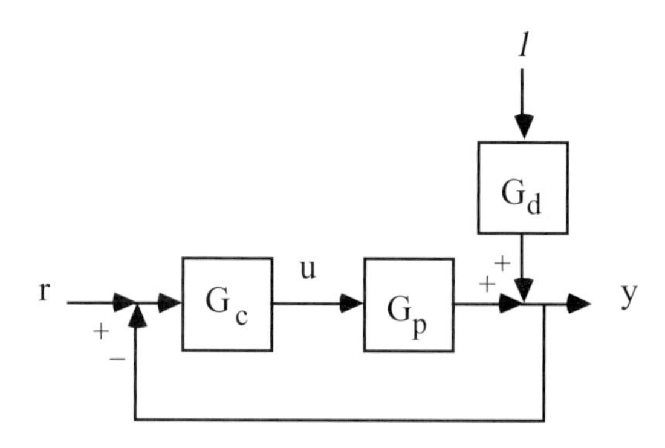

Figure 14–8 Multivariable block diagram.

Setpoints, outputs, manipulated inputs, and the disturbance input are

$$r = \begin{bmatrix} r_1 \\ r_2 \end{bmatrix}, \quad y = \begin{bmatrix} y_1 \\ y_2 \end{bmatrix}, \quad u = \begin{bmatrix} u_1 \\ u_2 \end{bmatrix}, \quad l = [l_1]$$

The controller, process, and disturbance matrices are

$$G_c(s) = \begin{bmatrix} g_{c11}(s) & g_{c12}(s) \\ g_{c21}(s) & g_{c22}(s) \end{bmatrix}, \quad G_p(s) = \begin{bmatrix} g_{11}(s) & g_{12}(s) \\ g_{21}(s) & g_{22}(s) \end{bmatrix}, \quad G_d(s) = \begin{bmatrix} g_{d1}(s) \\ g_{d2}(s) \end{bmatrix}$$

Notice that if SISO controllers are used, the controller transfer function matrix is

$$G_c(s) = \begin{bmatrix} g_{c1}(s) & 0 \\ 0 & g_{c2}(s) \end{bmatrix}$$

where $g_{c1}(s)$ and $g_{c2}(s)$ will usually be PID-type controllers. The relationship between inputs (both manipulated and disturbance) and outputs is

$$y(s) = G_p(s)u(s) + G_d(s)l(s)$$

Since the error signal is $e(s) = r(s) - y(s)$, we can derive (where I is the identity matrix)

$$[I + G_p(s)G_c(s)]y(s) = G_p(s)G_c(s)r(s) + G_d(s)l(s)$$

and using matrix inversion to solve for $y(s)$, we find

$$y(s) = [I + G_p(s)G_c(s)]^{-1}G_p(s)G_c(s)r(s) + [I + G_p(s)G_c(s)]^{-1}G_d(s)l(s) \qquad (14.20)$$

This last equation is the closed-loop transfer function relationship for multivariable systems. Notice that the order of multiplication is very important. In general, $G_P(s)G_c(s) \neq G_c(s)G_p(s)$. This is clear if, for example, G_p is a 4×3 matrix and G_c is a 3×4 matrix.

Recall that the SISO closed-loop transfer function is

$$y(s) = \frac{g_c(s)g_p(s)}{1 + g_c(s)g_p(s)} r(s) + \frac{g_d(s)}{1 + g_c(s)g_p(s)} l(s)$$

For SISO systems, the order of multiplication does not matter; for MIMO systems, it is crucial.

14.6 Decoupling

We have seen the problems with control-loop interaction; these occur because a manipulated input affects more than one controlled output. One approach to handling this problem is known as *decoupling*. The idea is to develop "synthetic" manipulated inputs that affect only one process output each. This approach is illustrated in Figure 14–9.

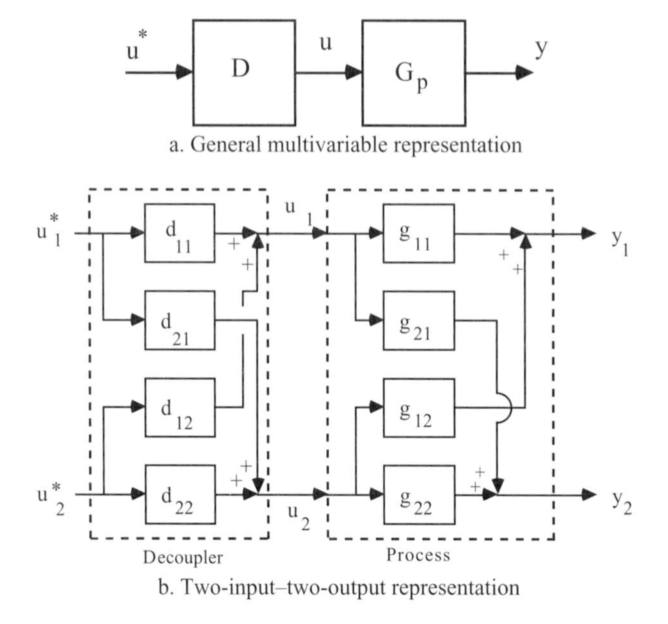

Figure 14–9 Decoupling control strategy—synthetic inputs and process outputs.

The relationship between the synthetic input vector and the process output vector is

$$y(s) = G_p(s)D(s)u^*(s) \tag{14.21}$$

For a two input–two output process,

$$\begin{bmatrix} y_1(s) \\ y_2(s) \end{bmatrix} = G_p(s)D(s)\begin{bmatrix} u_1^*(s) \\ u_2^*(s) \end{bmatrix} \tag{14.22}$$

where $G_p(s)D(s)$ is a 2×2 transfer function matrix. There are a number of possible choices for the "target" $G_p(s)D(s)$ matrix. Two common ones lead to ideal decoupling and simplified decoupling.

Ideal Decoupling

One possible choice for $G_p(s)D(s)$ is

$$G_p(s)D(s) = \begin{bmatrix} g_{11}(s) & 0 \\ 0 & g_{22}(s) \end{bmatrix} \tag{14.23}$$

Solving for $D(s)$, we find

$$D(s) = \tilde{G}_p^{-1}(s)\begin{bmatrix} \tilde{g}_{11}(s) & 0 \\ 0 & \tilde{g}_{22}(s) \end{bmatrix} \tag{14.24}$$

where we use the (~) notation to make it clear that these calculations are performed based on a process model. The relationship between the synthetic inputs and process outputs, $y(s) = G_P(s)D(s)u^*(s)$, that results is

$$\begin{bmatrix} y_1(s) \\ y_2(s) \end{bmatrix} = \begin{bmatrix} \tilde{g}_{11}(s) & 0 \\ 0 & \tilde{g}_{22}(s) \end{bmatrix}\begin{bmatrix} u_1^*(s) \\ u_2^*(s) \end{bmatrix} \tag{14.25}$$

and we see that each synthetic input affects only one process output (assuming a perfect model).

$$y_1(s) = g_{11}(s)u_1^*(s)$$
$$y_2(s) = g_{22}(s)u_2^*(s)$$

The main advantage is that independent SISO tuning parameters can be used for each control loop. A major disadvantage is that the resulting decoupler is the inverse of the process transfer function matrix. There are numerous problems with this approach, including the fact that the decoupler will be unstable if there are RHP transmission zeros associated with the process transfer function matrix. Also, this type of decoupler is known to be extremely sensitive to model error.

Simplified Decoupling

An alternative approach, known as simplified decoupling, is shown for a 2 × 2 example in Figure 14–10. Here, the decoupling matrix is restricted to the form

$$D(s) = \begin{bmatrix} 1 & d_{12}(s) \\ d_{21}(s) & 1 \end{bmatrix} \tag{14.26}$$

Here, we specify a decoupled response and the decoupler with the structure in Equation (14.26),

$$G_p(s)D(s) = \begin{bmatrix} g_{11}^*(s) & 0 \\ 0 & g_{22}^*(s) \end{bmatrix}$$
$$\begin{bmatrix} g_{11}(s) & g_{12}(s) \\ g_{21}(s) & g_{22}(s) \end{bmatrix}\begin{bmatrix} 1 & d_{12}(s) \\ d_{21}(s) & 1 \end{bmatrix} = \begin{bmatrix} g_{11}^*(s) & 0 \\ 0 & g_{22}^*(s) \end{bmatrix} \tag{14.27}$$

and we can solve four equations in four unknowns to find

$$\begin{aligned} d_{12}(s) &= -\frac{g_{12}(s)}{g_{11}(s)} \\ d_{21}(s) &= -\frac{g_{21}(s)}{g_{22}(s)} \\ g_{11}^{*}(s) &= g_{11}(s) - \frac{g_{12}(s)g_{21}(s)}{g_{22}(s)} \\ g_{22}^{*}(s) &= g_{22}(s) - \frac{g_{21}(s)g_{12}(s)}{g_{11}(s)} \end{aligned} \tag{14.28}$$

Notice that the decoupling elements (d_{12} and d_{21}) are the same as the feed-forward control design presented in Chapter 10. The effect of the synthetic inputs are being "fed-forward" to minimize control loop interaction. Similar to feed-forward control, some factorization of the decoupling elements must be performed to make certain that they are stable and physically realizable.

The feedback control diagram for simplified decoupling is shown in Figure 14–11. Whether ideal or simplified decoupling is being implemented, the result is a true multivariable strategy, as shown in Figure 14–12.

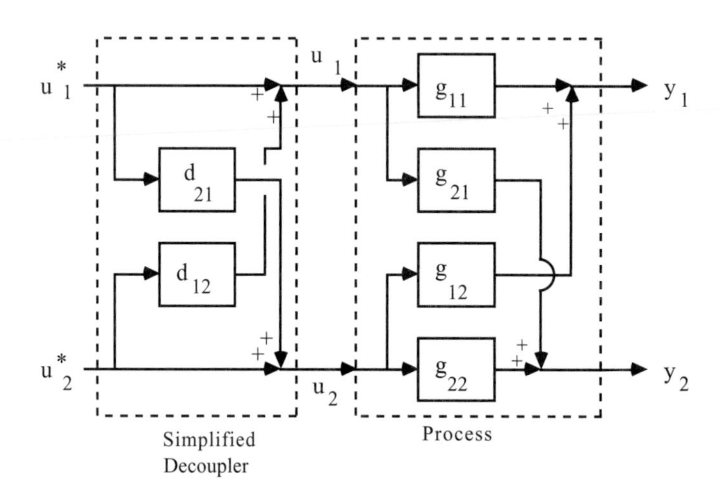

Figure 14–10 Simplified decoupling control strategy—synthetic inputs and process outputs.

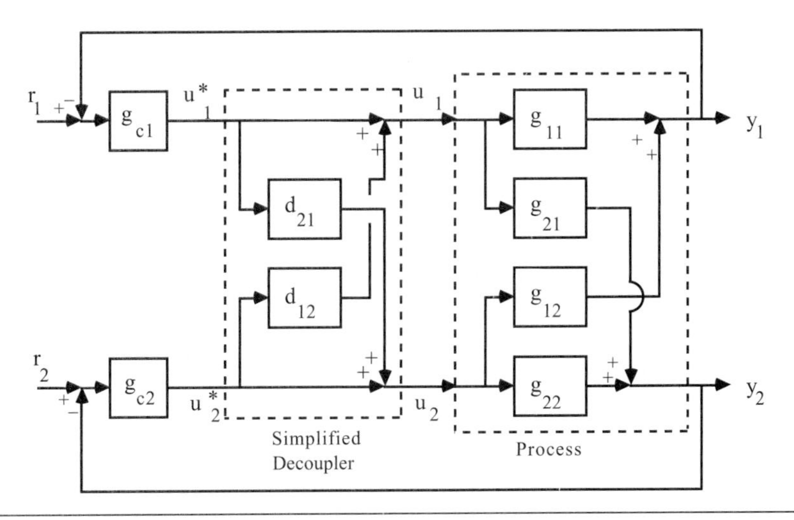

Figure 14–11 Feedback control using simplified decoupling.

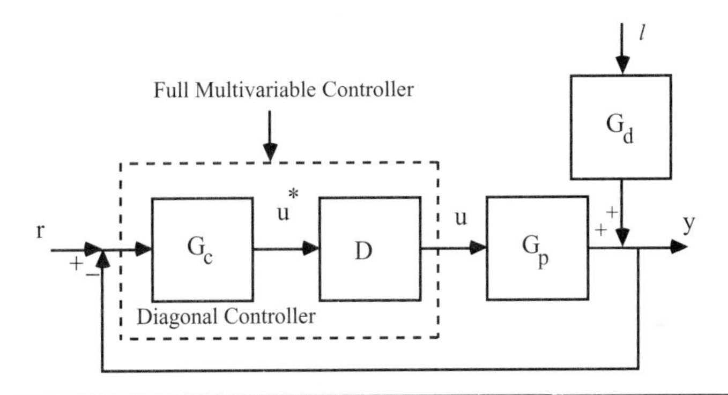

Figure 14–12 Illustration of multivariable control using decoupling.

Static Decoupling

In practice, static decoupling (based on steady-state gains) is used much more often than dynamic decoupling, since the dynamic version may not be physically realizable, or the dynamic parameters may be too uncertain. In this case, simply set $s = 0$ in the decoupling equations.

14.7 IMC

The IMC design procedure for MIMO systems is similar to the design procedure developed in Chapter 8 for SISO systems. The IMC block diagram is shown in the Figure 14–13, where the blocks now represent matrix transfer functions and the inputs and outputs are vectors. We first use an example to illustrate the multivariable IMC design procedure, then follow with the general procedure.

Example 14.1, continued

Here we consider *system A* from Example 14.1. Recall that system A had a RHP transmission zero ($z = 3\ \text{min}^{-1}$)

$$\tilde{G}_p(s) = \begin{bmatrix} \frac{s+3}{(s+1)(s+2)} & \frac{2}{(s+1)(s+2)} \\ \frac{3}{(s+1)(s+2)} & \frac{1}{(s+1)(s+2)} \end{bmatrix}$$

$$= \begin{bmatrix} \frac{1.5(0.333s+1)}{(s+1)(0.5s+1)} & \frac{1}{(s+1)(0.5s+1)} \\ \frac{1.5}{(s+1)(0.5s+1)} & \frac{0.5}{(s+1)(0.5s+1)} \end{bmatrix}$$

There are a number of ways to factor this matrix. The easiest way is to place the RHP transmission zero on the diagonal of the "bad (noninvertible) matrix."

$$\tilde{G}_{p+}(s) = \begin{bmatrix} \frac{-0.333s+1}{0.333s+1} & 0 \\ 0 & \frac{-0.333s+1}{0.333s+1} \end{bmatrix}$$

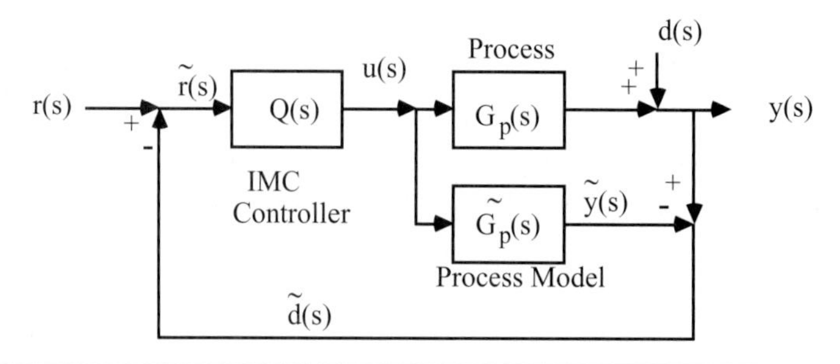

Figure 14–13 IMC.

Since $\tilde{G}_p = \tilde{G}_{p+}(s)\tilde{G}_{p-}(s)$, we can solve for the "good matrix" from $\tilde{G}_{p-}(s) = \tilde{G}_{p+}^{-1}(s)\tilde{G}_p$, and find that

$$\tilde{G}_p(s) = \underbrace{\begin{bmatrix} \frac{-0.333s+1}{0.333s+1} & 0 \\ 0 & \frac{-0.333s+1}{0.333s+1} \end{bmatrix}}_{\tilde{G}_{p+}} \bullet$$

$$\underbrace{\begin{bmatrix} \frac{1.5(0.333s+1)^2}{(-0.333s+1)(s+1)(0.5s+1)} & \frac{(0.333s+1)}{(-0.333s+1)(s+1)(0.5s+1)} \\ \frac{1.5(0.333s+1)}{(-0.333s+1)(s+1)(0.5s+1)} & \frac{0.5(0.333s+1)}{(-0.333s+1)(s+1)(0.5s+1)} \end{bmatrix}}_{\tilde{G}_{p-}}$$

Where the controller is

$$Q(s) = \begin{bmatrix} \frac{1.5(0.333s+1)^2}{(-0.333s+1)(s+1)(0.5s+1)} & \frac{(0.333s+1)}{(-0.333s+1)(s+1)(0.5s+1)} \\ \frac{1.5(0.333s+1)}{(-0.333s+1)(s+1)(0.5s+1)} & \frac{0.5(0.333s+1)}{(-0.333s+1)(s+1)(0.5s+1)} \end{bmatrix}^{-1} \bullet$$

$$\begin{bmatrix} \frac{1}{\lambda_1 s+1} & 0 \\ 0 & \frac{1}{(\lambda_2 s+1)^2} \end{bmatrix}$$

$$= \begin{bmatrix} \frac{-0.667(s+1)(0.5s+1)}{(\lambda_1 s+1)(0.333s+1)} & \frac{1.333(s+1)(0.5s+1)}{(\lambda_2 s+1)^2(0.333s+1)} \\ \frac{2(s+1)(0.5s+1)}{(\lambda_1 s+1)(0.333s+1)} & \frac{-2(0.333s+1)(s+1)(0.5s+1)}{(\lambda_2 s+1)^2(0.333s+1)} \end{bmatrix}$$

Assume a perfect model with no disturbances. For $\lambda_1 = \lambda_2 = 0.333$ min, in Figure 14–14 we find the response to a step setpoint change in output 1. Notice that the output response is decoupled, although both manipulated inputs are changed. Also, although the original $g_{11}(s)$ transfer function did not have a RHP zero, the multivariable system has inverse response behavior in the closed loop. It should be noted that, in general, the IMC filter factors (λ_1 and λ_2) should be tuned to have different values.

The reader should show that a setpoint change in output 2 also leads to inverse response behavior in output 2.

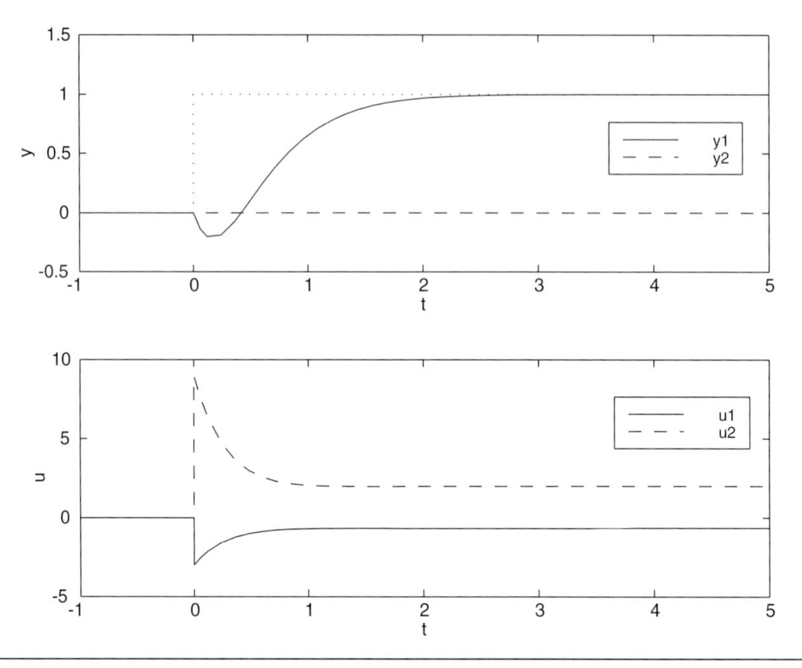

Figure 14–14 Response to a setpoint change in output 1, with a diagonal factorization matrix and $\lambda_1 = \lambda_2 = 0.333$ min.

A major disadvantage to the diagonal factorization is that inverse response appears in all output setpoint responses. It is also possible to perform a factorization that places all the inverse response behavior in one of the output variables, resulting in good closed-loop performance in the other outputs. See Holt and Morari (1985) for more details.

The general multivariable IMC procedure is as follows.

1. Factor the process model into invertible and noninvertible elements.

$$\tilde{G}_p = \tilde{G}_{p+}(s)\tilde{G}_{p-}(s) \tag{14.29}$$

The difficulty is that there are many ways to factor the process transfer function matrix. One way is to place the RHP transmission zero on the diagonal.

$$\tilde{G}_{p+}(s) = \begin{bmatrix} \frac{-\frac{1}{z}s+1}{\frac{1}{z}s+1} & 0 & \cdots & 0 \\ 0 & \frac{-\frac{1}{z}s+1}{\frac{1}{z}s+1} & 0 & 0 \\ \vdots & 0 & \ddots & \vdots \\ 0 & 0 & \cdots & \frac{-\frac{1}{z}s+1}{\frac{1}{z}s+1} \end{bmatrix} \tag{14.30}$$

Although this is perhaps the simplest method, it usually does not result in the best performance. Other factorization methods are beyond the scope of this text. The motivated reader should consult Morari and Zafiriou (1989) for more details.

2. Form the idealized controller.

$$\tilde{Q}(s) = \tilde{G}_{p-}^{-1}(s) \tag{14.31}$$

3. Add a filter to make all elements in the controller matrix proper.

$$Q(s) = \tilde{Q}(s)F(s) = \tilde{G}_{p-}^{-1}(s)F(s) \tag{14.32}$$

where $F(s)$ is normally a diagonal matrix.

$$F(s) = diag\left[\frac{1}{(\lambda_i s+1)^n}\right] \tag{14.33}$$

4. The filter factors are adjusted to vary the response and robustness characteristics. In the limit of a perfect model ($G_p(s) = \tilde{G}_p(s)$), the response will be

$$\begin{aligned} y(s) &= G_p(s)Q(s)r(s) = \underbrace{G_{p+}(s)G_{p-}}_{G_p(s)}\underbrace{\tilde{G}_{p-}^{-1}(s)F(s)}_{Q(s)}r(s) \\ &= G_{p+}(s)F(s)r(s) \end{aligned} \tag{14.34}$$

and where, as in the SISO case, the "bad stuff" must appear in the output response.

14.8 MATLAB tzero, svd, and LTI Functions

The MATLAB Control Toolbox has a routine (tzero) for calculating the transmission zeros of a multivariable model. MATLAB can also be used to perform a SVD analysis (using the function svd).

Transmission Zero Calculation

Here we illustrate the use of tzero by way of an example.

Example 4.2, continued

Operating Point 1 (minimum phase)

Consider first the minimum-phase operating point. First, set all of the individual matrix elements as follows:

```
g11 = tf([2.6],[62 1]);
g12 = tf([1.5],conv([62 1],[23 1]));
g21 = tf([1.4],conv([30 1],[90 1]));
g22 = tf([2.8],[90 1]);
```

Then define the transfer function matrix

```
G = [g11 g12;g21 g22]

Transfer function from input 1 to output...
          2.6
 #1:   --------
       62 s + 1

                 1.4
 #2:   --------------------
       2700 s^2 + 120 s + 1

Transfer function from input 2 to output...
                 1.5
 #1:   -------------------
       1426 s^2 + 85 s + 1
```

```
        2.8
 #2:  --------
      90 s + 1
```

The command to evaluate transmission zeros is

```
z = tzero(G)

z =
  -0.01733947114395
  -0.05947212305895
```

and we see that both zeros are in the left-half-plane, so the model inverse will be stable.

Operating Point 2 (nonminimum phase)

Consider now the nonminimum-phase operating point. Set all of the individual matrix elements as follows:

```
g11 = tf([1.5],[63 1]);
g12 = tf([2.5],conv([39 1],[63 1]));
g21 = tf([2.5],conv([56 1],[91 1]));
g22 = tf([1.6],[91 1]);
```

Then define the transfer function matrix

```
 G2 = [g11 g12;g21 g22]

Transfer function from input 1 to output...
        1.5
 #1:  --------
      63 s + 1

              2.5
 #2:  --------------------
      5096 s^2 + 147 s + 1

Transfer function from input 2 to output...
```

```
                2.5
 #1:   --------------------
       2457 s^2 + 102 s + 1

          1.6
 #2:   --------
       91 s + 1
```

and calculate the transmission zeros

```
 z2 = tzero(G2)

z2 =
   0.01300046557077
  -0.05649863406894
```

The RHP (positive) zero indicates that the matrix inverse will be unstable.

SVD

Here we illustrate the use of `svd` to perform a SVD, by way of an example.

Example 14.3, continued

The MATLAB `svd` function is straightforward to use. First, enter the scaled G matrix, then issue the `svd` command, which yields the following results:

```
» G_scaled = [1.667 1.667;0.5 -0.5];

» [U,S,V] = svd(G_scaled)

U =
    1.0000     0.0000
    0.0000    -1.0000

S =
    2.3575          0
         0     0.7071

V =
   0.7071    -0.7071
   0.7071     0.7071
```

LTI Objects in SIMULINK Block Diagrams

Here we illustrate the use of the LTI feature of the Control Toolbox to perform multivariable simulations. First, the process and controller objects are generated using the LTI `tf` (transfer function) command. Then the block diagram is constructed and the LTI objects from the Control Toolbox are placed in specific blocks on the diagram.

Example 14.1, continued

Here, we use Example 14.1, at the nonminimum phase (system A) operating point, for illustration. First, define the process using the LTI objects.

```
% define the process
g11 = tf([1 3],[1 3 2]);
g12 = tf([2],[1 3 2]);
g21 = tf([3],[1 3 2])
g22 = tf([1],[1 3 2]);
G = [g11 g12;g21 g22];
%
% define the controller (L1 and L2 are the tuning parameters)
% set the L1 and L2 values before simulating the .mdl file
q11 = tf(-(2/3)*[0.5 1.5 1],[(1/3)*L1 ((1/3)+L1) 1]);
q12 = tf((4/3)*[0.5 1.5 1],conv([L2^2 2*L2 1],[(1/3) 1]));
q21 = tf(2*[0.5 1.5 1],[(1/3)*L1 (1/3)+L1 1]);
q22 = tf(-2*conv([0.5 1.5 1],[(1/3) 1]),conv([L2^2 2*L2 1],
       [(1/3) 1]));
Q = [q11 q12;q21 q22];
```

Then, construct the `.mdl` diagram shown in Figure 14–15, where the multivariable (MV) IMC controller block contains the `Q` object, the plant block contains the *G* object and the model block is equal to the plant block for this (ideal) example.

Set the tuning parameter values (L1 and L2) before running the simulation. For a setpoint change in output 1, the responses are shown in Figure 14–14.

14.9 Summary

Transmission zeros are the multivariable equivalent of SISO transfer function zeros. RHP (positive) transmission zeros indicate that the process inverse will be unstable. A transfer

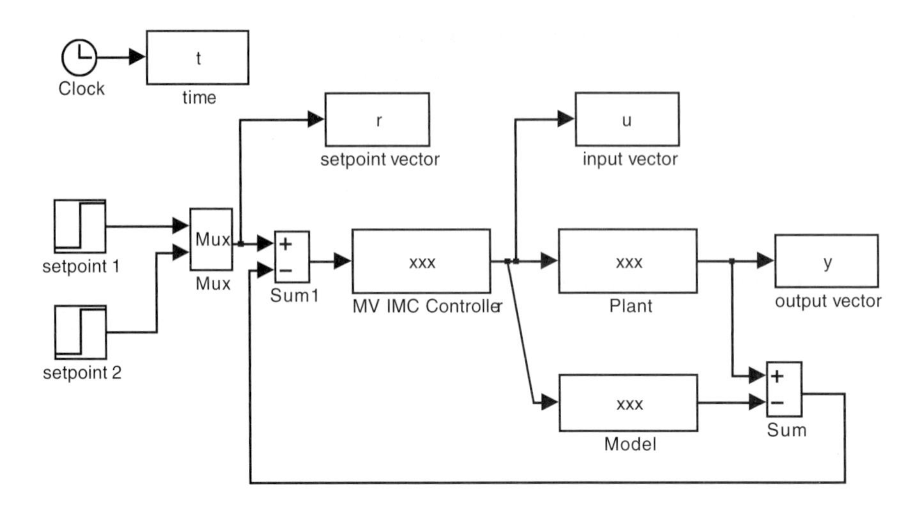

Figure 14–15 MV IMC block diagram.

function matrix may have no individual transfer functions with RHP zeros, and yet have a RHP matrix transmission zero. Similarly, a transfer function matrix may have an element with a RHP zero, and yet have no RHP matrix transmission zeros. Processes with RHP transmission zeros have inherent closed-loop performance limitations; often there is not a significant advantage to full multivariable control (using decoupling or MV IMC) of these processes.

Multivariable processes exhibit a directional sensitivity, where two different input vector directions could result in outputs with a significantly different output magnitude. Processes that are ill-conditioned (high condition number, or ratio of maximum to minimum singular values) will often be too sensitive for full multivariable control.

Processes with RHP transmission zeros have internal closed-loop performance limitations; often there is not a significant advantage to full multivariable control (using decoupling or MV IMC) of these processes.

References

The quadruple tank problem (Example 14.2) is presented in the following paper: Johansson, K. H., "The Quadruple-Tank Process. A Multivariable Laboratory Process with an Adjustable Zero," *IEEE Trans. Cont. Syst. Tech.*, **8**(3), 456–465 (2000).

Control performance limitations due to RHP transmission zeros is discussed in the following texts and papers:

Holt, B. R., and M., Morari "Design of Resilient Processing Plants—VI: The Effect of Right-Half-Plane Zeros on Dynamic Resilience," *Chem. Eng. Sci.*, **40**(1), 59–74 (1985).

Morari, M., and E. Zafiriou, *Robust Process Control*, Prentice Hall, Upper Saddle River, NJ (1989).

Skogestad, S., and I. Postlethwaite, *Multivariable Feedback Control: Analysis and Design*, Wiley, New York (1996).

A more detailed discussion of the operating window approach is presented in the following: Marlin, T. E., *Process Control: Designing Processes and Control Systems for Dynamic Performance*, 2nd ed., McGraw Hill, New York (2000).

A measure of controllability, based on the operating window approach, is presented in the following paper: Vinson, D. R., and C. Georgakis, "A New Measure of Process Output Controllability," *J. Process Control*, **10**, 185–194 (2000).

Student Exercises

1. Consider Example 14.2, the quadruple tank problem.

  a. Find the RGA for each operating point. Which variable pairing do you recommend for each operating point?

  b. Consider the nonminimum-phase (RHP transmission zeros) operating point (operating point 2) in Example 14.2. Design independent single-loop controllers for relatively fast closed-loop responses. (*Hint*: Approximate the transfer functions as first order and use PI tuning rules.) Then close both loops simultaneously using the independent SISO tuning. What do you find? How much detuning is necessary to obtain good performance with both loops closed?

2. Find the inverse of the process transfer function matrix for *system B* in Example 14.1.

3. Consider the mixing tank presented in Example 14.3.
Write the dynamic material and energy balances for this system. The material balance will yield the rate of change of height with respect to time, and the energy balance will yield the rate of change of temperature with respect to time. Linearize the model at the steady state to develop the input-output transfer function matrix.

4. Perform simulations on the mixing tank shown in Example 14.3.
   a. First, work with the scaled transfer function matrix and make step input changes in both the strongest and weakest directions. Compare the magnitudes of the output responses.
   b. Now, work with the dimensional variables. Convert the strongest and weakest input directions to dimensional inputs and compare the effect of step changes on the dimensional outputs.
5. Consider Example 14.1, controlled using the multivariable IMC procedure. Use the SIMULINK diagram and LTI objects shown in Section 14.8.3 to perform simulations for various tuning parameter (λ_1 and λ_2 values). Show that decreasing the tuning parameter values below 0.333 does not significantly speed up the response but does increase the manipulated input action.
6. Consider Example 14.1, controlled using the multivariable IMC procedure. Use the SIMULINK diagram and LTI objects shown in Section 14.8. but revise these for the minimum-phase (system B) operating point. Since there is not a RHP transmission zero, no factorization of the process transfer function is necessary. Perform simulations for various tuning parameter (λ_1 and λ_2) values. Show that decreasing the tuning parameter values to approach 0 leads to almost instantaneous control, at the expense of large manipulated variable action.
7. Consider Example 14.2 (quadruple tank problem). The modeling equations are presented in the appendix. Verify that the given steady-state values are correct for each operating point. Also, linearize and find the state space model and verify the transfer function matrix that was given for each operating point in Example 14.2.
8. Consider Example 14.2 (quadruple tank problem). Use MATLAB to calculate the transmission zeros for both operating points.
9. Consider Example 14.2 (quadruple tank problem). Design a multivariable IMC controller for the second (nonminimum phase) operating point. Use the standard factorization that places the RHP transmission zero on the diagonal terms of the "bad stuff" ($\tilde{G}_{p+}(s)$) matrix. Implement the closed-loop controller in SIMULINK and discuss the effect of the IMC filter factors on setpoint changes in each output (at different times). Consider the allowable changes in the process outputs and manipulated inputs, given the steady-state data shown in the appendix.
10. Consider Example 14.3 (mixing tank) with the transfer function matrix shown in Equation (14.11). Implement the simplified decoupling strategy. Compare this with static decoupling. Discuss the effect of controller tuning. For setpoint

changes of a reasonable magnitude, make certain that the manipulated input ranges are feasible.

11. Consider a process ("the shower problem") where a cold water stream at 15°C is mixed with a hot water stream of 50°C. The outputs are the total mixed stream flow rate and temperature. Assume steady-state flow rates of 0.19 and 0.13 liters/sec for the cold and hot streams respectively. The steady-state outputs are then 0.32 liters/sec and 29°C. Scale the inputs and outputs to cover appropriate ranges, find the scaled gain matrix and perform an SVD analysis. What is the condition number? What are the most sensitive input and output directions (provide a physical interpretation of this result)? For a rectangular input space where the cold water flow can range from 0 to 0.25 liters/second and the hot water flow can range from 0 to 0.19 liters/sec, construct the output operating window. Given those input flow limits, is it possible to operate over a mixed stream flow rate of 0.19 to 0.44 liters/sec, with a mixed stream temperature between 23 and 34°C?

Appendix 14.1

The nonlinear modeling equations for the quad tank process are presented by Johansson (2000).

$$\frac{dh_1}{dt} = -\frac{a_1}{A_1}\sqrt{2gh_1} + \frac{a_3}{A_1}\sqrt{2gh_3} + \frac{\gamma_1 k_1}{A_1} v_1$$

$$\frac{dh_2}{dt} = -\frac{a_2}{A_2}\sqrt{2gh_2} + \frac{a_4}{A_2}\sqrt{2gh_4} + \frac{\gamma_2 k_2}{A_2} v_2$$

$$\frac{dh_3}{dt} = -\frac{a_3}{A_3}\sqrt{2gh_3} + \frac{(1-\gamma_2)k_2}{A_3} v_2$$

$$\frac{dh_4}{dt} = -\frac{a_4}{A_4}\sqrt{2gh_4} + \frac{(1-\gamma_1)k_2}{A_4} v_1$$

where

A_i	=	cross-sectional area of tank i;
a_i	=	cross-sectional area of the outlet hole i;
h_i	=	water height of tank i;
k_i	=	pump coefficient i;
γ_i	=	flow splits;
g	=	gravitational constant

The parameter values are

$$
\begin{aligned}
A_1, A_3 &= 28 \text{ cm}^3 \\
A_2, A_4 &= 32 \text{ cm}^3 \\
a_1, a_3 &= 0.071 \text{ cm}^2 \\
a_2, a_4 &= 0.057 \text{ cm}^2 \\
\beta &= 0.5 \text{ V/cm} \\
g &= 981 \text{ cm/s}^2
\end{aligned}
$$

And the measured outputs are $y_i = \beta h_i$ (for i = 1,2). Note that the maximum tank height is 20 cm.

The following states, inputs and coefficients correspond to the two operating points:

Operating point 1

$$
\begin{aligned}
h_{1s}, h_{2s} &= 12.4, 12.7 \text{ cm} \\
h_{3s}, h_{4s} &= 1.8, 1.4 \text{ cm} \\
v_{1s}, v_{2s} &= 3.00, 3.00 \text{ cm} \\
k_1, k_2 &= 3.33, 3.35 \text{ cm}^3\text{/Vs} \\
\gamma_1, \gamma_2 &= 0.7, 0.6
\end{aligned}
$$

Operating point 2

$$
\begin{aligned}
h_{1s}, h_{2s} &= 12.6, 13.0 \text{ cm} \\
h_{3s}, h_{4s} &= 4.8, 4.9 \text{ cm} \\
v_{1s}, v_{2s} &= 3.15, 3.15 \text{ cm} \\
k_1, k_2 &= 3.14, 3.29 \text{ cm}^3\text{/Vs} \\
\gamma_1, \gamma_2 &= 0.43, 0.34
\end{aligned}
$$

CHAPTER 15

Plantwide Control

A process engineer is concerned about more than the control of a single loop or multiple loops associated with a single-unit operation. Of real concern is the operation of the entire plant. After studying this chapter, the reader should be able to do the following:

- Develop an appreciation for some of the complexities of plantwide control
- Understand the control strategies associated with equipment/unit operations that are part of a typical chemical process plant
- Understand the "snowball effect" and how to fix a stream flow in the recycle loop to avoid it
- Understand that recycle can significantly change the process dynamics
- Understand the control and optimization hierarchy—that important operation and control decisions at different corporate levels are made on different timescales

The major sections of this chapter are as follows:

15.1 Background

This textbook has emphasized controller design, primarily for SISO control loops. Much attention has been focused on control algorithms (such as PID or IMC) and tuning [adjusting the PID parameters, or the IMC "filter factor" (λ)]. Some attention has also been paid to control-loop interaction, particularly for two input–two output systems. Much of what a control or process engineer does, however, is not related to specific control algorithm and tuning details. A process engineer must be concerned about the operation of an entire process plant or operating unit, and not just individual control loops or even unit operations. The issues addressed by a process engineer are similar to those developed in Chapter 1, but on a much larger scale. At this point, the plant operating and control objectives are really inseparable, so we repeat from Chapter 1 the following topics on the development of a control strategy:

- What are the major operational objectives of the plant?
- What sensors should be paired with what manipulated inputs to form a control structure to achieve the major objectives?
- For any control valve or sensor failure, what steps will a process operator (and control system) need to take?

Much of this type of discussion occurs during the design [or "retrofit" (redesign)] stage of a process. Here, process engineers develop and review process flow sheets and process and instrumentation diagrams (P&ID). Typically, a detailed operational description of the proposed process and control strategy is developed, with consideration to all possible operating modes that can be selected by an operator, as well as to all equipment failure modes.

In this chapter, we cannot do justice to the scope of work that is needed to develop a control strategy and get an automation and control (including the plant!) strategy "up and running." We attempt to show some of the major issues in "plantwide control" by using some illustrative examples. In addition to showing examples of the many control loops required by even a simple process, we provide an overview of the various levels (hierarchy) involved in a chemical manufacturing organization.

We should note that the term *plantwide control* is somewhat of a misnomer. Process unit control is really a more proper term, since an entire chemical process manufacturing plant is usually composed of a large number of process units. Many of these process units contain reactors and separators with recycle. In addition, there are recycle flows between many of the process units, and the units draw from and contribute to many utility streams, such as steam and natural gas. Although process unit control is a better term, we will continue the tradition of using plantwide control as the term for regulation of a process unit.

15.2 Steady-State and Dynamic Effects of Recycle

A typical chemical process includes a chemical reactor to create products from reactants. Generally, the larger the reactor the greater the conversion of reactants to products; the main limit is imposed by equilibrium considerations. Rather than having a chemical reactor that is large enough to approach equilibrium conversion, it is usually more economical to have a smaller reactor, with a separator and a recycle stream to improve the overall conversion of reactants to products. If secondary reactions occur, or inert components can "build up," it is common to include a purge stream. Recycle systems are a source of challenging control problems, including the so-called "snowball" effect (sensitivity of recycle flow rates to disturbances) and the potentially long-time-scale dynamics, although the individual unit operations have reasonably short time scales.

Steady-State Effects of Recycle

To illustrate recycle control system issues and challenges, we begin with the simple example shown in Figure 15–1. The goal of this process is to convert the feed component A to product B. The make-up feedstream (component A) mixes with the recycle stream (also primarily component A) and is fed to the isothermal CSTR. A first-order reaction occurs; the reaction is incomplete, so the reactor product stream (which is fed to the distillation column) contains both components A and B. The distillation column separates this feedstream into two relatively pure product streams. The distillate product, containing primarily A, is recycled; the bottoms product contains primarily component B and is sent to storage for sale to a customer. Clearly, the bottoms product must meet certain specifications; here we assume a composition specification.

Since you have been focused primarily on SISO control systems, the control system may seem somewhat complex to you. The make-up feedstream is under flow control, and mixes with the recycle stream; this fixes the flow to the isothermal CSTR. The CSTR level controller manipulates the flow out of the CSTR to the distillation column. Now, consider the distillation column; see Module 13 for a more detailed discussion of distillation column control. A standard two-product column has a minimum of 3 controlled outputs (pressure, distillate receiver level, and bottoms level) and 5 manipulated inputs (cooling water flow, distillate product flow, bottoms flow, reflux flow, and steam-to-reboiler). In addition, the feed flow rate is an input that may either be a disturbance or manipulated input, depending on the control strategy. For this particular flow sheet we use the control variable pairing known as "conventional dual composition control," where the overhead (distillate) and bottoms products are measured and controlled. Each of the measured outputs has been paired with a manipulated input; there are no steady-state degrees of freedom. That is, if the column feed flow rate and composition are known, all other flows and compositions in the column can be calculated. This allows us to take a simplified view shown in Figure 15–2; it should be clear that, if the overhead and bottoms product stream compositions are fixed, then the overall material balance and the component balance on A can be used to solve for the distillate (recycle) and bottoms product flowrates.

Example 15.1 is used to illustrate an inherently steady-state problem (the "snowball effect") that exists with the proposed control strategy. This is followed by Example 15.2, which illustrates how recycle systems can appreciably change the dynamic characteristics of a process.

Example 15.1: A Simple Recycle Problem

Consider a single, first-order reaction

$$A \rightarrow B$$

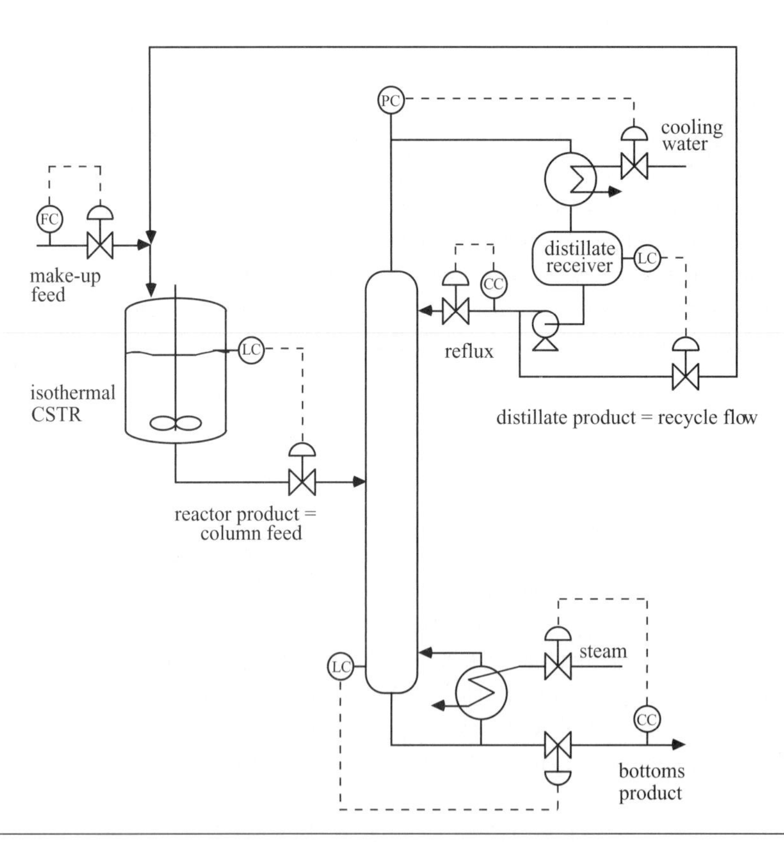

Figure 15–1 Recycle flow control and instrumentation diagram.

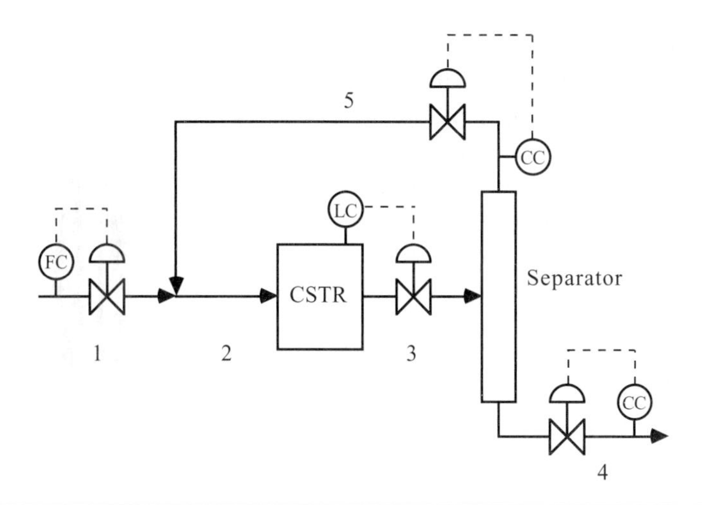

Figure 15–2 Simplified recycle flow control and instrumentation diagram. Only the loops directly affecting the external flows from the separator are shown.

and an idealized flow sheet composed of a CSTR and a "perfect separator." An example is shown in Figure 15–3; a basis of 100 moles/hour of the feedstream (stream 1) is used and the stream component flow rates are shown directly on the flow sheet. A reactor volume of 200 moles and a reaction rate constant of 0.6 hr^{-1} are assumed; this results in a 16% conversion of reactant A to product B in the CSTR. In addition, the separator perfectly separates component A from B. Because of the relatively low conversion, a 5:1 recycle/feed stream ratio is required.

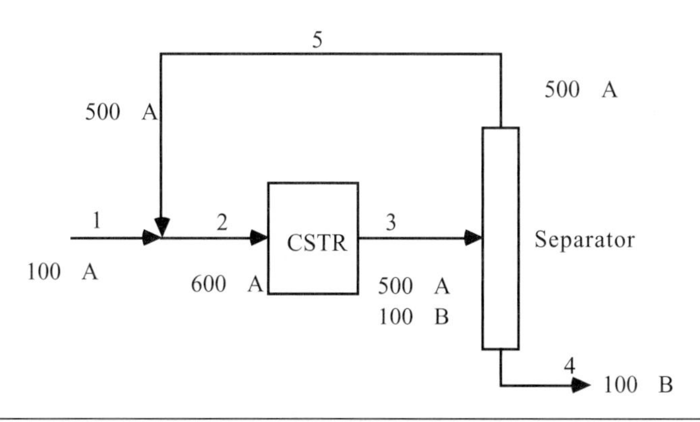

Figure 15–3 Idealized flow sheet example with reaction, separation, and recycle.

A surprising result is obtained if the fresh feed is increased from 100 moles/hour to 110 moles/hour, while maintaining the reactor volume at a constant level and assuming a perfect separator. The new material balance is shown in Figure 15–4.

Notice the tremendous change in the recycle flow rate. A 10% *increase* in feed flow rate has resulted in more than doubling the recycle flow rate (from 500 to 1210 moles/hr)! Similarly consider a 10% *decrease* in feed flow rate, which results in the stream flow rates shown in Figure 15–5. A 10 moles/time decrease (from the nominal amount of 100 from Figure 15–3) has led to a decrease in the recycle stream flow rate from 500 to 230 moles/hr. You are encouraged to work Exercise 1 to verify these results.

This extreme sensitivity to small disturbances, illustrated by Example 15.1, has been termed the "snowball effect" by Luyben (see *Plantwide Process Control* by Luyben et al., 1999). To minimize this effect, Luyben et al. suggest that one of the flows within the

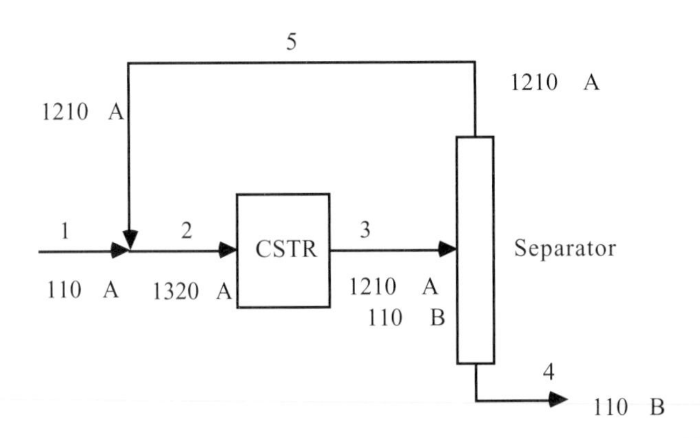

Figure 15–4 Idealized flow sheet example with a change in feed flow rate of 10%.

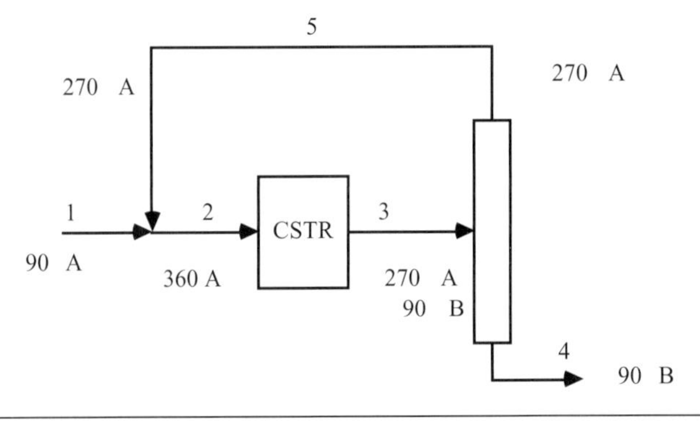

Figure 15–5 Idealized flow sheet example with a change in feed flow rate of -10%.

recycle system be fixed. For example, consider the strategy shown in Figure 15–6. Here, the flow rate of the mixed stream to the CSTR is maintained at a constant value by manipulating the feedstream flow. The more complete control instrumentation diagram is shown in Figure 15–7. Another alternative is to fix the flow rate of stream 3 (between the CSTR and column) by using a flow controller on stream 3, and to control the CSTR volume by manipulating the fresh make-up (stream 1) flow rate (see Exercise 2).

In this section we have covered a relatively simple process with recycle, where there were 7 control loops. A number of more complex flow sheet examples are shown later, in Section 15.5.

Dynamic Effects of Recycle

It is known that recycle systems can exhibit dynamic behavior that is significantly different from the equivalent process components without a recycle. The reason for this can be found by analyzing the simple, positive feedback block diagram in Figure 15–8; here, $g_F(s)$ and $g_R(s)$ represent the forward and recycle processes, respectively. Notice that the difference between this diagram and a standard feedback-control diagram is that recycle processes exhibit *positive feedback*. Recall that standard feedback control is also known as *negative feedback*.

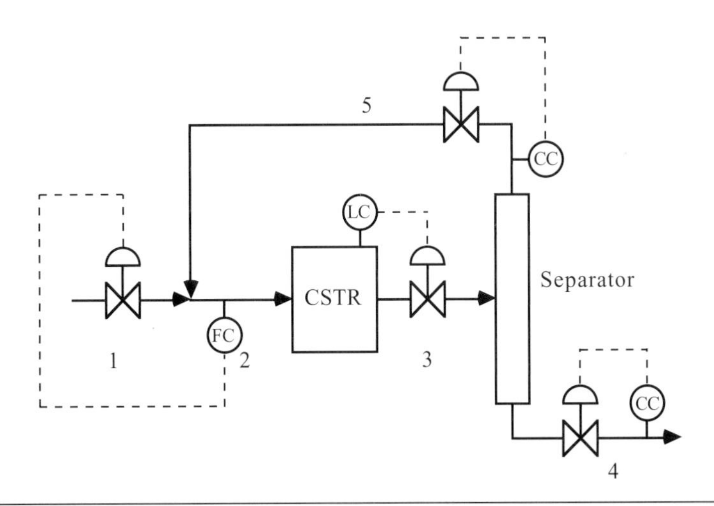

Figure 15–6 Simplified recycle flow control and instrumentation diagram, with the reactor feed flow rate (stream 2) fixed to minimize the "snowball effect."

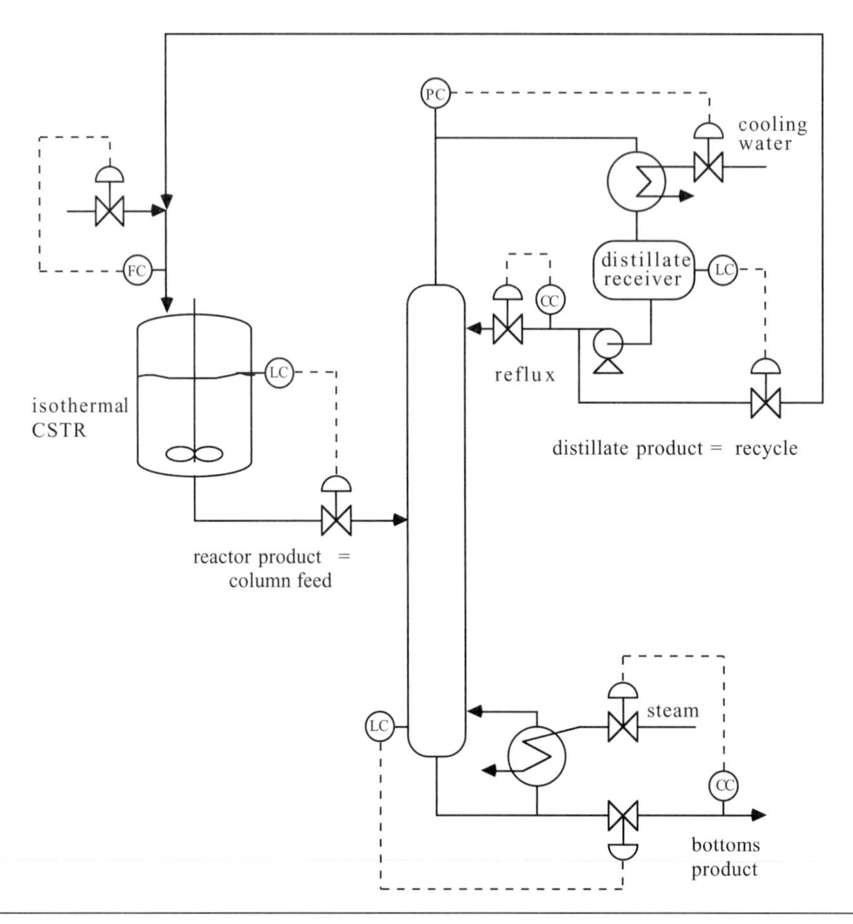

Figure 15–7 Recycle flow control and instrumentation diagram. The reactor feed flow is fixed to minimize the "snowball effect."

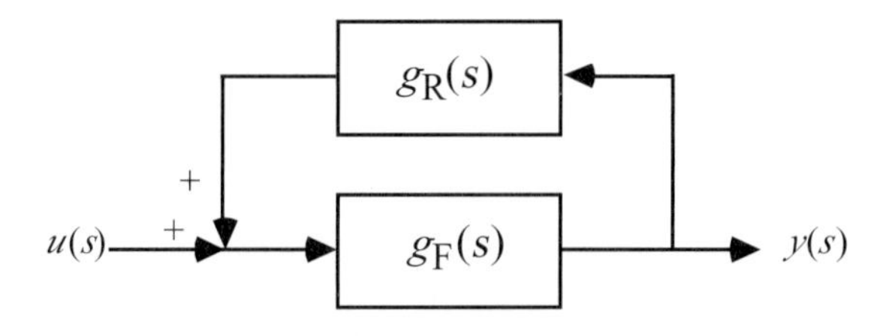

Figure 15–8 Illustration of recycle as positive feedback.

The reader should show that the closed-loop transfer function relationship between an input (u) and the output (y) is

$$y(s) = \frac{g_F(s)}{1 - g_F(s)g_R(s)} u(s) \tag{15.1}$$

It is easy to show that the closed-loop dynamics can be substantially longer than the individual time constants. This is an interesting result because it indicates that individual unit operations in a recycle system may be designed for relatively fast behavior, yet the dominant timescale of the entire recycle system can be substantially longer. This is illustrated by the following example.

Example 15.2: Dynamic Effect of Recycle

To illustrate the positive feedback effect of recycle, we use the following first-order transfer function to represent the forward process, where the time unit is minutes

$$g_F(s) = \frac{1}{s+1} \tag{15.2}$$

We also neglect any dynamic behavior of the recycle process and assume that it can be characterized as a constant gain (k_R) between 0 and 1. The resulting input-output relationship is [from Equations (15.1) and (15.2)]

$$y(s) = \frac{\left(\frac{1}{1-k_R}\right)}{\left(\frac{1}{1-k_R}\right)s + 1} u(s) \tag{15.3}$$

Notice that, as the recycle gain approaches 1, the effective gain and time constant increase dramatically. This effect is also shown in Figure 15–9, where a unit step input has been applied at t = 0 min. The response of the forward path ($k_R = 0$) is rapid with a magnitude of 1, while larger recycle gains cause slower responses with larger magnitude. This is an interesting result, since a recycle process with no dynamics has caused a tremendous increase in the time constant of the overall (forward + recycle path) system. For $k_R = 0.9$ with a forward time constant of 1 minute, the overall dynamic time constant is 10 minutes. In Example 15.6 in Section 15.6 a physically based recycle system with similar changes in time constant are shown.

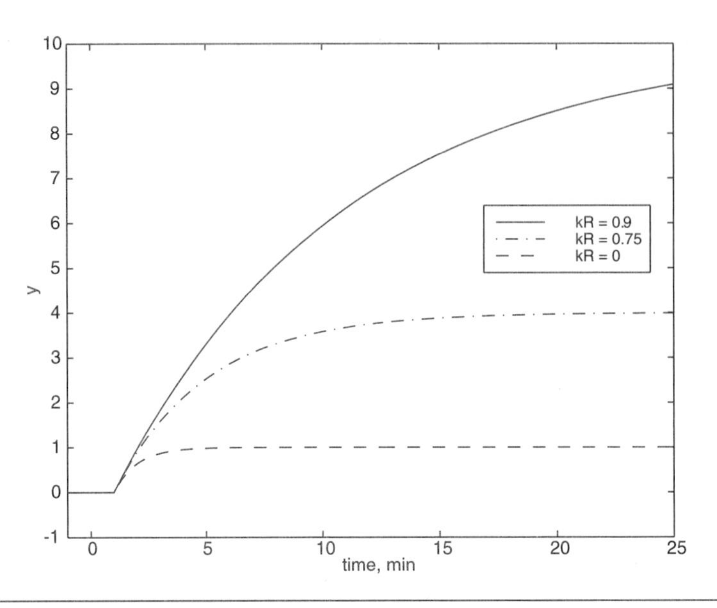

Figure 15–9 Effect of recycle gain on response to a step input on a first-order process.

15.3 Unit Operations Not Previously Covered

In this text, we have covered the control of a few unit operations but have not discussed control structure determination in detail. In this section, we first discuss the two major structures for maintaining material balance control, and then we discuss the control of some typical unit operations that occur in a process flow sheet.

Supply-Side vs. Demand-Side Control of Production

Material balance control can be accomplished using two approaches: supply side (direction of flow) and demand side (direction opposite to flow). A simple surge tank level example is shown in Figure 15–10. The level controller manipulates the flow out of the vessel in supply-side control, while the level controller manipulates the feed flow to the vessel in demand-side control.

Using supply-side control, a change in the flow rate setpoint propagates downstream as a disturbance to all units that follow. In demand-side control, a flow rate setpoint change propagates upstream to all units that precede the unit. Luyben (1999) has studied demand-side control structures for complex flow sheets and finds that they often add additional lags and create control problems, particularly when distillation columns are involved. This causes greater product variability, so the supply-side structure is generally recommended.

Compressor Control

It is more economical to condense a vapor stream and pump it than to vaporize a liquid stream and compress it. Gas compressors are often used on recycle streams composed of light components (hydrogen, methane, etc.). For these systems, it is not economical to condense the streams and pump them as liquids.

Compressors that operate at a constant speed are usually controlled using one of the strategies shown in Figure 15–11. Usually the "pinch" method will have lower operating costs than the "spillback" method. Variable speed control methods are shown in Figure 15–12 for an electric motor drive and for a steam turbine drive. The operating costs of variable speed compressors are generally lower than for fixed speed systems, with perhaps a slightly higher initial capital cost.

When compressors are used on recycle streams, Douglas (1988) generally recommends that they be operated "wide open"; that is, at a constant maximum flow rate. This is because the economic return from the increased overall yield usually outweighs the incremental operating cost of the compressor.

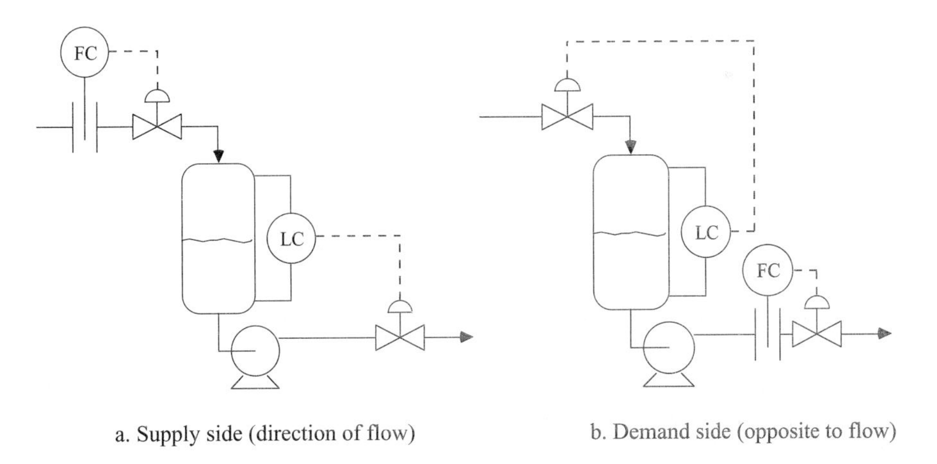

Figure 15–10 Control in direction of flow (supply side) vs. direction opposite to flow (demand side).

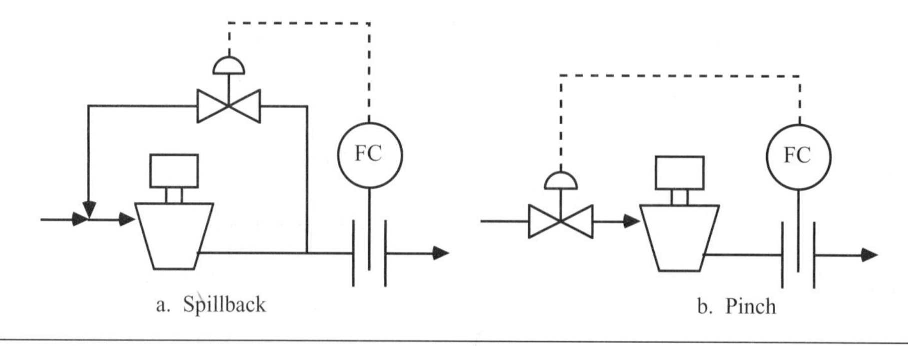

Figure 15–11 Constant speed control systems.

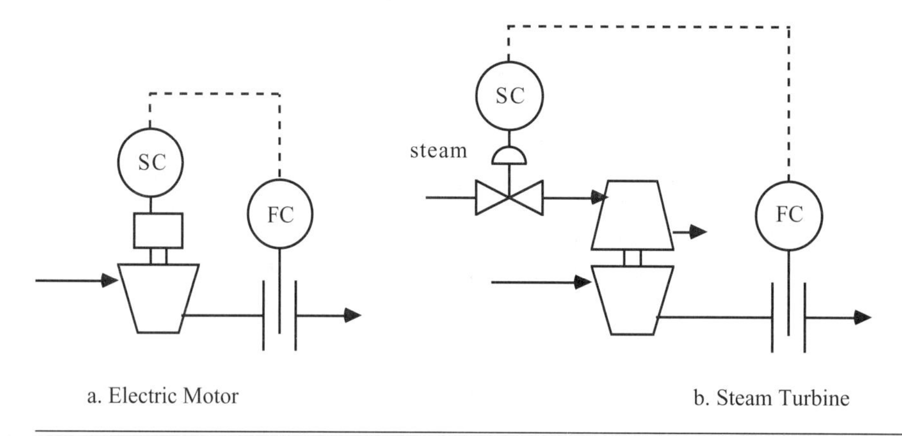

Figure 15–12 Variable speed control systems.

Heat Exchangers

Heat exchangers are used in a large number of temperature-control services, so here we focus on the regulation of the exit temperature of a cold process stream that is exchanging heat with a hot process stream. A common application would be a feed/effluent exchanger for a chemical reactor. Three possible strategies are shown in Figure 15–13. The preferred method will often be that shown in Figure 15–13c, since the dynamics will be much faster than for the other two strategies shown. A change in the cold bypass flow rate has an almost immediate effect on the cold stream outlet temperature.

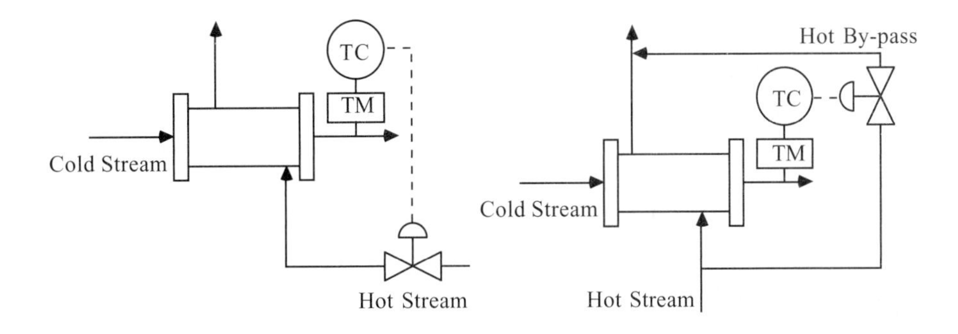

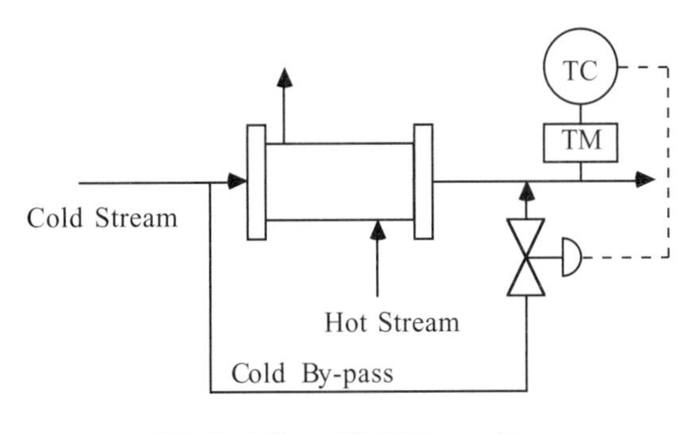

Figure 15–13 Alternative temperature control strategies for a heat exchanger.

Adiabatic Plug Flow Reactors

It is important to control the inlet temperature of adiabatic plug-flow reactors, since there is no method to add/remove energy from the reactor. For an exothermic reaction with a substantial heat effect, one might consider using the heat from the reactor effluent to preheat the feed to the reactor, as shown in Figure 15–14a. A problem with this strategy is that it is not very flexible; if the bypass valve saturates (becomes fully closed, for example), there is a loss of temperature control. Also, additional equipment will usually need to be added to start up the system. The alternative shown in Figure 15–14b leads to a more flexible design. Particularly at start up, the furnace allows the cold feed to be heated to a high enough temperature to initiate the reaction. The sacrifice is higher energy consumption; the relative increase depends on the steady-state ratio of the feed/effluent exchanger and furnace duties.

15.4 The Control and Optimization Hierarchy

Operating Levels

The actual operation of a chemical manufacturing company involves decisions and actions that occur at a number of levels and timescales, as shown in Figure 15–15. At the *corporate* level, allocation decisions are made. Market demand projections, raw material availability, and past operating costs are used to set long-term plans for corporate-wide optimization. These decisions are typically made on an infrequent basis—weekly, monthly, or quarterly. Often these decisions are made using linear programming (LP) techniques. This information is passed to the *plant* level.

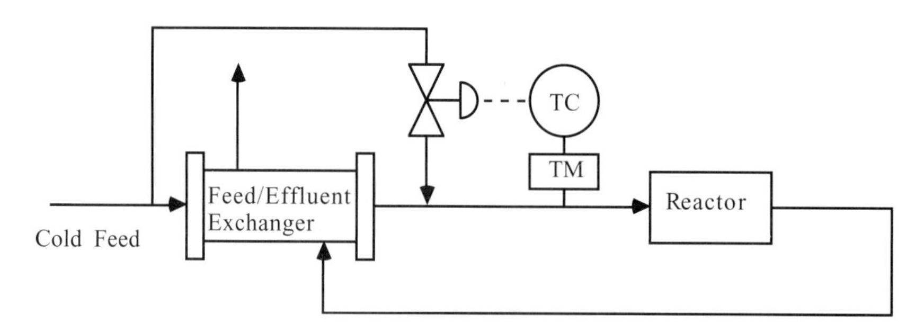

a. Reactor with Feed/Effluent Exchanger

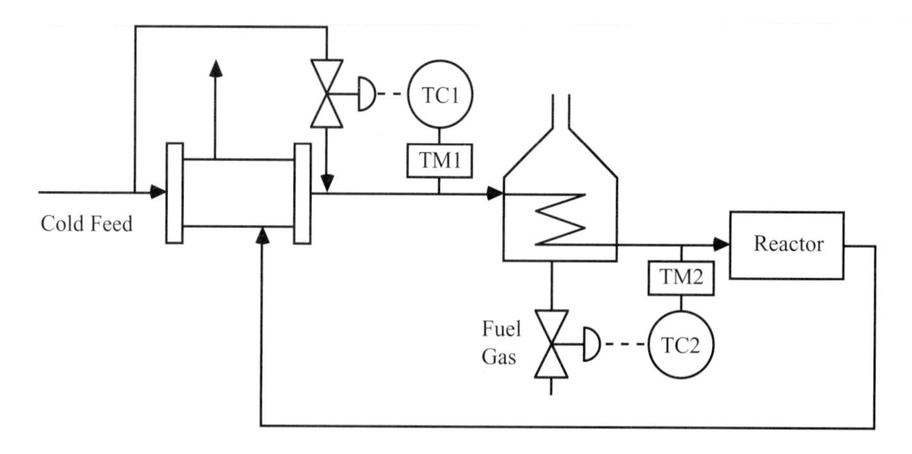

b. Reactor with Feed/Effluent Exchanger and a Furnace

Figure 15–14 Temperature control for an adiabatic plug flow reactor.

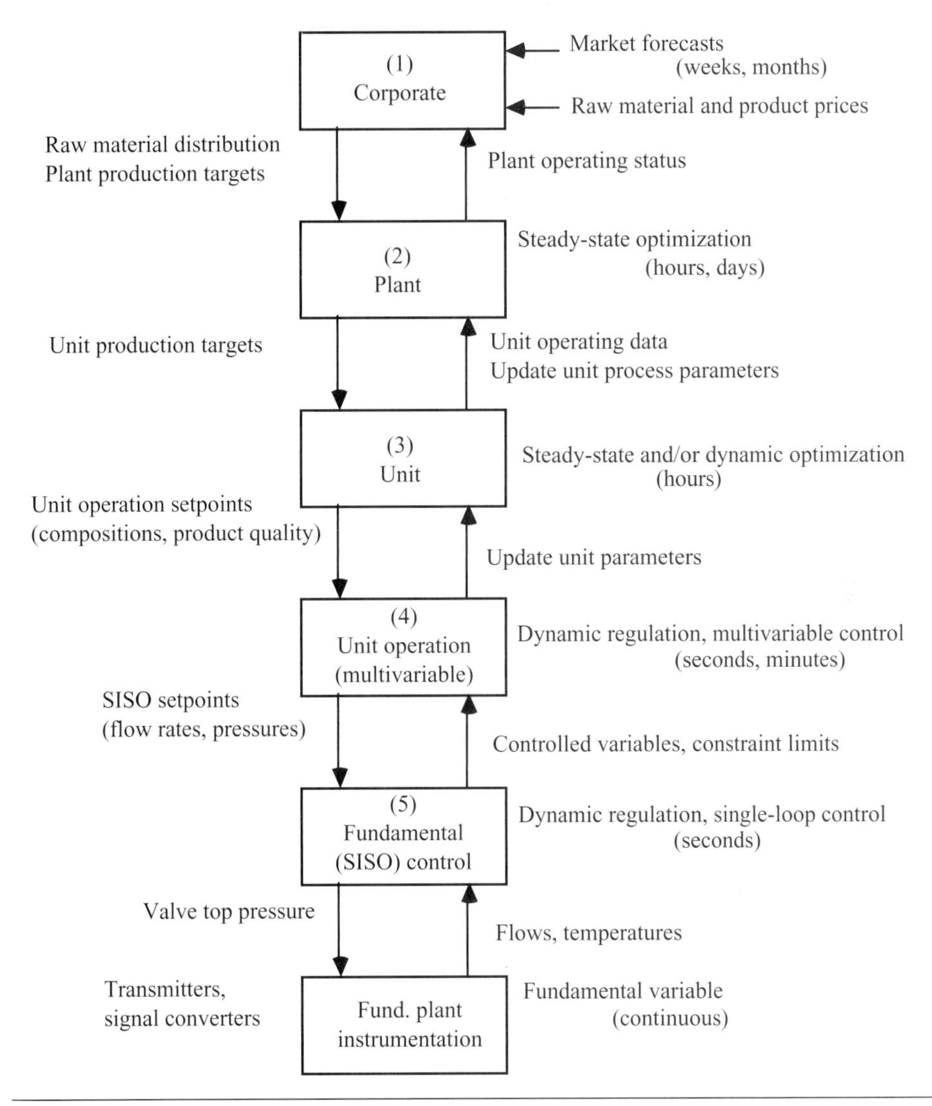

Figure 15–15 Corporate operating levels.

At the next level are plant decisions about allocation among various plant operating units. These may be made daily, or even more frequently, depending on the plant automation and control system. For example, steam may be generated by a number of different sources, with different costs, depending on relative energy and product values. In an oil refinery, different units make similar components for blending into gasoline (which is produced in many different grades), and these relative amounts may change depending on energy and product values, operating problems with certain units, and so forth.

The *operating unit* level takes information available on a daily basis and over the course of the day makes hour-by-hour changes, on average, to meet those goals. The *unit operation* level involves individual equipment, and desired setpoints may be changed on an hourly (or more frequent) basis. The regulation on this level occurs on the order of minutes.

The *fundamental control* level consists primarily of process flow or pressure controllers. The lowest level generally involves the manipulation of a control valve, or the reading of a sensor, where the time frame is on the order of 1 second or less. Notice that there is a feedback of information from the lower levels back up to the higher levels. The basic idea of feedback control occurs at the unit operation and fundamental control levels. Here, control algorithms are used to adjust manipulated inputs (controller outputs or control variables) to maintain process outputs (process variables) at desired values (setpoints).

Petroleum Refining Example

An example of the overall operating problem is illustrated in Figure 15–16, for a typical petroleum refining company. At the highest level, corporate management decides where to purchase crude and how to distribute the crude oil to the various refineries in the corporation. At the next level (level 2), refinery A takes the current and future crude delivery projections and gasoline production projections and determines the operating conditions for each process unit (level 3, e.g., the catalytic cracking unit) in the refinery. Setpoints for the unit operations (level 4, e.g., distillation) are determined at this point. The unit operations level determines the process flow rates, such as the distillate or reflux flow rates (level 5). These controllers then determine, for example, the pressure to the control valves to regulate various flow rates.

Discussion

For many plants, much of the feedback provided between different levels is performed manually. During the past decade, major corporations have moved toward automating more of these functions, and better integrating the decisions that are made at different levels. The broad term for this, at the plant level, is computer-integrated manufacturing (CIM). Developments in information technology and database management are making this task easier. Corporate-wide technology is known as supply chain management or enterprise optimization. This software links databases from various suppliers (of raw materials) with corporate scheduling, and with customer orders (of products).

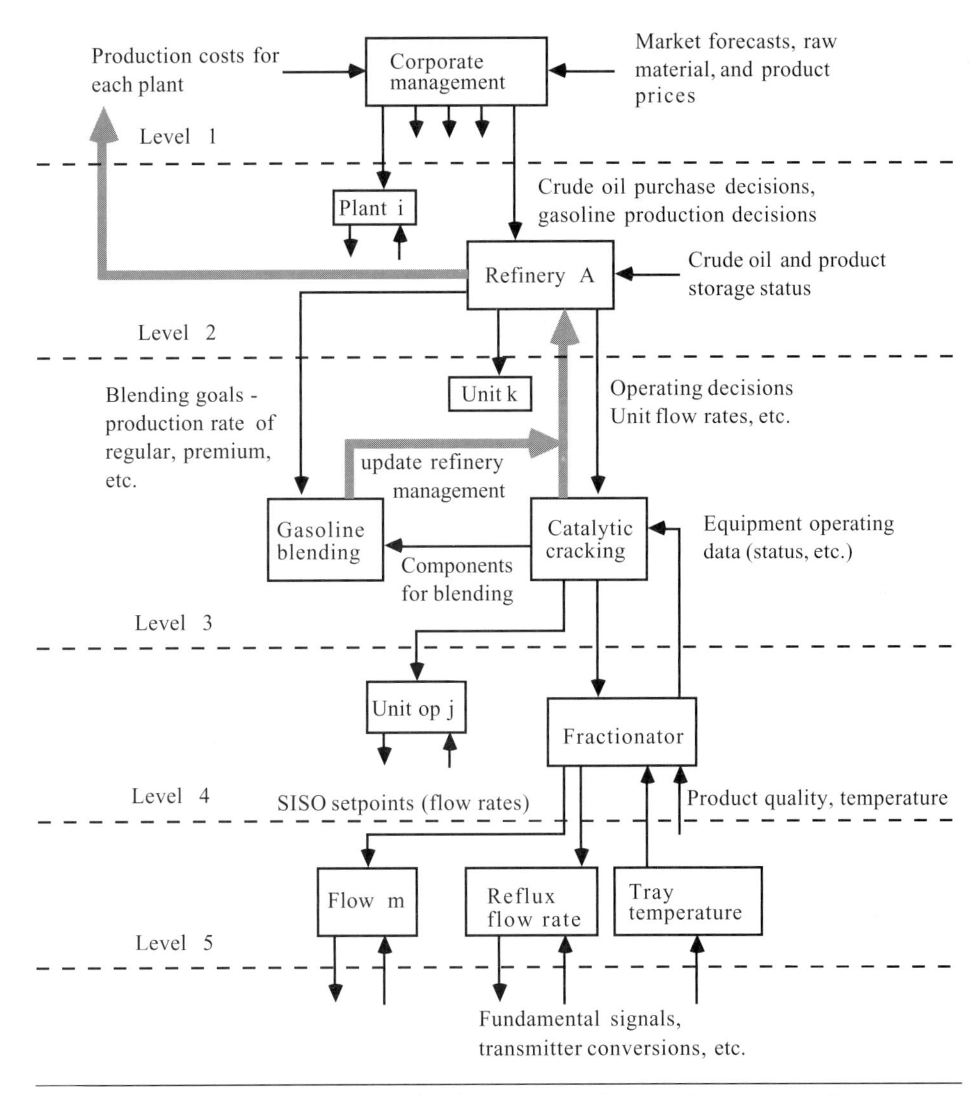

Figure 15–16 Corporate operating levels: petroleum refining example.

15.5 Further Plantwide Control Examples

Examples and challenges of control of processes with recycle were presented in Section 15.2. In this section we give the reader more experience with plantwide control strategies by studying two more examples: MPN and HDA.

Example 15.3: MPN Process and Instrumentation Diagram

Consider the process flow sheet shown in Figure 15–17. Here, methanol is reacted with acrylonitrile to form MPN. A slight excess (2%) of ACN is supplied and essentially all of the MeOH is reacted. There is no purge stream, so the excess ACN appears in the MPN product stream. Here the recycle stream is liquid-phase distillate stream.

The objective is to produce MPN at a specified rate, while satisfying product purity constraints. In Figure 15–17, there are nine control valves, representing nine degrees of freedom. Also shown are 13 measurements: seven flow, three level, two temperature, and one pressure. Clearly, there are not enough degrees of freedom to specify a setpoint value for every measured variable. Inventory variables, such as level and pressure, must be controlled; without feedback, these integrating variables could easily violate alarm limits. This means that four control valves must be used to control the three levels and the column pressure.

We consider it an important exercise for the reader to attempt to place nine control loops on Figure 15–17. Please do this before reading any further.

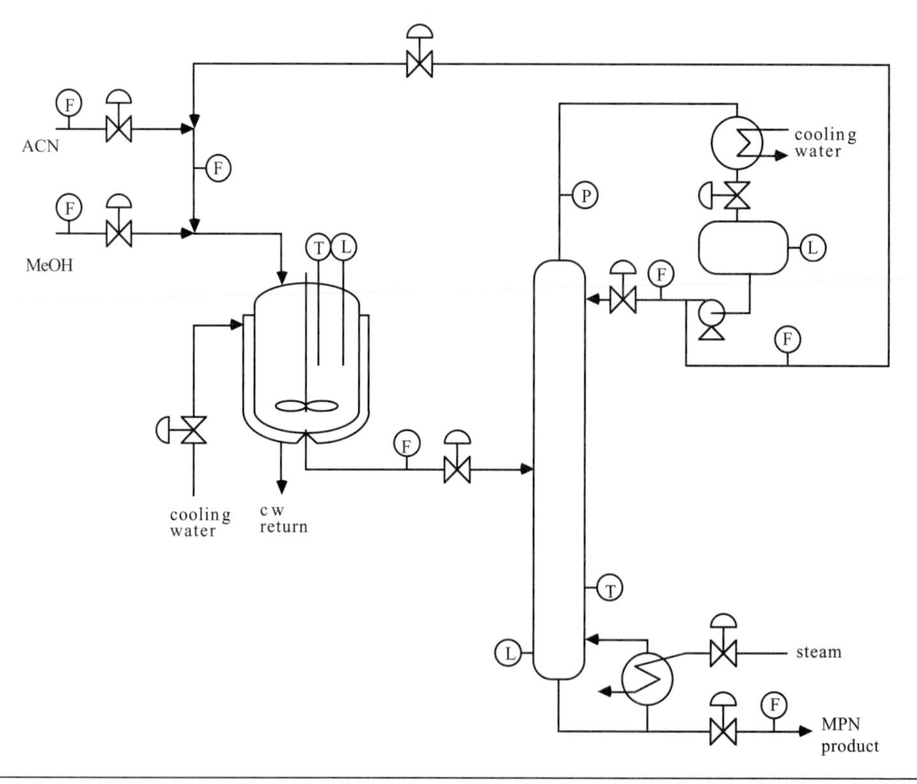

Figure 15–17 Preliminary instrumentation diagram for an MPN process (adapted from Zhan and Grassi, 1999).

A Possible Solution to Example 15.3

One possible solution is shown in Figure 15–18.

1. The production rate objective should be satisfied by the methanol flow rate controller, since there is essentially total conversion of the methanol to MPN.
2. The excess ACN is obtained by specifying the ACN flow rate to be 2% higher than the methanol flow rate.
3. Exothermic reactor temperatures must be controlled to assure stable reactor operation. Here, the reactor temperature is controlled by manipulating the cooling water valve.
4. The reactor level (related to residence time) is controlled by manipulating the reactor outlet flow valve. Notice that there are now only five more loops that can be specified, since there are a total of nine control valves. The final five loops are

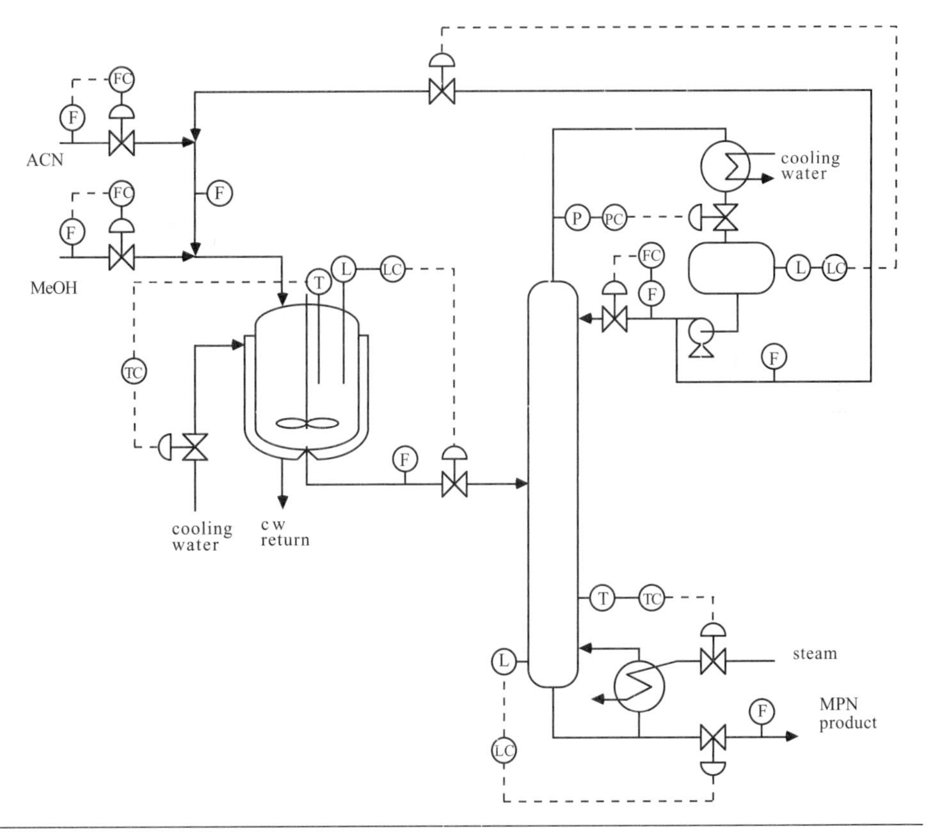

Figure 15–18 A typical P&ID for the MPN process.

associated with the distillation column; three of these are associated with inventory variables (distillate receiver level, bottoms level, column pressure).

5. A temperature in the bottom of the column manipulates the steam control valve. Because of equilibrium relationships, the temperature is related to the product purity.
6. The reflux stream is simply flow controlled. An alternative would be to control a tray temperature in the top of the column; this would be related to controlling the distillate purity. Since the distillate stream is recycled to the reactor, it is not considered important to maintain the stream composition at a given value.
7. The distillate receiver level is controlled by manipulating the distillate (or recycle) flow valve.
8. The column pressure is controlled by manipulating the condensate flow. This essentially changes the level of condensate in the overhead condenser, changing the effective heat transfer area, and therefore the condenser duty.
9. The column bottom level is maintained by manipulating the bottom control valve.

Problems with This Solution

The main problem with this solution is that the control structure exhibits a high sensitivity in the recycle flow rate to small disturbances. This so-called *snowball effect* was discussed in Section 15.2. Luyben et al. (1999) have recommended that at least one control loop in the recycle flow path be flow controlled in order to minimize the snowball effect.

Alternative Solution 1

An alternative solution [designated plantwide control structure 1 by Zhan and Grassi (1999)] is shown in Figure 15–19. Here, a flow controller is used to regulate the mixed stream (recycle + ACN) flow rate by manipulating the distillate (recycle) flow rate; this allows the control structure to minimize the snowball effect. The distillate level controller manipulates the fresh ACN flow control valve. The other seven loops are identical to the previous solution shown in Figure 15–18.

Other Alternative Solutions

There are some alternative solutions that include flow control of a recycle loop stream. For example (alternative 2), the reactor outlet stream can be flow controlled. The reactor level can then be controlled by manipulating the ACN feed-flow valve (see Exercise 3). In addition, a reactor concentration sensor can be used to improve control (again, see Exercise 3).

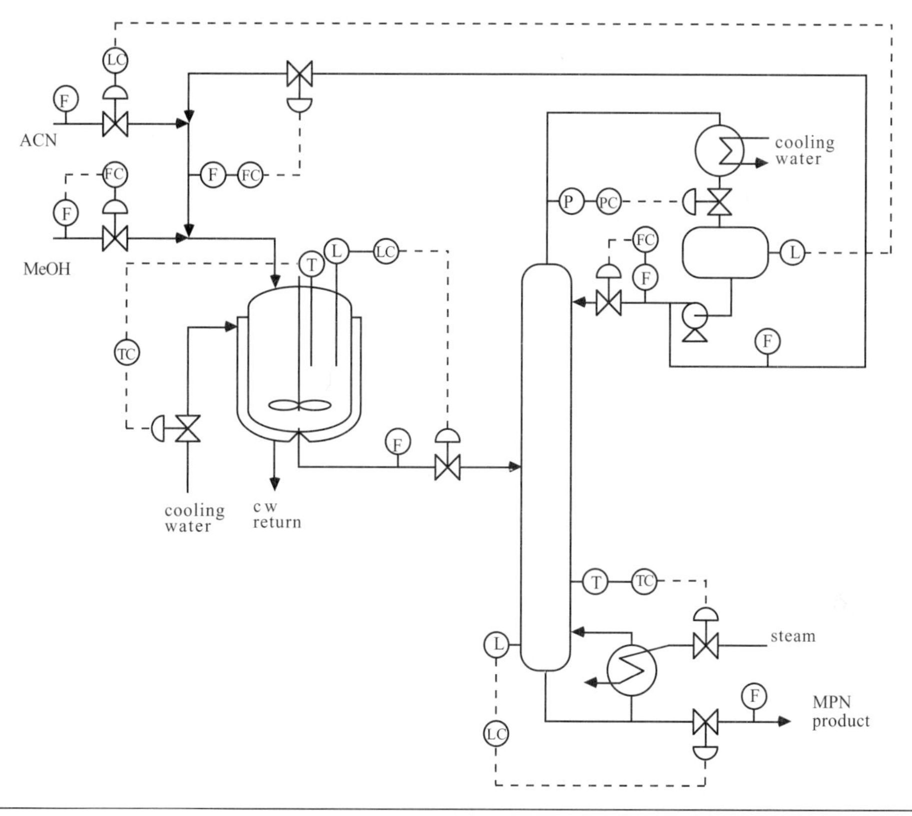

Figure 15–19 Plantwide control structure 1 developed by Zhan and Grassi (1999) for the MPN process.

Example 15.4: HDA Process

Toluene hydrodealkylation (HDA) is one of the primary processes for the production of benzene. A simplified process and instrumentation diagram for an HDA process is shown in Figure 15–20. For clarity, we have omitted a number of details, such as pumps on distillation reflux streams and furnace combustion air dampers.

The feed streams are high purity (99.98 mol%) toluene and hydrogen (96 mol% hydrogen, 4 mol% methane). The process operating objective is to produce a desired rate of benzene at a purity of 99.98 mol%. Because of coking considerations, a 5:1 hydrogen/aromatics ratio must be maintained at the reactor entrance. The reactor inlet pressure is to be maintained at just under 500 psig (the pressure of the hydrogen feed stream); the pressure drop between the feed stream mixing point and the flash drum is roughly 35 psi. The minimum reactor inlet temperature is 1150°F, while the maximum outlet temperature

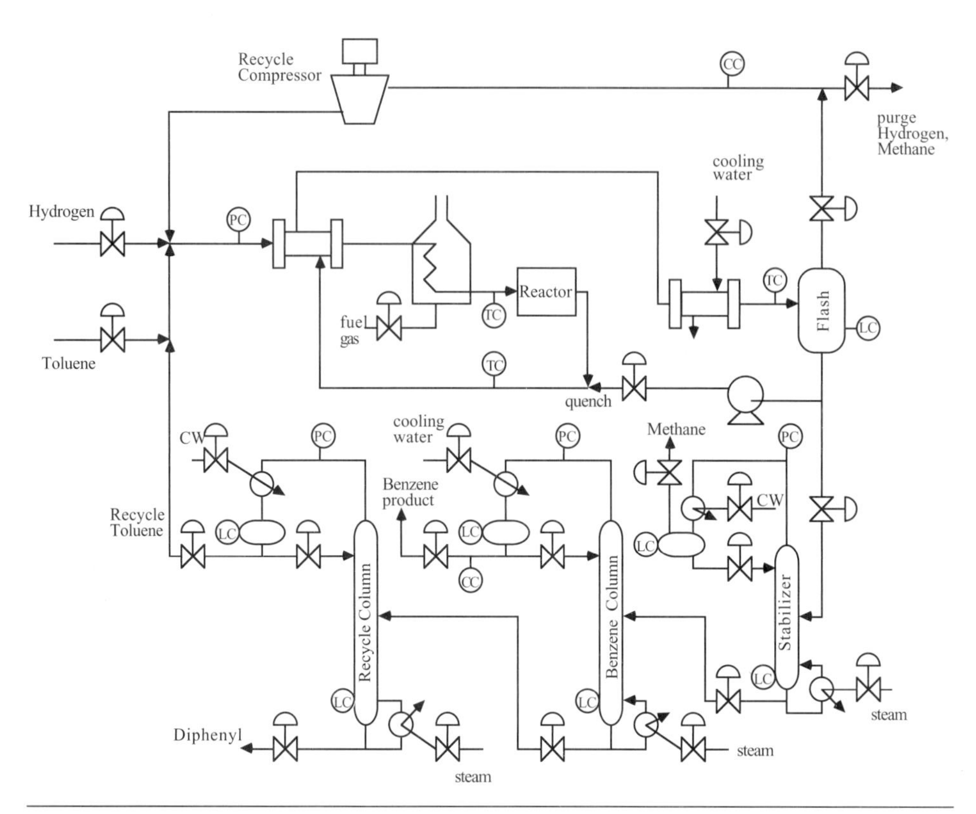

Figure 15–20 Preliminary P&ID for an HDA process.

is 1300°F. The reactor exit stream must be immediately quenched to 1150°F to minimize secondary reactions.

The primary reaction is an irreversible reaction of toluene and hydrogen to produce benzene and methane, while the secondary reaction is the reversible reaction of benzene to form diphenyl and hydrogen

$$C_7H_8 + H_2 \rightarrow C_6H_6 + CH_4$$
$$2C_6H_6 \leftrightarrow C_{12}H_{10} + H_2 \tag{15.4}$$

The selectivity decreases as the temperature increases. It is clear that a purge stream is necessary to eliminate methane from the process system.

We consider it an important exercise for the student to attempt to place control loops on Figure 15–20. *Please do this before analyzing the suggested solution shown in Figure 15–21.* Use the following suggestions to guide your control structure development.

- Luyben et al. (1999) recommend placing a flow control loop on a stream in the recycle path. For this process, they suggest placing the recycle toluene stream on flow control.
- Douglas (1988) recommends operating recycle compressors "wide open" because the value of recovered components is usually higher than the additional compressor power cost.
- A tray temperature in each column can be used to "infer" a product composition. Dual composition control for the columns is not necessary, and this petrochemical plant has had good success by placing reflux streams under flow control.

Notice that there are 23 control degrees of freedom, since there are 23 control valves. It is not necessary that all these valves be used for feedback control; one or more valves may be held at a constant valve position.

A Possible Solution to Example 15.4

One possible solution to the HDA problem is shown in Figure 15–21. This solution was developed by Luyben et al. (1999). The important control structure decisions to note are as follows:

- The total toluene flow (recycle + feed) is set to prevent the *snowball effect* (recall that it is recommended that one loop in a recycle system be placed on flow control). Notice that this flow also sets the production rate of benzene.
- The makeup (feed) toluene flow is manipulated by the recycle column distillate receiver level controller. Your intuition may suggest that there would be a significant lag between the feed flow, through the various unit operations, to the final distillation column. This is not the case, however, since the total toluene flow is regulated. Any change in toluene feed has an immediate effect on the recycle toluene flow and, hence, the distillate receiver level.
- The overhead valve from the flash drum is set wide open to allow maximum flow through the recycle compressor. The flash drum operating pressure is then dictated by pressure drop through the system, since the pressure is controlled at the reactor entrance.
- Since the diphenyl flow rate from the recycle column is so low, the level is controlled by the reboiler heat duty (steam to the reboiler).
- The benzene purity is maintained using a cascade strategy, where the output of the composition controller is the setpoint to a tray temperature controller. This strategy yields a faster rejection of feed disturbances and provides satisfactory composition control when the composition sensor fails or needs recalibration.

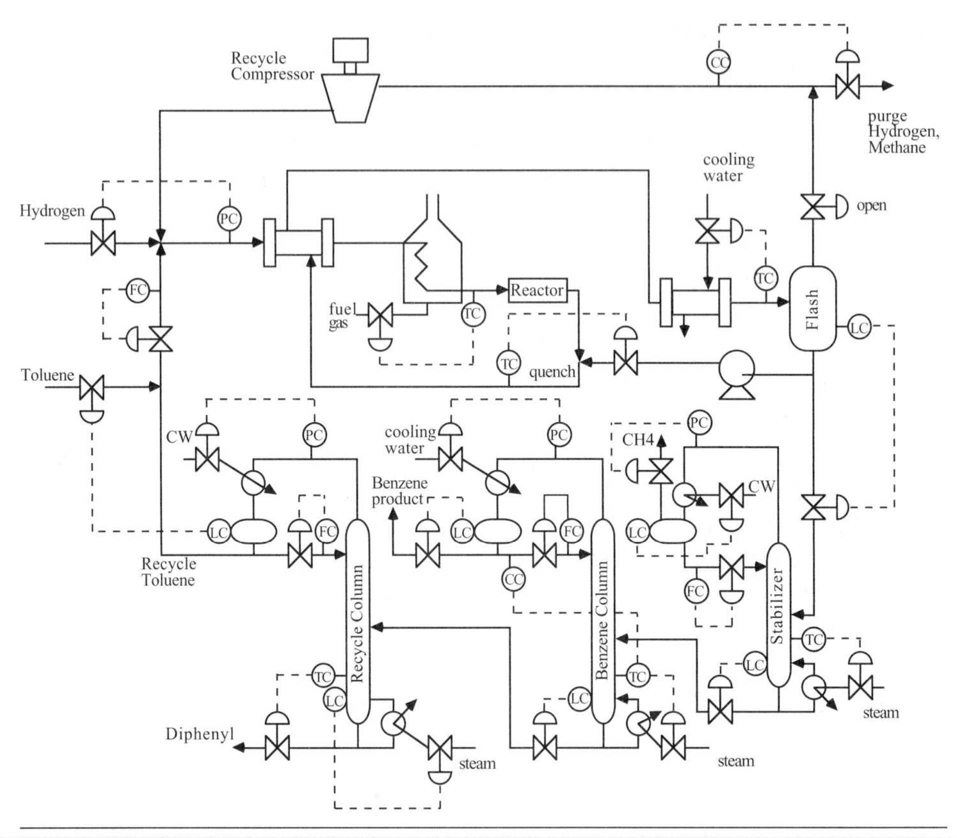

Figure 15–21 Plantwide control structure for an HDA process, as suggested by Luyben et al. (1999).

It should also be noted that only the basic control strategy is presented. In practice, ratio, feed forward, and cascade control will be added for improved control. Also, the control techniques presented in Section 15.3 can be used.

The Art of Process Engineering

At this point, you may be overwhelmed by the complexity of a plantwide control problem. Please realize that placing controllers on a process flow sheet is an "art," and different process engineers will often develop different control structures for the same process flow sheet. In practice, you will not develop these control strategies in isolation. At the process design (or retrofit) stage, a large number of engineers will be involved in the development of a control structure; this is truly a team effort, involving participants from the plant, the engineering design firm, and other contractors involved in automation. An operational description of the process automation system will evolve over many months of meetings. These discussions will lead to numerous iterations of the control structure. Experience during and after plant or control-system startup will normally lead to changes in this structure.

Although computer simulation packages are widely available, a process engineer often gains insight about operating and design decisions using simple "back of the envelope" calculations, as shown in the next example.

Example 15.5: HDA "Back of the Envelope" Material Balance

Much insight about the effect of design decisions can be obtained by rough "back of the envelope" calculations. Here we consider the HDA flow sheet shown in Figure 15–22. The following assumptions are made to simplify the analysis: (1) only the primary reaction (toluene + hydrogen = benzene + methane) occurs, (2) a single purge stream of hydrogen and methane, (3) a pure benzene product stream, (4) pure toluene product and recycle streams.

A basis of 1 mole of toluene make-up (stream 1) is used. Here we use n_i to represent a molar flow rate, where the subscript indicates the stream number. The reader should derive the following relationships (see Student Exercise 4).

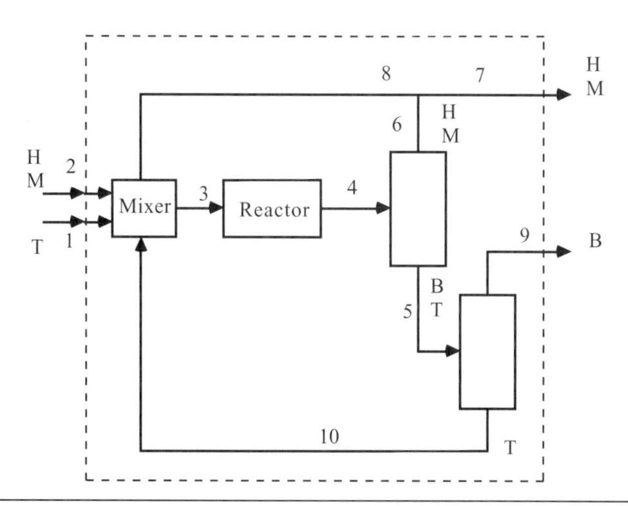

Figure 15–22 "Back of the envelope" material balance for the simplified HDA problem.

Toluene recycle (stream 10) flow: $n_{10} = (1-X_T)/X_T$, where X_T is the single-pass conversion of toluene in the reactor.

Hydrogen make-up (stream 2) flow: $n_2 = 1/(0.96-y_{7,H})$, where $y_{7,H}$ is the mole fraction of H_2 in the purge stream (stream 7).

Hydrogen recycle (stream 8) flow:

$$n_8 = \frac{1}{y_{7,H}} \cdot \left[\frac{5}{X_T} - \frac{0.96}{0.96 - y_{7,H}} \right] \tag{15.5}$$

Two of the major operating costs are the hydrogen make-up (raw material) and vapor recycle (stream 8, compressor costs). We can now quickly calculate the hydrogen make-up flow and vapor recycle as a function of the toluene conversion and hydrogen purge fraction. For example, consider a single pass toluene conversion of 0.7. The following table (Table 15–1) is quickly constructed; additional rows can be generated by considering other toluene conversions (see Student Exercise 4). Notice that there is a direct trade-off of hydrogen raw material costs with recycle compressor operating costs. Douglas (1988) finds that the incremental raw material cost is higher than the incremental recycle compression costs for these types of processes.

Consideration of the secondary reaction in Equation (15.4) results in the simplified flow sheet shown in Exercise 5.

15.6 Simulations

SIMULINK provides a nice modeling environment for simulating very simple process flow sheets. For rigorous process design and analysis, chemical process flow sheeting packages such as ASPEN PLUS and HYSYS should be used. Flow sheets can quickly be developed using standard "icons" that represent unit operations, such as reactors and separators. This next example illustrates the use of SIMULINK to solve a flow sheet similar to Example 15.1.

Table 15–1 H_2 Make-up and Vapor Recycle as a Function of H_2 Fraction in Purge Stream

H_2 purge frac, $y_{7,H}$	*Toluene conv., X_T*	*H_2 make-up, n_2*	*Recycle, n_8*
0.3	0.7	1.515	18.96
0.4	0.7	1.786	13.57
0.5	0.7	2.174	10.11
0.6	0.7	2.778	7.46

Example 15.6: A Simple Recycle System

Here we consider a slightly different steady state than that shown in Figure 15–3. In Figure 15–23 the make-up feedstream (stream 1) has 0.9 mole fraction A. The CSTR has a molar holdup of N = 200 moles and a reaction rate constant of k = 0.6 hr^{-1}. One of our objectives is to compare the responses of the flow sheets with and without a recycle stream, to disturbances in the fresh feed (stream 1) composition. A SIMULINK `.mdl` for the recycle system dynamics is shown in Figure 15–24, where the mixing point, the CSTR,

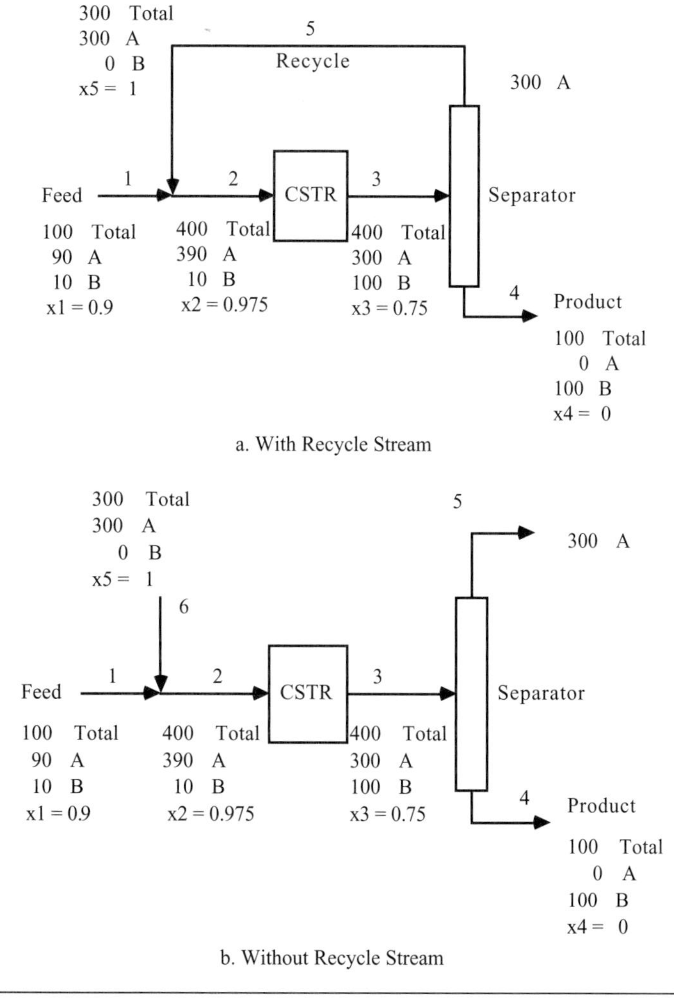

Figure 15–23 Flow sheet example. Steady-state flows shown.

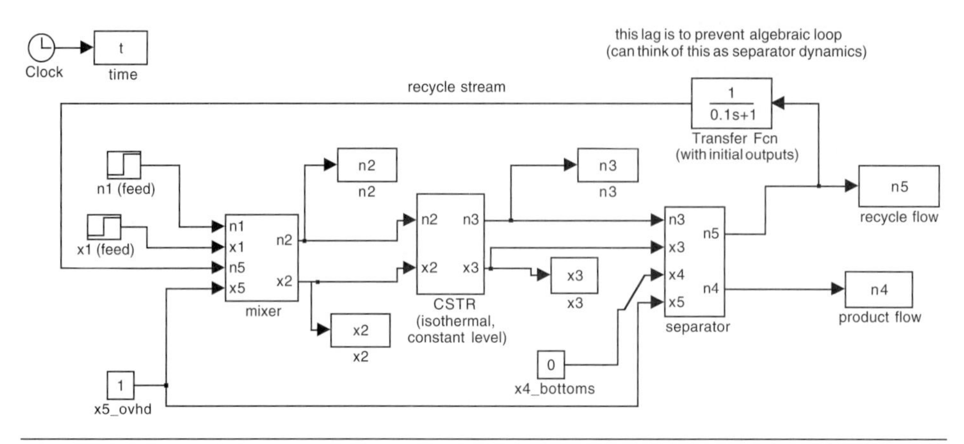

Figure 15–24 SIMULINK diagram (`.mdl`) for the dynamic flow sheet.

and the separator are "masked" functions. The CSTR is "unmasked" in Figure 15–25, to reveal the modeling equation

$$\frac{dx_3}{dt} = \frac{n_2}{N} \cdot (x_2 - x_3) - kx_3 \tag{15.6}$$

Consider a step change in the feed composition of A from 0.9 to 1.0 mole fraction. The results for the recycle system are compared with the "straight through" system without recycle in Figure 15–26. Notice that the recycle system has a tremendous change in the separator product stream (recycle flow) compared to the no-recycle case. As discussed in Section 15.2 (and Example 15.2), the recycle system has a large gain and time constant compared to the system without recycle.

15.7 Summary

This chapter has provided a brief overview of perhaps the most interesting and challenging aspects of process control. The intuitive placement of controllers on a process with recycle can often lead to extreme sensitivity to disturbances known as the snowball effect. This effect can be minimized by fixing the flow rate of one stream in the recycle loop.

Recycle can also significantly increase the dynamic timescale of the process. The specification of a plantwide control structure is based on years of operational experience with similar plants and will involve input from a large number of engineers and technicians. If you are interested in a much more detailed presentation of plantwide control, including many "rules of thumb" and other suggestions, the book by Luyben et al. (1999) is highly recommended.

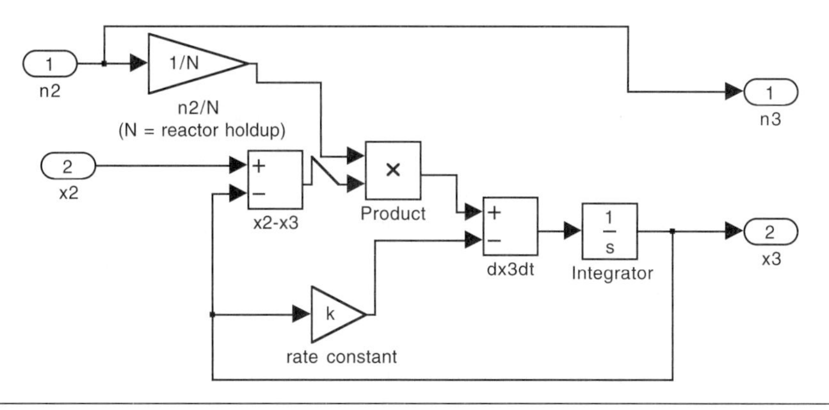

Figure 15–25 Unmasked view of the CSTR block in Figure 15–24.

Corporations have a large number of levels where decisions need to be made (Section 15.4). The various levels range from the long-term corporate purchasing and scheduling decisions at the top, to fundamental control instrumentation at the bottom, as shown in Figures 15–15 and 15–16.

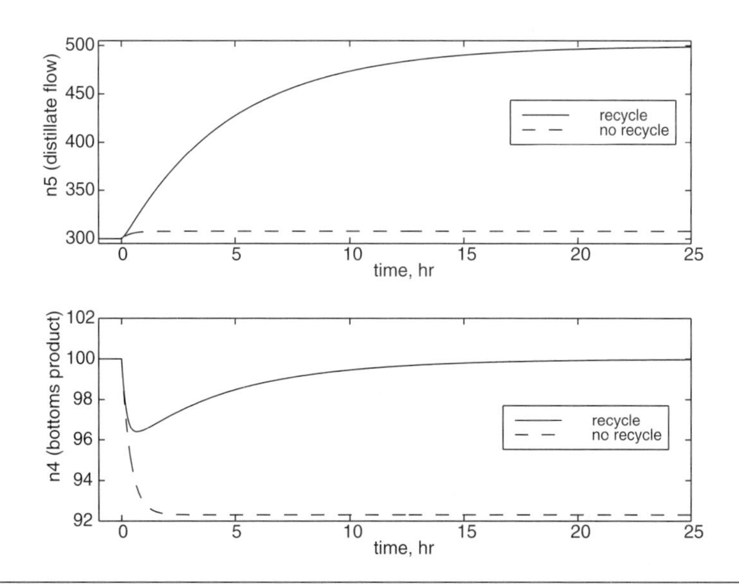

Figure 15–26 Response to a step disturbance in the feedstream composition of A (from 0.9 to 1.0) at t = 0 hr. Comparison of recycle with no-recycle systems. Top plot: distillate product flow rate (n_5). Bottom plot: flow rate of product B (n_4).

References

The MPN process is presented and discussed by in the following: Zhan, Q., and V. G. Grassi, "Dynamic Matrix Control of a Specialty Chemical Plant with Different Plantwide Control Structures," in *Proceedings of the 1999 American Control Conference*, pp. 632–636 (1999).

The best reference for plantwide control is the following monograph: Luyben, W. L., B. D. Tyréus, and M. L. Luyben, *Plantwide Process Control*, McGraw Hill, New York (1999).

Plantwide strategies are also discussed in the following papers and textbook:

Ponton, J. W., and D. M. Laing, "A Hierarchical Approach to the Design of Process Control Systems, *Trans. Inst. Chem. Eng.*, **71**(A), 181–188 (1993).

Stephanopoulos, G., *Chemical Process Control*, Prentice Hall, Englewood Cliffs, NJ (1984).

Problems with "on-demand" control structures are discussed in the following: W. L. Luyben, "Inherent Dynamic Problems with On-Demand Control Structures," *Ind. Eng. Chem. Res.*, **38**, 2315–2329 (1999).

An overview of process systems engineering is provided in the following encyclopedia article: Bequette, B. W., and L. P. Russo, "Process Design, Simulation, Optimization and Operation," in the *Encyclopedia of Physical Science and Technology*, 3rd ed. Vol. 2, pp. 751–766. (R. Matsubara, ed.), Academic, New York (2002).

The following book provides a nice treatment of the synthesis of chemical process flow sheets to meet design objectives. Once the process is designed, Douglas recommends that recycle streams be operated at a constant maximum flow rate, since the resulting yield improvements are worth the additional energy costs.

Douglas, J. M. *Conceptual Design of Chemical Processes*, McGraw Hill, New York (1988).

The Tennessee Eastman "challenge problem" (Exercise 6) was presented in the following paper: Downs, J. J., and E. F. Vogel, "A Plant-Wide Industrial Process Control Problem," *Comp. Chem. Engng.*, **17**, 245–255 (1995).

Student Exercises

1. Consider Example 15.1. Assume that the isothermal CSTR has perfect level control; that is, the volume is constant. The CSTR contains 200 moles, and the reaction rate constant is 0.6 hr^{-1}. Verify the results shown in Figures 15–3, 15–4 and 15–5.

2. Consider Example 15.1. Revise Figure 15–7 to obtain the following alternative control strategy. Draw the control instrumentation diagram obtained by fixing the flow rate of stream 3 (between the CSTR and column) by using a flow controller on stream 3, and to control the CSTR volume by manipulating the fresh make-up (stream 1) flow rate.
3. Consider the MPN example. Alternative 2 is shown here. Now, consider the use of a reactor concentration sensor. Let the reactor concentration controller manipulate the methanol feed flow rate. Sketch this control strategy on the diagram. Which controller setpoint is set to yield a desired production rate?

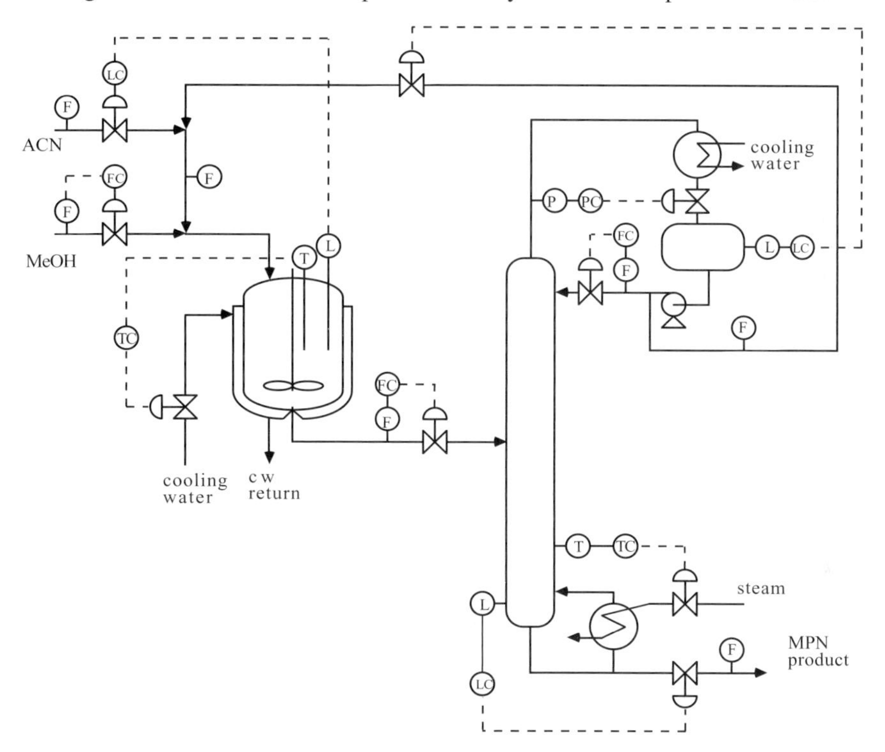

4. Consider the simplified flow sheet shown in Figure 15–22 (Example 15.5).
   a. From material balances find the relationships for all flows as a function of hydrogen fraction in the purge stream and the single-pass toluene conversion in the reactor.
   b. Construct a table similar to Table 15–1.
   c. Based solely on the current raw material costs (neglecting capital and energy costs) of hydrogen and toluene, can a profit be obtained with a hydrogen purge concentration of 0.5 mole fraction?

5. Consider the "back of the envelope" calculations for the HDA process (Example 15.5). In our first pass at material balances, we neglected the diphenyl formation. Now, consider the selectivity of conversion to toluene and include diphenyl material balances. The additional separator is shown in the flow sheet here.
   a. Find how the raw material, recycle, and benzene product flows change as a function of toluene conversion, selectivity (fraction of toluene converted to benzene vs. overall conversion of toluene) and hydrogen purge fraction. Construct a table similar to Table 15–1.
   b. Based solely on the current raw material costs (neglecting capital and energy costs) of hydrogen and toluene, can a profit be obtained with a hydrogen purge concentration of 0.5 mole fraction and a selectivity of 0.95?

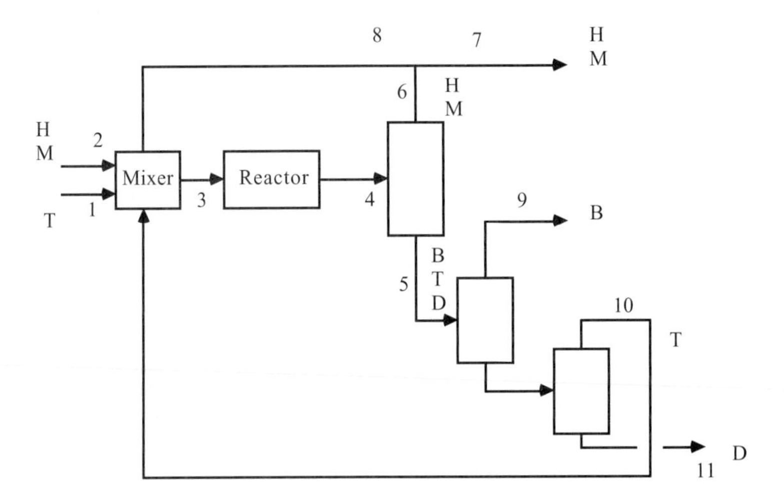

6. A Tennessee Eastman "challenge problem" was published by Downs and Vogel in 1993 (see the references section of this chapter). A fairly large number of academic studies have been performed on plantwide control problems. Conduct a literature review on this topic, read at least two articles, and compare the different control techniques proposed.
7. Most process plants have three steam header pressure levels, approximately 15 psig, 150 psig, and 600 psig. Discuss possible answers to the following:
   - What determines the pressure level needed for a particular unit operation?
   - How would you control the pressure of each of the steam headers?

- How would you determine the cost of various levels of steam? If some low pressure steam were being released to the atmosphere to maintain the steam balance (and the low pressure header at 15 psig), what should be the incremental cost of this steam to give an economic reward using it for heating needs?

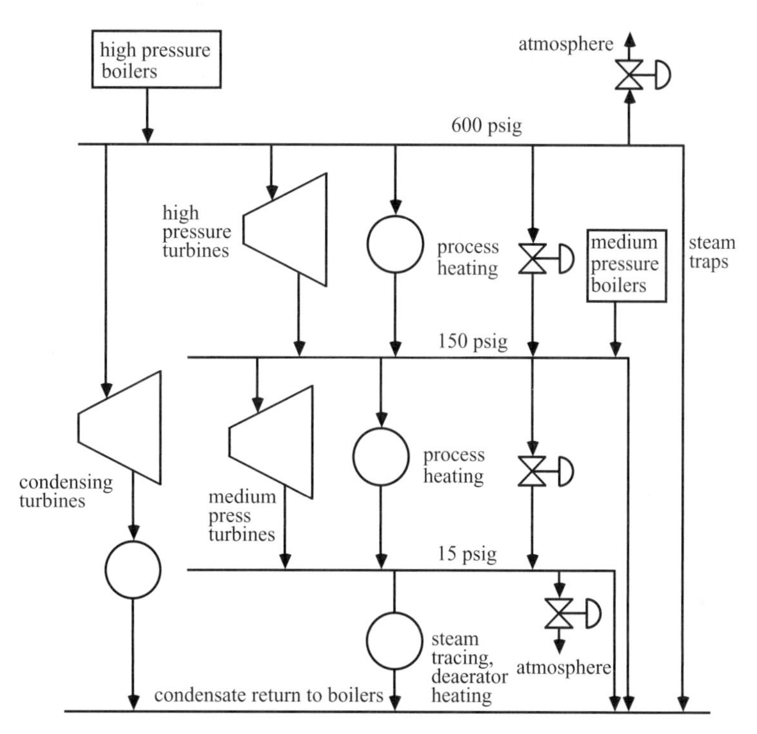

CHAPTER 16

Model Predictive Control

MPC is the class of advanced control techniques most widely applied in the process industries. A primary advantage to the approach is the explicit handling of constraints. In addition, the formulation for multivariable systems with time delays is straightforward.

After studying this chapter, the reader should understand the basic ideas behind SISO MPC, including the following:

- Dynamic matrix control (DMC)
- Tuning parameters such as prediction and control horizons and control weight
- The effect tuning parameters have on the control performance and stability
- The ability to explicitly include constraints and handle multivariable processes

The major sections of this chapter are as follows:

16.1 Motivation

Most of the techniques presented in this textbook have been based on continuous-time models. The controllers are normally implemented by microprocessors and, therefore, must be converted to discrete time before implementation. We have focused on continuous time because few changes in the techniques are needed for the discrete-time case. One advanced control technique that has solely had discrete-time application is *model predictive control* (MPC). MPC is by far the most commonly applied advanced control technique in the chemical process industry, so it is worth devoting a chapter to it. One major contributor to the success of MPC is the ability to handle constraints in an optimal fashion. The optimization-based procedure is intuitive and is also a natural way of handling multivariable systems.

Basic Description

The basic idea behind MPC is shown in Figure 16–1. At each time step, k, an optimization problem is solved, as illustrated on the top half of the figure. An objective function (usually quadratic) based on output predictions over a *prediction horizon* of P time steps is minimized by a selection of manipulated variable moves over a *control horizon* of M control moves. Although M moves are optimized, only the first move is implemented. After u_k is implemented, the measurement at the next time step, y_{k+1} is obtained, as shown in the bottom half of the figure. A correction for model error is performed, since the measured output y_{k+1} will, in general, not be equal to the model predicted value. A new optimization problem is then solved, again, over a prediction horizon of P steps by adjusting M control moves. This approach is also known as receding horizon control.

For clarity, we initially focus on unconstrained SISO systems. Several questions naturally arise. What type of objective function should be used for optimization? What type of model should be used to predict the output? How should the model be initialized to predict the future output values? What is the desired setpoint trajectory? How should constraints be implemented? Some of these topics are discussed in the next section.

16.2 Optimization Problem

The term optimization implies a best value for some type of performance criterion. This performance criterion is known as an objective function. Here, we first discuss possible objective functions, then possible process models that can be used for MPC.

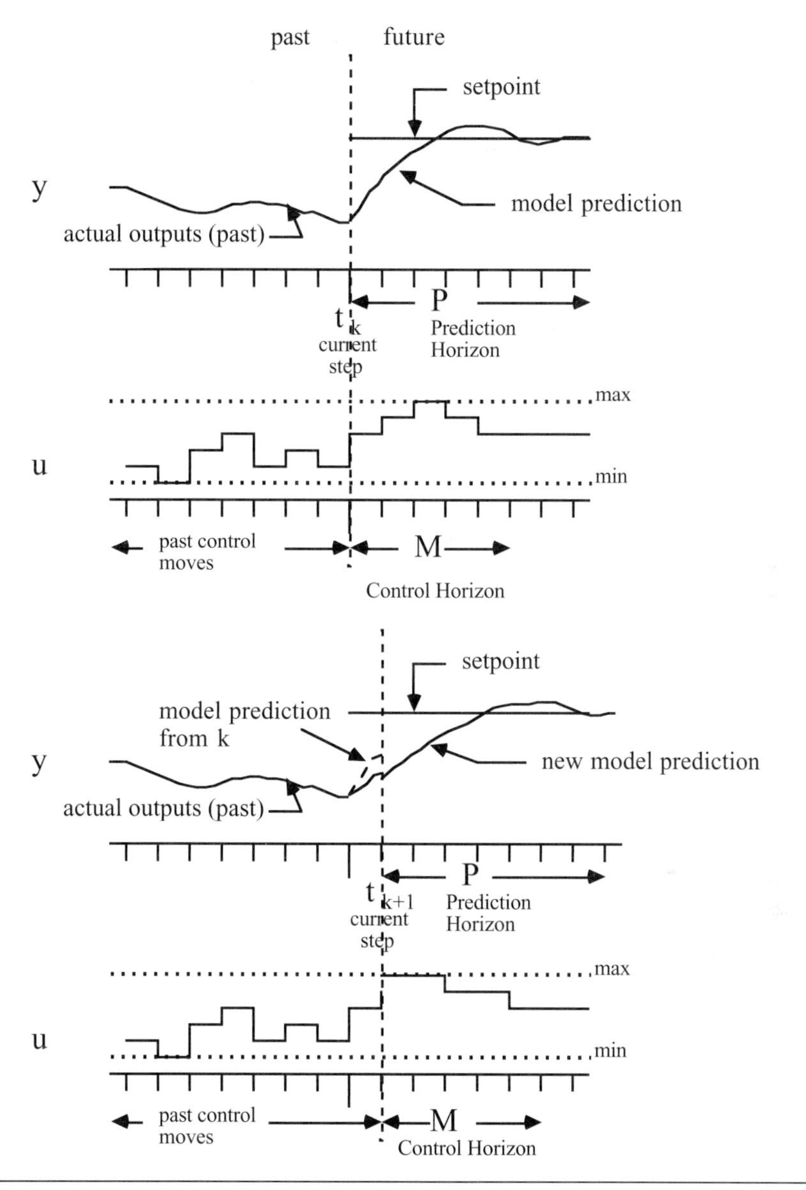

Figure 16–1 Basic concept of MPC.

Objective Functions

Here, there are several different choices for objective functions. The first one that comes to mind is a standard *least-squares* or "quadratic" objective function. The objective function is a "sum of squares" of the predicted errors (differences between the setpoints and the model-predicted outputs) and the control moves (changes in control action from step to step).

A quadratic objective function for a prediction horizon of 3 and a control horizon of 2 can be written

$$\Phi = (r_{k+1} - \hat{y}_{k+1})^2 + (r_{k+2} - \hat{y}_{k+2})^2 + (r_{k+3} - \hat{y}_{k+3})^2 + w\Delta u^2_{k} + w\Delta u^2_{k+1}$$

where $\hat{y}$ represents the model predicted output, r is the setpoint, Δu is the change in manipulated input from one sample time to the next, w is a weight for the changes in the manipulated input, and the subscripts indicate the sample time (k is the current sample time). For a prediction horizon of P and a control horizon of M, the least-squares objective function is written

$$\Phi = \sum_{i=1}^{P} \left(r_{k+i} - \hat{y}_{k+i}\right)^2 + w \sum_{i=0}^{M-1} \Delta u_{k+i}^2 \tag{16.1}$$

Another possible objective function is to simply take a sum of the *absolute values* of the predicted errors and control moves.

For a prediction horizon of 3 and a control horizon of 2, the absolute value objective function is

$$\Phi = |r_{k+1} - \hat{y}_{k+1}| + |r_{k+2} - \hat{y}_{k+2}| + |r_{k+3} - \hat{y}_{k+3}| + w|\Delta u_k| + w|\Delta u_{k+1}|$$

which has the following general form for a prediction horizon of P and a control horizon of M:

$$\Phi = \sum_{i=1}^{P} \left|r_{k+i} - \hat{y}_{k+i}\right| + w \sum_{i=0}^{M-1} \left|\Delta u_{k+i}\right| \tag{16.2}$$

The optimization problem solved is usually stated as a minimization of the objective function, obtained by adjusting the M control moves, subject to modeling equations (equality constraints), and constraints on the inputs and outputs.

$$\begin{aligned} &\min_{\Delta u_k, \ldots, \Delta u_{k+M-1}} \quad \Phi \\ &s.t. \ \text{modeling equations and constraints} \end{aligned} \tag{16.3}$$

Least-squares formulations are by far the most common objective functions in MPC. Least squares yields analytical solutions for unconstrained problems and penalizes larger errors (relatively) more than smaller errors. The absolute value objective function has been used in a few algorithms because a linear programming (LP) problem results. LPs are routinely solved in large-scale scheduling and allocation problems. For example, an oil company often uses an LP to decide how to distribute oil to various refineries and to decide how much and what product to produce at each plant. This topic was discussed in Chapter 15. The LP approach is not useful for model predictive control, because the manipulated variable moves often "hop" from one extreme constraint to another. For the rest of this chapter we use a quadratic (least squares) formulation for the objective function.

Models

Many different types of models are possible for calculating the predicted values of the process outputs, which are used in evaluating the objective function. Since the outputs are evaluated at discrete-time steps, it makes sense to use discrete models for the output prediction. Here, we review step and impulse response models (introduced in Chapter 4), both of which are used in common MPC algorithms.

Finite Step Response

FSR models are obtained by making a unit step input change to a process operating at steady state. The model coefficients are simply the output values at each time step, as shown in Figure 16–2. Here, s_i represents the step response coefficient for the ith sample time after the unit step input change. If a non-unit step change is made, the output is scaled accordingly.

The step response model is the vector of step response coefficients,

$$S = [s_1 \; s_2 \; s_3 \; s_4 \; s_5 \ldots s_N]^T \tag{16.4}$$

where the model length N is long enough so that the coefficient values are relatively constant (i.e., the process is close to a new steady state).

Finite Impulse Response

Another common form of model is a finite impulse response (FIR). Here, a unit pulse is applied to the manipulated input, and the model coefficients are simply the values of the outputs at each time step after the pulse input is applied. As shown in Figure 16–3, h_i represents the ith impulse response coefficient.

There is a direct relationship between step and impulse response models:

$$\begin{aligned} h_i &= s_i - s_{i-1} \\ s_i &= \sum_{j=1}^{i} h_j \end{aligned} \tag{16.5}$$

Figure 16–4 illustrates how impulse response coefficients can be obtained from step responses. The impulse response coefficients are simply the changes in the step response

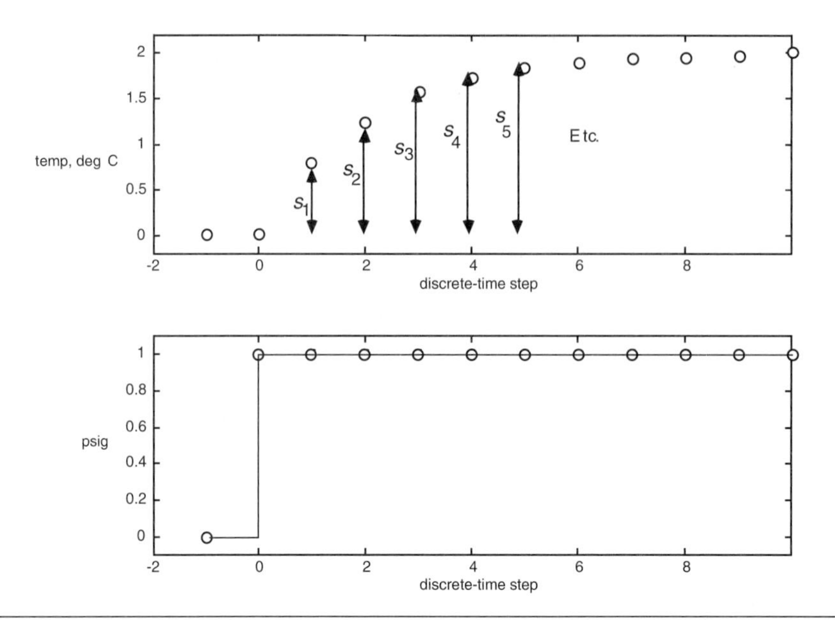

Figure 16–2 Illustration of step response parameter identification.

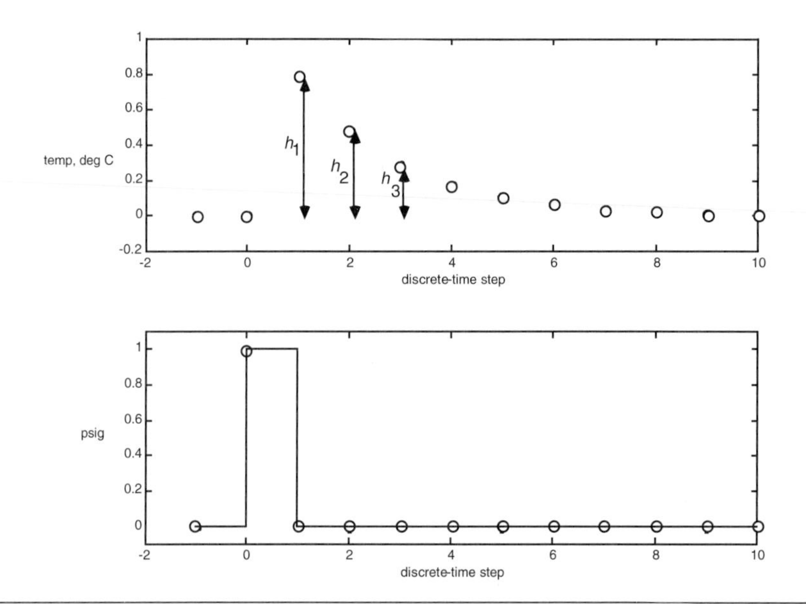

Figure 16–3 Illustration of impulse response parameter identification.

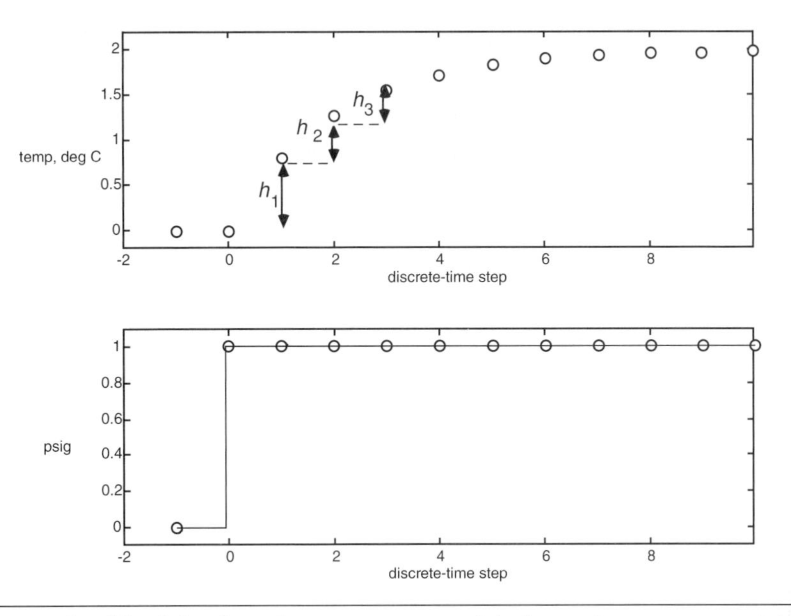

Figure 16–4 Obtaining impulse response models from step response data.

coefficient at each time step. Similarly, step response coefficients can be found from impulse responses; a step response coefficient is the sum of the impulse response coefficients to that point. It should be noted that there are two major limitations to step and impulse response models. They can only be used to represent open-loop stable processes, and they require a large number of parameters (model coefficients) compared to state space and transfer function models.

16.3 Dynamic Matrix Control

DMC was developed by Shell Oil Company in the 1960s and 1970s (see Cutler and Ramaker, 1980). It is based on a step response model, which has the form (see Appendix 16.1 for a derivation)

$$\hat{y}_k = s_1\Delta u_{k-1} + s_2\Delta u_{k-2} + \ldots + s_{N-1}\Delta u_{k-N+1} + s_N u_{k-\mathrm{N}} \tag{16.6}$$

which is written in the form

$$\hat{y}_k = \sum_{i=1}^{N-1} s_i \Delta u_{k-i} + s_N u_{k-N} \tag{16.7}$$

where $\hat{y}_k$ is the model prediction at time step k, and u_{k-N} is the manipulated input N steps in the past.

Note that the model-predicted output is unlikely to be equal to the actual measured output at time step k. The difference between the measured output (y_k) and the model prediction is called the additive disturbance.

$$d_k = y_k - \hat{y}_k \tag{16.8}$$

The "corrected prediction" is then equal to the actual measured output at step k,

$$\hat{y}_k = \hat{y}_k + d_k \tag{16.9}$$

Similarly, the corrected predicted output at the first time step in the future can be found from

$$\begin{aligned}
\hat{y}^c_{k+1} &= \hat{y}_{k+1} + \hat{d}_{k+1} \\
\hat{y}^c_{k+1} &= \sum_{i=1}^{N-1} s_i \Delta u_{k-i+1} + s_N u_{k-N+1} + \hat{d}_{k+1} \\
\hat{y}^c_{k+1} &= s_1 \Delta u_k + \sum_{i=2}^{N-1} s_i \Delta u_{k-i+1} + s_N u_{k-N+1} + \hat{d}_{k+1}
\end{aligned} \tag{16.10}$$

So, for the jth step into the future, we find

$$\begin{aligned}
\hat{y}^c_{k+j} &= \hat{y}_{k+j} + \hat{d}_{k+j} \\
\hat{y}^c_{k+j} &= \underbrace{\sum_{i=1}^{j} s_i \Delta u_{k-i+j}}_{\text{effect of future control moves}} + \underbrace{\sum_{i=j+1}^{N-1} s_i \Delta u_{k-i+j} + s_N u_{k-N+j}}_{\text{effect of past control moves}} + \underbrace{\hat{d}_{k+j}}_{\text{correction term}}
\end{aligned} \tag{16.11}$$

and we can separate the effects of past and future control moves

$$\begin{aligned}
\hat{y}^c_{k+j} = {} & s_1 \Delta u_{k+j-1} + s_2 \Delta u_{k+j-2} + \ldots + s_j \Delta u_k && \{\text{effect of current and future moves} \\
& + s_N u_{k-N+j} + s_{j+1} \Delta u_{k-1} + s_{j+2} \Delta u_{k-2} \\
& + \ldots + s_{N-1} \Delta u_{k-N+j+1} && \{\text{effect of past moves} \\
& + \hat{d}_{k+j} && \{\text{correction term}
\end{aligned} \tag{16.12}$$

The most common assumption is that the correction term is constant in the future (this is the "constant additive disturbance assumption"):

$$\hat{d}_{k+j} = \hat{d}_{k+j-1} = \ldots = d_k = y_k - \hat{y}_k \tag{16.13}$$

Also, realize that there are no control moves beyond the control horizon of M steps, so

$$\Delta u_{k+M} = \Delta u_{k+M+1} = \ldots = \Delta u_{k+P-1} = 0$$

In matrix-vector form, a prediction horizon of P steps and a control horizon of M steps, yields

$$\underbrace{\begin{bmatrix} \hat{y}^c_{k+1} \\ \hat{y}^c_{k+2} \\ \vdots \\ \hat{y}^c_{k+j} \\ \vdots \\ \hat{y}^c_{k+P} \end{bmatrix}}_{\substack{P\times 1 \\ \text{corrected output} \\ \text{predictions, } \hat{Y}^c}} = \underbrace{\begin{bmatrix} s_1 & 0 & 0 & \cdots & 0 & 0 \\ s_2 & s_1 & 0 & \cdots & 0 & 0 \\ \vdots & \vdots & & & & \\ s_j & s_{j-1} & s_{j-2} & \cdots & \cdots & s_{j-M+1} \\ \vdots & \vdots & \vdots & & & \vdots \\ s_P & s_{P-1} & s_{P-2} & \cdots & \cdots & s_{P-M+1} \end{bmatrix}}_{\substack{P\times M \\ \text{dynamic matrix, } S_f}} \underbrace{\begin{bmatrix} \Delta u_k \\ \Delta u_{k+1} \\ \vdots \\ \Delta u_{k+M-2} \\ \Delta u_{k+M-1} \end{bmatrix}}_{\substack{M\times 1 \\ \text{current and future} \\ \text{control moves, } \Delta u_f}}$$

$$+ \underbrace{\begin{bmatrix} s_2 & s_3 & s_4 & \cdots & s_{N-2} & s_{N-1} \\ s_3 & s_4 & s_5 & \cdots & s_{N-1} & 0 \\ \vdots & \vdots & & & 0 & 0 \\ s_{j+1} & s_{j+2} & \cdots & s_{N-1} & 0 & 0 \\ \vdots & \vdots & \vdots & & & \vdots \\ s_{P+1} & s_{P+2} & \cdots & 0 & \cdots & 0 \end{bmatrix}}_{\substack{P\times(N-2) \\ \text{matrix, } S_{past}}} \underbrace{\begin{bmatrix} \Delta u_{k-1} \\ \Delta u_{k-2} \\ \vdots \\ \Delta u_{k-N+3} \\ \Delta u_{k-N+2} \end{bmatrix}}_{\substack{(N-2)\times 1 \\ \text{past control} \\ \text{moves, } \Delta u_{past}}} \tag{16.14}$$

$$+ s_N \underbrace{\begin{bmatrix} u_{k-N+1} \\ u_{k-N+2} \\ \vdots \\ u_{k-N+P} \end{bmatrix}}_{\substack{P\times 1 \\ \text{past inputs, } u_P}} + \underbrace{\begin{bmatrix} d_{k+1} \\ d_{k+2} \\ \vdots \\ d_{k+P} \end{bmatrix}}_{\substack{P\times 1 \\ \text{predicted disturbances, } \hat{d}}}$$

which we write using matrix-vector notation

$$\underbrace{\hat{Y}^c}_{\substack{\text{corrected} \\ \text{predicted} \\ \text{outputs}}} = \underbrace{S_f \Delta u_f}_{\substack{\text{effect of} \\ \text{current and} \\ \text{future moves}}} + \underbrace{S_{past}\Delta u_{past} + s_N u_P}_{\text{effect of past moves}} + \underbrace{\hat{d}}_{\substack{\text{predicted} \\ \text{disturbances}}} \tag{16.15}$$

In Equation (16.15) the corrected-predicted output response is naturally composed of a “forced response” (contributions of the current and future control moves) and a “free response” (the output changes that are predicted if there are no future control moves). The difference between the setpoint trajectory, r, and the future predictions is

$$\underbrace{r - \hat{Y}^c}_{\substack{\text{corrected} \\ \text{predicted} \\ \text{error, } E^c}} = \underbrace{r - \left[S_{past}\Delta u_P + s_N u_P + \hat{d}\right]}_{\substack{\text{unforced error (if no current and future} \\ \text{control moves were made), } E}} - S_f \Delta u_f \tag{16.16}$$

which can be written

$$E^c = E - S_f \Delta u_f \tag{16.17}$$

where the future predicted errors are composed of "free response" (E) and "forced response" ($-S_f \Delta u_f$) contributions.

The least-squares objective function (16.1) is

$$\Phi = \sum_{i=1}^{P} \left(e_{k+i}^c\right)^2 + w \sum_{i=0}^{M-1} \left(\Delta u_{k+i}\right)^2 \tag{16.18}$$

Notice that the quadratic terms can written in matrix-vector form as

$$\sum_{i=1}^{P} \left(e_{k+i}^c\right)^2 = \begin{bmatrix} e_{k+1}^c & e_{k+2}^c & \cdots & e_{k+P}^c \end{bmatrix} \begin{bmatrix} e_{k+1}^c \\ e_{k+2}^c \\ \vdots \\ e_{k+P}^c \end{bmatrix} \tag{16.19}$$

$$= \left(E^c\right)^T E^c$$

and

$$w \sum_{i=0}^{M-1} \left(\Delta u_{k+i}\right)^2 = w \cdot \begin{bmatrix} \Delta u_k & \Delta u_{k+1} & \cdots & \Delta u_{k+M-1} \end{bmatrix} \begin{bmatrix} \Delta u_k \\ \Delta u_{k+1} \\ \vdots \\ \Delta u_{k+M-1} \end{bmatrix}$$

$$= \begin{bmatrix} \Delta u_k & \Delta u_{k+1} & \cdots & \Delta u_{k+M-1} \end{bmatrix} \begin{bmatrix} w & 0 & 0 & 0 \\ 0 & w & 0 & 0 \\ 0 & 0 & \ddots & 0 \\ 0 & 0 & 0 & w \end{bmatrix} \begin{bmatrix} \Delta u_k \\ \Delta u_{k+1} \\ \vdots \\ \Delta u_{k+M-1} \end{bmatrix} \tag{16.20}$$

$$= \Delta u_f^T W \Delta u_f$$

Therefore the objective function can be written in the form

$$\Phi = (E^c)^T E^c + (\Delta u_f)^T W \Delta u_f \tag{16.21}$$

subject to the modeling equation equality constraint (16.17)

$$E^c = E - S_f \Delta u_f \tag{16.17}$$

Substituting (16.17) into (16.21), the objective function can be written

$$\Phi = (E - S_f \Delta u_f)^T (E - S_f \Delta u_f + (\Delta u_f)^T W \Delta u_f \tag{16.22}$$

The solution for the minimization of this objective function is (see Appendix 16.2)

$$\Delta u_f = \underbrace{\left(S_f^T S_f + W\right)^{-1} S_f^T}_{K} \underbrace{E}_{\text{unforced errors}} \tag{16.23}$$

Notice that the current and future control move vector (Δu_f) is proportional to the unforced error vector (E). That is, a controller gain matrix, K, multiplies the unforced error vector (the future errors that would occur if there were no control move changes implemented).

Because only the current control move is actually implemented, we use the first row of the K matrix, and

$$\Delta u_k = K_1 E \tag{16.24}$$

where K_1 represents the first row of the K matrix, where $K = (S_f^T S_f + W)^{-1} S_f^T$

Perhaps it is worth summarizing the steps involved in implementing DMC on a process.

1. Develop a discrete step response model with length N (16.4), based on a sample time Δt.
2. Specify the prediction (P) and control (M) horizons. $N \geq P \geq M$
3. Specify the weighting on the control action ($w = 0$ if no weighting).
4. All calculations assume deviation variable form, so remember to convert to/from physical units.

The effect of all of these tuning parameters is now discussed for SISO systems.

Model-length and sample-time selection are not independent. The model length should be approximately the "settling time" of the process, that is, the time required to reach a new steady state after a step input change. For most systems, the model length will be roughly 50 coefficients. The sample time is usually on the order of one tenth the dominant time constant, so the model length is roughly the settling time of the process.

Prediction and control horizons differ in length. Usually, the prediction horizon is selected to be much longer than the control horizon. This is particularly true if the control weighting factor is selected to be zero. Usually, if the prediction horizon is much longer than the control horizon, the control system is less sensitive to model error. Often $P = 20$ or so, while $M = 1$–3.

Control weighting is often set to zero if the prediction horizon is much longer than the control horizon. As the control horizon is increased, the control moves tend to become more aggressive so a larger weight is needed to penalize the control moves.

Example 16.1: First-Order Process

Here we study the first-order process, where the time unit is minutes,

$$g_p(s) = \frac{1}{5s+1}$$

Ordinarily, we would select a sample time of roughly 0.5 minutes, and a model horizon of 50. Here we select a sample time of 1 minute to study the effect of other tuning parameters. The step response coefficients generated from this plant are shown in Figure 16–5. We can see that a model length of at least 20 should be used, since that is when the output is reasonably close to steady state.

The importance of prediction horizon is shown clearly in Figure 16–6. In both simulations a control horizon of 1 is used. A prediction horizon of 1 results in the setpoint being achieved in 1 time step, while a prediction horizon of 5 yields a much slower response. Notice, however, that $P = 1$ requires much more control action than $P = 5$. Although not shown here, the shorter prediction horizon is more sensitive to model uncertainty; these simulations have assumed a perfect model.

The importance of the model length is shown in Figure 16–7. Since a model length of 6 does not capture the complete dynamics of the process (see Figure 16–5), it effectively results in a model error and poor performance compared with a model length of 25.

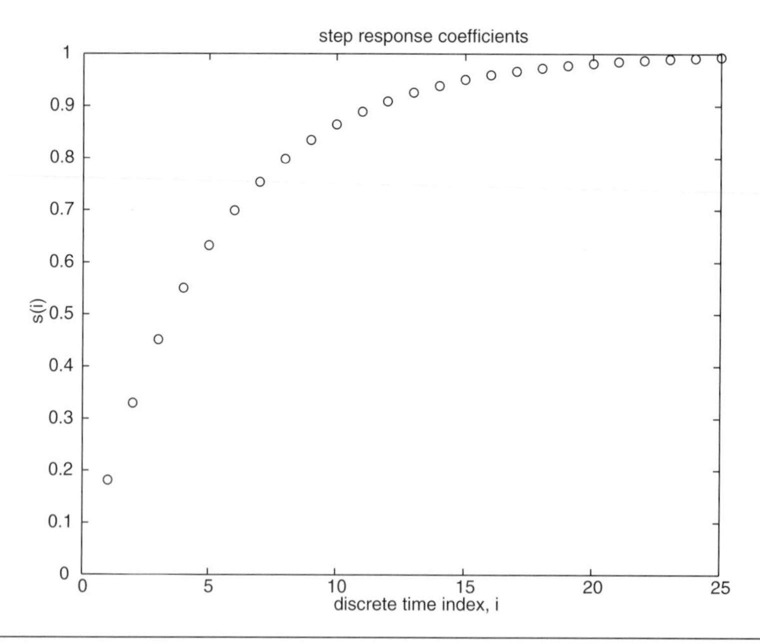

Figure 16–5 Step response coefficients for the first-order example. Sample time = 1 minute; model length, $N = 25$.

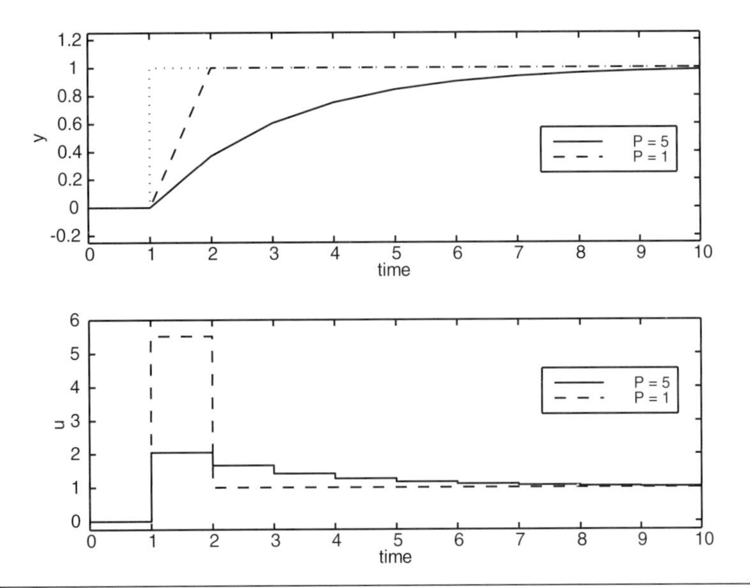

Figure 16–6 Closed-loop responses to a step setpoint change at t = 1 min, first-order example. Sample time = 1 minute; model length, $N = 25$; control horizon, $M = 1$; no weighting on input. Comparison of prediction horizons, $P = 5$ and $P = 1$.

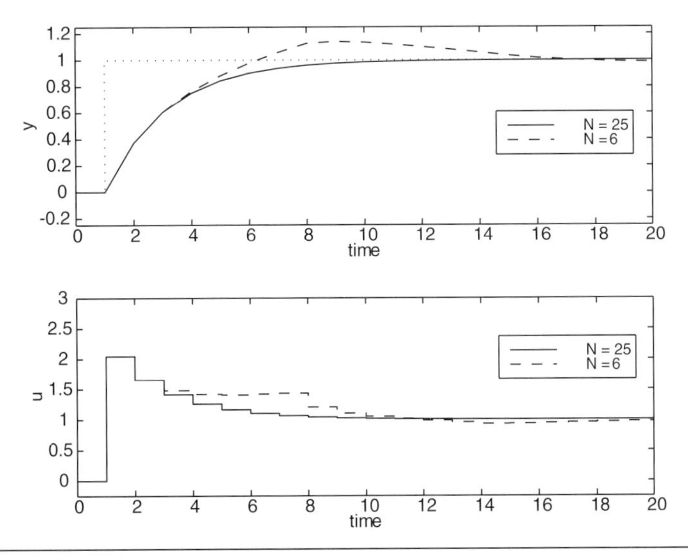

Figure 16–7 Closed-loop responses to a step setpoint change at t = 1 min, first-order example. Sample time = 1 minute; prediction horizon, $P = 5$; control horizon, $M = 1$; no weighting on input. Comparison of model lengths, $N = 25$ and $N = 6$.

From Equation (16.14), for $N = 6$, $P = 5$, $M = 1$, we find (see Figure 16–5 for the step response coefficients) the following dynamic (S_f) and past move (S_{past}) matrices.

$$S_f = \underbrace{\begin{bmatrix} s_1 \\ s_2 \\ s_3 \\ s_4 \\ s_5 \end{bmatrix}}_{\substack{P\times M(5\times 1)\\ \text{dynamic matrix}}} = \begin{bmatrix} 0.181 \\ 0.330 \\ 0.451 \\ 0.551 \\ 0.632 \end{bmatrix}$$

$$S_{past} = \underbrace{\begin{bmatrix} s_2 & s_3 & s_4 & s_5 \\ s_3 & s_4 & s_5 & 0 \\ s_4 & s_5 & 0 & 0 \\ s_5 & 0 & 0 & 0 \\ 0 & 0 & 0 & 0 \end{bmatrix}}_{P\times(N-2),(5\times 4)} = \begin{bmatrix} 0.330 & 0.451 & 0.551 & 0.632 \\ 0.451 & 0.551 & 0.632 & 0 \\ 0.551 & 0.632 & 0 & 0 \\ 0.632 & 0 & 0 & 0 \\ 0 & 0 & 0 & 0 \end{bmatrix}$$

also, selecting a control move weight of 0

$$W = 0$$

we calculate the feedback gain matrix from Equation (16.23)

$$K = (S_f^T S_f + W)^{-1}\, S_f^T$$

and $K_1 = K$ since the control horizon (M) is 1 (the reader should verify these numerical results using MATLAB), the feedback gain matrix is:

$$K_1 = [0.173 \quad 0.315 \quad 0.431 \quad 0.525 \quad 0.603]$$

in the simulations shown, the process is initially at steady state when the unit step setpoint change at t = 1 minute is made. Since deviation variables are used, the values of all previous manipulated inputs and outputs are 0. The unforced error vector, E, from Equation (16.16) is (we have transposed it to a row vector to save space)

$$E^T = r^T = [1 \quad 1 \quad 1 \quad 1 \quad 1]$$

so the first control move, after the setpoint change is made is

$$\Delta u_k = K_1 E = \left[0.173 \quad 0.315 \quad 0.431 \quad 0.525 \quad 0.603\right] \cdot \begin{bmatrix} 1 \\ 1 \\ 1 \\ 1 \\ 1 \end{bmatrix} = 2.05$$

which is consistent with the first control move shown in Figure 16–7.

Example 16.2: Van de Vusse Reactor

Consider the van de Vusse reactor problem covered in Module 5. The continuous state space model is

$$A = \begin{bmatrix} -2.4048 & 0 \\ 0.83333 & -2.2381 \end{bmatrix}, \quad B = \begin{bmatrix} 7 \\ -1.117 \end{bmatrix}$$

$$C = \begin{bmatrix} 0 & 1 \end{bmatrix}, \quad D = \begin{bmatrix} 0 \end{bmatrix}$$

where the measured state (output) is the concentration of the second component and the manipulated input is the dilution rate. The manipulated input-output process transfer function for the reactor is

$$g_p(s) = \frac{-1.1170s + 3.1472}{s^2 + 4.6429s + 5.3821} = \frac{0.5848(-0.3549s + 1)}{0.1828s^2 + 0.8627s + 1}$$

Step Response Model

The discrete step response model is shown in Figure 16–8, for a sample time of 0.1 minute. Notice that a model length of at least 35 is needed to capture the complete dynamic behavior.

Effect of Prediction Horizon

A sample time of 0.1 minute, a model length of $N = 50$, and a control horizon of $M = 1$ are used in the following simulations. For this particular example, the prediction horizon does not appear to have an appreciable effect, as shown in Figure 16–9; the setpoint tracking performance is roughly the same for prediction horizons of $P = 10$ and 25. There is a lower limit, however, to the length of the prediction horizon that can be used. A prediction horizon of 7 or less results in an unstable closed-loop system, as shown in Figure 16–10. This is not due to any model error, since we have assumed a perfect model in these simulations. If the prediction horizon is too short, the initial step response coefficients dominate. Since these are negative while the later coefficients are positive (corresponding to a positive process gain), the prediction is really in error. The effect is the same as using a PID controller with a controller gain that is the wrong sign.

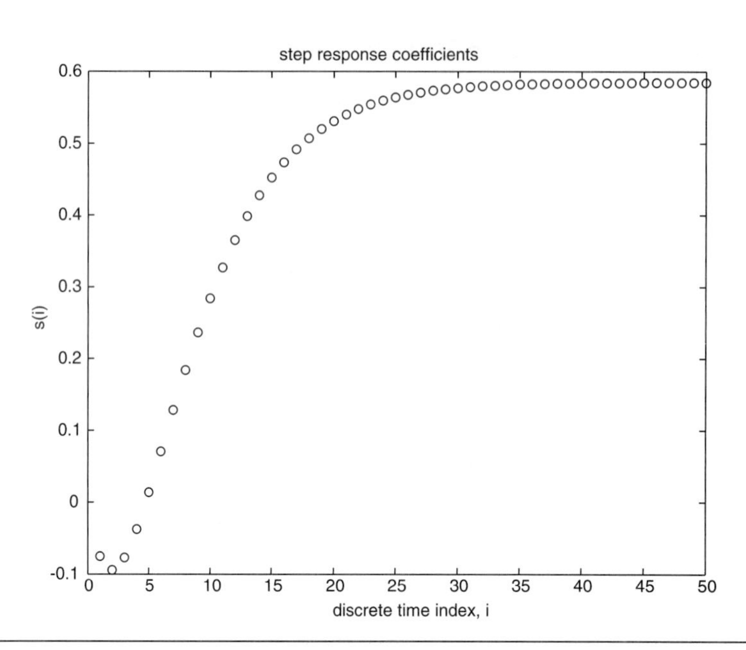

Figure 16–8 Step response coefficients for the van de Vusse example. Sample time = 0.1 minutes, model length, $N = 50$.

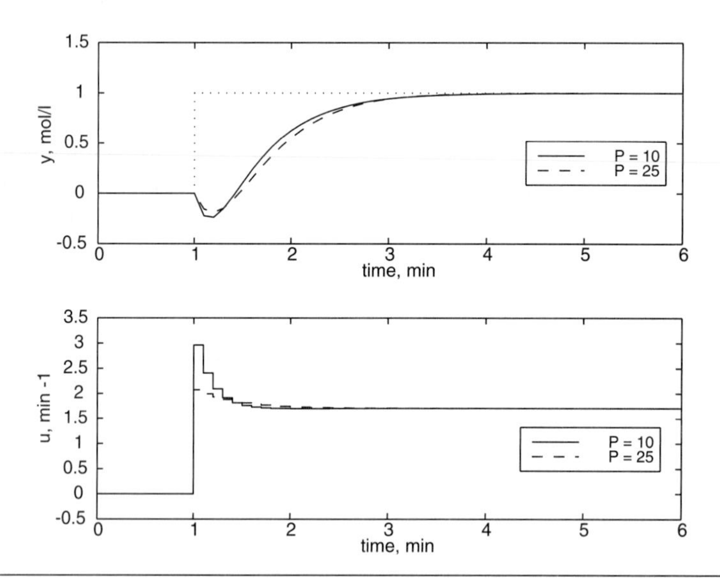

Figure 16–9 Closed-loop responses to a step setpoint change, van de Vusse example. Sample time = 0.1 minute; model length, $N = 50$; control horizon, $M = 1$; no weighting on input. Comparison of prediction horizons, $P = 10$ and $P = 25$.

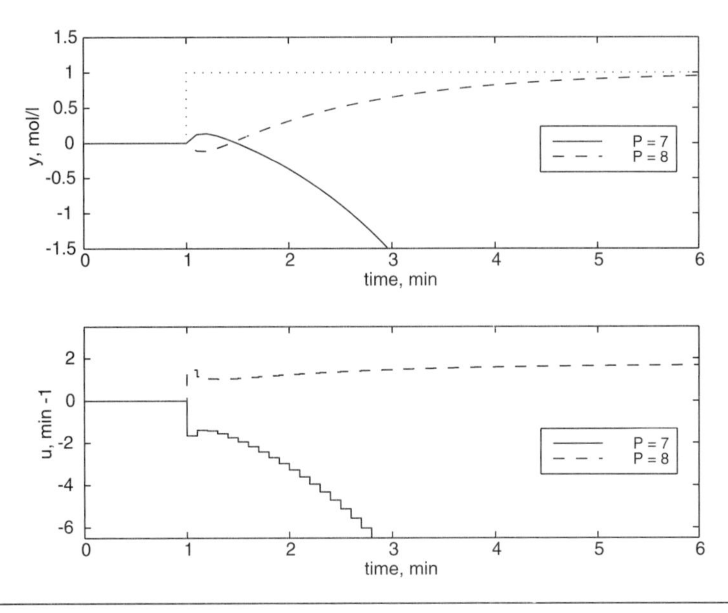

Figure 16–10 Closed-loop responses to a step setpoint change, van de Vusse example. Sample time = 0.1 minute; model length, $N = 50$; control horizon, $M = 1$; no weighting on input. Comparison of prediction horizons, $P = 7$ and $P = 8$.

16.4 Constraints and Multivariable Systems

The dynamic matrix control technique presented in the previous section is based on an unconstrained optimization of current and future control moves. The combination of a linear model and a quadratic objective function lead to an analytical solution for the control moves. In practice, constraints on manipulated inputs (control moves) can be very important. If Equation (16.23), which is the analytical solution to Equation (16.22), results in an infeasible control action (e.g., a violation of constraints, such as a flow rate greater than the maximum possible flow), then obviously the control moves must be "truncated" at the maximum or minimum values. Since the resulting truncated solutions may not be optimal if the control horizon is greater than 1, it is important to use a constrained optimization formulation for these problems. Fortunately, dynamic matrix control is easily formulated to explicitly handle constraints by using quadratic programming (QP); the method is known as Quadratic Dynamic Matrix Control (QDMC).

Quadratic DMC (QDMC)

QDMC considers constraints on the manipulated inputs. The input constraints can be of the following form

$$u_{\min} \le u_{k+i} \le u_{\max} \tag{16.25}$$

which is suitable for minimum and maximum flowrates, for example. In addition, velocity constraints that limit the magnitude of the control moves at each sample time have the following form

$$\Delta u_{\min} \le \Delta u_{k+i} \le \Delta u_{\max} \tag{16.26}$$

where ordinarily, $\Delta u_{\min} = -\Delta u_{\max}$. To use a standard quadratic program (QP), the constraints in (16.25) need to be written in terms of the control moves, Δu_{k+i}. Since the previously implemented control action (u_{k-1}) is known, we can write

$$\begin{aligned} u_k &= u_{k-1} + \Delta u_k \\ u_{k+1} &= u_{k-1} + \Delta u_k + \Delta u_{k+1} \end{aligned} \tag{16.27}$$

and so on. Since the manipulated input constraints are enforced over the control horizon of M steps, (16.25) and (16.27) yield

$$\begin{bmatrix} u_{\min} \\ u_{\min} \\ \vdots \\ u_{\min} \end{bmatrix} \le \begin{bmatrix} u_{k-1} \\ u_{k-1} \\ \vdots \\ u_{k-1} \end{bmatrix} + \begin{bmatrix} 1 & 0 & 0 & \cdots & 0 \\ 1 & 1 & 0 & \cdots & 0 \\ \vdots & \vdots & & & \\ 1 & 1 & 1 & \cdots & 1 \end{bmatrix} \begin{bmatrix} \Delta u_k \\ \Delta u_{k+1} \\ \vdots \\ \Delta u_{k+M-1} \end{bmatrix} \le \begin{bmatrix} u_{\max} \\ u_{\max} \\ \vdots \\ u_{\max} \end{bmatrix} \tag{16.28}$$

Most standard QP codes use a "one-sided" form

$$\begin{bmatrix} 1 & 0 & 0 & \cdots & 0 \\ 1 & 1 & 0 & \cdots & 0 \\ \vdots & \vdots & & & \\ 1 & 1 & 1 & \cdots & 1 \end{bmatrix} \begin{bmatrix} \Delta u_k \\ \Delta u_{k+1} \\ \vdots \\ \Delta u_{k+M-1} \end{bmatrix} \ge \begin{bmatrix} u_{\min} - u_{k-1} \\ u_{\min} - u_{k-1} \\ \vdots \\ u_{\min} - u_{k-1} \end{bmatrix} \tag{16.29a}$$

and

$$-\begin{bmatrix} 1 & 0 & 0 & \cdots & 0 \\ 1 & 1 & 0 & \cdots & 0 \\ \vdots & \vdots & & & \\ 1 & 1 & 1 & \cdots & 1 \end{bmatrix} \begin{bmatrix} \Delta u_k \\ \Delta u_{k+1} \\ \vdots \\ \Delta u_{k+M-1} \end{bmatrix} \ge \begin{bmatrix} u_{k-1} - u_{\max} \\ u_{k-1} - u_{\max} \\ \vdots \\ u_{k-1} - u_{\max} \end{bmatrix} \tag{16.29b}$$

which have the form $A\Delta u_f \ge b$.

The velocity constraints are implemented as bounds on the control moves

$$\begin{bmatrix} \Delta u_{\min} \\ \Delta u_{\min} \\ \vdots \\ \Delta u_{\min} \end{bmatrix} \le \begin{bmatrix} \Delta u_k \\ \Delta u_{k+1} \\ \vdots \\ \Delta u_{k+M-1} \end{bmatrix} \le \begin{bmatrix} \Delta u_{\max} \\ \Delta u_{\max} \\ \vdots \\ \Delta u_{\max} \end{bmatrix} \tag{16.30}$$

The majority of constrained MPC problems can be solved based on the input constraints considered above. For completeness, however, we also show how constraints on the process outputs can be included.

It may be desirable to force the predicted process outputs to be within a range of minimum and maximum values

$$y_{\min} \le \hat{y}^{c}_{k+i} \le y_{\max} \tag{16.31}$$

Here we first rewrite Equation (16.15) in the following form

$$\underbrace{\hat{Y}^{c}}_{\substack{\text{corrected}\\ \text{predicted}\\ \text{outputs}}} = \underbrace{S_f \Delta u_f}_{\substack{\text{effect of}\\ \text{current and}\\ \text{future moves}}} + \underbrace{S_{past}\Delta u_{past} + s_N u_P}_{\text{effect of past moves}} + \underbrace{\hat{d}}_{\substack{\text{predictive}\\ \text{disturbances}}} \tag{16.32}$$

$$\hat{Y}^{c} = S_f \Delta u_f + f$$

where f, the free response of the "corrected-predicted output" (if no current and future control moves are made) is

$$f = S_{past}\Delta u_{past} + s_N u_P + \hat{d} \tag{16.33}$$

so that (16.31) can be written

$$y_{\min} - f \le S_f \Delta u_f \le y_{\max} - f \tag{16.34}$$

or, in terms of one-sided inequalities

$$\begin{aligned} S_f \Delta u_f &\ge y_{\min} - f \\ -S_f \Delta u_f &\ge -y_{\max} + f \end{aligned} \tag{16.35}$$

Using shorthand matrix-vector notation, the quadratic programming problem is stated as

$$\min_{\Delta u_f} \; \Phi = \frac{1}{2} \cdot \Delta u_f^T H \Delta u_f + c^T \Delta u_f \tag{16.36}$$

$$\text{s.t.} \quad \begin{aligned} A\Delta u_f &\ge b \\ \Delta u_{\min} \le \Delta u_f &\le \Delta u_{\max} \end{aligned} \tag{16.37}$$

where the matrices and vectors in the objective function are

$$\begin{aligned} H &= S_f^T S_f + W \\ c^T &= E^T S_f \end{aligned} \tag{16.38}$$

and the inequality matrices, A and b, in (16.37) incorporate the matrices in (16.29) and (16.35), with appropriate dimensionality. For example, if there are no output constraints, then A and b are

$$A = \begin{bmatrix} 1 & 0 & 0 & \cdots & 0 \\ 1 & 1 & 0 & \cdots & 0 \\ \vdots & \vdots & & & \\ 1 & 1 & 1 & 1 & 1 \\ -1 & 0 & 0 & & 0 \\ -1 & -1 & 0 & & 0 \\ & & & & \\ -1 & -1 & -1 & -1 & -1 \end{bmatrix}$$

$$b = \begin{bmatrix} u_{\min} - u_{k-1} \\ u_{\min} - u_{k-1} \\ \vdots \\ u_{\min} - u_{k-1} \\ u_{k-1} - u_{\max} \\ u_{k-1} - u_{\max} \\ \vdots \\ u_{k-1} - u_{\max} \end{bmatrix} \tag{16.39}$$

Futher presentation is beyond the scope of this text. Example m-files for the study of constrained systems will be placed on the book web page.

Multivariable Systems

In this chapter we have focused on single input–single output processes. One reason that MPC has been so widely applied is that it naturally handles multivariable systems. The same basic ideas presented in this chapter hold for the multivariable case, but the formulation for step response models in a DMC framework is somewhat tedious. We recommend that you use the Model Predictive Control Toolbox to perform multivariable studies.

16.5 Other MPC Methods

MPC is a basic concept or idea that can be implemented in many ways, depending on models used and assumptions made (about disturbances and so on). With DMC, future output predictions were composed of two effects: the unforced (or free) response (effect of past control moves) and the forced response (effect of current and future control moves). Virtually all other model-based procedures use the same idea. For example, if state space models are used, the free response can be obtained simply by iteration from the current state values, assuming no input changes. Here we discuss a number of industrial techniques and current MPC research.

MPC with Industrial Applications

The previous section covered the most widely applied (in industry) MPC technique, DMC. There are a wide variety of other MPC techniques that have been used in industrial applications; only a few are discussed here. A nice overview of these approaches is provided by Qin and Badgwell (1997).

Model predictive heuristic control (MPHC) was developed in France and received initial applications in the French petroleum refining industry. The software was called IDCOM for "identification and command." This work is summarized by Richalet et al. (1978). A finite impulse response model is used, and the desired setpoint trajectory over the prediction horizon is first order.

IDCOM-M, developed by Setpoint, Inc. (now part of ASPEN Technology), differs from IDCOM in a number of ways. In addition to handling multivariable processes, it includes constraint handling, where hard constraints can be ranked in terms of priority. Also, rather than considering outputs over an entire prediction horizon, each output is controlled at a specific point in the future (termed the coincidence point).

Generalized predictive control (GPC) arose from adaptive control research, where model parameters are continuously updated to account for changing process dynamics. It is geared toward SISO processes and has had few major industrial applications. Clarke (1988) provides an overview of some applications.

MPC Research

Problems with Existing Techniques

Many of the "classical" MPC approaches used in industry have performance limitations, which we summarize here.

- *Model structure.* The finite step and impulse response models limit applications to open-loop stable processes and require many model coefficients to describe the response (indeed, even a first-order system with two parameters may require 50 or more step response coefficients). Integrating systems have been handled by formulating the derivative of an integrating output as the controlled output.
- *Disturbance assumption.* In Chapter 8 we found that the standard IMC design may not handle input disturbances very well, since it is formulated for output disturbances. Similarly, DMC uses a "constant output disturbance" assumption. This may not yield good performance if the real disturbance occurs at the plant input.
- *Finite horizons.* We found in the examples presented in Section 16.3 that control performance could deteriorate if the prediction or control horizons were not formulated correctly, even if the model was perfect.
- *Model type.* The step and impulse response models are all linear. For some processes (exothermic reactors) where the process operating conditions are changed frequently (different product specifications for each consumer, for example), a single linear model may not describe the dynamic behavior of the process over the wide range of conditions. Batch processes also operate over a wide range of conditions. For these systems, better control performance may be achieved if nonlinear models are used.

Solutions Being Developed

Much academic research has gone into developing approaches to solving these problems. It is now recognized that there are many advantages to using discrete *state space* models. State space models require fewer model parameters than step and impulse response models to describe process behavior. State space models also provide more flexibility for disturbance estimation. *State estimation* techniques can be easily formulated to handle input disturbances, for example. State estimation also allows unstable systems to be controlled. This is discussed by Muske and Rawlings (1993), for example.

Infinite horizon approaches can be used to guarantee closed-loop stability. Although it may seem impossible to solve an optimization problem over an infinite horizon, linear models have analytical solutions, allowing rapid computations to be performed. This topic is covered by Muske and Rawlings (1993).

Nonlinear models can be used to improve performance, compared with linear models. Reviews of these approaches are provided by Bequette (1991) and Henson (1998). An alternative is to include several linear models to approximate the nonlinear system over a range of operating conditions. The linear models can then be "scheduled" or "interpolated" so that the best linear model approximation to the nonlinear process is used at a particular operating condition. An application to drug infusion during surgery is discussed by Rao et al. (2001).

16.6 MATLAB

There is an MPC Toolbox available through Mathworks. It is recommended if you are going perform many MPC simulations, especially if you are working with multivariable systems.

For those without the MPC Toolbox, simple *m*-files that can be modified for problems of interest are presented in Appendix 16.3. It is suggested that you download these files from the course web page and use them for Problems 1 and 2. For each new problem, simply insert the new continuous state space model at the appropriate location in the `dmcsim.m` file.

16.7 Summary

MPC is the most widely applied advanced control technique in industry. Our presentation focused on DMC, which has achieved a great deal of success in the petroleum refining and petrochemicals industries. Most MPC techniques have been based on step or impulse response models, but there has been a recent trend toward state space models. New mathematical techniques are making it easier to develop state space models from plant data. Regardless of the model type, the same basic ideas are used. At the current time step, future process output predictions are based on two contributions: the *free or unforced response* (how the outputs will change if no further control moves are made) and the *forced response* (the effect of current and future control moves on the predicted outputs). In DMC, the free response is the effect of the past control moves and the additive correction term, while the forced response is the effect of the current and future (to be calculated) control moves.

We studied two simple examples to develop a basic understanding of how to tune SISO model predictive controllers. When using step or impulse models, it is important to make certain that the model length (N) is long enough to capture the steady-state change; although not discussed, it is important to "filter," or smooth, the step response data. We found that it is also important to make certain that the prediction horizon (P) is long enough to avoid stability problems, particularly for systems with inverse response behavior. Generally, control horizons (M) are much shorter than are prediction horizons, yielding more robust performance. A disadvantage of MPC is that there are many parameters (model length, prediction horizon, control horizon, manipulated input weighting, and even sample time) that affect the closed-loop performance.

The variables used in this chapter are as follows:

M	control horizon ($N \geq P \geq M$)
P	prediction horizon ($P \geq M$)
N	model length ($N \geq P$)
S_f	dynamic matrix
S	vector of step response coefficients
S_{past}	matrix, used for effect of past control moves
s_i	step response coefficient
Δt	sample time
u	manipulated input
Δu	change in manipulated input (control move)
w	weight applied to manipulated in the objective function (often 0 if $P >> M$)

The following nomenclature is common

DMC	dynamic matrix control
FIR	finite impulse response
FSR	finite step response
MPC	model predictive control

References and Relevant Literature

There are large number of papers that discuss many different aspects and formulations of MPC. The basic DMC approach is presented by Cutler and Ramaker (1980). An early academic paper analyzing DMC is presented by Marchetti et al. (1983). Garcia et al. (1989) review early theory and applications, while Morari and Lee (1999) summarize recent results and current research topics. The monograph by Camacho (1999) presents a number of techniques but focuses on GPC. The textbook by Maciejowski (2002) includes MATLAB files and a number of application case studies.

Camacho, E.F., and C. Bordons, *Model Predictive Control*, Springer-Verlag, London (1999).

Clarke, D.W., "Application of Generalized Predictive Control to Industrial Processes," *IEEE Control Syst. Mag.*, **April**, pp. 49–55 (1988).

Cutler, C.R., and B.L. Ramaker, "Dynamic Matrix Control—A Computer Control Algorithm," in *Proc. Joint Automatic Control Conference*, San Francisco, CA, Paper WP5-B (1980).

Garcia, C.E., and A.M. Morshedi, "Quadratic Programming Solution of Dynamic Matrix Control (QDMC)," *Chem. Eng. Commun.*, **46**, 73–87 (1986).

Garcia, C.E., D.M. Prett, and M. Morari, "Model Predictive Control: Theory and Practice—A Survey," *Automatica*, **25**, 335–348 (1989).

Maciejowski, J.M., *Predictive Control: with Constraints*, Prentice Hall, Harlow, England (2002).

Marchetti, J.L., D.A. Mellichamp, and D.E. Seborg, "Predictive Control Based on Discrete Convolution Models," *Ind. Eng. Chem. Proc. Des. Dev.*, **22**(3), 488–495 (1983).

Morari, M., and J.H. Lee, "Model Predictive Control: Past, Present and Future," *Comp. Chem. Eng.*, **23**, 667–682 (1999).

Muske, K., and J.B. Rawlings, "Model Predictive Control with Linear Models," *AIChE J.*, **39**(2), 262–287 (1993).

Qin, S.J., and T.A. Badgwell, "An Overview of Industrial Model Predictive Control Technology," *Proceedings of the Fifth Conference on Chemical Process Control,* AIChE Symp. Ser. 316, Vol. 93, pp. 232–256 (Kantor, J.C., C.E. Garcia, and B. Carnahan, Eds.), (1997).

Richalet, J.A., A. Rault, J.D. Testud, and J. Papon, "Model Predictive Heuristic Control: Applications to Industrial Processes," *Automatica*, **14**, 413–428 (1978).

Shah, S.L., "A Tutorial Introduction to Constrained Long Range Predictive Control," *Pulp Pap. Can.*, **96**(4), 57–63 (1995).

A focus on nonlinear systems is presented in the following papers and books.

Allgower, F., T.A. Badgwell, J.S. Qin, J.B. Rawlings, and S.J. Wright, "Nonlinear Predictive Control and Moving Horizon Estimation—An Introductory Overview," Advances in Control: Highlights of ECC '99, Ch. 12, pp. 391–449, (P.M. Frank, Ed.), Springer-Verlag, Berlin/New York (1999).

Bequette, B.W., "Nonlinear Control of Chemical Processes: A Review," *Ind. Eng. Chem. Res.*, **30**, 1391–1413 (1991).

Henson, M.A., "Nonlinear Model Predictive Control: Current Status and Future Directions," *Comp. Chem. Eng.*, **23**, 187–202 (1998).

Lee, J.H., and N.L. Ricker, "Extended Kalman Filter Based Nonlinear Model Predictive Control," *Ind. Eng. Chem. Res.*, **33**(6), 1530–1541 (1994).

Meadows, E.S., and J.B. Rawlings, "Model Predictive Control," in *Nonlinear Process Control*, pp. 233–310 (M.A. Henson and D.E. Seborg, Eds.), Prentice Hall, Upper Saddle River, NJ (1997).

Sistu, P.B., and B.W. Bequette, "Model Predictive Control of Processes with Input Multiplicities," *Chem. Eng. Sci.,* **50**(6), 921–936 (1995).

An application of a multiple model-based MPC approach is discussed in the following paper:

Rao, R., C.C. Palerm, B. Aufderheide, and B.W. Bequette, "Experimental Studies on Automated Regulation of Hemodynamic Variables," *IEEE Engineering in Medicine and Biology Magazine*, **20**(1), 24-38 (Jan/Feb, 2001).

Student Exercises

1. Reproduce the DMC simulations for the van de Vusse reactor shown in Figures 16–9 and 16–10 using the files shown in Appendix 16.3 (also, you may download these files from the book web page).
2. Run the DMC simulations for the van de Vusse reactor, for a prediction horizon of 10 (and a model length of 50). Vary the control horizon and discuss the effect; for example, if $M = 5$, the manipulated input resembles a "pulse" change. Also, try different weighting on the manipulated input and discuss the effect.
3. Consider a biochemical reactor operated at a stable operating condition (Module 7), with the state space model

$$A = \begin{bmatrix} 0 & 0.9056 \\ -0.75 & -2.5640 \end{bmatrix}, \; B = \begin{bmatrix} -1.5301 \\ 3.8255 \end{bmatrix}$$
$$C = \begin{bmatrix} 1 & 0 \end{bmatrix}, \; D = \begin{bmatrix} 0 \end{bmatrix}$$

   where the input is the dilution rate (units = min^{-1}) and the output is the biomass concentration (gmol/L). Modify the DMC simulation code presented in Appendix 16.3 to simulate the MPC applied to this system. Use a sample time of 0.05 hours and discuss the effect of model length, prediction horizon, and control horizon on the performance for setpoint changes of 0.02 gmol/L. Consider the effect of these parameters on the magnitude of manipulated input changes. What are your recommended tuning parameters if the maximum manipulated input change is 0.2 min^{-1}?
4. Consider a jacketed chemical reactor, where jacket temperature is the manipulated input and reactor temperature is the measured output. For the 100 ft^3 propylene glycol reactor presented in Module 8, the state space model is

$$A = \begin{bmatrix} -7.9909 & -0.013674 \\ 2922.9 & 4.5564 \end{bmatrix}, \; B = \begin{bmatrix} 0 \\ 1.4582 \end{bmatrix}$$
$$C = \begin{bmatrix} 0 & 1 \end{bmatrix}, \; D = \begin{bmatrix} 0 \end{bmatrix}$$

Modify the DMC simulation code presented in Appendix 16.3 to simulate MPC applied to this system. Use a sample time of 0.05 hours and discuss the effect of model length, prediction horizon, and control horizon (remember that $N > P > M$) on the performance for setpoint changes of 5°F. Consider the effect of the tuning parameters on the manipulated input changes (that is, is the change in jacket temperature reasonable?).

5. Consider a discrete state space model with the following form

$$x_{k+1} = \Phi x_k + \Gamma u_k$$
$$y_{k+1} = C x_{k+1}$$

Assume that the state vector is perfectly known at the current time step (k), the model is perfect, and there are no disturbances. Predict the output up to P time steps in the future, based on M current and future control moves (Δu_k, ..., Δu_{k+M-1}). Realize that $u_k = u_{k-1} + \Delta u_k$, etc. Show how the output prediction is naturally composed of free and forced response contributions.

Appendix 16.1: Derivation of the Step Response Formulation

From superposition of all previous inputs, the output at step k can be found from

$$y_k = \sum_{i=1}^{\infty} s_i \Delta u_{k-i} = s_1 \Delta u_{k-1} + \cdots + s_N \Delta u_{k-N} + s_{N+1} \Delta u_{k-N-1} + \cdots + s_{N+\infty} \Delta u_{k-\infty}$$

However, assuming that all step response coefficients greater than step N are equal to the value at step N (this is only true for stable systems with a long model horizon), we can write

$$y_k = \sum_{i=1}^{\infty} s_i \Delta u_{k-i} = s_1 \Delta u_{k-1} + \cdots + s_{N-1} \Delta u_{k-N+1} + s_N \Delta u_{k-N} + \cdots + s_N \Delta u_{k-\infty}$$

$$= s_1 \Delta u_{k-1} + \cdots + s_{N-1} \Delta u_{k-N+1} + s_N \underbrace{\left(\Delta u_{k-N} + \cdots + \Delta u_{k-\infty}\right)}_{u_{k-N}}$$

$$y_k = s_N u_{k-N} + \sum_{i=1}^{N-1} s_i \Delta u_{k-i}$$

Appendix 16.2: Derivation of the Least Squares Solution for Control Moves

The objective function is

$$\Phi = (E - S_f\Delta u_f)^T(E - S_f\Delta u_f) + (\Delta u_f)^T W\Delta u_f$$

Expanding the objective function, we find

$$\begin{aligned}\Phi &= E^TE - \Delta u_f^T S_f^T E - \Delta u_f^T S_f^T E + \Delta u_f^T S_f^T S_f \Delta u_f + \Delta u_f^T W\Delta u_f \\ &= -2\Delta u_f^T S_f^T E + \Delta u_f^T(S_f^T S_f + W)\Delta u_f\end{aligned}$$

Unconstrained System

The minimization with respect to the control move vector, when there are no constraints, is

$$\frac{\partial \Phi}{\partial \Delta u_f} = -2S_f^T E + 2\left(S_f^T S_f + W\right)\Delta u_f = 0$$

and solving for Δu_f we find

$$\Delta u_f = \underbrace{\left(S_f^T S_f + W\right)^{-1} S_f^T}_{K} \underbrace{E}_{\text{unforced errors}} \tag{16.23}$$

Constrained System

The objective function for a constrained system is no different than the unconstrained

$$\min_{\Delta u_f} \quad \Phi = \Delta u_f^T \underbrace{\left(S_f^T S_f + W\right)}_{H}\Delta u_f - \Delta u_f^T 2S_f^T E$$

but this is often written in the following form

$$\min_{\Delta u_f} \quad \Phi = \frac{1}{2}\cdot \Delta u_f^T H\Delta u_f + c^T\Delta u_f$$

and since

$$\Delta u_f^T S_f^T E = E^T S_f \Delta u_f$$

the objective function in the form of most quadratic programs is

$$\min_{\Delta u_f} \quad \Phi = \frac{1}{2}\cdot \Delta u_f^T \underbrace{\left(S_f^T S_f + W\right)}_{H}\Delta u_f \underbrace{-E^T S_f}_{c^T}\Delta u_f$$

and

$$H = S_f^T S_f + W$$
$$c^T = -E^T S_f$$

The constraints

$$A\Delta u_f \geq b$$
$$\Delta u_{\min} \leq \Delta u_f \leq \Delta u_{\max}$$

are shown in Section 16.4.

Appendix 16.3

Use the following files to reproduce the van de Vusse simulations shown in Figure 16–9. These are based on an unconstrained DMC strategy. Files for the constrained formulation, involving `qp` and `quadprog` from the `optimization toolbox`, are available on the book web page.

dmcsim.m

```
% dmcsim
%
% 26 Oct 00 - revised 4/5 Aug 01 for Chapter 16 - MPC
% b.w. bequette
% unconstrained DMC simulation, calls: smatgen.m and dmccalc.m
% currently contains van de vusse reactor model and plant
%
% MPC tuning and simulation parameters
  n = 50;         % model length
  p = 10;         % prediction horizon
  m =  1;         % control horizon
  weight = 0.0;   % weighting factor
  ysp = 1;        % setpoint change (from 0)
  timesp = 1;     % time of setpoint change
  delt =   0.1;   % sample time
  tfinal = 6;     % final simulation time
  noise  = 0;     % noise added to response coefficients
%
  t = 0:delt:tfinal;  % time vector
  kfinal = length(t); % number of time intervals
```

```
  ksp = fix(timesp/delt);
  r = [zeros(ksp,1);ones(kfinal-ksp,1)*ysp]; % setpoint vector
%
% ----- insert continuous model here -----------
% model (continuous state space form)
%
  a = [-2.4048 0;0.8333 -2.2381]; % a matrix - van de vusse
  b = [7; -1.117];                % b matrix - van de vusse
  c = [0 1];                      % c matrix - van de vusse
  d = 0;                          % d matrix - van de vusse
  sysc_mod = ss(a,b,c,d); % create LTI "object"
%
% ----- insert plant here -----------
% perfect model assumption (plant = model)
  ap = a;
  bp = b;
  cp = c;
  dp = d;
  sysc_plant = ss(ap,bp,cp,dp);
%
% discretize the plant with a sample time, delt
%
  sysd_plant = c2d(sysc_plant,delt)
  [phi,gamma,cd,dd] = ssdata(sysd_plant)
%
% evaluate discrete model step response coefficients
%
  [s] = step(sysc_mod,[delt:delt:n*delt]);
%
% generate dynamic matrices (both past and future)
%
  [Sf,Sp,Kmat] = smatgen(s,p,m,n,weight);
%
% plant initial conditions
  xinit = zeros(size(a,1),1);
  uinit = 0;
  yinit = 0;
% initialize input vector
  u     = ones(min(p,kfinal),1)*uinit;
```

```
%
  dup    = zeros(n-2,1);
  sn     = s(n);        % last step response coefficient
  x(:,1)= xinit;
  y(1)   = yinit;
  dist(1) = 0;
%
% set-up is done, start simulations
%
 for k = 1:kfinal;
%
 du(k) = dmccalc(Sp,Kmat,sn,dup,dist(k),r(k),u,k,n);
% perform control calculation
  if k > 1;
 u(k) = u(k-1)+du(k); % control input
  else
 u(k) = uinit + du(k);
  end
% plant equations
 x(:,k+1) = phi*x(:,k)+gamma*u(k);
 y(k+1) = cd*x(:,k+1);
% model prediction
  if k-n+1>0;
   ymod(k+1) = s(1)*du(k) + Sp(1,:)*dup + sn*u(k-n+1);
  else
   ymod(k+1) = s(1)*du(k) + Sp(1,:)*dup;
  end
% disturbance compensation
%
 dist(k+1) = y(k+1) - ymod(k+1);
% additive disturbance assumption
% put input change into vector of past control moves
 dup = [du(k);dup(1:n-3)];
 end
%
% stairs plotting for input (zero-order hold) and setpoint
%
  [tt,uu] = stairs(t,u);
  [ttr,rr] = stairs(t,r);
```

```
%
  figure(1)
  subplot(2,1,1)
  plot(ttr,rr,'--',t,y(1:length(t)))
  ylabel('y')
  xlabel('time')
  title('plant output')
  subplot(2,1,2)
  plot(tt,uu)
  ylabel('u')
  xlabel('time')
```

smatgen.m

```
function [Sf,Sp,Kmat] = smatgen(s,p,m,n,w)
%
% b.w. bequette
% 28 Sept 00, revised 2 Oct 00
% generates dynamic matrix and feedback gain matrix
% assumes s = step response column vector
% Sf   = Dynamic Matrix for future control moves (forced)
% Sp   = Matrix for past control moves (free)
% Kmat = DMC feedback gain matrix
% s    = step response coefficient vector
% p    = prediction horizon
% m    = control horizon
% n    = model horizon
% w    = weight on control input
%
% first, find the dynamic matrix
  for j = 1:m;
         Sf(:,j) = [zeros(j-1,1);s(1:p-j+1)];
  end
%
% now, find the matrix for past moves
%
  for i = 1:p;
         Sp(i,:) = [s(i+1:n-1)' zeros(1,i-1)];
  end
```

```
%
% find the feedback gain matrix, Kmat
%
  Kmat = inv(Sf'*Sf + w*eye(m))*Sf';
```

dmccalc.m

```
function [delu] = dmccalc(Sp,Kmat,sn,delup,d,r,u,k,n)
%
% for use with dmcsim.m
% b.w. bequette
% 2 oct 00
% calculate the optimum control move
%
% first, calculate uold = u(k-n+1)...u(k-n+p)
%
  [m,p] = size(Kmat);
    uold = zeros(p,1);
  for i = 1:p;
       if k-n+i>0;
       uold(i) = u(k-n+i);
       else
       uold(i) = 0;
       end
  end
  dvec  = d*ones(p,1);
  rvec   = r*ones(p,1);
  y_free = Sp*delup + sn*uold + dvec;
  e_free = rvec-y_free;
  delu = Kmat(1,:)*e_free;
```

CHAPTER 17

Summary

A lot of material has been covered in this textbook, yet there are many control-related topics that have not been covered. The purpose of this chapter is to provide a concise summary of the topics covered and to provide an overview of what the reader should expect in a process engineering position. This chapter should spark your memory and help you determine which topics require a more in-depth review.

After reading this chapter the reader should be able to:

- Recall most of the important topics covered in the textbook
- "Ace" a final examination in a dynamics and control course
- Have a feel for important practical issues in process engineering

The major sections of this chapter are as follows:

17.1 Overview of Topics Covered in This Textbook

In the next few pages we attempt to concisely review the topics covered in this textbook. This summary should assist you in preparing for a final examination in a process control course, for example. It also serves the purpose of providing a quick way of helping you determine what topics you would like to understand better. In the next section these topics are reviewed in the chronological order presented in the text. In the subsequent section a concise review of the material presented in the modules is presented.

Chapters

Chapter 1 provided the motivation for process control, while Chapter 2 introduced fundamental models. Chapter 3 presented dynamic behavior, with a focus on transfer functions. Chapter 4 covered the development of empirical models, including continuous time step responses, as well as the identification of parameters for discrete-time models. Chapter 5 provided an introduction to the analysis of closed-loop control strategies, introducing the idea of a block diagram. Chapter 6 presented PID controller tuning techniques, followed by frequency response analysis in Chapter 7. Many advanced control techniques rely on the use of a process model embedded in the control strategy; Chapter 8 covers the Internal Model Control strategy. Internal model controllers can often be rearranged to form a PID controller, as shown in Chapter 9. Improved disturbance rejection is the main motivation of the cascade and feedforward techniques presented in Chapter 10. Enhancements to PID control, including autotuning, integral windup protection, and nonlinear approaches are discussed in Chapter 11. Ratio, split-range and selective controllers are presented in Chapter 12. The interaction of multiple SISO loops, with the RGA as the main analysis tool, is developed in Chapter 13. In Chapter 14, right-half-plane transmission zeros indicate dynamic performance limitations; the singular value decomposition provides insight about possible steady-state performance limitations. The topic of plantwide control is the focus of Chapter 15, followed by model predictive control in Chapter 16. Specific topics are presented in the pages that follow.

Given a process, be able to do the following:

- Identify the control objective and the state, output, manipulated, and disturbance variables
- Develop the fundamental nonlinear model
- Solve for the steady state
- Linearize the nonlinear model about the steady state to find the state-space model
- Find the transfer functions relating each input to each output
- Convert to different sets of units

Given experimental step responses, the reader should be able to estimate parameters for low-order transfer function models that match the measured process output.

Given steady-state information, the reader should be able to do the following:

- Convert deviation variable results to physical variable results
- Convert physical variable results to deviation variables

Given a process transfer function, be able to

- Relate the gain-time constant form and the pole-zero form
- Relate poles and time constants
- Relate the location of a pole in the complex plane to the speed of response
- Consider a first-order + dead time transfer function (Why does the first-order Padé approximation for dead time provide a good approximation for the response to a step input? *Hint*: Think of RHP zeros.)
- Use the final- and initial-value theorems of Laplace transforms

Given a physical process with feedback control (a process instrumentation and control diagram), the reader should be able to do the following:

- Draw the block diagram for any type of feedback control (all variables should be clearly identified on the block diagram

For a given block diagram, the reader should be able to do the following:

- Derive the closed-loop transfer functions (for example, relate the setpoint to the output for a feedback control scheme)
- Relate the disturbance to the output for a feedback control scheme

For a given process transfer function, the reader should be able to do the following:

- Determine the offset due to proportional feedback control
- Perform a closed-loop Ziegler-Nichols (continuous oscillation) test
- Compare the performance of Ziegler-Nichols and Tyreus-Luyben suggested tuning parameter values
- Use the Routh stability criterion to determine if a closed-loop system is stable (the denominator of your closed-loop transfer function will usually be a polynomial with an order of 2–4

- Determine the range of tuning parameter values that will yield a stable feedback system if the controller tuning parameters are not given
- Use the *direct synthesis method* to find a control algorithm that yields a desired closed-loop response. Find the corresponding PID parameters if the controller has PID form (sometimes in series with lead-lag)
- Calculate the poles and discuss the response characteristics for a closed-loop transfer function that results in a first or second-order polynomial
- Discuss the response characteristics if you are given the closed-loop poles for a higher order closed-loop polynomial

Given a Bode diagram of $g_c(\omega)g_p(\omega)$, be able to do the following:

- Determine the gain and phase margins
- Use that information to calculate the amount of additional deadtime allowable before instability occurs

Given a Nyquist diagram of $g_c(\omega)g_p(\omega)$, be able to do the following:

- Determine if the closed loop system will be stable
- Understand the relationship between Bode and Nyquist plots (How is the information from a Bode plot translated to the Nyquist diagram? How is information of the Nyquist diagram translated to a Bode plot?)

Understand that the closed-loop Ziegler-Nichols method is equivalent to finding the proportional controller gain where the phase angle is -180° when the open-loop ($g_c{*}g_p$) amplitude ratio is one.

For a given transfer function the reader should be able to:

- Determine the magnitude of the output from the transfer function at low input frequencies and at high input frequencies
- Understand what is meant by the term "nonminimum phase"
- Be able to determine if a transfer function is proper, semi-proper, or improper

Be able to handle the complete IMC design procedure for a given process transfer function:

- Factor $g_p(s)$ into minimum-phase and nonminimum-phase elements {good [$g_{p-}(s)$] vs. bad [$g_{p+}(s)$] stuff}

- Invert the minimum phase elements to form the ideal IMC controller
- Cascade the ideal IMC controller with a filter that is high enough order to form a physically realizable IMC controller
- Determine what the closed-loop response will look like for an IMC scheme with a perfect model
- Understand the effect of λ, the IMC filter time constant, on the closed-loop response (How does λ impact the robustness? A good first guess for tuning λ is to use roughly one third to one half the dominant time constant, depending on the amount of time-delay and model uncertainty; note that λ has units of time)
- Know that the *IMC-based PID* procedure will give the same closed-loop results as the IMC strategy when there are no time delays (the behavior will be different when there are time delays because the IMC-based PID procedure uses a Padé approximation for the time delay)

Design a physically realizable feed-forward controller

Derive equivalent transfer functions to analyze a cascade-control strategy as a standard feedback control strategy

Use ARW techniques (controllers with integral action can exhibit "reset windup" when the manipulated input becomes constrained)

For multivariable systems, the reader should be able to do the following:

- Use the RGA for input-output pairing for MVSISO systems
- Renumber the inputs and outputs such that the favorable RGA pairings appear on the diagonal of the new gain matrix
- **Realize that you should *never* pair on a negative or zero relative gain**
- Find the structure of the control matrix that corresponds to the RGA pairing that has been selected
- Find the closed-loop transfer functions for any multivariable block diagram
- Realize that the order of multiplication for matrices is critical
- Implement steady-state decoupling
- Calculate multivariable transmission zeros
- Use a singular value decomposition to understand directional effects
- Implement MV IMC, using a diagonal factorization

For a system with more than two inputs and two outputs, the reader should know the following:

- To check the RGAs of all reduced-order structures (this is to make certain that when a control loop is out of service, the reduced order system is not failure sensitive)

Place control loops on a process flow sheet (one heuristic is that one stream in a recycle loop should be under flow control)

Understand the basic idea behind MPC. For the step response-based technique of DMC, understand

- How the model length (N) is determined
- Are longer prediction horizons (P) more robust than shorter ones?
- Are shorter control horizons (M) more robust than longer ones?

Modules

The modules in the final section of the text provide detailed application examples to illustrate the techniques presented in the Chapters. Module 1 reviews MATLAB, while Module 2 covers SIMULINK. Numerical integration of ordinary differential equations using MATLAB is presented in Module 3, while Module 4 presents useful functions available in the Control System Toolbox. An isothermal CSTR with a series/parallel reaction structure (the van de Vusse reaction) is studied in Module 5. Frequency response techniques are used to analyze the classic first-order + dead time model in Module 6. Biochemical reactors and the classic exothermic CSTR are convered in Modules 7 and 8, respectively. Steam and surge drum level control problems differ significantly, as presented in Modules 9 and 10. Batch reactors are studied in Module 11, followed by biomedical systems in Module 12. Linear and nonlinear effects in distillation are discussed in Module 13. A set of case studies is summarized in Module 14; these are particularly useful for open-ended final projects in a typical process control course. Flow and digital control, Modules 15 and 16, complete the text.

17.2 Process Engineering in Practice

There is a limit to the amount of material that can be covered in a single textbook with reasonable length and depth of coverage. In this text we have chosen topics that we consider to form a firm basis for further study in automation and control. Hopefully you now have a firm foundation for further study on automation and control-related topics.

Topics Not Covered

Important topics not addressed in this textbook to a significant extent include the following:

- Statistical process control (SPC): SPC is used by process engineers and operators to determine if a process is producing a product that is meeting specified quality targets. SPC techniques attempt to differentiate between normal statistical deviations in product quality and deviations due to operating problems (these may be control related)
- Advanced digital control techniques: An introduction to the most important advanced control technique, model predictive control (MPC), was presented in Chapter 16. There are a large number of important topics in MPC that were not presented. These include the use of different types of models (linear step response models were covered), different model updating techniques (output additive disturbance was covered), and different objective functions (least squares over a finite prediction horizon was covered). Nonlinear models and adaptive linear models can also be used.
- Inferential control: Very often the most important process variable (product quality or property) cannot be measured on-line. Other measurements must be used to "infer" the quality variable; one example is to use tray temperature in distillation to infer the distillate purity.
- Noise filtering: In practice, measurements are corrupted by noise. Often "filters" are used to "average" recent measurements to obtain a better estimate of the actual value.
- Specific control system hardware and software: We have chosen not to cover specific hardware or software for automation projects in detail. Although important in the actual implementation of a control project, we feel that the role of the process engineer is to specify control objectives, determine the proper measured variables, etc. Instrumentation specialists and other project engineers will be involved with many of the detailed specifications necessary to complete an automation project.

Art and Philosophy of Process Engineering

The majority of students majoring in chemical engineering become process engineers in their first industrial position. A typical responsibility is to provide technical support for an entire operating unit (termed a "plant" in Chapter 15). This includes troubleshooting (determining why a product is suddenly off-specification, for example), and capital project justification (designing new equipment to reduce energy consumption, minimize waste, maximize profitability, etc.), among numerous other things. One of the first things that you will do is review process flow diagrams (to understand the basic process flows) and process and instrumentation diagrams (P&IDs, to understand the basic control strategy). You will spend a lot of time tracing pipes through piperacks, to understand physically the layout of the process unit. You will also spend a lot of time talking to process operators, shift foremen, and the unit manager to better understand basic operating procedures. An important part of process operation, naturally, is the automation and control strategy. Each process plant will have an engineer whose primary responsibility is the basic maintenance of the computer control system. Most often a distributed control system (DCS) is used. As a process engineer for a particular unit, you will be expected to achieve some level of knowledge of the control system.

The point of the previous paragraph is that you, as a process engineer, do not operate in isolation. There are many people involved in the operation of a chemical process, no matter which industry you are working in. There is an entire history to the operation of a particular process, and you will not be expected to be an expert overnight. You need to be willing, however, to listen and learn from the experiences of others. At the same time, you need to realize that not everything you will be told will be technically correct and that some information is biased (due to management and labor relation issues, etc.). You will need to achieve a delicate balance between believing everything you are told and being so skeptical that you believe nothing.

So, will you be deriving equations, manipulating block diagrams, and using the Routh stability criterion on a daily basis? In all probability, no. But hopefully this textbook has given you some basic knowledge of dynamics and control that you can use in a broad, qualitative sense. Without explicitly trying, you will probably begin to think in terms of inputs, outputs, and objectives. You will begin to develop a feel for the order of magnitude of an effect of an input change on an output. You will naturally perform "back of the envelope" calculations based on material and energy balances to understand these relationships (for example, if the steam flow to an exchanger changes by a certain amount, how much do you expect the process fluid temperature to change?).

Eventually, as you learn more about the process, you may conceive ideas to improve the control of the process. There is no better way to develop better control strategies than to achieve a deeper knowledge of the process. The best "loop-tuner" in the world will not achieve good control loop performance if a bad input-output pairing is made to begin with.

Rules of Process Operations

We have all heard of Murphy's law—"if something can go wrong, it will go wrong." Similarly with process operations, it is not a question of *if a control element will fail*, but rather, *when will the control element fail*? Safe processes are designed with the knowledge that most moving equipment will eventually fail. Also, sensors will fail or need to be recalibrated. As we noted in Chapter 1, all control valves are designed with a bypass line. When the control valve fails, it can be taken out of service while the flow is regulated manually through the bypass valve. Similarly, virtually all pumps have spare pumps placed "in parallel" with them. When one pump fails (detected as a loss of pressure), the other pump automatically "kicks on" before the process is affected. This also allows preventive maintenance to be performed on all pieces of moving equipment.

It should also be noted that the probability of a failure or technical problem occurring between 8 am and 5 pm on Monday–Friday is roughly 27% (you do the math), so 73% of failures and technical problems occur outside the normal working hours of a process engineer. If the technical problem is critical to continued process operation (remember, many of these operating units are producing several $ million/day of product), and you are the process engineer "on-call," you will likely be called in to help solve the problem. This is usually exciting (sometimes you may wish for less excitement in your life), sometimes with almost an "emergency room" sense of urgency as you and many others try and troubleshoot/solve the problem.

17.3 Suggested Further Reading

A number of literature sources were suggested in Chapter 1. Now that you have a good basic background in process control, you may wish to obtain a deeper or more applied knowledge by studying the following sources.

For books with many practical tips and present numerous application examples, read the following:

Shinskey, F. G. *Feedback Controllers for the Process Industries*, McGraw Hill, New York (1994).

Shinskey, F. G., *Process Control Systems. Application, Design and Tuning*, 4th ed., McGraw Hill, New York (1996).

The plantwide process control book by Luyben et al. provides numerous unit operation and entire flow-sheet examples.

Luyben, W. L., B. D. Tyréus, and M. L. Luyben, *Plantwide Process Control*, McGraw Hill, New York (1999).

For a more detailed coverage of linear systems analysis and an introduction to nonlinear systems behavior and analysis, read the following textbook:

Bequette, B. W., *Process Dynamics: Modeling, Analysis and Simulation*, Prentice Hall, Upper Saddle River, NJ (1998).

For a nice introduction to robust control system design for multivariable systems, read the following:

Skogestad, S., and I. Postlethwaite, *Multivariable Feedback Control: Analysis and Design*, Wiley, New York (1996).

This textbook focuses on chemical process control. For a discussion of the differences between process control and other control fields (both practice and education), see the following article:

Bequette, B. W., and B. A. Ogunnaike, "Chemical Process Control Education and Practice," *IEEE Control Syst. Mag.*, **21**(2), 10–17 (2001).

17.4 Notation

A	Jacobian	state space—relates states to the state derivatives
B	input matrix	state space—relates inputs to state derivatives
C	output matrix	state space—relates states to outputs
D		state space—direct transmittance from inputs to outputs
e	error	setpoint—measured process output
l	load (disturbance) input	
p	pole	roots of the denominator polynomial of $g_p(s)$; equal to the eigenvalues of A
r	setpoint	desired value of the output
s	Laplace domain variable	when used in frequency-response techniques, set $s = j\omega$
x	state variable	
y	output variable	measured output (also known as the controlled variable or process variable)
u	manipulated input	also known as the controller output

This tutorial provides a brief overview of essential MATLAB commands that are entered in the command window. *You will learn this material more quickly if you use MATLAB interactively as you are reviewing this tutorial.* The MATLAB commands will be shown in the following font style:

```
Monaco font
```

The prompt for a user input is shown by the double arrow (»). MATLAB has an extensive on-line help facility. For example, type `help pi` at the prompt,

```
»  help pi
   PI      3.1415926535897....
   PI = 4*atan(1) = imag(log(-1)) = 3.1415926535897....
```

so we see that MATLAB has the number π "built in." Notice that although MATLAB reports the function in uppercase letters, lowercase is actually used when entering the function in the command window.

M1.2 Matrix Operations

Matrix Notation Review

Matrices are ordered arrays of one or more dimensions. A one-dimensional matrix that can be either a column or a row is known as a vector. In this text, we normally use subscripts to denote the element numbers. Consider the following vectors of length 3, where c is a column vector and r is a row vector.

$$c = \begin{bmatrix} c_1 \\ c_2 \\ c_3 \end{bmatrix} = \begin{bmatrix} 7 \\ 1 \\ 4 \end{bmatrix}, r = \begin{bmatrix} r_1 & r_2 & r_3 \end{bmatrix} = \begin{bmatrix} 9 & 5 & 2 \end{bmatrix}$$

Transposing each vector results in the following, where the superscript 'T' represents a transpose:

$$c^T = \begin{bmatrix} 7 & 1 & 4 \end{bmatrix}, r^T = \begin{bmatrix} 9 \\ 5 \\ 2 \end{bmatrix}$$

A 3 × 2 matrix has three rows and two columns, as shown by the matrix *G*:

$$G = \begin{bmatrix} g_{11} & g_{12} \\ g_{21} & g_{22} \\ g_{31} & g_{32} \end{bmatrix} = \begin{bmatrix} 9 & 7 \\ 3 & 5 \\ 1 & 8 \end{bmatrix}$$

Notice that the *ij* subscript refers to the element in the *i*th row and *j*th column.

Dimensional consistency is very important in matrix operations. Consider the matrix operation

$$F = GH$$

To be dimensionally consistent, the number of columns in *G* must be equal to the number of rows in *H*. When considering matrix multiplication, it is convenient to write the dimensions underneath the matrices. For a *G* matrix with *mg* rows and *ng* columns, and an *H* matrix with *nh* columns, then the rest of the dimensions must be

$$\underset{(mg,nh)}{F} = \underset{(mg,ng)}{G}\;\underset{(ng,nh)}{H}$$

The *ij*th element of *F* is found in the following fashion:

$$f_{i,j} = g_{i,1} \cdot h_{1,j} + g_{i,2} \cdot h_{2,j} + \ldots + g_{i,ng} \cdot h_{ng,j}$$

For the following *F* and *G* matrices,

$$G = \begin{bmatrix} 9 & 7 \\ 3 & 5 \\ 1 & 8 \end{bmatrix}, H = \begin{bmatrix} 1 & 3 & 2 & 0 \\ 2 & 5 & 7 & 4 \end{bmatrix}$$

you should be able to calculate that

$$F = \begin{bmatrix} 23 & 62 & 67 & 28 \\ 13 & 34 & 41 & 20 \\ 17 & 43 & 58 & 32 \end{bmatrix}$$

MATLAB Matrix Operations

The basic entity in MATLAB is a rectangular matrix; the entries can be real or complex. Commas or spaces are used to delineate the separate values in a matrix. Consider the following vector, x (recall that a vector is simply a matrix with only one row or column), with six elements

```
» x = [1,3,5,7,9,11]
x =
    1    3    5    7    9    11
```

For this example, we would use

```
» axis([0  50  0  4]);
```

Multiple curves can be placed on the same plot in the following fashion (*perform this*).

```
» plot(t,4*exp(-0.1*t),t,t.*exp(-0.1*t),'--')
```

The subplot command can be used to make multiple plots, as illustrated in Figure M1–3.

```
» subplot(2,1,1), plot(t,4*exp(-0.1*t)), xlabel('t'),ylabel('y1')
» subplot(2,1,2), plot(t,t.*exp(-0.1*t)), xlabel('t'),ylabel('y2')
```

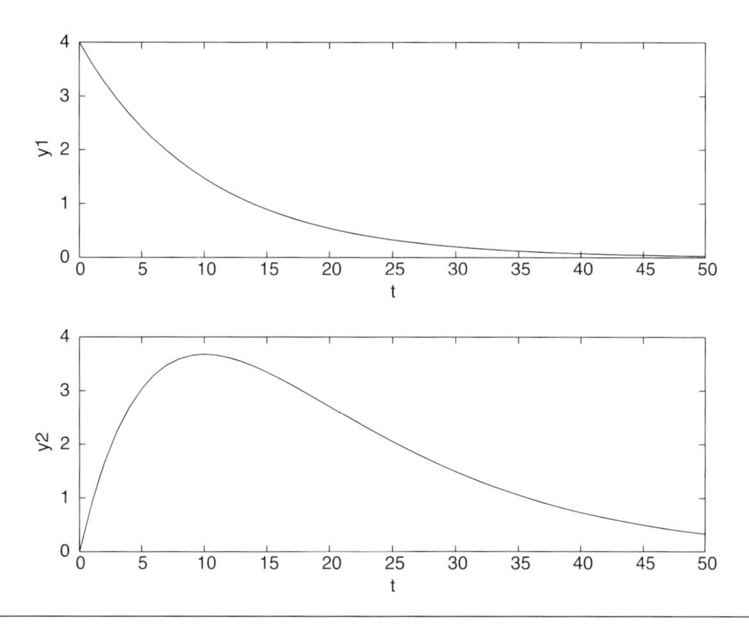

Figure M1–3 Use of subplot feature.

Here, subplot(i,j,k) means that there are i "rows" of figures, j "columns" of figures, and the current plot is the kth figure (counting left to right and top to bottom). The following is an example of a four plot sequence shown in Figure M1–4:

```
» t = 0:1:60;
» subplot(2,2,1),plot(t,4*exp(-0.1*t)), xlabel('t')
» subplot(2,2,2),plot(t,t.*exp(-0.1*t)), xlabel('t')
» subplot(2,2,3),plot(t,sin(.25*t)), xlabel('t')
» subplot(2,2,4),plot(t,cos(.25*t)), xlabel('t')
```

To return to single plots, simply enter subplot(1,1,1).

M1.6 More Matrix Stuff

A matrix can be constructed from two or more vectors. If we wish to create a matrix v which consists of two columns, the first column containing the vector x (in column form)

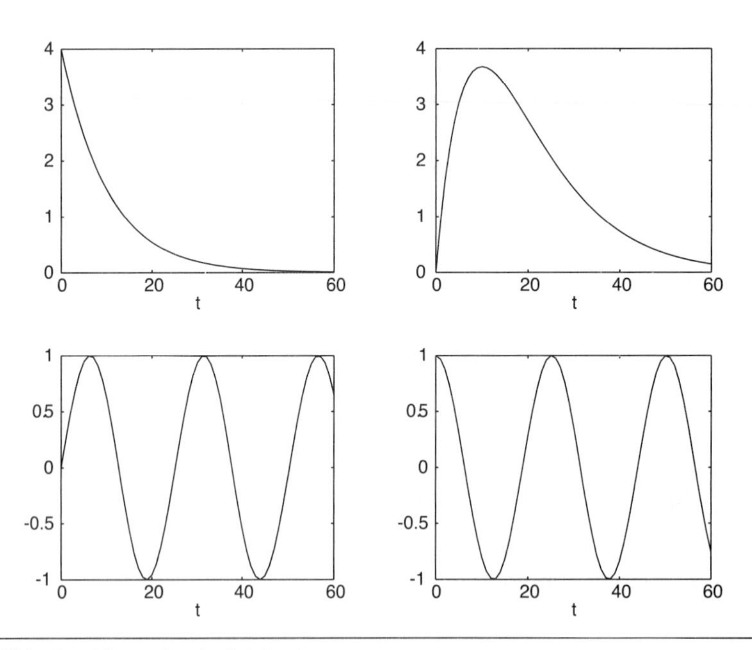

Figure M1–4 Use of subplot feature.

and the second column containing the vector z (in column form), we can use the following (where x and z are previously generated row vectors):

```
» v = [x',z']
v =
     1     5
     3    10
     5    15
     7    20
     9    25
    11    30
```

If we wished to look at the first column of v, we could use (where : indicates all rows, while the 1 indicates the first column)

```
» v(:,1)
ans =
     1
     3
     5
     7
     9
    11
```

If we wished to look at the second column of v, we could use

```
» v(:,2)
ans =
     5
    10
    15
    20
    25
    30
```

And we can construct the same plot as before by using the following plot command ('--' gives a dashed line):

```
» plot(v(:,1),v(:,2),'--')
```

M1.7 For Loops

A for loop in MATLAB is similar to a DO Loop in FORTRAN. The main difference is that the FORTRAN DO loop must have an integer index variable; for does not have this restriction. An example of a for loop that is virtually identical to a DO loop is

```
» for k = 1:5001;
  t(k) = (k-1)*0.01;
  y(k) = sin(t(k));
end
```

Another way of implementing the same loop is to increment t from 0 to 50 in intervals of 0.01:

```
» k = 0
» for t = 0:0.01:50;
  k = k + 1;
  y(k) = sin(t);
end
```

The developers of MATLAB highly recommend that you use the vectorized version of the above for loops:

```
t = 0:0.01:50;
y = sin(t);
```

since the computation time for this method is over 200 times faster than the nonvectorized methods.

M1.8 m-Files

Thus far we have shown the interactive features of MATLAB by entering one command at a time. One reason that MATLAB is powerful is that it is a language, and programs of MATLAB code can be saved for later use. There are two ways of generating your own MATLAB code: script files and function routines.

Script Files

A script file is simply a sequence of commands that could have been entered interactively in the MATLAB command window. When the sequence is long or must be performed a number of times, it is much easier to generate a script file.

The following example is for the so-called quadratic map (population model)

$$x_{k+1} = \alpha \cdot x_k \cdot (1 - x_k)$$

where x_k represents the value of the population (dimensionless) at the kth time step. We have titled the file `popmod.m`. Use the MATLAB text editor to generate this file and save it in a working directory as `popmod.m`.

```
% popmod.m
% population model, script file example
%
  clear x,k
  n       = input('input final time step ');
  alpha   = input('input alpha ');
  xinit   = input('input initial population ');
  x(1)    = xinit;
  time(1)= 0;
  for  k = 2:n+1;
    time(k)  = k-1;
    x(k)     = alpha*x(k-1)*(1-x(k-1));
  end
  plot(time,x)
%  end of script file example
```

Notice that we have used the MATLAB input function to prompt the user for data. Also note that a percent sign (%) may be used to put comments in a script or function file. Any text after a % is ignored by the program.

The file is run by simply entering

```
 » popmod
```

in the MATLAB command window.

Function Routines

A more powerful way of solving problems is to write MATLAB function routines. Function routines are similar to subroutines in FORTRAN. Consider the previous example.

```
  function [time,x] = pmod(alpha,xinit,n)
% population model example, pmod.m
  clear time; clear x; clear k;
  x(1)    = xinit;
  time(1)= 0;
  for  k = 2:n+1;
    time(k) = k-1;
    x(k)     = alpha*x(k-1)*(1-x(k-1));
  end
%  end of function file example
```

where we generate and save a file titled `pmod.m`. We can now "run" this function routine (using `alpha` = 2.8, `xinit` = 0.1, `n` = 30) by typing the following in the MATLAB command window:

```
» [tstep,xpop]=pmod(2.8,0.1,30);
» plot(tstep,xpop)
```

This function routine can also be called by other function routines. This feature leads to "structured programming"; structured programs are easy to follow and debug.

Commonly Used MATLAB Functions

MATLAB has many built-in function routines that you will use throughout this text. The most commonly used routines are `fzero`, `fsolve`, and `ode45`, used to solve a single nonlinear algebraic equation, multiple nonlinear algebraic equations, and a set of nonlinear differential equations, respectively. The optimization toolbox is required to use `fzero` and `fsolve`.

Single Nonlinear Algebraic Equation

`fzero` is used to solve a single nonlinear algebraic equation. As an example, consider

$$x^2-2x-3 = 0$$

The first step is to generate a function file, which we will call fcn (any name can be used). For this simple example, the function file can consist of two lines,

```
function y = fcn(x)
y = x^2 - 2*x -3;
```

and can be saved as fcn.m. The user must enter the following command in the command window, where 0 is the initial guess. Notice that fzero calls the fcn routine, using an initial guess for the solution,

```
»  y = fzero(@fcn,0)
```

and the following answer is returned

```
y =  -1
```

We know that there will be two solutions, since this is a quadratic equation. Indeed, an initial guess of x = 2 yields the following result:

```
» y = fzero(@fcn,2)
  y =   3.0000
```

Since the function is a polynomial, we can use a function developed for polynomial solutions. The MATLAB function roots finds the solution to the quadratic equation, using

```
»  roots([1 -2 -3])
   ans =
        3
       -1
```

which yields the two solutions expected.

M1.9 Summary of Commonly Used Commands

axis	axis limits for plots
clear	removes all variables from workspace
diary	save the text of a MATLAB session
end	end of loop
exp	exponential function
figure(n)	generates a figure window
for	generates loop structure
format	output display format
function	user generated function
gtext	place text on a plot
help	help function
hold	holds current plot and allows new plot to be placed on current plot
if	conditional test
length	length of a vector
lookfor	keyword search on help variables
plot	plots vectors
size	size of the array (rows, columns)
subplot	multiple plots in a figure window
who	view variables in workspace
whos	view variables in workspace, with more detail (size, etc.)
*	matrix multiplication
'	transpose
;	suppress printing (also end of row, when used in matrices)
.*	element-by-element multiplication
./	element-by-element division
:	denotes a column or row in a matrix; also creates a vector
%	placed before comment statements

M1.10 Frequently Used MATLAB Functions

Function	***Use***	
eig	eigenvalues, eigenvectors	
fsolve	solve algebraic equations	Optimization toolbox
fzero	solve a single algebraic equation	Optimization toolbox
impulse	impulse response	Control toolbox
ode45	integrate set of ordinary differential equations	
polyfit	least-squares fit of a polynomial	
ss2tf	convert state space to transfer function model	Control toolbox
step	step response	Control toolbox
tf2ss	convert transfer function to state-space model	Control toolbox

Additional Exercises

1. Plot the following three curves on a single plot and a multiple plots (using the `subplot` command): 2 cos(t), sin(t), and cos(t)+sin(t). Use a time period such that two or three peaks occur for each curve. Use solid, dashed, and + symbols for the different curves. Use roughly 25–50 points for each curve.
2. **a.** Calculate the rank, determinant, and matrix inverse of the following matrices (use `help rank`, `help det`, and `help inv`):

$$A = \begin{bmatrix} 1 & 2 & 1 \\ -1 & -2 & -1 \\ 2 & 4 & 2 \end{bmatrix}$$

$$B = \begin{bmatrix} 1 & 2 & 1 \\ -1 & 4 & -1 \\ 2 & 4 & 2 \end{bmatrix}$$

$$C = \begin{bmatrix} 1 & 2 & 1 \\ -1 & 4 & -1 \\ 2 & 4 & 5 \end{bmatrix}$$

3. Find $C \cdot C^{-1}$ where

$$C = \begin{bmatrix} 1 & 2 & 1 \\ -1 & 4 & -1 \\ 2 & 4 & 5 \end{bmatrix}$$

4. Calculate $\mathbf{x}^T\mathbf{x}$, and calculate $\mathbf{x}\mathbf{x}^T$ where

$$x = \begin{bmatrix} 1 \\ 2 \\ 3 \\ 4 \end{bmatrix}$$

5. Find the eigenvalues of the matrix

$$D = \begin{bmatrix} -1 & 0 & 0 & 2 \\ 1 & -2 & 0 & 6 \\ 1 & 3 & -1 & 8 \\ 0 & 0 & 0 & -2 \end{bmatrix}$$

6. Find the solutions to the equation $f(x) = 3x^3 + x^2 + 5x - 6 = 0$. Use `roots` and `fzero`.

7. Integrate the equations, from $t = 0$ to $t = 5$

$$\frac{dx_1}{dt} = -x_1 + x_2$$
$$\frac{dx_2}{dt} = -x_2$$

with the initial condition $x_1(0) = x_2(0) = 1$. Use `ode45` (Module 3) and plot your results.

Introduction to SIMULINK

Although the standard MATLAB package is useful for linear systems analysis, SIMULINK is far more useful for control system simulation. SIMULINK enables the rapid construction and simulation of control block diagrams. The goal of the tutorial is to introduce the use of SIMULINK for control-system simulation. The version available at the time of writing of this textbook is SIMULINK 5.0, part of Release 13 (including MATLAB 6.5) from MATHWORKS. The version that you are using can be obtained by entering `ver` in the MATLAB command window.

The easiest way to learn how to use SIMULINK is to implement each step of the tutorial, rather than simply reading it. The basic steps to using SIMULINK are independent of the platform (PC, MAC, UNIX, ...).

The sections of this module are as follows:

M2.1 Background
M2.2 Open-Loop Simulations
M2.3 Feedback-Control Simulations
M2.4 Developing Alternative Controller Icons
M2.5 Summary

M2.1 Background

The first step is to start up MATLAB on the machine you are using. In the Launch Pad window of the MATLAB desktop, select SIMULINK and then the SIMULINK Library Browser. A number of options are listed, as shown in Figure M2–1 for SIMULINK 4 (Release 12); SIMULINK 5 (Release 13) has a number of additional options. Notice that `Continuous` has been highlighted; this will provide a list of continuous function blocks available. Selecting `Continuous` will provide the list of blocks shown in Figure M2–2. The ones that we often use are `Transfer Fcn` and `State-Space`.

Selecting the `Sources` icon yields the library shown in Figure M2–3. The most commonly used sources are `Clock` (which is used to generate a time vector) and `Step` (which generates a step input).

The `Sinks` icon from Figure M2–1 can be selected to reveal the set of sinks icons shown in Figure M2–4. The one that we use most often is the `To Workspace` icon. A variable passed to this icon is written to a vector in the MATLAB workspace. The default data method (obtained by "double-clicking" on the icon after it is placed in a model worspace) should be changed from "`structure`" to "`array`" in order to save data in an appropriate form for plotting.

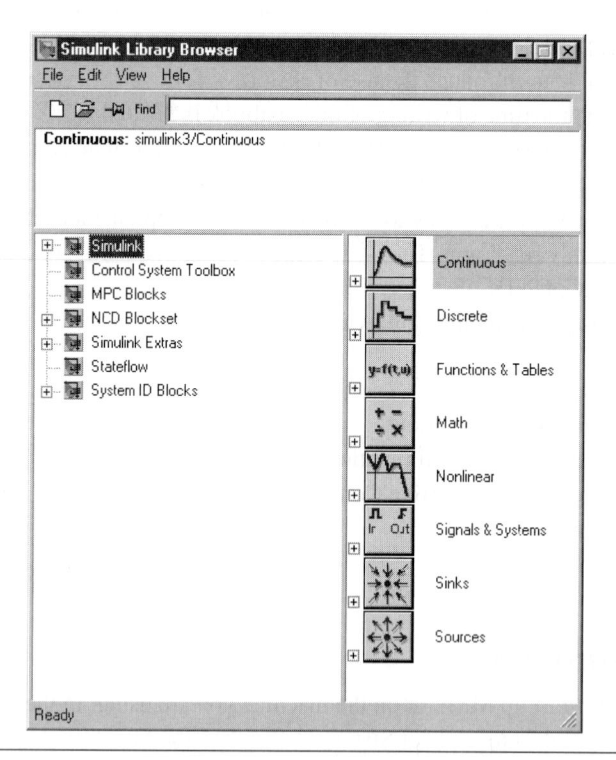

Figure M2–1 SIMULINK Library Browser (Version 4.1, Release 12.1).

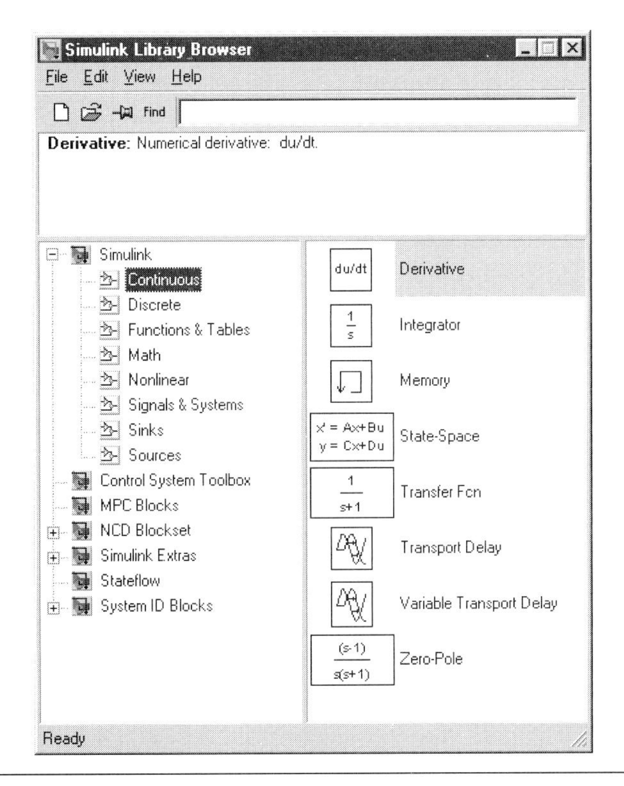

Figure M2–2 SIMULINK Continuous blocks.

M2.2 Open-Loop Simulations

You now have enough information to generate an open-loop simulation. The `Clock`, `simout`, `step`, and `Transfer Fcn` blocks can be dragged to a model (`.mdl`) workspace, as shown in Figure M2–5a. Renaming the blocks and variables, and connecting the blocks, results in the model shown in Figure M2–5b. The transfer function studied is the Van de Vusse reactor (see Module 5).

The s-polynomials in the process transfer function were entered by double clicking on the `transfer function` icon and entering the coefficients for the numerator and denominator polynomials. Notice also that the default step (used for the step input change) is to step from a value of 0 to a value of 1 at $t = 1$. These default values can be changed by double clicking the `step` icon. The simulation parameters can be changed by going to the `Simulation` "pull-down" menu and modifying the `stop time` (default = 10) or the `integration solver method` (default = `ode45`).

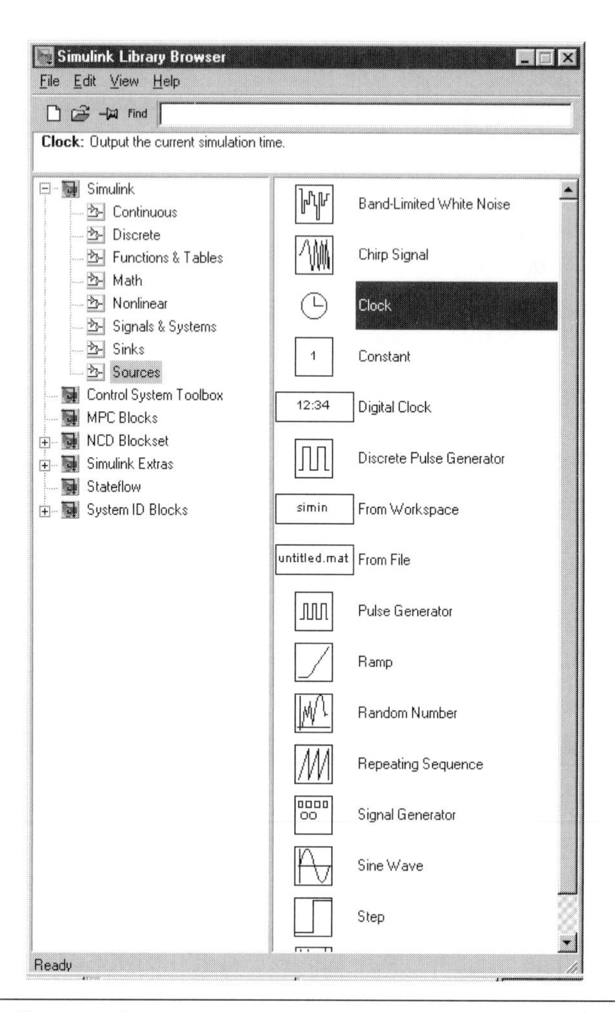

Figure M2–3 SIMULINK Sources.

The reader should generate simulations and observe the "inverse response" behavior of the output with respect to a step input change. Use the `subplot` command to place the process output (y) on the top plot, and the manipulated input (u) on the bottom plot. *Perform this now*. If desired, change the default simulation stop time by selecting the parameters "pull down" menu.

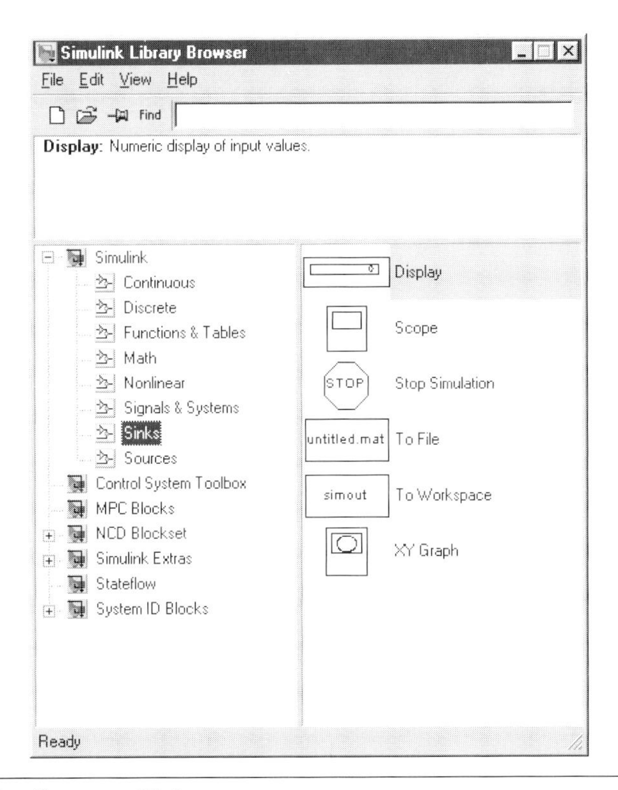

Figure M2–4 SIMULINK Sinks.

M2.3 Feedback-Control Simulations

The `Math` icon from Figure M2–2 can be selected, resulting in the functions shown in Figure M2–6. Additional icons can be found by selecting the `Simulink Extras` icon shown in Figure M2–1. Selecting the `Additional Linear` icon from this group yields the set of icons shown in Figure M2–7. The most useful icon here is the `PID Controller`. Any icon can be "dragged" into the `untitled` model workspace. In Figure M2–8, we show the preliminary stage of the construction of a control block diagram, where icons have been dragged from their respective libraries into the `untitled` model workspace.

The labels (names below each icon) can easily be changed. The default parameters for each icon are changed by double clicking the icon and entering new parameter values. Also, connections can be made between the outputs of one icon and inputs of another. Figure M2–8b shows how the icons from Figure M2–8a have been changed and linked together to form a feedback-control block diagram.

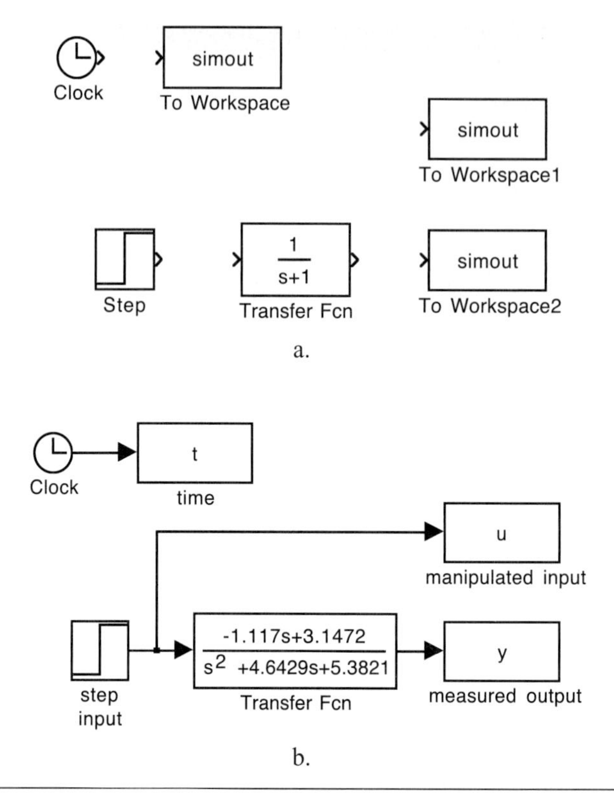

Figure M2–5 Development of an open-loop simulation. (a) Placement of function blocks. (b) Renaming and connection of blocks.

It should be noted that the form of the PID control law used by the SIMULINK `PID Controller` icon is not the typical form that we use as process-control engineers. The form can be found by double clicking the icon to reveal the controller transfer function representation

$$g_c(s) = P + \frac{I}{s} + Ds$$

while we normally deal with the following PID structure:

$$g_c(s) = k_c\left(1 + \frac{1}{\tau_I s} + \tau_D s\right)$$

So the numerical values for the following parameters must be entered in the SIMULINK PID controller:

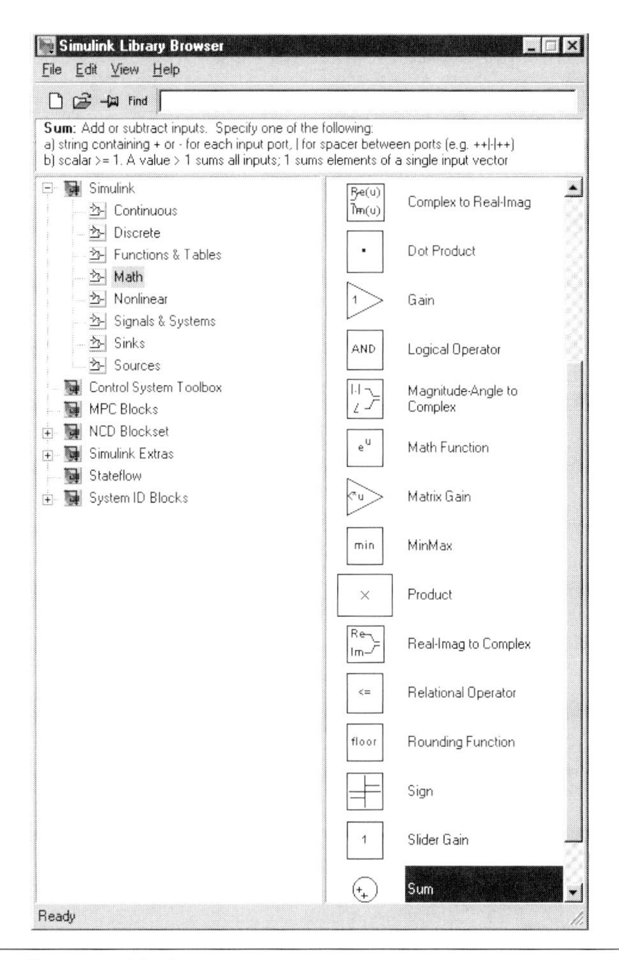

Figure M2–6 SIMULINK Math.

$$P = k_c \quad I = k_c/\tau_I \quad D = k_c\tau_D$$

The *s*-polynomials in the process transfer function were entered by double clicking on the transfer function and entering the coefficients for the numerator and denominator polynomials. Notice also that the default step (used for the step setpoint change) is to step from a value of 0 to a value of 1 at $t = 1$. These default values can be changed by double clicking the `step` icon. The simulation parameters can be changed by going to the `Simulation` "pull-down" menu and modifying the start time (default = 0), stop time (default = 10) or the integration solver method (default = ode45).

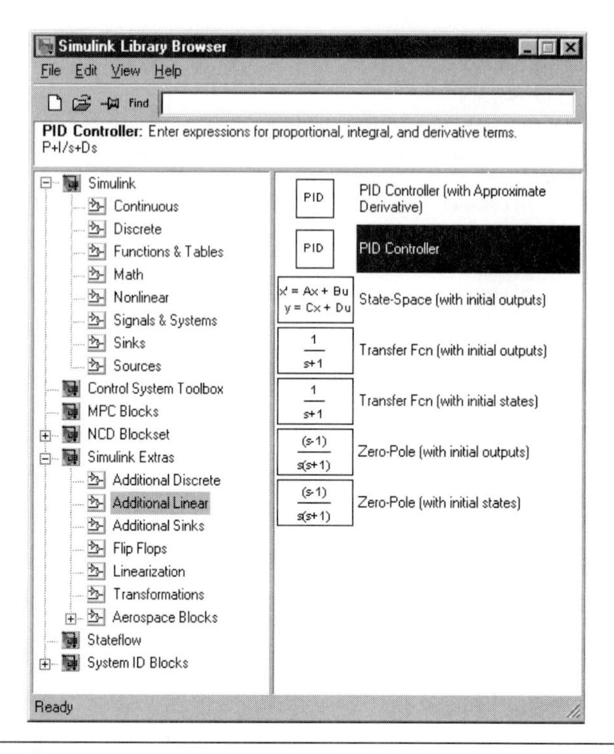

Figure M2–7 SIMULINK Additional Linear.

The controller tuning parameters of $k_c = 1.89$ and $\tau_I = 1.23$ are used by entering $P = 1.89$ and $I = 1.89/1.23$ in the default `PID Controller` block. The following plot commands were used to generate Figure M2–9:

```
» subplot(2,1,1),plot(t,r,'--',t,y)
» xlabel('t (min)')
» ylabel('y (mol/l)')
» subplot(2,1,2),plot(t,u)
» xlabel('t (min)')
» ylabel('u (min^-1)')
```

The curves in Figure M2–9 could be made smoother by selecting the '`refine`' box in the Simulation pull-down menu. Selecting `refine = 2` will plot two points for each integration step.

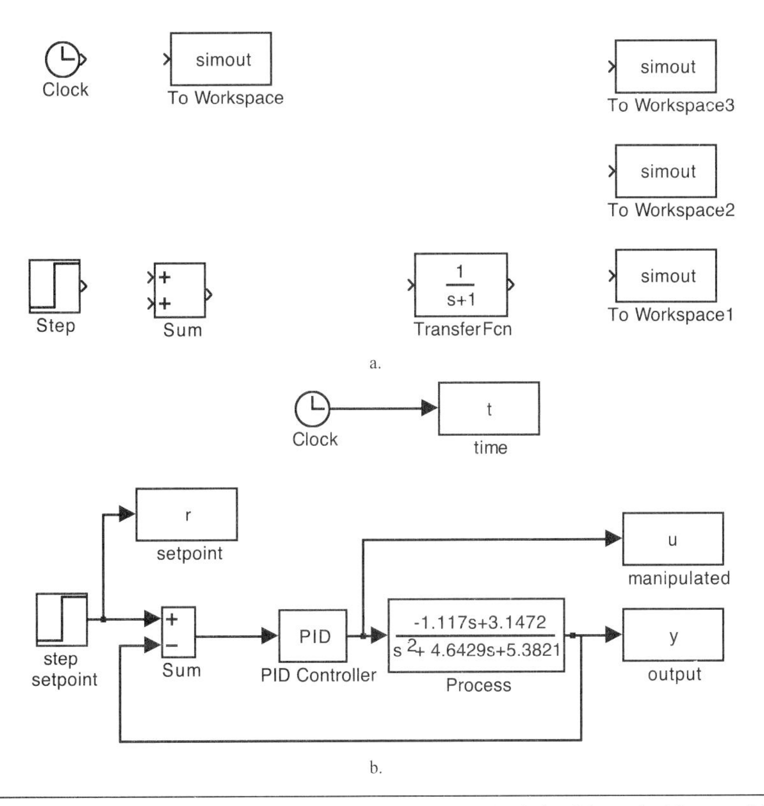

Figure M2–8 Block diagram for feedback control of the Van de Vusse CSTR. (a) Preliminary block diagram. (b) Completed block diagram, with name and parameter changes.

This brief tutorial has gotten you started in the world of SIMULINK-based control block diagram simulation. You may now easily compare the effect of different tuning parameters or different formulations of a PID controller ("ideal" vs. "real," for example).

Let's say you generated responses for a set of tuning parameters that we call case 1 for convenience. You could generate time, input, and output vectors for this case by setting

```
» t1 = t;
» y1 = y;
» u1 = u;
```

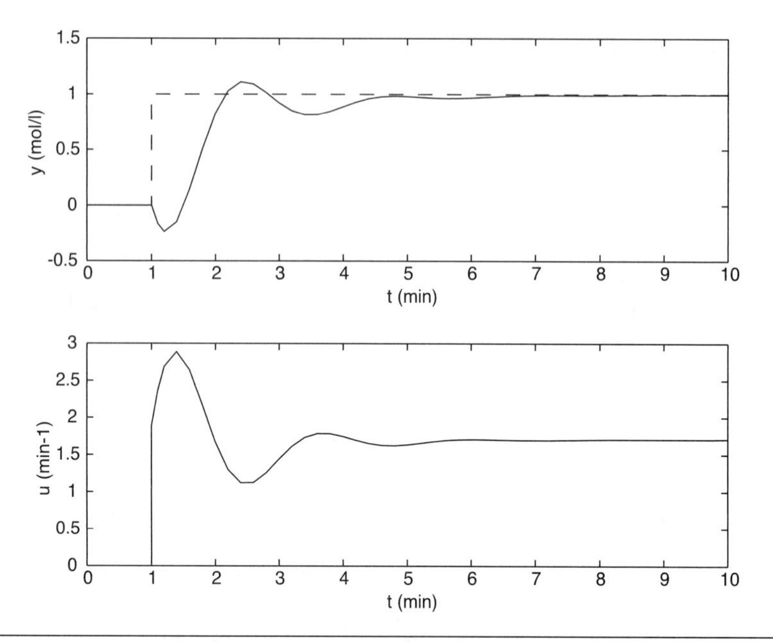

Figure M2–9 Measured output and manipulated input responses to a unit step setpoint change.

after running the case 1 values. You could then enter case 2 values, run another simulation, and create new vectors:

```
» t2 = t;
» y2 = y;
» u2 = u;
```

Then, compare case 1 and case 2 results by

```
» subplot(2,1,1),plot(t1,y1,t2,y2,'--')
» xlabel('t (min)')
» ylabel('y (mol/l)')
» subplot(2,1,2),plot(t1,u1,t2,u2,'--')
» xlabel('t (min)')
» ylabel('u (min^-1)')
```

which automatically plots case 1 as a solid line and case 2 as a dashed line.

Similarly, you could modify the controller type by placing a transfer function block for the controller and using a "real PID" transfer function (this only differs when there is derivative action).

Other Commonly Used Icons

Often you will want to simulate the behavior of systems that have time delays. The `Transport Delay` icon can be selected from the `Continuous` library shown in Figure M2–2. The transport delay icon is shown in Figure M2–10. Our experience is that simulations can become somewhat "flaky" if 0 is entered for a transport delay. We recommend that you remove the transport delay block for simulations where no time delay is involved.

Manipulated variables are often constrained to between minimum (0 flow, for example) and maximum (fully open valve) values. A `saturation` icon from the `Discontinuities` library can be used to simulate this behavior. The saturation icon is shown in Figure M2–11.

Actuators (valves) and sensors (measurement devices) often have additional dynamic lags that can be simulated by transfer functions. These can be placed on the block diagram in the same fashion that a transfer function was used to represent the process earlier.

It should be noted that icons can be "flipped" or "rotated" by selecting the icon and going to the format "pull-down" menu and selecting `Flip Block` or `Rotate Block`. The block diagram of Figure M2–8 has been extended to include the saturation element and transport delay, as shown in Figure M2–12.

The default data method for the "`to workspace`" blocks (`r,t,u,y` in Figure M2–12) must be changed from "`structure`" to "`array`" in order to save data in an appropriate form for plotting.

Figure M2–10 Transport Delay icon.

Figure M2–11 Saturation element.

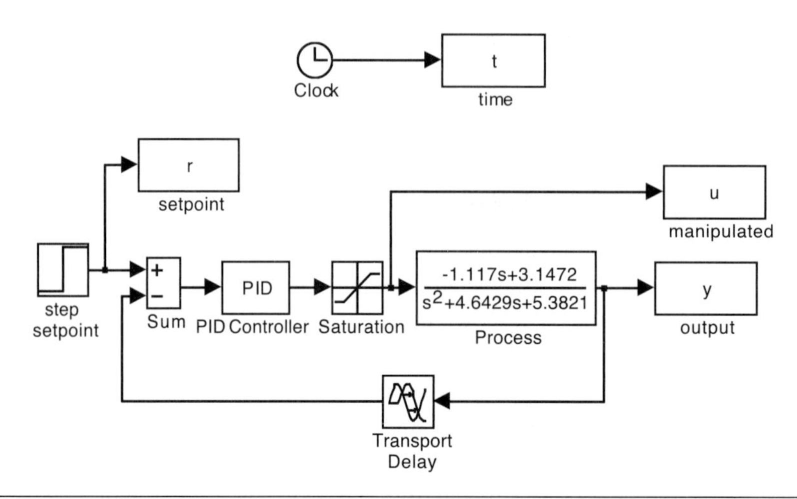

Figure M2–12 Block diagram with saturation and time-delay elements.

M2.4 Developing Alternative Controller Icons

It was noted earlier that the default SIMULINK PID controller block uses a form different from that used by most process engineers. It is easy to generate new PID controller blocks as shown below. The default PID controller icon is shown in Figure M2–13a. This is "unmasked" (by clicking the icon and selecting 'Look under mask' from the Edit pull-down menu) to yield the diagram shown in Figure M2–13b. Again, this has the form

$$g_c(s) = P + \frac{I}{s} + Ds$$

while we normally prefer the following PID structure:

$$g_c(s) = k_c\left(1 + \frac{1}{\tau_I s} + \tau_D s\right)$$

Of course, the two algorithms are related by

$$P = k_c \quad I = k_c/\tau_I \quad D = k_c\tau_D$$

but it would be much less confusing to work with our standard form.

We have generated a new ideal analog PID, as shown in Figure M2–14a. Notice that there are two inputs to the controller, the setpoint (r) and the measured output (y), rather than just the error signal that is the input to the default SIMULINK PID controller. Our new implementation is unmasked in Figure M2–14b to reveal that

$$g_c(s) = k_c\left(1 + \frac{1}{\tau_I s} + \tau_D s\right)$$

is the algorithm used.

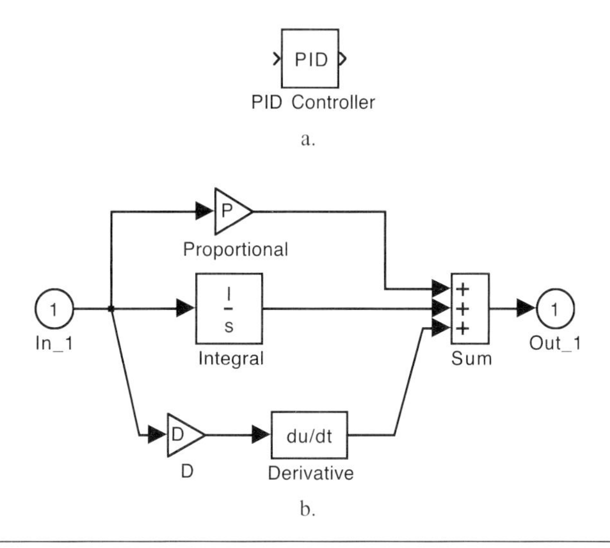

Figure M2–13 Default ideal PID controller. (a) Block. (b) Block *unmasked*.

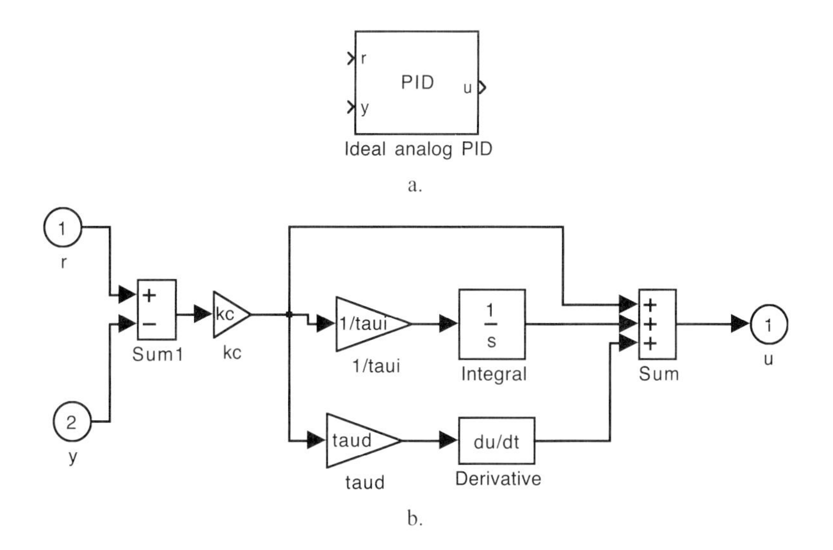

Figure M2–14 Preferred implementation of ideal PID controller. (a) Block. (b) Unmasked block.

M2.5 Summary

SIMULINK is a very powerful block diagram simulation language. Simple simulations, including the majority of those used as examples in this textbook, can be set up rapidly (in a matter of minutes). The goal of this module was to provide enough of an introduction to get you started on the development of open- and closed-loop simulations. With experience, the development of these simulations will become second nature. It is recommended that you perform the simulations shown in this module, as well as the additional exercises, to rapidly acquire these simulation skills.

Additional Exercises

1. Compare step responses of the transfer function and state space models for the van de Vusse reactor.
   a. Transfer function: Perform the open-loop simulation for a step input change from 0 to 1 at $t = 1$, as shown in Figure M2–5.
   b. State space: Replace the transfer function block shown in Figure M2–5 with a state space block. Enter the following matrices in the MATLAB command window:

$$A = \begin{bmatrix} -2.4048 & 0 \\ 0.83333 & -2.2381 \end{bmatrix} \qquad B = \begin{bmatrix} 7 \\ -1.117 \end{bmatrix}$$

$$C = \begin{bmatrix} 0 & 1 \end{bmatrix} \qquad D = \begin{bmatrix} 0 \end{bmatrix}$$

   Show that the resulting step responses of the transfer function and state space models are identical.

2. Consider the closed-loop SIMULINK diagram for the van de Vusse reactor (Figure M2–8). Replace the default PID controller with a "`Real PID`" controller, as shown below in both the masked and unmasked versions. How do your results compare with ones obtained using the default PID controller?

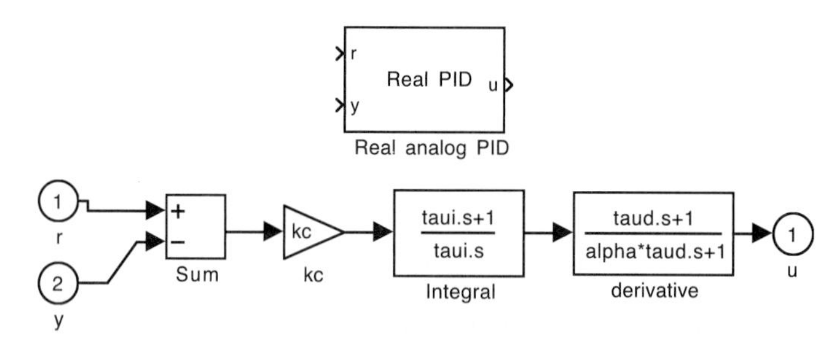

3. The following control algorithm uses a filtered derivative of the process output (see Chapter 5). Develop a SIMULINK implementation of this algorithm, where $e(s) = r(s) - y(s)$.

$$u(s) = k_c\left(\frac{\tau_I s + 1}{\tau_I s}\right)e(s) - k_c\,\frac{(\tau_D s)}{(\tau_F s + 1)}\,y(s)$$

MODULE 3

Ordinary Differential Equations

MATLAB provides several powerful approaches to integrating sets of initial-value, ordinary differential equations. In this module, we will discuss two approaches: using functions called by various `ode` routines from a MATLAB command window or script file, and using functions called by a block from a SIMULINK diagram.

After reviewing this module, the reader should be able to do the following:

- Write a basic function file containing differential equations to be integrated by MATLAB `ode` routines
- Write a more advanced function file, with parameters based through an argument list and containing differential equations to be integrated by MATLAB `ode` routines
- Write a function (more specifically, an S-function) file containing differential equations to be integrated by SIMULINK
- Generate the SIMULINK block diagram containing an S-function block

The sections of this module are as follows:

M3.1 MATLAB ode—Basic

MATLAB has several different routines for numerical integration; all use the same basic call statements:

```
[t,x]= ode45('xprime',tspan,x0)   % nonstiff, medium order
[t,x]= ode23('xprime',tspan,x0)   % nonstiff, low order
[t,x]= ode113('xprime',tspan,x0)  % nonstiff, variable order
[t,x]= ode15s('xprime',tspan,x0)  % stiff, variable order
[t,x]= ode23s('xprime',tspan,x0)  % stiff, low order
[t,x]= ode23t('xprime',tspan,x0)  % moderately stiff, trapezoidal
[t,x]= ode23tb('xprime',tspan,x0) % stiff, low order
```

`ode45` uses fourth-order and `ode23` uses second-order Runge-Kutta integration. All routines use a *variable integration step size* (Δt is not constant). The integration step size is adjusted by the routine to provide the necessary accuracy, without taking too much computation time.

To use these routines, the user must first generate an *m*-file to evaluate the state variable derivatives. The m-file must be named '`xprime.m`', and contain the following statement in the first line of the file

```
function xdot = xprime(t,x);
```

where '`xprime`' is the name of the function routine (usually selected as something meaningfully related to the system of equations), '`xdot`' is the vector of time derivatives of the states, '`t`' is time, and '`x`' is the vector of states. '`xdot`' must be generated as a column vector.

The command given to perform the integration is

```
[t,x]= solver('xprime',tspan,x0)
```

where

`solver`	is any of the integration routines (`ode45`, etc.)
`xprime`	is a string variable containing the name of the m-file for the state derivatives
`tspan`	is the vector of time values (row vector)
`x0`	is the initial condition vector for the state variables (column vector).

The arrays that are returned are

`t`	a (column) vector of time
`x`	an array of state variables as a function of time (column 1 is state 1, etc.).

For example, if the time vector has 50 elements and there are three state variables, then the state variable vector has the 50 rows and three columns. After the integration is performed, if the student wishes to plot all three variables as a function of time, she/he simply enters

```
plot(t,x)
```

If you only want to plot the second state variable, then the command `plot(t,x(2,:))` is given.

Example M3.1: Van de Vusse Reaction

Consider the following set of differential equations that describe the van de Vusse reaction scheme in an isothermal, CSTR.

$$\frac{dC_A}{dt} = \frac{F}{V}\left(C_{Af} - C_A\right) - k_1 C_A - k_3 C_A^2$$

$$\frac{dC_B}{dt} = -\frac{F}{V} C_B + k_1 C_A - k_2 C_B$$

The parameter values are

$$k_1 = 5/6\text{min}^{-1} \qquad k_2 = 5/3\text{min}^{-1} \qquad k_3 = 1/6\text{mol/liter}\cdot\text{min}$$

The input values used in the following simulation are

$$F/V = 4/7\text{min}^{-1} \qquad C_{Af} = 10\text{mol/liter}$$

The differential equations are placed in a file named `vdv_ode.m`

```
  function xdot = vdv_ode(t,x);
%
% Solves the two differential equations modeling
% the van de vusse reaction
% scheme in an isothermal CSTR. The states are the concentration
% of A and B in the reactor.
%
```

```
% [t,x] = ode45(vdv_ode,[0 5],x0)
% integrates from t = 0 to t = 5 min, with initial conditions
% ca0 = x0(1) and cb0 = x0(2), and x0 is a column vector
% 16 Jan 99
% b.w. bequette
%
% since the states are passed to this routine in the x vector,
% convert to natural notation

  ca = x(1);
  cb = x(2);

% the parameters are:
  k1 = 5/6;     % rate constant for A-->B (min^-1)
  k2 = 5/3;     % rate constant for B-->C (min^-1)
  k3 = 1/6;     % rate constant for 2A-->D (mol/(l min))
% the input values are:

  fov = 4/7;    % dilution rate (min^-1)
  caf = 10;     % mol/l

% the modeling equations are:
  dcadt = fov*(caf-ca) - k1*ca -k3*ca*ca;
  dcbdt = -fov*cb + k1*ca - k2*cb;

% now, create the column vector of state derivatives

  xdot = [dcadt;dcbdt];

% end of file
```

In the MATLAB command window, enter the initial conditions and run `ode45`

```
» x0 = [2;1.117]
» [t,x] = ode45('vdv_ode',[0 5],x0);
» subplot(2,1,1),plot(t,x(:,1)), xlabel('t'), ylabel('ca')
» subplot(2,1,2),plot(t,x(:,2)), xlabel('t'), ylabel('cb')
```

The plots are shown in Figure M3–1. Notice that the system converges to the steady-state values of $C_A=3$, $C_B=1.117$.

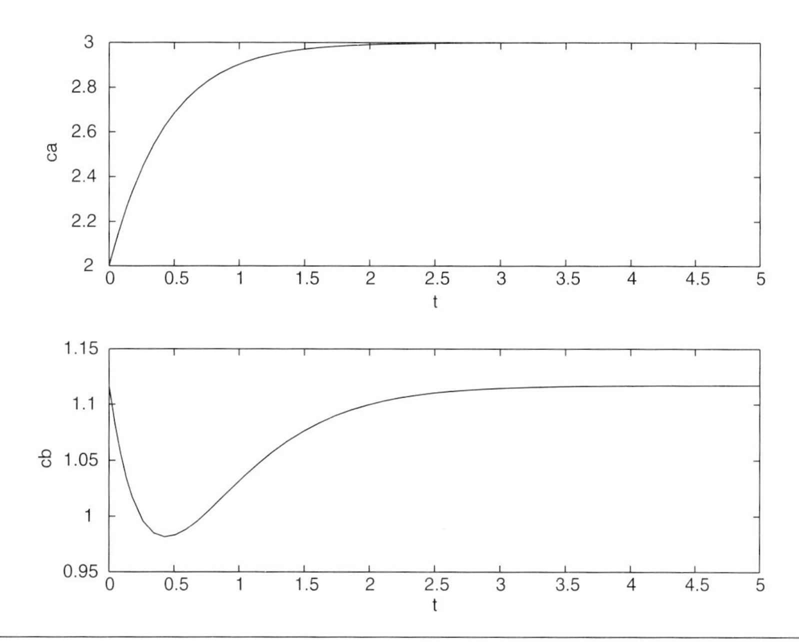

Figure M3–1 Transient response to initial conditions perturbed from the steady-state values, $C_A(0) = 2$, $C_B(0) = 1.117$.

M3.2 MATLAB ode—Options

The MATLAB integration routines have additional options (such as changing the default tolerances for integration) that can be set by the user. Also, parameters for the differential equations may be passed through the argument list to the function m-file.

```
[t,x]= ode45('xprime',tspan,x0,options,P1,P2,...)
[t,x]= ode23('xprime',tspan,x0,options,P1,P2,...)
[t,x]= ode113('xprime',tspan,x0,options,P1,P2,...)
[t,x]= ode15s('xprime',tspan,x0,options,P1,P2,...)
[t,x]= ode23s('xprime',tspan,x0,options,P1,P2,...)
[t,x]= ode23t('xprime',tspan,x0,options,P1,P2,...)
[t,x]= ode23tb('xprime',tspan,x0,options,P1,P2,...)
```

To use the default options (generally recommended) and pass through parameter values, simply enter `[ ]` for the options element in the argument list. For example, create an `'xprime.m'` m-file for the differential equation

```
function xdot = xprime(t,x,P1,P2,...)
...
```

and enter

```
[t,x]= ode45('xprime',tspan,x0,[ ],P1,P2,...);
```

in the MATLAB command window.

A typical application might be to be able to change the magnitude of an input variable. For example, consider the van de Vusse reactor shown in Example M3.1. The first line of the vdv_ode.m file could be modified in the fashion

```
function xdot = vdv_ode(t,x,delstep);
```

where `delstep` represents the change in the manipulated input. The file is further modified by changing the following line

```
fov = 4/7 + delstep;    % dilution rate (min^-1)
```

so that `delstep` represents a change from the nominal input value of 4/7 min^{-1}. If the magnitude of the step change is 0.1, then the line entered in the MATLAB command window is

```
[t,x]= ode45('vdv_ode',[0 5],x0,[ ],0.1)
```

M3.3 SIMULINK sfun (.mdl Files)

One of the powers of SIMULINK is the ability to construct block diagrams that provide an intuitive approach to simulating feedback-control systems. Often we use linear transfer functions to simulate the behavior of linear systems. A block titled "`sfun`" can obtained from the "`nonlinear`" library in SIMULINK and placed in a SIMULINK diagram. When simulations are performed, the equations in the sfun m-file are solved simultaneously with other blocks in the block diagram. The `sfun` file must be given a name with a '`.m`' extension, while the block diagram is given a '`.mdl`' extension.

```
b =
                    u1
         x1          7
         x2     -1.117

c =
                    x1          x2
         y1          0           1

d =
                    u1
         y1          0

Continuous-time system.
```

The poles and zeros can be calculated using the `pole` and `tzero` commands.

```
» pole(vdv_ss)

ans =
   -2.2381
   -2.4048

» tzero(vdv_ss)

ans =
   2.8173
```

Clearly, the system is stable (the poles of the transfer function = the eigenvalues of the A matrix = negative, indicating stability) and contains a RHP zero, indicating an inverse response.

Transfer Function

The transfer function model is

$$g(s) = \frac{-1.117s + 3.1472}{s^2 + 4.6429s + 5.3821} = \frac{0.5848(-0.3549s + 1)}{0.1858s^2 + 0.8627s + 1}$$

The numerator and denominator polynomials are entered:

```
»  num = [-1.117,3.1472];
»  den = [1,4.6429,5.3821];
```

The transfer function object command is entered:

```
»  vdv_tf = tf(num,den)
```

The result is displayed:

```
Transfer function:
  -1.117 s + 3.147
---------------------
s^2 + 4.643 s + 5.382

» pole(vdv_tf)

ans =
   -2.4048
   -2.2381

» tzero(vdv_tf)

ans =
   2.8173
```

Again, the RHP zero indicates an inverse response.

Zero-Pole Gain

This structure is not often used by process engineers. It is similar to the transfer function form, but the polynomials have been factored to clearly show the zeros and poles. An example with one zero and two poles is

$$g(s) = \frac{k(s - z_1)}{(s - p_1)(s - p_2)}$$

Notice that the gain term, k, is not the same as the process gain and is definitely a source of confusion.

Consider the transfer function with the following zero-pole-gain form (you should be able to write the previous transfer function in this form):

$$g(s) = \frac{-1.117(s - 2.817)}{(s + 2.238)(s + 2.405)}$$

The MATLAB command, followed by the result displayed, is

```
» vdv_zpk = zpk(2.817,[-2.238 -2.405],-1.117)

Zero/pole/gain:
 -1.117 (s-2.817)
--------------------
(s+2.238) (s+2.405)
```

Converting Between Model Types

A model in any one of the standard forms can easily be converted to another standard form.

Converting from State Space Form

As an example, assume that the state space form has been entered as

```
» a = [-2.4048,0;0.8333,-2.2381];
» b = [7;-1.117];
» c = [0,1];
» d = [0];
» vdv_ss = ss(a,b,c,d)
```

The transfer function form can found from the state space form using

```
» vdv_tf = tf(vdv_ss)
```

and the MATLAB command window displays the following:

```
Transfer function:
  -1.117 s + 3.147
----------------------
s^2 + 4.643 s + 5.382
```

Also, the zero-pole-gain form can be found from the state space form using

```
» vdv_zpk = zpk(vdv_ss)

Zero/pole/gain:
 -1.117 (s-2.817)
-------------------
(s+2.238) (s+2.405)
```

An alternative approach is to use `ss2tf` or `ss2zp`,

```
[num,den] = ss2tf(a,b,c,d,iu)
[z,p,k]   = ss2zp(a,b,c,d,iu)
```

where `iu` is the *i*th input variable (2, if you want the transfer functions relating input 2 to all of the outputs). If there is more than one output, then num will be a matrix of elements, where each row contains the coefficients of the polynomial associated with that output (row 1 is the numerator polynomial for output 1, etc.).

Converting from Transfer Function to State Space Form

Assume that the transfer function form has been created:

```
» num = [-1.117,3.1472];
» den = [1,4.6429,5.3821];
» vdv_tf = tf(num,den);
```

The state space form can be found by using

```
»vdv_ss1 = ss(vdv_tf)

a =
                    x1          x2
         x1    -4.6429     -1.3455
         x2          4           0

b =
                    u1
```

```
          x1             1
          x2             0

c =

                        x1           x2
          y1        -1.117       0.7868

d =

                        u1
          y1             0

Continuous-time system.
```

Recall that there are many state space models that yield the same transfer function model. The state space realization of a transfer function, used by MATLAB, is known as the *controllable canonical* form. Notice that this form in this example (`vdv_ss1`) is different than the original state space form (`vdv_ss`). You should show that either state space form results in the same transfer function.

Multiple Inputs and/or Outputs

The above examples have illustrated LTI objects that are SISO. The extension to MIMO is straightforward for the state space model (simply increase the dimensions of the matrices appropriately.

For transfer functions, the easiest way to form a MIMO transfer function matrix is to concatenate the individual matrix elements. Consider

$$G(s) = \begin{bmatrix} \frac{2}{3s+1} & \frac{-1}{4s+1} \\ \frac{-1}{2s+1} & \frac{5}{6s+1} \end{bmatrix}$$

Enter all the individual transfer functions, then specify the elements in the transfer function matrix:

```
» g11 = tf([2],[3 1]);
» g12 = tf([-1],[4 1]);
» g21 = tf([-1],[2 1]);
» g22 = tf([5],[6 1]);
» G = [g11 g12;g21 g22]
```

MATLAB returns the following:

```
Transfer function from input 1 to output...
         2
 #1:  -------
      3 s + 1

        -1
 #2:  -------
      2 s + 1

Transfer function from input 2 to output...
        -1
 #1:  -------
      4 s + 1

         5
 #2:  -------
      6 s + 1
```

Input Time Delays

Most chemical processes have significant time delays that can cause control problems. The LTI objects assume that the time delay is on the inputs. This means that each column of a transfer function matrix, for example, is assumed to have the same time delay. Consider the previous two input–two output transfer function matrix, with time delays of 1.5 and 2.5 time units on input 1 and 2, respectively:

$$G(s) = \begin{bmatrix} \dfrac{2e^{-1.5s}}{3s+1} & \dfrac{-1e^{-2.5s}}{4s+1} \\ \dfrac{-1e^{-1.5s}}{2s+1} & \dfrac{5e^{-2.5s}}{6s+1} \end{bmatrix}$$

Since the delay-free transfer function has already been generated, we can add input delays using the following:

```
» set(G,'InputDelay',[1.5 2.5])
```

The new transfer function properties can be found using

```
» G

Transfer function from input 1 to output...
          2
 #1:  -------
      3 s + 1

         -1
 #2:  -------
      2 s + 1

Transfer function from input 2 to output...
         -1
 #1:  -------
      4 s + 1

          5
 #2:  -------
      6 s + 1

Input delays (listed by channel): 1.5  2.5
```

Delays to state-space models can be handled in a similar fashion.

M4.2 Forming Discrete-Time Models

The formation of discrete-time models is the same as for continuous-time models, except that a sample time should also be specified. This sample time is simply the next argument in the function call statement,

```
sys = ss(a,b,c,d,Ts)  % discrete state space
sys = tf(num,den,Ts)  % discrete transfer function
sys = zpk(z,p,k,Ts) % discrete zero-pole-gain
```

where `Ts` is the sample time.

Discrete State Space Models

A discrete-time state space model has the form

$$x(k+1) = Ax(k) + Bu(k)$$
$$y(k) = Cx(k) + Du(k)$$

where k represents the discrete-time step index. Consider the discrete state space model

$$A = \begin{bmatrix} 0.78625 & 0 \\ 0.06607 & 0.79947 \end{bmatrix}, B = \begin{bmatrix} 0.62219 \\ -0.07506 \end{bmatrix}, C = \begin{bmatrix} 0 & 1 \end{bmatrix}, D = \begin{bmatrix} 0 \end{bmatrix}$$

with a sample time of $\Delta t = 0.1$ minutes. Please note that, if you need to convert a continuous state space model to discrete, see Section M4.3.

After the *a, b, c* and *d* matrices are entered, the state space object can be created:

```
» vdv_ssd = ss(a,b,c,d,0.1)

a =

                    x1            x2
        x1     0.78625             0
        x2     0.06607       0.79947

b =

                    u1
        x1     0.62219
        x2    -0.07506

c =

                    x1            x2
        y1           0             1

d =

                    u1
        y1           0

Sampling time: 0.1
Discrete-time system.
```

If you have a discrete-time model but do not know the sample time, enter –1 for `Ts`.

```
»vdv_ssd1 = ss(a,b,c,d,-1)

a =
                    x1          x2
          x1   0.78625           0
          x2   0.06607     0.79947


b =
                    u1
          x1   0.62219
          x2  -0.07506


c =
                    x1          x2
          y1         0           1


d =
                    u1
          y1         0

Sampling time: unspecified
Discrete-time system.
```

The poles and zeros are found using the same commands as for continuous systems.

```
» pole(vdv_ssd)
ans =
    0.7995
    0.7863
» tzero(vdv_ssd)
ans =
   1.3339
```

The poles are inside the unit circle, indicating stability. The zero is outside the unit circle, indicating that the model inverse is unstable.

Discrete Transfer Function

The discrete transfer function model is

$$g(z) = \frac{-0.07506z + 0.1001}{z^2 - 1.586z + 0.6286}$$

which represents the following input-output model

$$y(k+1) = 1.586y(k) - 0.6286y(k-1) - 0.07506u(k) + 0.1001u(k-1)$$

The transfer function object is created

```
» vdv_tfd = tf([-0.07506 0.1001],[1 -1.586 0.6286],0.1)

Transfer function:
 -0.07506 z + 0.1001
----------------------
z^2 - 1.586 z + 0.6286

Sampling time: 0.1

» pole(vdv_tfd)
ans =
     0.7995
     0.7863

» tzero(vdv_tfd)
ans =
   1.3339
```

Again, the poles are inside the unit circle, indicating stability. The zero is outside the unit circle, indicating that the model inverse is unstable.

Discrete Filter Form

It is common to use the backward shift notation for discrete chemical process models. There is an additional Control Toolbox LTI object specifically for this.

```
sys = filt(num,den,Ts)  % discrete transfer function, z^-1
```

Consider the transfer function in backward shift form:

$$g(z) = \frac{0.1564z^{-1} + 0.2408z^{-2}}{1 - 0.3513z^{-1} + 0.0307z^{-2}}$$

The MATLAB command and results are

```
» vdv_ff = filt([0 0.1564 0.2408],[1 -0.3513 0.0307],0.1)

Transfer function:
  0.1564 z^-1 + 0.2408 z^-2
-----------------------------
1 - 0.3513 z^-1 + 0.0307 z^-2

Sampling time: 0.1
```

Note that with the discrete filter form, it is critical to have the leading zero element in the numerator polynomial.

Converting Between Discrete Model Types

A model in any one of the standard discrete forms can easily be converted to another standard discrete form.

Converting from State Space Form to Transfer Function Form

As an example, assume that the discrete state space form has been entered as `vdv_ssd`. The discrete transfer function is found using `tf`:

```
» vdv_tfd1 = tf(vdv_ssd)

Transfer function:
 -0.07506 z + 0.1001
----------------------
z^2 - 1.586 z + 0.6286

Sampling time: 0.1
```

M4.3 Converting Continuous Models to Discrete

The Control Toolbox can also easily convert continuous models to discrete, using the `c2d` command.

```
sysd = c2d(sysc,Ts,method)    % continuous to discrete
```

where `sysc` is the LTI object for a continuous model, `sysd` is the discrete object to be created, `Ts` is the sample time, and `method` is the type of hold placed on the input variable. We always assume a zero-order hold (`'zoh'`); this is also the default if no entry is made for `method`.

Consider the continuous state space model for the van de Vusse reactor, `vdv_ss`. It can be converted to the discrete-time model using the following MATLAB commands:

```
» vdv_ssd = c2d(vdv_ss,0.1,'zoh')

a =
                       x1            x2
          x1      0.78625             0
          x2     0.066067       0.79947

b =
                       u1
          x1      0.62219
          x2    -0.075061

c =
                       x1            x2
          y1            0             1

d =
                       u1
          y1            0

Sampling time: 0.1
Discrete-time system.
```

M4.4 Converting Discrete Models to Continuous

Discrete-time models are often developed based on the sampled inputs and outputs, using parameter estimation. Control system design, however, is usually performed based on continuous models. It is important, then, to be able to convert discrete models to continuous. Fortunately, the Control Toolbox makes this easy via the `d2c` command.

```
sys = d2c(sysd, method)  % discrete to continuous
```

where `sysd` is the LTI object for a discrete model, `sysc` is the continuous object to be created, and `method` is the type of hold placed on the input variable. We always assume that a zero-order hold (`'zoh'`) is used for the discrete model; this is also the default if no entry is made for `method`.

Consider the discrete transfer function model for the van de Vusse reactor, `vdv_tfd`. It can be converted to the continuous-time model using the following MATLAB command:

```
» vdv_tfc = d2c(vdv_tfd,'zoh')

Transfer function:
  -1.117 s + 3.147
---------------------
s^2 + 4.643 s + 5.382
```

M4.5 Step and Impulse Responses

Step responses of LTI objects can be determined or compared in a number of different ways. The following command compares the step responses of the continuous and discrete-time van de Vusse models previously entered in the MATLAB command window.

```
» step(vdv_tfc,vdv_tfd)
```

The resulting response is shown in Figure M4–1. Notice that a zero-order hold has been applied to the discrete step response. Also, note that the step command assumes that the time unit is seconds, which is not normally the case for chemical process systems.

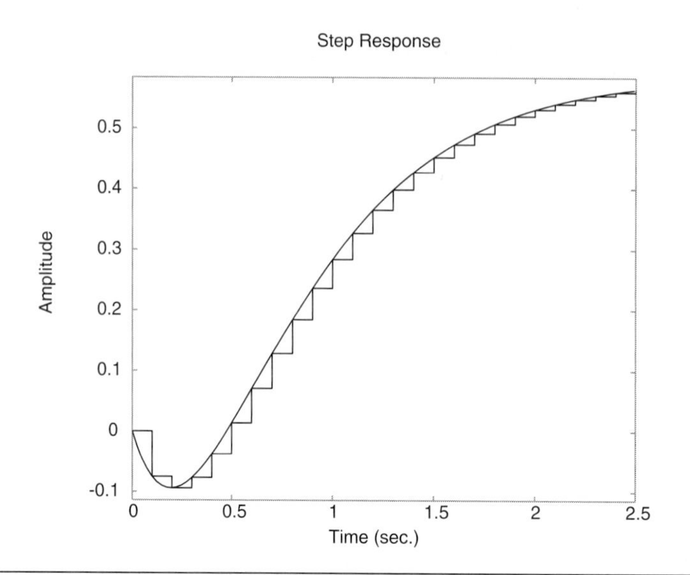

Figure M4–1 Step responses of continuous and discrete van de Vusse reactor model. Illustration of zero-order hold. Notice the default labeling of a time unit of seconds.

Alternatively, the following sequence of commands can be used to generate Figure M4–2:

```
» [yc,tc] = step(vdv_tfc);
» [yd,td] = step(vdv_tfd);
» plot(tc,yc,td,yd,'o')
» xlabel('time')
» ylabel('y')
» title('step response of van de Vusse reactor models')
» legend('continuous','discrete')
```

Similarly, the impulse responses shown in Figure M4–3 are generated using

```
» impulse(vdv_tfc,vdv_tfd)
```

Notice that the tremendous difference in the impulse response for the continuous and discrete-time models is due to the pulse implementation on the discrete-time model.

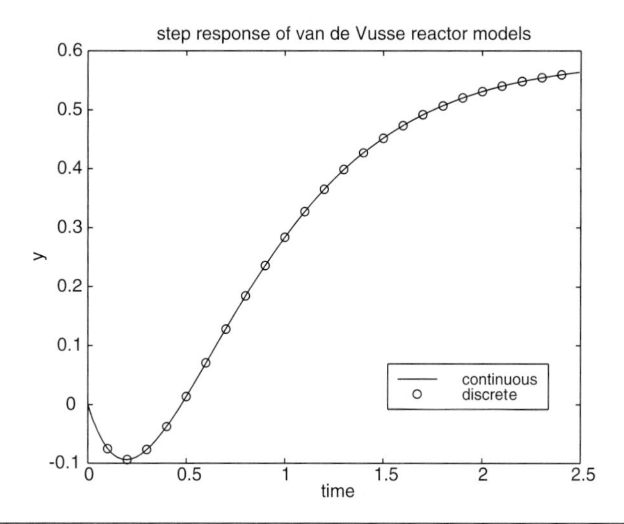

Figure M4–2 Step responses of continuous and discrete van de Vusse reactor model.

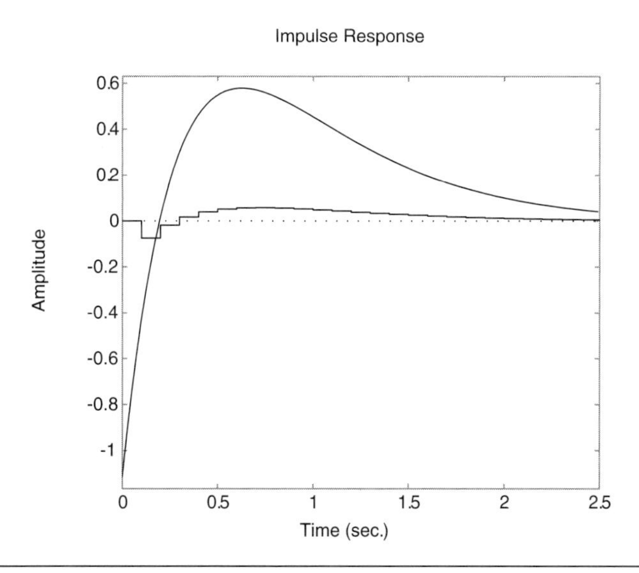

Figure M4–3 Impulse responses of continuous and discrete van de Vusse reactor model. Notice the default labeling with a time unit of seconds.

M4.6 Summary

The MATLAB Control Toolbox can be used to create LTI objects. These objectives can then easily be converted to other forms. Conversion from continuous to discrete is particularly useful when implementing a digital control algorithm. The commands used include the following:

```
sys    =  ss(a,b,c,d)          % state space form
sys    =  tf(num,den)          % transfer function form
sys    =  zpk(z,p,k)           % zero-pole-gain form
sysd   =  filt(num,den,Ts)     % backward shift form
sysd   =  c2d(sys,Ts,'zoh'     % continuous to discrete
sys    =  d2c(sysd,'zoh')      % discrete to continuous
pi     =  pole(sys)            % poles
zi     =  tzero(sys)           % zeros
[y,t]  =  step(sys)            % step response
[y,t]  =  impulse(sys)         % impulse response
```

In addition, the following do not require the Control Toolbox:

```
[num,den] = ss2tf(a,b,c,d,iu)  % state space to transfer function
[a,b,c,d] = tf2ss(a,b,c,d)     % transfer function to state space
```

Examples for the most important commands can be found in Section M4.1.

Reference

More details on converting between model types can be found in the following: Bequette, B. W., *Process Dynamics: Modeling, Analysis and Simulation*, Prentice Hall, Upper Saddle River, NJ (1998).

Additional Exercises

1. Consider the continuous state space model, where the time unit is minutes

$$\dot{x} = \begin{bmatrix} -1 & 0 & 0 \\ 1 & -1 & 0 \\ 0 & 1 & -1 \end{bmatrix} x + \begin{bmatrix} 1 \\ 0 \\ 0 \end{bmatrix} u$$

$$y = \begin{bmatrix} 0 & 0 & 1 \end{bmatrix} x + 0u$$

a. Find the eigenvalues of the A matrix.

b. Find the continuous-time transfer function and calculate the poles and zeros. Is the inverse of this transfer function stable?

c. Find the discrete-time transfer function for a sample time of 0.1 minutes. What do you notice about the number of zeros? Is the inverse of this transfer function stable for this sample time? You should increase the sample time to 1.8399 minutes to find that the transmission zero shifts to the right-half-plane.

d. Compare the step responses of the continuous model with the discrete models.

e. Find the "filter" form (backward shift notation) of the discrete transfer function for sample times of 0.1 and 1.

2. Show that the following three state space models result in the same transfer function model.

a. Model 1

$$\dot{x} = \begin{bmatrix} 0 & 1 \\ -1 & -1 \end{bmatrix} x + \begin{bmatrix} 0 \\ 1 \end{bmatrix} u$$
$$y = \begin{bmatrix} 1 & 0 \end{bmatrix} x + 0u$$

b. Model 2

$$\dot{x} = \begin{bmatrix} -1 & -1 \\ 1 & 0 \end{bmatrix} x + \begin{bmatrix} 1 \\ 0 \end{bmatrix} u$$
$$y = \begin{bmatrix} 0 & 1 \end{bmatrix} x + 0u$$

c. Model 3

$$\dot{x} = \begin{bmatrix} -0.5 & 0.86603 \\ -0.86603 & -0.5 \end{bmatrix} x + \begin{bmatrix} -1.633 \\ 0 \end{bmatrix} u$$
$$y = \begin{bmatrix} 0 & 0.70711 \end{bmatrix} x + 0u$$

Isothermal Chemical Reactor

There are a number of important concepts that are presented in this module.

- The nonlinear concept of input multiplicity
- The challenge of inverse response behavior
- The Ziegler-Nichols closed-loop oscillation tuning method
- IMC

This module contains too many topics to cover interactively during a typical 2-hour session. It is recommended that the reader (or instructor) select a topic of particular interest. The two different controller design sections (M5.4 and M5.5) can be studied independently. The approximate time needed and the focus (in parentheses) of each section of this module are as follows:

M5.1 Background
M5.2 Model (Chapter 2)
M5.3 Steady-State and Dynamic Behavior (Chapter 3)—1–1.5 Hours
M5.4 Classical Feedback Control (Chapters 5 and 6)—1.5 Hours
M5.5 Internal Model Control (Chapter 8)—3 Hours

M5.1 Background

The reactant conversion in a chemical reactor is a function of the residence time or its inverse, the space velocity. For an isothermal CSTR, the product concentration can be controlled by manipulating the feed flow rate, which changes the residence time (for a constant volume reactor). The feedback strategy is shown in the schematic diagram shown in Figure M5–1.

Here, we consider a series-parallel reaction of the following form (known as the van de Vusse reaction scheme):

$$A \xrightarrow{k_1} B \xrightarrow{k_2} C$$

$$2A \xrightarrow{k_3} D$$

The desired product is the component B, the intermediate component in the series reaction. In this module, we find interesting steady-state and dynamic behavior that can occur with this reaction scheme. Klatt and Engell (1998) note that the production of cyclopentenol from cyclopentadiene is based on such a reaction scheme (where A = cyclopentadiene, B = cyclopentenol, C = cyclopentanediol, and D = dicyclopentadiene).

M5.2 Model (Chapter 2)

The molar rate of formation (per unit volume) of each component is

$$r_A = -k_1 C_A - k_3 C_A^2$$
$$r_B = k_1 C_A - k_2 C_B$$
$$r_C = k_2 C_B$$
$$r_D = \frac{1}{2} k_3 C_A^2$$

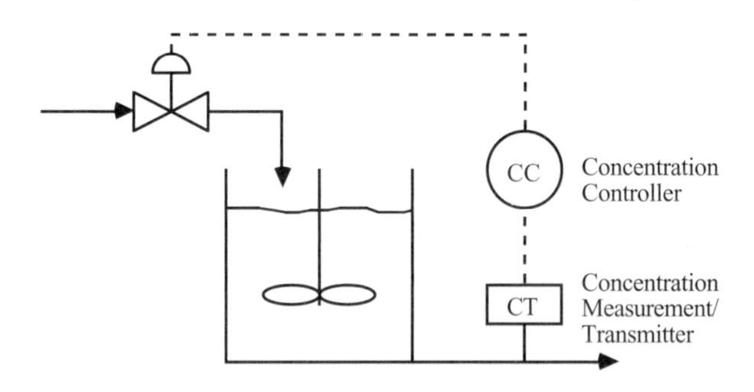

Figure M5–1 Feedback strategy for the isothermal reactor.

Assuming a constant volume reactor, it is easy to derive the following equations (do this):

$$\frac{dC_A}{dt} = \frac{F}{V}\left(C_{Af} - C_A\right) - k_1 C_A - k_3 C_A^2$$

$$\frac{dC_B}{dt} = -\frac{F}{V}C_B + k_1 C_A - k_2 C_B$$

$$\frac{dC_C}{dt} = -\frac{F}{V}C_C + k_2 C_B$$

$$\frac{dC_D}{dt} = -\frac{F}{V}C_D + \frac{1}{2}k_3 C_A^2$$

Notice that the first two equations do not depend on the concentration of components C or D. Since we are only concerned about the concentration of component B, we only need to solve the first two equations:

$$\frac{dC_A}{dt} = \frac{F}{V}\left(C_{Af} - C_A\right) - k_1 C_A - k_3 C_A^2$$

$$\frac{dC_B}{dt} = -\frac{F}{V}C_B + k_1 C_A - k_2 C_B$$

M5.3 Steady-State and Dynamic Behavior (Chapter 3)

At steady state, the component A balance yields a quadratic expression,

$$-k_3 C_{As}^2 + \left(-k_1 - \frac{F_s}{V}\right)C_{As} + \frac{F_s}{V}C_{Afs} = 0$$

where the subscript s is used to indicate the steady state value. Solving this quadratic and using the positive root (clearly, there cannot be negative concentrations), we find

$$C_{As} = \frac{-\left(k_1 + \frac{F_s}{V}\right)}{2k_3} + \frac{\sqrt{\left(k_1 + \frac{F_s}{V}\right)^2 + 4k_3\frac{F_s}{V}C_{Afs}}}{2k_3}$$

and solving for the steady-state concentration of B,

$$C_{Bs} = \frac{k_1 C_{As}}{\frac{F_s}{V} + k_2}$$

Using these two expressions, we can find the steady-state input-output curve relating C_{Bs} and F_s/V. Here we consider the dilution rate (F/V) to be the manipulated input. The main advantage is that it is independent of scale (reactor volume). Whether the reactor is 1 or 10,000 liters, the same dilution rate will yield the same concentrations (assuming the feed stream is the same composition).

For the particular reaction under consideration, the rate constants are

$$k_1 = 50 \text{ hr}^{-1} = 5/6 \ \text{min}^{-1}$$

$$k_2 = 100 \text{ hr}^{-1} = 5/3 \ \text{min}^{-1}$$

$$k_3 = 10 \ \frac{\text{mol}}{\text{liter} \cdot \text{hr}} = \frac{1}{6} \frac{\text{mol}}{\text{liter} \cdot \text{min}}$$

and the steady-state feed concentration is $C_{Afs} = 10$ gmol/liter.

Steady-State Input-Output Curve

Notice that this process has a nonlinear relationship between the steady-state dilution rate (F/V) and the steady-state concentration of B (Figure M5–2). There exists a maximum concentration of B that can be achieved. It is interesting to note that the reactor can not be controlled at this maximum point because the process gain is zero. For a given desired value of the concentration of B (as long as it is less than the maximum possible value), there are two dilution rates that can achieve the concentration. This is known as input multiplicity.

In this example, a steady-state concentration of B of 1.117 gmol/liter can be obtained with either $F_s/V = 0.5714 \text{ min}^{-1}$ (case 1) or $F_s/V = 2.8744 \text{ min}^{-1}$ (case 2). Notice that the process gain (slope of the steady-state input-output curve) is positive for case 1 and negative for case 2. For controller design, it is clearly important to know whether you are operating on the left-hand side (e.g., case 1) or right-hand side (e.g., case 2) of the "peak," since the sign of the controller gain must be the same sign as the process gain. Also notice that a steady-state concentration greater than 1.266 gmol/liter cannot be achieved regardless of the controller used.

Linear Analysis

The linear state space model is

$$\dot{x} = Ax + Bu$$
$$y = Cx + Du$$

where the states, inputs, and output are in deviation variable form. The first input (dilution rate) is manipulated and the second (feed concentration of A) is a disturbance input.

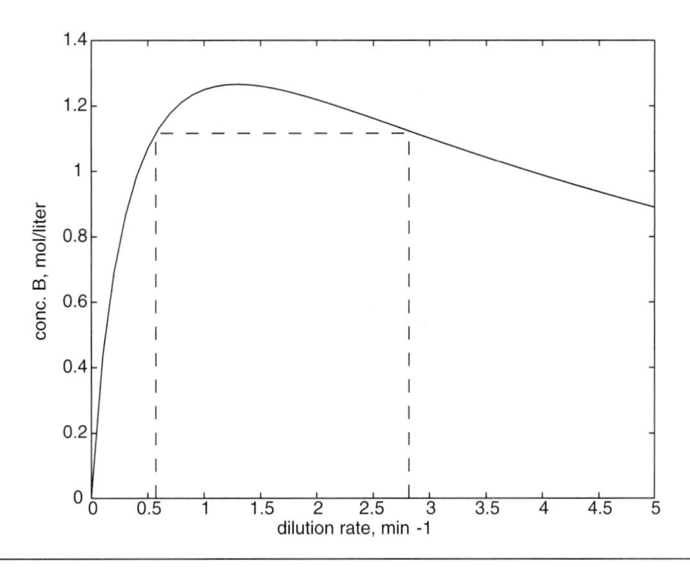

Figure M5–2 Steady-state input-output curve. The dotted lines indicate two possible flow rates to achieve a concentration of 1.117 gmol of *B*/liter.

$$x = \begin{bmatrix} x_1 \\ x_2 \end{bmatrix} = \begin{bmatrix} C_A - C_{As} \\ C_B - C_{Bs} \end{bmatrix}$$

$$u = \begin{bmatrix} F/V - F_s/V \\ C_{Af} - C_{Afs} \end{bmatrix}$$

$$y = x_2 = \left[C_B - C_{Bs} \right]$$

You should linearize the two modeling equations at steady-state solution to find the following state space matrices:

$$A = \begin{bmatrix} -\frac{F_s}{V} - k_1 - 2k_3 C_{As} & 0 \\ k_1 & -\frac{F_s}{V} - k_2 \end{bmatrix}$$

$$B = \begin{bmatrix} C_{Afs} - C_{As} & \frac{F_s}{V} \\ -C_{Bs} & 0 \end{bmatrix}$$

$$C = \begin{bmatrix} 0 & 1 \end{bmatrix}$$

$$D = \begin{bmatrix} 0 & 0 \end{bmatrix}$$

Case 1—Operation on the "Left-Hand Side" of the Peak Concentration

Based on the steady-state operating point of $C_{As} = 3$ gmol/liter, $C_{Bs} = 1.117$ gmol/liter, and $F_s/V = 0.5714\ \text{min}^{-1}$, the state space model is

$$A = \begin{bmatrix} -2.4048 & 0 \\ 0.83333 & -2.2381 \end{bmatrix}$$

$$B = \begin{bmatrix} 7 & 0.5714 \\ -1.117 & 0 \end{bmatrix}$$

$$C = \begin{bmatrix} 0 & 1 \end{bmatrix}$$

$$D = \begin{bmatrix} 0 & 0 \end{bmatrix}$$

The manipulated input-output process transfer function for the reactor is [you can find this analytically, using $G(s) = C(sI-A)^{-1}B$, or numerically, using Module 4]

$$g_p(s) = \frac{-1.1170s + 3.1472}{s^2 + 4.6429s + 5.3821} = \frac{0.5848(-0.3549s + 1)}{0.1828s^2 + 0.8627s + 1}$$

and the disturbance input-output transfer function is

$$g_d(s) = \frac{0.4762}{s^2 + 4.6429s + 5.3821} = \frac{0.0885}{0.1828s^2 + 0.8627s + 1}$$

The responses of the linear and nonlinear models to a step change in F/V of $0.1\ \text{min}^{-1}$ are compared in Figure M5–3.

Note: A motivated reader may wish to demonstrate that the linear and nonlinear models have the same response for a step change of $0.01\ \text{min}^{-1}$, while the differences in response to a step change of $0.5\ \text{min}^{-1}$ are substantial. Function files for integration of the nonlinear model are presented in Module 3. The focus of the current module is on responses of the linear model.

M5.4 Classical Feedback Control (Chapters 5 and 6)

Closed-Loop Performance Requirements

It is desirable to be able to make setpoint changes of ± 0.1 gmol/liter. Also, the manipulated dilution rate should not change by more than $0.4\ \text{min}^{-1}$, and there should be less than 20% overshoot. Here, we assume that the process is represented by the transfer function

$$g_p(s) = \frac{-1.1170s + 3.1472}{s^2 + 4.6429s + 5.3821} = \frac{0.5848(-0.3549s + 1)}{0.1828s^2 + 0.8627s + 1}$$

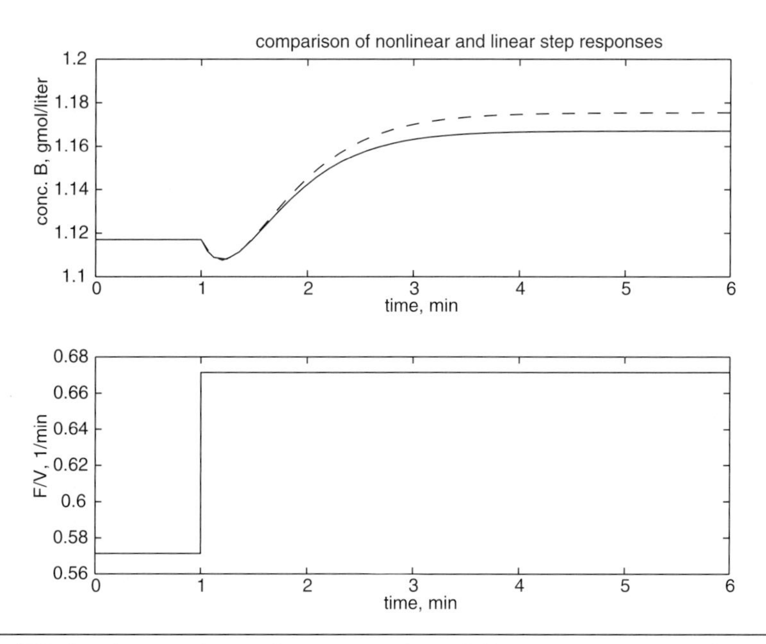

Figure M5–3 Comparison of concentration of B response to step change in F/V of 0.1 min^{-1}. Nonlinear (solid) vs. linear (dashed).

Problem 1. Based on the transfer function given, answer the following questions:

a. For a desired concentration change of 0.1 gmol/liter, what is the expected steady-state manipulated dilution rate change? If the volume is 1 liter, what is the expected steady-state flow-rate change? If the volume is 1000 gallons, what is the expected steady-state flow-rate change?

b. To keep the manipulated input dilution rate change less than 0.4 min^{-1}, what is the maximum concentration setpoint change that can be made?

c. What are the units of the controller proportional gain, k_c?

Problem 2. Generate the SIMULINK block diagram shown in Figure M5–4. Recall that the default SIMULINK controller block requires parameters in the decoupled (k_c,k_I,k_D) form,

$$u(t) = k_c e(t) + k_I \int_o^t e(\sigma)d\sigma + k_D \frac{de}{dt}$$

while we use the following form in this text:

$$u(t) = k_c \left[e(t) + \frac{1}{\tau_I} \int_o^t e(\sigma)d\sigma + \tau_D \frac{de}{dt} \right]$$

If the default SIMULINK PID controller block is used, then the tuning parameters for the integral and derivative modes are

$$k_I = k_c/\tau_I \qquad k_D = k_c\tau_D$$

Note that variables can be specified on the input window for the PID controller block, so that numerical values can simply be entered in the MATLAB command window before each new simulation.

Perform some initial simulations with arbitrary PID parameter values. Use a setpoint change of 0.1 at $t = 1$ minute. Adjust the tuning parameters to obtain an "acceptable" closed-loop response, so that the concentration overshoot and maximum input flow-rate changes are satisfied.

Problem 3. Consider P-only control and a step setpoint change of 0.1 gmol/liter at $t = 1$ minute.

a. Find the offset when $k_c = 1$ is used. Is this the amount expected? Show your calculation.

b. Discuss the behavior as k_c is increased. At what value of k_c does the system appear to go unstable? Is this consistent with your analytical results from the Routh stability criterion? Show those results here.

Note that the system solver (integration code), although accurate, may generate points that yield a "coarse" plot. You can make the curves appear smoother by selecting the `parameters` pull-down menu and changing `refine` from 1 to 2 or 4. This generates more points for plotting purposes.

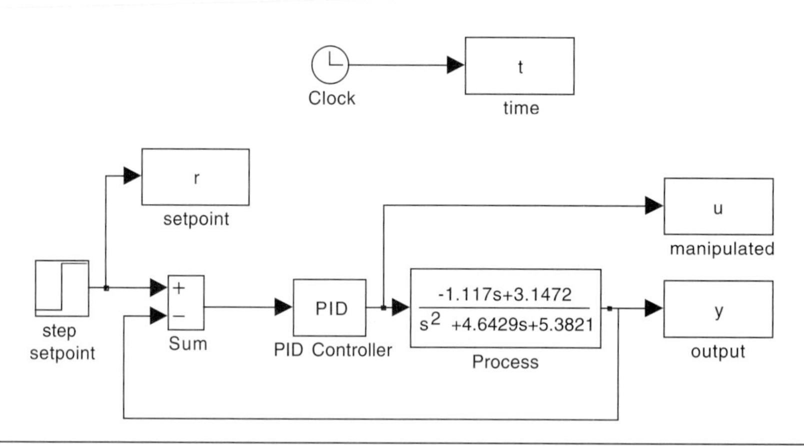

Figure M5–4 SIMULINK block diagram.

a. Find the two steady-state flow rates that yield a concentration of $B = 0.9$ gmol/liter.

b. Find the process transfer function for the lower flow rate that yields $C_B = 0.9$ gmol/liter.

Answer: $$g_p(s) = \frac{-0.9s + 1.447}{s^2 + 1.933s + 0.9231} = \frac{1.567(-0.622s + 1)}{(0.931s + 1)(1.163s + 1)}$$

c. What are the poles and zeros of this transfer function?

MODULE 6

First-Order + Time-Delay Processes

The most commonly used model to describe the dynamics of chemical processes is first order + time delay. In this module, we relate the stability of feedback control of these processes to frequency-response analysis. The important principles that we wish to illustrate are as follows:

- Gain margin is a measure of gain uncertainty that can be tolerated
- Phase margin is a measure of the time-delay uncertainty that can be tolerated
- The "cross-over frequency" is related to the ultimate period from the closed-loop Ziegler-Nichols approach
- Ziegler-Nichols recommended tuning values do not provide adequate gain and phase margins
- IMC-based PI is easier to tune for robustness than are the other approaches

The sections of this module are as follows:

It is assumed that the reader has the MATLAB Control Toolbox, which has the Bode and Nyquist functions.

M6.1 Motivation

Since many chemical processes are modeled as first order + time delay (at least for controller design), it is important to understand the effect of control structure (IMC, P, PI, PID) and tuning parameters on the stability of the nominal (perfect model) and uncertain (actual) closed-loop system. In this module, we use the following transfer function to represent the model

$$\tilde{g}_p(s) = \frac{e^{-5s}}{10s+1}$$

For illustrative purposes, we assume that the units of the gain are in degrees Celsius per percent and that minutes are the time units.

M6.2 Closed-Loop Time-Domain Simulation

Here, we begin by using the SIMULINK diagram shown in Figure M6–1.

The system is on the verge of instability when $k_c = 3.8$ %/°C, as shown in Figure M6–2; the ultimate proportional gain is then $k_{cu} = 3.8$ %/°C. The period of oscillation (ultimate period, P_u) is approximately 17.2 minutes. Both of these values can be obtained from frequency-response analysis, as shown in Section M6.3. Notice that the manipulated variable action is constant from $t = 1$ to $t = 6$ minutes. This is due to the initial "proportional kick." The manipulated input does not change until the process output is observed to change at $t = 6$ minutes (remember there is a 5-minute time delay). If we used a controller with integral action, the manipulated input would have kept increasing during this interval.

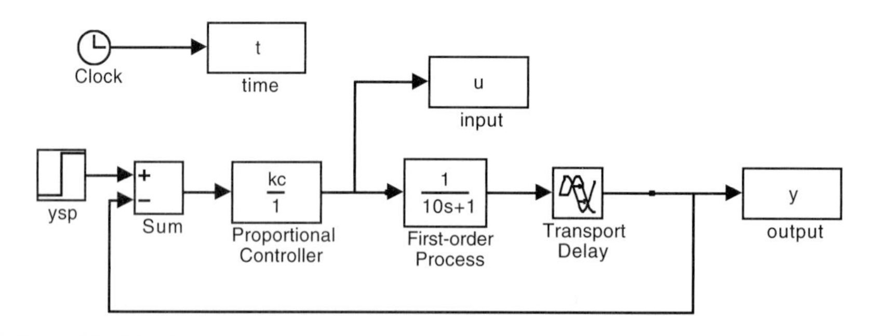

Figure M6–1 SIMULINK diagram. For best results, choose a stiff integrator, such as `ode15s` (stiff/NDF), from the "`parameters`" pull-down menu.

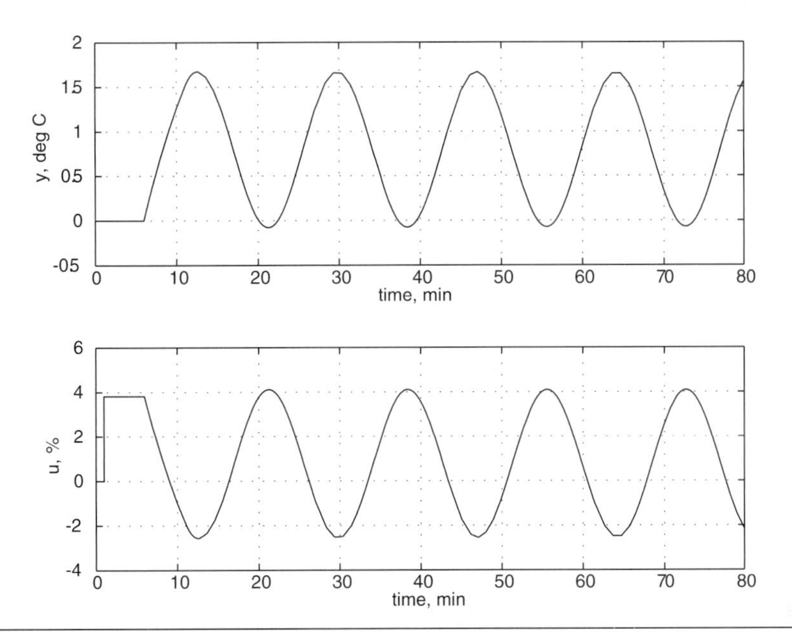

Figure M6–2 Closed-loop response for a setpoint change of 1°C at time = 1 minute, with k_c = 3.8 %/°C. The period of oscillation (P_u, peak-to-peak time) is 17.2 minutes.

M6.3 Bode Analysis

Here, we enter the first-order + time-delay model in the MATLAB command window.

```
» fodt = tf([1],[10 1],'InputDelay',5)

Transfer function:
   1
--------
10 s + 1

Input delay: 5
```

Now, let us perform a Bode analysis assuming a proportional controller gain of 1.

```
» kc = 1;
» [mag,phase,w] = bode(kc*fodt);
» subplot(2,1,1), loglog(w,squeeze(mag))
```

```
» subplot(2,1,2), semilogx(w,squeeze(phase),[0.01 1],
  [-180 -180],'--')
```

The resulting Bode plot, with the –180° line drawn for convenience, is shown in Figure M6–3. Notice that the crossover frequency is around 0.36 rad/min, where the amplitude ratio is approximately 0.26. The gain margin is then 1/0.26 = 3.8. This means that the controller gain could be increased to roughly 3.8 before the process goes unstable. This is consistent with the finding of $k_c = 3.8$ for instability, shown in the previous section (time-domain simulation).

The gain and phase margins can be calculated using the following MATLAB function:

```
» [Gm,Pm,Wco,Wpm] = imargin(squeeze(mag),squeeze(phase),w)

Gm = 3.80093442825351
Pm = Inf
Wco = 0.36655875121677
Wpm = NaN
```

where ω_{co} is the crossover frequency (where $\phi = -180°$) and ω_{pm} is the frequency where AR = 1.

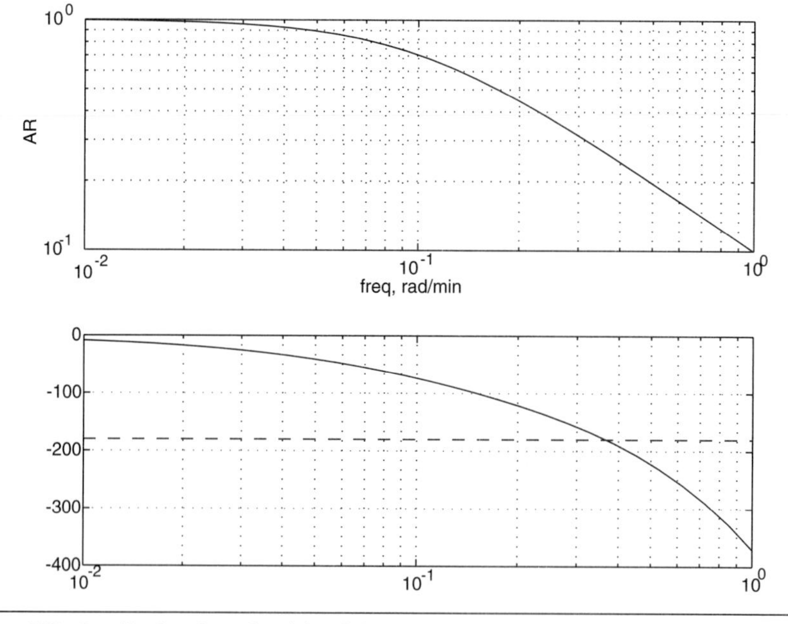

Figure M6–3 Bode plot of $g_c(s)g_p(s)$ for $k_c = 1$.

Gain Margin

The gain margin of 3.8 indicates that the controller gain can be increased from 1 to 3.8 before the process becomes unstable—again, this is consistent with the previous results. Also notice that the phase margin is infinity. This is because the amplitude ratio is always less than 1 (it is exactly 1 only at 0 frequency, when the proportional gain is 1).

Crossover Frequency/Ultimate Period

The crossover frequency is 0.366 rad/minute. Converting this to cycles/minute, we find a frequency of 0.0583 cycles/minute, which corresponds to a period of 1/0.0583 = 17.16 minutes. This is exactly the value found in the time-domain simulations.

Nyquist Diagram

The Nyquist plot (Figure M6–4) can be generated by using the following commands:

```
» [re,im,w] = nyquist(kc*fodt);
» plot(squeeze(re),squeeze(im))
```

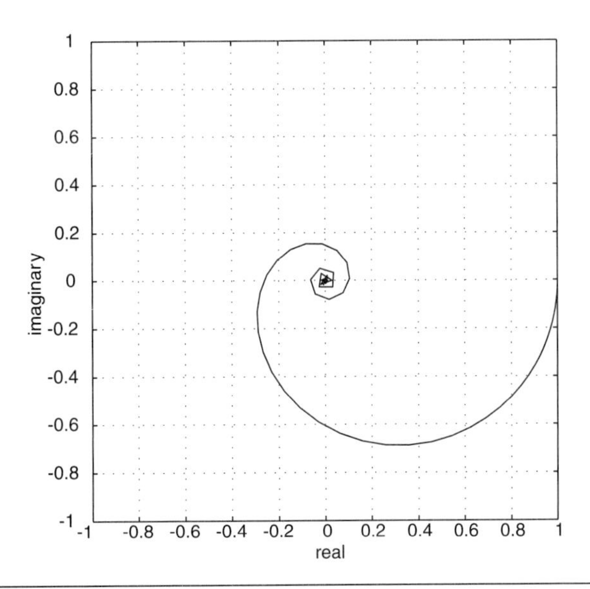

Figure M6–4 Nyquist plot of $g_c(s)g_p(s)$ for $k_c = 1$.

Problem 1. Use a controller proportional gain of 3.8 and perform a Bode analysis to show that the amplitude ratio is 1 at the crossover frequency. What is the phase margin and frequency when AR = 1 (for the controller proportional gain of 3.8)?

Solution

The Bode and Nyquist plots for $g_c(s)g_p(s)$ with $k_c = 3.8$ are shown in Figures M6–5 and M6–6. Notice that the phase margin is 1 (the AR = 1 at the crossover frequency) and the phase margin is 0 (the phase angle is -180° when the AR = 1).

This is also shown by the gain and phase margin analysis from MATLAB:

```
» kc = 3.8;
» [mag,phase,w] = bode(kc*fodt);
» [Gm,Pm,Wco,Wpm] = imargin(squeeze(mag),squeeze(phase),w)

Gm  = 1.00024590217198
Pm  = 0.03106191271067
Wco = 0.36655875121677
Wpm = 0.36646150971664
```

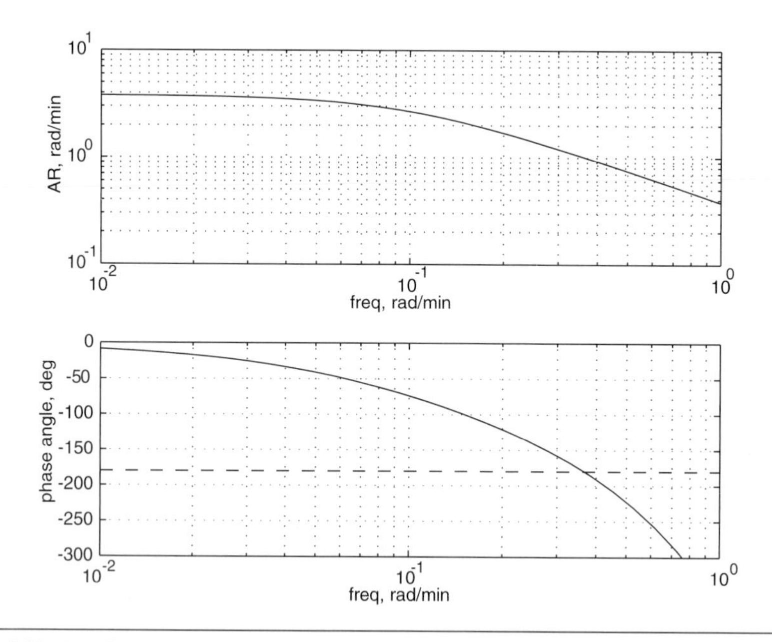

Figure M6–5 Bode plot of $g_c(s)g_p(s)$ for $k_c = 3.8$. Notice that the gain margin is 1 and the phase margin is 0°, indicating that the system is on the verge of instability.

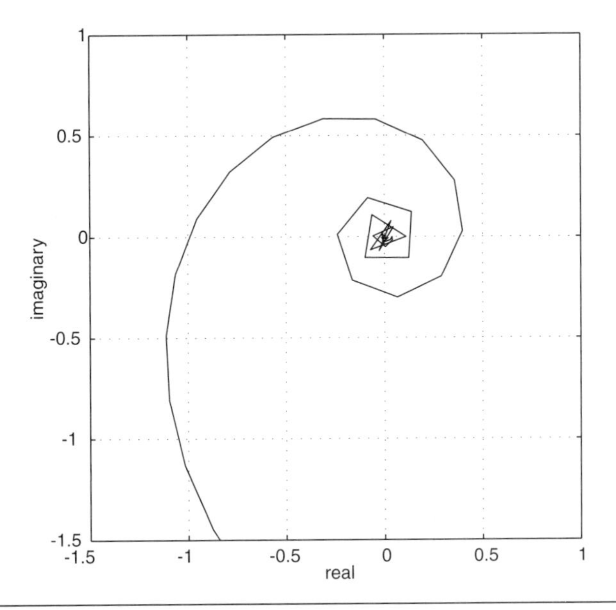

Figure M6–6 Nyquist plot of $g_c(s)g_p(s)$ for $k_c = 3.8$. The Nyquist curve passes through the critical point (–1,0), indicating the system is on the verge of instability.

Notice that the gain margin is effectively 1, the phase margin is effectively 0, and the crossover and phase margin frequencies converge to 0.366 rad/min.

M6.4 Ziegler-Nichols Tuning

The tuning parameters recommended by Ziegler and Nichols (1942) were shown in Table 6–1. Generally, these values are not very robust. The objective of this section is for you to calculate the values of the gain and phase margins for each of these recommended values. Tyreus and Luyben modified the suggested Ziegler-Nichols parameters for increased robustness (see Luyben and Luyben, 1997); these recommended values were shown in Table 6–2.

Recall that the P-only controller tuning recommended by Ziegler and Nichols (1942) results in a controller that has a gain margin of 2 (since the proportional gain is one half of the ultimate gain). This is also shown by the following MATLAB analysis.

```
» kc = 3.8/2;
» [mag,phase,w] = bode(kc*fodt);
```

```
» [Gm,Pm,Wco,Wpm] = imargin(squeeze(mag),squeeze(phase),w)

Gm  =  2.00049180434395
Pm  = 75.49091926803467
Wco =  0.36655875121677
Wpm =  0.16127245556240
```

The gain margin is 2 and the phase margin is 75°, indicating that a reasonable amount of uncertainty can be tolerated before the closed-loop system goes unstable. The closed-loop responses (measured output and manipulated input) for $k_c = 1.9$ are shown in Figure M6–7.

Problem 2. Find the gain and phase margins when Ziegler-Nichols PI settings are used. Compare these with the gain and phase margins of the Tyreus-Luyben recommended PI settings. Compare the closed-loop responses (time domain) for both of these controller settings; include plots of the manipulated input action. *Hint*: Create a transfer function for

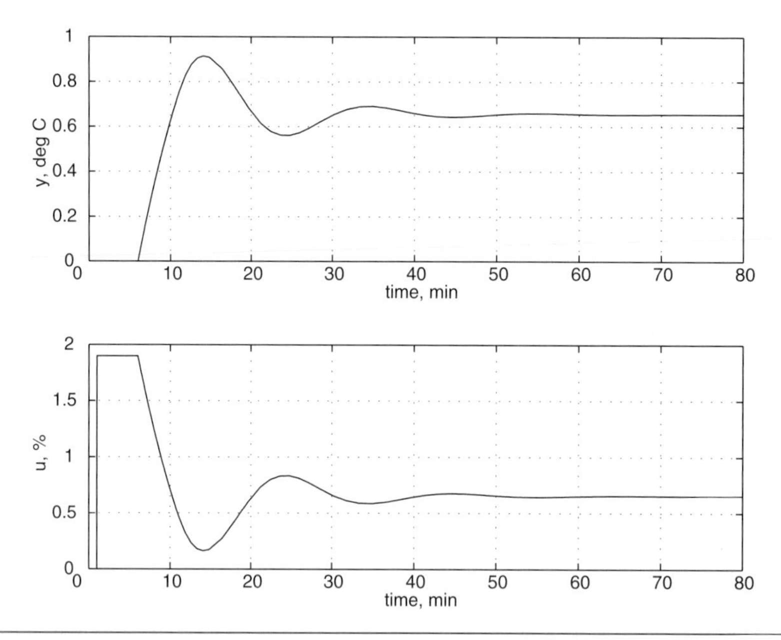

Figure M6–7 Closed-loop response for a setpoint change of 1 at time = 1 minute, with $k_c = 1.9$ (Ziegler-Nichols recommended P-only setting). Notice the offset due to the P-only controller.

the PI controller and multiply it with the process transfer function. Recall that a PI controller transfer function is

$$g_c(s) = \frac{k_c(\tau_I s + 1)}{\tau_I s}$$

The Ziegler-Nichols recommended parameters for a PI controller are then

$$g_c(s) = \frac{1.71(14.3s + 1)}{14.3s} = \frac{24.45s + 1.71}{14.3s}$$

The commands to MATLAB are then

```
» gc = tf([24.45 1.71],[14.3 0])

Transfer function:
24.45 s + 1
-----------
  14.3 s

» [mag,phase,w] = bode(gc*fodt);
» [Gm,Pm,Wco,Wpm] = imargin(squeeze(mag),squeeze(phase),w)
```

Thus far, we have used the gain margin to determine either how much the controller gain can be increased before instability occurs, or how much the process gain can increase before instability occurs. Next, we illustrate how the phase margin is related to time-delay uncertainty that can be tolerated.

Problem 3. Use the Ziegler-Nichols recommended P-only value of $k_c = 0.5\,k_{cu} = 3.8/2 = 1.9$. What is the phase margin for this value of controller gain? What is the frequency associated with this phase margin? Calculate the amount of additional time delay that can be tolerated from

$$\Delta\theta = \frac{\Delta\phi}{-\omega_{pm}} \cdot \frac{2\pi}{360°} = \_\_\_\_\_\_\_\_\_\_$$

Add this time delay to the process model and show that the new phase margin is now 0. Show that the time-domain closed-loop simulations have an oscillation with a frequency of ω_{pm}. Include plots of the manipulated input action.

M6.5 IMC-Based PID Control

Recall from Chapter 9 that Padé approximations for dead time could be used to develop a IMC-based PI and PID controllers for first-order + time-delay systems. Here, we compare the gain and phase margins for an IMC-based PI controller (using the "improved" settings) for various values of λ.

The improved IMC-based PI settings are as follows:

$$k_c = \frac{\tau_p + 0.5\theta}{k_p \lambda}$$

$$\tau_I = \tau_p + 0.5\theta$$

Problem 4. For the IMC-based PI (improved) settings, find the value of λ that leads to a gain margin of 2.5. Also, find the value of λ that leads to a phase margin of 75°. Compare the time-domain closed-loop responses for these two systems (always include plots of the manipulated inputs). *Hint*: Recall that the improved IMC-based PI settings for first-order + time-delays systems are

$$k_c = \frac{(\tau_p + 0.5\theta)}{k_p \lambda}, \quad \tau_I = \tau_p + 0.5\theta$$

For $\lambda = 10$, we find

```
» gc = tf([1.25*12.5 1.25],[12.5 0])

Transfer function:
15.62 s + 1.25
--------------
    12.5 s

» [mag,phase,w] = bode(gc*fodt);
» [Gm,Pm,Wco,Wpm] = imargin(squeeze(mag),squeeze(phase),w)

Gm  =  2.63261419901939
Pm  = 63.19455846288847
Wco =  0.32379172211747
Wpm =  0.11488259758305
```

Here, we see that the gain margin requirement is satisfied. Increase λ until the phase margin is satisfied.

Problem 5. Compare the Ziegler-Nichols PI, Tyreus-Luyben PI and IMC-based improved PI (with $\lambda = 10$ minutes) control performance in time-domain simulations.

M6.6 Summary

Frequency-response techniques have been used to study the robustness of a simple first-order + time-delay process model. The gain margin is a measure of gain uncertainty that can be tolerated, while phase margin is a measure of dead time uncertainty that can be tolerated. An uncertain time constant affects both the gain margin and phase margin.

References

The original Ziegler-Nichols parameters are developed in the following paper: Ziegler, J. G., and N. B. Nichols, "Optimum Settings for Automatic Controllers," *Trans. ASME*, **64**, 759–768 (1942).

The Tyreus-Luyben recommended values are presented in Chapter 3 of the following textbook: Luyben, M. L. and W. L. Luyben, *Essentials of Process Control*, McGraw Hill, New York (1997).

Additional Exercises

1. The transfer function for the van de Vusse reactor is given below; the input variable is dilution rate (F/V, min^{-1}) and the output is concentration of component B (mol/liter). Find the gain and phase margins for the Ziegler-Nichols recommended P, PI, and PID parameters. Show the resulting closed-loop time-domain responses to a unit setpoint change.

$$g_p(s) = \frac{-1.1170s + 3.1472}{s^2 + 4.6429s + 5.3821} = \frac{0.5848(-0.3549s + 1)}{0.1858s^2 + 0.8638s + 1}$$

2. The IMC-based PID design procedure for the van de Vusse reactor (Additional Exercise 1) results in a PID with first-order lag controller. Find how the gain and phase margins vary as a function of the IMC filter factor, λ. Vary λ from 0.1 minutes to 1 minute. What is the recommended value of λ if the minimum gain and phase margins are 3 and 75°, respectively?

Appendix M6.1

Answers to Problems 2, 3 and 4. The reader should show how these results are obtained and generate the resulting closed-loop time domain plots.

Problem 2: Ziegler-Nichols PI: Gm = 1.97, Pm = 53, Wco = 0.33, Wpm = 0.16. Tyreus-Luyben: Gm = 3.09, Pm = 102, Wco = 0.351, Wpm = 0.0746.

Problem 3: Gm = 2.00, Pm = 75.49, Wco = 0.366, Wpm = 0.161. Additional time delay = 8.18 minutes.

Problem 4: For $\lambda = 9.5$, $k_c = 1.315$, Gm = 2.50. Increase λ to obtain a Pm of at least 75°.

Biochemical Reactors

The objective of this module is to illustrate a practical problem where the IMC-based PID procedure can be used. The sections of the module are as follows:

M7.1 Background
M7.2 Steady-State and Dynamic Behavior
M7.3 Stable Steady-State Operating Point
M7.4 Unstable Steady-State Operating Point
M7.5 SIMULINK Model File

The procedure for a *stable process* will be used in Section M7.3, while the procedure for an *unstable process* will be used in Section M7.4. In both cases, use the SIMULINK model file presented in Section M7.5.

Although you will be using a linear controller, you will be implementing this controller on a nonlinear process. The inputs and outputs to/from this process are in physical variables, while the controller design procedure is based on deviation variables.

M7.1 Background

Biochemical reactors are used in a wide variety of processes, from waste treatment to alcohol fermentation. Biomass (cells) consume substrate (sugar or waste chemicals) and produce more cells. A typical control and instrumentation diagram, with biomass concentration as the measured output, is shown in Figure M7–1.

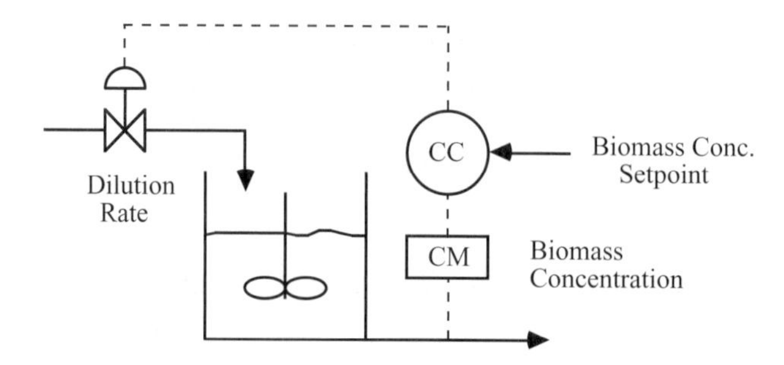

Figure M7–1 Simplified control and instrumentation diagram for a biochemical reactor.

Model

The modeling equations for a bioreactor are

$$\frac{dx_1}{dt} = (\mu - D)x_1$$
$$\frac{dx_2}{dt} = D(x_{2f} - x_2) - \frac{\mu x_1}{Y} \tag{M7.1}$$

where the state variables are x_1 = biomass (cell) concentration = mass of cells/volume, and x_2 = substrate concentration = mass of substrate/volume.

The manipulated input is D = dilution rate = F/V = volumetric flow rate/reactor volume, and the disturbance input is x_{2f} = substrate feed concentration.

Two possible expressions for the specific growth rate are monod and substrate inhibition kinetics, which include

$$\mu = \frac{\mu_{max} x_2}{k_m + x_2} \qquad \text{Monod}$$

$$\mu = \frac{\mu_{max} x_2}{k_m + x_2 + k_1 x_2^2} \qquad \text{Substrate Inhibition}$$

Notice that the monod-specific growth rate model is a subset of the substrate inhibition model ($k_1 = 0$).

Scale Up

One motivation for working with the dilution rate as the manipulated input is that the resulting dynamic model is independent of scale. A reactor volume of 1 liter with a flow rate of 0.3 liters/hour has the same dynamic behavior as a reactor volume of 1000 liters with a flow rate of 300 liters/hour. Thus, a small-scale (laboratory or pilot plant) reactor can be used to predict the behavior of a production-scale reactor.

M7.2 Steady-State and Dynamic Behavior

We use the following parameters for a substrate inhibition model for this control study:

$$\begin{aligned} \mu_{max} &= 0.53\text{hr}^{-1} \\ k_m &= 0.12\text{g/liter} \\ k_1 &= 0.4545\text{liter/g} \\ Y &= 0.4 \end{aligned}$$

The steady-state dilution rate is $D_s = 0.3 \text{ hr}^{-1}$ (and the residence time is 3.33 hours) and the feed substrate concentration is $x_{2fs} = 4.0$ g/liter.

Steady-State Conditions

The nonlinear process has the following three steady-state solutions (operating points) to Equation M7.1 for a dilution rate of 0.3 hr^{-1}:

Steady state	*Biomass concentration*	*Substrate concentration*	*stability*
Equilibrium 1—wash out	$x_{1s} = 0$	$x_{2s} = 4.0$	stable
Equilibrium 2—nontrivial	$x_{1s} = 0.995103$	$x_{2s} = 1.512243$	unstable
Equilibrium 3—nontrivial	$x_{1s} = 1.530163$	$x_{2s} = 0.174593$	stable

Notice that we definitely do not want to operate a equilibrium point 1. Here there is no reaction occurring because the cells have been washed out of the reactor. The outlet substrate concentration is the same as the inlet substrate concentration under these conditions.

Linear Model

The state-space model matrices are

$$A = \begin{bmatrix} \mu_s - D_s & x_{1s}\mu'_s \\ -\dfrac{\mu_s}{Y} & -D_s - \dfrac{\mu'_s x_{1s}}{Y} \end{bmatrix} \qquad B = \begin{bmatrix} -x_{1s} \\ x_{2fs} - x_{2s} \end{bmatrix}$$

where the partial derivative of the specific growth rate with respect to the substrate concentration is

$$\mu'_s = \frac{\partial \mu}{\partial x_{2s}} = \frac{\mu_{\max} k_m}{\left(k_m + x_{2s}\right)^2}$$

and where dilution rate is the manipulated input. Different control strategies have been used to control continuous biochemical reactors. One is based on measuring the biomass concentration and manipulating the dilution rate. Another is based on measuring the substrate concentration and manipulating the dilution rate.

In the following two problems, the biomass concentration is the measured output and the dilution rate is the manipulated input. You will be performing your control simulations using the nonlinear process, so the inputs and outputs from the process are in physical variable form, while the linear controller design is based on the use of deviation variables. Assume that the dilution rate is physically constrained between 0 and 0.6 hr^{-1}.

M7.3 Stable Steady-State Operating Point

Design an IMC-based PID controller to control the bioreactor at equilibrium point 3—the *stable* nontrivial point. The steady state (also use this as the initial condition for your simulations) is

$$x(0) = \begin{bmatrix} 1.530163 \\ 0.174593 \end{bmatrix}$$

At this operating point, the state space model is

$$A = \begin{bmatrix} 0 & 0.9056 \\ -0.75 & -2.5640 \end{bmatrix} \qquad B = \begin{bmatrix} -1.5301 \\ 3.8255 \end{bmatrix}$$

$$C = \begin{bmatrix} 1 & 0 \end{bmatrix} \qquad D = \begin{bmatrix} 0 \end{bmatrix}$$

Use MATLAB to find that the eigenvalues are –0.3 and –2.2640 hr^{-1}, so the system is stable and the IMC-based PID method for stable systems can be used (see Table 9–1). Find the transfer function relating the dilution rate to the biomass concentration and use this for controller design. You may wish to use the MATLAB function `ss2tf` to find the process transfer function

$$g_p(s) = \frac{-1.5302s - 0.4590}{s^2 + 2.564s + 0.6792}$$

After placing the process model in gain and time constant form and recognizing pole-zero cancellation, you should find

$$g_p(s) = \frac{-0.6758}{0.4417s + 1}$$

Notice that the time constant of 0.4417 hr is significantly shorter than the residence time (3.33 hr). Also notice from Table 9–1, that the IMC-based controller is a PI controller.

a. Show how the response to a small setpoint change varies with λ. Suggested setpoint changes are from 1.53016 to 1.52 and from 1.53016 to 1.54 g/liter.

b. For a particular value of λ, show how the magnitude of the setpoint change affects your response. This is where the nonlinearity comes into play. Try changing from the steady-state value of 1.53016 to 1.0, 1.4, 1.6, and 2.5 g/liter. You will find problems with control saturation and for larger setpoint changes.

c. Often measurements cannot be made instantaneously, and there will be a transport delay associated with the measurement. Use the transport delay function in SIMULINK. Use $\theta = 0.25$ hour and discuss how the time-delay affects your choice of λ.

d. Consider now disturbance rejection. Do not make a setpoint change but do make a step change in substrate feed concentration (x_{2f}) from 4 to 3.5 at $t = 1$ hour. Compare disturbance rejection results with the PI controller developed in row 2 of Table 9–1.

M7.4 Unstable Steady-State Operating Point

Design an IMC-based PID controller to control the bioreactor at equilibrium point 2—the *unstable* nontrivial point. The steady state (also use this as the initial condition for your simulations) is

$$x(0) = \begin{bmatrix} 0.995103 \\ 1.512243 \end{bmatrix}$$

At this point, the state space model is

$$A = \begin{bmatrix} 0 & -0.0679 \\ -0.7500 & -0.1302 \end{bmatrix} \qquad B = \begin{bmatrix} -0.9951 \\ 2.4878 \end{bmatrix}$$

$$C = \begin{bmatrix} 1 & 0 \end{bmatrix} \qquad D = \begin{bmatrix} 0 \end{bmatrix}$$

Use MATLAB to find that the eigenvalues are –0.3 and 0.169836 hr^{-1}, so the system is unstable and the IMC-based PID method for unstable systems must be used (Table 9–3). Find the transfer function relating the dilution rate to the biomass concentration and use this for controller design. You may wish to use the MATLAB function `ss2tf` to find the process transfer function

$$g_p(s) = \frac{-0.9951s - 0.2985}{s^2 + 0.1302s - 0.0509}$$

After placing the process model in gain and time constant form (and cancelling common poles and zeros), you should find

$$g_p(s) = \frac{5.8644}{-5.888s + 1}$$

That is, the transfer function has a RHP pole at 0.1698 hr^{-1} which is consistent with the state space model. Notice that we can use the first entry in Table 9–3, which is a PI controller.

a. Show how the response to a small setpoint change varies with λ (show explicitly how the PID tuning parameters vary with λ). I suggest setpoint changes from 0.995103 to 0.985 and 1.005 g/liter.

b. For a particular value of λ, show how the magnitude of the setpoint change affects your response. This is where the nonlinearity comes into play. Try changing from the steady-state value of 0.995013 to 0.5, 0.75, 1.5, and 2.0 g/liter.

c. Use the transport delay block in SIMULINK. Start with $\theta = 0.25$ hour and discuss how the time delay affects your choice of λ.

d. Consider now disturbance rejection. Do not make a setpoint change but do make a step change in substract feed concentration (x_{2f}) from 4 to 3.5 gmol/liter at $t = 1$ hour.

M7.5 SIMULINK Model File

In order to save time on constructing your block diagram, copy the block for the bioreactor from the textbook web page (see Figure M7–2)

```
biofbc.mdl
```

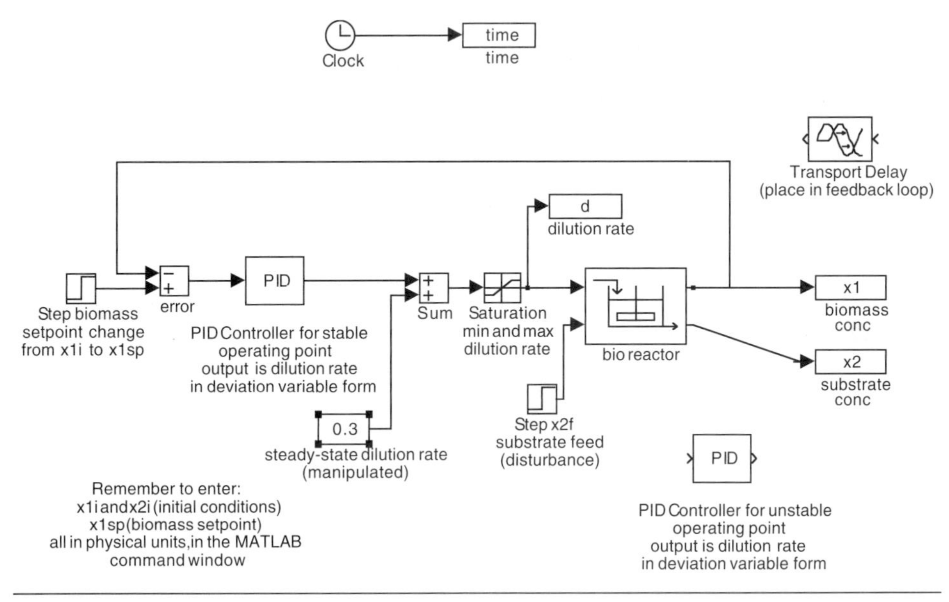

Figure M7–2 SIMULINK block diagram for bioreactor.

I recommend that you make your proportional and integral terms in the PID controller a function of lambda (λ). Then, before each simulation run, you merely have to type in a new lambda in the MATLAB window, rather than typing in new PI parameters. Always realize that the integral term in the default SIMULINK PID block is our k_c/τ_I.

Notice that the SIMULINK diagram contains two extra blocks. The transport delay block can be inserted in the feedback loop to illustrate the effects of a measurement time delay. The PID controller for the open-loop unstable point can be inserted when Additional Exercise 2 is being solved.

The following must be entered into the MATLAB command window: initial conditions, $x1$i and $x2$i; setpoint for biomass concentration, $x1$sp; final simulation time, tfinal; substrate feed concentration, x2f; and the IMC filter factor, lambda.

- The transfer function parameters are used by the PID controller block. The IMC-based PID parameters are different depending on whether the process is open-loop stable (Additional Exercise 1) or open-loop unstable (Additional Exercise 2). This is also seen by comparing Tables 9–1 and 9–3.
- Notice that a saturation block is used so that the dilution rate is bounded between 0.0 and 0.6. This is consistent with "real-world" constraints on manipulated variable action.

- The typical rule-of-thumb for λ for open-loop stable systems is roughly one third to one half of the dominant time constant. There are no good rules-of-thumb for open-loop unstable systems.

When performing your simulations, plot the dilution rate to see if it is hitting a constraint.

Reference

The parameters used in this module were presented in the following paper: Agrawal, P., and H. C. Lim, "Analyses of Various Control Schemes for Continuous Bioreactors," *Adv. Biochem. Eng. Biotechnol.*, **30**, 61–90 (1984).

Additional Exercises

1. Data for a specific growth rate coefficient as a function of substrate concentration for a biochemical reactor are shown below

x_2, g/liter	0	0.1	0.25	0.5	0.75	1	1.5	3	5
μ, hr^{-1}	0	0.38	0.54	0.63	0.66	0.68	0.70	0.73	0.74

   a. Estimate the parameter values for a Monod-specific growth rate model.
   b. The production rate of cells is Dx_1. Find the steady-state value of the dilution rate that maximizes the production rate of cells. The substrate feed concentration is 5 g/liter.
   c. Find the steady-state concentration of biomass and substrate at this dilution rate.
   d. Find the linear state space model at this dilution rate, with dilution rate and substrate feed concentration as the input variables. Also, find the transfer function relating dilution rate to biomass concentration.
   e. Simulate the responses (using the nonlinear dynamic model) of the concentration of biomass and substrate to step increases and decreases of 10% in the dilution rate (changes are from the dilution rate found in question b.). Compare these results with those of the linear system (remember to convert deviation variables back to physical variables).

2. Literature search: Perform a literature review (using Current Contents, EI Compendex, or some other professional search package, and NOT an Internet-based search engine) on the control of fermentation processes. Select one or two articles that are most interesting to you and discuss the importance of this particular process. Example: What is a typical annual production rate?

MODULE 8

CSTR

Unlike many of the other modules, this module is too lengthy to be performed in a single 1- to 2-hour session. Each instructor or reader is likely to have a different focus. Much of the initial material is related to modeling and interesting nonlinear aspects. If the focus is on controller design, the initial material can be skimmed before Problem M8.4 is completed. After studying this module the reader should be able to understand:

- The effect of scale-up on the steady-state and dynamic characteristics of a jacketed chemical reactor. Further, understand how these characteristics affect the control system design and operating performance.
- These reactors can have input-output operating curves that are nonlinear; that is, the "process gain" is not constant.
- Larger reactors have a reduced capability of removing heat compared to smaller reactors. Often, a larger reactor is operated at an unstable point, while the smaller reactor (with the same residence time) is stable.

The major sections of this module are

M8.1 Background
M8.2 Simplified Modeling Equations
M8.3 Example Chemical Process—Propylene Glycol Production
M8.4 Effect of Reactor Scale
M8.5 Further Study: Detailed Model
M8.6 Other Considerations
M8.9 Summary

M8.1 Background

Chemical reactors often have significant heat effects, so it is important to be able to add or remove heat from them. In a jacketed CSTR (continuous stirred-tank reactor) the heat is added or removed by virtue of the temperature difference between a jacket fluid and the reactor fluid. Often, the heat transfer fluid is pumped through agitation nozzles that circulate the fluid through the jacket at a high velocity; an example recirculating jacket system is shown in Figure M8–1. Notice that cascade control is shown, with a primary controller (reactor temperature) and a secondary controller (jacket temperature). The output of the reactor temperature controller is the jacket temperature setpoint. The output of the jacket temperature controller is the valve position for the jacket make-up valve (usually it is assumed that this is directly related to the flowrate of the jacket make-up fluid). The coolant can be chilled water, but is often a heat transfer fluid, such as an ethylene glycol/water mixture or a proprietary mixture of hydrocarbons.

A different cascade control strategy is shown in Figure M8–2. In this cascade control stategy, the output of the reactor temperature controller is the setpoint for jacket *inlet* temperature. The output of the jacket inlet temperature controller is the valve position of the coolant make-up valve. The first cascade control strategy (Figure M8–1) is the most commonly used reactor temperature control strategy.

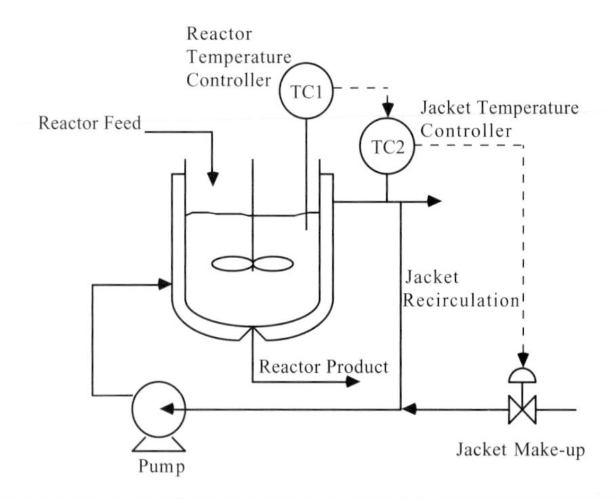

Figure M8–1 Cascade control, with jacket outlet temperature as the secondary control variable.

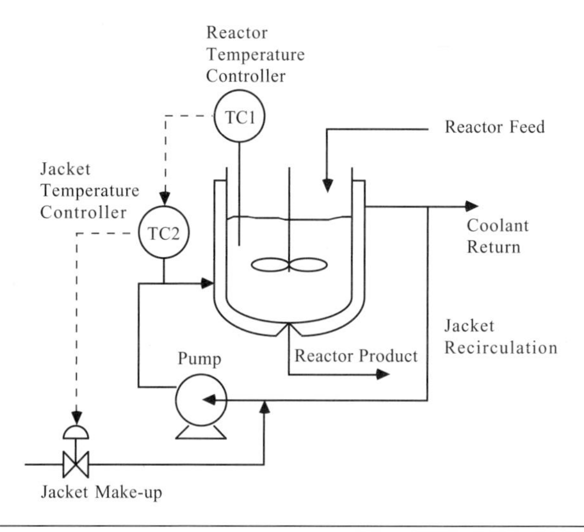

Figure M8–2 Cascade control, with jacket inlet temperature as the secondary control variable.

M8.2 Simplified Modeling Equations

Here we make the following assumptions

1. Perfect mixing in reactor and jacket
2. Constant volume reactor and jacket
3. Constant parameter values

In addition, to develop a simplified model, we assume that the *jacket temperature* can be directly (and nearly instantaneously) manipulated. This assumption is very good if a boiling heat transfer fluid is used, for example; changing the pressure on the jacket side would result in near instantaneous change in jacket temperature (note that a recirculating jacket system would not be used in this scenario). Even for the recirculating heat transfer system, the assumption of the jacket temperature being directly manipulated can be good if the jacket dynamics are rapid compared to the reactor dynamics.

Overall Reactor Material Balance

The reactor overall material balance equation is

$$\frac{dV\rho}{dt} = F_{in}\rho_{in} - F_{out}\rho$$

where V is the constant liquid reactor volume, ρ is the density of reactor fluid and ρ_{in} is the density of the inlet stream. F_{in} and F_{out} are the flow rates of the inlet and outlet streams. Assuming a constant density ($\rho_{in} = \rho$) and volume, it is easy to show that $F_{out} = F_{in} = F$; this will be used in the component balance.

Balance on Component A

Here we consider, for simplicity, the simple reaction $A \rightarrow B$. The balance on component A is

$$V\frac{dC_A}{dt} = FC_{Af} - FC_A - Vr_A \tag{M8.1}$$

where C_A is the concentration of component A in the reactor and r_A is the rate of reaction per unit volume. The Arrhenius expression is normally used for the rate of reaction. A first-order reaction results in the following

$$r_A = k_0 \exp\left(\frac{-E_a}{RT}\right)C_A \tag{M8.2}$$

where k_0 is the frequency factor, E_a is the activation energy, R is the ideal gas constant, and T is the reactor temperature on an absolute scale (R, Rankine or K, Kelvin).

Reactor Energy Balance

The reactor energy balance, assuming constant volume, heat capacity (c_p) and density (ρ), and neglecting changes in the kinetic and potential energy is

$$\underbrace{V\rho c_p \frac{dT}{dt}}_{accumulation} = \underbrace{F\rho c_p\left(T_f - T\right)}_{energy\ in-out\ by\ flow} + \underbrace{(-\Delta H)Vr_A}_{\substack{heat\ of\ reaction\\ contribution}} - \underbrace{UA\left(T - T_j\right)}_{\substack{heat\ transferred\\ to\ jacket}} \tag{M8.3}$$

where $-\Delta H$ is the heat of reaction, U is the heat transfer coefficient, A is the heat transfer area, T_f is the feed temperature, and T_j is the jacket temperature.

State Variable Form of the Equations

We can write Equations (M8.1–M8.3) in the standard state variable form

$$\frac{dC_A}{dt} = f_1(C_A, T) = \frac{F}{V}\left(C_{Af} - C_A\right) - k_0 \exp\left(\frac{-E_a}{RT}\right)C_A \tag{M8.4a}$$

$$\frac{dT}{dt} = f_2(C_A, T) = \frac{F}{V}\left(T_f - T\right) + \frac{(-\Delta H)}{\rho c_p} k_0 \exp\left(\frac{-E_a}{RT}\right)C_A - \frac{UA}{V\rho c_p}\left(T - T_j\right) \tag{M8.4b}$$

Steady-State Solution

The steady-state solution is obtained when the two state derivatives are set equal to zero

$$f_1(C_A, T) = \frac{dC_A}{dt} = 0 = \frac{F}{V}(C_{Af} - C_A) - k_0 \exp\left(\frac{-E_a}{RT}\right) C_A \quad \text{(M8.5a)}$$

$$f_2(C_A, T) = \frac{dT}{dt} = 0 = \frac{F}{V}(T_f - T) + \frac{(-\Delta H)}{\rho c_p} k_0 \exp\left(\frac{-E_a}{RT}\right) C_A - \frac{UA}{V\rho c_p}(T - T_j) \quad \text{(M8.5b)}$$

Here, all of the inputs and parameters must be specified, allowing the two equations to be solved for steady-state values of the two states. For these equations, a nonlinear technique, such as Newton Raphson, would need to be used. It should be noted, however, that there is an easier selection of the two *unknowns* to solve for. For a given steady-state temperature, solve Equation (M8.5a) for the steady-state concentration. Then, solve Equation (M8.5b) for the value of the jacket temperature (perform this in the following exercise).

Problem M8.1 Steady-State Concentration and Jacket Temperature for a Specified Reactor Temperature

a. Solve for the steady-state concentration as a function of the specified steady-state temperature, from Equation (M8.5a). Use a subscript s to indicate a steady-state value.

Result:

$$C_{As} = \frac{(F_s/V)C_{Afs}}{(F_s/V) + k_o \exp(-E_a/RT_s)} \quad \text{(M8.6a)}$$

b. Solve for the steady-state jacket temperature as a function of the steady-state reactor temperature and concentration from Equation (M8.5b).

Result:

$$T_{js} = T_s + \left[F_s \rho c_p (T_s - T_{fs}) - (-\Delta H) V k_o \exp(-E_a/RT_s) C_{As}\right] \cdot \frac{1}{UA} \quad \text{(M8.6b)}$$

Alternatively, one could specify the desired concentration, solve for the required reactor temperature (from M8.5a), and then solve for the required jacket temperature (from M8.5b). Again, this assumes that the residence time and all other parameters are fixed.

Linearization

The goal of the linearization procedure is to find a model with the form

$$\dot{x} = Ax + Bu$$
$$y = Cx + Du$$

Here we define the states, output, and inputs in deviation variable form

$$x = \begin{bmatrix} x_1 \\ x_2 \end{bmatrix} = \begin{bmatrix} C_A - C_{As} \\ T - T_s \end{bmatrix}$$

$$y = T - T_s$$

$$u = \begin{bmatrix} u_1 \\ u_2 \\ u_3 \\ u_4 \end{bmatrix} = \begin{bmatrix} T_j - T_{js} \\ T_f - T_{fs} \\ C_{Af} - C_{Afs} \\ F - F_s \end{bmatrix}$$

Notice that the first input (jacket temperature, T_j) is manipulated, while the last three inputs are disturbances.

The Jacobian matrix is

$$A = \begin{bmatrix} a_{11} & a_{12} \\ a_{21} & a_{22} \end{bmatrix} = \begin{bmatrix} \frac{\partial f_1}{\partial x_1} & \frac{\partial f_1}{\partial x_2} \\ \frac{\partial f_2}{\partial x_1} & \frac{\partial f_2}{\partial x_2} \end{bmatrix} = \begin{bmatrix} -\frac{F}{V} - k_s & -C_{As}k_s' \\ \frac{(-\Delta H)}{\rho c_p} k_s & -\frac{F}{V} - \frac{UA}{V\rho c_p} + \frac{(-\Delta H)}{\rho c_p} C_{As}k_s' \end{bmatrix} \quad \text{(M8.7)}$$

where $k_s = k_0 \exp(-E_a/RT_s)$ and $k_s' = \partial k_s/\partial T_s = k_0 \exp(-E_a/RT_s)(E_a/RT_s^2) = k_s(E_a/RT_s^2)$.

The first column (related to the manipulated input, T_j) of the B matrix is

$$B_1 = \begin{bmatrix} b_{11} \\ b_{21} \end{bmatrix} = \begin{bmatrix} \partial f_1/\partial u_1 \\ \partial f_2/\partial u_1 \end{bmatrix} = \begin{bmatrix} 0 \\ UA/V\rho c_p \end{bmatrix} \quad \text{(M8.8)}$$

The output matrix is, since we consider temperature (second state) to be measured,

$$C = \begin{bmatrix} 0 & 1 \end{bmatrix}$$

Also, the input is not measured, so D = 0.

M8.3 Example Chemical Process—Propylene Glycol Production

Roughly 1.3 billion pounds of propylene glycol are produced per year. It has a wide variety of uses, including: anti-freeze applications, including aircraft deicing; a solvent for a number of drugs; moisturizers; and as artificial smoke or fog, for fire-fighting training or theatrical productions. Propylene glycol is produced by the hydrolysis of propylene oxide with sulfuric acid as a catalyst

$$CH_2 - O - CH - CH_3 + H_2O \rightarrow CH_2OH - CH_2O - CH_3$$

Water is supplied in excess, so the reaction is first-order in propylene oxide concentration. The rate of reaction of propylene oxide (component A) is first-order

$$r_A = -k_0 \exp(-E_a/RT)C_A$$

Parameter Values

This system has the following activation energy, frequency factor, and heat of reaction values

$$E_a = 32{,}400 \text{Btu/lbmol}$$
$$k_0 = 16.96 \times 10^{12} \text{ hr}^{-1}$$
$$-\Delta H = 39000 \text{Btu/lbmol PO}$$

The other parameters are

$$U = 75 \text{ Btu/hr ft}^2 \text{°F}$$
$$\rho c_p = 53.25 \text{Btu/ft}^3 \text{°F}$$
$$R = 1.987 \text{Btu/lbmol °F}$$

Assume that the reactor is to be operated with the following residence time, feed concentration, and feed temperature

$$V/F = 15 \text{ minutes} = 0.25 \text{ hr}$$
$$C_{Af} = 0.132 \text{ lbmol/ft}^3$$
$$T_f = 60\text{°F} = 519.67\text{R}$$

Also assume the reactor is designed as a vertical cylinder with a height/diameter ratio of 2:1, that complete heat transfer area coverage occurs when the reactor is 75% full, and that the reactor is operated at 85% of the design volume.

M8.4 Effect of Reactor Scale

In this section we study the effect of reactor size on the steady-state and dynamic behavior of the reactor. Each reactor is designed with a residence time of 15 minutes, with a desired conversion of 50% (C_A = 0.066 lbmol/ft^3), resulting in a reactor temperature of 101.1°F.

Reactor Scale

Consider now the following reactors, operated at 85% of the design volume.

Reactor Volume (nominal)	100 ft^3	5000 gallon (668 ft^3)
Heat transfer area	88 ft^2	309 ft^2
Diameter	4 ft	7.5 ft
Operating Volume, V	85 ft^3	500 ft^3
Operating Flowrate, F	340 ft^3/hr	2000 ft^3/hr
UA	6600 Btu/hr °F	23200 Btu/hr °F

Problem M8.2 Steady-State Jacket Temperatures

Find the steady-state *jacket* temperatures for the 100 ft^3 and 5000 gallon reactors.

Solution: Substituting the parameter values and steady-state concentration and temperature into Equations (M8.6a) and (M8.6b), the resulting jacket temperatures are 80°F (100 ft^3) and 65°F (5000 gallon).

Steady-State (Nonlinear) Results

Expanding on Problem M8.1 we can find the jacket temperature as a function of reactor temperature, and solve for a number of reactor temperatures to construct a steady-state input-output plot. This is shown in Figure M8–3 for different reactor sizes. The '+' symbol denotes the design operating point for each reactor. The jacket and reactor temperatures at this design point are consistent with the results in Problem M8.2.

There are several interesting things to note:

1. All of the input output curves are *nonlinear*. Recall that a process gain is change in output/change in input, or the slope of the input-output curve. A linear system has a constant gain and therefore a constant slope.
2. The larger (5000 gallon and 1000 ft^3) reactors exhibit *multiple steady-state* behavior. That is, for a certain jacket temperature there is the possibility of three reactor temperatures. For these larger reactors the intermediate temperature (desired) operating point is unstable. This will be shown in the linear stability analysis that follows.
3. For a given reactor temperature, a larger reactor requires a lower jacket temperature. This is because a larger reactor has a reduced heat transfer area/volume (A/V) ratio and a reduced ability to transfer heat.

Linear Open-Loop Results

The steady-state operating point is $C_{As} = 0.066$, $T_s = 101.1$°F.

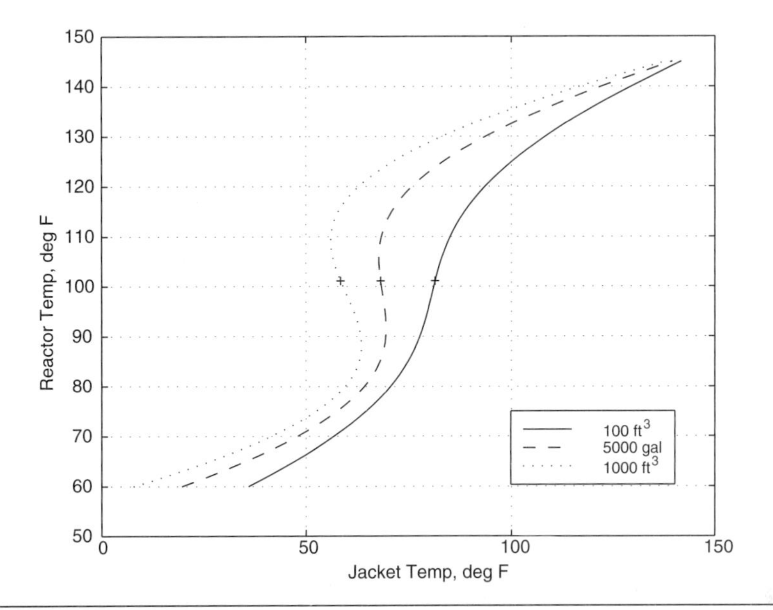

Figure M8–3 Steady-state input-output plot of reactor temperature as a function of jacket temperature. Notice that the larger reactors exhibit multiple steady-state behavior (this is also known as hysteresis). The desired steady-state operating point is denoted by '+'.

a. For the 100-ft³ reactor the state space model is (note that the time unit is hours)

$$A = \begin{bmatrix} -7.9909 & -0.013674 \\ 2922.9 & 4.5564 \end{bmatrix} \quad B = \begin{bmatrix} 0 \\ 1.4582 \end{bmatrix}$$

$$C = \begin{bmatrix} 0 & 1 \end{bmatrix} \quad D = 0$$

and you should be able to show (analytically, or using MATLAB) that the process transfer function is

$$g_p(s) = \frac{1.458s + 11.65}{s^2 + 3.434s + 3.557} = \frac{0.4099s + 3275}{0.2811s^2 + 0.9664s + 1} \tag{M8.9}$$

where the process gain is $k_p = 3.275°F/°F$ and the time unit is hours. You will want to factor this into the gain-time constant form for controller design.

b. For the 5000-gallon reactor the state space model is

$$A = \begin{bmatrix} -7.9909 & -0.013674 \\ 2922.9 & 5.1432 \end{bmatrix} \quad B = \begin{bmatrix} 0 \\ 0.87136 \end{bmatrix}$$

$$C = \begin{bmatrix} 0 & 1 \end{bmatrix} \quad D = 0$$

and you should be able to show (analytically, or using MATLAB) that the process transfer function is

$$g_p(s) = \frac{0.8714s + 6.963}{s^2 + 2.848s - 1.132} \tag{M8.10}$$

You will want to factor this into the gain-time constant form for controller design.

Problem M8.3 Open-Loop Stability

Calculate the poles and zeros of the process transfer function for both size reactors. You should find that the 100-ft^3 reactor is stable, while the 5000-gallon (668 ft^3) reactor is unstable.

Since the 5000-gallon reactor is open-loop unstable, it can only be operated at the design point of 101.1°F using a closed-loop control system.

Linear Closed-Loop Results

Recall that the closed-loop transfer function relating the setpoint to the output is

$$y(s) = \frac{g_c(s)g_p(s)}{1 + g_c(s)g_p(s)} \cdot r(s)$$

where the stability is determined by the roots of the denominator polynomial. The Routh stability criterion (chapter 5) can be used to determine the ranges of the controller tuning parameters that will assure a closed-loop stable system. See Additional Exercise 4 at the end of the module.

Problem M8.4 Closed-Loop Simulations

Develop a SIMULINK block diagram to perform closed-loop simulations for the two reactor sizes (100 ft^3 and 5000 gal). For simplicity, use the linear process transfer functions (M8.8 and M8.9) to represent the respective reactor. Include two additional first-order transfer functions with time constants of 0.1 hour to account for unmodeled mixing and jacket dynamics (although these are included in your simulation studies, neglect them for any model-based controller design procedure). Here any of your favorite control system design techniques can be used: (i) Ziegler-Nichols, (ii) Tyreus-Luyben, (iii) IMC, (iv) IMC-based PID, (v) "ad-hoc" iterative tuning.

In the following we use the *IMC-based PID* procedure for both the 100-ft^3 (stable, Table 9–1) and the 5000-gallon (unstable, Table 9–3) reactors. The IMC filter factor (λ) is adjusted to obtain a satisfactory closed-loop response to a step setpoint change in reactor temperature (including a consideration of the action of the manipulated jacket temperature). For the unstable reactor a setpoint filter (see Figure 9–8) should be used for improved response.

Make plots of both the reactor temperature (output) response and the manipulated jacket temperature (input). Since the linear simulations are in deviation variable form, add the steady-state values to the deviation variables to plot the physical values. Your responses for a 5°F setpoint change should be similar to Figure M8–4.

M8.5 For Further Study: Detailed Model

A more detailed model of a CSTR includes the effect of cooling jacket dynamics. The three modeling equations are

$$\frac{dC_A}{dt} = f_1(C_A, T, T_j) = \frac{F}{V}(C_{Af} - C_A) - k_0 \exp\left(\frac{-\Delta E}{RT}\right) C_A \qquad \text{(M8.11a)}$$

$$\frac{dT}{dt} = f_2(C_A, T, T_j) = \frac{F}{V}(T_f - T) + \frac{(-\Delta H)}{\rho c_p} k_0 \exp\left(\frac{-\Delta E}{RT}\right) C_A - \frac{UA}{V \rho c_p}(T - T_j) \qquad \text{(M8.11b)}$$

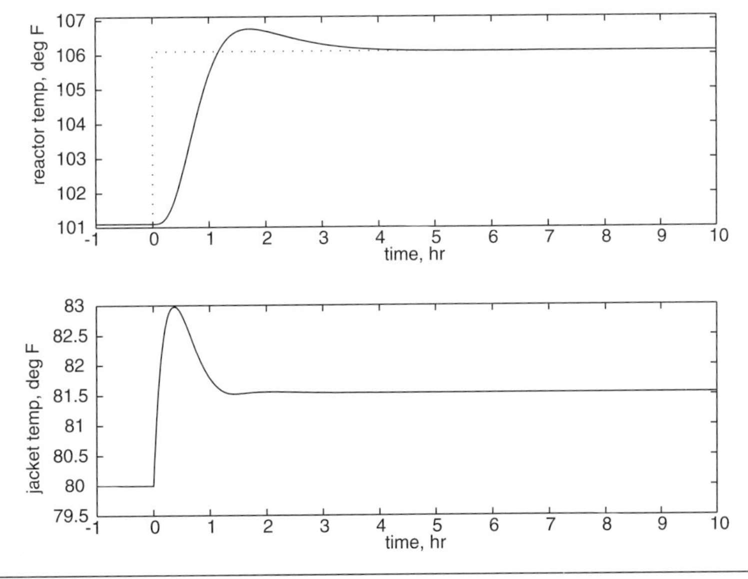

Figure M8–4 Closed-loop responses for reactor temperature setpoint changes of +5°F. Linear simulations for 100 ft³ reactor. IMC-based PID with λ = 0.5 hr.

$$\frac{dT_j}{dt} = f_3(C_A, T, T_j) = \frac{F_j}{V_j}(T_{jin} - T_j) + \frac{UA}{V_j \rho_j c_{pj}}(T - T_j) \quad \text{(M8.11c)}$$

where F_j is the flow rate through the jacket, and T_{jin} is the jacket inlet temperature. The variable definition should be clear from Figure M8–5. Notice that the manipulated input is the jacket make-up flow rate (F_{jf}), which does not appear in the above modeling equations.

A static energy balance can be used to find the jacket inlet temperature as a function of the jacket make-up flow rate (F_{jf})

$$T_{jin} = T_j + \frac{F_{jf}}{F_j}(T_{jf} - T_j) \quad \text{(M8.11d)}$$

When (M8.11d) is substituted into (M8.11c), the following equations result

$$\frac{dC_A}{dt} = f_1(C_A, T, T_j) = \frac{F}{V}(C_{Af} - C_A) - k_0 \exp\left(\frac{-\Delta E}{RT}\right) C_A \quad \text{(M8.12a)}$$

$$\frac{dT}{dt} = f_2(C_A, T, T_j) = \frac{F}{V}(T_f - T) + \frac{(-\Delta H)}{\rho c_p} k_0 \exp\left(\frac{-\Delta E}{RT}\right) C_A - \frac{UA}{V \rho c_p}(T - T_j) \quad \text{(M8.12b)}$$

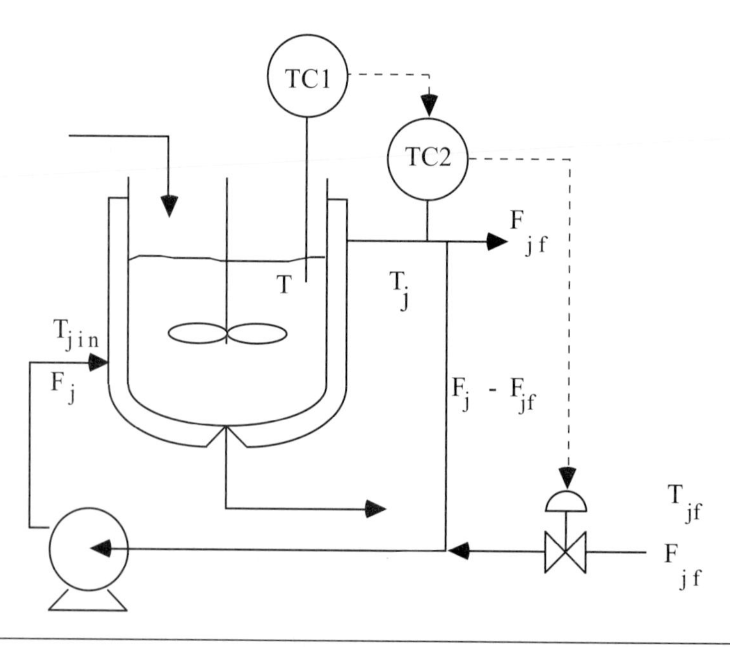

Figure M8–5 CSTR with recirculating jacket. Definition of variables.

$$\frac{dT_j}{dt} = f_3\left(C_A, T, T_j\right) = \frac{F_{jf}}{V_j}\left(T_{jf} - T_j\right) + \frac{UA}{V_j\rho_j c_{pj}}\left(T - T_j\right) \tag{M8.12c}$$

which is the same set of equations that would be derived from a reactor with a "once-through" jacket, shown in Figure M8–6.

This is an interesting result. The model for a CSTR with a "once-through" jacket is the same as that for a recirculating jacket, assuming a static energy balance around the mixing point. This is one reason that it is common to see Figure M8–6 used to represent a CSTR when Figure M8–5 is the more common configuration in industry. It should be noted that the assumption of a constant heat-transfer coefficient is much better for Figure M8–5 than M8–6, since the jacket velocity will stay relatively constant for this system.

Linear Open-Loop Results

Here we again study the propylene glycol reactor presented in Section M8.3, with the following additional jacket parameters

$$V_j/V = 0.25$$
$$T_{jf} = 0°F$$
$$\rho_j c_{pj} = 55.6\ Btu/ft^3{}°F$$

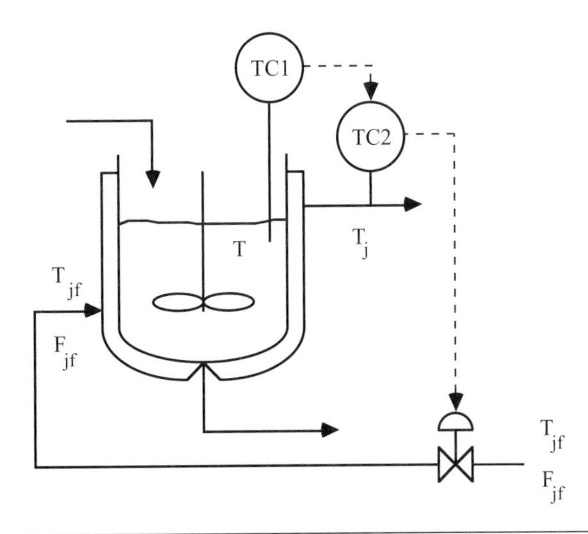

Figure M8–6 CSTR with "once-through" jacket. Definition of variables.

For the 100-ft³ reactor the state space model is (see Appendix M8.1 for the linearization results)

$$A = \begin{bmatrix} -7.9909 & -0.013674 & 0 \\ 2922.9 & 4.5564 & 1.4582 \\ 0 & 4.7482 & -5.8977 \end{bmatrix} \qquad B = \begin{bmatrix} 0 \\ 0 \\ -3.2558 \end{bmatrix}$$
$$C = \begin{bmatrix} 0 & 1 & 0 \\ 0 & 0 & 1 \end{bmatrix} \qquad D = \begin{bmatrix} 0 \\ 0 \end{bmatrix} \qquad \text{(M8.13)}$$

where $x = [C_A - C_{As} \quad T - T_s \quad T_j - T_{js}]^T$, $u = F_{jf} - F_{jfs}$, $y = [T - T_s \quad T_j - T_{js}]^T$

You should be able to show (using MATLAB) that the process transfer function relating the jacket flow rate (u_1) to the reactor temperature (y_1) is

$$g_{p11}(s) = \frac{-4.747s - 37.94}{s^3 + 9.332s^2 + 16.89s - 34.45} \qquad \text{(M8.14)}$$

which has an unstable pole.

Nonlinear Open-Loop Results

There is also multiple steady-state behavior as shown in Figure M8–7. This is an interesting result, because the 100-ft³ reactor was stable when the simplified model was used, and there was only one operating point. The simplified model essentially assumes perfect control of the jacket temperature. This means that the jacket temperature controller stabilizes the process.

Linear Closed-Loop Results

Since the CSTR is open-loop unstable, a feedback controller must be designed to stabilize the reactor operation. In the next problem, the Routh stability criterion is used to find the range of controller gains that stabilize the CSTR.

Problem M8.5 A Stabilizing P-only Controller for the 100-ft³ Reactor

Use the Routh stability criterion to determine the range of controller gains, for a P-only controller, that stabilize the reactor (based on the transfer function M8.14). Based on simulation results, what value of gain do you recommend for the P-only controller?

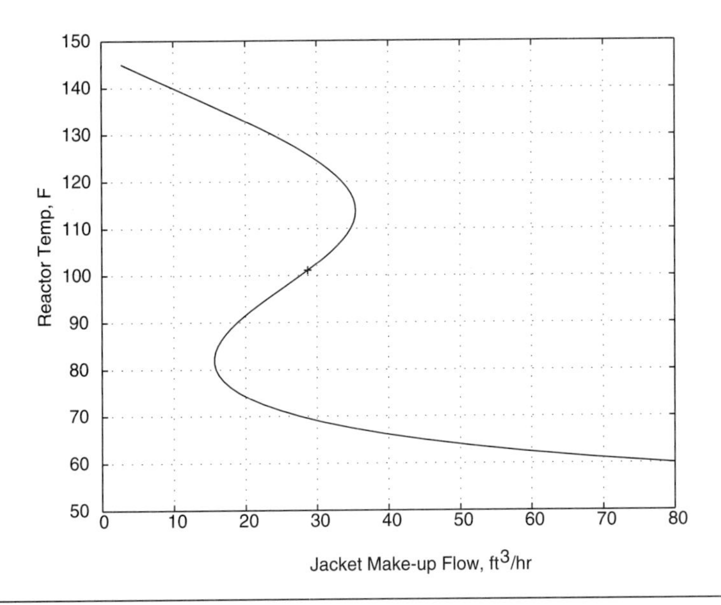

Figure M8–7 Steady-state input-output plot of reactor temperature as a function of jacket flow rate for the 100-ft^3 reactor. The desired steady-state operating point is denoted by '+'.

Problem M8.6 Cascade Control of the 100-ft^3 Reactor

Develop a SIMULINK diagram of a cascade controller CSTR, where a state space representation [Equation (M8.12)] is used for the reactor. First, tune the secondary loop to stabilize the process. Then, tune the primary loop for the desired setpoint response behavior.

M8.6 Other Considerations

Nonlinear Behavior

Although general nonlinear models were developed in this module, and some nonlinear steady-state results were presented, dynamic simulations were based on linear models. Although the responses on the physical (nonlinear) system will be similar to the linear results for small changes around the designed operating point, the nonlinear behavior can be significantly different for large changes.

Download the nonlinear SIMULINK/MATLAB files from the book web page and perform closed-loop simulations. Show that responses of the nonlinear system are similar to the linear simulations, for small setpoint changes. For a sequence of larger setpoint changes, show that the nonlinear system begins to deviate significantly from the linear system.

Split-Range Control

Thus far we have only considered the case where coolant was circulated through the jacket at a lower temperature than the reactor vessel, to remove energy due to the exothermic reaction. During reactor start-up, however, it may be necessary to add energy to the reactor to bring the temperature up to a point where the reaction can be initiated. In this case a warm fluid must be circulated through the jacket. This requires a split-range controller as shown in Figures M8–8 and M8–9. Notice that the cold fluid is fail-open, while the hot fluid is fail-closed.

M8.7 Summary

Exothermic CSTRs can exhibit interesting, and challenging, nonlinear behavior. Even a simple two-state model, with jacket temperature manipulated and the reactor temperature as the (controlled) output variable, can have multiple steady-state behavior. For reactors with three steady states, the intermediate steady-state is unstable, requiring a feedback controller for stable operation.

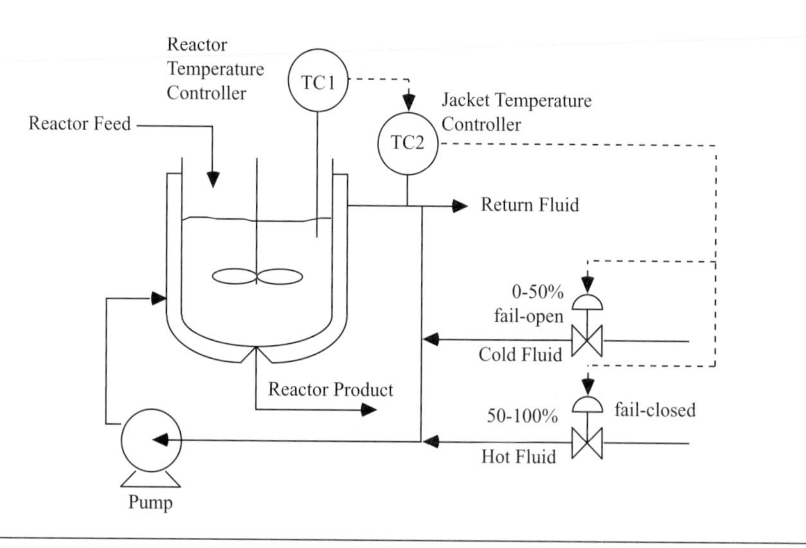

Figure M8–8 Split-range strategy for secondary controller.

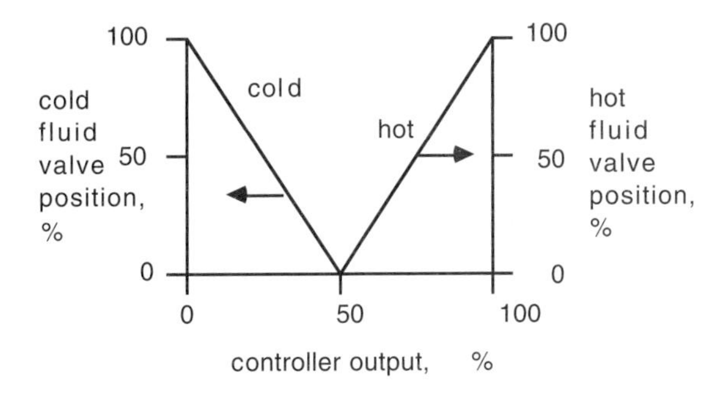

Figure M8–9 Split-range controller action.

CSTRs often have recirculating jacket fluid heat-transfer systems to improve the heat removal capability. For tighter control, and particularly for start-up and transition control, it is important to incorporate a split-range strategy; here, the jacket temperature controller output manipulates either a hot fluid valve or a cold fluid valve, depending on the range of the controller output. For most reactors it is desirable for the cold valve to fail open and the hot valve to fail closed.

References

The Fogler text uses the propylene glycol example for the analysis of an adiabatic CSTR. The Seider et al. text uses it in the context of simulation-based process design. The original experimental results for the kinetic rate expressions are provided in Furusawa et al. An analysis of a two-state CSTR is presented in the *Process Dynamics* text by Bequette. The article by Lipták provides an excellent overview of both batch and continuous reactor control.

Bequette, B. W., *Process Dynamics: Modeling, Analysis and Simulation*, Prentice Hall, New York (1998).

Fogler, H. S., *Elements of Chemical Reaction Engineering*, 2nd ed., Prentice Hall, New York (1992).

Furusawa, T., H. Nishimura, and T. Miyauchi, "Experimental Study of a Bistable Continuous Stirred-Tank Reactor, *J. Chem. Eng. Jpn.*, **2**(1), 95–100 (1969).

Lipták, B. G., "Controlling and Optimizing Chemical Reactors," *Chemical Engineering Mag.*, pp. 69–81 (May 26, 1986).

Seider, W. D., J. D. Seader, and D. R. Lewin, *Process Design Principles*, Wiley, New York (1999).

Additional Exercises

1. Perform a search on propylene glycol. What is the worldwide demand? What is the current sales price? Enumerate some of the uses of this product.
2. Consider the simple CSTR model used in Section M8.4. Transfer functions for the 100-ft^3 and 5000-gallon reactor were given.
   a. Find the heat transfer area, diameter, and UA for a 1000-ft^3 reactor (with an operating volume of 850 ft^3), designed with the same residence time and height/diameter ratio as the other reactors. Also, find the required jacket temperature to achieve the reactor temperature of 101.1°F.
   b. Find the state space model and calculate the eigenvalues for the 1000-ft^3 reactor. Find the transfer function and calculate the poles. You should find that this reactor is open-loop unstable. Now, find the range of proportional gains that will yield a stable closed-loop system under proportional-only control.
3. Consider the propylene glycol example
   a. For a desired propylene oxide conversion of 50% at a residence time of 0.25 hr, show that the reactor operating temperature is 101.1°F.
   b. For a desired propylene oxide conversion of 50% and a feed concentration of 0.132 lbmol/ft^3, find the production rate of propylene glycol for two different feed flow rates: (i) 340 ft^3//hr, and (ii) 2000 ft^3/hr. Assume that the plant operates for 8400 hours/year. If propylene glycol sells for $0.70/lb, what are the expected annual revenues (before operating costs are considered) for each flowrate?
   c. Assume complete liquid coverage of the heat transfer area when the reactor is 75% full. What is the heat transfer area for a 100-ft^3 reactor? Remember to include both the "bottom" and "side" of the reactor.
   d. Again, assume complete liquid coverage of the heat transfer area when the reactor is 75% full. What is the heat transfer area for a 5000-gallon (668 ft^3) reactor? Remember to include both the "bottom" and "side" of the reactor.
   e. Comment on the A/V ratio as the volume of a reactor increases. For a specified reactor conversion (and temperature) and residence time, how does the jacket temperature change as a function of reactor volume?
4. Propylene glycol example, closed-loop behavior.
   Perform the following for both the 100-ft^3 and 5000-gallon (668 ft^3) reactors.
   a. For a PI controller, with an integral time of 1 hour, find the range of controller gains for closed-loop stability using the Routh stability criterion.
   b. Develop a SIMULINK block diagram to perform closed-loop simulations for the two reactor sizes. For simplicity, use the linear process transfer functions to represent the reactor. Include two additional first-order transfer

functions with time constants of 0.1 hour (6 minutes) to account for unmodeled mixing and jacket dynamics. Select tuning parameters within the range of those that you obtained in part **a**. Adjust the parameters until you are satisfied with closed-loop response to a step setpoint change.

Make plots of both the reactor temperature response and the manipulated jacket temperature.

5. Consider the following CSTR parameters

$$E_a = 11{,}843\text{kcal/kgmol}$$
$$k_0 = 14{,}825*3600\text{hr}^{-1}$$
$$-\Delta H = 5215 \text{ kcal/kgmol}$$
$$UA/V = 250\text{kcal/(m}^{3\circ}C\text{ hr)}$$
$$\rho c_p = 500 \text{ kcal/(m}^{3\circ}C)$$
$$R = 1.987\text{kcal/kgmol}^\circ C$$

Assume that the reactor is to be operated with the following steady-state inputs:

$$V/F = 1 \text{ hr}$$
$$C_{Af} = 10 \text{ kgmol/m}^3$$
$$T_f = 25^\circ\text{C}$$
$$T_j = 25^\circ\text{C}$$

Solve for the steady-state concentration and temperature and find the corresponding state space model.

Appendix M8.1

For Equations (M8.12) we develop a linear state space model, with the states, outputs, input, and disturbances as the following deviation variables:

$$x = \begin{bmatrix} x_1 \\ x_2 \\ x_3 \end{bmatrix} = \begin{bmatrix} C_A - C_{As} \\ T - T_s \\ T_j - T_{js} \end{bmatrix}$$

$$y = \begin{bmatrix} y_1 \\ y_2 \end{bmatrix} = \begin{bmatrix} T - T_s \\ T_j - T_{js} \end{bmatrix}$$

$$u = F_{jf} - F_{jfs}$$

$$l = \begin{bmatrix} l_1 \\ l_2 \\ l_3 \\ l_4 \end{bmatrix} = \begin{bmatrix} T_f - T_{fs} \\ T_{jf} - T_{jfs} \\ C_{Af} - C_{Afs} \\ F - F_s \end{bmatrix}$$

The Jacobian matrix elements are

$$A = \begin{bmatrix} a_{11} & a_{12} & a_{13} \\ a_{21} & a_{22} & a_{23} \\ a_{31} & a_{32} & a_{33} \end{bmatrix} = \begin{bmatrix} -\dfrac{F_s}{V} - k_s & -C_{As}k_s' & 0 \\ \dfrac{(-\Delta H)}{\rho c_p}k_s & -\dfrac{F}{V} - \dfrac{UA}{V\rho c_p} + \dfrac{(-\Delta H)}{\rho c_p}C_{As}k_s' & \dfrac{UA}{V\rho c_p} \\ 0 & \dfrac{UA}{V_j\rho_j c_{pj}} & -\left(\dfrac{F_{jfs}}{V_j} + \dfrac{UA}{V_j\rho_j c_{pj}}\right) \end{bmatrix}$$

where $k_s' = \dfrac{\partial k_s}{\partial T_s} = k_0 \exp(-E_a/RT_s)(E_a/RT_s^2) = k_s(E_a/RT_s^2)$

The first column of the B matrix is (this column is related to input 1, the jacket feed flow rate)

$$B = \begin{bmatrix} b_{11} \\ b_{21} \\ b_{31} \end{bmatrix} = \begin{bmatrix} \partial f_1/\partial u_1 \\ \partial f_2/\partial u_1 \\ \partial f_3/\partial u_1 \end{bmatrix} = \begin{bmatrix} 0 \\ 0 \\ \dfrac{\left(T_{jfs} - T_{js}\right)}{V_j} \end{bmatrix}$$

If both the reactor temperature (state 2) and the jacket temperature (state 3) are measured for a cascade control structure, the measured output matrix is

$$C = \begin{bmatrix} c_{11} & c_{12} & c_{13} \\ c_{21} & c_{22} & c_{23} \end{bmatrix} = \begin{bmatrix} 0 & 1 & 0 \\ 0 & 0 & 1 \end{bmatrix}$$

Steam Drum Level

Steam utility systems are prevalent in almost all manufacturing complexes, as well as in many other commercial systems (including hospitals, universities, etc.). Since these systems are so important and have been in existence for many years, the control strategies are reasonably advanced. In this learning module, the instructors and students can choose their preferred control system design technique for an example steam drum system.

The sections of this module are as follows:

M9.1 Background

Virtually every manufacturing plant has a steam utility system. The heart of the steam system is the boiler (in general, there are a number of boilers in a plant). One of the important control loops on a steam boiler is the steam drum level controller, which is shown schematically in Figure M9–1. An actual boiler control problem consists of many other control loops, including steam pressure control.

The control of steam drum level is tougher than most other level control problems because there is inverse response combined with a ramp in the level when a step increase in feed-water flow rate is made. A typical step response is shown in Figure M9–2. The

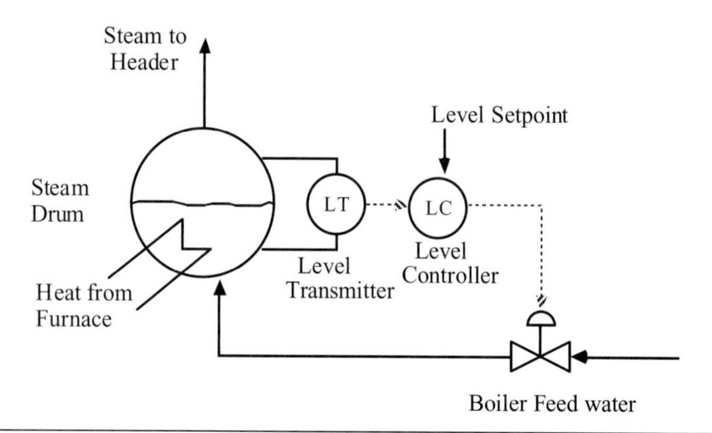

Figure M9–1 Control instrumentation diagram for boiler level.

ramp nature (integrator) is because if there an increase in the incoming water with no corresponding increase in outlet steam flow, then the level must continually increase. The inverse response occurs because the incoming cold water causes some of the bubbles in the steam drum to collapse, causing the level to drop slightly. However, since the incoming flow rate is higher that the outgoing flow rate, the level must eventually start rising.

This is an example of a system where constraints on the process output (steam drum level) are important. If the liquid level gets too high, liquid can enter the superheater section, expanding rapidly and causing pipe rupture. If the liquid level gets too low, then water is no longer in the pipes in the radiation section below the steam drum, causing the pipes to get too hot and fail.

M9.2 Process Model

The relationship between feed-water flow rate and drum level for the process studied in this module is

$$g_p(s) = \frac{0.25(-s+1)}{s(2s+1)} \tag{M9.1}$$

Assume that the valve transfer function is

$$g_v(s) = \frac{1}{0.15s+1} \tag{M9.2}$$

Appendix M9.1: SIMULINK Diagram for Feed-Forward/ Feedback Control of Steam Drum Level

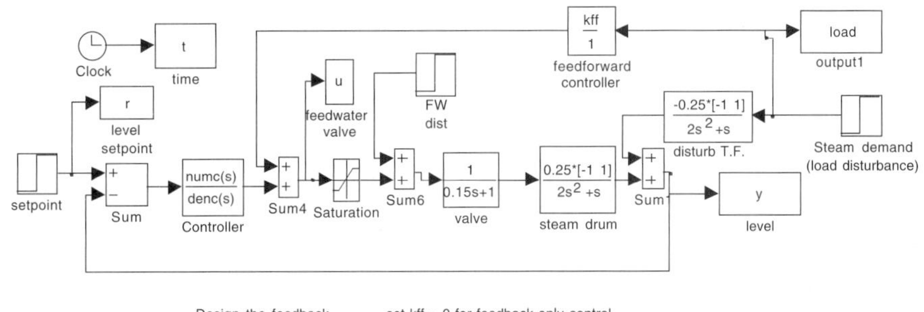

Appendix M9.2: SIMULINK Diagram for 3-Mode Control of Steam Drum Level

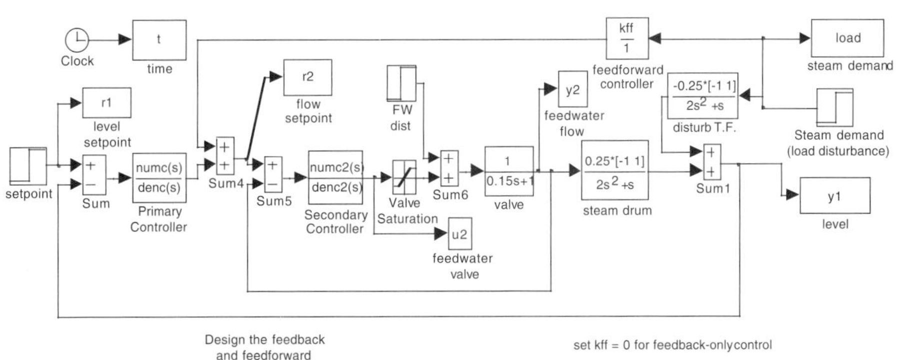

MODULE 10

Surge Vessel Level Control

Surge vessels are used to help reduce the effect of flow rate variations between interconnected process units. It is not necessary to maintain tight level control in a surge vessel, resulting in a significantly different control objective compared to that of the steam drum level controller of Module 9.

The sections of this module are as follows:

M10.1 Background

Much of this text has focused on tracking setpoint changes, where the goal has often been to design a controller to yield a first-order type of response to a step setpoint change. There are many processes, however, where disturbance rejection (regulatory control) is more important than setpoint tracking, because setpoints may not be changed frequently. Consider the surge drum level control system shown in Figure M10–1. The purpose of the surge vessel is to smooth variations in the flow from process 1 and maintain a relatively constant flow rate to process 2. The level can vary substantially from the setpoint, as long as the vessel does not overflow or go dry. The main objective is to vary the manipulated flow rate (the outlet flow from the vessel) as little as possible, while satisfying level constraints.

The goal of this module is to develop guidelines for tuning level controllers for surge vessels. The performance specifications are significantly different than for the steam drum level systems studied in the previous module.

M10.2 Process Model

Here, we develop a process model for the vessel in Figure M10–1a, which is a vertically oriented cylinder that can be assumed to have a constant cross-sectional area. The development of a model for the vessel in Figure M10–1b is left as an exercise for the reader. A material balance around the vessel, assuming constant density and cross-sectional area (A), yields

$$\frac{dh}{dt} = \frac{1}{A}\left(F_1 - F_2\right) \tag{M10.1}$$

The deviation variables for the manipulated input, load disturbance, and measured output are

$$\begin{aligned} u &= F_2 - F_s, \\ l &= F_1 - F_s, \\ y &= h - h_s \end{aligned} \tag{M10.2}$$

The process and load disturbance transfer functions are then

$$g_p(s) = -g_d(s) = -\frac{1}{As} = \frac{k}{s} \tag{M10.3}$$

Notice that this is an integrating process, since the pole = 0.

M10.3 Controller Design

Here we consider a P-only controller, since offset is less important than minimizing manipulated variable movement. The question is:

What value of the proportional gain should be used?

It seems reasonable to select a value for the proportional gain that will just assure that constraints will not be violated for the worst-case expected disturbance. Assume that level alarms are set at 20% and 80% of the volume of the vessel. Let ΔH represent the magnitude of the maximum allowable deviation of the tank height from setpoint, and let ΔL represent the maximum magnitude of a step disturbance.

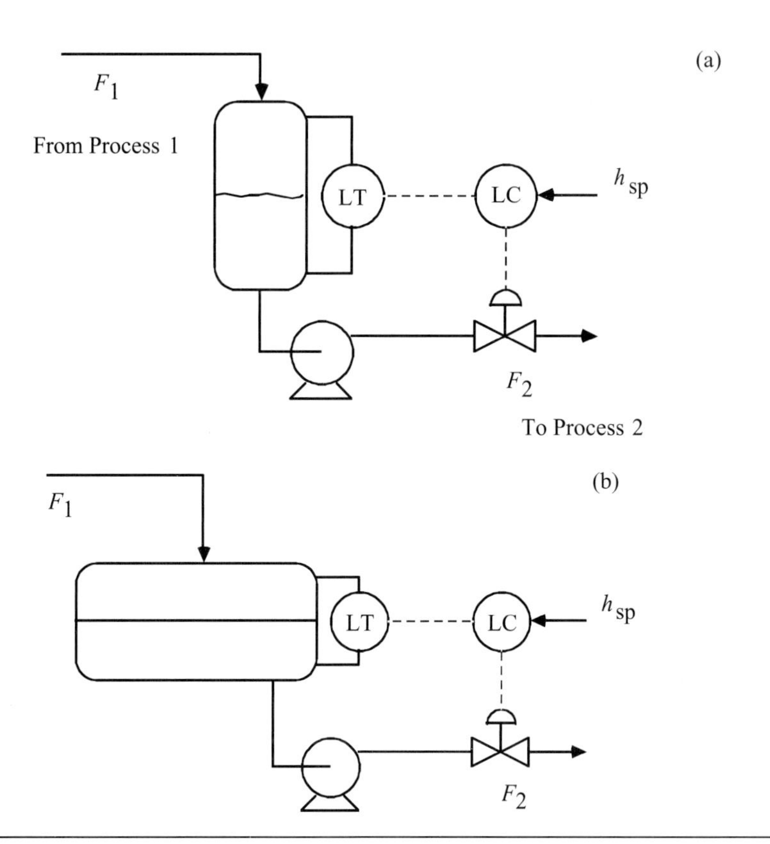

Figure M10–1 Surge vessel level control. (a) Vertical orientation. (b) Horizontal orientation.

Proportional Gain

Assuming that the flow rate out is proportional to the deviation from setpoint,

$$F_2 = F_s + k_c e = F_s + k_c(h - h_{sp}) = F_s + k_c(-\Delta H) \qquad \text{(M10.4)}$$

where k_c is a negative number since the flow rate must increase if the tank height is above the setpoint value. The step change in the load disturbance is ΔL,

$$F_1 = F_s + \Delta L$$

and the value at the new steady-state after a step disturbance is [from Equation (M10.1)]

$$0 = \frac{1}{A}\left(F_s + L - F_s + k_c \Delta H\right)$$

which yields

$$k_c = -\Delta L/\Delta H \tag{M10.5}$$

Nonlinear Proportional Gain

Since small deviations from setpoint are not critical, it may be desirable to have a controller gain that varies as a function of the magnitude of the error. A possible formulation is

$$k_c = k_{c0}|e| \tag{M10.6}$$

Again, the value of the k_{c0} term can be determined from the proportional gain term in Equation (M10.5),

$$k_{c0} = \frac{k_c}{|e|} = \frac{k_c}{\Delta H} = -\frac{\Delta L}{\Delta H^2} \tag{M10.7}$$

M10.4 Numerical Example

Consider a surge vessel that has a diameter of 1 m, and a maximum height of 2 m, with low- and high-level alarms set at 20% and 80%. Also, assume that the height setpoint is 1 meter, and the steady-state flow rate is 0.5 m^3/min. The constant cross-sectional area is then 0.785 m^2. Since the steady-state flow rate is 0.5 m^3/min, assume that the inlet flow rate can vary between 0 and 1 m^3/min.

You should show that the process and load disturbance transfer functions are

$$g_p(s) = -g_d(s) = \frac{-1.274}{s}$$

Since the physical variable minimum and maximum alarm heights are 0.4 and 1.6 m, respectively, in terms of deviation variables, the minimum and maximum heights are

$$y_{\min} = -0.6m$$
$$y_{\max} = 0.6m$$

So, $\Delta H = 0.6$ m. The minimum and maximum manipulated variable values are

$$u_{\min} = -0.5m^3/\text{min}$$
$$u_{\max} = 0.5m^3/\text{min}$$

So, $\Delta L = 0.5$ m^3/min. From Equations (M10.5) and (M10.7), we find k_c = -0.833 m^2/min and k_{c0} = -1.389 m/min.

Step Disturbances

Compare the step responses of the linear and nonlinear controllers to a small step disturbance of 0.05 m^3/min and a large step disturbance of 0.5 m^3/min. For a small step change in the load disturbance (inlet flow rate), the linear and nonlinear controllers are shown in Figure M10–2. Notice that the manipulated input changes more slowly for the nonlinear controller. The minor disadvantage is that the offset is larger for the nonlinear controller, as designed. This is not a real problem, as the offset is small compared with the capacity of the vessel.

Responses to a large step in the inlet flow rate are compared in Figure M10–3. For this case, these responses are similar and there is not much incentive to use the nonlinear strategy.

Sinusoidal Disturbances

Consider a sinusoidal disturbance with an amplitude of 0.05 m^3/min and a frequency of 1/min. The linear and nonlinear strategies are compared in Figure M10–4. Clearly, there is an incentive for nonlinear control, since the manipulated variable does not change nearly as much for the nonlinear case as for the linear case.

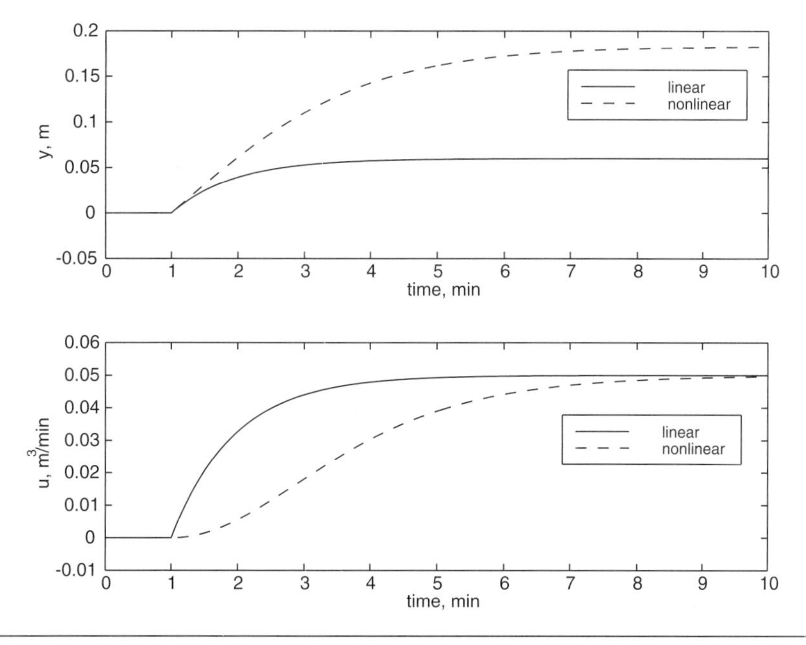

Figure M10–2 Response to a small step in inlet flow rate (0.05 m^3/min).

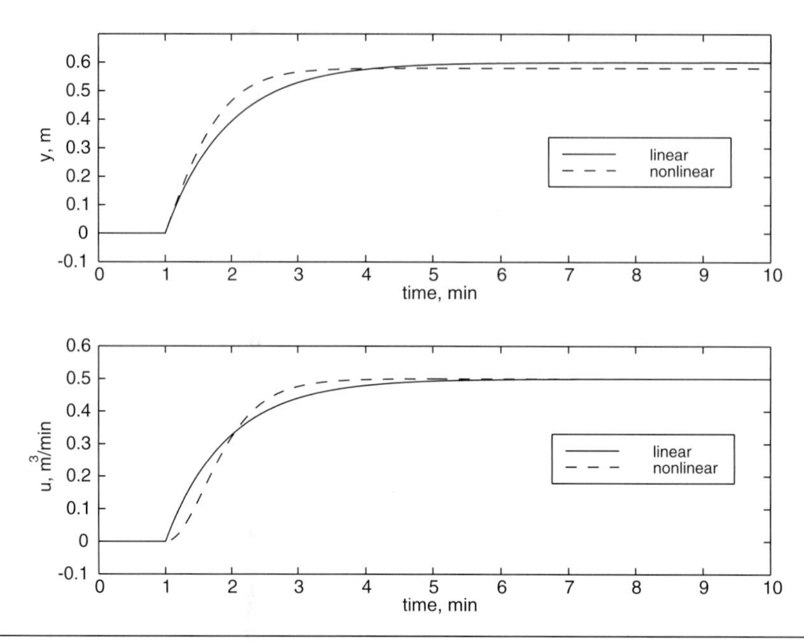

Figure M10–3 Response to a large step in inlet flow rate (0.5 m^3/min).

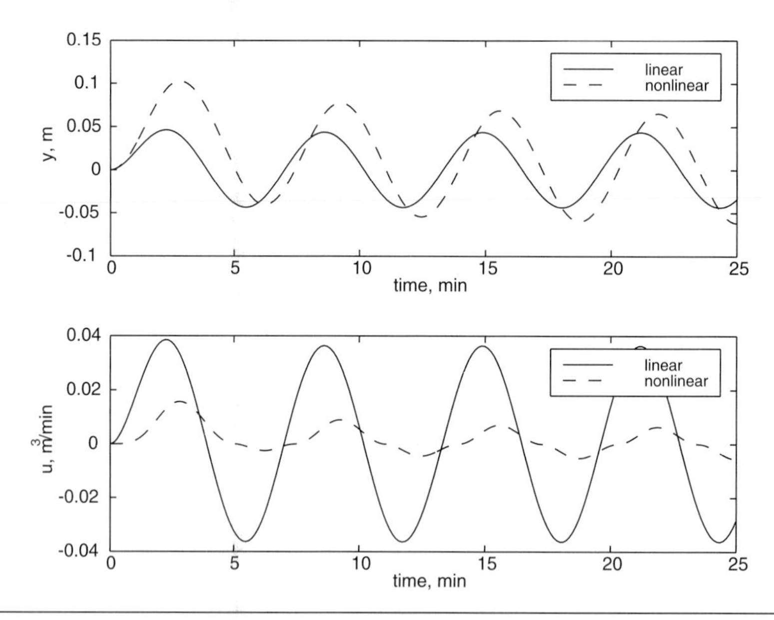

Figure M10–4 Response to a small amplitude (0.05 m^3/min) sinusoidal inlet flow rate disturbance.

The responses to large-magnitude sinusoidal disturbance with an amplitude of 0.5 m^3/min and a frequency of 1/min is shown in Figure M10–5. In this case, there is not much incentive for the nonlinear control strategy.

You should show that a high-frequency disturbance is rejected very well by the nonlinear strategy.

Additional Simulations

Consider the large-magnitude sinusoidal disturbance shown in Figure M10–5. Show that doubling the amplitude of the disturbance leads to saturation of the manipulated variable, and to violation of the high- and low-level alarms.

M10.5 Summary

For surge vessels it is generally more important to allow levels to "float" in order to minimize flow rate variations. The approach shown in this module uses a gain that is proportional

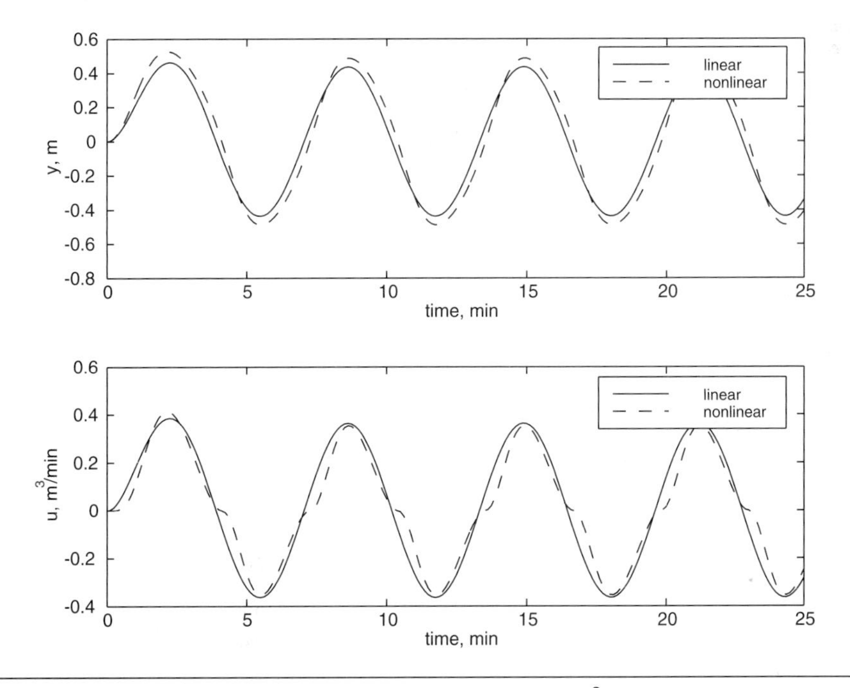

Figure M10–5 Response to a large-amplitude (0.5 m^3/min) sinusoidal inlet flow rate disturbance.

to the absolute value of the error; this is effectively a nonlinear controller, where the control action is a function of the square of the error. Nonlinear control was shown to lead to reduced control action compared with simple P-only control.

Reference

Nonlinear level controllers with integral action are discussed in the following:

Marlin, T. E., *Process Control: Designing Processes and Control Systems for Dynamic Performance*, 2nd ed. McGraw Hill, New York (2000).

Additional Exercises

1. Develop a nonlinear model for the rate of change of tank height for the horizontal vessel orientation shown in Figure M10–1b. Consider a surge vessel that has a diameter of 1 m and a length of 2 m, with low- and high-level alarms set at 20% and 80%. Also, assume that the height setpoint is 0.5 m, and the steady-state flow rate is 0.5 m^3/min. The steady-state flow rate is 0.5 m^3/min, so assume that the inlet flow rate can vary between 0 and 1 m^3/min. Design a proportional controller for this system. For the same disturbances given in the example, compare the performance of the horizontal tank with that of the vertical drum.
2. Implement a so-called gap controller with the following form

$$k_c = k_{c\,\min} \quad \text{for } |e| \le |e_{break}|$$
$$k_c = k_{c\,\max} \quad \text{for } |e| > |e_{break}|$$

   on the level control problems discussed in this chapter. Revise the SIMULINK diagram shown in Figure M10–6 for your simulations. Discuss your selection of k_{cmin}, k_{cmax}, and e_{break}.
3. Consider a more general form of controller than Equation (M10.6). Perform simulations using a controller with the form

$$k_c = k_{c0} + b|e(t)|$$

   Discuss the effect of the two tuning parameters (k_{c0} and b) on the response to different amplitude step and sign disturbances.

Appendix M10.1: The SIMULINK Block Diagram

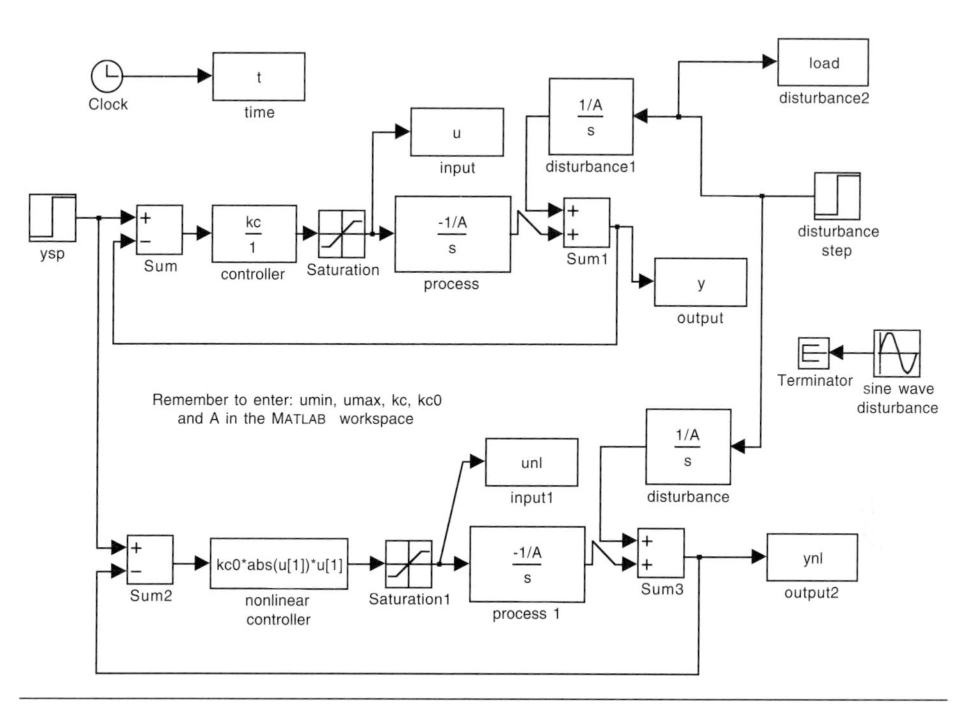

Figure M10–6 SIMULINK diagram for surge level control simulations.

MODULE 11

Batch Reactor

Batch chemical reactors are commonly used in the specialty chemical and pharmaceutical industries. The focus of this textbook has been on continuous processes that operate at some nominal steady-state value. Batch processes are dynamic by nature and usually have a time-varying setpoint profile. Some of the challenges of batch reactor control are outlined in this module.

After studying this module, the reader should be able to do the following:

- Understand the effect of scale-up on the dynamic characteristics of a jacketed batch chemical reactor. Further, understand how these characteristics affect the control-system design and operating performance.
- Understand some common control configurations and controller tuning techniques. Tuning for ramp vs. step setpoint changes, and tuning for disturbance rejection.

M11.1 Background

Jacketed batch chemical reactors are often used in the pharmaceutical and specialty chemicals industries. Very often these reactors need to have both "heating" and "cooling" modes. That is, the temperature of a reactor is often ramped up from the ambient reactor charge conditions (near 20°C) to a temperature where the reaction begins to "take off." If the reaction is exothermic, the heat released through the reaction must be removed by circulating a cold fluid through the jacket.

A common strategy is shown in Figure M11–1, where both hot and cold jacket feed streams are available. Here we have shown a split-range jacket temperature control strategy. When the jacket temperature controller output is between 0 and 50%, the cold glycol valve is open. When the jacket temperature controller output is between 50 and 100%, the hot glycol valve is open. The split-range behavior is shown in Figure M11–2. Notice that the cold glycol valve is fail-closed, while the hot glycol valve is fail-open.

An alternative approach is for the reactor temperature controller output to be the setpoint to a jacket inlet temperature controller, as shown in Figure M11–3. In Section M11–2 we will focus on the strategy shown in Figure M11–1. Simplifying assumptions will be used to ease the design of the reactor temperature controller, and to better understand the effect of reactor scale.

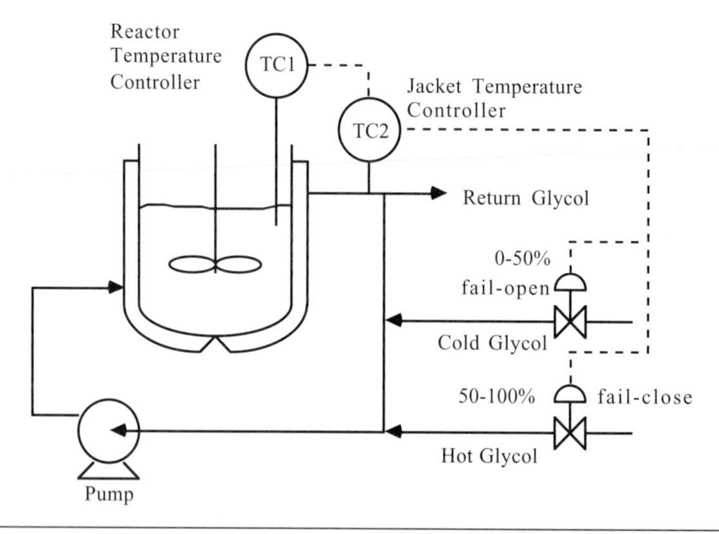

Figure M11–1 Batch reactor temperature control. The jacket temperature controller has a split-range output, where the cold glycol valve is open during "cooling mode" and the hot glycol valve is open during "heating mode."

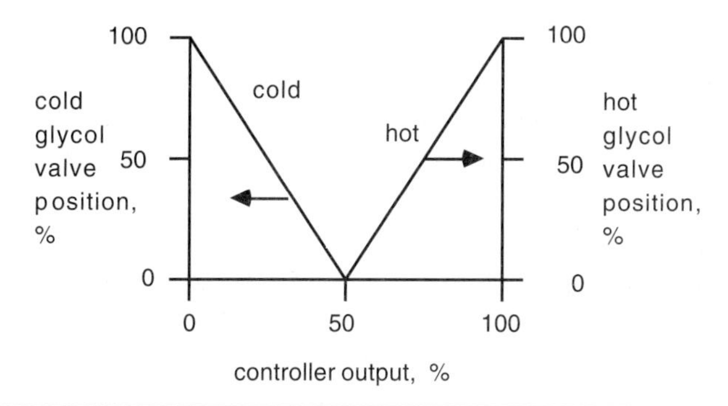

Figure M11–2 Depiction of the split-range controller action.

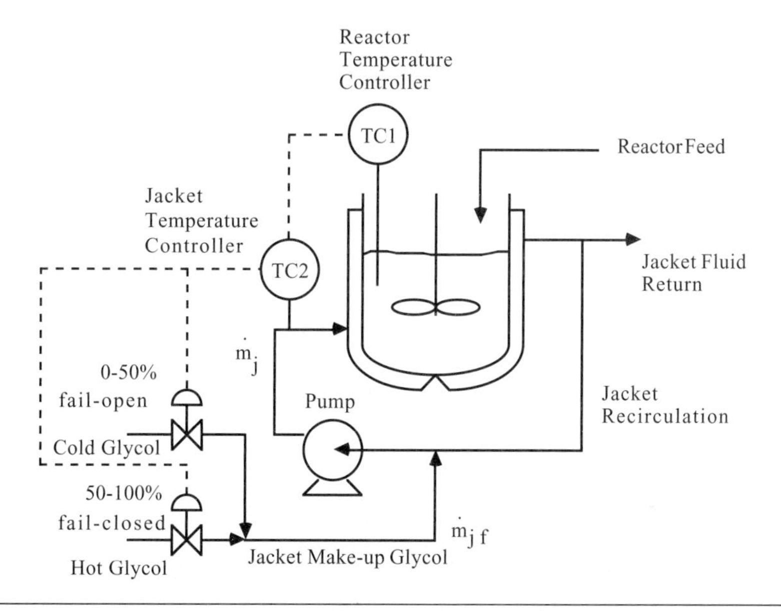

Figure M11–3 Batch reactor temperature control. The jacket inlet temperature controller has a split-range output, where the cold glycol valve is open during "cooling mode" and the hot glycol valve is open during "heating mode."

M11.2 Batch Model 1: Jacket Temperature Manipulated

In this section we focus on the development of a mathematical model for the reactor temperature dynamics; we will assume that the jacket temperature is directly manipulated. A number of other assumptions will be made in the development of a mathematical model to describe the dynamic batch reactor behavior. First of all, we will assume that the reactor is perfectly mixed. Also, for simplicity we lump all reaction heat effects into a single heat generation term, q_{gen}. Physical parameters, such as density and the heat transfer coefficient, are assumed to be constant.

$$\left(mc_p\right)_r \frac{dT}{dt} = -UA\left(T - T_j\right) + q_{gen} \tag{M11.1}$$

where U = heat transfer coefficient, A = area for heat transfer.

Notice that the accumulation term includes the effect of the "thermal mass" of the reactor wall

$$(mc_p)_r = (mc_p)_m + (mc_p)_f \tag{M11.2}$$

where $(mc_p)_m$ is the "thermal mass" (mass * heat capacity, including the metal as well as the glass lining) of the reactor wall and any other inert components (agitator, baffles, etc.). Also, $(mc_p)_f$ is the contribution of the reactor fluid ($V\rho c_p$). For small (laboratory) scale reactors the contribution of the reactor wall and other inert components can be significant, while the relative contribution is insignificant for large scale reactors. Equation (M11.1) can be rewritten

$$\frac{dT}{dt} = \left[-UA\left(T - T_j\right) + q_{gen}\right] / \left(mc_p\right)_r \tag{M11.3}$$

The fundamental model shown in (M11.3) is linear. The more familiar state-space form is

$$\frac{dT}{dt} = -\frac{UA}{\left(mc_p\right)_r} T + \frac{UA}{\left(mc_p\right)_r} T_j + \frac{1}{\left(mc_p\right)_r} q_{gen} \tag{M11.4a}$$

Or, using deviation variable form

$$\frac{dx}{dt} = -\frac{UA}{\left(mc_p\right)_r} x + \frac{UA}{\left(mc_p\right)_r} u + \frac{1}{\left(mc_p\right)_r} l \tag{M11.4b}$$

where reactor temperature is the state (and output), jacket temperature is the manipulated input, and heat generation is the load disturbance.

The resulting input-output transfer function models relating the manipulated and disturbance inputs to the measured reactor output are first-order (see Additional Exercise 1)

$$y(s) = \frac{k_p}{\tau_p s + 1} u(s) + \frac{k_d}{\tau_p s + 1} l(s)$$

$$k_p = 1$$

$$\tau_p = \frac{(mc_p)_r}{UA} \qquad \text{(M11.5)}$$

$$k_d = \frac{1}{UA}$$

Effect of Scale (Size)

Notice that the time constant is the "thermal mass" divided by the UA. If both of these terms increased proportionally with the reactor volume, then studies conducted in a laboratory would be directly applicable to large-scale production vessels. The problem is that the mass of the fluid increases linearly with reactor volume, but the heat-transfer area does not. As a result, the time constant of the reactor increases as the reactor gets larger. Essentially, on a per-volume basis, the ability to transfer heat goes down as the volume increases.

In the following studies we will consider three different reactors: (i) laboratory scale (1 liter), (ii) pilot-plant scale (0.5 m^3), and (iii) manufacturing scale (10 m^3). We will assume that the heat-transfer coefficient is the same for each of these reactors, but that the heat transfer area changes. The heat transfer coefficient used is

$$U = \frac{0.1 \text{ kcal}}{\text{s m}^2 \text{ °C}} = \frac{0.4184 \text{ kW}}{\text{m}^2 \text{ °C}}$$

Assume that the reactor is a vertical cylinder with a height/diameter ratio of 2, and that a heating/cooling jacket fully covers the bottom and sides of the reactor. This yields the results shown in Table M11–1.

Quasi-Steady-State Behavior

If the heat-generation term is constant, then there exists a steady-state temperature difference between the reactor and jacket such that the heat flows are balanced. From Equation (M11.1) we see

$$UA(T - T_j) = q_{gen}$$

Table M11–1 Reactor Characteristics

V	0.001 m^3	0.5 m^3	10 m^3
A	0.0523 m^2	3.295 m^2	24.28 m^2
UA	0.0219 kW/°C	1.379 kW/°C	10.16 kW/°C

Problem M11.1 Effect of Reactor Scale on Heat Transfer Capability

For each scale of reactor, determine the heat flow, on a per-volume basis, if the temperature difference between the reactor and jacket is 20°C. How much more "efficient" at transferring heat is a 1 liter reactor compared to a 10-m^3 reactor?

Problem M11.2 Effect of Reactor Scale on Process Time Constant

For each scale of reactor, find the process time constant (τ_p) assuming that the process fluid is water and neglecting the thermal effect of the reactor mass. For controller tuning, we suggest that you use an initial λ value of 1/4 τ_p for future simulations.

IMC-Based Design

In the Internal Model Control (IMC) design procedure used throughout most of this text we have focused on the response to step setpoint changes. IMC can also be designed for ramp setpoint changes or for the rejection of process input disturbances. What differs about the design is the specification of the filter. For each case the filter is shown below.

Step setpoint change $$f(s) = \frac{1}{(\lambda s + 1)^n}$$

Ramp setpoint change $$f(s) = \frac{n\lambda s + 1}{(\lambda s + 1)^n}$$

Input load disturbance $$f(s) = \frac{\gamma s + 1}{(\lambda s + 1)^n}$$

where n is chosen to make the internal model controller physically realizable. For a first-order process, you should verify the following results.

Design	IMC, q(s)	IMC-based PID, $g_c(s)$
Step setpoint	$\frac{1}{k_p} \cdot \frac{\tau_p s + 1}{\lambda s + 1}$	$\left(\frac{\tau_p}{k_p \lambda}\right) \cdot \frac{\tau_p s + 1}{\tau_p s}$ (see Table 9–1A)
Ramp setpoint	$\frac{1}{k_p} \cdot \frac{\tau_p s + 1}{\lambda s + 1} \cdot \frac{2\lambda s + 1}{\lambda s + 1}$	$\left(\frac{2\tau_p}{k_p \lambda}\right) \cdot \frac{\tau_p s + 1}{\tau_p s} \cdot \frac{2\lambda s + 1}{2\lambda s}$
Input disturbance	$\frac{1}{k_p} \cdot \frac{\tau_p s + 1}{\lambda s + 1} \cdot \frac{\gamma s + 1}{\lambda s + 1}$	(see Table 9–1B)

It is very important to use antireset windup (ARW) strategies when using PID controllers on batch reactors, since the manipulated inputs are often constrained. For the following problems assume that the jacket temperature is constrained to be between (–10°C and 150°C). For each of the following problems use the corresponding controller design.

Problem M11.3 IMC-Based PI Control: Step Setpoint Tracking

Here we require that the reactor temperature setpoint be changed from 20°C to 50°C. Compare responses to step setpoint changes, for each size reactor. Discuss the practical limitations to the expected dynamic performance for each size reactor, especially considering the constraints on jacket temperature.

Problem M11.4 IMC-Based PI Control: Ramp Setpoint Tracking

Here we require that the reactor temperature setpoint be changed from 20°C to 50°C. Compare responses to ramp setpoint changes, for each size reactor. Discuss the practical limitations to the expected dynamic performance for each size reactor, especially considering the constraints on jacket temperature. Also, what are practical limitations on the ramp time as a function of scale? For example, perhaps a 10-minute ramp time is fine for the laboratory scale reactor; is it feasible for the larger scale reactors?

Problem M11.5 IMC-Based PI Control: Disturbance Rejection

After the reactor has been heated to 50°C, assume that a reaction heat flow of 0.05 kW/liter (50 kW/m^3) is initiated. Compare the closed-loop performance as a function of reactor scale; compare the maximum deviation from the desired value of 50°C for each scale reactor.

M11.3 Batch Model 2: Jacket Inlet Temperature Manipulated

In this section we focus on the development of a mathematical model for the reactor and jacket; we will assume that the jacket inlet temperature is manipulated, and the jacket flowrate is constant. This model is appropriate for the strategy shown in Figure M11–3. A number of other assumptions will be made in the development of a mathematical model to describe the dynamic batch reactor behavior. First of all, we will assume that the reactor is perfectly mixed. Also, for simplicity we lump all reaction heat effects into a single heat-generation term, q_{gen}. Similarly, the jacket is perfectly mixed; often, agitation nozzles are used to create a high "swirl velocity" in the jacket, so this assumption will be fine on the time scale of interest. Physical parameters, such as density and the heat transfer coefficient, are assumed to be constant. The dynamics of the recirculation heat-transfer system are neglected in this model. Energy balances on the reactor and jacket yield the following equations:

$$\begin{aligned}(mc_p)_r \frac{dT}{dt} &= -UA(T - T_j) + q_{gen} \\ (mc_p)_j \frac{dT_j}{dt} &= \dot{m}_j c_{pj}(T_{jin} - T_j) + UA(T - T_j)\end{aligned} \qquad \text{(M11.6)}$$

where the accumulation terms include the effect of any "thermal mass" of the reactor and jacket walls. For example, the reactor energy balance includes the term

$$(mc_p)_r = (mc_p)_m + (mc_p)_f$$

where $(mc_p)_m$ is the "thermal mass" (mass * heat capacity, including the metal as well as the glass lining) of the reactor wall and any other inert components (agitator, baffles, etc.). Also, $(mc_p)_f = V\rho c_{pf}$ is the contribution of the reactor fluid. The term $\dot{m}_j$ represents the mass flow rate of the jacket fluid.

Equation (M11.6) can be rewritten

$$\begin{aligned}\frac{dT}{dt} &= \left[-UA(T - T_j) + q_{gen}\right] / (mc_p)_r \\ \frac{dT_j}{dt} &= \left[\dot{m}_j c_{pj}(T_{jin} - T_j) + UA(T - T_j)\right] / (mc_p)_j\end{aligned} \qquad \text{(M11.7)}$$

The linear state space model is

$$\begin{bmatrix} \frac{dT}{dt} \\ \frac{dT_j}{dt} \end{bmatrix} = \begin{bmatrix} -\frac{UA}{(mc_p)_r} & \frac{UA}{(mc_p)_r} \\ \frac{UA}{(mc_p)_j} & -\frac{(\dot{m}_j c_{pj} + UA)}{(mc_p)_j} \end{bmatrix} \begin{bmatrix} T \\ T_j \end{bmatrix} + \begin{bmatrix} 0 \\ \frac{\dot{m}_j c_{pj}}{(mc_p)_j} \end{bmatrix} T_{jin} + \begin{bmatrix} 1/(mc_p)_r \\ 0 \end{bmatrix} q_{gen} \qquad \text{(M11.8)}$$

$$T = \begin{bmatrix} 1 & 0 \end{bmatrix} \begin{bmatrix} T \\ T_j \end{bmatrix}$$

The resulting input-output transfer function models relating the manipulated and disturbance inputs to the measured reactor output are (see Additional Exercise 2)

$$T(s) = \frac{k_p}{\tau^2 s^2 + 2\zeta\tau s + 1} T_{jin}(s) + \frac{k_d(\tau_{nd} s + 1)}{\tau^2 s^2 + 2\zeta\tau s + 1} q_{rxn}(s)$$

$$k_p = 1$$

$$\tau^2 = \frac{(mc_p)_j (mc_p)_r}{(\dot{m}_j c_{pj}) UA} \tag{M11.9}$$

$$2\zeta\tau = \frac{UA\left[(mc_p)_j + (mc_p)_r\right] + (\dot{m}_j c_{pj})(mc_p)_r}{(\dot{m}_j c_{pj}) UA}$$

Problem M11.6 Effect of Reactor Scale on Process Parameters

For each scale of reactor (Table M11–1), find the process parameters assuming that the process and jacket fluid is water and neglecting the thermal effect of the non-fluid components of the reactor and jacket. Assume that the jacket volume is 1/4 of the reactor volume.

IMC-Based PID Tuning Parameters

Using the IMC-based PID tuning rules, with a desired first-order closed-loop response with a time constant of λ, we find

$$k_c = \frac{2\zeta\tau}{k_p \lambda}$$

$$\tau_I = 2\zeta\tau \tag{M11.10}$$

$$\tau_D = \frac{\tau^2}{2\zeta\tau}$$

We can also use the following order of magnitude analysis to reduce the expressions

$$\dot{m}_j c_{pj} >> (mc_p)_r \approx (mc_p)_j$$

so

$$\tau_I = 2\zeta\tau \approx \frac{(mc_p)_r}{UA}$$

$$\tau_D = \frac{\tau^2}{2\zeta\tau} \approx \frac{(mc_p)_j}{\dot{m}_j c_{pj}} \tag{M11.11}$$

$$k_c = \frac{2\zeta\tau}{k_p \lambda} = \frac{\tau_I}{\lambda} \approx \frac{(mc_p)_r / UA}{\lambda}$$

Also, as in Section M11.2, let the closed-loop time constant be roughly 1/4 of the open-loop time constant, so the resulting controller proportional gain is

$$k_c = 4$$

and the proportional band is then 25%. For a PI controller, we can start with the same parameter values, but set the derivative time constant to 0.

Problem M11.7 IMC-Based PI Control: Step Setpoint Tracking

Here we require that the reactor temperature setpoint be changed from 20°C to 50°C. Compare responses to step setpoint changes, for each size reactor. Discuss the practical limitations to the expected dynamic performance for each size reactor, especially considering the constraints on jacket inlet temperature (–10 to 150°C).

M11.4 Batch Model 3: Cascade Control

The process model, if the jacket make-up flow rate is the manipulated input, is

$$\begin{aligned} \left(mc_p\right)_r \frac{dT}{dt} &= -UA\left(T - T_j\right) + q_{gen} \\ \left(mc_p\right)_j \frac{dT_j}{dt} &= \dot{m}_{jfcold} c_{pj}\left(T_{jfcold} - T_j\right) + \dot{m}_{jfhot} c_{pj}\left(T_{jfhot} - T_j\right) + UA\left(T - T_j\right) \end{aligned} \tag{M11.12}$$

As written, this assumes that the jacket is a "once through" flow and that there are cold and hot jacket feed streams. It can also be shown that a jacket recirculating heat-transfer system has the same model equations, where the jacket mass flow rate is simply the make-up flow to the recirculating jacket heat-transfer system [compare Equations (M8.11) and (M8.12) for a CSTR]. Here a split-range control strategy would be used, with either the cold or hot fluid stream admitted to the jacket, depending on the split-range controller output.

Linearizing Equation (M11.12), we find the following state space model matrices

$$A = \begin{bmatrix} \dfrac{-UA}{\left(mc_p\right)_r} & \dfrac{UA}{\left(mc_p\right)_r} \\ \dfrac{UA}{\left(mc_p\right)_j} & -\dfrac{\left(\dot{m}_{jfcold} + \dot{m}_{jfhot}\right)c_{pjf}}{\left(mc_p\right)_j} - \dfrac{UA}{\left(mc_p\right)_j} \end{bmatrix}$$

$$B = \begin{bmatrix} 0 & 0 & \dfrac{1}{\left(mc_p\right)_r} \\ \dfrac{T_{jfcold} - T_{js}}{\left(mc_p\right)_j} & \dfrac{T_{jfhot} - T_{js}}{\left(mc_p\right)_j} & 0 \end{bmatrix}$$

where inputs 1 and 2 (corresponding to columns 1 and 2 of the B matrix) are the cold and hot jacket feedstreams, respectively. The third input is the disturbance (reaction heat flow), q_{gen}. Since the measured outputs are the reactor and jacket temperature,

$$C = \begin{bmatrix} 1 & 0 \\ 0 & 1 \end{bmatrix}$$

$$D = \begin{bmatrix} 0 & 0 & 0 \\ 0 & 0 & 0 \end{bmatrix}$$

Problem M11.8 Simulation Model for Intermediate Scale Reactor

For the 0.5 m^3 reactor, assuming that both the reactor and jacket fluids are water, and the jacket volume is 25% of the reactor volume (and neglecting the thermal mass of the non-fluid components), find the state space model. For the initial steady-state assume that there is no reaction heat flow, so the cold and hot fluid flow rates are zero initially. What do you notice about the eigenvalues of the A matrix? Find the transfer functions relating the manipulated inputs to the outputs; what do you notice about one of the poles? Observe the responses to step changes in the manipulated inputs. Are these realistic? (*Hint:* You should find that the responses are integrating in nature; this is expected for the initial responses, but not for the long-term behavior.)

Problem M11.9 Cascade Control with Split-Range Jacket Temperature Controller (this problem is better for a longer-term project)

Develop a jacket temperature controller that has a split-range output (see Chapter 12 for an example SIMULINK model for a stirred-tank heater). The controller output should be similar to Figure M11–2. Design the jacket temperature (secondary) controller using any method that you desire. Next, design the reactor temperature (primary) controller.

Perform simulations where the reactor temperature setpoint is changed from 20°C to 50°C. Compare responses to step setpoint changes, for each size reactor. Discuss the practical limitations to the expected dynamic performance for each size reactor, especially considering the constraints on jacket inlet temperature (–10 to 150°C).

It is recommended that both the primary and secondary controllers include antireset windup (Chapter 11) strategies.

M11.5 Summary

This module has assumed very simple models for batch reactors, where heat effects (due to an exothermic reaction, for example) are captured by a simple heat generation term.

This resulted in low-order models that demonstrated clearly the importance of process scale. A smaller reactor has a larger heat transfer area to volume ratio, making it much more efficient at transferring heat compared to a larger reactor. That is, smaller temperature differences between the reactor and jacket are needed to remove heat energy from the reactor. Smaller reactors then are less prone to suffer from problems with jacket temperature constraints. In addition, the smaller reactors have smaller process time constants, allowing better closed-loop temperature control (faster closed-loop responses). Certainly, more complicated models can be developed, which include reactor rate terms and semi-batch reactor operation (see Additional Exercise 4). More comprehensive examples will be placed on the textbook web page; you are encouraged to download these files for advanced study.

Reference

The following article provides an excellent overview of both batch and continuous reactor control:

Lipták, B. G., "Controlling and Optimizing Chemical Reactors," *Chem. Eng. Mag.*, pp. 69–81 (May 26, 1986).

Additional Exercises

1. Find the transfer function relationships shown in Equation (M11.5).
2. Find the transfer function relationships shown in Equation (M11.9).
3. Pfaudler is a major manufacturer of reaction vessels. Search the Pfaudler web page to find 100-gallon and 1500-gallon reactors. Do the heat transfer areas vary with volume in the way that you would expect?
4. All of the examples in this module have included a heat generation term, for simplicity. This resulted in simple, low-order models. Develop a more comprehensive model of a semi-batch reactor (flow in, but not out), with a simple first-order irreversible reaction ($A \rightarrow B$). Develop the modeling equations, including the following states: V (liquid volume), N_A (total amount of component A in the reactor), T (reactor temperature), and T_j (jacket temperature). Allow the heat-transfer area to vary as a function of the reactor volume. Include a steady-state energy balance on the recirculating jacket fluid heat-transfer system, and assume that a split-range controller manipulates either the cold or hot glycol flow rate.

MODULE 12

Biomedical Systems

There are a number of important applications of feedback control for biomedical systems. In this module we present a number of examples and suggest that instructors or students pick one of the topics for each interactive learning session. The goals are to learn more about modeling and control by applying the basic concepts to important problems in biomedical systems.

The sections of this module are as follows:

M12.1 Overview

Physiological systems are composed of a large number of feedback control loops. Examples include the baroreceptor reflex that regulates blood pressure, and the synthesis of insulin and glucagon by the pancreas to regulate blood glucose. The focus of the examples presented in this module is on exogenous feedback regulation using devices external to the human body.

There is a rich history of automated control applications in biomedicine. Initial studies on feedback control of anesthesia were performed by Bickford at the Mayo Clinic in the 1950s. Several different anesthetics were studied as the manipulated input, while the measured variable was the integrated rectified amplitude of the EEG. The control strategy did not gain acceptance because it was not clear what feature of the EEG signal should be used as an indicator of the depth of anesthesia.

Modeling biomedical processes can be quite challenging. For physiological systems, drug material balances can be written based on a compartmental model; that is the body or tissues are "lumped" into "volumes" in the same fashion that stirred tanks are used to represent chemical process behavior.

M12.2 Pharmacokinetic Models

Pharmacokinetics is the study of the dynamic behavior of drug concentrations in blood and tissues. Compartmental models are widely used to conceptually describe this behavior. One common 3-compartment model is shown in Figure M12–1.

Compartment 1 is often used to represent blood plasma. The drug is assumed to be directly infused to compartment 1. The drug can then diffuse between compartment 1 and compartments 2 and 3, based on the transfer parameters, k_{ij}. For example, k_{12} represents the kinetics for transfer of the drug from compartment 1 to compartment 2. Compartment 2 can be thought of as well-perfused tissues, such as muscle and brain, while compartment 3 represents residual tissues and bone.

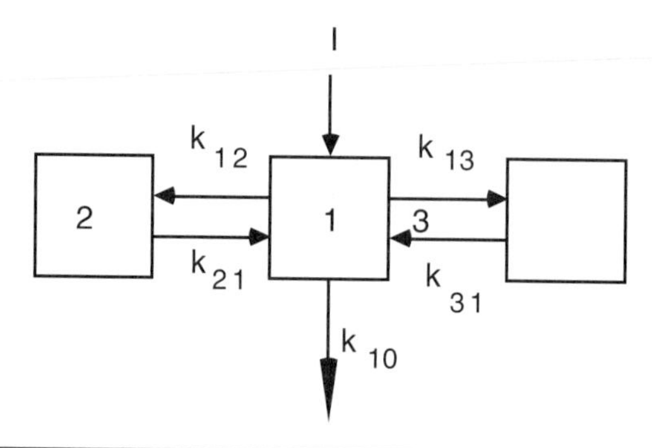

Figure M12–1 Three-compartment model.

Modeling Equations

If we let I represent the mass infusion rate (mass/unit time) of the drug, the overall material balance can be written as

$$\begin{aligned}\frac{dM_1}{dt} &= -(k_{12}+k_{13}+k_{10})M_1 + k_{21}M_2 + k_{31}M_3 + I \\ \frac{dM_2}{dt} &= k_{12}M_1 - k_{21}M_2 \\ \frac{dM_3}{dt} &= k_{13}M_1 - k_{31}M_3\end{aligned} \tag{M12.1}$$

This has the form of a state space model

$$\begin{bmatrix}\dot{M}_1\\ \dot{M}_2\\ \dot{M}_3\end{bmatrix} = \begin{bmatrix}-(k_{12}+k_{13}+k_{10}) & k_{21} & k_{31}\\ k_{12} & -k_{12} & 0\\ k_{13} & 0 & -k_{31}\end{bmatrix}\begin{bmatrix}M_1\\ M_2\\ M_3\end{bmatrix} + \begin{bmatrix}1\\0\\0\end{bmatrix} I \tag{M12.2}$$

One is normally concerned more about drug concentrations rather than the total amount of drug. The blood plasma concentration can be represented by $C_1 = M_1/V_1$. The state space model for the measured output is then

$$C_1 = \begin{bmatrix}\frac{1}{V_1} & 0 & 0\end{bmatrix}\begin{bmatrix}M_1\\ M_2\\ M_3\end{bmatrix} + [0]I \tag{M12.3}$$

Alternatively, the states can be formulated as concentrations, resulting in the following model:

$$\begin{aligned}\begin{bmatrix}\dot{C}_1\\ \dot{C}_2\\ \dot{C}_3\end{bmatrix} &= \begin{bmatrix}-(k_{12}+k_{13}+k_{10}) & k_{21} & k_{31}\\ k_{12} & -k_{12} & 0\\ k_{13} & 0 & -k_{31}\end{bmatrix}\begin{bmatrix}C_1\\ C_2\\ C_3\end{bmatrix} + \begin{bmatrix}\frac{1}{V_1}\\0\\0\end{bmatrix} I \\ C_1 &= \begin{bmatrix}1 & 0 & 0\end{bmatrix}\begin{bmatrix}C_1\\ C_2\\ C_3\end{bmatrix}\end{aligned} \tag{M12.4}$$

M12.3 Intravenous Delivery of Anesthetic Drugs

Traditionally, most anesthetics have been vapor-phase, such as isoflurane. In recent years there has been a move toward the intravenous (IV) delivery of drugs for anesthesia. One of the more popular IV anesthetics is propofol. For a particular population of subjects, the following kinetic values have been determined: $k_{10} = 0.152$, $k_{12} = 0.207$, $k_{13} = 0.040$, $k_{21} = 0.092$, and $k_{31} = 0.0048$, all with units of min^{-1}. Also, assume that the volume of the blood plasma compartment is $V_1 = 12$ liters.

1. What is the required steady-state propofol infusion rate to maintain a desired blood concentration of 2 *μg/ml*?
2. Develop the state space model, based on Equations (M12.2) and (M12.3 or M12.4), assuming that the plasma volume is 12 liters. What are the units of the manipulated input (propofol infusion rate) and measured output (blood concentration)?
3. Develop the process transfer function model, if the manipulated drug infusion rate has units of mg/hr, and the blood concentration has units of *μg/ml.*
4. Develop a feedback-control strategy to control blood concentration by manipulating the propofol infusion rate. Make a setpoint change from 1 to 2 *μg/ml.* Are the closed-loop response and drug infusion rate feasible?

M12.4 Blood Glucose Control in Diabetic Patients

The human body has many innate feedback-control loops. For example, blood glucose is regulated by the production of insulin by the pancreas. When food is consumed and broken down by the digestive system, the blood glucose level rises, stimulating insulin production. Cells use insulin to break down the glucose. People with diabetes have a reduced capability of producing insulin. Type I diabetes mellitus patients cannot produce any insulin and must administer insulin shots several times a day to help regulate their blood glucose level. A typical patient is then serving as a control system. Some actions are feedforward in nature, such as the administration of an insulin shot to coincide with meal consumption. Other actions are feedback in nature, such as when the patient changes dosages based on blood glucose measurements, obtained from "finger pricks" and analysis of glucose strips. In the long-term, hyperglycemia (high blood sugar) can lead to blindness and cardiovascular problems. In the short-term, hypoglycemia (low blood sugar) can lead to fainting or a diabetic coma.

Clearly, there is a tremendous motivation for the development of closed-loop insulin delivery systems. Current technology involves external pumps with insulin reservoirs that can deliver insulin continuously, rather than having "pulses" due to shots administered several times a day. With an external infusion pump, the patient must still perform tests to monitor blood glucose levels.

There is a major research effort to develop implantable glucose sensors and insulin infusion pumps. An implanted sensor will measure glucose and send a telemetry signal to a small, wristwatch-sized controller that can send another telemetry signal to a small implanted insulin pump. This closed-loop system essentially acts as an artificial pancreas, yielding a lifestyle that differs little from a nondiabetic person on a day-to-day basis.

Nonlinear Model

One of the more widely used models of the effect of insulin infusion and glucose (meal) inputs on the blood glucose concentration is known as the Bergman "minimal model" (Bergman et al., 1981). This is described by three differential equations:

$$\begin{aligned}
\frac{dG}{dt} &= -p_1 G - X(G + G_b) + \frac{G_{meal}}{V_1} \\
\frac{dX}{dt} &= -p_2 X + p_3 I \\
\frac{dI}{dt} &= -n(I + I_b) + \frac{U}{V_1}
\end{aligned} \tag{M12.5}$$

where G and I represent the deviation in blood glucose and insulin concentrations, respectively. Also, X is proportional to the insulin concentration in a "remote" compartment. The inputs are G_{meal}, a meal disturbance input of glucose, and U, the manipulated insulin infusion rate. The parameters include p_1, p_2, p_3, n, and V_1 (which represents the blood volume). Other parameters are G_b and I_b, the "basal" (baseline or steady state) values of blood glucose and insulin concentration. These values can be used to determine the basal infusion rate of insulin necessary to maintain a steady state.

Linear Model

A linear state space model can be developed for use in control-system design. Here, define the state, input, and output variables (in deviation form) as

$$\begin{bmatrix} x_1 \\ x_2 \\ x_3 \end{bmatrix} = \begin{bmatrix} G \\ X \\ I \end{bmatrix} \qquad \begin{bmatrix} u_1 \\ u_2 \end{bmatrix} = \begin{bmatrix} U - U_b \\ G_{meal} - 0 \end{bmatrix} \qquad y = G \tag{M12.6}$$

where the first input is manipulated (insulin infusion) and the second input represents a meal glucose disturbance. The state space model is

$$A = \begin{bmatrix} -p_1 & -G_b & 0 \\ 0 & -p_2 & p_3 \\ 0 & 0 & -n \end{bmatrix} \qquad B = \begin{bmatrix} 0 & 1/V_1 \\ 0 & 0 \\ 1/V_1 & 0 \end{bmatrix} \tag{M12.7}$$

$$C = \begin{bmatrix} 1 & 0 & 0 \end{bmatrix} \qquad D = \begin{bmatrix} 0 & 0 \end{bmatrix}$$

Example Set of Parameters

Consider a diabetic that is modeled using the following set of parameters (Lynch and Bequette, 2001):

$$\begin{aligned} G_b &= 4.5 \text{ mmol/liter} \\ I_b &= 4.5 \text{ mU/liter} \\ V_1 &= 12 \text{ liters} \\ p_1 &= 0 \text{ min}^{-1} \\ p_2 &= 0.025 \text{ min}^{-1} \\ p_3 &= 0.000013 \text{ mU/liter} \\ n &= 5/54 \text{ min}^{-1} \end{aligned}$$

It is necessary to be careful with units. Since the concentrations are in mmol/liter, and the glucose disturbance has units of grams, we must apply a conversion factor of 5.5556 mmol/g to the G_{meal} term. Solving for the steady states, we find that the basal insulin infusion rate must be U_b = 16.667 mU/min. For these parameters, the resulting state space model is

$$A = \begin{bmatrix} 0 & -4.5 & 0 \\ 0 & -0.025 & 0.000013 \\ 0 & 0 & -5/54 \end{bmatrix} \qquad B = \begin{bmatrix} 0 & 0.4630 \\ 0 & 0 \\ 1/12 & 0 \end{bmatrix} \tag{M12.8}$$

In the United States it is more common to work with glucose concentration units of mg/deciliter rather than mmol/liter. Since the molecular weight of glucose is 180 g/mol, we must multiply the glucose state (mmol/liter) by 18 to obtain the measured glucose output (mg/deciliter). This yields the following state-output relationship

$$C = \begin{bmatrix} 18 & 0 & 0 \end{bmatrix} \qquad D = \begin{bmatrix} 0 & 0 \end{bmatrix} \tag{M12.9}$$

You should find that the process transfer function is

$$g_p(s) = \frac{-3.79}{(40s+1)(10.8s+1)s} \tag{M12.10}$$

Due to pole/zero cancellation, the disturbance transfer function is

$$g_d(s) = \frac{8.334}{s} \tag{M12.11}$$

In practice, the glucose does not directly enter the blood stream. The processing in the gut before entering the blood can be modeled as a first-order transfer function with a 20-minute time constant. This means that the disturbance transfer function, including the lag in the gut, is

$$g_d(s) = \frac{8.334}{s(20s+1)} \tag{M12.12}$$

Also, unlike many chemical process disturbances, a meal disturbance is best modeled as a pulse. Assume that a 50 g glucose meal is consumed over a 15-minute period; the pulse then has a magnitude of 3.333 g/minute for a duration of 15 minutes.

Desired Control Performance

The steady-state glucose concentration of 4.5 mmol/liter corresponds to a measured glucose concentration of 81 mg/deciliter. It is important that a diabetic maintain her/his blood glucose concentration above 70 mg/deciliter (to avoid short-term problems with hypoglycemia, such as fainting). Notice that the insulin infusion rate (manipulated input) cannot go below 0; it is critical that you consider this constraint in your simulations.

A number of control strategies could be used. One suggestion is to simplify the process transfer function (M12.10) to the following form

$$g_p(s) = \frac{k_p}{(\tau_p s+1)s} \tag{M12.13}$$

and design an IMC-based PD or PID controller. Also, since the diabetic knows when they are consuming a meal, some type of feed-forward action can be used.

Perform disturbance rejection simulations and discuss your results. What feedback-only design (and tuning) do you recommend? What are the minimum and maximum blood glucose values that occur after a pulse meal consumption of 50 g glucose over a 15-minute period? Does your strategy satisfy constraints on the insulin infusion rate? How much better is the performance if feed-forward control is included? Compare the resulting blood glucose profiles with those found in the literature.

M12.5 Blood Pressure Control in Post-Operative Patients

Health care is an important area that can benefit from control systems technology. For example, after undergoing open-chest surgery (such as cardiac bypass or lung surgery), many patients have a higher than desired blood pressure; one reason that a lower blood pressure is needed is to reduce bleeding from sutures. Typically, a critical care nurse will set up a continuous infusion of the drug sodium nitroprusside to reduce the patient's blood pressure. The nurse must make frequent checks of blood pressure and change the drug infusion rate to maintain the desired blood pressure. In a sense, the nurse is serving as a feedback controller. There is a clear motivation to develop an automated feedback control system, where a blood pressure sensor sends a signal to a blood pressure controller, which adjusts the speed of a drug infusion pump. In addition to providing tighter regulation of blood pressure, this strategy frees up the nurse to spend more time monitoring patients for other problems.

A 70-kg patient exhibits the following input-output behavior for the effect of sodium nitroprusside (SNP) on mean arterial pressure (MAP):

$$g_p(s) = \frac{-1.0e^{-0.5s}}{0.67s+1}$$

where the pressure unit is mm Hg, the drug flow units are ml/hr and the time unit is minutes. The initial MAP (with no SNP infused) is 175 mm Hg. The desired MAP setpoint is 100 ± 15 mm Hg, and the maximum allowable (short-term) SNP infusion rate is 150 ml/hr. The desired settling time (time for the MAP to remain within limits) is 5 minutes.

1. Can a proportional-only controller be designed to yield an offset of ± 15 mm Hg while meeting the infusion rate constraint?
2. Design an internal model controller for this system. What is the minimum λ value that can be used and still satisfy the drug infusion limit? What is the maximum λ value that can be used and still satisfy the settling time limit?
3. Design an IMC-based PI controller for this system, using the "improved" parameters (Table 9.2). Compare the closed-loop performance with a "pure" IMC strategy (developed in problem 2).

M12.6 Critical Care Patients

Critical care patients have often suffered a "disturbance" to the normal operation of their physiological system; this disturbance could have been generated by surgery or some sort of trauma (e.g., a heart attack). A responsibility of the critical care physician is to maintain certain patient outputs within an acceptable operating range. Two important outputs to be maintained are mean arterial pressure (MAP) and cardiac output (CO). During surgery the anesthesiologist will infuse several drugs into the patient in order to control these states close to the desired values. A conceptual diagram is shown in Figure M12–2.

The goal of this control system design is to manipulate the flow rate of two drugs, dopamine (DPM) and sodium nitroprusside (SNP), to maintain the two outputs at their desired setpoints. A successful implementation of such a strategy allows the anesthesiologist to spend more time monitoring other patient states, such as "depth of anesthesia."

A simplified model representing the input-output behavior for a particular patient is

$$\begin{bmatrix} y_1(s) \\ y_2(s) \end{bmatrix} = \begin{bmatrix} \dfrac{-6e^{-0.75s}}{0.67s+1} & \dfrac{3e^{-s}}{2s+1} \\ \dfrac{12e^{-0.75s}}{0.67s+1} & \dfrac{5e^{-s}}{5s+1} \end{bmatrix} \begin{bmatrix} u_1(s) \\ u_2(s) \end{bmatrix} \tag{M12.14}$$

where inputs 1 and 2 are SNP and DPM (ml/hr), and outputs 1 and 2 are MAP (mmHg) and CO (liters/min). The time constants have units of minutes.

1. Based on the RGA (Chapter 13), what are the natural pairings? Ordinarily, one would control MAP by manipulating SNP and control CO by manipulating SNP. Does the RGA rule out this strategy? Why or why not?

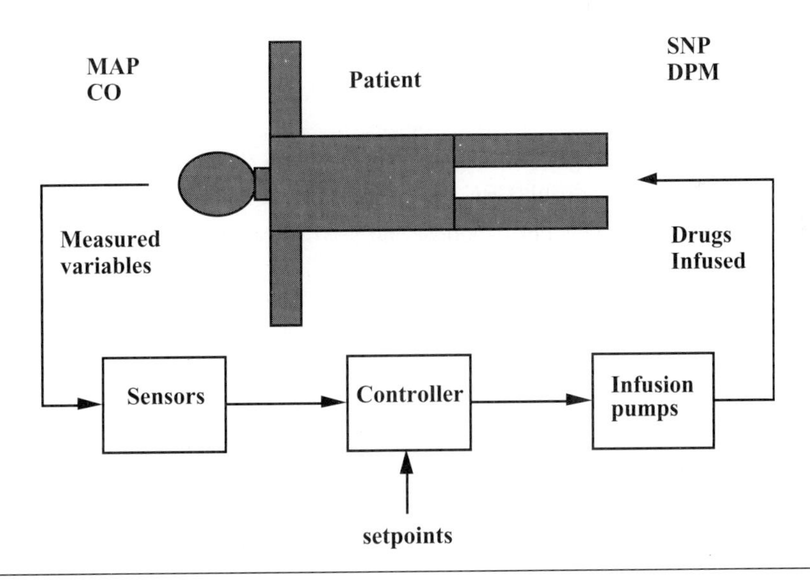

Figure M12–2 Drug infusion control.

2. Consider the transfer function matrix. If the nominal steady-state drug infusion rates are 0, can the cardiac output (CO) be decreased by infusing these drugs? Why or why not?
3. Perform an SVD analysis (Chapter 14), assuming that the transfer function matrix (M12.14) is properly scaled. What are the strongest and weakest output directions for this process?
4. Assume that the initial steady states for MAP and CO are 120 mm Hg and 4 liters/min, respectively. Design controllers and perform simulations for step setpoint changes to 100 mm Hg and 4.5 liters/min. Discuss controller performance and the different behavior if setpoint changes are made individually versus simultaneously.

M12.7 Summary

Automation and control in biomedicine is an important area of research, with many potential long-term health care benefits. Sensors, the difficulty of developing a reliable model, and "fuzzy" performance specifications all create difficulties in feedback-control design. For each of the examples presented in this module, it is recommended that you perform a literature review to understand the current state of the art.

References

A nice textbook on this general topic is:

Northrop, R. B., *Endogenous and Exogenous Regulation and Control of Physiological Systems*, Chapman and Hall/CRC, Boca Raton, FL (2000).

For examples of biomedical control applications, see the following special issue:

Bequette, B. W., and F. J. Doyle III (Eds.), *IEEE Engineering in Medicine and Biology Magazine*, Special issue on automation in biomedicine, Jan/Feb (2001).

A comparison of sets of propofol pharmacokinetic parameters is presented in:

Vuyk, J., F. H. M. Engbers, A. G. O. Burm, A. A. Vletter and J. G. Bovill, "Performance of Computer-Controlled Infusion of Propofol: An Evaluation of Five Parmacokinetic Parameter Sets," *Anesth. Analg.*, 81, 1275–1282 (1995).

Control of blood glucose in diabetics is discussed in the following papers:

Bellazzi, R., G. Nucci, and C. Cobelli, "The Subcutaneous Route to Insulin-Dependent Diabetes Therapy," *IEEE Engineering in Medicine and Biology Magazine*, 20(1), 54–64 (2001).

Parker, R. S., F. J. Doyle III, and N. A. Peppas, "The Intravenous Route to Blood Glucose Control," *IEEE Engineering in Medicine and Biology Magazine*, 20(1), 65–73 (2001).

The Bergman "minimal model" is discussed in the following paper:

Bergman, R. N., L. S. Philips, and C. Cobelli, "Physiological Evaluation of Factors Controlling Glucose Tolerance in Man," J. Clin. Invest., 68, 1456–1467 (1981).

The parameters for the diabetes model are presented in the following paper:

Lynch, S. M., and B. W. Bequette, "Estimation-based Model Predictive Control of Blood Glucose in Type I Diabetics: A Simulation Study," in Proceedings of the 27th Northeast Bioengineering Conference, Storrs, CT, pp. 79–80 (2001).

A review of blood pressure control is presented by:

Isaka, S., and A. V. Sebald, "Control Strategies for Arterial Blood Pressure Regulation," *IEEE Trans. Biomed. Eng.*, 40(4), 353–363 (1993).

An experimental study of the simultaneous control of blood pressure and cardiac output is presented in the paper:

Rao, R., C. C. Palerm, B. Aufderheide, and B. W. Bequette, "Experimental Studies on Automated Regulation of Hemodynamic Variables," *IEEE Engineering in Medicine and Biology Magazine*, 20(1), 24–38 (Jan/Feb, 2001).

Additional Exercises

1. Much research has been conducted on the development of a measure of anesthetic depth. A device that measured the bispectral index (BIS) is on the market and has been used in some feedback-control studies. Perform a literature search and discuss at least two papers that present feedback-control results based on a BIS measurement and the manipulation of either an intravenous (such as propofol) or vapor-phase (such as isoflurane) anesthetic.
2. Ventricular Assist Devices (VAD) help a weakened left ventricle to supply more "pumping action" to the blood. Perform a literature search and discuss at least two papers presenting VAD control studies.

Distillation Control

Distillation remains the most common method of separating chemical components from a mixture. In this module, we provide a simple example of a binary distillation column that has behavior illustrative of more complex columns. The goal is to better understand the dynamic behavior of multivariable, interacting systems. After studying this module the reader should be able to do the following:

- Understand the different control loops associated with distillation operation
- Understand the importance of input "direction" when making step input changes
- Understand how an input "direction" affects the magnitude and "direction" of the resulting output change
- Interpret SVD results in terms of these sensitive directions
- Consider failure sensitivity when designing controllers

The sections of this module are as follows:

M13.1 Description of Distillation Control

A schematic diagram of a typical binary distillation column control strategy is shown in Figure M13–1. Notice that there are five measured variables (pressure, distillate receiver level, base level, distillate composition, and bottoms composition) and five manipulated variables (condenser heat duty, reflux flow rate, distillate flow rate, bottoms flow rate, and reboiler heat duty). In addition, if the feed to the column is regulated, then there is an additional measurement (feed flow rate) and manipulated input (also feed flow rate or valve position). Normally, it is assumed that the pressure is controlled by manipulating the condenser heat duty (or the flow of cooling water to the condenser), that the distillate receiver level is controlled by manipulating the distillate product flow rate, and that the base level is controlled by manipulating the bottoms product flow rate. It is also assumed that these loops are tightly tuned, and the process outputs are "perfectly controlled" at their setpoint values.

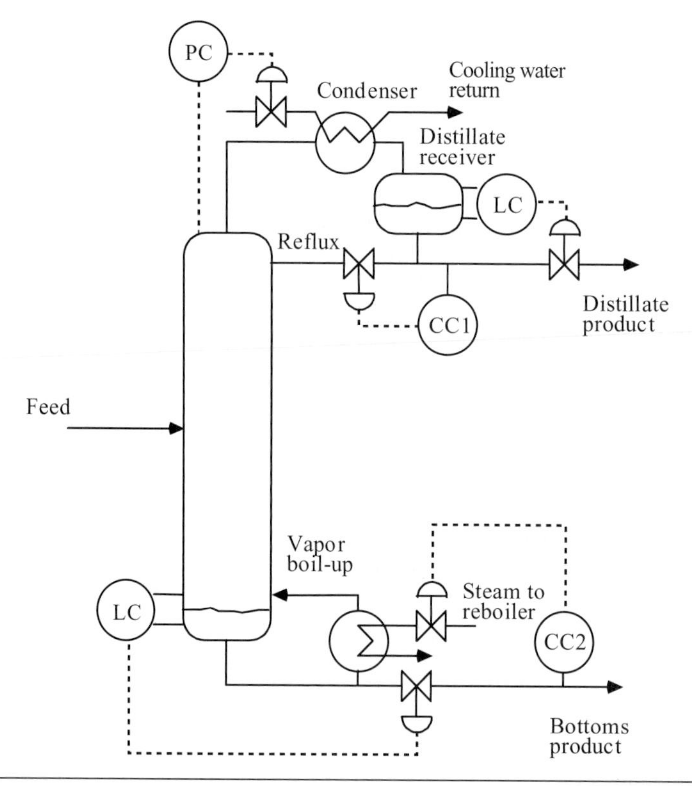

Figure M13–1 Dual composition control of a distillation column.

For a number of reasons, composition sensors may not be installed or many not be in service. The distillate and bottoms streams compositions can be inferred by measuring the temperatures of trays near each end of the column, as shown in Figure M13–2. The tray temperature measurement dynamics are typically more rapid than the product stream composition measurements, so tighter disturbance rejection may be possible with this strategy. For convenience, we assume throughout the rest of this module that composition measurements are available. The analysis and design techniques presented in this module can easily be used for the temperature-control system shown in Figure M13–2.

Simplified Model

The column studied is a 41-stage binary separation. For ease of modeling, it is assumed that the manipulated inputs are reflux flow rate and vapor boil-up rate. In practice, the

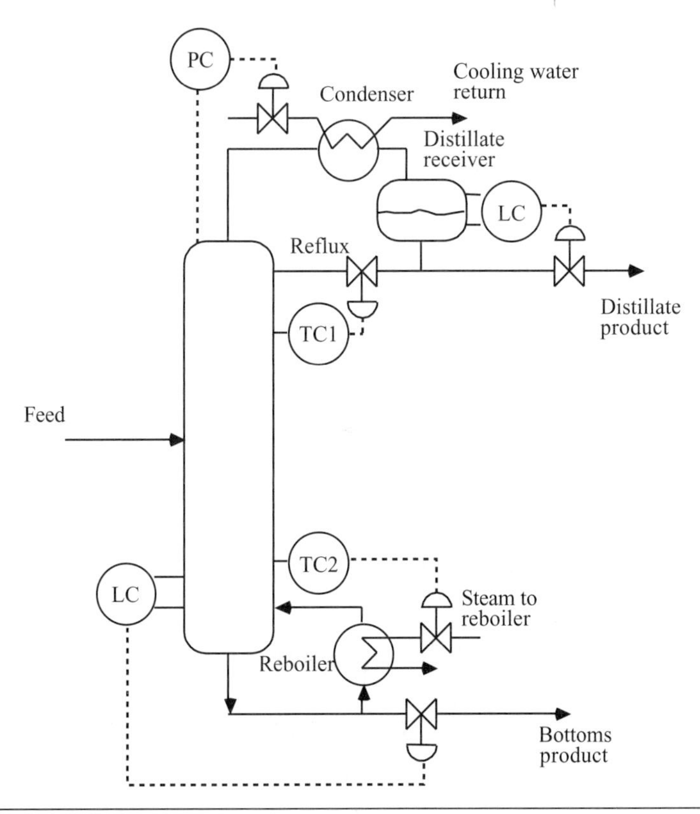

Figure M13–2 Dual temperature control of a distillation column.

steam flow rate to the reboiler would be manipulated, but it is related to the vapor boil-up by the heats of vaporization of the bottoms stream and the steam.

An input-output transfer function model for the dual composition control strategy is

$$\begin{bmatrix} x_D(s) \\ x_B(s) \end{bmatrix} = \begin{bmatrix} \dfrac{0.878}{75s+1} & \dfrac{-0.864}{75s+1} \\ \dfrac{1.082}{75s+1} & \dfrac{-1.096}{75s+1} \end{bmatrix} \begin{bmatrix} L(s) \\ V(s) \end{bmatrix} + \begin{bmatrix} \dfrac{0.394}{75s+1} & \dfrac{0.881}{75s+1} \\ \dfrac{0.586}{75s+1} & \dfrac{1.119}{75s+1} \end{bmatrix} \begin{bmatrix} F(s) \\ z_F(s) \end{bmatrix} \tag{M13.1}$$

The outputs are the distillate (x_D) and bottoms (x_B) composition, in mole fraction of light component. The manipulated inputs are the reflux (L) and vapor boil-up (V) rates, in kilomoles per minute. The disturbance inputs are feed flow rate (F, kmol/minute) and feed light component mole fraction (zF). The unit of time is minutes. The steady-state compositions are 0.99 and 0.01 mole fraction light component for the distillate and bottoms product streams, respectively. The steady-state manipulated inputs (reflux and vapor flow rates) are 2.706 and 3.206 kmol/minute, respectively. The feed to the column is 1 kmol/minute, with a feed composition of 0.5 mole fraction light component and a feed quality of 1.

M13.2 Open-Loop Behavior

Response to Reflux Change

Here, we consider open-loop step changes in the manipulated inputs, with additional time delays of 2 minutes in the concentration measurements. The responses to step changes of ± 0.01 kmol/minute in reflux at $t = 10$ minutes, while keeping vapor boil-up constant, are shown in Figure M13–3. The response to a positive reflux flow change are shown as solid lines, while responses to a negative reflux change are shown as dashed lines; also, we have chosen to plot mole percent rather than mole fraction (the unit used in the transfer function matrix). The responses are symmetric because we are using a linear model. Naturally, the bottoms composition cannot decrease below 0, as obtained with a step decrease in reflux flow; this illustrates a clear limitation to the use of a linear model.

Response to Vapor Boil-Up Change

The reader should generate responses to step changes of ± 0.01 kmol/minute in the vapor boil-up rate (V) at $t = 10$ minutes (see Additional Exercise 1). Compare and contrast these responses with those of the reflux changes. Are these changes consistent with the transfer function matrix?

Response to Simultaneous Changes in Reflux and Vapor Boil-Up

It should be noted that the effect of the input "direction" is very important. Contrast the response to reflux (+0.005 kmol/minute) and vapor (-0.005 kmol/minute) changes that are in the "opposite" directions, with simultaneous positive (+0.005 kmol/minute) changes in reflux and vapor. The responses shown in Figure M13–4 are two orders of magnitude larger than in Figure M13–5, although the magnitude of the input changes are the same. The SVD analysis shown in Section M13.6 is used to explain these results. Remember that this is a linear model, based on the nominal steady-state conditions. These are not nonlinear effects; rather, they illustrate the important of "direction" in multivariable processes, even when the processes are linear.

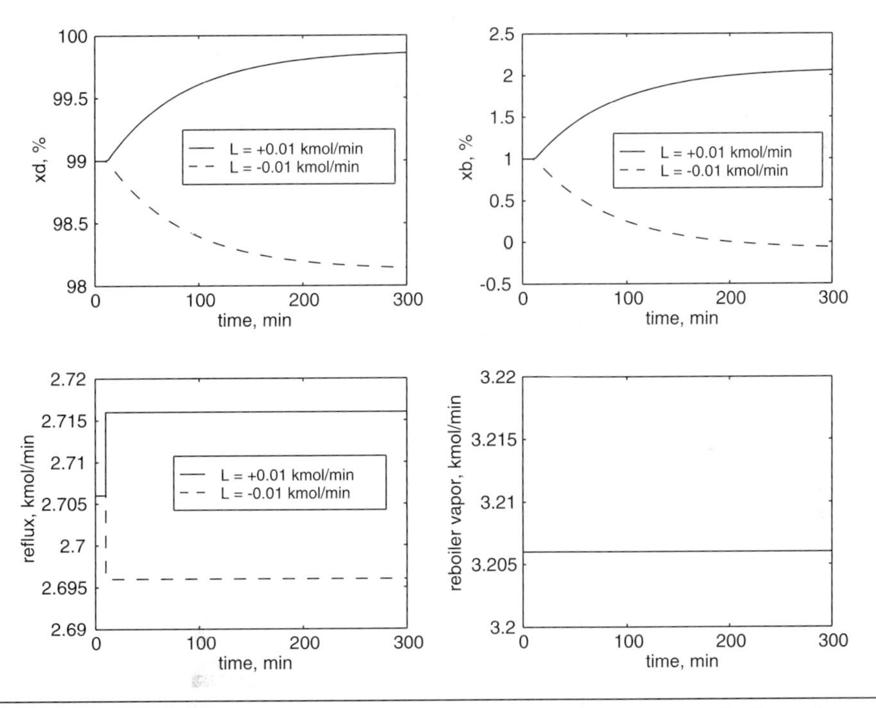

Figure M13–3 Open-loop step response. Change of ±0.01 kmol/minute in reflux flow rate.

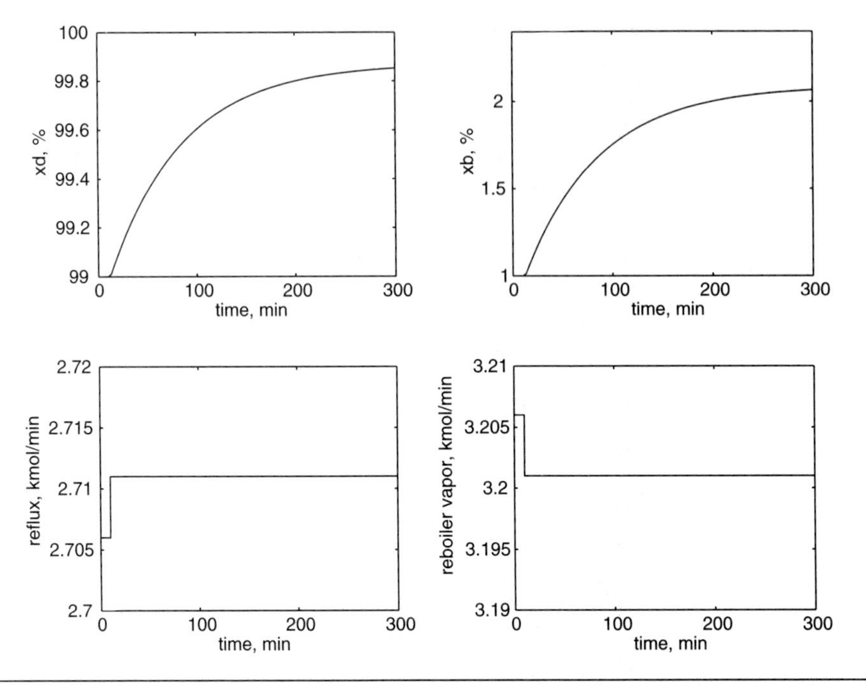

Figure M13–4 Open-loop response to a simultaneous +0.005 kmol/minute changes in reflux and –0.005 kmol/minute in vapor.

M13.3 SISO Control

The IMC-based PID procedure was used to design the x_D-L (loop 1) and x_B-V (loop 2) controllers, in each case assuming the "other loop" was open. Time delays were neglected in the control system design, resulting in PI controllers

$$k_{c1} = \frac{75}{0.878\lambda_1}, \frac{\text{kmol/min}}{\text{mol fraction}} \tau_{I1} = 75 \text{ min}$$

$$k_{c2} = \frac{75}{-1.096\lambda_2}, \frac{\text{kmol/min}}{\text{mol fraction}} \tau_{I2} = 75 \text{ min}$$

A first-order closed-loop response is achieved for each of these controllers operated independently (only one loop closed at a time). *The reader should verify, through simulation of the independent controllers, that IMC filter factors of 25 minutes lead to closed-loop time constants of roughly 25 minutes* (see Additional Exercise 2).

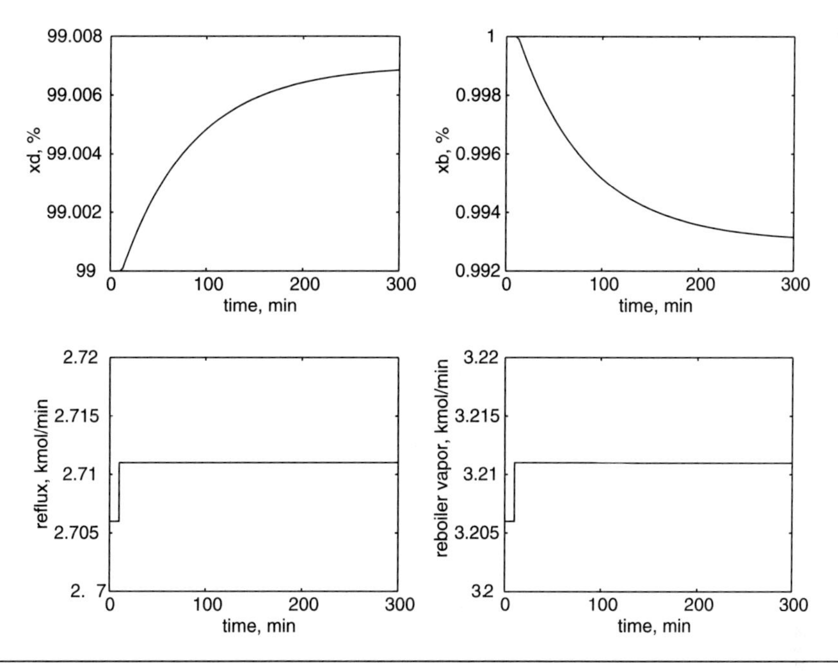

Figure M13–5 Open-loop response to simultaneous +0.005 kmol/minute changes in reflux and vapor.

M13.4 RGA Analysis

The RGA (Chapter 13) for this system is

$$\Lambda = \begin{bmatrix} 35 & -34 \\ -34 & 35 \end{bmatrix}$$

This RGA indicates that the x_D-L (distillate composition-reflux flow) and x_B-V (bottoms composition-reboiler vapor flow) input-output pairings are the only possible pairings, since pairing on a negative relative gain would lead to a failure-sensitive system. It also indicates that this system is sensitive to model uncertainty and that significant tuning parameter changes must be made (typically, the magnitude of the proportional gains must be increased) so that both loops can be closed simultaneously with good performance. Very often, with high relative gain systems, tuning parameters that work well with both loops closed will be unstable if one of the loops is opened (either by an operator or by the failure or saturation of a valve). With high relative gain systems, one must often decide to simply keep one loop under control and sacrifice the control of the less important output variable.

M13.5 Multiple SISO Controllers

Setpoint Changes

In Section M13.3, the composition control loops were tuned independently using the IMC-based PI technique, neglecting time delays. The control system was simulated assuming a 2-minute time delay on all measurements. The IMC filter factors (λ_1, λ_2) were both set to 25 minutes, one third of the process time constant. Here, we show the responses for both loops closed simultaneously. The response for a setpoint change of 1 mol% (0.01 mole fraction) in the distillate composition is shown in Figure M13–6. Note that the controllers are highly interacting and the response time is extremely slow (the outputs are still far from the setpoint after 1000 minutes).

It is intriguing that simultaneous setpoint changes of 0.625 mol% in the distillate and 0.78 mol% in bottoms composition lead to much more rapid closed-loop responses, as shown in Figure M13–7. Notice here that the time scale is 100 minutes, and that the desired setpoint has been achieved in 55 minutes.

Without any other information, one might assume that nonlinearities could be used to explain the difference in the system behavior displayed in Figures M13–6 and M13–7.

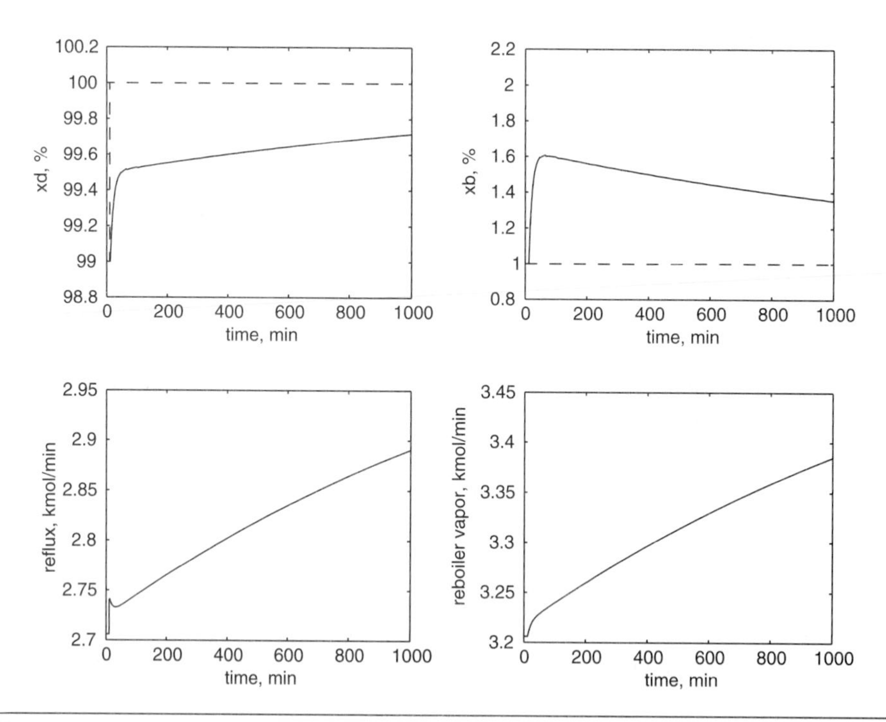

Figure M13–6 Setpoint change of $r = [1,0]$, mol% light comp. Note the long timescale.

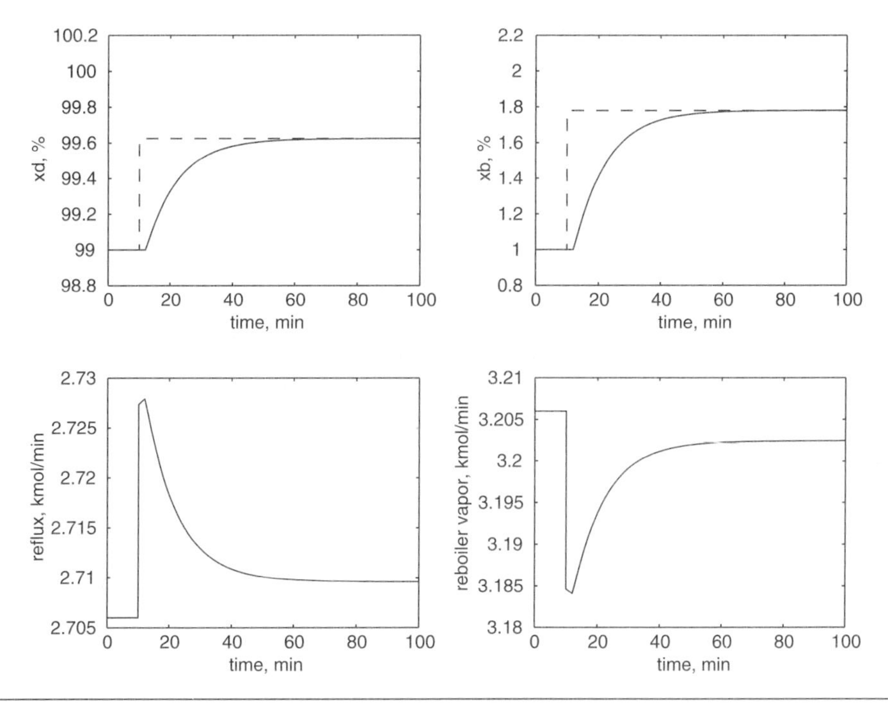

Figure M13–7 Setpoint change of $r = [0.625, 0.78]$, mol% light comp. Note the relatively short timescale compared to Figure M13–6.

These simulations, however, are based on the linear transfer function model. The differences observed are due to the "directional sensitivity" of multivariable systems. Singular value analysis will be used in Section M13.6 to better understand this behavior.

The reader should tune the controllers for tighter performance for the setpoint change in the "slow" direction (Figure M13–8). Decrease λ_1 and λ_2 (increase the magnitude of the proportional gains) and observe the behavior (Additional Exercise 3). Also, once tighter control has been achieved with both loops closed, consider the effect of opening one loop. You should observe that the other loop goes unstable (Additional Exercise 4).

Disturbance Rejection

The focus of the previous sections has been on setpoint changes. It is also very important for any control strategy to reject feed flowrate and feed composition disturbances. In Additional Exercise 9 you have the opportunity to perform simulations and discuss results for disturbance rejection.

M13.6 Singular Value Analysis

The SVD can be used to predict directional sensitivity of a process. The process gain matrix is decomposed into three matrices,

$$G = U\Sigma V^T$$

where U is the left singular vector matrix, Σ the diagonal matrix of singular values, ordered, and V the right singular vector matrix. The left and right singular vector matrices are both orthonormal matrices; that is, each column of the matrix is orthogonal to all other columns and the columns each are unit length. The diagonal singular value matrix is ordered so that the largest singular value is in the (1,1) position. *Note that the standard notation for SVD is to use U to represent the left singular vector matrix. Please do not confuse this with the u vector commonly used to represent the vector of manipulated inputs.*

When performing a singular value analysis, it is important to scale the inputs and outputs to cover the same range. For this system, we assume that we desire the output compositions to vary by only ± 0.01 mole fraction. Also, we assume that the inputs vary by ± 0.5 kmol/minute. The scaled process gain matrix is then

$$G_{scaled} = \underbrace{\begin{bmatrix} 100 & 0 \\ 0 & 100 \end{bmatrix}}_{\text{output scaling}} \underbrace{\begin{bmatrix} 0.878 & -0.864 \\ 1.082 & -1.096 \end{bmatrix}}_{\text{gain matrix}} \underbrace{\begin{bmatrix} 0.5 & 0 \\ 0 & 0.5 \end{bmatrix}}_{\text{input scaling}}$$

The MATLAB SVD analysis of this system is shown below.

MATLAB SVD Analysis

```
» g = [0.878 -0.864;1.082 -1.096];

» gs = [100 0;0 100]*g*[0.5 0;0 0.5]

gs =

   43.9000  -43.2000
   54.1000  -54.8000

» [u,s,v] = svd(gs)

u =

    0.6246   -0.7809
    0.7809    0.6246
```

```
s =

   98.6043         0
         0    0.6957


v =

    0.7066   -0.7077
   -0.7077   -0.7066

» cond(gs)

ans =   141.7320
```

The SVD of the scaled gain matrix is

$$G_{scaled} = \underbrace{\begin{bmatrix} 0.62 & -0.78 \\ 0.78 & 0.62 \end{bmatrix}}_{\text{left singular vector matrix}} \underbrace{\begin{bmatrix} 98.6 & 0 \\ 0 & 0.696 \end{bmatrix}}_{\text{singular value matrix}} \underbrace{\begin{bmatrix} 0.70 & -0.70 \\ -0.70 & -0.70 \end{bmatrix}^T}_{\text{right singular vector matrix}}$$

The condition number is

$$\frac{\sigma_{max}}{\sigma_{min}} = \frac{98.6}{0.696} = 142$$

which indicates that this is an ill-conditioned system. The first column of the left singular vector matrix indicates that the most sensitive output direction is a simultaneous change in the distillate and bottoms composition in the same direction. The first column of the right singular vector matrix indicates that the strongest input direction is to change reflux and vapor boil-up by the same magnitude, but in different directions (increase reflux and decrease vapor boil-up, or vice versa). Physically, this is because these types of changes have a greater effect on the overall material balance around the column.

We can view these SVD results to understand both open-loop and closed-loop effects.

Open-loop. An input in the most sensitive direction will have a large-magnitude effect on the outputs. This means that an input in the most sensitive direction has a high "gain" effect on the output; that is, a small input change causes a large output change. Similarly, an input in the least sensitive direction (column 2 of the V matrix) has very little effect on the output. These effects are illustrated by the open-loop responses in Figures M13–4 and M13–5. Figure M13–4 is a forcing in the most sensitive input direction, while Figure M13–5 is a forcing in the least sensitive direction.

Closed-loop. A desired output change in the most sensitive output direction will require an input in the most sensitive input direction, and these are associated with the largest singular value. This means that a setpoint change in the most sensitive output direction will not require as large a magnitude input change as a setpoint change in the least sensitive output direction. It should be noted that the response shown in Figure M13–7 is based a setpoint change in the most sensitive output direction, which requires little manipulated variable action. The reader should show that a setpoint change in the weakest direction will yield a slow response, similar to Figure M13–6.

M13.7 Nonlinear Effects

The results presented in previous sections are purely linear phenomena. In addition, distillation columns have important nonlinear effects. For example, even an infinite reflux rate (requiring an infinite vapor boil-up rate) will yield a mole fraction of the light component in the distillate that is less than 1.0. If the steady-state value of the distillate composition is 0.99, it can increase by less than 0.01, yet it can decrease to almost as low as the feed composition of 0.5. The range for increasing purity is then much smaller than the range for decreasing purity.

Figures M13–8 and M13–9 show small open-loop changes in the manipulated inputs. The reader should perform additional small open-loop changes to verify that the outputs do not change much when the purity is increasing but can change tremendously when the purity is decreasing (Additional Exercise 5).

Because of nonlinear effects, a controller designed at setpoint may not operate well when the setpoint is significantly changed. The reader is encouraged to perform SISO control simulations (Additional Exercise 6) and multiple loop simulations (Additional Exercise 7).

M13.8 Other Issues in Distillation Column Control

Distillation is probably the most widely studied multiloop control problem, with many papers and books published on the topic. In this chapter, we have focused on "conventional control," where reflux flow rate is manipulated to control distillate composition, and vapor boil-up (related to reboiler duty) is manipulated to control bottoms composition. The pressure and level loops are assumed to be perfectly controlled. An alternative control structure, often used on high-purity columns, is known as material balance control. In material balance control, distillate flow rate is manipulated to control distillate composition. (See Additional Exercise 8 for a study using material balance control.) In practice,

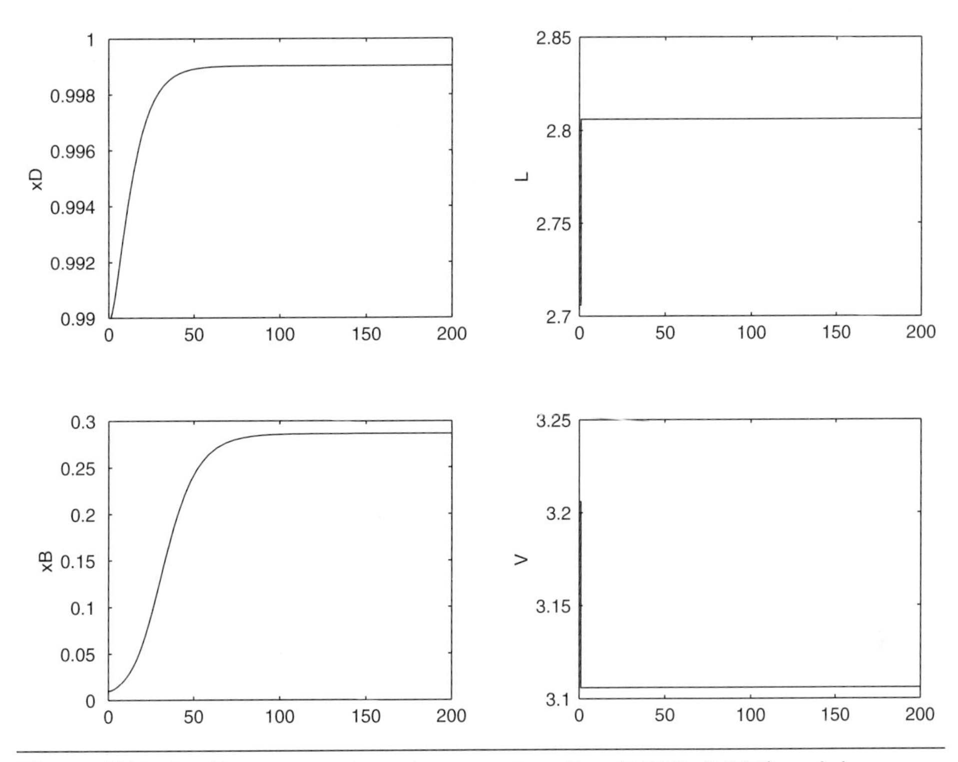

Figure M13–8 Responses to an increase in reflux (2.706–2.806) and decrease in vapor (3.206–3.106) rates.

for this strategy to work, the distillate level must be very tightly controlled by manipulation of the reflux flow rate.

In this chapter we have assumed that the pressure and level loops are perfectly controlled. In Figures M13–1 and M13–2 we used a simplified representation for the pressure control loop, where the pressure was controlled by manipulating the condenser cooling water flow rate. In practice, there are many different ways to control column pressure. (See the references section for articles discussing various pressure control methods.)

The multivariable interactions, particularly in high-purity distillation, make it very difficult to control the distillate and bottoms compositions simultaneously. More often, the important product is under feedback control, while the less important product is manually controlled. Alternatively, the controller for the most important product could be tightly tuned, while the other composition loop is loosely tuned.

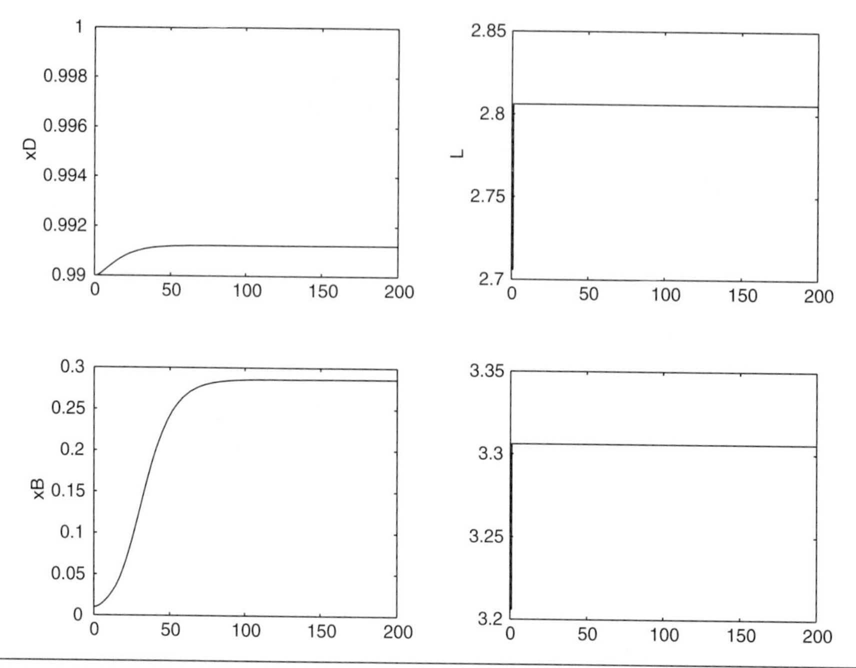

Figure M13–9 Responses to a increase in reflux (2.706–2.806) and vapor (3.206–3.306) rates.

M13.9 Summary

In this chapter, we have focused on "conventional control" of distillation columns, where reflux is manipulated to control distillate composition and reboiler vapor boil-up rate is manipulated to control bottoms composition. The RGA indicates that this is the only satisfactory pairing for this strategy. We have found that this process is very sensitive to the multivariable *direction* of the manipulated inputs (open loop), or the setpoints (closed loop). The SVD is a useful mathematical tool for finding the strongest and weakest directions for a multivariable process.

References

The distillation model is presented in Bequette (1998), based on a column studied by Morari and Zafiriou (1989) and Skogestad and Postlethwaite (1996).

Bequette, B. W., *Process Dynamics. Modeling, Analysis and Simulation*, Prentice Hall, Upper Saddle River, NJ (1998).

Morari, M., and E. Zafiriou, *Robust Process Control*, Prentice Hall, Englewood Cliffs, NJ (1989).

Skogestad, S., and I. Postlethwaite, *Multivariable Feedback Design*, Wiley, New York (1996).

Sigurd Skogestad has published a nice series of articles that discuss the dynamics and control of distillation columns. The following review paper can be found at his web site (www.kjemi.unit.no/~skoge): Skogestad, S., "Dynamics and Control of Distillation Columns. A Tutorial Introduction," paper presented at Distillation and Absorption '97, Maastricht, Netherlands (8–10 September, 1997).

Other papers of interest include the followng:

Skogestad, S., and M. Morari, "Understanding the Dynamic Behavior of Distillation Columns," *Ind. Eng. Chem. Res.*, **27**(10), 1848–1862 (1988).

Skogestad, S., and M. Morari, "Control Configuration Selection for Distillation Columns," *AIChE J.*, **33**(10), 1620–1635 (1987).

For a discussion of the many practical issues in distillation control, see the following:

Shinskey, F. G. *Distillation Control*, 2nd ed., McGraw-Hill, New York (1984).

There are many ways to regulate the pressure in a distillation column. Two nice articles are as follows:

Chin, T. G., "Guide to Distillation Pressure Control Methods," *Hydrocarbon Processing*, **86**(10), 145–153 (1979).

Sloley, A. W., "Effectively Control Column Pressure," *Chem. Eng. Prog.*, **97**(1), 38–48 (2001).

Additional Exercises

1. Generate responses to step changes of ± 0.01 kmol/minute in the vapor boil-up rate (V) at $t = 10$ minutes. Compare and contrast these responses with those of the reflux changes shown in Figure M13–3. Are these changes consistent with the transfer function matrix?
2. Consider the SISO controllers. The reader should verify through simulation that the independent controllers with IMC filter factors of 25 minutes lead to closed-loop time constants of roughly 25 minutes.
   a. First, close the x_D-L loop and make a setpoint change of 0.005 mole fraction.
   b. Next, close the x_B-V loop (with x_D-L open) and make a setpoint change of -0.005 mole fraction. In each case, show the change of both compositions.
3. Consider the case where both SISO controllers are closed simultaneously. Tune the controllers tighter for the setpoint change in the "slow" direction (Figure M13–8). Decrease λ_1 and λ_2 (increase the magnitude of the proportional gains) and observe the behavior.

4. Using the tuning parameters found in Additional Exercise 3, study the response to a setpoint change if one of the loops is "opened." You should find that the other loop goes unstable or has oscillatory performance.
5. Now perform open-loop step changes on the nonlinear model. Show that the purity never changes much when it is increasing but can exhibit large changes when decreasing.
6. Perform closed-loop SISO studies on the nonlinear system. For small setpoint changes (say, from 0.990 to 0.991 on the distillate composition), the results are similar to the linear case. For large setpoint changes (say, from 0.990 to 0.999), the differences can become more substantial.
7. Perform closed-loop studies on the nonlinear system with both loops closed. For small setpoint changes in the most sensitive directions (say, from 0.9900 to 0.9906 on distillate and from 0.0100 to 0.0108 on bottoms), the results are similar to the linear case.
8. Consider a so-called *material balance* control strategy, where distillate flow is manipulated to control distillate composition. Assume the same scalings for the inputs and outputs as for the conventional control strategy. Perform an SVD analysis on the scaled gain matrix to determine the most and least sensitive setpoint change directions. Compare and contrast closed-loop simulations with those for the conventional control strategy. The transfer function model for a material balance control strategy (where D is the distillate flow rate) is

$$\begin{bmatrix} x_D(s) \\ x_B(s) \end{bmatrix} = \begin{bmatrix} \dfrac{-0.878}{75s+1} & \dfrac{0.014}{75s+1} \\ \dfrac{-1.082}{75s+1} & \dfrac{-0.014}{75s+1} \end{bmatrix} \begin{bmatrix} D(s) \\ V(s) \end{bmatrix}$$

9. The focus of the previous problems has been setpoint tracking. In this problem consider changes of $\pm$ 0.1 kmol/min in the feed flow rate and $\pm$ 0.05 mole fraction in the feed composition. Base your simulations on the matrix transfer function model presented in Section M13.1. Compare and contrast closed-loop performance when only single loops are closed vs. both loops closed.

Appendix M13.1

The SIMULINK .mdl diagram for the open-loop transfer function model is shown in Figure M13–10.

The nonlinear model of the open-loop system is shown in Figure M13–11.

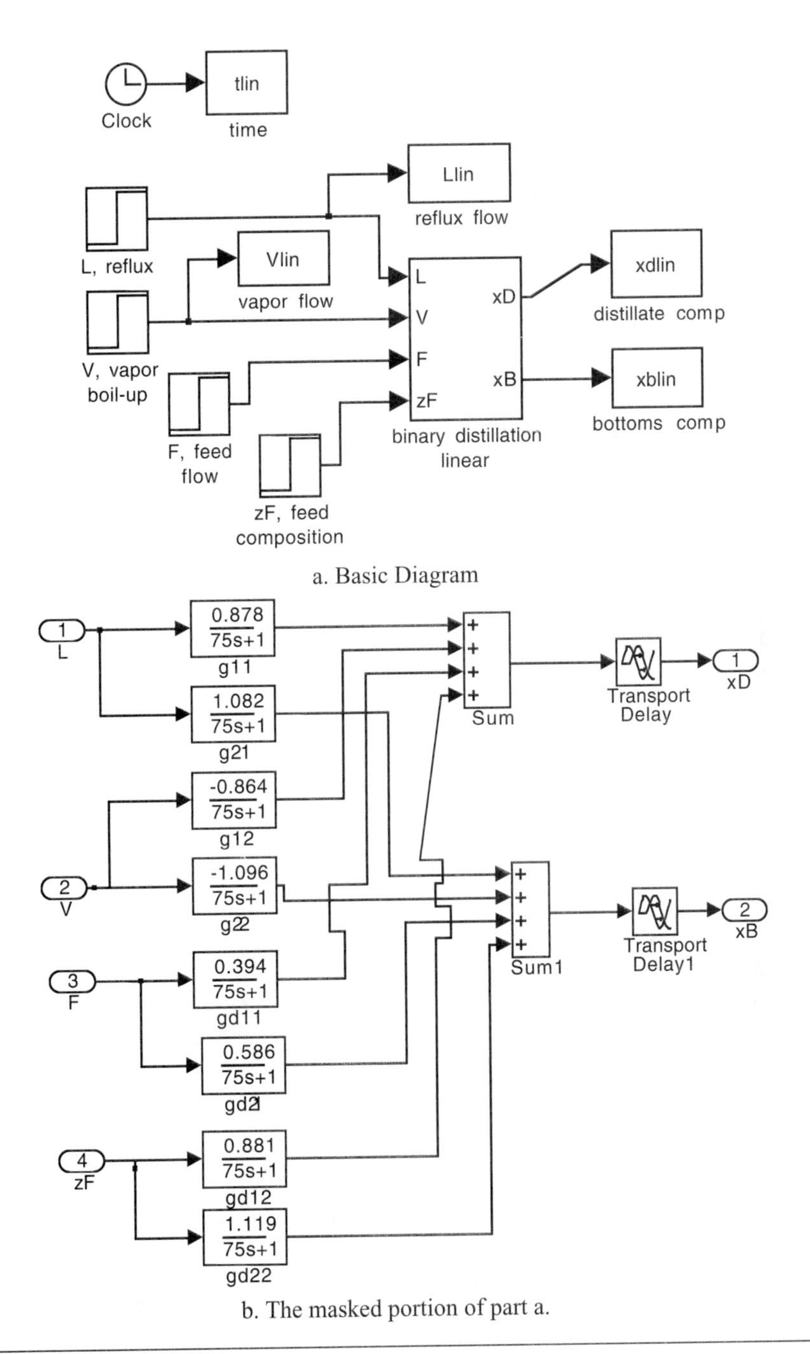

Figure M13–10 SIMULINK diagram for linear model. `dist_OLtf.mdl`.

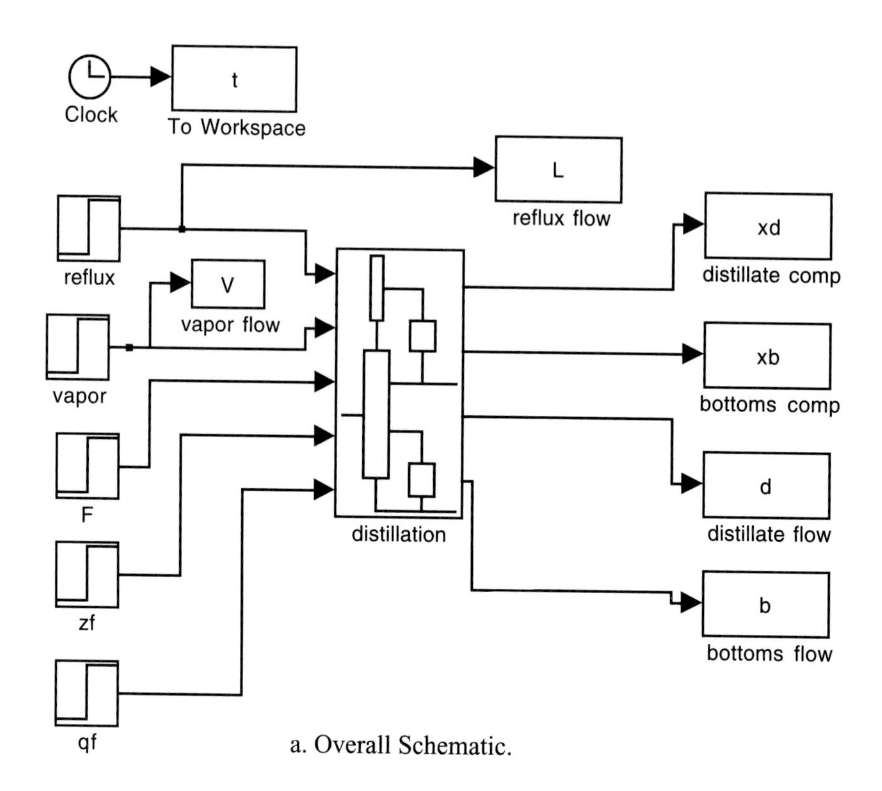

a. Overall Schematic.

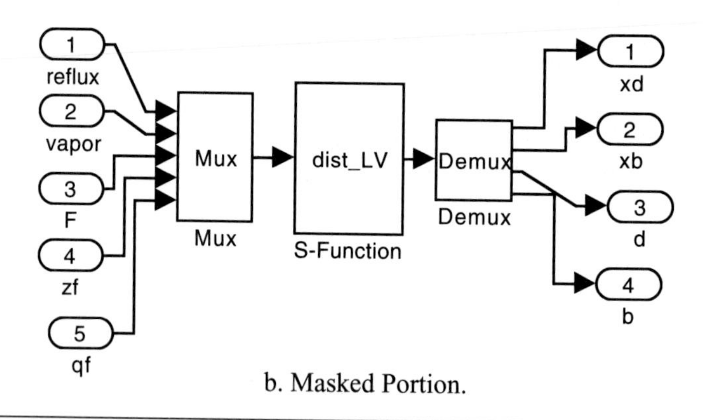

b. Masked Portion.

Figure M13–11 SIMULINK diagram for open-loop nonlinear system. `dist_blk.mdl`.

MODULE 14

Case Study Problems

The purpose of this module is to present some realistic multivariable process control problems that require the integration of a number of different concepts and techniques presented in the text. SIMULINK files for each of these processes can be downloaded from the text web page. The various case studies are as follows:

A partial solution for the drug infusion case study is presented on the textbook web page.

M14.1 Background

The development and implementation of an industrial control strategy is complex and involves a number of steps. A goal of this module is to present a number of interesting important process control problems that require the synthesis of concepts presented in this textbook. One use of this module is for a special project studied during the final few weeks of a typical undergraduate process control course. Students, often working in small groups, select one of the challenge problems discussed in this module. Here we discuss some of the critical steps of the control system development.

It is important to consider the control objectives for a particular process; in addition, it one should consider previously developed strategies and understand how successful they were/are. Sometimes this information is available from a currently operating plant in the corporation. In other cases, a literature search can provide important background material. The next step is to develop a process model, either from first principles, or from testing an existing process. In this module we assume that the step testing approach is being used; since we do not have direct access to an operating plant, SIMULINK modules are used to mimic plant behavior; these can be downloaded from the textbook web page. Models can then be used to develop SISO control loops, and to understand/predict possible multivariable interaction problems when all loops are closed. If the closed-loop performance of multiple SISO loops is not satisfactory, then multivariable techniques can be developed.

When working on these problems, use the following steps and answer the following questions:

- What types of control strategies and objectives have been used on similar processes?
- What types of input tests should be performed to develop models for control system design? Do the responses of the models adequately match the actual plant responses?
- Do the orders of magnitudes of the model gains, time constants, and time delays make physical sense?
- If using independent SISO control loops (MVSISO), what pairings are suggested by the RGA analysis?
- If an SISO controller is tuned based on the other loops being open, does the tuning need to change when all the other loops are closed?
- Should a decoupling (steady-state or dynamic) type of control strategy be used for real (MIMO) multivariable control? What about multivariable IMC?
- Can you think of any other measured or manipulated variables that could be used?
- What are the important process disturbances?

When tuning the independent SISO loops (assuming the other loops are open), it is probably best to use a procedure that you are familiar with, such as IMC-based PI. You will generally use a closed-loop time constant (λ, the IMC filter factor) of roughly one half of the open-loop time constant (if this satisfies the minimum requirements for λ based on the dead time).

When testing controller performance for these multivariable processes, it is probably best to initially make a step change in one setpoint while keeping all other setpoints constant (for example, change r_1 without changing r_2). Do this for each loop. Compare how a setpoint change in one output variable affects the other output variables. Also consider the manipulated variable responses.

M14.2 Reactive Ion Etcher

An important process in semiconductor device manufacturing is plasma etching (or reactive ion etching). In this unit operation, reactive ions are used to selectively remove (etch) layers of solid films on a wafer. A major objective of a control system is to achieve relative etch rates (angstroms/second) by manipulating the process inputs. Currently, etch rate measurements are not readily available, so it is assumed that by measuring and controlling other variables, good control of the etch rate can be obtained.

Two variables important to control are the voltage bias of the plasma and the fluorine concentration, since they ultimately determine the etch rate. Two important manipulated variables are the RF power and the outlet valve position (throttle). A schematic of a reactive ion etcher is shown in Figure M14–1.

A reactive ion etcher has power and throttle position inputs (P and T) and voltage (V_{bias}) flourine concentration (F) as outputs.

$$u(s) = \begin{bmatrix} P(s) \\ T(s) \end{bmatrix} \qquad y(s) = \begin{bmatrix} V_{bias}(s) \\ F(s) \end{bmatrix}$$

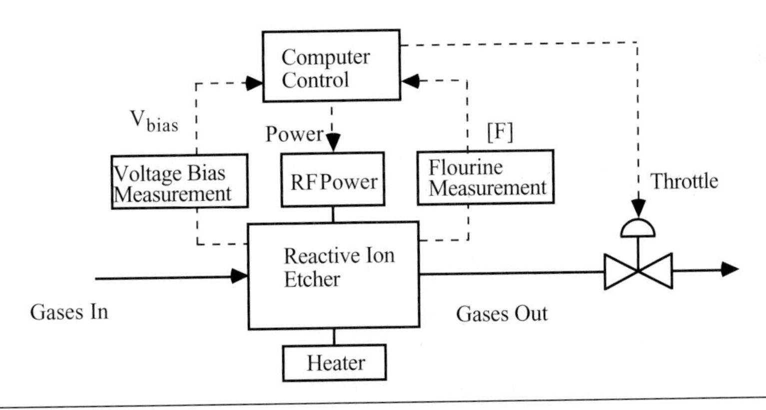

Figure M14–1 Reactive ion etcher.

Questions to Answer

Are fundamental (first principles) models of plasma etchers easy to develop?

Will fundamental models or input/output tests (i.e., step tests) normally be used for control system design?

The nominal and range of operating conditions for this system are as follows:

- Inputs: power 1000 W, range from 0 to 2000 W; throttle, 50%, range from 0 to 100%.
- Outputs: voltage, 250 V; fluorine, 50% of range.

Disturbances of up to ±25 W in power are possible.

A SIMULINK block diagram for a reactive ion etcher is shown in Figure M14–2. The inputs and outputs of the etcher are in physical (not deviation) variables.

M14.3 Rotary Lime Kiln Temperature Control

Lime kilns are used in the paper industry to convert lime mud to lime. A schematic control instrumentation diagram for a lime kiln is shown in Figure M14–3. Lime mud enters the "back" or the "cold" end, while the lime product exits the "front" or "hot" end. The kiln rotates at approximately one revolution per minute and is inclined so that the mud flows from back to front. The quality of the lime depends on the temperature profile along the kiln. Typical temperatures are 2250°F at the front end and 425°F at the rear end, with nominal damper positions of 50% of range. The inputs are fuel gas flow rate and the damper position, and the outputs are the front and rear temperatures.

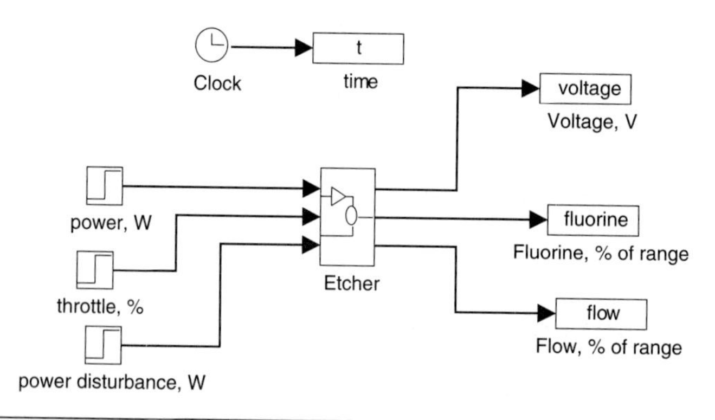

Figure M14–2 SIMULINK diagram for reactive ion etcher.

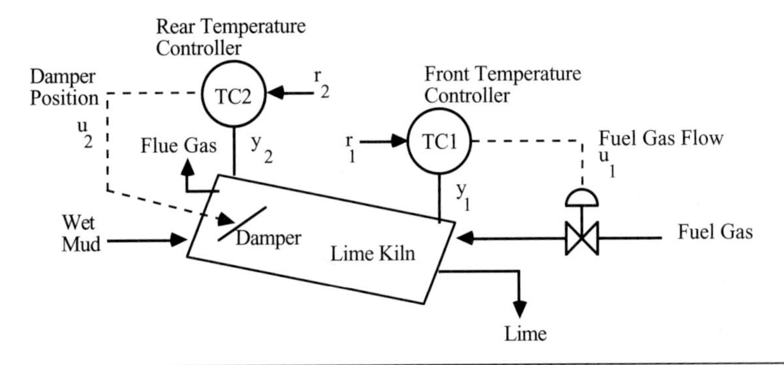

Figure M14–3 Rotary lime kiln temperature control. The manipulated variables are fuel gas flow rate (u_1) and damper position (u_2). The measured variables are front temperature (y_1) and rear temperature (y_2).

M14.4 Fluidized Catalytic Cracking Unit

A fluidized catalytic cracking unit (FCCU) is used to produce the majority of gasoline in an oil refinery. A simplified process schematic and instrumentation diagram is shown in Figure M14–4. Heavy oil feedstock from the crude unit is mixed with recirculating catalyst and reacted in a riser tube. The catalyst cracks the oil to form lighter hydrocarbons. The lighter hydrocarbon products are separated from the catalyst in the reactor (which is actually a separator; it is called a reactor for historical purposes). The spent catalysis (which contains a great deal of carbon) is sent to the regenerator, where partial combustion is used to remove the carbon from the catalyst. Regenerated catalyst (manipulated variable) is recirculated back to mix with the inlet feed oil from the crude unit. The important measured variables are reactor (separator) temperature (T_1) and the regenerator gas temperature (T_{cy}). The manipulated variables are the catalyst recirculation rate (F_s) and regenerator air rate (F_a).

The steady-state input values are 294 kg/s (F_s), 25.35 kg/s (F_a) and the output values are 776.9°F (T_1) and 998.1 (T_{cy}), the regenerator temperature (T_{rg}) is 965.4°F. Assume that the manipulated inputs can vary by ± 50% of the nominal steady-state values.

M14.5 Anaerobic Sludge Digester

Anaerobic sludge digesters are used to degrade compounds (in waste water that is high in suspended solids) to carbon dioxide and water. A schematic of an anaerobic digester is shown in Figure M14–5.

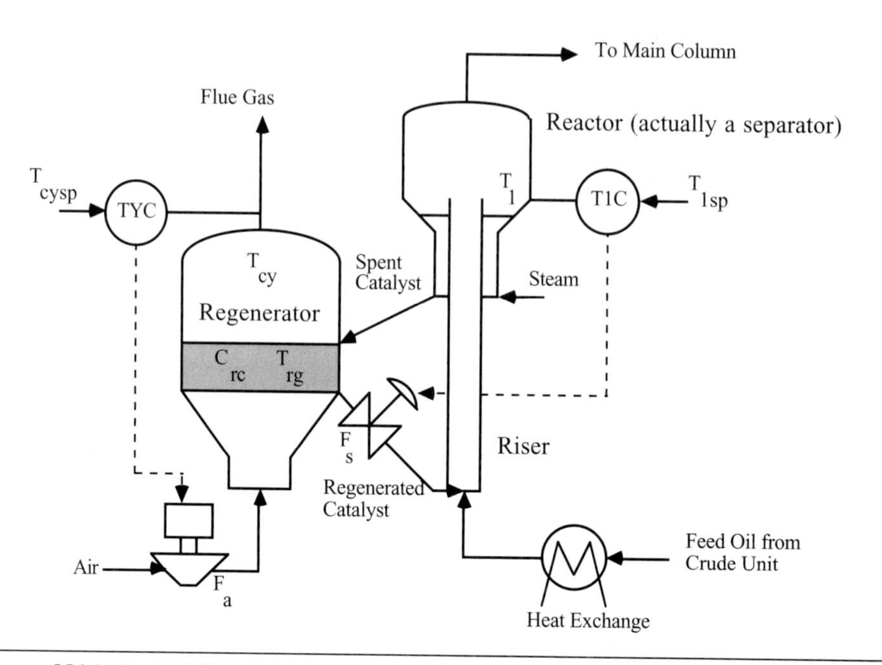

Figure M14–4 FCCU control strategy. F_s and F_a are manipulated. T_1 and T_{cy} are measured.

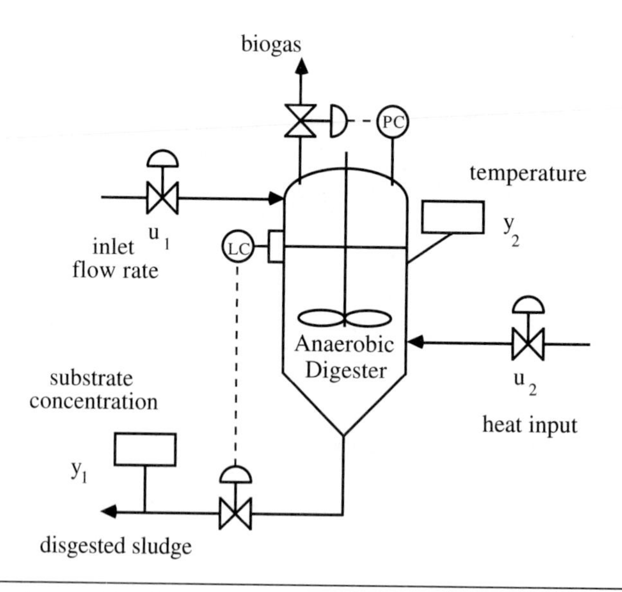

Figure M14–5 Anaerobic digester. The level and pressure controllers are assumed to be perfect.

The time unit is days, and the inputs and outputs have been scaled to cover the same ranges. The outputs and inputs are y_1 = substrate concentration, y_2 = temperature, u_1 = inlet flow rate, and u_2 = heat addition rate (°C/day). The steady-state inlet flow rate is 300 m^3/day, the heat addition rate is 10°C /day. The steady-state temperature is 55°C and substrate concentration is 300 mg/liter.

Besides considering setpoint responses, assume that disturbances in the inlet flow rate occur.

M14.6 Drug Infusion System

Critical care patients have often suffered a "disturbance" to the normal operation of their physiological system; this disturbance could have been generated by surgery or some sort of trauma (e.g., a heart attack). A responsibility of the critical care physician is to maintain certain patient outputs within an acceptable operating range. Two important outputs to be maintained are mean arterial pressure (*MAP*) and cardiac output (*CO*). Often the anesthesiologist will infuse several drugs into the patient in order to control these states close to the desired values. A conceptual diagram is shown in Figure M14–6.

The goal of this control system design is to manipulate the flow rate of two drugs, dopamine (*DPM*) and sodium nitroprusside (*SNP*), to maintain the two outputs (*MAP* and

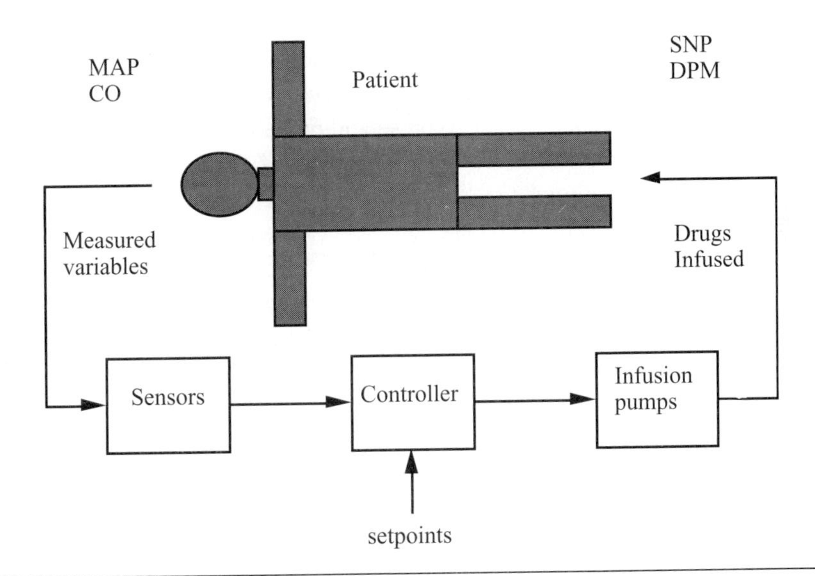

Figure M14–6 Drug infusion control.

CO) at their desired setpoints. A successful implementation of such a strategy allows the anesthesiologist to spend more time monitoring other patient states, such as "depth of anesthesia."

The steady-state (initial) drug infusion rates are both 0, while the steady-state (initial) mean arterial pressure is 120 mmHg and cardiac output is 130 ml/(kg min). A partial solution to this problem is presented on the textbook web page. The maximum flow rates for both drugs is 10 mg/(kg min).

M14.7 Suggested Case Study Schedule

It is suggested that these case studies be assigned during the last 4–5 weeks of a typical one-semester course. The suggested time frame is realistic enough to allow for the study of other modules and for short homework assignments during this period.

1. Literature review and motivation for problem	1 week
2. Model development (step tests)	2 weeks
3. SISO controller design	3 weeks
4. MVSISO controller design	4 weeks
5. Oral presentation	4–5 weeks
6. Written report	4–5 weeks

Each problem (unit operation) will have an "advisor" who will answer questions that you have regarding the problem and possible solutions techniques. After you turn in part 1, you will be assigned an advisor. You should probably plan to meet with the advisor once per week, during office hours, during the project period.

An example partial solution to a biomedical engineering control problem (drug infusion control) is presented on the web page.

1. *Literature review and motivation for problem (due after 1 week)*
Write a short memo (two pages or fewer), giving the following information:
- The example that you have decided to study.
- The importance of the unit operation to the particular industry. What are typical sizes or production rates?
- A literature search with references related to control of the unit operation. *A web-based search (using Google or some other search engine) is unacceptable.* We are interested in sources published in books, journals, magazines,

or conference proceedings. Appropriate databases include Current Contents and the EI Compendex.

The SIMULINK files for each problem are available on the textbook web page. You will apply step inputs to identify the process transfer functions for the next phase of the project.

2. *Model development (due after 2 weeks)*
   Write a short memo (one page or under) summarizing your modeling results. Attach the following information to the memo:
   - The transfer function matrix relating each input to each output
   - Comparison of open-loop step tests with model step responses (superimpose the model and plant responses)
3. *SISO controller design (due after 3 weeks)*
   Write a short report summarizing the control system design for each independent SISO controller. Please include the following:
   - A description of the design procedure (IMC-based PID, Ziegler-Nichols, direct synthesis, etc.) and the tuning parameters used.
   - Closed-loop responses for each independent loop. It is appropriate to demonstrate the effect of different values for the tuning parameters.
   - How does a setpoint change in a controlled loop affect the uncontrolled loop?
4. *MVSISO controller design (due after 4 weeks)*
   - The RGA analysis: Based on the RGA, suggest how the variables should be paired in a MVSISO control structure (i.e., y_1 paired with u_1 and y_2 paired with u_2).
   - The suggested PI or PID tuning parameters that you would use if the loops did not interact: For example, what tuning parameters would you recommend for loop 1 if loop 2 was open; you may use IMC-based PI or PID, Cohen-Coon, Ziegler-Nichols, Direct Synthesis, or frequency-response-based tuning (satisfying certain gain and phase margins).
   - A couple of plots: One for the setpoint response of loop 1, with loop 2 open; another for the setpoint response of loop 2, with loop 1 open.
   - Do you suspect that the control loops must be detuned if all loops are closed? Use the RGA to assist you with this.
   - Are there dynamic reasons that you might want not pair outputs and inputs based on the RGA?
   - What are the practical limits to the magnitude of setpoint changes?

5. *Oral presentation (performed after 4–5 weeks)*
   Prepare a short (fewer than 15 minutes) oral presentation with overhead transparencies. Each group member should make an equal contribution. The problem advisor and at least one other consultant will observe the presentation and ask questions.
6. *Final Report (due after 4–5 weeks)*
   Write a short final report including the following:
   a. A title page, giving a list of the group members. Include a one paragraph *abstract* that summarizes the results presented in the report.
   b. A short *background* section describing the problem. Include literature references.
   c. Summarize the model developed in the preliminary reports.
   d. A short section with the preliminary *pairing* selection based on RGA analysis.
   e. A *single-loop* results section giving the setpoint responses of each control loop separately (i.e., setpoint response for loop 1 assuming loop 2 is open).
   f. A *MVSISO* results section for the setpoint responses of each control loop. In this, you will show the results of a setpoint change in loop 1 with loop 2 closed. Did you need to detune either loop?
   g. Disturbance rejection results. If there are no disturbance transfer functions, assume disturbances to the inputs.
   h. A *recommendations* section. Do you feel that MVSISO control is sufficient? Do you feel that a multivariable technique such as decoupling should be used? Do you feel that the system needs to be studied further using more advanced techniques?
   i. The previous reports (with the advisor comments) placed in the appendix of the report.

M14.8 Summary

We have presented module of case study example problems. The simulation files can be downloaded from the textbook web page, allowing a complete, realistic control system study. This includes model development, SISO controller design, MVSISO design (including retuning of controllers and consideration of failure tolerance). Additional case study problems will be added to the web page.

Additional Exercises

For each of the case studies, the following additional studies can be performed, depending on the interests of the students and instructor:

Implement antireset windup techniques (Chapter 11)

Implement digital control (Module 16)

Provide SVD and operability analysis (Chapter 14)

Implement decoupling (Chapter 14)

Implement multivariable IMC (Chapter 14)

Implement multivariable MPC using the MPC Toolbox

MODULE 15

Flow Control

The objective of this module is to discuss some of the practical issues in flow control. After studying this module, the reader should be able to do the following:

- Determine the process gains for typical pieces of instrumentation
- Understand how to determine the proportional band of a controller
- Discuss the flow characteristics of equal percentage, quick-opening, and linear control valves and understand when each should be used
- Realize that control valves should be specified as fail-open or fail-closed

The sections of this module are as follows:

M15.1 Motivating Example

Consider the control schematic diagram for the flow control loop shown in Figure M15–1. In practice, this loop contains a number of components, as detailed in Figure M15–2.

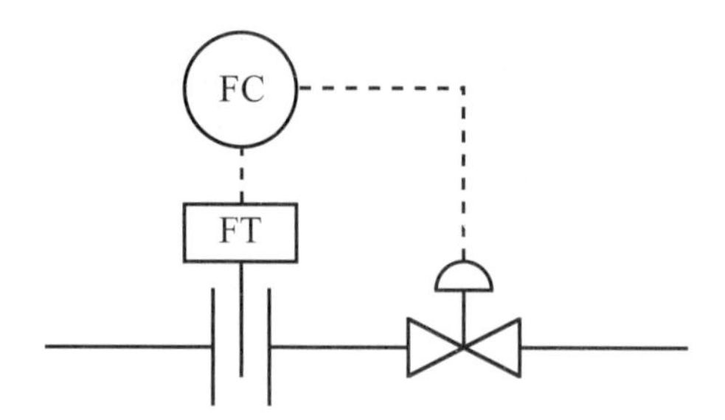

Figure M15–1 Schematic instrumentation diagram for a flow controller.

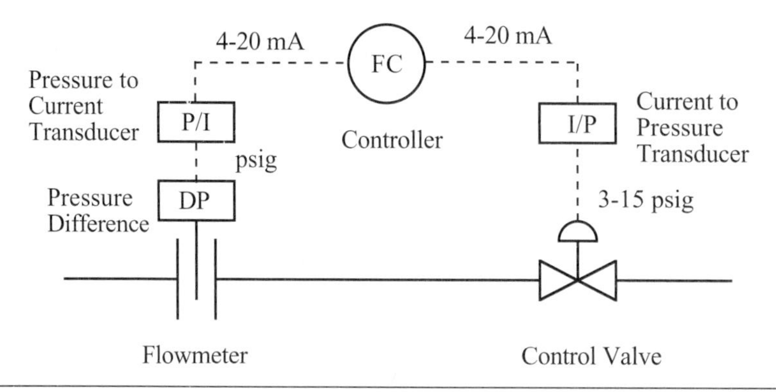

Figure M15–2 More detailed instrumentation diagram for a flow controller.

We can see that this loop consists of the following components.

- Flowmeter—often an orifice plate meter. The volumetric flow rate is proportional to the square root of the pressure drop across the orifice plate
- DP (differential pressure) cell—measures the pressure drop across the orifice plate
- P/I (pressure to current) transducer—converts the pressure signal to a 4- to 20-mA current signal
- Controller—the input to the controller is a 4- to 20-mA current signal. The output of the controller is a 4- to 20-mA current signal
- I/P (current to pressure) transducer—converts the 4- to 20-mA signal to a 3- to 15-psig pressure signal
- Control valve actuator—translates the 3- to 15-psig pressure signal to a valve position

Notice that this control loop is based on having an electronic analog controller. In practice, there are still a number of pneumatic analog controllers, where the input and output signals are pressure signals (typically 3–15 psig). Also, more of the "field equipment" (flow transmitters, etc.) are microprocessor-based, so often the signals to and from the controller are digital in nature.

Now we further analyze each component of instrumentation in terms of its transfer function.

Flowmeter

We can think of the input to the flowmeter as the actual flow rate of the process fluid. The output of the flowmeter is the pressure drop across the orifice plate. Here, we determine the gain of the flowmeter by simply considering the minimum and maximum flows and how they relate to the pressure drop

$$k_{fm} = \frac{\text{change in pressure drop}}{\text{change in flow rate}}$$

Consider a flowmeter where the flow rate varies from 0 to 5 gpm, while the pressure drop varies over a range of 0–2 psig. The flowmeter gain is then

$$k_{fm} = \frac{2 \text{ psig}}{5 \text{ gpm}}$$

The dynamics of flowmeters are so fast compared with the process dynamics that we can normally neglect the time constant and assume that the transfer function is represented as a static gain. Also, in practice the gain is *not constant* because the pressure drop is a nonlinear function of the flow rate. This is discussed in depth in Section M15.2.

DP Cell

The DP cell measures the pressure drop across the orifice plate. We can think of the gain as being part of the flowmeter gain. The dynamics of differential pressure cells are so fast compared with the process dynamics that we can normally neglect the time constant and assume that the transfer function is represented as a static gain.

Pressure to Current Transducer

The input to the P/I transducer is the pressure drop across the orifice plate. Assume that input to the P/I transducer is scaled so that 0–2 psig is 0–100% of the input range. The range of the output is 4–20 mA, so the gain for the P/I transducer is

$$k_{P/I} = \frac{20-4 \text{ mA}}{2-0 \text{ psig}} = 8 \text{ mA/psig}$$

Again, the dynamics are so fast that they are normally neglected.

Controller—Concept of Proportional Band

The input to the controller ranges from 4 to 20 mA and the output of the controller ranges from 4 to 20 mA. This might lead us to believe that the controller gain is one. We must recognize, however, that the controller gain, k_c, is a tuning parameter. Remember that the proportional gain does not act on the actual input to the controller; it acts on the error [the difference between setpoint and the process measured variable (input to the controller)].

$$k_c = \frac{\text{change in controller output}}{\text{change in error}}$$

Often a controller uses *proportional band* (PB) as a tuning parameter, rather than proportional gain. The PB is defined as the range of error that causes the controller output to change by the full range (100%). We see that relationship can be written

$$\text{PB} = \frac{\Delta e}{\Delta u} \cdot 100 \tag{M15.1}$$

For example, if a change in error (Δe) of 10% causes a 50% change in controller output (Δu), then the PB is 20.

We recognize that the controller gain we are familiar with can be written

$$\Delta u = k_c \Delta e \tag{M15.2}$$

Notice also that Equation (M15.1) can be written

$$\Delta u = \frac{100}{\text{PB}} \cdot \Delta e \tag{M15.3}$$

Comparing Equations (M15.3) and (M15.2), we see that

$$k_c = \frac{100}{\text{PB}}$$

which leads to the following definition of PB:

$$\text{PB} = \frac{100}{k_c} \tag{M15.4}$$

We have always used a simplified representation for a controller for analysis purposes. In practice, a controller performs the computations shown in Figure M15–3.

We see that the controller transfer function operates on signals that are scaled from 0 to 100% of range. The following scaling factors are used:

- S_{ei} converts electronic signal to physical units (e.g., gpm/mA)
- S_i converts engineering units to percentage of range of measured variables
- S_o converts percentage of range of controller output to control signal

Current-to-Pressure Transducer

The input to the I/P transducer is a 4- to 20-mA signal and the output is a 3- to 15-psig signal. The I/P gain is then

$$k_{I/P} = \frac{15-3\text{ psig}}{20-4\text{ mA}} = 0.75\text{ psig/mA}$$

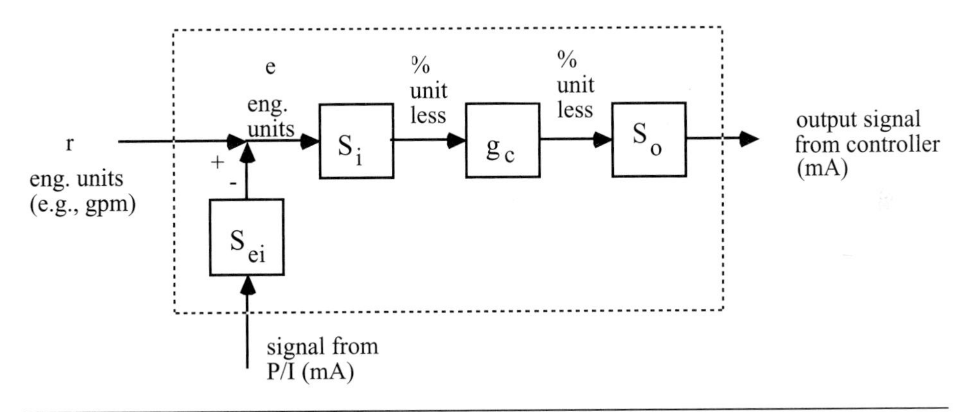

Figure M15–3 Schematic of computations and scaling performed by a controller.

Control Valve

Control valves will be covered in more depth in Section M15.3. Consider here the two cases: a fail-closed valve and a fail-open valve.

Fail-Closed Valve

If the pressure signal to the valve changed, this valve would close, hence the term fail-closed. These valves are also called *air-to-open*. Consider a case where flow rate through the valve is 5 gpm when it is fully open (15-psig signal). Assuming that the valve is fully closed at 3 psig (or less), the valve gain is

$$k_v = \frac{5-0 \text{ gpm}}{15-3\text{psig}} = 0.4167 \text{ gpm/psig}$$

Fail-Open Valve

If the pressure signal to the valve changed, this valve would open, hence the term fail-open. These valves are also called *air-to-close*. Consider a case where the flow rate through the valve is 5 gpm when it is fully open (3-psig signal). Assuming that the valve is fully closed at 15 psig (or more), the valve gain is

$$k_v = \frac{5-0 \text{ gpm}}{3-15\text{psig}} = -0.4167 \text{ gpm/psig}$$

One of the most important things that a process engineer can do is to specify whether a valve is air-to-open or air-to-close. For example, if the control valve was manipulating the cooling water flow to a nuclear reactor, then a fail-open valve should be specified.

Valve dynamics can become significant for larger valves. A laboratory valve may have a time constant of perhaps 1–2 seconds, a valve on a 3-inch process pipe may have a time constant of 3–10 seconds, and a valve on the Alaska pipeline may have a time constant of minutes.

M15.2 Flowmeters

An orifice plate meter is the most common flow measurement device. A constricting orifice is placed between two flanges, as shown in Figure M15–4. As the fluid "speeds up" through the orifice, the pressure drops. The volumetric flow rate is related to the pressure drop across the orifice. As shown in the motivating example, a DP cell is typically used to measure the pressure drop; here, we use a manometer for illustrative purposes.

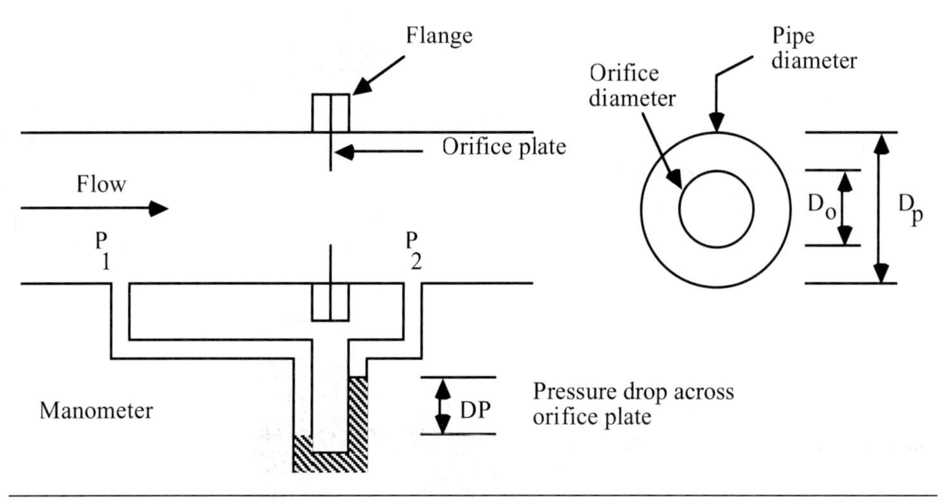

Figure M15–4 Orifice meter.

The equation for flow through an orifice is (McCabe and Smith, 1976)

$$v_o = \frac{C_o}{\sqrt{1-\beta^4}} \cdot \sqrt{\frac{2g_c(P_1 - P_2)}{\rho}} \tag{M15.5}$$

where v_o is the average velocity through the orifice;

β is the ratio of orifice to pipe diameters, D_o/D_p;

g_c is Newton's law gravitational constant, 32.174 ft $lb_m/lb_f s^2$; ρ is the fluid density;

C_o is the orifice coefficient (typically 0.61); and

P_1,P_2 are the upstream and downstream pressures.

The volumetric flow rate, F, is

$$F = v_o \cdot \frac{\rho D_o^2}{4} \tag{M15.6}$$

so the volumetric flow rate [combining Equations (M15.5) and (M15.6)] is

$$F = \frac{\rho D_o^2}{4} \cdot \frac{C_o}{\sqrt{1-\beta^4}} \cdot \sqrt{\frac{2g_c(P_1 - P_2)}{\rho}} \tag{M15.7}$$

which can be written

$$F = C_1\sqrt{P_1 - P_2} \tag{M15.8}$$

where

$$C_1 = \frac{\rho D_o^2}{4} \cdot \frac{C_o}{\sqrt{1-\beta^4}} \cdot \sqrt{\frac{2g_c}{\rho}} \quad \text{(M15.9)}$$

Notice that C_1 will be a constant for a given system. Notice also that an orifice plate has a nonlinear input-output relationship. Consider the input to be the volumetric flow rate. The output is the differential pressure. After rearranging Equation (M15.8) we find Equation (M15.10) and notice that the pressure drop is a function of the square of the flow rate,

$$P_1 - P_2 = \frac{F^2}{C_1} \quad \text{(M15.10)}$$

This means that the orifice plate gain (pressure drop change/flow rate change) increases with increasing flow. The orifice plate gain is

$$k_{op} = \frac{\partial(P_1 - P_2)}{\partial F} = \frac{2F}{C_1} \quad \text{(M15.11)}$$

and we see that if the flow rate doubles, then the orifice plate gain doubles. If a controller is tuned to obtain good control when the flow rate is low, there is a chance that the control loop will become unstable when the flow rate is high, because of the increase in overall control loop gain, which is due to the increasing orifice plate gain. A common way of handling this varying gain problem is to use a *square root extractor*.

Square Root Extractor

Assume that the input to a square root extractor is the pressure drop signal. The output is a 4- to 20-mA signal that is proportional to the square root of the pressure drop. Let P_{op} represent the signal entering the square root extractor and I represent the signal out of the square root extractor. The input-output relationship is then

$$I = 4 + C_2\sqrt{P_1 - P_2} \quad \text{(M15.12)}$$

where C_2 is a constant. The gain of the square root extractor is

$$k_{sr} = \frac{\partial I}{\partial(P_1 - P_2)} = \frac{C_2}{2} \cdot \frac{1}{\sqrt{P_1 - P_2}} \quad \text{(M15.13)}$$

Notice that the square root extractor gain is inversely proportional to the square root of the pressure drop. The purpose of a square root extractor can be seen by combining Equations (M15.12) and (M15.10) to yield

$$I = 4 + C_2\sqrt{\frac{F^2}{C_1}} = 4 + \frac{C_2}{\sqrt{C_1}} \cdot F = 4 + C_3 F \quad \text{(M15.14)}$$

The gain between flow rate and the signal out of the square root extractor is

$$\frac{\partial I}{\partial F} = C_3$$

where $C_3 = C_2/\sqrt{C_1}$. That is, we have a signal that is now directly proportional to the flow rate.

M15.3 Control Valves

The relationship for flow through a valve is

$$F = C_v f(x)\sqrt{\frac{\Delta P_v}{s.g.}} \tag{M15.15}$$

where F is the volumetric flow rate, C_v is the valve coefficient, x is the fraction of valve opening, ΔP_v is the pressure drop across the valve, s.g. is the specific gravity of the fluid, and $f(x)$ is the flow characteristic.

Three common valve characteristics are linear, equal-percentage, and quick-opening. For a linear valve,

$$f(x) = x \tag{M15.16}$$

For an equal-percentage valve,

$$f(x) = \alpha^{x-1} \tag{M15.17}$$

For a quick-opening valve

$$f(x) = \sqrt{x} \tag{M15.18}$$

The three characteristics are compared in Figure M15–5.

At this point it is reasonable to ask why we would ever want a valve characteristic that is not linear. After all, it seems reasonable that we would want the increase in flow to be linearly related to the increase in valve opening.

Think about the orifice plate nonlinearity. If a square root extractor is not used, then the orifice plate gain is low at low flow rates and high at high flow rates [see Equation (M15.11)]. Notice that the quick-opening valve has a high gain at low flow rates and a low gain at high flow rates, exactly the opposite of the orifice plate. Since the flowmeter and control valve are both elements in a control loop, the overall gain of the loop may be relatively unchanged by combining the nonlinearity of the direct-acting valve with the nonlinearity of the orifice plate flowmeter. This is probably less important in this day and age because square root extractors or computer computation of the actual flow rate can

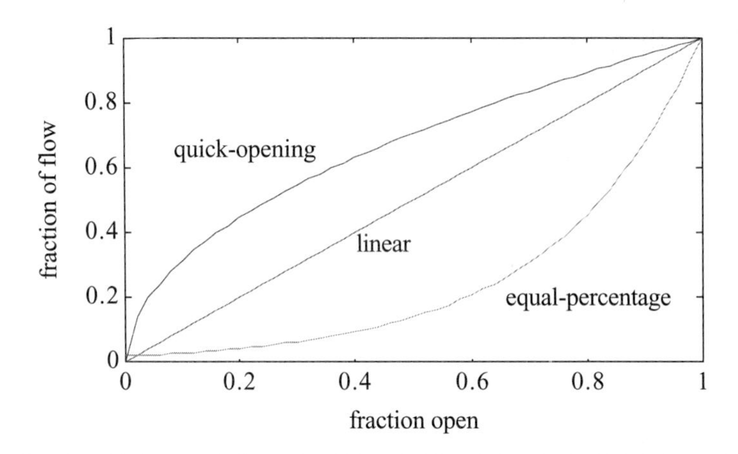

Figure M15–5 Flow characteristics of control valves. $\alpha = 50$ for equal-percentage valve.

compensate for the orifice plate nonlinearity. A quick-opening valve is then no longer necessary. We discuss this because a number of older plants still have old analog equipment, which may not have nonlinear compensation (e.g., square root extraction) for the orifice plate. Perhaps a process was originally designed with the direct-acting valve and an orifice plate flowmeter with square root extraction. If a recent "retrofit" (installation of new equipment) to change the control instrumentation involved adding nonlinear compensation (square root extraction, etc.) but did involve the replacement of the control valve, then the overall loop gain could change tremendously as a function of the flow rate. These thoughts may help you when you are troubleshooting a control problem.

Notice that the curves in Figure M15–5 are based on a constant pressure drop across the valve. This may be reasonable, for example, if the pressure drop through a piping system was negligible and all the pressure drop was due to the valve. In this case, a linear valve is desirable because there is a linear relationship between the valve position and the actual flow rate.

There are many systems, however, where the pressure drop due to piping is a significant portion of the pressure drop through the system and the pressure drop through the valve is not constant, and thus it may be desirable to use a nonlinear valve. This is shown more clearly in the next section.

M15.4 Pumping and Piping Systems

Consider the pumping and piping system shown in Figure M15–6. The pump is taking suction on the surge tank and pumping a liquid through a chemical process system (perhaps a heat exchanger and a reactor).

A typical pump head curve is shown in Figure M15–7. The pump head is a direct measure of the pressure increase between the suction and discharge of the pump that can be obtained as a function of the flow rate through the pump.

$$\Delta P_p = P_1 - P_0 = \frac{g}{g_c} \cdot \rho h_p$$

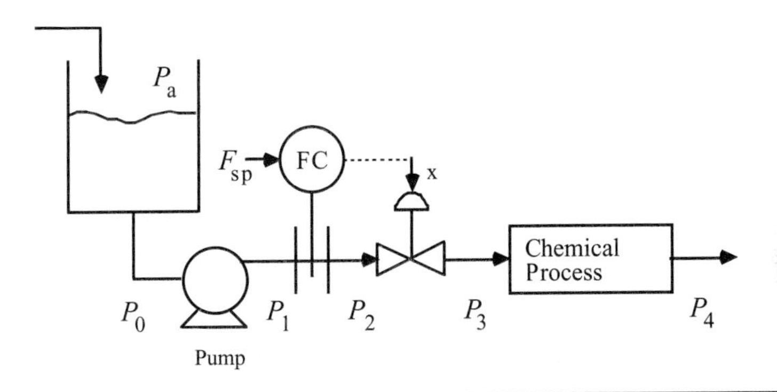

Figure M15–6 Process flowsheet for a pumping and piping system.

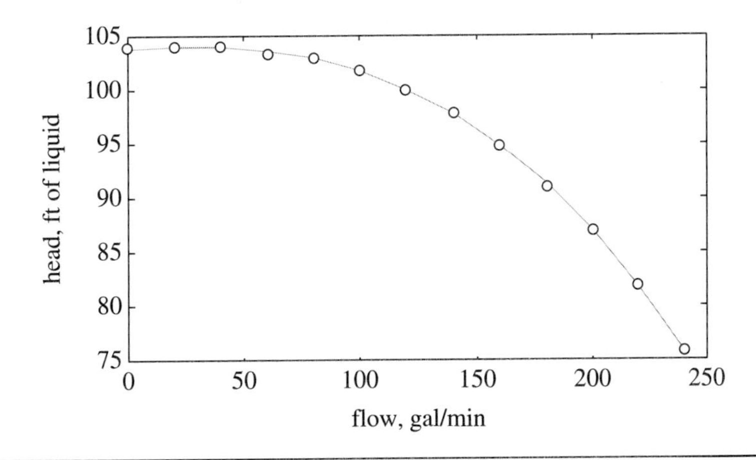

Figure M15–7 Pump head curve for a typical centrifugal pump [see Figure 6-7 from Perry and Green (1984)].

where h_p is the pump head in feet. The process will operate at the intersection of the pump head and system head curves, as shown in the following discussion.

Assume that the pressures P_a and P_4 are known. P_a is the ambient (or atmospheric) pressure and P_4 is the pressure at some known point downstream. If the height of liquid in the tank is known, then P_0 can be calculated for any given flow rate as

$$P_o = P_a + \rho \frac{g}{g_c} h - \Delta P_{sp} \tag{M15.19}$$

where h is the height of fluid between the tank level and the suction side of the pump and ΔP_{sp} is the pressure loss due to friction in the suction piping. The pressure drop is generally proportional to the square of the volumetric flow rate.

The pump discharge pressure, P_1, can be determined either using P_0 and the *pump head curve* as a function of flow rate,

$$P_1 = P_0 + \Delta P_p \tag{M15.20}$$

or using the pressure drop through the discharge piping system and the final pressure, P_4.

Using the second method, we can basically determine the system head as a combination of the pressure drop through the chemical process (ΔP_{cp}) and across the orifice plate (ΔP_{op}).

$$\Delta P_{sys} = \Delta P_{cp} + \Delta P_{op} \tag{M15.21}$$

The pump discharge pressure can be found as

$$P_1 = P_4 + \Delta P_{sys} + \Delta P_v \tag{M15.22}$$

Obviously, the problem solution is where P_1 from Equation (M15.20) is equal to P_1 from Equation (M15.22).

Numerical Example

Consider the pump head curve shown in Figure M15–7. Assume that the maximum flow rate is 240 gpm. The pump head at this flow rate is 76 feet of water. Now, let's make some assumptions about the other pressures in the system. First, we know that P_1 calculated from Equation (M15.20) is equal to P_1 from Equation (M15.22), or

$$P_0 + \Delta P_p = P_4 + \Delta P_{sys} + \Delta P_v \tag{M15.23}$$

For this particular example, assume that $P_4 = P_0$, so

$$\Delta P_p = \Delta P_{sys} + \Delta P_v \tag{M15.24}$$

At the maximum flow rate of 240 gpm, assume that the system head is

$$\Delta P_{sys} = 66 \text{ feet}$$

so the valve pressure drop must be

$$\Delta P_v = 10 \text{ feet}$$

We can easily determine the system head at any other flow rate (F_2) as

$$\Delta P_{sys,2} = \Delta P_{sys,1} \cdot \left(\frac{F_2}{F_1}\right)^2 \quad \text{(M15.25)}$$

That is, $\Delta P_{sys,2} = 66\left(\frac{F}{240}\right)^2$.

The system head is compared with the pump head curve in Figure M15–8. Notice that in practice, the pressure drop can be predicted by knowing the pipe diameter, the number of elbows and fitting, the pipe roughness, and so forth, and by calculating an overall friction factor. The pressure drop is a function of the square of the flow rate, so we are assuming that the pressure drop has been calculated or measured at a particular flow rate and we simply use the ratio of the square of the flow rates to determine the pressure drop at any other flow rate. The difference between the pump head and the system head is the pressure drop across the valve.

The relationship for flow through a valve is

$$F = C_v f(x)\sqrt{\frac{\Delta P_v}{s.g.}} \quad \text{(M15.26)}$$

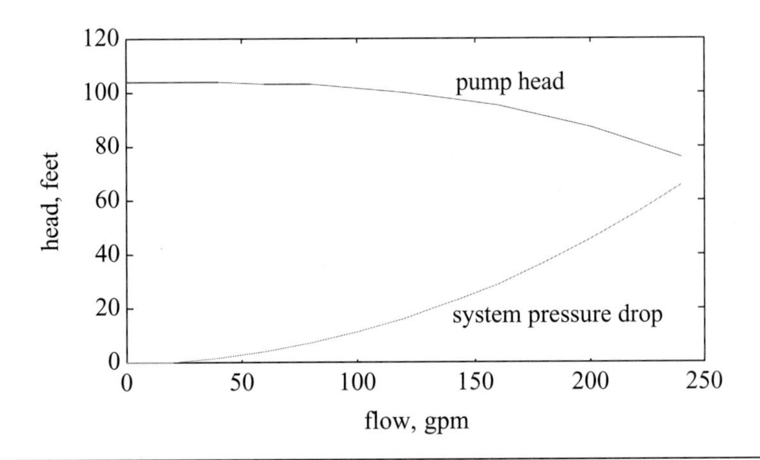

Figure M15–8 Pump head and system head curves.

Now assuming that the valve is wide open at the maximum flow rate of 240 gpm (x = f(x) = 1) and the fluid is water, we can calculate the valve coefficient as

$$C_v = \frac{F}{\sqrt{\Delta P_v}} = \frac{240\text{gpm}}{\sqrt{10\text{ft}}} = 75.9\,\text{gpm}/\sqrt{\text{ft}}$$

Notice that the unit of pressure drop that we are using is feet of water.

Then, for a given flow rate and valve pressure drop, we can calculate the fraction that the valve must be open. That is,

$$f(x) = \frac{F}{C_v\sqrt{\Delta P_v}}$$

And for a linear valve, the fraction that the valve is open is

$$x = f(x) = \frac{F}{C_v\sqrt{\Delta P_v}}$$

and for an equal-percentage valve, the fraction that the value is open is

$$x = 1 + \frac{\ln(f(x))}{\ln(a)} = 1 + \frac{\ln\left(\dfrac{F}{C_v\sqrt{\Delta P_v}}\right)}{\ln(a)}$$

In each case, $\Delta P_v = \Delta P_p - \Delta P_{sys}$. These relationships are shown in Figure M15–9.

Notice that the linear valve has a more linear characteristic at low flow rates, while the equal-percentage valve has a more linear characteristic over a wider range of flow

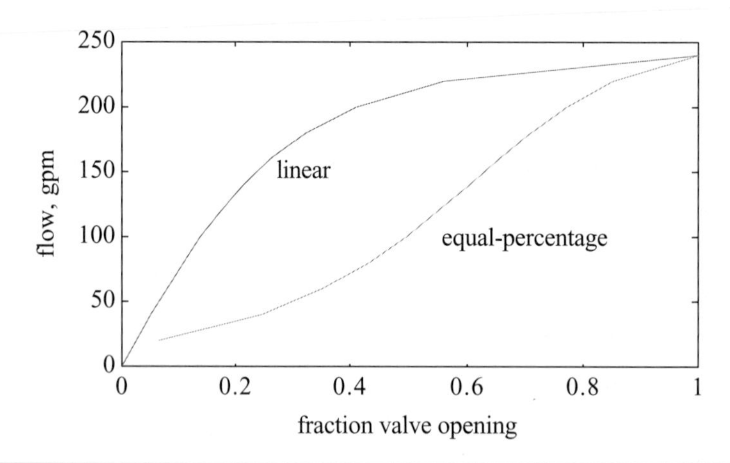

Figure M15–9 Comparison of installed characteristics of control valves.

rates, and in the higher flow rates in particular. Notice also that the gain of the linear valve is high at low valve opening and low at high valve openings. The equal-percentage valve has a gain that is roughly the same at high and low valve openings and has a maximum gain that is roughly twice the minimum gain. The linear valve has a maximum gain that is roughly 10 times the minimum gain. If a control system using the linear valve was tuned at a high flow rate (valve gain low, controller gain high), the control system might go unstable when operated at low flow rates. This problem is less likely to occur with the equal-percentage valve.

M15.5 Summary

We have seen that a direct-acting valve might be used to compensate for the orifice plate nonlinearity if no square root extractor is used. We also noted that a linear valve should be used if the pressure drop across the valve is constant (say, the upstream and downstream headers are maintained at constant pressure, and there is negligible pressure drop through the piping). We notice that an equal-percentage valve should be used if a substantial amount of the pressure drop at high flow rates is due to piping, while most of the pressure drop at low flow rates is across the valve.

Many industrial controllers use proportional band rather than proportional gain. You can still use any of the techniques presented in this text to find the proportional gain, and simply convert to PB using $PB = 100/k_c$.

References

McCabe, W. L., and J. C. Smith, *Unit Operations of Chemical Engineering*, 3rd ed., McGraw-Hill, New York (1976).

Perry, R. H., and D. Green (ed.), *Perry's Chemical Engineering Handbook*, 6th ed., pp. 6–8, New York: McGraw Hill (1984).

Additional Exercises

1. Consider the control valve diagram presented in Figure M15–5 and Equations (M15.16)–(M15.18). Find the gain between the fraction of valve opening, x, and the output of the orifice plate, ΔP, as a function of flow rate for each of the valves shown in Figure M15–5. Assume that the sum of the pressure drop

across the valve and the orifice plate is constant. Assume that the constant pressure drop is 30 psig, and that at maximum flow, three quarters of the pressure drop is across the valve and one quarter is across the orifice plate. The gain calculation is shown conceptually below.

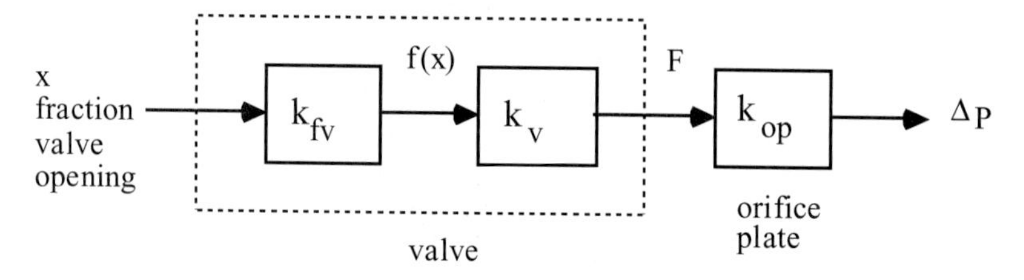

2. Instrumentation Search. Select one of the following measurement devices and use Internet resources to learn more about it. Determine what types of signals are input to or output from the device. For flow meters, what range of flow rates can be handled by a particular flowmeter model?
   a. Vortex-shedding flowmeters
   b. Orifice-plate flowmeters
   c. Mass flowmeters
   d. Thermocouple-based temperature measurements
   e. Differential pressure (delta P) measurements
   f. Control valves
   g. pH

Digital Control

This module provides a tutorial overview of digital control system analysis and design. After studying this module, the reader should understand the following:

- Discrete and continuous controller performance is similar for short sample times
- The discrete equivalent to a continuous-system RHP zero is a zero outside the unit circle
- A discrete control system is stable if the poles of the discrete closed-loop characteristic equation are inside the unit circle
- The discrete IMC procedure factors zeros outside the unit circle and negative zeros inside the unit circle into the "bad stuff"
- A continuous process transfer function with no RHP zeros may have an equivalent discrete transfer function with a zero outside the unit circle, if the *relative order* (difference in denominator and numerator orders) is 3 or higher.

The major sections of this module are as follows:

M16.1 Background

The focus of this textbook has been on continuous systems although, in practice, most controllers are implemented digitally. Fortunately, the sample time is usually small compared with the process dynamics, so continuous systems theory is usually sufficient.

Processes are best described as continuous systems (with differential equation models), with inputs that are applied continuously and outputs that are available continuously. With digital controllers, however, measurements are made at discrete intervals of time and control moves are made at discrete intervals of time. Discrete-time models were developed in Chapter 4 and the conversion of continuous time to discrete time models (using LTI objects in MATLAB) were presented in Module 4.

Recall that continuous systems are stable if their poles are negative [in the left-half plane (LHP)]; also, process models can only be inverted to form stable controllers if the models have zeros in the LHP. Models with RHP zeros must be factored into "good" and "bad" stuff; only the portion with LHP (good stuff) zeros can be inverted to form a controller. For continuous systems, then, the focus is on the LHP vs. the RHP. Discrete-time systems are stable if their poles are "inside the unit circle," that is, with a magnitude <1. If a discrete-time process model has a zero outside the unit circle (magnitude >1), it cannot be inverted to form a stable controller; the zero must be factored into the bad stuff before the good stuff is inverted to form a stable controller.

Stable continuous process models will always have stable discrete process transfer functions. That is, if all poles of a continuous process model are negative, then the discrete process model will have all discrete poles inside the unit circle. It is possible, however, for a continuous process with no RHP zeros to have a discrete zero that is outside the unit circle. This is possible for continuous transfer functions that have a *relative order* (difference between the denominator and numerator polynomial orders) of 3 or higher. See Additional Exercise 4 for an example of this problem.

The purpose of this interactive module is to obtain a better understanding of the effect of discretization on control system performance. We focus initially on PID control, then study IMC.

M16.2 PID Controllers

Consider the continuous (analog) form of an ideal PID controller,

$$u(t) = u_0 + k_c\left[e(t) + \frac{1}{\tau_I}\int_0^t e(t)dt + \tau_D\frac{de}{dt}\right] \tag{M16.1}$$

We use a constant sample time, Δt, and the index k to represent the value of the continuous time signal at discrete step k. For example, the error at discrete sample time k is equal to the continuous system value at time $t = t_k$.

$$e(k) = e(t_k) \tag{M16.2}$$

We approximate the integral term

$$\begin{aligned}\int_0^{t_k} e(t)dt &\approx e(t_1)\cdot\Delta t + e(t_2)\cdot\Delta t + \cdots + e(t_k)\cdot\Delta t \\ &\approx \sum_{i=1}^{k} e(t_i)\Delta t = \sum_{i=1}^{k} e(i)\Delta t\end{aligned} \tag{M16.3}$$

and the derivative term is approximated using backward finite differences:

$$\frac{de(t_k)}{dt} \approx \frac{e(k)-e(k-1)}{\Delta t} \tag{M16.4}$$

Substituting Equations (M16.2)–(M16.4) into Equation (M16.1), we find

$$u(k) = u_0 + k_c\left[e(k) + \frac{\Delta t}{\tau_I}\sum_{i=0}^{k} e(i) + \frac{\tau_D}{\Delta t}\left(e(k)-e(k-1)\right)\right] \tag{M16.5}$$

Notice that the manipulated variable action is calculated at discrete intervals. The most common way of implementing the discrete control actions on a continuous process is to assume *a zero-order hold*. A zero-order hold means that the control action is held constant between the sample times, and that the manipulated variable changes instantly at each sample time.

Equation (M16.5) is known as the position form of the discrete PID controller equation. The "velocity form" can be found by subtracting the position form at step $k - 1$ from that at step k:

$$u(k) = u(k-1) + k_c\left[\left(1 + \frac{\Delta t}{\tau_I} + \frac{\tau_D}{\Delta t}\right)e(k) + \left(-1 - \frac{2\tau_D}{\Delta t}\right)e(k-1) + \frac{\tau_D}{\Delta t}e(k-2)\right] \tag{M16.6}$$

One of the major advantages to the velocity form of Equation (M16.6) is that it is naturally "antireset windup" (see Chapter 11 for a discussion of reset windup and procedures to prevent it). If the manipulated input saturates, the position form of Equation (M16.5) has an integral term that continues to accumulate, causing a potential windup problem.

Most of the PID control system design techniques covered earlier in this text (direct synthesis, IMC-based PID, etc.) should yield successful tuning parameters for the discrete PID version, if the sample time is sufficiently small. Generally, the sample time should be

0.01 to 0.1 times the dominant time constant. In practice, most digital control loops are sampled several times per second, which is far more frequently than is actually needed for most process systems.

Z-Transform Representation of PID Control

In the same way that Laplace transforms are used to characterize continuous-time systems, Z-transforms are used to represent the input-output relationships for discrete-time systems. Consider the discrete PID algorithm in Equation (M16.6). This can be written as

$$u(k) - u(k-1) = b_0 e(k) + b_1 e(k-1) + b_2 e(k-2) \qquad \text{(M16.7)}$$

where $b_0 = k_c\left(1 + \frac{\Delta t}{\tau_I} + \frac{\tau_D}{\Delta t}\right)$, $b_1 = -k_c\left(1 + \frac{2\tau_D}{\Delta t}\right)$, $b_2 = \frac{k_c \tau_D}{\Delta t}$.

In an analogy to continuous-time systems, where we let $u(s)$ represent the Laplace transform of $u(t)$, we let $u(z)$ represent the Z-transform of $u(k)$. It is common to refer to z^{-1} as the backward shift operator:

$$\begin{aligned} Z[u(k)] &= u(z) \\ Z[u(k-1)] &= z^{-1}u(z) \\ Z[u(k-2)] &= z^{-2}u(z) \end{aligned}$$

handling e(k), etc., similarly, we can write Equation (M16.7) as

$$(1 - z^{-1})u(z) = (b_0 + b_1 z^{-1} + b_2 z^{-2})e(z)$$

which can be written

$$u(z) = \frac{\left(b_0 + b_1 z^{-1} + b_2 z^{-2}\right)}{1 - z^{-1}} e(z) \qquad \text{(M16.8)}$$

We can also multiply the numerator and denominator polynomials by z^2/z^2, to write

$$u(z) = \frac{\left(b_0 z^2 + b_1 z + b_2\right)}{z^2 - z} e(z) \qquad \text{(M16.9)}$$

Note that we can view this as an input-output block

A discrete closed-loop block diagram is similar to that of a continuous block diagram. The closed-loop relationship for a setpoint change is

$$y(z) = \frac{g_c(z)g_p(z)}{1+g_c(z)g_p(z)} \cdot r(z) \tag{M16.10}$$

M16.3 Stability Analysis for Digital Control Systems

For a standard discrete feedback control system, the closed-loop system is stable if the poles of the closed-loop transfer function are all inside the unit circle. The closed-loop transfer function is

$$g_{CL}(z) = \frac{g_c(z)g_p(z)}{1+g_c(z)g_p(z)}$$

so the roots of

$$1 + g_c(z)g_p(z) = 0$$

must be less than 1 in magnitude.

Example M16.1: Stability of a Discrete Control System

Consider a first-order process, where the output is temperature (°C), the manipulated input is valve position (%), and the time unit is minutes.

$$g_p(s) = \frac{1}{10s+1}$$

If discretized with a sample time of 1 minute, the discrete-time process (the MATLAB Control Toolbox can be used to convert continuous to discrete models—see Module 4, Section M4.3) is

$$g_p(z) = \frac{0.0952z^{-1}}{1-0.9048z^{-1}} = \frac{0.0952}{z-0.9048}$$

With a P-only controller, we must check the roots of

$$1+\frac{0.0952k_c}{z-0.9048} = 0$$

and solving for the root, we find

$$z = \frac{0.9048}{1-0.0952k_c}$$

where the magnitude of the root is less than 1 for $-1<k_c<20.01$. Since the process gain is positive, only positive values of the proportional gain make sense. This is result is verified in Figure M16–1, where the discrete P-only controller is unstable, while the continuous P-controller is stable with only a small amount of offset; $k_c = 20.01$ for both the continuous and discrete P controllers.

Problem M16.1 *(see the* SIMULINK `.mdl` *file in Appendix M16.1)*

Consider the first-order process given in Example M16.1. For a discrete PI controller, with $\tau_I = 10$ minutes, find the proportional gain that just makes the closed-loop system unstable. Verify your analytical result via simulation; show plots similar to Figure M16–1.

M16.4 Performance of Digital Control Systems

The performance of a digital PID controller is rarely significantly different from that of a continuous PID controller, as long as the sample time is relatively short compared with the dominant process time constant and the desired closed-loop response time. A rule of thumb for digital control systems in general is that the sample time should be less than one tenth

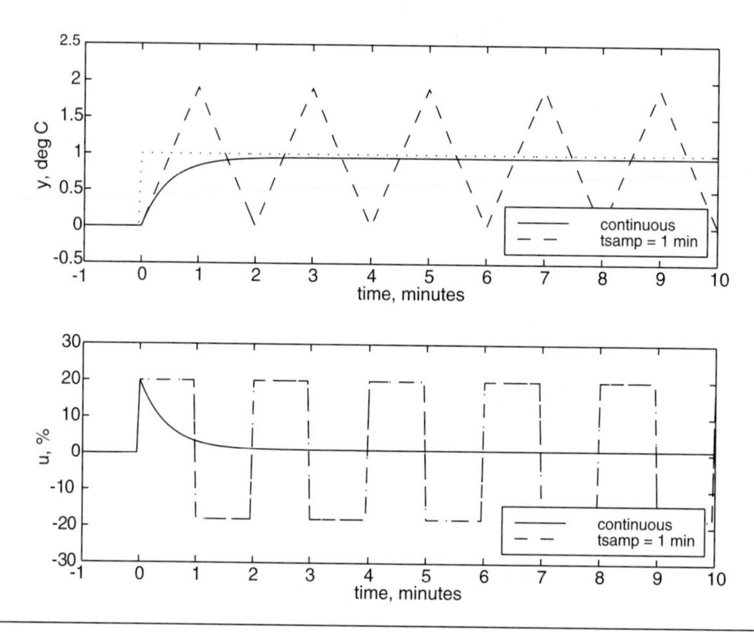

Figure M16–1 Comparison of setpoint responses for continuous and discrete P-only controllers on the first-order process. $k_c = 20.01$ %/°C.

the dominant time constant. In practice, digital PID controllers have a sample time much shorter than that. Many distributed control systems have a sample rate that is several times per second, so the sample time is less than 1 second. This makes sense for flow control loops but is much shorter than necessary for other loops. The main disadvantages to sampling too fast are that (i) more frequent computations than necessary are being performed, (ii) numerical round-off could be a problem if only small control moves are being made at each sample time, and (iii) there could be increased measurement noise sensitivity.

Example M16.2: Effect of Sample Time on PI Control Performance

Consider the first-order + time-delay process

$$g_p(s) = \frac{e^{-5s}}{10s+1}$$

where the output is temperature (°C), the manipulated input is valve position (%), and the time unit is minutes. In Module 6 we used this example to compare the robustness of different tuning procedures for PID-type controllers. The IMC-based "improved PI" parameters (see Chapter 9), with $\lambda = 7.5$ minutes, yields $k_c = 1.67\%/°C$ and $\tau_I = 12.5$ minutes for an analog PI controller. In Figure M16–2 we compare analog PI control with discrete

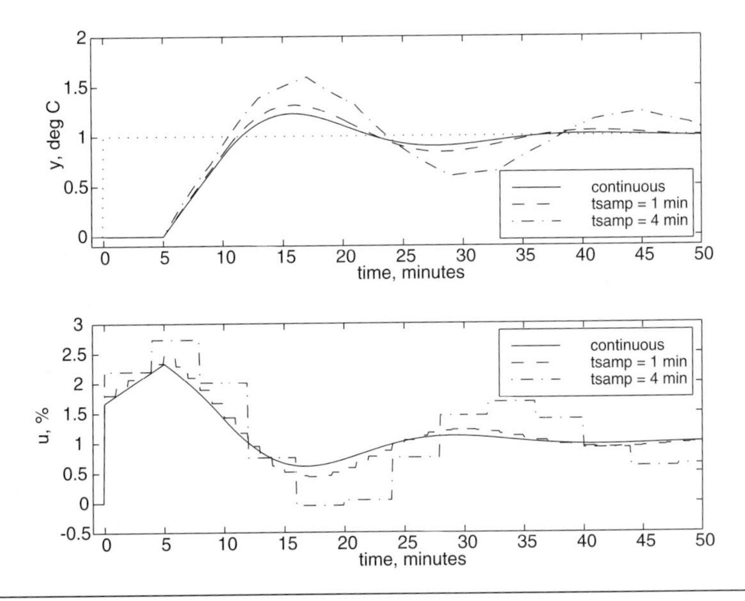

Figure M16–2 Comparison of setpoint responses for different sample times. IMC-based PI controller tuning used ($\lambda = 7.5$ minutes).

PI control for two sample times (1 and 4 minutes). Clearly, the control performance degrades as the sample time increases.

Problem M16.2 *(see the* SIMULINK `.mdl` *file in Appendix M16.1)*

Perform simulations with different sample times and compare the performance of the digital controller with the continuous controller while keeping the PID tuning parameters the same as shown in the example.

a. Is there much difference in performance when a sample time of 0.25 minutes is used?

b. Does the discrete control system go unstable for any sample times greater than 4 minutes? You may wish to increase the integration final time for this simulation.

M16.5 Discrete IMC

The design of discrete IMCs is similar to that of the continuous IMC design procedure detailed in Chapter 8. The primary difference is that the factorization is performed on a discrete-time (Z-domain) model.

Factorize into good stuff and bad stuff the discrete model

$$\tilde{g}_p(z) = \tilde{g}_{p-}(z)\tilde{g}_{p+}(z) \qquad \text{(M16.11)}$$

where $\tilde{g}_{p+}(z)$ contains zeros outside the unit circle and zeros inside the unit circle that negative. The reason for this is that negative zeros inside the unit circle cause oscillatory poles in the controller if not removed. Make certain that the numerator of $\tilde{g}_{p-}(z)$ is one order less than the denominator polynomial. Also, form an "all-pass" in $\tilde{g}_{p+}(z)$ by including a pole at $1/z_i$ for each z_i outside the unit circle. Make certain that "gain" (found by setting $z = 1$) of the "bad stuff" is equal to one, while the gain of the "good stuff" is equal to the gain of the unfactored transfer function.

The design procedure is then analogous to that developed for continuous IMC in Chapter 8. The controller is the inverse of the "good stuff" times a discrete filter to make the controller physically realizable (numerator and denominator polynomials are the same order).

$$q(z) = \tilde{g}_{p-}^{-1}(z) \cdot f(z) \qquad \text{(M16.12)}$$

where $f(z)$ has the following form

$$f(z) = \frac{(1-\alpha)z^{-1}}{1-\alpha z^{-1}} = \frac{(1-\alpha)}{z-\alpha} \qquad \text{(M16.13)}$$

where α is a discrete tuning parameter. It is related to the continuous IMC filter factor by

$$\alpha = e^{-\Delta t/\lambda} \quad \text{(M16.14)}$$

where Δt is the sample time. The discrete IMC procedure is slightly more complex than the continuous-time case for a couple of reasons. First of all, the continuous system parameters such as time constants are much easier to interpret than the discrete-time parameters; discrete parameters are a function of the sample time of the digital system. Also, it is easier to keep mental track of gains in continuous-time, since setting $s = 0$ normally eliminates most terms in a transfer function. In discrete-time, setting $z = 1$ results in the sum of a number of coefficients, and is simply not as natural to handle.

The following example presents a step-by-step application of the discrete IMC procedure.

Example M16.3

Consider the van de Vusse reactor model studied in Module 5; this was also used as an example for the conversion between model types in Module 4. The continuous transfer function is

$$\tilde{g}_p(s) = \frac{-1.1170s + 3.1472}{s^2 + 4.6429s + 5.3821} = \frac{0.5848(-0.3549s + 1)}{0.1828s^2 + 0.8627s + 1}$$

The discrete-time model (for a sample time of 0.1 minutes) is

$$\tilde{g}_p(z) = \frac{-0.07506z + 0.1001}{z^2 - 1.586z + 0.6286} = \frac{-0.07506z^{-1} + 0.1001z^{-2}}{1 - 1.586z^{-1} + 0.6286z^{-2}}$$

This can be written in discrete pole-zero form as

$$\tilde{g}_p(z) = \frac{-0.075061(z - 1.334)}{(z - 0.7995)(z - 0.7863)}$$

and it is easy to verify that the gain (found by setting $z = 1$) is 0.5851, which is consistent with the continuous time model (found by setting $s = 0$).

Now, the first step of the factorization procedure is to remove the zeros that are outside the unit circle (there is a zero at 1.334), as well as the negative zeros inside the unit circle (there are none in this particular problem)

$$\begin{aligned}\tilde{g}_p(z) &= \tilde{g}_{p-}(z)\tilde{g}_{p+}(z) \\ &= \frac{-0.75061}{(z - 0.7995)(z - 0.7863)} \cdot (z - 1.334)\end{aligned}$$

The next step is to form the "all-pass"

$$\tilde{g}_p(z) = \tilde{g}_{p-}(z)\tilde{g}_{p+}(z)$$

$$= \frac{-0.75061\left(z - \frac{1}{1.334}\right)}{(z-0.7995)(z-0.7863)} \cdot \frac{(z-1.334)}{\left(z - \frac{1}{1.334}\right)}$$

Next, make certain that the gain of the "bad stuff" is 1 and that the gain of the "good stuff" is the same as the gain of the original transfer function

$$\tilde{g}_p(z) = \tilde{g}_{p-}(z)\tilde{g}_{p+}(z)$$

$$= \frac{-0.75061\left(z - \frac{1}{1.334}\right)(1-1.334)}{(z-0.7995)(z-0.7863)\left(1 - \frac{1}{1.334}\right)} \cdot \frac{(z-1.334)\left(1 - \frac{1}{1.334}\right)}{\left(z - \frac{1}{1.334}\right)(1-1.334)}$$

$$= \frac{0.1001(z-0.7496)}{(z-0.7995)(z-0.7863)} \cdot \frac{(-0.7496)(z-1.334)}{z-0.7496}$$

The IMC is obtained by inverting the good stuff

$$\tilde{q}(z) = \frac{(z-0.7995)(z-0.7863)}{(0.1001)(z-0.7496)}$$

and using a first-order filter to make it realizable:

$$q(z) = \frac{(z-0.7995)(z-0.7863)}{(0.1001)(z-0.7496)} \cdot \frac{(1-\alpha)}{(z-\alpha)}$$

where the discrete tuning parameter, α, is related to the continuous tuning parameter, λ, by

$$\alpha = \exp(-\Delta t/\lambda) = \exp(-0.1/\lambda)$$

Here, we compare the responses of the discrete control system with the continuous system. For this sample time and choice of $\lambda = 0.2$ min, there is virtually no difference, as shown in Figure M16–3. The block diagram and *m*-file used for this simulation is presented in Appendix M16.2.

Problem M16.3

Choose Exercise 2, 3, or 4 from the Additional Exercises at the end of this module for further study of the IMC design procedure. Exercise 2 is a straightforward first-order example where there are no particular performance limitations. Exercise 3 illustrates the importance of factoring out negative zeros inside the unit circle. Exercise 4 is an example

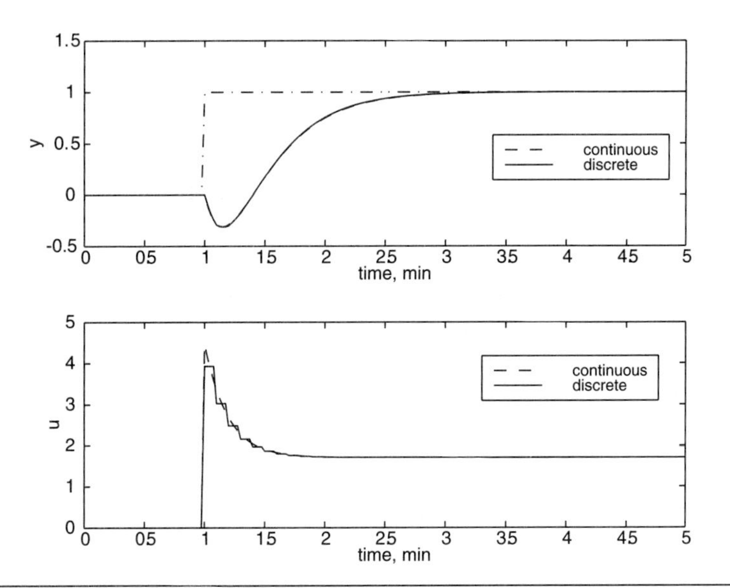

Figure M16–3 Comparison of continuous and discrete IMC for $\lambda = 0.2$ minutes. Discrete sample time, $\Delta t = 0.1$ minutes.

where the continuous-time transfer function is third-order, but minimum-phase (no RHP zeros or time delays) so there is no limitation to achievable IMC performance. The discrete-time model, however, has a zero outside the unit circle. This poses substantial performance limitations for the discrete-time system.

M16.6 Summary

The unit circle plays the same role in discrete-time systems that the real line does in continuous-time systems. A zero outside the unit circle causes the same performance limitations that an RHP zero causes for continuous systems. It is possible, however, for a continuous-time system with no RHP zero to have a discrete zero that is outside the unit circle, if the continuous-time system has a relative order of 3 or higher.

Discrete-time models with zeros outside the unit circle must be factored into the bad stuff for discrete IMC controller design. This procedure is similar to continuous-time models, where RHP zeros must be factored into the bad stuff. The main difference between continuous and discrete systems is that even zeros than are inside the unit circle, but negative, must be factored out; if they are not, the controller will often be too oscillatory for acceptable performance.

References

In this textbook we have used the Control Toolbox to convert continuous-time models to discrete time models (assuming a zero-order hold). For more details on how to perform this conversion analytically, see the following textbooks:

Stephanopoulos, G., *Chemical Process Control*, Prentice Hall, Upper Saddle River, NJ (1984).

Seborg, D. E., T. F. Edgar, and D. A. Mellichamp, *Process Dynamics and Control*, Wiley, New York (1989).

The basic discrete IMC design procedure is covered (at an advanced level using different notation than in our textbook) in the following:

Zafiriou, E., and M. Morari, "Digital Controllers for SISO Systems: A Review and a New Algorithm," *Int. J. Cont.*, **42**(4), 855–876 (1985).

Morari, M., and E. Zafiriou, *Robust Process Control*, Prentice Hall, Englewood Cliffs, NJ (1989).

Additional Exercises

1. Consider the van de Vusse reactor, which has the following state space model at the nonminimum-phase operating point:

$$\dot{x} = Ax + Bu$$
$$y = Cx + Du$$

where

$$A = \begin{bmatrix} -2.4048 & 0 \\ 0.8333 & -2.2381 \end{bmatrix}, \quad B = \begin{bmatrix} 7 \\ -1.117 \end{bmatrix}, \quad C = \begin{bmatrix} 0 & 1 \end{bmatrix}, \quad D = \begin{bmatrix} 0 \end{bmatrix}$$

Find the sample time that will bring the zero (discrete time) that is outside the unit circle to the unit circle. Compare the discrete-step response for this sample time with the continuous system step response.

2. Consider the first-order process (where the time unit is minutes and the gain is °C/%)

$$g_p(s) = \frac{1}{10s+1}$$

If discretized with a sample time of 1 minute, the discrete-time model is

$$\tilde{g}_p(z) = \frac{0.0952z^{-1}}{1 - 0.9048z^{-1}}$$

Design continuous and discrete IMC controllers and compare performances on a unit step setpoint change. Study various values of the IMC tuning parameter, λ.

3. Consider the second-order process (where the time unit is minutes and the gain is %/%)

$$g_p(s) = \frac{1}{(10s+1)(25s+1)}$$

If discretized with a sample time of 3 minutes, the discrete-time model is

$$\tilde{g}_p(z) = \frac{0.0157z^{-1} + 0.0136z^{-2}}{1 - 1.6277z^{-1} + 0.65702z^{-2}}$$

Consider factorizations for the discrete IMC design when (a) the negative zero inside is not factored out, and (b) the negative zero inside the unit circle is factored out. Compare the two digital controllers and continuous IMC performances on a unit step setpoint change. Study various values of the IMC tuning parameter, λ.

4. Consider the third-order process (where the time unit is minutes, and the gain is %/%)

$$g_p(s) = \frac{1}{(s+1)^3}$$

which has no zeros (that is, it is minimum phase). If the model is perfect, we know that a continuous-time controller can be tuned arbitrarily tightly.

If the continuous model is discretized with a sample time of 1 minute, the discrete-time model (in factored form) is

$$\tilde{g}_p(z) = \frac{0.0803(z+1.7990)(z+0.1238)}{(z-0.3679)^3}$$

Notice that one of the process zeros is at –1.7790 (outside the unit circle); if the inverse of the model is used for controller design, an unstable controller results. It is interesting that the continuous-time model has no RHP zero, yet the discretized model has a zero outside the unit circle.

Consider factorizations for the discrete IMC design when (a) the zero at –1.7990 is not factored out, and (b) the zero at –1.7990 is factored out. Compare the two digital controllers and continuous IMC performances on a unit step setpoint change. Study various values of the IMC tuning parameter, λ.

Appendix M16.1: SIMULINK .mdl File for Example M16.2

The .mdl figure Figure M16–4 can be used to simulate Example M16.2. To use this for Example M16.1, simply remove the "transport delay" blocks.

Appendix M16.2: SIMULINK .m and .mdl Files for Example M16.3

Use the following m-file to run simulations based on the cd_IMC.mdl SIMULINK model, and make plots. See Figure M16–5.

```
% Run continuous and discrete IMC simulations
% For use with the simulink diagram: cd_IMC.mdl
%
% B.W. Bequette
```

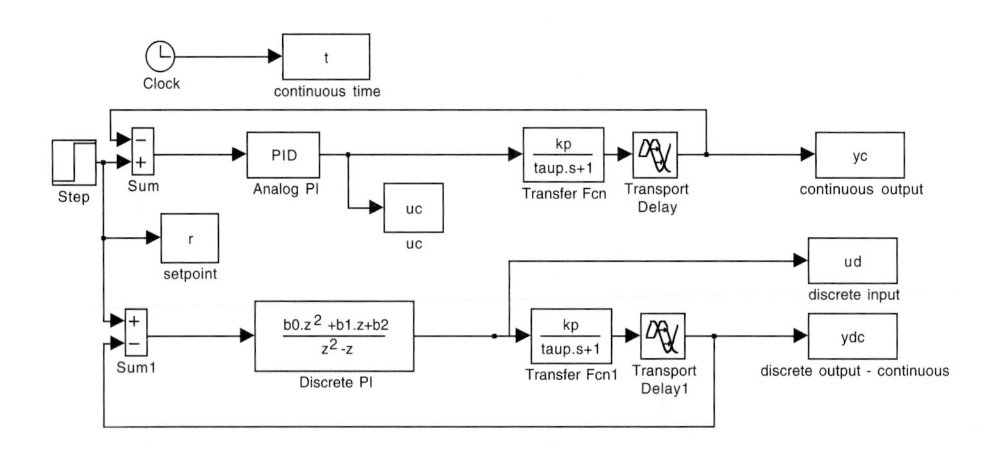

Figure M16–4 .mdl figure.

```
% 14 Sept 00
%
% plant values are (notice the other models are embedded in the
% simulink diagram

  a = [-2.4048 0;0.8333 -2.2381];
  b = [7; -1.117];
  c = [0 1];
  d = 0;
%
% first enter the value of lambda desired in the command window
%
  alpha = exp(-0.1/lambda);
%
  tfinal = 5; % final simulation time
  sim('cd_IMC')
%
  figure(3)
  title('comparison of discrete and continuous IMC')
  subplot(2,1,1),plot(t,yimc,'--',t,yd,t,r,'-.')
  legend('continuous','discrete')
  xlabel('time, min')
  ylabel('y')
  subplot(2,1,2),plot(t,uimc,'--',t,ud)
  legend('continuous','discrete')
  ylabel('u')
  xlabel('time, min')
```

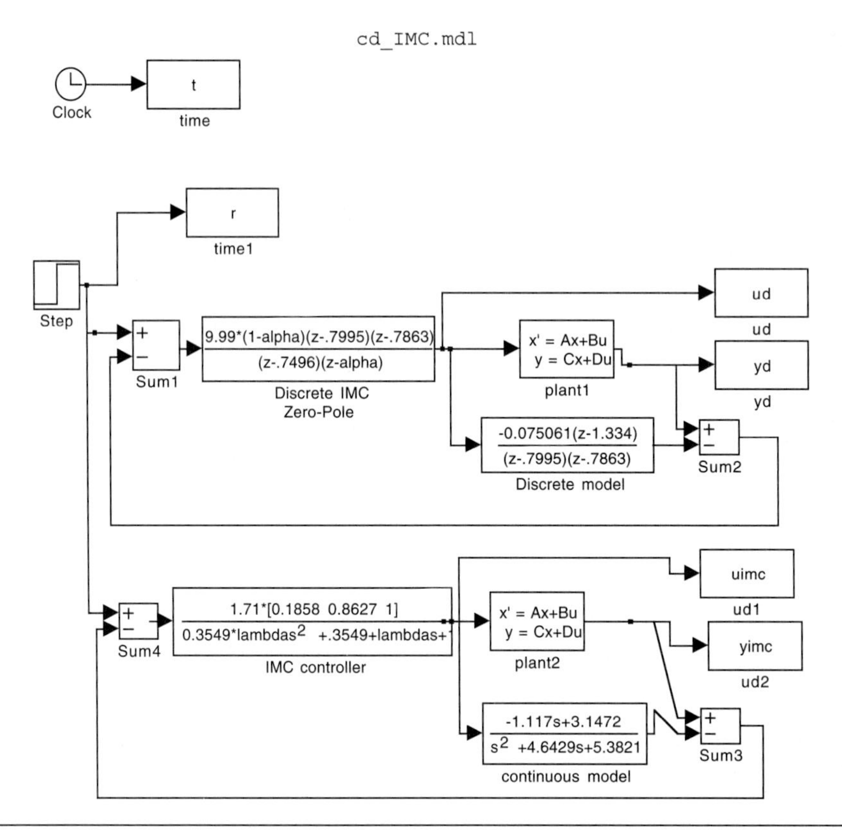

Figure M16–5 Comparison of discrete and continuous IMC.

Index